AF517453

Sustainable Development of
Dryland Agriculture
in India

Other Related Books

Advancements in Micronutrient Research	A. Hemantaranjan
Advances in Plant Physiology (Vol. 18)	A. Hemantaranjan
Agriculture and Fertilizers	O.C. Bockman
Agrotechnology for Dryland Farming (2nd Revised Ed.)	A. M. Dhopte
Challenges and Strategies of Dryland Agriculture	S.C. Rao
Climate Change Adaptation Strategies in Agriculture and Allied Sectors	GSHLV Prasad Rao
Crop Production in Salt-Affected Soil	I.C. Gupta
Dryland Technology 2nd Edition	ML Jat
Farm Mechanization for Production	Dhirendra Khare
Fertilizer and Plant Nutrition Guide	Fao
Fundamentals af Agronomy	R.L. Arya
Fundamentals of Organic Farming and Gardening	Alexander Mcgregor
Glimpse on General Agriculture	P. Laxman Rao
Introduction to Crops of India 2Nd Ed	N.R. Das
Kharpatvar Niyantran (Weed Management) (Hindi)	J.P. Tiwari
Krishi Jagat: Jet Evam Krishi Parikshak Pariksha Upoagi (Hindi)	M.L. Jat
Manual on Fundamentals of Agronomy	L. K. Jain
Nanofertilizers: Challenges and Prospects	J.C. Tarafdar
Organic Farming Theory & Practice	S.P. Palaniappan
Salt Affected Soils: Reclamation and Management	S.K. Gupta
Sasay Vigyan Ka Moolbhut Gyan (Fundamental Knowledge of Agronomy) (Hindi)	Ram Pyare
Seed Technology, 2Nd Ed.	Dhirendra Khare
Soil Conservation Technical Handbook	D. H. Hicks
Soil Guide A Handbook for Uniderstanding and Managing Agricultural Soils	Geoff Moore
Sustainable Agriculture	R.S. Meena
Sustainable Agriculture: from Common Principles to Common Practice	Fritz J. Hani
The Organic Farming Manual	A. L. Hansen
The Soil-Plant System in Relation to Inorganic Nutrition	M. Fried

Sustainable Development of Dryland Agriculture in India

Editor

R.P. Singh

Director (Retd.) CRIDA, Hyderabad

61, Sardar Club Scheme,

Jodhpur - 342 001, India

Published by:
SCIENTIFIC PUBLISHERS (INDIA)
Jodhpur -

5-A, New Pali Road
P.O. Box 91
Jodhpur - 342 001 INDIA
Tel.: +91-291-2433323
E-mail: info@scientificpub.com

Print 2022

ISBN: 978-81-7233-405-5 (HB)
eISBN: 9789386102225

Visit the Scienfific Publishers (India) website at
http://www.scientificpubonline.com
Printed in India

FOREWORD

Improved dryland agriculture holds the key to sustainable food and livelihood security for over 50% of our population. The new paradigm of dryland agriculture calls for concurrent attention to the principles of ecology, economics, equity and employment. Land and water management are important components of sustainable farming practices in dryland areas. This book is therefore a timely publication since it brings together a vast array of information on all aspects related to the improvement of the productivity, profitability, stability and sustainability of dryland farming systems. We owe a debt of gratitude to Dr. R.P. Singh for the trouble he has taken to get such an useful book prepared.

M.S. Swaminathan

FOREWORD

Improved dryland agriculture holds the key to sustainable food and livelihood security for over 50% of our population. The new paradigm of dryland agriculture calls for concurrent attention to the principles of ecology, economics, equity and employment. Land and water management are important components of sustainable farming practices in dryland areas. This book is therefore a timely publication since it brings together a vast array of information on all aspects related to the improvement of the productivity, profitability, stability and sustainability of dryland farming systems. We owe a debt of gratitude to Dr. R.P. Singh for the trouble he has taken to get such an useful book prepared.

M.S. Swaminathan

PREFACE

This is the second book in the series on sustainable development, the first being 'Sustainable Development of The Indian Arid Zone—A Research Perspective' (Eds. R.P. Singh and Surendra Singh). The present book is devoted to sustainable development of dryland agriculture-the Indian experience. There are quite a few publications now available on dryland agriculture, but this book has two distinct features viz, 1. it gives coverage to almost all the important topics prescribed in the syllabus on dryland agriculture for post-graduate students of Indian agricultural universities, and 2. each chapter is written by a scientist having first hand knowledge of the subject, backed by many years of field experience.

The first three chapters deal with historical perspective of dryland agricultural research in India; dryland farming-problems, past progress, and future prospects. Chapters 4 and 5 give information on natural resources-climate and soils, and deal with soil degradation-related constrainsts in Indian rainfed zones. Soil salinity with special reference to drylands is dealt exhaustively in chapter 6.

All aspects pertaining to soil and water conservation, soil fertility management and fertiliser use, agronomic aspects of plant nutrient management, and integrated nutrient management, are discussed in chapters 7,8,9, and 10, respectively. Besides giving principles and practices, a comprehensive review is provided in each of the chapters mentioned above.

Water is the life line of all activities and progress in dry areas. Inorder to manage scarce water resources, it is necessary to have an understanding of moisture retention by soils and its availability, water losses and their control, ways of improving water use efficiency (WUE), water harvesting procedures, and principles and practices of protective irrigation. Chapters 11 to 15 fulfill this requirement. Chapters 16 and 17 are exclusively devoted to tillage, and improved agricultural implements for drylands.

Climate, soils, soil and water conservation, soil fertility management, curtailing water losses, improving water use efficiency, water harvesting, protective irrigation (life saving), tillage and use of improved agricultural implements, all aim at increasing productivity of crops grown in dry areas. Research experience gained on crop production aspects, efficient crops and cropping systems, and their efficient management, is set out in chapters 18 to 20. Aberrant weather situations and agricultural drought being very common in dry regions, crop management options are available to meet each situation and the same are given in chapter 21. Of late, significant advances have been made in oilseed and pulse crops production in the country, a very comprehensive treatment of which is provided in chapters 23 and 24 of this book. Oilseed and pulse crops are the predominant dryland crops and hence a special treatment has been given to these crops.

Not all drylands are suitable for arable cropping. Degraded and marginal lands predominate in dry areas. The most feasible, viable, and eco-friendly way to manage such soils is through alternate land use systems, including agro-forestry and pastures. The management options for such lands are given in chapters 24 to 26.

Weeds and diseases take a heavy toll. Changing scenarios of weeds and their management in dryland crops and cropping systems, and principal diseases of dryland crops and their management are discussed in details in chapters 27 and 28, respectively.

Post-harvest technology is important in the context of heavy losses that occur after crop harvest and during processing. The ways and means, and the modern machinery to be used, are described in chapter 29.

Wastelands and their rehabilitation assume importance in view of the dimensions of the problem and the potentialities they hold for production of fodder, fibre, timber and food. The technologies available for rehabilitation of wastelands are set out in chapter 30.

It is now well recognized at all hands that natural resources are best managed on watershed basis, more so in dry areas. Principles and practices relating to watershed management are given in chapter 31, together with exprience gained over the years in this field. Technological feasibility and economic viability should go together, if technology has to make a dent on production and productivity fronts. Economics of recommended dryland practices is presented in chapter 32. Inview of the special features and characteristics of dryland areas, extension strategy has to be different than irrigated areas and because of this, chapter 33 has its own importance.

To cap it all, sustainability of dryland agriculture is of paramount importance. The concepts, dryland practices and systems that can ensure sustainability are given in chapter 34, along with some important policy issues.

It is hoped that the principles, practices, and experiences contained in various chapters of this book will be found useful by students, researchers, and development workers. Suggestions for improvement of contents are most welcome.

Thanks are due to all the authors who have contributed chapters to this book, thereby sharing their experiences with the readers.

(R.P. Singh)
Director (Retd.) CRIDA, Hyderabad
61, Sardar Club Scheme,
Jodhpur—342001
INDIA

CONTRIBUTORS

1. Dr. R.P. Singh
61, Sardar Club Scheme
JODHPUR - 342 001

2. Dr. B.R. Hegde
Chief Scientist
AICRPDA
GKVK Campus
BANGALORE - 560 065

3. Dr. S.M. Virmani
Principal Agroclimatologist
International Crops Research Institute for the Semi-Arid Tropics
PATANCHERU PO - 502 324 A.P.

4. Dr. J.C. Katyal
Director
Central Research Institute for Dryland Agriculture
Santoshnagar, Saidabad Post
HYDERABAD - 500 659, A.P.

5. Dr. N.T.Singh
Director (Retd.)
C/o Central Soil Salinity Research Institute
Karnal - 132 001
HARYANA

6. Dr. Prabhakar Pathak
Principal Scientist
International crops Research Institute for Semi-Arid Tropics
PATANCHERU PO, A.P. - 502 324

7. Dr. S.K. Das
Project Coordinator (R)
Central Research Institute for Dryland Agriculture
Saidabad Post
HYDERABAD - 500 659

8. Dr. R.K. Aggarwal
Principal Scientist
Central Arid Zone Research Institute
JODHPUR - 342 003

9. Dr. Ram K. Gupta
C/o Chief Scientist
Dryland Research Centre (AICRPDA)
College of Agriculture
INDORE - 452 001
(M.P.)

10. Dr. J.P. Gupta
Principal Scientist (Resource Mgmt.)
Central Arid Zone Research Institute
JODHPUR - 342 003

11. Dr. C.J. Itnal
Chief Scientist
UAS - Regional Agricultural Research Station
BIJAPUR - 586 101
KARNATAKA

12. Dr. V.M. Mayande
Scientist
Central Research Institute for Dryland Agriculture,
Saidabad Post
HYDERABAD - 500 659, A.P.

13. Dr. N.K. Umrani,
Associate Dean
Mahatma Phule Krishi Vishwa Vidalaya
RAHURI - 413 712
(Maharashtra) Distt : Ahmednagar

14. Dr. M.V.R. Prasad
Director
Directorate of Oilseeds Research
Rajendra Nagar
HYDERABAD - 500 030 A.P.

15. Dr. D. Kumar
Scientist
Division of Arable Cropping
Central Arid Zone Research Institute
JODHPUR - 342 003

16. Dr. Md. Osman
Scientist
Central Research Institute for Dryland Agriculture
Saidabad Post
HYDERABAD - 500 659

17. Dr. R. Deb Roy
Director
National Research Centre for Agroforestry
Pahuj Dam, Gwalior Road
JHANSI - 248 001 UP

18. Dr. K.C. Singh,
Senior Scientist
Central Arid Zone Research Institute
JODHPUR - 342 003

19. Dr. A. Ramakrishna
Principal Scientist (Agronomy)
International Crops Research Institute for Semi - Arid Tropics
Patancheru Post - 502 324
ANDHRA PRADESH

20. Dr. S.K. Singh
Asstt. Professor (PP)
SKUAST - Regional Agrl. Research Station
R.S.PURA, Jammu-Tawi - 181 102
(J & K)

21. Dr. T.P. Ojha
Deputy Director General (Retd.)
C/o Dr. I.P. Abrol
Deputy Director General (SA & AF)
Indian Council of Agricultural Research
Krishi Bhawan
NEW DELHI - 110 001

22. Dr. Sriniwas Sharma,
Principal Scientist
Division of Resource Management
Central Research Institute for Dryland Agriculture
Saidabad Post
HYDERABAD - 500 659, A.P.

23. Dr. C. Prasad
Director
National Academy of Agricultural Research Management
Rajendra Nagar
HYDERABAD - 500 030 A.P.

24. Dr. Surendra Singh
Principal Scientist
Central Arid Zone Research Institute
JODHPUR - 342 003

25. DR. Y. V. R. Reddy
Principal Scientist
Central Research Institute for Dryland Agriculture
Saidabad Post
HYDERABAD - 500 659

CONTENTS

1

DRYLAND AGRICULTURAL RESEARCH IN INDIA — A HISTORICAL PERSPECTIVE

R.P. SINGH

INTRODUCTION

Dryland agricultural research in India has been a late starter. Serious efforts in this direction should have been made in the eighteenth century when the country faced a series of famines. Even the Famine Commission, appointed in 1880, in the face of frequent scarcities and famines, did not make a categorical recommendation for initiation of research for increasing and stabilizing the productivity of rainfed crops. A few protective irrigation projects were set up in Madras, Bombay and the erstwhile states of Mysore and Hyderabad, for the benefit of dry areas as a result of the commission's recommendations. This did not take the country any far.

A realization that the amount and distribution of rainfall tilts the balance of food production in India came as late as 1923 when the first systematic scientific approach to tackle the problem of dry farming areas was first made. Thanks to the vision of Dr. H.H. Mann, Director of Agriculture of the erstwhile Bombay state and Mr. C.V. Mehta, Minister of Agriculture, that research on dry farming for the scarcity tracts of Bombay state, was initiated. The work was entrusted to Mr. V.A. Tamhane, Soil Physicist, who after working on a small plot at Manjri Farm, near Pune, passed on the work to Mr. N.V. Kanitkar in 1926.

After working for a few years, Mr. Kanitkar realized that the work was vast and complex and confining research to one place/station was not of much significance. The need for a broad based research project on dryland agriculture was considered imperative by him. A comprehensive scheme of research for the solution of the problems of dry areas was drawn up by Mr. Kanitkar, which was supported by Mr. H.H. Mann and Dr. W. Burns. The scheme, sponsored and supported by Government of Bombay, got the financial support of the Imperial Council of Agricultural Research.

Dryland agricultural research in this country has two distinct phases from historical view point. One, relates to the pre-green revolution era, and the other, to the post-green revolution period. Some of the salient features of each are given here under:

PRE-GREEN REVOLUTION ERA

India witnessed green revolution in the sixties, when high yielding, short-statured, and fertilizer responsive varieties/hybrids of cereals made a significant dent on food production in irrigated areas. However, as pointed out earlier, a comprehensive project on dryland farming was launched by the Imperial (now Indian) Council of Agricultural Research at five centres, viz., Sholapur and Bijapur in 1933, Hagari and Raichur in 1934, and Rohtak in 1935. The research programme at these stations had a total life span of about 10 years, the programme having terminated in 1943-44. The work carried out at five experimental stations consisted of analysis of rainfall and other climatic factors, physico-chemical characteristics of the soils in the dry areas, physiological studies of important millets grown in the dry tracts, and agronomy or soil management aspects of crop production. As a result of these research efforts made under agronomy/soil management studies, a series of useful dry farming practices, commonly known as the Bombay Dry Farming Practices, the Hyderabad Dry Farming Practices, and the Madras Dry Farming Practices were evolved. The common features of these practices were: contour bunding of fields for prevention of run-off and erosion, deep ploughing once in three years for better water intake and storage, use of FYM to supply plant nutrients once in three years in ploughed up fields, use of low seedrate and wider spacing, interculturing of crops for efficient use of limited moisture, and adoption of mixed cropping and crop rotations. The basic ingredient of these practices lay in soil and moisture conservation measures. Some other important findings of the studies made during 1933-34 to 1943-44 were :

1. A vegetal cover on land led to reduction in run-off, the groundnut crop effectively reduced the runoff by 50 per cent.
2. Deeper cultivation showed some reduction in runoff as compared to shallow cultivation.
3. Special practices like scooping showed nearly 50 per cent reduction in runoff.
4. All rainfalls exceeding 5 cm in a day cause runoff.
5. Bare fallow resulted in heavy erosion, the soil loss ranging from 45 t/ha at Sholapur to 54 t/ha at Manjri.
6. The soil removed by erosion was richer in all plant food ingredients than the original soil.
7. Evaporation accounts for the largest proportion of dissipation of water received from rainfall and absorbed by the soil. The second significant loss was by way of underground drainage - the drainage loss (water and nutrients) being more in shallow soils as compared to deep soils. Bare fallow lands lost more water by drainage than cropped lands.
8. Amongst the treatments which were found to minimize the loss of soil moisture, bunding, listing or scooping are most important. Preparatory tillage or ploughing was found to be more effective than harrowing in increasing the moisture in surface layer in majority of soils. Repeated harrowings or interculturing forms soil mulch and led to conservation of more moisture in the sub-surface layers than in unstirred soils. Manuring did not influence the moisture content of the soil. However, application of manure increased yields because of higher availability of moisture.
9. Fallowing resulted in increased soil moisture in majority of soils which have sufficient depth for conserving soil moisture at least in a normal season.
10. No evidence was available to support the theory of capillary rise of soil moisture under field conditions of the scarcity tracts of India.

However, the practices listed above were able to increase the yields of dryland crops by 15 to 20 per cent only. This did not enthuse the farmers to adopt them as the pay off was marginal only. The main constraint was the absence of suitable biological material responsive to management and inputs.

It was unfortunate that research on dry farming was discontinued in 1943-44, just after a decade of its initiation. Somebody could have realized that it is not easy to tackle the problems of dry areas - vast, varied, and complex as they are. Dryland research is much more complicated and challenging than carrying out research in irrigated areas. Soil erosion, among other factors, poses a serious threat to increasing productivity in dry areas. It was perhaps in this context that the Indian Council of Agricultural Research established eight Soil Conservation Research Centres in 1954. It was yet another attempt to improve the productivity of drylands. The mandate of these research centres was, however, limited. Nonetheless, the research achievements clearly brought out the need for new biological material for increasing the productivity of drylands. Climatological analysis which was first attempted in the Deccan post rainy season brought recognition of the fact that only the varieties that mature within a period of 100 to 120 days give high yields on drylands with limited moisture availability. The fact remains that Soil Conservation Research Centres addressed themselves to the local problems relating to runoff and soil loss but did not pay attention to the problems of dry farming *per se*. A coordinated effort to study the climate, soils, crops, animals, trees, and agricultural implements, and all that which goes to make dry farming a wholesome unit, was conspicuous by its absence. Thus, dryland research continued to be a low priority area till 1970.

POST-GREEN REVOLUTION ERA

Although the era of planned development began as early as 1950-51, the dry farming research did not attract the attention of the planners till the increasing socio-economic imbalances between irrigated and dryland agriculture sectors became serious as a consequence of green revolution beginning in 1965-66. It was the Fourth Plan (1969-74) document which emphasized the urgent need for creating conditions whereby the dryland farmers could participate meaningfully in the agricultural development process. In recognition of this need, the Indian Council of Agricultural Research (ICAR) thought of preparing a comprehensive research project on dryland agriculture. A handsome amount of Rs 200 million was earmarked for dryland research and development in the Fourth Plan. An All India Coordinated Research Project for Dryland Agriculture (AICRPDA) was thus born. A unique feature of this coordinated project was that Integrated Dryland Development Projects, started simultaneously in 1970, were linked with the research centres to give back-up support and for testing the technology in the farmers' fields. Since the technology was in the making, it couldn't be tested on a large scale on farmers' fields, although demonstrations for a few dry farming practices were laid out.

ALL INDIA COORDINATED RESEARCH PROJECT FOR DRYLAND AGRICULTURE

The AICRPDA is a multi-disciplinary project encompassing the disciplines of Agronomy, Soil Science, Plant Breeding and Agricultural Engineering. The project was launched in 1970 at 23 cooperating centres located in the 8 major agro-climatic regions of the country, representing all major soil types and cropping patterns. The financial allocation was Rs 14.75 million for the remaining part of the Fourth Plan period. The Government of Canada came forward to assist the Government of India in its efforts to increasing and stabilizing production and productivity in the dry areas.

The Coordinating Cell of the project was located at Hyderabad, Andhra Pradesh. To begin with, there were 15 main centres and 8 sub-centres. In 1984, 5 out of 8 sub-centres were upgraded to the level of main centres. An additional special centre (twenty fourth) was set up at the Indian Agricultural Research Institute, New Delhi, with the hope that it will carry out path-breaking research in the fields of Plant Physiology and Agro-meteorology. Later, it was merged with the Coordinating Cell of the project. In 1985, Dehradun centre was dropped.

Operational Research Programme

The first Operational Research Programme on Dryland Agriculture was launched at Indore (Madhya Pradesh) with U.K. collaboration. It was augmented with the addition of four more CIDA (Canadian International Development Agency) sponsored ORPs in 1976. In 1984, four additional ORPs were added to the list. Currently, nine Dryland Operational Research Projects are operating at Indore, Bangalore, Hyderabad, Sholapur, Ranchi, Hoshiarpur, Anantpur, Hisar and Arjia (Rajasthan).

Training in Dryland Agriculture

With the initiation of Dryland ORPs in 1976, the farmers training became institutionalized in 1977 with the setting up of a Krishi Vigyan Kendra (KVK) with facilities located at Hayatnagar Research Farm. In 1978, the Trainers' Training Centre (TTC), established as a part of AICRPDA, offered its first course to extension personnel involved in farmers' training. In 1979, a specific transfer of technology programme, Lab to Land, was added.

Watershed Development Programme

The year 1983 saw the beginning of yet another programme for development of dryland agriculture on watershed basis. The work began in 47 model watersheds, the average size of each being 700 to 800 hectares. A sum of Rs 25 lakhs for a period of five years was earmarked by the Government of India for development of each watershed on scientific basis. The programme consisted of soil and water conservation measures, crop management, and alternate land use systems. Of the 47 model watersheds, 30 watersheds were attached to AICRPDA, and later with CRIDA, for technical backstop, including training of personnel and monitoring of periodical progress.

CENTRAL RESEARCH INSTITUTE FOR DRYLAND AGRICULTURE

In recognition of commendable research work done by the scientists of AICRPDA and in view of the national urgency and importance of developing more stable and remunerative farming systems for drylands of the country, the Indian Council of Agricultural Research elevated the status of the Directorate of AICRPDA to that of a full fledged institute, christened as CRIDA. On April 19, 1985, the institute was formally inaugurated. The main purpose of establishing CRIDA was to carry out 'lead research' in the area of dryland agriculture, leaving location-specific problems and their solutions to AICRPDA.

Collaborative Research in Dryland Agriculture

Indo-Canadian Dryland Research

The period 1970 to 1987 saw the beginning and end of Indo-Canada collaboration in dryland research. The project (AICRPDA) was headed by the Project Director-cum-Assistant Director General on the Indian side. As the name suggests, the Canadian Scientists seconded by Agriculture Canada with Canadian Principal Advisor as head of the team, had an advisory role to play. The bilateral collaboration between the two Governments was signed on August 25, 1970. The important components of the agreement were: provision of Canadian experts in the priority areas of research, supply of equipment for both laboratory and field research, and training of Indian scientists in Canada and third countries.

This collaboration had three distinct phases, viz., phase I (1970-75), phase II (1976-80), and phase III (1981-87).

Phase I (1970-75) : During the period, a service laboratory for soil chemical analysis was established at the headquarters. Main centres were provided with equipment for soil physical analysis. Nine main centres were equipped with tillage-seeding-fertilizer application implements and machinery.

Research conducted during the first phase (1970-75) confined itself mainly to the identification of efficient crops and varieties for different regions, crop geometry and optimum plant populations, optimal time for planting crops, fertiliser use for dryland crops, basic soil problems in water intake and storage, weed control, tillage and seeding practices. The most fruitful areas of research relate to dates of sowing, fertilization, crop substitution and introduction. For this purpose, Type Experiments were initiated in the field of soils and agronomy, crop sciences and agricultural engineering with a view to generate some basic information on the above aspects. In all, eight experiments in soils and agronomy were started. Assessment of breeding material, testing of pre-release materials and crop introduction were the experiments included under crop sciences. Agricultural engineering experiments consisted of studies on increasing the infiltration and moisture storage, tillage practices, vertical mulching, sowing and fertilization, water harvesting and minimal irrigation.

Phase II (1976-80) : During the later part of the programme, i.e. the second phase (1976-80), the emphasis shifted from type experiments to location specific projects. An attempt was made to synthesize 'individuals' into an 'integrated whole'. A systems approach was thus brought to bear on the programme.

The emphasis was on developing cropping systems (both intercrop and sequential) that could ensure stable optimal yields and maximize profits. Full exploitation of the agro-climatic resources was the main objective. Towards further refinement of the systems, identification of genotypes, manipulation of sowing and harvesting dates, plant populations and fertiliser use received attention. Pulses and oilseed crops formed important components of the cropping systems. Stable and economically viable intercrop systems with potentials of higher productivity, income and land use efficiency were evolved. Suitable double crop systems were also evolved. These improvements provided additional employment opportunities and much better nutritional status.

Efficient management of inputs continued to receive emphasis. A new concept that fertiliser use should be tailored to the available soil moisture at least in post rainy season areas was evolved. Precision seeding and placement of fertilizers in relation to moisture in the seeding zone received better appreciation. Bullock drawn seed-cum-fertilizer drills or attachments were designed and developed.

Water harvesting and recycling was yet another area which received emphasis during the period. Strategies for aberrant weather situations were further refined. Year-round tillage as a means to control weeds and conserve soil moisture was crystallized into a concept of great significance to drylands.

Research in the areas of agricultural economics and pasture management received impetus. Agro-economic research was initiated at 16 centres. Pasture research was taken up at 5 centres. Alternate land uses for marginal lands started receiving attention.

Phase III (1981-87) : Two broad areas of research were identified for major thrust in the eighties. They were: (i) alternate land use systems for maximizing biomass production and income, and (ii) water harvesting and reuse for enhancing crop production and stabilization of income.

Refinement of cropping systems, increasing fertiliser-use efficiency, management of natural resources (soil, water, climate, etc.) on a watershed basis and watershed hydrology were yet other areas which received attention. CRIDA and AICRPDA scientists assisted the State Governments and State Agricultural Universities in the selection of sites, preparation of master plans, training of personnel, monitoring and evaluation, and providing technical back-stop to 30 model watersheds. CRIDA thus emerged as the pioneer institution in providing an effective linkage between research and development in the area of dryland agriculture.

ICAR-ICRISAT COLLABORATIVE RESEARCH PROJECTS

The Project Director-cum-Assistant Director General (Dry Farming) was identified to act as Member-Secretary of the ICAR-ICRISAT Policy Committee meetings held annually since the beginning of the project in 1970. A meaningful collaboration between CRIDA and ICRISAT on one to one scientist basis started from 1986 onwards in areas of biological nitrogen fixation, agroforestry systems, land configurations for soil and moisture conservation and economics of watershed technology. The Resource Management Programme of ICRISAT, in main, collaborated with the on-going research programme of CRIDA. CRIDA and ICRISAT also joined hands in management of dryland agriculture on watershed basis in two of the 30 ICAR model watersheds, monitored by CRIDA.

Additional collaboration was initiated in 1989-90 on evaluating the vegetative barriers, studying the behaviour of phosphorus on vertisols and investigations on perennial pigeon pea.

INDO-US COLLABORATION IN DRYLAND AGRICULTURAL RESEARCH

The Indo-US collaboration started in 1988, under the agreement of Indo-US Joint Sub-Commission on Agriculture. The Central Research Institute for Dryland Agriculture, Hyderabad; Central Arid Zone Research Institute, Jodhpur; and Central Soil and Water Conservation Research & Training Institute, Dehradun, are collaborating with United States Department of Agriculture in areas of enhancement of fertilizer use efficiency, germplasm enhancement in castor for drought tolerance and in groundnut for higher N_2 fixation, crop simulation modelling, and farm mechanization. Few agricultural universities having cooperating centres of dryland agricultural research are also participating in this programme. Comprehensive joint research projects have been drawn up and a mutually agreed programme, having specific task-frame and time-frame, is under implementation. The emphasis under this programme is on better understanding of the various processes that govern productivity of crops under dryland conditions and carry out 'lead research' for finding solutions to some of the vexed problems evading solution so far.

Currently, CRIDA-US collaborative research projects, finalized in November 1989, comprise of five projects, viz., (i) Enhancing fertiliser use efficiency in conjunction with residue management in dryland crops and cropping systems, (ii) Germplasm enhancement for drought tolerance-evaluation of castor *genotypes* for drought tolerance and yield, (iii) Evaluation of genotypic variability in legumes under drought for higher N fixation, NHI, N-use efficiency and yield, (iv) Crop simulation modelling in drylands, (v) Development of suitable systems/implements for efficient use of available draft power for maximization of resources.

Resume of Current Research Activities

Besides other objectives, CRIDA is mainly contributing to the understanding of crop growth and yield processes more especially under moisture stress conditions and development of strategies for improving productivity and sustainability of crop production under dryland farming. Currently, CRIDA is engaged in evaluation of alternate intercropping systems, forage production under drylands, interaction of arable crops with tree component in agri-silviculture system, possible utilization of organic N sources for castor. Basic studies on the effects of water stress and rewatering on physiological processes in dryland crops are in progress. Studies on nitrogen assimilation in rainfed sorghum and on crop simulation modelling are being continued.

In the arena of resource management, studies on agro-climatological characterization of dry farming regions, crop-weather relationships, sustainability of the recommended practices, interaction between nutrients and water use, residual effects of tillage and fertiliser application, etc. are being continued, besides carrying out experiments on agroforestry systems with emphasis on quantifying the effects of tree component on crops. Economics of different land use systems in dryland farming is currently receiving attention.

EPILOGUE

From the above historical perspective, it would be evident that the history of dryland agricultural research in India is full of significant events and achievements. During a short span of two decades (1970-1990) of systematic research, AICRPDA and CRIDA have made a noteworthy place for themselves on the scroll of agricultural research in this country. Sound foundations have been laid. Future success will depend on skilful management and capitalization of natural resources, scientific talents, and collaborative research potentialities.

2

DRYLAND FARMING : PAST PROGRESS AND FUTURE PROSPECTS

B.R. HEGDE

DEFINITION OF DRYLANDS

Dry farming as a practice antedates history, but the usage of the term in its present form and meaning probably started in Utah in 1863. Dr. John A Widtsoe, one of the earliest to take up to dry farming research defines the term as "the profitable production of useful crops without irrigation on lands that receive annually a rainfall of 20" or less "this he says 'may be extended to 30" with unfavourable distribution of rainfall". According to the definition given in the Encyclopedia Britannica "Dry farming is farming without irrigation in areas where the moisture supply is the chief limitation to crop production". Dry farming and Dryland Farming have been used synonymously to indicate similar farming situations.

It is important to distinguish between dryland agriculture in particular and rainfed agriculture in general. Clearly they both exclude irrigation. The emphasis in rainfed agriculture is often on disposal of excess water, maximum crop yields, high level of fertilizer inputs and water erosion constraints. In dryland agriculture, emphasis is on water conservation, sustainable crop yields, limited inputs for soil fertility maintenance, and both wind and water erosion constraints. Drylands generally receive less than 750 mm of annual rainfall in arid and semi-arid regions. However, in the present Indian context, drylands mean unirrigated rainfed areas.

After careful consideration of several definitions, Steiner et al. [1988] consider aridity index concept of the United Nations Conference on Desertification based on the balance between precipitation (P) and evapotranspiration (ETP) to be appropriate for wide scale adoption. According to this definition the areas with P/ETP ratio between 0.03 and 0.20 are arid and areas with the ratio between 0.2 to 0.5 are semi-arid. Another 'influential' definition is based on 'growing period concept' of the FAO. Areas having a growing period between 1 and 74 days are classified as arid and those with a growing period between 75 and 119 days are semi-arid. The growing period is the number of days during a year when precipitation exceeds half the potential evapotranspiration, plus a period to use an assumed 100 mm of water from excess precipitation (or less, if not available) stored in the soil profile.

ICAR while establishing the dryland centres in different agroclimatic zones of the country in 1970, used the simple formula of Thornthwaite [1955] for estimating the moisture deficiency index.

$$\text{Moisture Deficiency Index} = \frac{P-PE}{PE} \times 100$$

where, P = Precipitation and
PE = Potential evaporation.

According to this classification the limit for semi-arid climates is given by the values of the aridity index varying from -33.3 to -66.7.

HISTORY OF DRYLAND FARMING

Though crop production under dryland conditions was forced on the people from 1850 onwards, the scientific attention was first received in USA by about 1863. Some emigrants in Utah had turned upon their fields the alkali water from Malad Creek and naturally the crops failed. In desperation they took up to cultivation of wheat on drylands. From that day, dry farming has been an established practice in that portion of the great salt lake city. The agricultural research in the USA picked up from 1862 with the Morill bill making it possible to establish the agricultural colleges and in 1887 Hatch act providing for establishing agricultural experimental stations in all the states. Colorado State initiated dry farming experiments and established a substation at Cheyenne Wells in 1894.

The later part of 19th century and the early part of 20th century saw a tremendous impetus on dryland farming in USA and Canada. Early writers like Hardy W. Campbell (Soil Culture Manual, 1904), William McDonald (Text Book of Dry Farming 1909), John A. Widtsoe (Text book on dry farming, 1911), inspired by the works of Jethro Tull in England advocated repeated cultivation for moisture conservation. In fact tillage was considered to be manure. In the words of Widtsoe, 'Conservation of soil moisture depends upon vigorous, unremitting, continuous stirring of the top soil. Cultivation, cultivation and more cultivation! must be the war cry of the dryland farmer who battles against the water thirst of an arid climate!

The dry farming in USA and Canada was synonymous with wheat-fallow system. The years between 1909 and 1916 had favourable precipitation with high price for wheat. Under the extra stimulation of propaganda by rail roads, which had large land holdings and needed settlers to develop trade along their line and by real estate dealers and with easy credit available, large areas were broken up and planted to wheat. No attention to conservation systems was given and people with no previous experience took to dry farming. Campbell's soil culture manual was an instruction book on dry farming during these good precipitation years.

From 1917 onwards upto 1923 there was a succession of dry years and many unqualified dry farmers went broke. The major cause for the failure was considered to be recommending one specific system of dry farming than diversification.

During the first and the second world wars, a period of wet years, especially in the plains, coupled with an unlimited market and unusually high prices for wheat led again to big expansions of dry farming, much of it by 'Suitcase farmers' who did not live on their land. The practice of deep ploughing followed by frequent intercultivation, during the dry years, led to the rapid decline of soil organic matter, deterioration of soil structure, accelerated soil erosion and often complete crop failures. The grain bowl of USA turned to 'dust bowl' by 1932. Farmers were ill equipped to deal with this harsher environment. At this time the US Government responded with its land utilization programme, under which the most severely eroded land was purchased by the USDA and removed from production. The USDA's Soil Conservation Service instituted technical programme, which covered the land with drought resistant cover crops and returned to grazing subsequently.

History of Dry Farming in India

Dry farming was followed in India from time immemorial, but scientific study was not initiated till 1923. A.K.Y.N. Aiyer [1917] listed the important practices followed by the farmers and found that the practice was in conformity with that of Campbell's teachings. Field bunding, fall ploughing, frequent intercultivations, drill sowing and growing drought resistant crops like *ragi, jowar, bajra, haraka* are some of the important practices listed by him.

Frequent famines prompted the Government of India to appoint a Famine Commission in 1880 which made among other recommendations to establish protective irrigation projects for the benefit of dry and scanty tracts. It also recommended the revival of the department of agriculture of the Government of India and establishment of the department in all the provinces.

Research efforts made during 1930, were mainly in terms of improving crop yields. Kanitkar et al. [1968] summarised the results of these efforts as Madras/Bombay/Rohtak dry farming methods which included:

— Bunding with bund former to conserve soil and water.

— Deep ploughing once in 3 years for better intake and storage of rain water.

— Use of farm yard manure (FYM) to supply plant nutrients.

— Contour cultivation and contour sowing.

— Use of low seed rate.

— Intercultivation for weed and evaporation control.

These practices gave 15-20 per cent increase in yield over a base yield of 200 to 400 kg/ha. In other words, the actual yield increase varied from 30 to 80 kg/ha. This marginal increase did not enthuse the dryland farmers to adopt the practices.

By the mid 1950s, the emphasis was shifted to soil management. Soil Conservation research and training centres were established in eight locations. The programmes focused on contour bunding. However, such an effort sometimes gave negative results, particularly in vertisols. Water congestion and washout were the main problems. Even when yield increases were observed, they were again not more than 15-20 per cent over the base yield of 200-400 kg/ha.

The importance of shorter duration crops to fit into the moisture availability period was recognized during 1960s. And it was during the mid-1960s that the answer was available through hybrids and high yielding varieties of crops, particularly coarse cereals. These high yielding hybrids and varieties were not only responsive to fertilizers but also to management.

Through the establishment of All India Coordinited Research Project for Dryland Agriculture (AICRPDA) in 1970, the emphasis was shifted to multidisciplinary approach to tackle the problems from several angles. Similar efforts were initiated at the International Crops Research Institute for the Semi-arid Tropics (ICRISAT) at Hyderabad during 1972.

Important events in the development of dryland farming in India

1920 Scarcity tract development given importance by the Royal Commission on Agriculture.

1923 Establishing Dryland Research Station at Manjri

1933 Research Stations established at Bijapur and Solapur

1934 Research Stations established at Hagari and Raichur

1935 Research Station established at Rohtak (Punjab)

1942 Bombay Land Development act passed

1944 Monograph on dry farming in India by N.V. Kanitkar

1953 Establishing Central Soil Conservation Board

1955 Dry Farming demonstration Centres started

1970 23 Research Centres established under AICRPDA

1972 Establishment of ICRISAT

1976 Establishment of Dryland Operational Research Projects

1983 Starting of 47 model watersheds under ICAR

1984 Initiation of World Bank assisted Watershed Development Programmes in four States. Establishing Dryland Development Board in Karnataka

1985 Birth of Central Research Institute for Dryland Agriculture at Hyderabad

1986 Launching of NWDPRA programmes by Government of India in 15 states.

TECHNOLOGICAL ADVANCES

The basic problems of dryland farming and remedies are listed by Widtsoe [1911] in his book on dry farming:

— Storage of all the available rain water in the soil.

— Retention of the moisture in the soil until it is used by the crops

— Prevention of direct evaporation of soil moisture during the growing season

— Regulation of the amount of water drawn from the soil by the plants

— Choice of suitable crops and varieties.

— Application of suitable crop treatments.

Modern research has further helped to understand the problems more thoroughly and the studies in the AICRPDA and ICRISAT have come out with new recommendations.

— Offseason and contour tillage in light textured red soils (Alfisols and related soils) and optimised minimum tillage in self-ploughing vertisols.

— Early sowing to fully utilize the moisture availability period and make use of the mineralized nutrients. Pest avoidance is another advantage of early sowing.

— Crop residue incorporation to improve the nutrient status and moisture retention capacity of soils.

— Crop and varietal selection based on the moisture availability period.

— Contingent crop planning to meet the vagaries of weather.

— Moderate use of fertilizers.

— Use of organic and soil mulches to avoid rapid evaporation of soil moisture.

— Adoption of modern tools for sowing, intercultivation, harvesting and post-harvest operations.

— Adoption of optimum plant population (not low plant population), timely weed control and need-based plant protection measures.

— *In situ* moisture conservation, water harvesting and recycling.

— Adoption of location specific intercropping and multiple cropping practices.

— Adoption of suitable agro-forestry and agri-horti systems especially in marginal lands. Vegetative methods of erosion control using grasses like *Vetiveria zizanioides* in place of earthen bunds.

Recent efforts are for stabilizing the production at a fairly high level rather than maximizing the yield at 'any cost'.

WATERSHED DEVELOPMENT — A HOLISTIC APPROACH

Rainfall is the main source of surface water, and its conservation is essential for successful crop production on drylands. As water moves down the slope, from ridge to the valley, its management also should start at the ridge and extend to the valley. Both arable and non-arable areas are to be considered together as the management of water in one area will influence the other. Forest, pasture land, crop land and wastelands should be treated as interlinked units of a hydrological entity. Watershed approach in dryland development is essential from yet another angle of effective resource utilization. Wastelands and denuded forest areas have a tremendous untapped potential for developing horticulture, pasture land and forest trees to provide fruit, fodder and fuel. Establishing vegetative cover to such over-exploited areas will go a long way in reducing the soil erosion, extending the life of reservoirs and restoring the ecological balance.

The concept of dryland development on watershed basis materialized with the identification of 47 model watersheds by the ICAR for resource management during 1983, followed by four world bank assisted watershed development programmes in Karnataka, Maharashtra, Andhra Pradesh and Madhya Pradesh. During the same year, the Government of Karnataka initiated watershed development programmes in all the 19 districts managed by Dryland Development Boards. The convincing success in these watersheds led to the acceptance of the principle by the Government of India to implement all dryland development programmes on watershed basis. During the year 1986, the Government of India launched the National Watershed Development Programme for Rainfed Areas in 99 districts in 15 states at a total cost of Rs 239 crores with the cost to be shared equally by the central and state governments. The programme is accepted for implementation on a larger scale in the eighth five year plan.

The basic objectives of integrated watershed development are :

i) To conserve the basic resources of soil, rain water and vegetation,

ii) To achieve higher biomass production both in arable and non-arable areas,

iii) To impart stability to crop yields through proper rain water management, restructuring of cropping patterns and ensuring proper land use,

iv) To enhance the income of individuals through adoption of alternative enterprises

v) To restore ecological balance.

Keeping the above objectives in view the identified components of watershed development are:

— Treatment of arable as well as non-arable area for effective *in situ* moisture conservation.

— Implementation of programmes like afforestation, horticulture, agroforestry, pasture land development, animal husbandry and aquaculture in stored water.

— Identification of a sound crop production system and its implementation through the involvement of development departments and input agencies.

— Developing suitable infrastructural facility and people's organizations to maintain the developed resources.

The conceived objectives and the adopted approaches in watershed development programmes indicate a NEW FARMING SYSTEM integrating several components which lead to sustainable agriculture.

It is very interesting to note that several agencies like DANIDA, German Development Bank and World Bank have come forward to assist the watershed development programmes in the state.

FUTURE PROSPECTS

Much of the increase in food production in the recent past is estimated to be mainly due to irrigated areas. Since 1950, the extent of irrigated land in the world has increased from 94 million ha to about 220 million ha. During the 1980s, however, the rate of irrigation development has dropped materially and is presently less than 1 per cent per year, whereas the world population is increasing at 1.7 per cent per annum. Irrigated land accounts for 18 per cent of the cultivated land but produces 33 per cent of the food. The cost of irrigation and drainage development of new systems averages over $5,000 per hectare and can be as high as $10,000. Therefore, it is inevitable that in future the additional food has to come mainly from the drylands [Unger et al., 1988].

In India only 43 m. ha of land out of the cultivated area of 143 m. ha was irrigated by 1986-87, leaving the remaining 100 m. ha as rainfed. According to experts, even when the ultimate irrigation potential is reached, 55 per cent of the net sown area will be still, rainfed. The contribution of rainfed agriculture in India is about 45 per cent of the total foodgrain, 75 per cent of oilseeds, 90 per cent of pulses and about 70 per cent of cotton. By the end of the century the contribution of drylands will have to be 60 per cent if India is to provide adequate food to 1000 million people. To achieve this target, tremendous efforts both in the development and research fronts are essential.

(a) *In situ moisture conservation* : Newer and acceptable methods for *in situ* moisture conservation will have to be developed instead of entirely depending on bunding programmes for moisture conservation. Vegetative barriers on contour cannot only be economically viable alternatives to earthen bunds for soil conservation but can also act as effective key lines for contour cultivation. Identification of locally available grasses and suitable methods of their establishment need to be studied on priority basis.

(b)*Nutrient management* : Eventhough, the consumption of fertilizers in the country has increased to 12.2 million tons by the year 1992-93, about 80 per cent of it is used in crops under irrigated conditions. However, there is a good response to the application of fertilizers under dryland conditions which is as high as 23 kg per kg of nutrient under favourable situations. One of the sure ways to increase the yields under dryland conditions is to step up the consumption of fertilizers at least to about 50 kg total nutrients per ha.

Soil organic matter : Increasing and maintaining soil organic matter becomes almost impossible especially under tropical conditions when the crop residue is needed both for feeding the animals and for fuel purposes. Under extreme conditions the soil organic matter content drops to the point where the soil becomes useless and soil erosion becomes rampant. This is the final stage in land degradation.

Biomass production : Fuel, fodder and fibre can be easily produced from the waste lands in the arid and semi-arid conditions. With the per capita cultivable area falling to 0.15 ha by 2000 A.D. there is no alternative but to harness the eroded waste lands for biomass production. Identifying suitable vegetation and the methods of their establishment should be a priority item in the future of our dryland research. Covering the wastelands with vegetation will be inevitable to reduce the soil erosion.

Water harvesting and recycling : Water harvesting is one of the most promising and technically feasible recommendation even in low rainfall areas with 15-20 per cent of the total rainfall lost as runoff. At the same time the methods now suggested for seepage control, lifting and irrigation are often uneconomical from the simple calculations of cost and benefit. Research in the directions of cost effective sealants and manually operated lifting devices need to be given proper attention. Acceptable methods of harnessing wind and solar energy should receive priority.

Agricultural implements : When returns are taken into consideration, the labour in drylands is costly. Time and labour saving devices for land preparation, sowing and intercutivation can improve the crop yields by achieving the much talked timeliness and precision under dryland situations. Implements operated by human labour or single bullock are likely to be more acceptable to the resource poor dryland farmer.

Risk minimization : A major problem in dryland areas is that a single technological adoption may not increase the yield, and although total package improvements may be economically beneficial over a cycle of years, it may not be advantageous in low rainfall years. Therefore, the credit policy in the drought prone areas should be to recover the loan over a cycle of good and bad years but not during unfavourable years. The cyclical credit policy of NABARD adopted in Chitravathy Watershed on experimental basis should be extended to other watersheds.

Sustainability concept : The intrinsic land value should always appreciate to sustain the required productivity. We cannot afford depreciation of natural resources. The scientists and the Government must develop criteria to detect the rate of land degradation to evaluate the impact of degradation on productivity and to determine the population supporting capacity of their land [Steiner et al., 1988]. Implementation of policies that will ensure that lands are not expected to produce beyond their capacity is a major long term challenge in drylands.

REFERENCES

Steiner, J.L., Day, J.C., Papendick, R.I., Meyer, R.E. and Bertrand, A.R. 1988. Improving and sustaining productivity in dryland regions of developing countries. *Advances in Soil Science* 8: 80-122.

Unger, P.W., Jordan, W.R., Sneed, T.V. and Jensen, R.W. 1988. In: *Preface to the Proceedings of the International Conference on Dryland Farming* held at Amarillo/Bushland, Texas, USA, August 1988.

3

PROBLEMS AND PROSPECTS OF DRYLAND AGRICULTURE IN INDIA

R.P. SINGH

INTRODUCTION

The problems relating to dryland agriculture in India are more numerous, varied and intricate as compared to the temperate drylands. Tropical drylands are different from temperate drylands in three respects :

1. Tropical drylands have much less carry-over of moisture from one season to other.
2. The intensity of rainfall is very high, leading to severe soil erosion.
3. The atmospheric evaporative demand in tropics is several times more than that of temperate drylands.

Three-fourths of our arable land (143 m ha) is rainfed. Food production is tied up with amount and distribution of rainfall. More than 90 per cent of the area under sorghum, groundnut, and pulses is rainfed. In case of maize and chick-pea; 82 to 85 per cent area is rainfed. Even 78 per cent of cotton area is rainfed. In case of rapeseed/mustard, about 65.8 per cent of the area is rainfed. Interestingly, but not surprisingly, 61.7, 44.0, and 35.0 per cent area under rice, barley and wheat, respectively, is rainfed. In terms of production, rainfed agriculture accounts for more than 40 per cent of the total food grains production, nearly 75 per cent of all oilseeds, 90 per cent of the pulses and 70 per cent of cotton. With the current pace of irrigation development, it is assumed that the gross irrigated area is likely to increase to 75 x 10^6 ha by the year 2000. This means that more than 55 per cent of the gross cropped area will continue to be farmed under rainfed conditions.

CHARACTERISTICS OF RAINFED FARMING

Rainfed farming in India is practised under a wide range of climate and soil conditions, the main features of which are discussed below.

Rainfall : Based on average annual rainfall, the country can be divided into three zones : low (<750 mm), medium (750-1150 mm) and high (>1150 mm) rainfall zones. The rainfed area is nearly equally distributed in the above three zones.

Nearly 80 per cent of annual rainfall is received during the rainy season, June to September. There is lot of variation in the amount of rainfall in arid and semi-arid regions. For instance, coefficient of variation varies from 20 per cent at Hyderabad receiving an annual rainfall of 792 mm to 43 per cent at Jodhpur receiving an annual rainfall of 382 mm. There is considerable variation in the quantum of rainfall from place to place. The coefficient of variation increases with the decrease in annual rainfall.

The total rainfall, its distribution, potential evapotranspiration, and soil water storage govern the length of the growing season at a particular location. The growing period varies from 60 to 300 days, depending on the factors noted above.

The amount of rainfall, but more importantly it's distribution, is more critical for drylands. They govern the productivity of rainfed crops in India. The main types of weather aberrations encountered are (1) late onset of monsoon, (2) early withdrawal of monsoon, and (3) long gaps in rains.

MAJOR SOIL GROUPS ENCOUNTERED IN RAINFED AREAS

Besides the climatic features, different soil types are encountered in dryland areas with problems of varied nature. The alluvial soils occupy the largest area. The problems of dryland agriculture are not so acute in these soils as they are in black and red soils. The severest problems are encountered in vertisols and alfisols and related soils. Black (vertisols) and red (alfisols and related soils) are mostly distributed in central and south India. The coastal areas have alfisols, laterite and lateritic soils.

Vertisols : Vertisols (black soils) occupy about 22 per cent of the geographical area. Annual precipitation ranges from 500 to 1500 mm. These soils are characterised by high clay content (30-70%), and have high available water holding capacity. These soils are highly erodible, and produce large amounts of runoff. Provision of surface drainage is essential in medium and high rainfall areas for improved productivity. The soils are low in nitrogen and often in available phosphorus.

Alfisols : Alfisols are commonly termed red soils. These soils occupy about 20 per cent of the geographical area. Annual rainfall ranges from 750 to 2000 mm. These soils are characterized by moderate (10 to 20%) clay and low organic matter content, poor nutrient status, particularly with respect to nitrogen, phosphorus, sulphur, and calcium. Have low water retention capacity due to shallow depths. Frequently have a compact subsoil, are prone to erosion and crusting, and produce large volumes of runoff.

Aridisols : Aridisols are found in low rainfall regions. These soils are characterized by low clay content, low water retentivity, and poor soil fertility. Prospects of increased crop productivity are relatively low. These soils are most suited to alternate land use systems.

Sub-montane soils : Sub-montane soils are spread in the hill and foothill regions of India. These areas receive annual rainfall varying from 750 to 2000 mm. These soils are highly prone to erosion and runoff loss. Adoption of appropriate soil and water conservation measures is a must.

Inceptisols and Entisols : Inceptisols and entisols commonly termed alluvial soils occupy nearly 21 per cent of the geographical area. These soils are very deep, of variable texture, and generally deficient in nitrogen and available phosphorus. The soils possess high agronomic potential.

Soil erosion is a serious problem in dryland areas. Out of 175 million hectares of land subjected to various processes of land degradation, 150 m ha suffer from water and wind erosion. About 100 m ha are severely eroded. Besides the loss of soil, there is heavy loss of nutrients. India is annually losing 8.4 m t of N P K through erosion. This loss of soil and nutrients could be saved if soil erosion is checked.

CROPS GROWN IN RAINFED AREAS AND THEIR PRODUCTIVITY

In Table 1 are presented the area and productivity of the principal rainfed crops grown in India.

Maharashtra, Rajasthan, Madhya Pradesh, Gujarat, Uttar Pradesh, Karnataka, and Tamil Nadu are the principal states growing most of the rainfed crops. The productivity of cereals (coarse grains) ranges from 0.522 to 1.325 Mg/ha, and that of pulses and oilseed crops varies from 0.540 to 0.650 Mg/ha, which is quite low.

Besides the above, there are several socio-economic problems in dry farming areas, which include small and fragmented holdings, poor resource-base of dryland farmers and severe unemployment faced by them during most part of the year. Poor infrastructural facilities are yet other problems encountered in rainfed regions.

The essential constrains that limit crop production in dryland areas are moisture stress and nutrient stress.

Table 1. Area, productivity, and major growing regions of principal rainfed crops [Directorate of Economics and Statistics, 1986].

Crops	Area		Productivity	Principal growing areas
	ha x 10^6	%[a]	Mg/ha	
Sorghum [*Sorghum bicolor* (L.) Moench]	15.79	(3.9)	0.690	Maharashtra 42.0[b], Karnataka 12.9 Madhya Pradesh 12.5, Andhra Pradesh 10.7
Pearl millet [*Pennisetum glaucum* (L.) Leeble]	10.69	(4.7)	0.522	Rajasthan 44.5, Maharashtra 15.9 Gujarat 12.3, Uttar Pradesh 8.1
Maize [*Zea mays* L.]	5.88	(16.7)	1.325	Uttar Pradesh 21.4, Rajasthan 16.5 Madhya Pradesh 14.1, Bihar 11.4
Pulses including chickpea [*Cicer arietinum*] and pigeonpea (*Cajanus cajan Millsp.*)]	23.82	(7.2)	0.540	Madhya Pradesh 20.7, Rajasthan 16.3 Uttar Pradesh 13.3, Maharashtra 12.0
Oilseeds including groundnut [*Arachis* sp.), rapeseed (*Brassica napus*) and mustard (*Brassica* sp.)]	18.87	(16.7)	0.650	Madhya Pradesh 14.5, Gujarat 12.5 Maharashtra 12.0, Andhra Pradesh 11.7 Uttar Pradesh 10.3, Rajasthan 9.2 Karnataka 8.6, Tamil Nadu 7.0
Cotton (*Gossypium* sp.)	7.58	(29.4)	0.175	Maharashtra 36.3, Gujarat 18.5 Karnataka 9.9

[a]Percent irrigated area [b]Percent of the total area occupied
Source : I.P. Abrol [1988]

In view of the diversity of the problems, an integrated approach has to be adopted to resolve them. The specific approaches however will depend upon each situation. Efficient resource management is the key to increasing the productivity of these areas. The principal approaches should include water resources management which encompasses cropping strategy based on rainfall analysis and moisture availability periods, *in-situ* moisture conservation technologies for optimum utilization of rainfall, and rain water harvesting and use for crops at critical stress periods. Soil and water conservation measures lay the foundation for good crop management. Improved crop production technologies include nutrient management, seeding, use of improved farm tools and implements, effective and timely control of weeds, diseases and pests, etc. Management of natural resources (soil, water, vegetation, etc.) is best done on watershed basis - a strategy well conceived and received at all hands. Alternate land use systems are some of the viable options for marginal and submarginal soils. All the above noted strategies are discussed below :

WATER RESOURCES MANAGEMENT

A brief resume of the research and development work carried out in India on water resources development in rainfed regions is discussed under three heads :

1. Cropping strategy, based on rainfall analysis and moisture availability periods.
2. *In-situ* moisture conservation technologies for optimum utilization of rainfall.
3. Rain water harvesting and use for crops at critical stress periods.

CROP PLANNING AND CLIMATE ANALYSIS

Crop production on drylands depends upon the amount and distribution of rainfall. With the initiation of the dryland research, historical rainfall data of twenty-three representative locations in the country were analyzed with regard to: (a) the dates of onset and termination of the rainy season and its variability, and (b) the distribution of rainfall within the rainy season.

The information so generated was fitted into the process of selection of efficient crops, varieties, and cropping systems. Later, the agroclimatological studies helped dryland research workers to: (a) identify suitable crops and varieties for early, normal, and late commencement of sowing rains; (b) identify and match intercropping systems with the rainfall pattern; (c) determine optimum sowing periods for different crops and cropping systems; and (d) assess the amount of inevitable runoff available for water harvesting and recycling in arid, semiarid, and subhumid regions of the country. All the data developed on efficient crops and cropping systems have been translated into packages of agronomic practices which are being disseminated by the state governments.

Weekly water balance computations have been carried out to determine the water availability periods for crop growth. Based on this information, regions that are suited for monocropping, intercropping, and double cropping have been identified. The date of commencement of sowing rains was observed to exercise considerable influence on the water availability periods for crop growth in dry farming tracts. Therefore, the drought vulnerability of rainfed crops was assessed, reflecting dates of commencement of rainy seasons, and threshold values of moisture availability for obtaining above-average yields were identified. In order to explain the productivity patterns of rainfed crops in relation to various dates of commencement of the rainy season, crop/weather models were developed using the concept of water requirement satisfaction of the crop during its growth cycle (water requirement satisfaction indicates the amount of soil water available for the crop in a given season in relation to the water needed for its optimum growth).

Both long-term and short-term strategies for mitigating the effects of droughts in dry farming areas have been worked out and documented. A contingent crop production strategy has been worked out as a result of these experiments and has become a set of guidelines for managing weather aberrations in rainfed farming.

Rainfed/dryland farmers generally grow long-duration local varieties of crops that do not exactly fit into the growing season. Under such circumstances, introducing a suitable crop or cultivar that fits into a particular water availability period not only minimizes the risks but also results in a net yield enhancement. Based on the research conducted by AICRPDA, a number of cropping systems have been identified that effectively utilize the rainfall resources of a given region and provide a stable income over the years.

In order to get the best out of the seasonal rainfall, it is necessary to provide a vegetal cover during the *kharif* season. Secondly, the long duration local varieties of *kharif* crops should be replaced with short-duration high yielding improved varieties. Thirdly, recommended doses of fertilizer have to be used. Fourthly, the crop must be sown at the right time according to weather and available soil moisture, and establish a good stand. Fifthly, a watch out for insect/pest attack and use of plant protection measures when necessary, will have to be undertaken. Last, but not the least, the crop has to be harvested at physiological maturity.

Choice of crops and cropping systems will depend on soil depth and available moisture. For instance, in a shallow soil with an available moisture of 10 cm, it is possible to have a sole crop only. In medium soils with a moisture availability of 15 cm, a sole crop of longer duration or intercropping could be possible. In deep soils with a moisture availability of 20 cm, double cropping is possible. In areas where *rabi* crops are grown on conserved soil moisture, the choice of crops and varieties will depend on the soil moisture stored in the soil profile at the time of sowing. For instance, with a soil moisture storage of 300 mm, wheat, pea, chickpea, could be taken in the North Western part of India. When the moisture drops down to 200 to 300 mm, *desi* wheat, barley, lentil, chickpea could be possible. Rapeseed and mustard is a good choice in soils having a moisture storage range of 100 to 150 mm at sowing time. When the moisture storage is low (50 to 75 cm), *taramira* (*Eruca sativa*) is the obvious choice.

To match growing seasons with limited water availability periods high yielding short duration crop varieties have been evolved. Examples of successful introduction and spread of new varieties/crops include the increase in area under sorghum hybrids from about 1.7 x 10^6 ha in 1970-71 to about 5.0 x 10^6 ha in 1982-83 with average yields increasing from 0.21 to 0.7 Mg/ha. In western Madhya Pradesh, the black soils that previously were fallowed during *kharif*, more than 1 million ha were planted to soybean in 1985-86 compared with only 25,000 ha 10 years earlier. Similarly, average yield of finger millets grown on 1 million ha in Karnataka increased from 0.9 Mg/ha during 1970-76 to about 1.3 Mg/ha during 1977-83 [Abrol, 1988].

Mixed cropping involving growing more than one crop on the same field at one time or in an overlapping sequence is a widespread traditional practice. Intercropping involving sowing of different crops in alternating rows/groups of rows reduces risk of total crop failure and significantly increases the cropping intensity. Intercropping research has focussed on developing systems for increased productivity of grain legumes and oilseeds without adversely affecting the production of the cereal components. Production technologies involving defining of base and companion crop combinations, plant population, fertilizer application, etc. have been standardized for intercropping systems suitable for different agroclimatic regions. Farmers are increasingly adopting recommended intercropping systems, important among them are sorghum and pigeonpea, sorghum and groundnut, groundnut and pigeonpea, sunflower (*Helianthus annuus* L.) and groundnut, and soybean (*Glycine max* L.) and cotton. Pigeonpea is the most favoured companion crop since it provides fuel in addition to the valued pulse. Thus, a number of finger millet varieties of different durations are now available to match the duration of the growing season due to late onset of rains.

IN SITU MOISTURE CONSERVATION

The second approach is to utilize the water where it falls by means of appropriate land treatments and *in-situ* (i.e. at the same place) moisture conservation practices.

Some of these include off-season tillage, mulching, dead furrows, keyline cultivation, compartmental bunding, graded border strips, and interplot water harvesting. Off-season tillage helps the rainwater to enter into the soil profile more effectively and, in addition, helps in weed control. Such a practice in alfisols (i.e. highly leached red brown soils) of the Telengana region in the state of Andhra Pradesh increased the sorghum yield by 43 per cent. Off-season tillage, however, is not suggested for aridisols (i.e. sandy soils of the desert regions) as this would accelerate wind erosion. Deep ploughing (i.e. > 22 cm), once every two or three years, also helps to increase crop yield in soils with a hard subsoil below the plough layer.

Mulching is a well-known practice for minimizing the movement of moisture from the soil into the atmosphere. Poor infiltration of rainwater is a problem in vertisols (i.e. heavy black soils), which often leads to water stagnation. Vertical mulching (e.g. placing sorghum stubbles in trenches 40 cm deep, 15 cm wide, and protruding 10 cm above the ground level) has been found to enhance available soil moisture by 4-5 cm and to increase the grain yield of sorghum by 40-50 per cent in vertisols at several locations in India. The making of "idle furrows", also called "dead furrows" at each 3.6 m interval helps in improving water retention. Such treatment at Anatapur in South India was found to increase the grain yield of groundnut by about 10 per cent.

In alfisols, crops such as sorghum and pigeonpea can be grown as intercrops on slopes, and ridges are made later that help in controlling soil and water erosion. Such simple practices enhanced the seed yields by 15-20 per cent, as compared to sowing on a flat surface. Broad bed and furrows (BBF) is another system eminently suited to vertisols. The basic objective of this system is to drain out excess water. A broad bed of 100 cm alternated with a 50 cm shallow furrow having a depth of 15 cm is useful in improving the moisture regime and draining excess water in places where water stagnation is a problem. This situation arises in high rainfall vertisol areas. ICRISAT conducted extensive experiments on this practice. On-farm trials at Akola in India showed that sorghum yield can be increased by 25 per cent as compared to flat sowing. Graded border strips of 130 x 11 m with a 0.1 per cent grade are effective in deep alfisols with medium rainfall. Finger millet yield was increased by 42 per cent as a result of this treatment at Bangalore in South India.

Ridge-and-furrow *in-situ* water harvesting is yet another promising system for the inceptisols (i.e. soils with poor horizon development as in alluvial soils) of subhumid areas. Maize is planted on the ridges and rice in furrows, thus meeting the varying water requirements of crops. The water requirement of rice is met by the water that is harvested into the furrows. Yield increases of about 18-21 per cent for both crops can be obtained by this system, as compared to flat sowing.

The interrow water harvesting system is a practice suitable for light textured soils found in the arid region. Significant yield advantages have been recorded with the pearl millet crop by adopting this practice, in combination with the recycling of the water for supplemental irrigation.

RUNOFF HARVESTING AND RECYCLING

Rainwater can be utilized by collecting in small dug out ponds excess runoff during high-intensity rainfalls and then recycling it for supplemental irrigation at critical stages of crop growth. This water also can be used for pre-sowing irrigation of a *rabi* crop in potential double-cropping regions. A *rabi* crop is a crop grown during the post-rainy season (i.e. October-March) on stored soil moisture. Extensive studies have been undertaken on these aspects both in arid and semiarid areas. Experiments on catchment size and slope, size of the pond, effective sealants, water lifting devices, and choice of crops have been conducted both in alfisol and vertisol regions.

Existing water harvesting systems suffer from the fact that the donor catchment (i.e. actual area that is contributing the runoff) does not benefit from the runoff water. In a modified system with small dugout ponds, the water can be used for the donor catchment. Studies have shown that appropriate pond size varies with rainfall, ranging from 200 cu m to 3,000 cu m. Seepage losses pose a major problem in these dugouts. Though natural silting minimizes seepage losses in large tanks, it causes a reduction in storage capacity in small dugouts. Among the sealants evaluated, cement plus soil in a 1:8 ratio was found to be reasonably efficient and cost effective. But an ideal sealant has not been found yet that is both efficacious and cost effective.

For lifting water (i.e. taking out water from the pond and adding it to the cropped area), experience has shown that for small ponds of 200 cu m, manual methods are more suitable. Only for big ponds of 3,000 cu m is it feasible to use mechanical pumps.

There are several benefits accruing out of supplemental irrigation from stored water. These are discussed below: A number of experiments have been conducted on the criteria for using stored rainwater. For a number of crops, e.g. sorghum, pearl millet, cowpea, and tobacco, it has been found that maximum payoffs in terms of yields can be obtained if irrigation is applied at flowering. For post-rainy-season crops such as wheat and barley, the crown root initiation stage was found to be the most appropriate. Experiences at Bellary in South India have shown that a single application of 10 cm of water increased the yield of cotton by 76 per cent in a subnormal year. In an ICRISAT experiment the sorghum yield in a vertisol region was 2,570 kilograms/hectare (kg/ha) without supplemental irrigation, whereas a single irrigation of 5 cm from runoff water boosted the yields to 3.570 kg/ha. Even greater benefits were apparent by supplemental irrigation in the aridisol regions of North West India. Experiments conducted at Jodhpur on light textured sandy loams showed that an irrigation of 5 cm to a pearl millet crop at a critical stage increased the yield by 35 per cent. Application of a modest dose of fertilizers further enhanced the yield.

The effects of supplemental irrigation have been studied on a number of short- and long-duration crops in both the *kharif* season (i.e. rainy season, June-September) and the *rabi* season (i.e. post-rainy season, October-March). Short-duration, shallow-rooted crops like sorghum and pearl millet responded better than deep-rooted crops like castor. The latter crops needed relatively more water for a significant yield response. Scheduling of irrigation by the moisture depletion approach was also attempted for wheat in the post-rainy season. Irrigation at 50 per cent depletion in two instalments gave the best results. As compared to grain crops, higher economic returns for vegetable crops such as chillies, brinjal, and beans can be obtained if they receive the water from a dugout pond. The principle is to sow the crops in the late rainy season and supply water at the reproductive stage.

The cost of initial construction of a farm pond is high for a water harvesting system, and it is one of the principal constraints in the spread of this technology. Planting a horticultural crop such as *ber* (*Zizyphus mauritiana*) helps in quick repayment. Recently, efforts have been made to introduce pisciculture to optimize the returns.

Among the water application methods tried, an irrigating alternate furrow system was found to be more effective than use of the furrow method and flooding. Despite the clear demonstration of benefits of *in-situ* moisture conservation and runoff recycling in both the experimental plots and the operational research projects, these practices have not yet been applied on a wide scale by the farmers. There are many constraints, both financial and technological, but the most critical constraint is the fact that farm holdings in India are very small and fragmented. The average size of an operating farm is about 1.8 hectares, which limits the application of conservation methods. Therefore, it is now realized that water conservation development projects should be undertaken on a watershed basis. The impact of water resource development can be better seen when improved practices are adopted in a watershed as a whole (i.e. a unit of land and a drainage area contributing runoff water to a common collecting point).

SOIL AND WATER CONSERVATION MEASURES

There are quite a few options in the area of soil and moisture conservation. In most of the situations, contour cultivation and contour sowing will serve as good measures for soil and water conservation. In other situations where slope is more than 1.5 per cent, mechanical measures like contour bunding in low rainfall areas will be appropriate. In deep black soils where water stagnation could be a problem, ridge and furrow, or broad bed and furrow (BBF) land configurations will be desirable.

In soils having a hard pan or compact layer, deep ploughing once in 3 to 4 years may have to be resorted to for ensuring better infiltration of water and root growth.

Incorporation of crop residues, use of surface mulches, vertical mulching, impounding of runoff water in a dug out pond etc., are other soil and moisture conservation measures. In arid areas, setting up of shelter belts/wind breaks is an effective measure of soil conservation.

It is now established that contour farming parallel to graded bunds/key lines helps in conserving both soil and water. For *in-situ* moisture conservation, furrows are made with a plough after every row, 3-4 weeks after sowing - an operation which is incidental to interculture done in dryland crops. Experience gained in the Operational Research Project at Hyderabad has revealed an additional advantage of about 15 per cent in the yield of sorghum and castor, on account of furrowing as a part of interculture operation done on contour. The practice is easily acceptable by small farmers as it does not involve any additional cost.

As a first step, simple agronomical practices like off-season tillage across the slope, key line cultivation,flat sowing and ridging later, strip cropping, intercropping system have to be adopted. These practices can be easily adopted without much additional cost. Later, mechanical structures may be introduced after gaining full confidence of the farmers. The emphasis should be on *in-situ* moisture conservation measures.

A technique long recommended and widely adopted for soil and water conservation in India involves construction of contour or graded earthen bunds in fields of upto 6 to 8 per cent slope. There is an increasing realization that this technique has several drawbacks; high initial cost, loss of land, maintenance difficulties, and short-term water logging resulting in reduced yields in case of contour bunds and reduced impounding in case of graded bunds. Planting vegetation along selected contour lines at a suitable interval, on the other hand, has several advantages; once established these serve to retain soil and reduce runoff, thus increasing *in-situ* conservation. Several plant species are being tested for planting as vegetative bunds.

Other soil management practices that have been developed for specific situations include compaction of highly permeable soils of low retentivity, chiselling soils having layers of high mechanical impedance at shallow depths, raised/sunken bed technology for slowly permeable flat soils for regions receiving more than 1000 mm seasonal rainfall, etc.

In the past, mechanical structures were considered to be the panacea for all soil erosion problems. We never thought of vegetative barriers which could be as efficacious as mechanical structures. Where the slope is steep and the gradient is high, there is no substitute for mechanical structures. There are situations where the soil erosion problems are not that severe and the gradient in the waterway is not that high, vegetative barriers could be established with advantage. At times, we may have to combine the mechanical drop structures with vegetative barriers. The whole approach will depend on the situation.

IMPROVED CROP PRODUCTION TECHNOLOGIES

Besides efficacious rain and soil management techniques soil and water conservation practices, choice of suitable cultivars matching the rainfall pattern, nutrient management, tillage and seeding are other important aspects to be reckoned with. These are discussed hereunder:

Nutrient Management : Nitrogen is the key element in dryland crop production. In order to induce early seedling vigour, it is necessary to give moderate quantities of N and P_2O_5 at the time of sowing. Nutrition induced vigour is able to mitigate the short-term stress of drought.

There are several N management problems encountered in rainfed areas. Besides all rainfed soils being deficient in N, heavy loss of N takes place. Above all, the recovery of applied N is very low. Since there is high cost of fertilizer N involved, the dryland farmers are often not enthused to use this nutrient to the extent required. Moreover, there is a risk involved in fertilizer use in sub normal/drought years.

Earlier, it was believed that fertilizer use increases water demand considerably. Experience has now shown that a fertilized crop yields 100-150 per cent higher over unfertilized crop, but requires only 20-25 per cent additional moisture. Fertilizer use in fact increases moisture use efficiency through increased root development, and increased production of vegetation.

It has been found that the response of crop to applied N will depend on available moisture and the crop/variety/ cropping system. In alfisols, the maximum response of applied N to sole and intercrop is obtained at 40 kg N/ha. In case of vertisol, the response of the sole crop is higher than the intercrop systems to applied N is not the same in alfisols and vertisols.

In a coarse textured soils like alfisols and related soils it is important to build organic matter in the soil. Continuous application of maize residues at Bangalore (alfisols), resulted in an yield of 23.5 q/ha over the treatment receiving no crop residues (18.8 q/ha). Maize residues were applied at the rate of 4 t/ha. Basal dose of 50 kg N and 50 kg P_2O_5/ha was given to maize crop grown after incorporation of maize residues.

Split application of nitrogen for *kharif* crops and deep placement for *rabi* crops has been found profitable. In acidic soils, liming has been found beneficial, particularly for pulses and groundnut. Application of sulphur (through gypsum) led to increased yields in case of oilseed crops, especially groundnuts. Considering the risk and heavy investment involved an integrated nutrient supply system with emphasis on recycling of farm water, inclusion of legumes in the cropping systems, use of efficient rhizobial strains, is called for.

Tillage : Off-season tillage is an important component of the dryland technology formulated for alfisols and related soils. The soil has to be ploughed soon after the harvesting of the *kharif* crop when there is still some moisture left in the soil profile. Any delay in off-season tillage will lead to high energy use and costs. There are other advantages like moisture conservation, weed control, etc. accruing out of summer tillage.

Seeding : Early seeding is extremely important in dryland areas. There are several advantages, important of which are: better moisture conservation, good seedling vigour and longer growing season. It is now established that wider rows have got distinct advantage in terms of energy saving over narrow rows. For instance, the labour hours per ha required for wider rows were 560.9 as against 449.3 in case of narrow rows. The row spacing did not make much difference in respect of yield.

Dry sowing is also considered appropriate under some dryland situations. The most important advantage of dry sowing is that it gives an early start to the crop. Dry seeding is possible in case of bold seed grains like castorbean, groundnut, pigeonpea, sorghum, etc. In case of small size grain crops like pearlmillet, fingermillet, sesamum, etc. dry seeding has some limitations. Moreover, it has limitations in soils where crust formation is common. Ranchi experience has shown that Gora rice and groundnut can safely be sown 6-10 days before the onset of monsoon. Similarly, cotton could be sown a week in advance of onset of monsoon at Indore. ICRISAT experience has shown that crops should be dry seeded in the week when there is about 70 per cent probability of getting 20 mm of rainfall. However, it is not an economic proposition if weed control cost is taken into consideration.

Borne out of experience, two crucial factors identified for success in dryland agriculture are timeliness and precision. Timeliness of seed bed preparation, seeding, weed control, top dressing and plant protection are extremely important in dryland agriculture. In case of cereals like corn and sorghum, each week's delay in sowing time after the optimum time of sowing will drastically reduce the yield. Similarly, weed control in case of short duration crops like greengram, blackgram and cowpea (within 20-25 days) and long duration crops like sorghum (within 35 days) and castor (within 45 days) is important.

Weeds take a heavy toll of moisture and nutrients. It has been estimated that 1000 kg dry matter of weeds take away about 30 kg of available nitrogen per ha. Further, they rob away 8-10 cm of water. If we can save 30 kg of N and 8-10 cm of water, it could give much relief in terms of input management.

Precision with regard to rate and depth of seeding is crucial. Productivity in dryland areas is poor because of sub optimal population. It is necessary that every attempt is made to obtain a good and timely crop stand establishment. For doing so, it is necessary that the rate and depth of seeding, and fertiliser placement are properly monitored. The seeds and fertilizers have to be placed in the moist zone, an operation which requires precise metering of seeds and an efficient fertilizer placement device. A good seed-cum-fertilizer drill could do this job efficiently. Unfortunately, the seed-cum-fertilizer drills which are really efficient are costly. Ones that are cheap are inefficient and serve no useful purpose.

ALTERNATE LAND USE SYSTEMS

Not all lands are suitable for crop production. Pasture and agroforestry systems could be tried with success on many a marginal and submarginal lands. Various agroforestry options are now available, principal of which are : (1) silvopastoral management system, (2) agro-horti system, (3) alley cropping, etc.

Promising tree species have been identified for each of these systems. *Leucaena leucocephala* and *Stylosanthes hamata* combination comes up very well in shallow alfisols. Another promising system is *Acacia albida* with an undergrowth of sorghum. *Ber* (*Zizyphus mauritiana*) + grain legume combination holds great prospects for drylands.

WATERSHED DEVELOPMENT PROGRAMME

The principal components of a watershed development programme are : (a) improvement of water resources, (b) *in-situ* soil and water conservation, (c) increased cropping intensity, and (d) alternate land-use systems for efficient use of lands. Alternate land-use systems refer to all land-use practices other than growing regular arable crops. These include agroforestry, horticulture, pastures, and tree farming.

The Government of India in its Seventh Five-Year Plan had launched a national development programme of rainfed agriculture on a watershed basis. By 1984-85 work was launched on 4,400 micro watersheds covering an area of 4.2 million hectares. Forty-seven model watersheds are being monitored by the Indian Council of Agricultural Research (ICAR) with field-level implementation by the state governments. The initial results have been quite encouraging. At Mittemari in Karnataka State in South India there has been significant improvement over a three-year period in crop yields, water resources development, and in the socioeconomic status of the people in the watershed, as compared to the outlying villages. At Tejpura in Uttar Pradesh there has been a marked improvement in the water table following the implementation of soil conservation programmes. Significant improvement in crop yields and cropping intensity also has been recorded. In some watersheds it has now become possible to plant a double crop. During the last few years watershed management projects have won national productivity awards for significant contributions to farm productivity in rainfed regions.

Of late, nongovernmental organizations also are taking an interest in the concept of watersheds as the organizational units for long-term development of rainfed regions. These organizations are voluntary or social work agencies, financed by public donations and trusts, or partly supported by the government. The Government of India trains hundred of personnel annually to work in watershed development programmes. Village development plans prepared by banks and other credit institutions are now being established on a watershed basis. Since there is an acute shortage of fodder and fuel wood in rural India, their supply has become a component of alternate land-use systems that have been introduced on a watershed basis.

Many of these water resource management technologies are being disseminated as a result of the operational research projects involving watersheds. However, unless a total package is adopted, individual components by themselves will not contribute to visible improvements in crop yields. The watershed concept, it is hoped, will go a long way in persuading farmers about the importance of comprehensive watershed resource management.

Considering the severe limitations on the future expansion of irrigation and the fact that groundwater exploitation cannot continue endlessly, there is little choice other than the efficient utilization of rainwater for a country like India, where the population is heavily dependent on agriculture.

INDIA'S EXPERIENCES IN RAINFED AGRICULTURE : LESSONS FOR OTHER COUNTRIES

Many developing countries in West and South Asia, the Middle East, and Africa have problems of rainfed agriculture somewhat similar to India's. The technologies and experiences of India can be beneficially utilized by these countries.

For example, the technologies of water harvesting in sandy soils of arid and semiarid regions can be used with appropriate modifications in the nations of West Asia that have an arid climate and light textured soils. In fact, technical cooperation in water resource management already is being undertaken by the Central Arid Zone Research Institute (CAZRI) in Jodhpur, India, to train experts in many countries in arid land management to combat desertification.

Land degradation, low productivity, erratic rainfall, and problem soils are constraints shared by India and a number of African nations. India over the years has gained rich experience in soil and water management technologies applicable to semiarid and subhumid zones. Of importance is the Indian experience in watershed development for rainfed agriculture, in particular our development of methodologies for surveying and planning watersheds, monitoring hydrological changes, and identifying the various constraints. This experience can be useful for rainfed regions of Africa.

India also has developed a series of alternate land-use systems including agroforestry, agrihorticulture (i.e. planting of fruit trees), silvipasture (i.e. grass and legume coverage), and alley cropping (i.e. tree plantings as hedges), with extensive research conducted over the past eight years. Some of these practices can be productively adapted in African countries that have low rainfall, adverse climatic conditions, marginal lands, and a high-density human and livestock population.

References

Abrol, I.P. Dryland Farming - The Indian Experience 1988. In: *Challenges in Dryland Agriculture - A Global Perspective. Proc. of Int. Conf. on Dryland Farming*, August 15-19, Amarillo/Bushland, Texas, USA, 50-53.

Singh, R.P. Crop Production under Rainfed Conditions in India. Paper presented at the FAO Regional Expert consultation on Rainfed Agriculture, Bangkok, Thailand, 29 November-2 December 1988.

Singh, R.P. 1990. Dryland/Rainfed Agriculture and Water Resources Management Research and Development in India. In: *Sharing Innovation : Global Perspectives on Food, Agriculture and Rural Development* (ed) Neil G. Kotler. Papers and Proceedings of a Colloquium organized by the Smithsonian Institution, USA, 94-105.

Singh, R.P. 1992. Problems and Prospects of Dryland Agriculture in India. In: *Land and Soils* (eds) T.N. Khoshoo and B.L. Deekshatulu. Indian National Science Academy, Har-Anand Publication, New Delhi, 266-278.

4

AGRICULTURAL CLIMATE OF INDIA

S.M. VIRMANI

INTRODUCTION

It is perhaps not surprising that weather continues to have a dominant influence on the fate of agricultural production in India, because there are a large number of weather-sensitive agricultural activities that control production and quality of food and fruits we grow. Therefore the study of weather (short-term weather phenomena) and climate (long-term weather) find application in an ever-widening array of land uses and agricultural practices. This is particularly the case in regions where dryland cropping systems are prevalent.

AGRICULTURAL CLIMATE: DEFINITION AND SCOPE

What is agricultural climate? It is a systematic study of the climate of a place or a region in relation to agriculture. It has both temporal (or time) and spatial (or space) dimensions. By temporal, we mean the changes in weather that occur on day to day or on month to month basis during a year. As an example the daily weather will generally show that temperature is higher during the day compared to the night. This day-night temperature difference is greater in winter than in summer months, or in cooler places compared to warmer places. The temporal dimension of the climate also means that the climate is dynamic. It changes from day to day and as such gives rise to seasons. The concept of seasonality is very important in managing agricultural climatic resources for crop production. The spatial dimension means that the climatic conditions vary from place and place. Therefore for agricultural planning, to be useful, climatic assessment must cover a fair number of locations.

OVERALL SETTING OF INDIAN AGRICULTURAL CLIMATE

India is located 8° to 36° North of the equator and between 68° to 96° East longitude. It is a tropical country. The tropic of cancer, which passes through the middle of the country, divides it into two distinct climates: in true tropical climate in the South where all the 12 months of the year have mean daily temperature exceeding 20°C; and in the North where a sub-tropical climate prevails. In sub-tropics during the winter months, it is cool to cold. Frosts occur sometime during the months of December and January. Three main climatic zones of India based on temperature are shown in Figure 1.

Some areas in the Northern India have a temperate climate. Here it snows during the winter months and freezing temperatures may extend to two months or more during the year.

Rainfall in India varies considerably. In some parts of the Thar desert, located in Western Rajasthan, the annual rainfall is as low as 100 mm; while in the East, the annual rainfall may be as high as 10,000 mm or more. In some areas it may rain for a month or two during the year, in others as many as 11 months may be rainy.

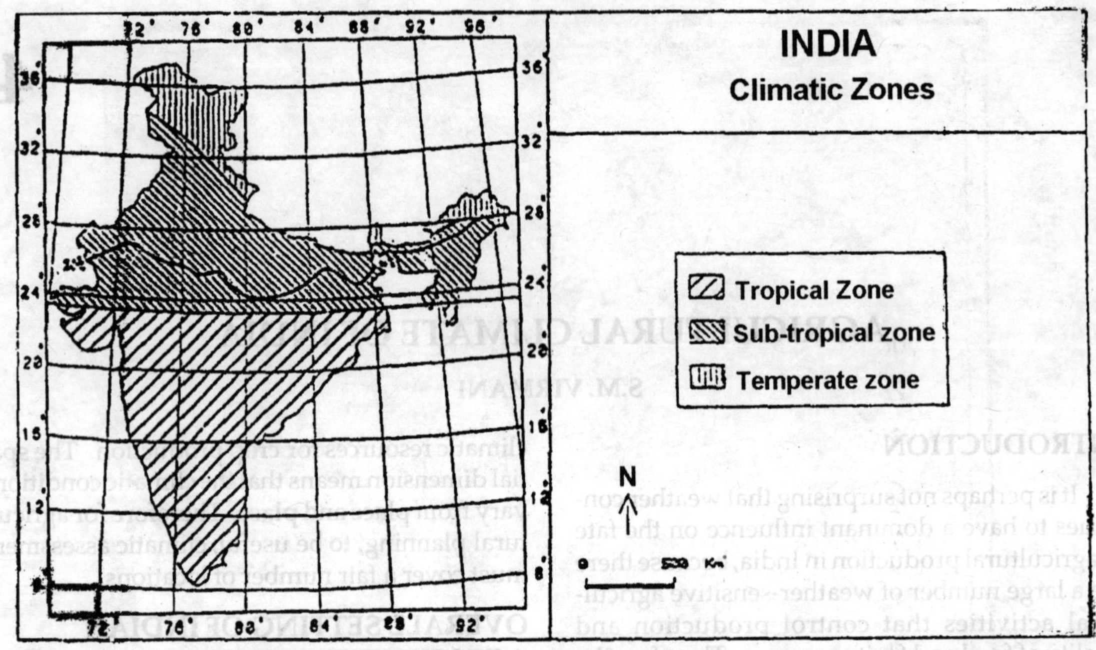

Fig. 1. India : Climate zones based on temperature

RAINFALL OVER INDIA

India is a monsoonal country. In the tropical and sub-tropical regions, almost 80% of the total annual rainfall is received during the monsoon rainy season. There are two types of monsoons received in India. During the main rainy season extending from June/July to September/October, the rains we receive are called as South-West monsoons. They are called thus, because the rainy season sets in first in the South of India and rains progress gradually towards the West of the country. Figure 2 shows this point clearly. At Trivandrum in Kerala State, the monsoon breaks around June 1; at Hyderabad in Andhra Pradesh around June 5; at Bombay around June 10; and at Jaipur or Jodhpur towards end of June or early July. The normal dates of onset are just averages, in actual terms there may be a week or 10 days delay or earliness. These are just guide dates for agricultural operations or crop calendars. The actual dates vary from year to year.

Figure 3 shows the normal dates of withdrawal of South-West monsoon; it also shows the average dates of withdrawal of the second monsoons called North-East (NE for short) monsoons. NE monsoons occur due to cyclonic disturbances in the Bay of Bengal. Their normal date of onset is between November 1 and November 15. These rains withdraw by about end of January and occur in areas located below 10° N latitude in India.

Thus in India, all areas located between 20° North of equator to 40° North, the rainfall is unimodal and almost all of the annual rainfall is received due to South-West monsoons. In South India, the areas located below 10° North, the rainfall is bimodal or it has two peaks - one peak during the South-West rainy season and the second peak in NE monsoon.

Figure 4 shows monthly distribution of rainfall for some selected locations of India. The rainy season extends to about 2 to 4 months across most of agricultural regions in Northern and North-West India; it extends to 5 months in Peninsular India (eg. Hyderabad), and is of a much longer duration in some Southern areas (eg. Bangalore or Trivandrum).

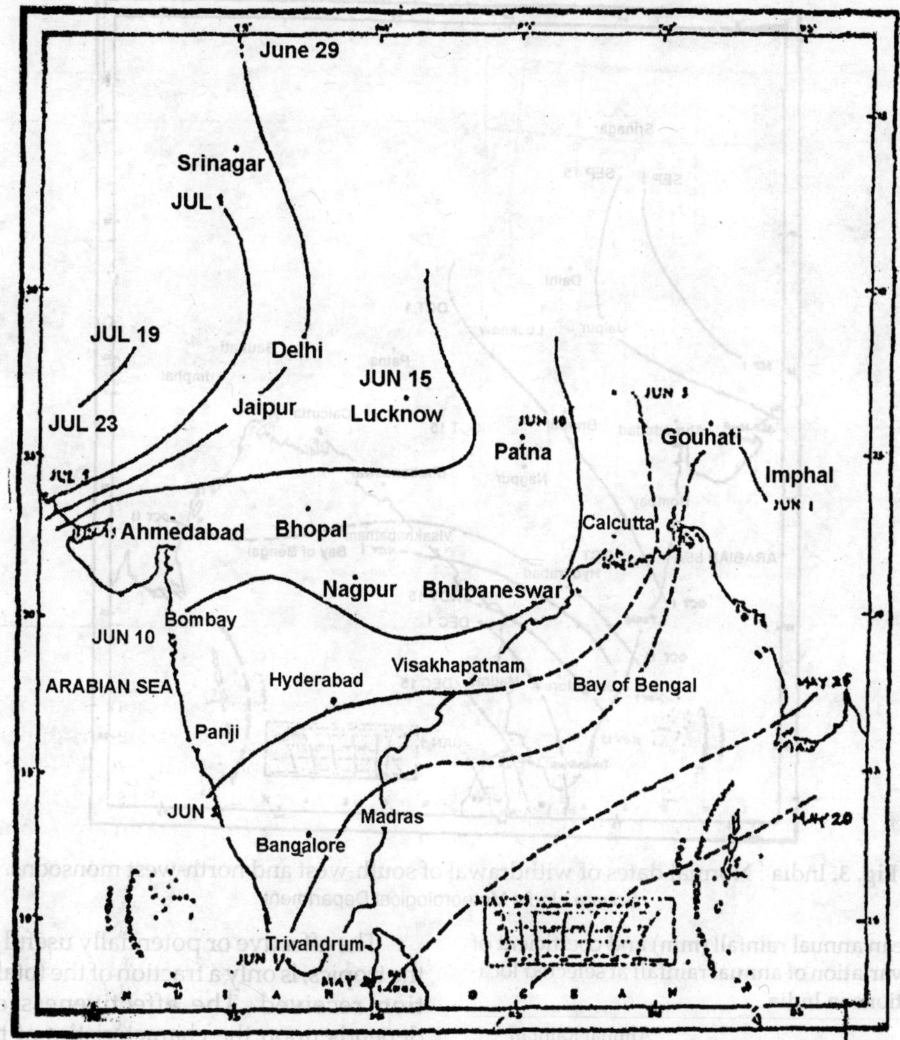

Fig. 2. India : Normal dates of onset of south-west monsoon.
Source : India Meteorological Department

COMMON FEATURES OF MONSOON RAINS OF AGRICULTURAL IMPORTANCE

The tropical monsoons occur in torrential intensity during short periods; there is a great variation in the total quantity of rainfall received at any location from year to year, and a significant difference in quantity of rainfall is measured over relatively short distances. In addition more than 50% of the total annual rainfall may be received in 5-20 rainy days. Thus there is a substantial loss of water due to surface runoff. This process also accelerates soil erosion.

Table 1 shows the annual rainfall and the associated coefficient of variation at some selected locations. Generally higher the rainfall, less is the coefficient of variation. This means that the stability of the seasonal rainfall and thus the growing season during monsoon is much less in areas with lower rainfall (eg. Jodhpur) compared to areas with a greater total rainfall (eg. Ranchi) in India.

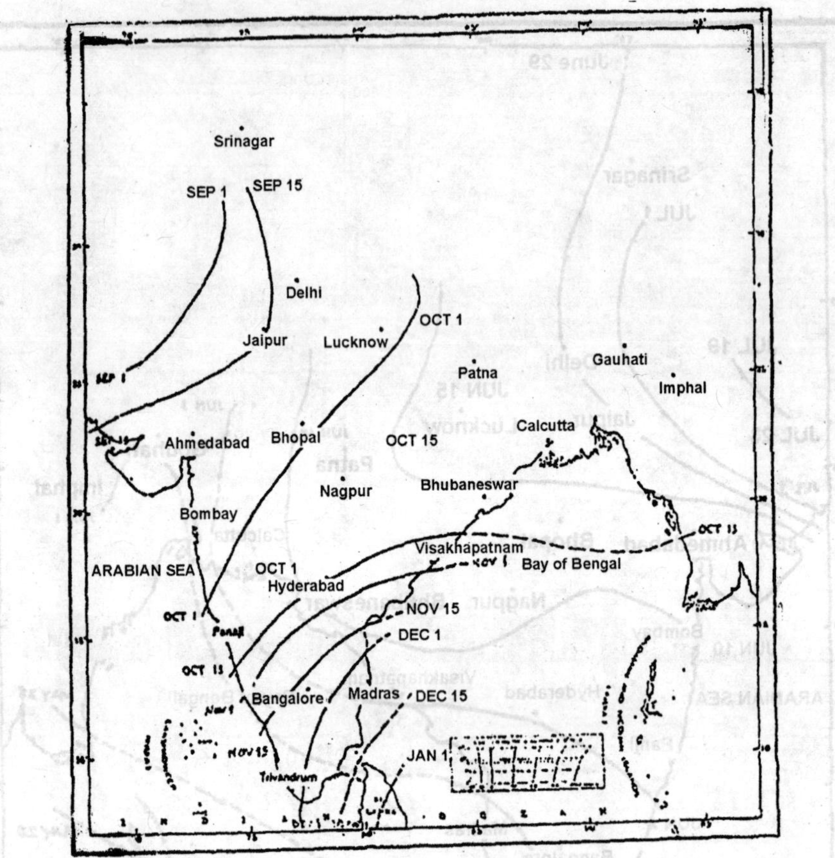

Fig. 3. India : Normal dates of withdrawal of south-west and north-west monsoons.
Source : India Meteorological Department

Table 1. Mean annual rainfall (mm) and coefficient of variation of annual rainfall at selected locations in India.

Location	Annual Rainfall	
	Mean (mm)	C.V. (%)
Hisar-Haryana	451	31
Anantapur-Andhra Pradesh	565	30
Ludhiana-Punjab	661	50
Sholapur-Maharashtra	708	28
Kovilpatti-Tamil Nadu	749	27
Anand-Gujarat	869	40
Bangalore-Karnataka	888	20
Faizabad-Uttar Pradesh (Plains)	1052	27
Ranichauri-Uttar Pradesh (Hills)	1264	22
Jabalpur-Madhya Pradesh	1308	24
Ranchi-Bihar	1338	16
Mohanpur-West Bengal	1389	19

The effective or potentially useful rainfall, in the tropics, is only a fraction of the total precipitation received. The effectiveness of rainfall depends upon the characteristics of the terrain, intensity and amount of rainfall, soil type, the soil/land surface conditions, and the vegetative cover. In general higher the slope of the land, a higher proportion of the rainfall goes as surface runoff and therefore reduced effectivity. Similarly higher the intensity of the rain (particularly if it exceeds the initial soil intake rate), the more the surface runoff and consequently lesser is the effectivity. Following conditions are conducive to higher rainfall effectivity: loamy soils, soils that are well ploughed or where the surface is rough, and lands that have adequate vegetative cover. In order to increase rainfall effectivity, efforts should be made to keep the land in a condition

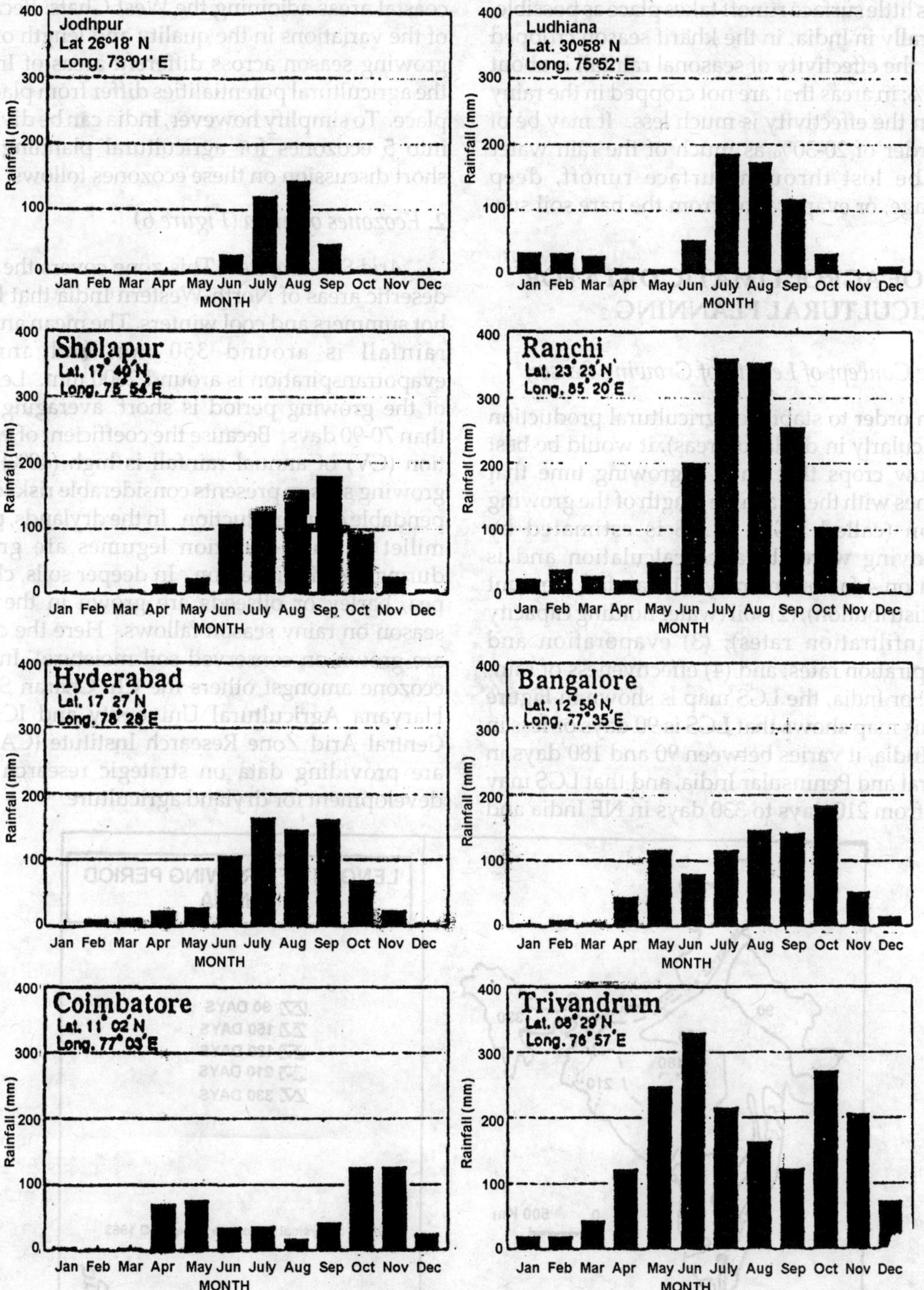

Fig. 4. India : Distribution of monthly rainfall at selected locations

Data Source : India Meteorological Department, 1931–1940 normals of climate. Climatological Tables, IMD

that as little surface runoff takes place as possible. Generally in India, in the kharif season cropped areas, the effectivity of seasonal rainfall is about 40-60%; in areas that are not cropped in the rainy season the effectivity is much less. It may be of the order of 20-30% as much of the rain water may be lost through surface runoff, deep drainage, or evaporation from the bare soil surface.

USE OF AGROCLIMATIC DATA FOR AGRICULTURAL PLANNING

1. *The Concept of Length of Growing Season*

In order to stabilize agricultural production (particularly in dryland areas), it would be best to grow crops that have a growing time that matches with the available length of the growing season (called LGS). LGS is estimated by employing water balance calculation and is based on 4 factors namely; (1) rainfall (amount and distribution); (2) soil (water holding capacity and infiltration rates); (3) evaporation and transpiration rates; and (4) effectiveness of rainfall. For India, the LGS map is shown in Figure 5. This map shows that LGS is 90 days or less in NW India, it varies between 90 and 180 days in Central and Peninsular India, and that LGS may vary from 210 days to 330 days in NE India and coastal areas adjoining the West Ghats. Because of the variations in the quality and length of the growing season across different areas of India, the agricultural potentialities differ from place to place. To simplify however, India can be divided into 5 ecozones for agricultural planning. A short discussion on these ecozones follows.

2. *Ecozones of India (Figure 6)*

Arid Sub-Tropics : This zone covers the arid desertic areas of North-Western India that have hot summers and cool winters. The mean annual rainfall is around 350 mm and annual evapotranspiration is around 2000 mm. Length of the growing period is short, averaging less than 70-90 days. Because the coefficient of variation (CV) of annual rainfall is high (40%), the growing season presents considerable risk to dependable crop production. In the drylands, pearl millet or short-duration legumes are grown during the kharif season. In deeper soils, chickpea, barley or oilseeds are grown in the rabi season on rainy season fallows. Here the crops are grown on conserved soil moisture. In this ecozone amongst others the Ch. Charan Singh Haryana Agricultural University and ICAR's Central Arid Zone Research Institute (CAZRI) are providing data on strategic research and development for dryland agriculture.

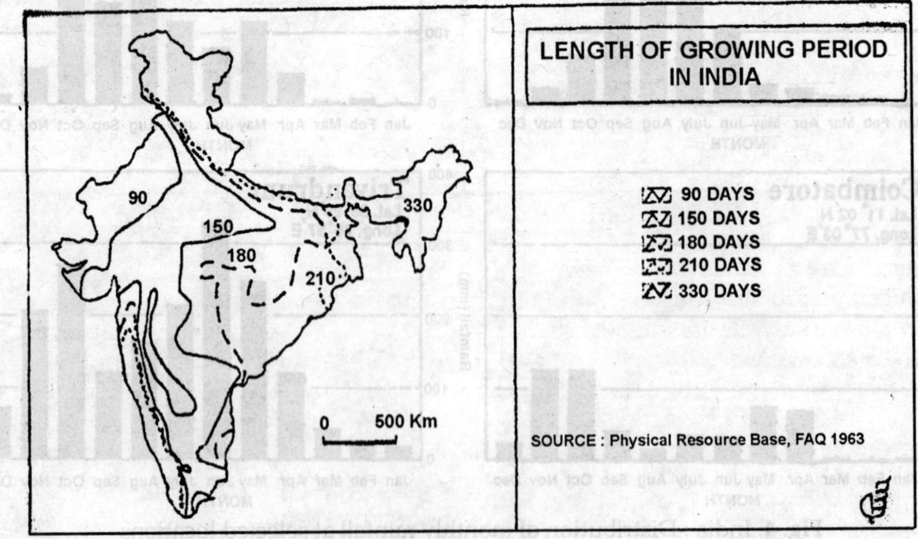

Fig. 5. India : (LGS) map

Semi-Arid Sub-Tropics : This ecozone covers large areas in North and Central India. Here the average annual rainfall is 900 mm and evapotranspiration about 1800 mm. The CV of annual rainfall is 20-30%. The length of growing period is 120-180 days or more. The area is suitable for double cropping (2 crops in a year sequentially). In drylands, maize, cotton, sorghum and rice (in areas with 1200 mm rainfall) and grain legumes are commonly grown in the rainy season. In the post rainy season, chickpea, oilseeds and cereals such as oats, barley, wheat are grown. Leading research institutions located in this ecozone are the State Agricultural Universities of Punjab, Uttar Pradesh, Madhya Pradesh, Gujarat, Orissa; and the Indian Agricultural Research Institute at New Delhi.

Semi-Arid Tropics : This zone comprises the Peninsular Indian region. It can be divided into two sub-zones : the relatively undependable rainfall areas with 500-750 mm. An example of undependable rainfall areas is Sholapur in Maharashtra, and that of dependable rainfall areas is Hyderabad in Andhra Pradesh. In the undependable rainfall areas, rainfall in June, July and August is erratic. The date of onset of the rainy season varies considerably, consequently kharif cropping is more risky. In the relatively dependable rainfall zone, the onset of rainy season is well-defined and rains are also well distributed. The reliability of dryland crop production in this region is relatively high and it presents a great potential for development. Major dryland research stations in this ecoregion are : Srikrushinagar, Sholapur. The ICAR's Central Research Institute for Dryland Agriculture (CRIDA) and the International Crops Research Institute for the Semi-Arid Tropics (ICRISAT) are supporting basic and strategic research for this ecozone.

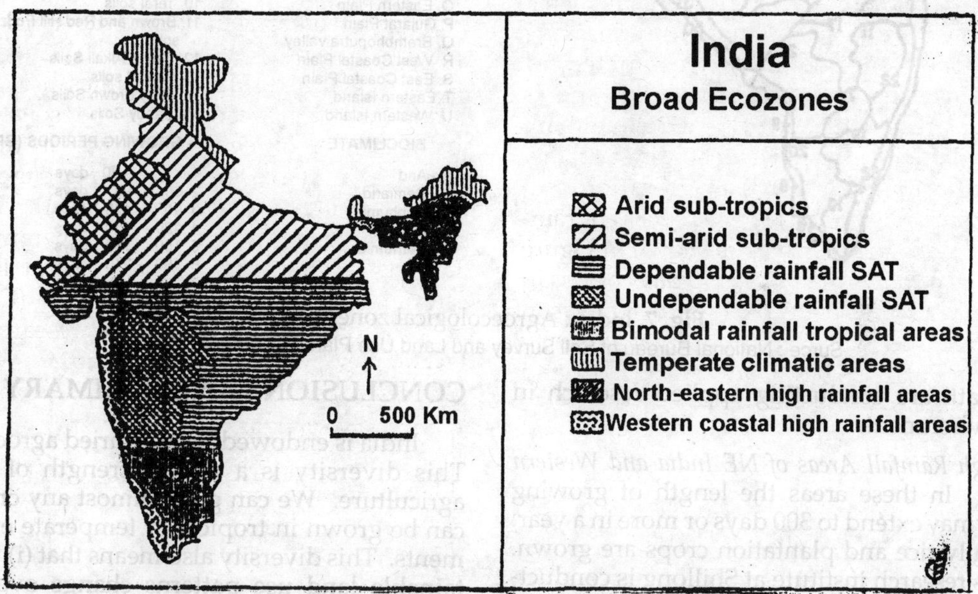

Fig. 6. India : Ecozones map

Sub-Humid Tropics : within the semi-arid tropiec the high rainfall (1200 mm) areas are characterized by a large surplus of water during the rainy season. The State Agricultural Universities at Faizabad, Jabalpur, Ranchi and Mohanpur are located here. Upland/lowland rice is the main crop during rainy season. One ICAR Institute, the Central Rice Research Institute at Cuttack is working actively on rice-based farming systems in this ecozone.

Bimodal Rainfall Tropical Areas : This zone consists of large tracts of Southern India in Southern Andhra Pradesh, Karnataka, Tamil Nadu and Kerala (excluding Western Ghats).

The mean annual rainfall is around 700 mm but since this is received during two seasons, there is a general scarcity of water in both the seasons. Drought is fairly common. Dryland Research Stations located at Anantapur, Bangalore and Kovilpatti are conducting applied research in this environment.

High Rainfall Areas of NE India and Western Ghats : In these areas the length of growing season may extend to 300 days or more in a year. Generally rice and plantation crops are grown. ICAR's research institute at Shillong is conducting studies on problems associated with the sustainability of hill agriculture practised in North-Eastern India, and there are a number of plantation crops research institutions located in Kerala (eg. Rubber Research Institute).

AGROECOLOGICAL ZONES MAP OF INDIA

The ICAR's National Bureau of Soil Survey and Land Use Planning has further sub-divided for the broad ecozones of India, into 54 zones based on physiography, soils, bioclimate, and growing periods. A map showing agroecological zones is shown in Figure 7 and can be used as a basis for regional/local planning.

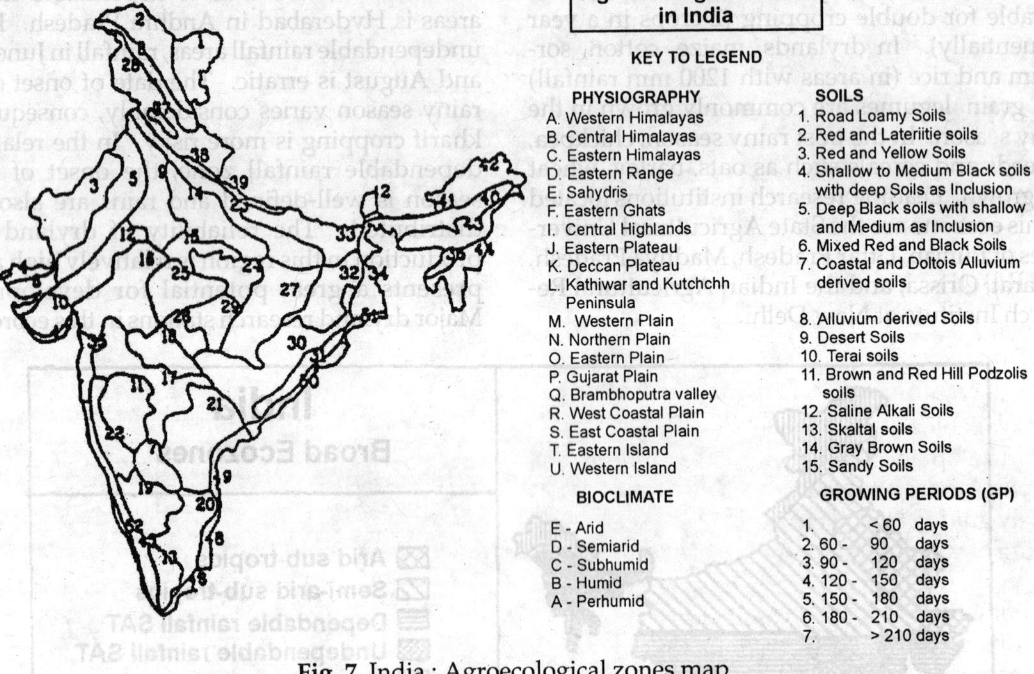

Fig. 7. India : Agroecological zones map
Surce : National Bureau of Soil Survey and Land Use Planning, Nagpur, 1991

CONCLUSIONS AND SUMMARY

India is endowed with a varied agroclimate. This diversity is a major strength of Indian agriculture. We can grow almost any crop that can be grown in tropical or temperate environments. This diversity also means that (i) the sustainable land use patterns change over short distances; (ii) no one package of agrotechnology can work across the country; and (iii) science-based agriculture will have to be supported at a fairly large number of research sites to effectively cover the ecological diversity. We must first fully understand the soil and agro-climatic resources, develop improved systems of farming suited to different agroecological zones and work with the development agencies and the farmers to fine-tune the technologies to their local conditions. This would lead to ecologically friendly and sustainable agriculture - the sole aim of all of agricultural research and development efforts.

5

SOILS INFORMATION AND SOIL DEGRADATION-RELATED CONSTRAINTS IN INDIAN RAINFED ZONES

J.C. KATYAL and K.P.R. VITTAL

INTRODUCTION

In the world and in particular, in our country India, the soil resources identification, and characterization are in infancy. As such a few data is only available for the soils under rainfed agriculture. The uplands being infested with multiple complex-slopes, the genesis of toposequences are heavily guided by soil erosion. Therefore, a continuum of soil orders occur on the sloppy terrain whose spatial variability is a difficult task to monitor with present available organisational set up. Taxonomic soil orders are being unique in their nature and existence, the characteristics are not affected by variations in the land utilization type. Hereunder an effort is made to describe the soil properties, their occurrences, soil-related constraints, evaluation of land utilization types and use of geographical information system for projecting land use policy with soil taxonomy backdrop. Familiarity with soil taxonomy [USDA 1978] is expected from the reader in this presentation.

CLIMATE AND SOILS

Nearly 80 per cent of the Indian land surface experiences arid and seasonally arid conditions. The amount of annual rainfall ranges between 1 to 14 m. A major portion of the total rainfall is restricted to three or four wet spells associated with fierce storms during intense rainy season. Violent rain storms promote runoff resulting in soil erosion during the rainy season. Pre-rainy season summer months are characterised by high day temperatures (40 to 45°C) and strong winds (10 to 50 kmph). Prevalence of these conditions contribute to wind erosion in soil left barren due to over-grazing. On the contrary, humid regions are subjected to severe water erosion due to floods. These effect the soil characteristics which, in turn, affect the land use. Henceforth is the need for the soils information.

Soil degradation - a quiet crisis, is engulfing the productivity of land all over the world. The resiliency of soil is finite and soil degradation cannot be reversed. Although similar in latitudes to northern hemisphere, process, factors and rate of soil degradation differ in southern hemisphere [Lal, 1987]. It was Khana in 600 AD, who recognized the process of soil erosion mediated degradation in 'Circa'. Soil degradation impairs soil structure, impoverishes soil of valuable organic matter, encourages fall in soil fertility and leads to miserable soil biology [Pande et al., 1988]. For most part of the hot plains and plateau, organic matter levels are poor because of high temperature related oxidation. With that soils are exceedingly low in nitrogen and humus levels, primarily because of climatic control and secondarily due to over exploitation of land [Jenny, 1980]. Due to subsistence agriculture and poverty, in our country almost all of the land is put to annual cropping irrespective of land capability. Distribution of the soils, the climatic characteristics and crops are described in Table 1. These physiographic regions are

Table 1. Extent of soil orders, their characteristics, climatic patterns, estimated crop growth and crops of the region

Taxonomy	Characterstics	Area (m ha and %)	Climate	Precipitation (mm)	PET (mm)	Water Deficit (mm)	Growing Period (days)	Land Utilisation type	Physiography
Cryorthents Cryothids	Shallow, sandy to loamy, skeletal soils with low AWC	15.6 (4.7%)	Cold Arid	< 150	< 800	< 300	<90	Millets and Wheat (in patches)	Western Himalayas
Camborthids, Calciorthids, Torripsamments Natrargids	Deep, sandy and coarse loamy, aline and medium AWC	29.6 (9.0%)	Hot Arid	< 300	1500-1900	1500-1800	< 90	Millets and pulses	Western Plain and Kutch Peninsula
Chromusterts, Rhodustalfs, and clayey, black Ustropepts	Deep, loamy red soils and clayey black soils with medium and high AWC	4.7 (1.4%)	Hot Arid	400-500	1800-1900	1500-1600	< 90	Millets, cotton and Oilseeds	Deccan Plateau
Ustochrept, stipsamments, Haplustalafs, Ustorthents, Ustifluvents	Deep, loam alluvium derived soils with medium, AWC and shallow to medium, sandy to loamy, gray brown soils with low to medium AWC	32.9 (10.0%)	Hot semi -arid	400-800	1400-1900	700-1000	90-150	Millets, wheat Pulses and maize	Northern Plain & Central High lands including parts of Gujarat Plains
Ustochrepts Ustropepts, Chromusterts, Ustorthents	Medium and deep, fine loamy and clayey, black soils with medium and high AWC	18.4 (5.6%)	Hot semi arid-	600-900	1600-2000	800-1200	90-150	Millets, Wheat and pulses	Central (Malwa) Highlands and Kathiawar Peninsula
Ustropepts, Ustorthents, Chromusterts	Shallow and medium loamy to clayey, black soils with low and meidum AWC; inclusion of deep, clayey, black soils, with high AWC	33.0 (10.0%)	Hot semi - arid	600-1000	1600-1800	800-1000	90-150	Millets, Cotton, and Pulses	Deccan Plateau

Table Contd...

Taxonomy	Characterstics	Area (m ha and %)	Climate	Precipit-ation (mm)	PET (mm)	Water Deficit (mm)	Growing Period (days)	Land Utilisation type	Physio-graphy
Ustropepts, Rhodustalfs, Plinthustalfs, Ustrothents	Medium, red loamy soils with low and medium AWC and deep clayey, black soils with medium and high AWC	20.8 (6.3%)	Hot semi - arid	600-1000	1600-1700	700-800	90-150	Millets, Rice and Oilseeds	Deccan Plateau & Eastern-ghats
Ustropepts, Haplustalfs, Rhodustalfs,	Shallow to deep, loamy, red soils with low to medium AWC	22.7 (6.9%)	Hot semi - arid	600-1000	1300-1600	400-700	90-150	Millets, Pulses and oilseeds	Eastern-ghats (TN uplands) & Deccan Plateau
Ustochrepts, Haplustalfs, Ustifluvents, Ustorthents	Deep, loamy, alluvium derived soils with meidum and high AWC	12.2 (3.7%)	Hot Sub-humid	1000-1200	1400-1800	500-700	150-180	Rice, Wheat Pulses and sugarcane	Northern Plain
Ustochrepts, Chromusterts,	Medium and deep, clayey black soils with medium AWC	8.2 (2.5%)	Hot sub-humid	1000-1500	1300-1500	600-700	150-180	Wheat, Pulses and millets	Central High lands (Malwa & Bundelkh-and)
Ustochrepts, Chromusterts, Ustorthents, Haplustalfs	Medium, loamy, red soils and medium deep clayey, black soils with medium and high AWC	13.7 (4.2 %)	Hot sub-humid	1000-1500	1500-1600	500-700	150-180	Wheat, Millets Cotton and Rice	Deccan Plateau & Central High lands (Bundel-khand)
Ustochrepts, Haplutalfs, Plinthustalfs, Rhodustalfs	Medium and deep, loamy red and yellow soils with medium AWC	13.2 (4.0%)	Hot sub-humid	1200--1600	1400-1500	500-700	150-180	Rice, millets and pulses	Eastern Plateau (Chhatisg-arh region)
Ustochrepts, Ustropepts, Haplustalfs, Plinthustalfs, Rhodustalfs	Shallow and medium, loamy, red soils and deep, loamy, lateritic soils with low and medium AWC	27.8 (8.5%)	Hot sub-humid	1000-1600	1400-1700	500-700	150-180	Rice, pulses and millets	Eastern (Chhota Nagpur) Plateau & Easterng-hats

Table Contd...

Taxonomy	Characterstics	Area (m ha and %)	Climate	Precipit-ation (mm)	PET (mm)	Water Deficit (mm)	Growing Period (days)	Land Utilisation type	Physio-graphy
Eutrochrepts, Udifluvents	Deep loamy, alluvium-derived soils with medium and high AWC	9.3 (2.8%)	Hot sub-humid	1400-1600	1300-1500	400-500	180-210	Rice, wheat and sugarcane	Eastern Plain
Eutrocherpts, Hapludalfs, Dystr-ochrepts,	Shallow and medium, loamy, forest and pod-zolic brown soils with low and medium AWC	17.7 (5.4%)	Warm sub-humid and humid	1600-200	800-1300	300-500	180-210(+)	Wheat, millets, maize and rice	Western Himalayas
Haplaguepts Hapludults, Udifluvents	Deep, loamy, alluvim medium and high AWC	11.7 (4.0%)	Hot per-humid	1400-	1000	100-	>210	Rice, jute and plantation crops	Assam and Bengal Plains
Haplumbrepts, Dystr-ochrepts, Hapludalfs, Hapludults	Shallow and medium brown red hill soils with low AWC	8.0 (2.4%)	Warm perhumi d	2000 4000	<1000	<100	>210	Rice and millets	Eastern Himalays
Dystrochreptes Hapludults, Rhodustalfs, Hapludalfs	Shallow and meidum, loamy red, yellow, lateritic soil with low AWC	10.7 (3.3.%)	Warm perhumi d	1600-2600	1000-1100	100-150	>210	Forest and rice in patches	North Eastern Hills
Ustropepts, Haplaguepts, Ustifluvents	Deep, loamy, coastal and deltaic alluvial soils with medium and high AWC	8.1 (2.5%)	Hot sub-humid	1200-1600	1200-1400	400-500	150->210	Rice and millets	Eastern Coastal Plain
Eutropepts, Haplustalfs, Hapludults, Dystraochrepts Udifluvents	Shallow and medium, loamy red and lateritic soils with low AWC	10.1 (3.0%)	Hot humid per humid	2000-3200	1400-1600	300-400	>210	Rice, tapioca, coconut and millets	Westernghats and Coastal Plain
Eutropepts Dystroppets, Hapludalfs, Udipsamments	Medium and deep, red loamy and sandy soils with very low to medium AWC	0.8 (03.%)	Hot per humid	1600-3000	1400-1600	300-400	>210	Forest coconut and rice	Islands of Andaman and Nicobar and Lakshaweep

Amended from Sehgal et al. 1990.

recognized as agro-ecological regions [Sehgal et al., 1990].

TOPOSEQUENCES IN DEGRADED LANDS

Soil erosion is a natural process of land formation and soil genesis resulting in potential loss of crop production. In alfisols of South India, the soil development by weathering is highly negligible and cannot offset the effects of erosion [Bronger and Bruhn, 1988].

In drylands of India, the changes in solum properties as a resultant of erosion are well felt. Occurrence of toposequence is a common phenomena in slopy lands. In undulating landscape as is under Hoshiarpur (Punjab), Dehradun (Uttar Pradesh) and Rakh-Dhiansar (Jammu and Kashmir) sub-mountane region, along with increasing soil depth the topsoil texture varies through loamy sands, sandy loam, loam, silt loam, to clay loam. Water stagnation in the valley is common in this region. In Solapur (Maharashtra) the toposequence is common with reddish brown soil on interfluve-ridge and black soil in foot valley.

In Rewa and Indore (Madhya Pradesh), light mixed red and black soil on hill slopes - shoulder and deep black clays in valleys are common. In Varanasi (Uttar Pradesh) alluvial soils, an inundated low land is common. In the oxisol region of Ranchi (Bihar) and Bhubaneswar (Orissa), red lateritic interfluve and shoulder areas and yellowish sandy clays in lower foot areas in regular. In Jhansi (Uttar Pradesh) and Hyderabad (Andhra Pradesh) soils are reddish alfisols on interfluve and linear region while in foot vertic soils to deep vertisols are common. In vertisols of Akola (Maharashtra) and aridisols of Hisar (Haryana) the depth of occurrence of calcium carbonate concretion layer is dependent upon the erosion that has affected the solum on the physiographic position on landscape [AICRPDA, 1983 a]. Major soil orders vary in piedmont plains flat topped with hill slopes and floods plains [Challa and Gaikwad, 1986] which is due to the confounded effect of erosion and soil development.

A benchmark soil project was initiated by the Indian Council of Agriculture Research in 1980. The chief objective of this project was to prepare an inventory of the soil potential of India. In all, 78 benchmark soils were identified which represent the soils of India. Since then 57 of these benchmark soils have been described in terms of agro-climatic situations in India including elements of their morphological and physico-chemical characteristics, geology and physiography, and soil classification units [Murthy et al., 1982]. Characteristics of typical soil series are presented in Table 2. There is a high variation in soil properties across the country [Sehgal et al., 1990].

The benchmark soils belonged to eight soil orders represented by Inceptisols (15), Alfisols (15), Vertisols (12), Aridisols (5), Entisols (6), Oxisols (2), Mollisols (1) and Ultisols (1). Figures in parentheses are number of soils. Soils (12) with an udic/aquic moisture regime represented largely by Inceptisols, occurred in the humid and sub-humid regions. Those with an ustic moisture regime (38) comprised mainly of vertisols and alfisols and belonged to the semi-arid tropical climate. Aridisols and entisols constitute the soils (7) with an aridic/torric moisture regime and are the soils of arid zones. Thus the majority of the soils (45) had an ustic to aridic moisture regime [Katyal and Sharma, 1991]. This situation was not unexpected because 75 per cent of the cultivated land in India is covered by the semi-arid tropics [Randhawa and Venkateswarlu, 1980].

With the exception of oxisols, alfisols with oxic characteristics, mollisols and ultisols were alkaline in reaction (mean pH 7.42). In general, soils with aridic to ustic moisture regimes had higher pH values than those with udic to aquic moisture regime. The majority of the soils was low to poor in organic matter content (mean organic matter per cent 0.87) and was calcareous (mean lime per cent 2.9). As the moisture regime changed from aquic to aridic, the organic matter levels declined while the lime content increased. A high lime content was a factor favouring higher pH of soils with ustic to aridic moisture regimes as was evident from a positive coeffi-

Table 2. Soil order, climate and soil fertility status of dryland research centres

Major soil order, climate	pH (1:2) electrical conductivity (mmhos/cm)	Organic carbon (%)	Phosphate Potash (kg/ha)	Probable deficiency of other nutrients	Research Centre
Alfisol					
Arid (AWC 5–6 cm)	6.0–7.0 0.03–0.05	0.25–0.30	10–15 150–200	Zinc	Anantpur
Semi-arid (AWC 5–13 cm)	5.5–6.5 0.05–0.20	0.50–0.75	8–12 180–250	Sulphur, Calcium, Zinc, Boron, Molybdenum	Hyderabad, Bangalore
Aridisol					
Arid (AWC 5–9 cm)	7.5–8.0 0.15–0.20	0.15–0.22	12–18 280–400	Zinc	Jodhpur, Hissar
Semi-arid (AWC 10–15 cm)	7.5–8.0 0.05–0.08	0.250–0.30	20–35 125–200		Dantiwada
Inceptisol					
Arid (AWC 5–6 cm)	7.5–8.0 0.75–0.90	0.20–0.38	15–25 250–350		Agra, Hoshiarpur
Dry sub-humid (AWC 20–25 cm)	7.0–7.5 0.05–0.30	0.10–0.25	15–35 80–150		Varanasi Rakh-Dhiansar
Oxisol					
Moist sub-humid (AWC 12–24 cm)	5.0–6.0 0.02–0.30	0.30–0.45	10–25 200–350	Calcium	Bhubaneswar, Ranchi
Vertic inceptisol					
Arid (AWC 9–11 cm)	7.5–8.0 0.15–0.20	0.50–0.70	25–30 350–425		Rajkot
Semi-arid (AWC 10–22 cm)	7.5–8.5 0.25–0.40	0.40–0.60	8–15 500–650	Sulphur, Calcium	Jhansi, Kovilpatti
Vertisol					
Semi-arid (AWC 18–40 cm)	7.8–8.5 0.20–0.35	0.20–0.35	20–40 500–800		Akola, Solapur, Indore, Bijapur, Udaipur
Dry sub-humid (AWC 12–15 cm)	6.5–7.0 0.15–0.19	0.10–0.20	10–15 450–500		Rewa

Climate classification is based on Thornthwaite; AWC is available water capacity
Adopted from Vittal 1986
For Centerwise (AICRPDA) data refer Sharma and Das 1992

cient of correlation between pH and lime [Katyal et al., 1982]. With the soil moisture regime falling to aridic or toric the clay content of soils decreased sharply. The presence of sand rich entisols and aridisols in this moisture regime soil groups contributed largely to the fall in clay content [Katyal and Sharma, 1991].

SOIL FERTILITY STATUS

Dryland agriculture is extensive and is practised under varied soil and climatic conditions. The soil characteristics and associated climatic conditions of dryland regions are presented in Tables 3 and 4. The soils are usually low in organic matter, low in available phosphorus and medium in potassium. The soil pH is almost neutral. Except in vertisols, the cation exchange capacity is lesser than 10 meq/100 g soil. In the halplustalf of Bangalore, phosphorus fixation is related to clay content. Available phosphorus (%) is given by 100 (2x clay content %). The formation of compact sub-surface layers like clay accumulations constrain the available rooting space, which in turn, reduces full utilization of the solum depth. As the sub-soil recharge from rainfall is reduced, its utilisation also is constrained [Vittal, 1986].

A report on available nitrogen status of Indian soils made by Ghosh and Hassan [1980] showed that 63 per cent of the total districts surveyed in India are low in available nitrogen, while 33 per cent fall under medium fertility class. Nitrogen content in these soils varies widely from 0.01 to over 1 per cent. The soils of India have very low nitrogen content [Ramamoorthy et al., 1971; Ramamoorthy and Velayutham, 1976; Prasad and Subbiah, 1982; Prasad and Thomas, 1982]. The effect of soil types on total nitrogen content was marked by the climatic conditions prevailing in India. Muttatkar and Roy Chaudhuri [1959] reported that nitrogen content in both black and alluvial soils increased with rainfall and decreased with increase in temperature. Nitrogen and carbon losses were most pronounced in areas having dry climates with long history of human occupation [Jenny and Roy Chaudhuri, 1960]. Low level of soil nitrogen was attributed primarily to climate and cropping systems that were in vogue. They further observed that high temperatures and low rainfall in the western and central plateau and high temperates in north-eastern tract and eastern region limit nitrogen accumulation in soil. Nitrogen content is relatively higher in hilly regions of north east and north-west India.

The data on district level P fertility status show that P deficiency is widespread in irrigated as well as unirrigated areas [Kanwar et al., 1984; Tandon, 1987]. An analysis of data from 24 districts which account for 50 per cent of the total sorghum production showed that the mean available P was low in 15 and medium in 9 districts. Based on nutrient surveys of crops it was found that groundnut crop was deficient in P in 98 per cent cases, fingermillet in 94 per cent, sorghum in 24-52 per cent, maize in 28 per cent and pearlmillet in 6 per cent cases [Singh and Venkateswarlu, 1985].

Soil deficiencies of potassium in the unirrigated semiarid tropics are important only in certain coarse-textured soils, in some red soils, and at high yield levels obtained without K application for a period of time [AICRPDA, 1983 b, 1988; Venkateswarlu, 1987]. Some possible areas of K deficiencies are soils of the Anantapur and Karimnagar area in Andhra Pradesh [Venkateswarlu, 1984], red soils of the Bangalore area [AICRPDA, 1983 a], and several alluvial soils in Uttar Pradesh [Ghosh and Hasan, 1976]. Soil test ratings may not always agree with the K status of standing crops. In Kurnool and West Godavari districts of Andhra Pradesh, rated as medium to high in available K, 29-60 per cent of the groundnut crop samples were reported to be deficient in K [Subba Rao, 1975].

Although S deficiencies occur to a varying degree in 90 districts, the available database is grossly inadequate to delineate S - deficient areas [Takkar, 1988; Tandon, 1986]. The following assessments are particularly relevant for the semi arid tropics.

- On an average, 17 per cent of the soils in Gujarat State are considered to be deficient in sulfur [Patel, 1988].

Table 3. Characteristics of some typical soils

Soil taxonomy	Clay (%)	pH (1 : 2 .5)	EC-1 dS m (1 : 2.5)	OC (%)	$CaCo_3$ (%)	CEC	Ext. aridity (cmol (p+kg)	Ext. Fe (%)	BS (%)	ESP (%)
Typic Cryorthent *	8.1	8.7	0.5	0.06	2.0	0.8		–	100	–
Typic Cryorthid	19.2	8.1	0.2	0.85	13.7	69	–	–	95	–
Lithic Camborthid	24.6	7.8	–	0.42	Tr	74	–	–	91	
Typic Torripsamment	6.7	8.2	–	0.08	7.6	2.4	–	–	92	
Typic Natrargid	47.4	8.4	4.6	0.17	2.5	27.0			100	28
Typic Pellustert	67.4	8.0	–	0.63	3.4	71.3			94	5
Typic Rhodustalf	24.8	6.3	–	0.18	–	13.1	–		57	–
Typic Ustochrept	17.7	7.9	0.6	0.19	1.1	9.2		0.70	95	–
Typic Natrustalf	30.5	9.7	3.0 (ECe)	0.17	5.7	13.5	–	–	100	84
Typic Ustipsamment	61.	7.8	0.2	00.5	–	6.0	–		98	11
Typic Chromustert	55.8	8.1	0.1	0.30	4.7	49.7	–	–	98	9
Vertic Ustochrept	63.2	7.2	0.2	0.54	Tr	61.9	–		98	
Lithic Ustorthent	16.7	7.9	<0.2	0.15	30.0	29.8	–	–	100	
Vertic Ustropept	39.0	7.8	0.4	0.35	34.0	40.0	–	–	100	
Typic Chromustert	70.2	8.4	0.5	0.49	1.6	61.7	–		100	9
Typic Pellustert	61.9	9.2	0.27	0.39	6.9	59.8	–		100	14.0
Udic Rhodustalf	32.7	7.0	–	0.60	–	17.2	0.16	2.1	99	–
Oxic Paleustalf	45.3	6.3	–	0.45	–	8.8	0.44	3.8	81	–
Typic Haplustalf	21.8	8.5	–	0.76	5.4	23.4		1.3	85	–
Typic Eutrochrept	22.7	7.3	0.4	0.12	–	9.8	–		89	–
Typic Hapludoll	17.5	7.1	–	0.72	Tr	8.6	–	0.5	88	–
Typic Chromustert	63.9	7.6	0.1	0.27	2.5	49.9	–		98	
Entic Chromustert	47.9	8.0	<0.1	0.28	Tr	50.1	–		98	6.5
Udic Chromustert	73.4	8.1	0.2	0.41	6.2	63.7	–	–	100	7.0

Table Contd...

Table Contd...

Soil taxonomy (Subgroup)	Clay (%)	pH (1:2.5)	EC-1 dS m (1:2.5)	OC (%)	CaCo (%)	CEC	Ext. acidity (cmol (p+kg)	Ext. Fe (%)	BS (%)	ESP (%)
Udic Haplustalf	38.3	7.3	0.1	0.65	Tr	25.5	–	–	78	9.7
Typic Haplustalf	32.0	6.0	<0.2	0.20	–	12.6	–		90	–
Udic Rhodustalf	30.9	6.2	–	0.22	–	10.4	–	–	60	–
Typic Paleustalf	35.0	6.9	<0.2	0.15	–	14.7	–	–	89	–
Ultic Paleustalf	33.8	6.3	–	0.11	–	11.0	–	2.2	81	–
Typic Hapludult	17.8	4.8	–	0.43	–	8.3	7.9	1.3	32	–
Typic Plinthustalf	34.2	6.0	–	0.38	–	17.5	–	3.0	59	–
Typic Ustifluvent	13.1	8.9	0.24	0.23	6.0	11.3	–	–	100	
Aquic Haplaquept	39.4	8.1	0.33	0.33	–	–	–	–	100	–
Typic Haplustlf	24.0	8.1	0.7	0.23	–	13.7	–	–	96	–
Mollic Haplaquept	20.5	6.9	–	0.52	1.2	5.5	–	1.2	97	–
Mollic Hapludalf	>9.0	0.5	0.1	0.45	–	5.6	–	–	64	–
Typic Dystrochrept	24.3	4.5	–	0.53	–	14.5	–	–	55	–
Aeric Fluvaquent	23.6	6.6	–	0.44	–	14.8	–	0.6	92	–
Aeric Haplaquept	9.2	5.2	–	0.31	–	4.3	–	–	40	–
Typic Hapludalf	18.8	5.2	–	0.72	–	12.6	–	–	52	–
Ultic Hapludalf	32.9	4.4	–	0.64	–	16.1	–	–	54	–
Vertic Halaquept	41.9	7.6	1.8	0.23	–	20.1	–	0.49	100	30.0
Udorthentic Pellustert	54.6	9.4	0.78	0.21	9.1	52.2	–	–	100	39.2
Oxic Dystropept	52.4	4.8	–	0.50	–	4.8	3.1	3.4	18	–
Typic Haplorthox	44.8	5.6	–	0.45	–	2.6	0.2	5.5	43	–
Typic Dyystrochrept	42.7	6.2	0.07	0.56	–	14.0	–	–	50	–
Typic Tropofluvent	37.2	5.6	0.04	0.25	–	10.0	–	–	60	–
Fluventic Sufaquent	33.2	3.6	6.3	3.9	–	9.0	–	–	45	–

Amended from Sehgal et al. 1990

Table 4. Micronutrient contents and some important properties of 57 benchmark soils

Soil order		DTPA-extractable (mg/kg)				Total (mg/kg)			Fe (%)	pH	Lime (%)	Organic matter	Clay (%)
		Zn	Cu	Mn	Fe	Zn	Cu	Mn					
Udic - acquic moisture regime													
Entisols	(1)	0.29	5.33	8.1	44.1	87.6	57.5	388	3.50	8.07	2.40	1.5	52
Inceptisols	(6)	0.81	2.61	40.4	30.4	53.4	26.4	276	2.71	7.07	1.80	1.2	24
Alfisols	(3)	0.55	2.96	50.2	56.8	40.9	24.0	187	1.86	5.77	0.78	0.7	27
Mollisols	(1)	1.86	1.99	47.3	59.0	30.5	30.3	388	3.18	6.87	1.65	2.1	34
Oxisols	(1)	1.40	2.92	155.4	16.3	74.3	54.4	800	4.61	6.14	0.80	2.6	43
Ustic moisture regime													
Entisols	(3)	0.61	0.56	17.9	18.4	40.1	19.4	275	1.82	7.58	0.88	1.1	7
Inceptisols	(9)	0.46	2.05	24.6	18.4	64.5	38.6	687	3.25	7.61	3.04	0.8	28
Alfisols	(12)	0.56	1.37	26.6	21.3	45.8	34.6	437	2.59	7.11	1.88	0.8	17
Vertisols	(12)	0.42	1.49	12.4	9.1	63.0	63.0	731	3.95	8.06	5.51	0.7	53
Ultisols	(1)	0.28	0.49	40.1	17.5	43.1	22.5	400	5.37	5.37	0.20	0.9	14
Oxisols	(1)	0.31	0.54	18.1	19.3	70.3	58.1	405	6.32	5.32	0.25	2.3	27
Aridic/torric moisture regime													
Entisols	(2)	0.25	0.51	7.8	7.8	38.4	17.2	431	2.47	8.44	4.55	0.3	8
Aridisols	(5)	0.38	1.28	12.70	9.6	61.4	50.7	814	4.52	8.33	4.10	0.5	17
Overall range		0.12 to 2.80	0.15 to 5.33	4.0 to 102.0	3.4 to 68.1	20 to 97	11 to 148	38 to 1941	1.3 to 8.0	4.16 to 10.50	6.20 to 8.90	0.10 to 2.00	3 to 64
Mean		0.54	1.71	26.0	20.3	55	41	537	3.3	7.42	2.92	0.87	28
SDm +		0.07	0.61	3.7	2.2	3	3	45	0.2	0.18	0.32	0.07	2

Data are means of various soil orders occurring within a soil moisture regime. Figures in paranthese are number of soils.
pH is in 1 : 2 water suspensions.
Adopted from Katyal and Sharma 1991.

- Sulfur deficiencies are considered to be limiting crop production in many parts of Karnataka [Badiger and Shivraj, 1988].
- In Madhya Pradesh, results of 1900 soil analyses from 12 districts were deficient in sulfur [Shinde, 1984].
- In Maharashtra, very meagre information is available on the sulfur status of soils under different climatic conditions [Karle, 1985].

Clay content in fact predominantly controls the micronutrient retention and availability. Variation in pH, lime, organic matter and clay contents with the changing soil moisture regime exerted a strong influence on the micronutrient distribution.

The mean data on some soil properties which can influence the distribution of the DTPA - extractable and total contents of micronutrient cations viz. Zn, Cu, Mn and Fe are presented in Table 4 [Katyal et al., 1982; Katyal and Sharma, 1991].

The currently followed critical limits of Zn availability (<0.6 mg Zn/kg soil by DTPA soil test) showed that 43 out of 57 benchmark soils were deficient in Zn. Similar to these findings a likelihood of a widespread Zn deficiency in other Indian soils was reported by Katyal et al. [1985]. Parallel to Zn availability levels, concentration of deficient soils varied among moisture regimes. Only 42 per cent of the soils with an aquic-udic moisture regime were categorised as deficient, while 84 to 96 per cent of the soils under the influence of an ustic to aridic moisture regime were Zn deficient. Among the soil orders, the maximum number of Zn deficient soils belonged to alfisols (14 out of 125), vertisols (10 out of 12), inceptisols (11 out of 16), and aridisols (4 out of 5). Acid, leached ultisols, (DTPA-Zn 0.2 mg/kg soil) will be Zn deficient.

Compared to Zn the deficiency of remaining micronutrients (Cu, Mn and Fe) in benchmark soils was limited. Only one benchmark soil showed values below the critical availability limits each of Cu (<0.2 mg Cu/kg soil) (Oxic Dystropepet) and Fe (<4.5 mg Fe/kg soil) (Vertic Ustropept). None of the soils tested less than the critical limit of Mn availability (<2.0 mg Mn/kg soil). On an overall basis, N and P are invariably limiting, due to poor abundance micronutrients are likely to constrain crop production more often in soils with ustic to aridic moisture regimes [Katyal and Sharma, 1991].

DEGRADATION INDUCED SOIL CONSTRAINTS

Degradation of soil over time occurs due to water, wind, management, ground water table, etc. resulting in paramount damage adversely affecting productivity. An Indian scenario is described below [Katyal et al., 1992].

Erosion of top soil : In India, the states of Gujarat, Maharashtra, Andhra Pradesh, Tamil Nadu, and Kerala are the worst sufferers of erosion. It is estimated that out of 175 m ha of degraded land (VI Plan Document), water erosion alone breaks 90 m ha [Yadav, 1986]. In the process, nearly 6,600 mt of soil is busted and carried to the ocean annually [Brown, 1984; Dhruvanarayana and Sastry, 1985]. Of the total area affected by erosion, sheet erosion plays havoc with about 72 m ha of red soils comprising of alfisols, ultisols and oxisols (4-10 t/ha/year), and 88 m ha of black soils representing vertic soils and vertisols (11-43 t/ha/year). Gully erosion is severe on 4 m ha (>33 t/ha/year), while hill side erosion afflicts 13 ha (>80 t/ha/year) [Dhruvanarayana and Sastry, 1985]. Extensive and deep gullies have reined massive area in Uttar Pradesh, Himachal Pradesh, Madhya Pradesh, Rajasthan, Gujarat, Maharashtra, Punjab, Bihar, Tamil Nadu and West Bengal [Singh and Wali, 1962; Patnaik, 1975; Lal and Banerji, 1974].

Hard setting : Excessive wetness induced degradation is an outcome of its adverse effect on certain physical and chemical characteristics. For instance, loss of soil strength due to wetness encourages soil erosion and facilitates hard setting. Wetness triggers anoxic soil environment. As a consequence, chemical and biological reduction under anaerobic soil conditions diminishes or enhances the availability of some essential plant nutrients. Zinc is a typical example in that its availability falls with anaerobiosis. On the other hand, solubility of oxidation-reduction sensitive nutrients like Fe and Mn can rise to potentially toxic levels [Fausey and Lal, 1990]. In India flooded lands are 6 m ha in extent [Yadav, 1986].

Mechanised agriculture, more common to western world, has been recognised to create quick soil degradation due to hard setting. Data are also available that suggest that hard setting can take place without the use of powered machinery and inappropriate management. Hardsetting of cultivated soil usually involves slumping which is a process of compaction without the application of external load. In a previously cultivated or loosed soil, hardsetting involves the collapse of some or all of the artificially/newly aggregated structure during and after wetting and hardsetting without restructuring during drying. Poor aeration, impeded infiltration and accelerated runoff and erosion contribute to this situation. Soils having - (i) cementing material such as high sesquioxides, (ii) low organic matter (2%), (iii) textures ranging from loamy sands to sandy clays, and (iv) dominance of hydrous micas and/or kaolinite type clay minerals [Mullins et al., 1990] are more vulnerable to hardsetting. Hardsetting soils impose following agronomic limitations :

- delay in seeding and tilling until soil is sufficiently wet
- poor tilth for seed bed
- crusting to prevent emergence
- constrained root growth
- poor growth of direct drilled crops

Chalka (alfisols) soil of Telangana region in Andhra Pradesh [Rastogi, 1991] are a typical example of hardsetting soils [El-Swaify et al., 1985]. Oxisols, alfisols and ultisols of Karnataka, Orissa, and Bihar and particularly those with rainfed agriculture also experience this problem. Treatment with waste organic materials [Gupta et al., 1984; Reddy et al., 1985] has been shown to minimize hardsetting and increase crop yields. In the hardsetting alfisols of Telangana, annual erosion rates ranged between 1-5 t/ha/year in cultivated fields. In contrast, with grass cover hardsetting was reduced and erosion rate fell to 0.5 t/ha/year [Sharma et al., 1988].

Wind erosion : Accelerated wind erosion and desertification are largely a result of man made problems. Aridization mediated wind erosion is closely linked with land misuse. Interventions like removal of vegetation cover, introduction of intensive agriculture, uncontrolled and excessive grazing, and large unprotected fields devoid of protective vegetative caps accelerate wind erosion [Sdasyuk, 1986]. In arid zone of India, an area of 0.5 m ha surveyed by Landsat Thematic Mapper sub-sense showed that 42 per cent area is affected by wind erosion [Raina et al., 1991]. About 58 per cent of western Rajasthan is covered by moving and semi-stabilised sand dunes [Yadav, 1986], where sand migration is about 5000 years old [Vishnu Mittre, 1977]. In Punjab, Haryana, Gujarat and Rajasthan, desertic conditions range 31.7 m ha. Available information suggest that cropping on marginal lands increased during 70's and 80's by 54 per cent. Coincidently, during the same period an equivalent decrease in pasture area took place. According to Mann [1982], 72 per cent of the area in the desertic Rajasthan is highly vulnerable, and 19 m ha of the area is moderately susceptible to degradation.

Nutrient exhaustion : The data collected under All India Coordinated Research Project on Long Term Fertilizer Experiments over a decade of intensive cropping pointed to a heavy drain on the soil nutrient pool. Soil test values confirmed a declining trend in soil fertility with time in entisols, inceptisols, vertisols and alfisols [Abrol and Katyal, 1990]. Alfisols, ultisols and oxisols with low cation exchange capacity exhibited higher nutrient depletion sensitivity due to intensive agriculture.

Irrigation : The direct effect of mismanagement of irrigated lands (irrigation without drainage) sets in motion degradation through salinity/alkalinity and water-logging. Reduction in effective cultivated area and productivity are the ultimate effects of irrigation. Also, faulty water management practices and irrigation cause seepage loss and impedance of drainage. Rapid rise on water table is the result that renders soil liable to waterlogging and secondary salinization. This phenomenon is of extensive occurrence in command areas [Joshi, 1987]. A measurable area in western and south western parts of Haryana and Punjab is endangered by rising saline groundwater table. At Hisar, the groundwater table in 1967 was perched at 16.7 m depth. By 1982, it galloped to within 2 m from the surface. The area affected by soil salinity and alkalinity is about 8 m ha [Yadav, 1986]. The coastal salt lands are highly erosive. Their erodability increased with depth indicating that the soils were susceptible to check bank and gully erosion in Maharashtra [Joshi and Karekar, 1985].

SOIL CONSTRAINTS FOR PRODUCTION

Regular soil erosion increases a magnitude of soil related constraints to production. The constraints are as follows:

Physical

- reduced rooting depth
- loss of soil water storage capacity
- crusting
- soil compaction
- hardening of plinthite
- soil colour
- soil albedo
- loss of clay and colloids
- influences soil tilth and consistency
- exposure of sub-soil
- rills and gullies
- changes in micro-relief

- mechanization is a problem
- more soil variability
- critical nature of wetness requirements for tillage

Chemical

- low CEC
- deficiency of N, P, K, Zn, S, etc.
- toxicants like Al, Mn
- high soil acidity

Biological

- biomass
- activity of microfauna like earthworms, etc.

Soil constraints are almost similar in tropics (Table 5) and semi arid tropics (Table 6).

The direct and indirect effects are as follows:

Direct

- decrease in soil depth
- decrease in organic matter content
- increased salinity, alkalinity, or acidification
- desertification
- increased soil loss

Indirect

- declining yields
- change in biological population
- increased incidence of crop failure
- increased surface stoniness
- increase surface crusting
- reduced quality of ground water and aquifers

The effect of soil properties on land use patterns are presented in Table 7.

Reduction in topsoil depth reduced yields in Dehradun [Yadav, 1984] and Hyderabad [Vittal et al., 1990 a, b and c, 1991] (Table 8).

EVALUATION OF SOIL CONSTRAINTS IN VARIOUS SOIL ORDERS

Many systems of land and soil evaluation are in use for deciding land uses. Land capability classification of the Soil Conservation Service of the U.S. Department of Agriculture [Klingebiel and Montgomery, 1961] has been adopted widely. Frame work of land evaluation [FAO, 1983] is also in use. Raj Kumar and Saxena [1985] have evaluated the benchmark soils of India using

Table 5. Soil-related constraints in the rainfed tropics

Constraint	Oxisols	Ultisols	Alfisols	Inceptisols	Vertisols
Erosion	2	2	3	3	3
Compaction	3	3	3	2	2
Crusting	1	1	3	3	2
Drought	2	2	3	1	2
Shallow effective rooting depth	3	3	3	1	2
Trafficability	1	1	1	1	3
Supraoptimal soil temperature	2	2	3	3	2
Chemical/ nutritional	3	3	2	1	2

3 = severe; 2 = moderate; 1 = slight constraint   Adopted from Lal 1987

Table 6. Soil-related constraints in the rainfed semi-arid tropics

Order	Great-group	Major constraints
Alfisols	Plinthustalfs	Soil volume, rooting, Ms
	Natrustalfs	Sodium problems
	Paleustalfs	Erosion, Ms
	Kandiustlfs	Erosion, nutrient supply, Ms
	Kanhaplutalfs	Erosion, nutrient supply, Ms
	Rhodustalfs	Erosion, P fix, Ms
	Haplustalfs	Erosion, Ms
Entisols	Ustifluvents	Ms
	Ustorthents	Ms
	Ustipsamments	Extreme Ms
Inceptisols	Ustochrepts	Ms, slope
	Ustropepts	Ms, slope
Mollisols	Natrustolls	Sodium
	Paleustolls	Ms
	Calciustolls	P fix, Ms
	Argiustolls	Ms
	Haplustolls	Ms
Oxisols	Acrustos	Ms, P fix, low nutrients
	Eutrustox	Ms, P fix,
	Haplustox	Ms, P fix, low nutrients
Ultisols	Plainthustults	Ms, low perm, acidity
	Paleustults	Ms, acidity
	Kandiustults	Ms, aicidity, low charge
	Kanhaplustults	Ms, acidity, low charge
	Rhodustults	Ms, acidity, P fixation
	Haplustults	Ms, acidity
Vertisols	Chromusterts	Tillage, Ms
	Pellusterts	Tillage, Ms
Andisols	Haplustands	Ms, P fix,

Ms = Moisture stress ; fix = fixation; perm = permeability   Adapted from Eswaran 1992

Table 7. Importance of soil-related characteristics on 3 types of land uses

Characteristic	Agriculture	Improved Pasture	Commercial forestry
Soil depth	XX		
Ground-water depth	XX	X	X
Slope	XX	X	X
Erodibility	XX	X	X
Topsoil texture	X		
Surface structure	XX	X	X
Bearing strength		XX	X
Viscosity		X	
Permeability	X	X	
Shrink-swell properties	X	X	X
Sub-surface texture	X		
Sub-surface chemistry	X	X	X

XX = major importance   X = ancillary importance   Amended from Swindale 1991

Table 8. Effect of loss of topsoil on crop yields

Crop		Loss in crop yield	
Sorghum	(1)	138	kg/ha/cm
Castorbeann	(1)	51	kg/ha/cm
Pearlmillet	(1)	12-17	kg/ha/cm
Pigeonpea	(2)	21.50	kg/ha/cm
Sunflower	(3)	35	kg/ha/cm
Cowpea	(3)	65	kg/ha/cm
Cenchrus ciliaris (green)	(3)	11-31	kg/ha/cm
Maize	(4)	21	t/hatopsoil
Wheat	(4)	8	t/ha/top soil

1. Vittal et al 1990 a 2. Vittal et al 1991 3. Vittal et al 1990c 4. Yadav et al 1984

land evaluation method proposed by Riquier et al. [1970]. Presently soil constraints for production as arrived through revised Storie Index Soil Rating method [Storie, 1978] for 63 bench mark soils [Sohan Lal, 1989]. The index is a function of four major factors, viz. rating on the basis of characteristics of physical profile, rating on the basis of surface texture, rating on the basis of slope and rating on the basis of drainage, alkali condition, nutrient (fertility) level, acidity, erosion and micro-relief. The index rating in per cent has been calculated by multiplication of fractions of the above four. The soils are graded into 6 groups as : 1-excellent, 2-good, 3-fair, 4-poor, 5-very poor, 6-non agricultural. Average Storie index soil ratings and grades computed for different great groups are presented in Table 9 and effects are briefly described below.

Alfisols: Soils of fine silty mixed mesic family of Typic Haplustalfs, fine loamy mixed hyperthermic family of Plinthustalfs clayey skeletal mixed isohyperthermic family of Udic Rhodustalfs occurring on nearly level to gently sloping lands are rated grade 2. Soils of fine mixed hyperthermic family of Aeric and Vertic Ocharqualfs, fine loamy mixed hyperthermic family of Typic Hapludalfs and Ustic Paleustalfs and fine mixed hyperthermic family of Typic Hapludalfs and Ultic Paleustalfs and fine mixed isohyperthermic family of Oxic Haplustalfs, Typic and Oxic Paleustalfs are grouped in grade 3. These soils have relatively more limitations in drainage, fertility status, acidity and setting of the soils. The soils of fine mixed hyperthermic family of Typic Haplustalfs are grouped in grade 4 and these soils have more drainage problems and limitations due to low fertility status and setting.

The soils having the limitations in setting and surface texture soils occurring on steep slopes are grouped in grade 5 while salt affected soils fall under grade 6. The soils of grade 2, 3 and 4 can be used for agricultural crops while the soils of grade 5 may be brought under cultivation by terracing and adopting suitable soil and water conservation measures. The soils of grade 6 may be reclaimed with application of amendments and provision of drainage before bringing under cultivation.

Aridisols : Soils are grouped in grade 4 and productivity of these soils may be increased by introduction of irrigation or in-situ water conservation measures and addition of organic manures. Natrargids and Salorthids have been grouped in grade 6 as soils are affected by coastal salinity and sodicity. Their location renders them difficult for reclamation and cultivation on sustained basis. However, adaptable salt tolerant plant species may be introduced on selective basis.

Entisols : Deep soils developed in alluvium (Fluvents) qualify for grade 1. The soils of Usti- and Torripsamments qualify for grades 3 to 4 and the productivity of these may be increased by addition of organic manures, fertilizers and following water use efficiency techniques. The soils of Aquents suborder qualify for grade 3 and is infested with the problem of the drainage. These soils are suited to upland rice depending upon the suitability of the climate. Lithic and skeletal soils occurring on steep slopes qualify for grade 6.

Table 9 Average rating of productivity constraints in 63 bench mark soils

Order	Great-group		Physical	Topsoil texture	Slope	Drainage alkali/ fertility	Overall (Storie rating)	Land capability classificati on
Alfisol								
	Ochraqualf	(3)	1	1	1	3	3	IIIW
	Hapludalf	(3)	1	1	1	4	4	IVe
	Haplustalf	(3)	2	1	1	2	3	II c,s-IVe
	Natrustalf	(2)	1	1	1	6	6	IIIs
	Paleustalf	(4)	1	1	1	2	3	IIe,s-IIIc,e
	Plinthustalf	(1)	1	1	1	1	2	IIs
	Rhodusalf	(2)	2	2	1	2	3	IIIs
Aridisol								
	Natrargid	(1)	1	2	1	6	6	IVs
	Paleargid	(1)	1	2	1	3	4	IIIe
	Caciorthid	(1)	3	1	1	2	4	IVc
	Cambarthid	(1)	1	2	1	6	6	IVs
	Salorthid	(1)	1	2	1	6	6	IVs
Entisol								
	Fluvaquent	(1)	1	1	1	3	3	IIIw
	Torrifluvent	(1)	1	1	1	1	1	IVc
	Udorthent	(1)	4	2	4	4	6	VIe
	Ustorthent	(1)	4	2	4	4	6	VIe
	Torripsamment	(1)	1	1	1	3	4	VIc
	Ustipsamment	(2)	1	2	1	2	3	IIIc
Inceptisol								
	Halaquept	(2)	1	1	1	6	6	IIw-IIIs
	Halaquept	(1)	1	1	1	3	3	IIIw
	Dystrochrept	(1)	1	1	1	3	3	–
	Eutrochrept	(3)	1	1	1	2	4	IIs-IIIw
	Ustocherept	(6)	1	1	1	1	2	IIc, e,s,-IIIs
	Dystropept	(1)	2	3	1	2	4	IIIs
	Ustropept	(2)	2	2	1	3	4	IIIs
Mollisol								
	Hapludoll	(1)	1	1	1	1	1	I
Oxisol								
	Etrthox	(1)	1	2	1	3	4	–
	Haplorthox	(1)	1	2	1	2	4	IIIs
Ultisol								
	Haplustult	(1)	3	1	1	2	4	IIIe
Vertisol								
	Chromustert	(9)	1	2	1	2	3	IIs-IIIe,s
	Pellustert	(4)	1	2	1	3	4	IIIs

– Rating suitable for production is given as below :

Excellent	1	80–100%	Poor	4	20–39%
Good	2	60–79%	Very poor	5	10–19%
Fair	3	40–59%	Not useful for agriculture		< 9%

Figures in parantheses are number of soil series and capability classification was as per Klingelbiel and Montogomery 1991. Storie index soil rating method was as per Storie 1978.

Amended from Sohan Lal 1989.

Inceptisols : Soils of Fluventic and Typic Eutrocherepts, and Typic and Udic Ustochrepts are grouped in grades 1 to 2. The soils are good in fertility status and occur in plains and river terraces. The soils of Haplaquepts due to drainage problems, Typic Dystrochrepts due to low fertility status and acidity and Vertic Ustocherpts due to clay texture are grouped in grade 3. These soils may be used for cultivation by providing suitable drainage conditions and application of manures and fertilizers. The soils of Oxic Dystropepts and Vertic Ustropepts and Eutropepts are grouped in grade 4 due to severe acidity and poor drainage, respectively. These soils may also be used successfully by adopting the measures indicated for grade 3 soils but may need amendments in Oxic Dystripepts. Due to depth limitation and clayey texture, the soils of the Lithic Vertic Ustochrepts are put in grade 5 while the salt affected soils, Halaquepts have been grouped in grade 6. The grade 5 soils may be used for pasture development while in grade 6 soils are adaptable to only salt tolerant plant species.

Mollisols : Soils occurring in terai region of Uttar Pradesh of coarse-loamy textural family class qualify for grade 1 which are highly productive.

Oxisols and Ultisols : Eutrorhox, Haplorthox and Haplustults which are relatively poor in fertility status qualify for grade 4. These soils are slightly acidic in nature and are expected to perform better with lime amendment.

Vertisols : Soils of Chromusterts and Pellusterts great groups belonging to fine and very fine textural family classes are grouped in grade 3 to 4. The main limitations of these soils are their susceptibility to erosion, poor tilth, high clay content and poor drainage conditions. The fertility of these soils is medium but management is difficult. With proper management, improved water intake, with vertical mulching and tillage, addition of organic residues, and other agronomic, like dry seeding, soil and water conservation practices like broadbed and furrow, compartmental bunding etc. good yields of climatically suitable crops can be obtained on sustained basis. Pellusterts great group are graded 5 as the soils are affected by sodicity especially in subsurface. Upland crops like rice may be preferred.

Most of the soils fall under grades 1, 2 and 3 which are good for agriculture. Grade 4 use is limited due to very fine or very coarse in surface texture with drainage problems, occurrence on sand dunes, skeletal nature of the soils, and soil depth limitations. Moderately deep soils occurring on steep slopes fall under grade 5 while shallow soils on very steep slopes with erosion hazards, and salt-affected soils are grouped under grade 6.

We believe that the above presented case study is only an example of preliminary exercise. Any effort in this direction needs the evaluation of constraints identified being tested on on-farm scale for monitoring the realities. This is necessitated by location specific problems in upland / rainfed agriculture.

USE OF GIS WITH SOILS INFORMATION FOR PROJECTIONS

From a knowledge of topography and the distribution of soils and their properties within an agro-ecological region and from knowledge of soil distribution, of depth to the water-table and soil research data we can prepare maps using Geographical Information System (GIS) for different uses. With added information from crop production research we can obtain maps or tables on crop responses to water and nutrients and the economics of production. From these derived maps and tables we can develop predictions for a variety of land uses.

The growing availability of computer based Geographical Information Systems (GIS) is the main reason for believing that all forms of information about soils and land can be utilized in land use planning. Hitherto unutilized interdisciplinary information is being put to use. Typical GIS products include area and distribution maps of any required commodity or attribute and multiattribute combinations on a single map. Earlier versions of GIS that were not user friendly have been changed and the limitations now lie largely in the skill and imagination of the users of GIS and the availability of sufficient, locationally referenced data. Some of the popular programmes are given below.

Michigan State University [Schultink, 1986] developed the Comprehensive Resource Information and Evaluation System (CRIES) which encompasses a systematic resource analysis approach to evaluate public and prime benefits derived from alternate land use options and policy scenarios.

The Australian Resource Information System (ARIS) has been used to describe extensive soil-vegetation region [Cocks and Walker, 1987] by employing geographically coded data. ARIS is now being used to develop prospects on the distribution of these and other natural resources which would be otherwise unavailable.

The Integrated Land and Watershed Management Information System (ILWIS) [Meijerink et al., 1988] developed by the International Institute for Aerospace Survey and Earth Sciences of the Netherlands has been used for land use zoning.

ARC/INFO [ESRI, 1990] is used to assess and zonate the biotic and abiotic stresses related to crops in the world's semi-arid tropics. This can be used to prioritize research problems identify research domains and assist in transferring improved technologies.

However, the process of creating a GIS data base is tedious and costly for developing and under developed countries. Inputting and outputting the spatial data requires time, patience and precision. But the costs associated with creating the data base are quickly justified by the value of the output. Work is in progress to link crop models (CERES/DSSAT) with GIS by interfacing. Because the system is dynamically interactive any number of scenarios to evaluate alternate land utilization types and their impact on the environment can be estimated.

Acknowledgements

Information is liberally borrowed from the works of several authors and reviews made by eminent well informed soils information managers. The sources of data is provided in the text itself. However, take the opportunity to thank all the workers of the past and present, without whose untiring endeavours the information on soils and constraints imposed due to their degradation would not have been recorded at least to this primary level.

REFERENCES

Abrol, I.P. and Katyal, J.C. 1990. In: *Sustainable Agriculture*, Vol.2, Indian Society of Agronomy, pp. 235-263.

AICRPDA, 1983 a. Highlights of Dryland Research in Red Soils of Karnataka (1971-1983). University of Agricultural Sciences, Bangalore, p. 100.

AICRPDA, 1983 b. Improved Agronomic Practices for Dryland Crops in India. All India Coordinated Research Project for Dryland Agriculture, Hyderabad, p. 136.

AICRPDA, 1988. Annual Report 1987-88. All India Coordinated Research Project for Dryland Agriculture, UAS, Bangalore.

Badiger, M.K. and Shivraj, B. 1988. Present status and future strategies in sulphur research needs in Karnataka. In: *Sulphur in Indian Agriculture*. Proc. TSI-FAI Symposium, PD II/4/1-8.

Bronger, A. and Bruhn, N. 1988. Proceedings. International Conference Dryland Farming. Amarillo/Bushland, Texas, USA, pp. 688-690.

Brown, L.R. 1984. *Journal of Soil and Water Conservation* 36: 162-165, 255-260.

Challa, O. and Gaikwad, S.T. 1986. *Journal of Indian Society of Soil Science* 34: 543-550.

Cocks, K.D. and Walker, P.A. 1987. *Applied Geography* 7: 17-27.

Dhruvanarayana, V.V. and Sastry, G. 1985. In: *Soil Erosion and Conservation* (eds) S.A. El-Swaify, A. Moldenhauer and A. Lo. Soil Conservation Society of America, Ankey, Iowa.

El-Swaify, S.A., Pathak, P., Rego, T.J. and Singh, S. 1985. *Advances in Soil Science* 1: 1-64.

ESRI 1990. Environmental Systems Research Institute, Redlands, California, USA.

Eswaran, H. 1992. *Journal of Indian Society of Soil Science* 40: 6-24.

FAO 1983. Guidelines: Land evaluation for rainfed agriculture, Rome.

Fausey, N.R. and Lal, R. 1990. *Advances in Soil Science* 11: 173-186.

Ghosh, A.B. and Hasan, R. 1976. Available potassium status of Indian soils, Bull. 10. Indian Society of Soil Science. New Delhi.

Ghosh, A.B. and Hasan, R. 1980. *Fertilizer News* 25(11): 19-24.

Gupta, R.P., Kumar, S. and Singh, T. 1984. Report of All India Coordinated Research Project on Improvement of Soil Physical Condition to Increase Agricultural Production of Problematic Areas. ICAR, New Delhi.

Jenny, H. 1980. The soil resource : origin and behaviour. *Ecological Studies* 37, Springer-Verlag, New York, p. 377.

Jenny, H. and Roy Chaudhuri, S.P. 1960. Effect of climate and cultivation on nitrogen and organic matter resources in Indian soils. ICAR, New Delhi.

Joshi, P.K. 1987. *Indian Journal of Agricultural Economics* 42: 416-423.

Joshi, R.G. and Karekar, S.B. 1985. *Current Agriculture* 9(1-2): 81-85.

Kanwar, J.S., Rego, T.J. and Sertharama, N. 1984. *Fertilizer News* 29(1): 42-52.

Karle, B.G. 1985. Sulphur nutrition in Maharashtra Agriculture. In: *Proceedings, National Seminar on Sulphur in Agriculture*, Coimbatore, pp. 55-68.

Katyal, J.C. and Sharma, B.D. 1991. *Geoderma* 49: 165-179.

Katyal, J.C., Sharma, B.D. and Mehta, S.K. 1982. Annual Report. All India Coordinated Scheme of Micronutrients in Soil and Plants. ICAR, New Delhi, p. 67.

Katyal, J.C., Doshi, S.P. and Malhotra, P.K. 1985. Use of cluster analysis for classification of benchmark soil samples from India in different micronutrient availability groups. *Journal of Agricultural Sciences Cambridge* 104: 421-424.

Katyal, J.C., Vittal, K.P.R. and Ramana Rao, B.V. 1992. An Indian Account of Soil Degradation. *XII National Symposium on Resource Management for Sustained Crop Production, Bikaner, Rajasthan*. Indian Society of Agronomy, pp. 45-57.

Klingebiel, A.A. and Montgomery, P.H. 1961. Handbook, USDA, p. 210.

Lal, R. 1987. In: *Management of acid tropical soils for sustainable agriculture. Proceedings of IBSRAM Workshop, Bangkok* (eds) P.A. Sanchez, E.R. Stoner and E. Pushparajh. No.2, 167-177, IITA.

Lal, V.B. and Banerji, C. 1974. Man's impact on erosion in the rural environment : The Indian experience. IAHS Publication No.13, pp. 46-52.

Mann, H.S. 1982. In: *Desertification and Soils Policy*. 12th International Soil Science Congress Symposium paper II pp. 44-55, New Delhi, India.

Meizerink, A.M.J., Valenzuela, C.R. and Stewart, A. 1988. Scientific Status Report on ILWIS Project, International Technical Communication.

Mullins, C.E., Mac Leod, D.A., North Cote, K.H., Tisdall, J.M. and Young, I.M.E. 1990. *Advances in Soil Science* 11: 38-108.

Murthy, R.S., Hirekerur, L.R., Deshpande, S.B. and Venkat Rao, B.V. 1982. Bench Mark Soils of India. NBSS&LUP, Nagpur.

Muttatkar, V.K. and Roy Chaudhuri, S.P. 1959. *Journal of Indian Society of Soil Science* 7: 255-262.

Pande, M.C., Joshi, R.C. and Alok Khare 1988. *Journal of Tropical Forestry* 4: 380-386.

Patel, M.S. 1988. In: *Sulphur in Indian Agriculture. Proceedings, TSI-FAI Symposium*, PD II/1/1-4.

Patnaik, N. 1975. *Indian Farming* 24: 7-9.

Prasad, R. and Subbiah, B.V. 1982. *Fertilizer News* 27: 27-42.

Prasad, R. and Thomas, J. 1982. Review of Soil Research in India. Part I-P-309. 12th International Congress of Soil Science, New Delhi.

Raina, P., Joshi, D.C. and Kolarkar, A.S. 1991. *Soil Use and Management* 7: 47-52.

Raj Kumar and Saxena, R.K. 1985. *Journal of Indian Society of Soil Science* 33: 596.

Ramamoorthy, B., Randhawa, N.S. and Zende, G.K. 1971. In: *Review of Soil Research in India*. Indian Society of Soil Science Publication, New Delhi, p. 77.

Ramamoorthy, B. and Velayutham, M. 1976. In: *Soil Fertility Theory and Practices*. ICAR, New Delhi, p. 156.

Randhawa, N.S. and Venkateswarlu, J. 1980. Indian experience in semi-arid tropics : Prospects and Retrospects. In: *Proceedings, International Symposium on Development and Transfer of Technology of Rainfed Agriculture and the SAT Farmer*. ICRISAT, Patancheru, A.P., pp. 207-220.

Rastogi, B.K. 1991. In: *Agroclimatic zone specific research* (ed) S.P. Gosh. ICAR, New Delhi, pp. 31-60.

Reddy, D.S., Raju, A.P. and Reddy, M.S. 1985. Consolidated report of the Coordinated Soil Physics Project, Hyderabad centre (1967-1983). APAU, Hyderabad.

Riquier, J., Bramao, D., Luis and Cornet, J.P. 1970. AEL TESR 7016, FAO, Rome.

Schultink, G. 1986. Microcomputer based GIS Technology for Resource Assessment and Rural Development Planning.

Sehgal, J.L., Mandal, D.K., Mandal, C. and Vadivelu, S. 1990. *Agroecological regions of India*. NBSS Technical Bulletin 24, NBSS&LUP, Nagpur.

Sdasyuk, G.V. 1986. In: *Arid land development and combat against desertification : An integrated approach*. GKNT, Moscow, pp. 87-93.

Shinde, D.A. 1984. Agronomy of Sulphur in Madhya Pradesh. College of Agriculture, Indore.

Sharma, K.L. and Das, S.K. 1992. Nitrogen and phosphorus management in dryland cropping systems. In: *Dryland Agriculture in India - State of Art of Research* (eds) L.L. Somani, K.P.R. Vittal and B. Venkateswarlu. Scientific Publishers, Jodhpur, pp. 305-350.

Sharma, S., Das, S.K. and Gupta, P.D. and Srivastava, N.N. 1988. *Indian Journal of Dryland Agriculture Research and Development* 3: 43-48.

Singh, R.P. and Venkateswarlu, J. 1985. *Fertiliser News* 30(4): 43-55.

Singh, J.S. and Wali, M.K. 1962. In: *Proceedings of National Academy of Sciences* B32: 118, 124 R.Zh. 1963.

Sohan Lal 1989. *Journal of Indian Society of Soil Science* 37: 78-86.

Storie, R.E. 1978. Storie Index Soil Rating. Special Publication 3203. Division of Agricultural Sciences, University of California, USA.

Subba Rao, I.V. 1975. Nutritional disorders of crops in Andhra Pradesh. Technical Bulletin 1, APAU, Hyderabad.

Swindale, L.D. 1991. *Journal of Indian Society of Soil Science* 39: 14-31.

Takkar, P.N. 1988. Sulphur status of Indian soils. In: *Sulphur in Indian Agriculture*, TSI-FAI Symp SI/2/1-31.

Tandon, H.L.S. 1986. Sulphur Research and Agricultural Production in India. 2nd ed. Fertilizer Development and Consultation Organization, New Delhi.

Tandon, H.L.S. 1987. *Fertilizer News* 26(6): 25-34.

USDA 1978. Soil Taxonomy. Handbook #436. NBSS&LUP, Nagpur, p. 754.

Venkateswarlu, J. 1984. Nutrient management in semi-arid red soils. Project Bull. 8, Part B. All India Coordinated Research Project for Dryland Agriculture, Hyderabad.

Venkateswarlu, J. 1987. Efficient resource management systems for drylands of India. *Advances in Soil Science* 7: 165-221.

Vishnu-Mittre 1977. In: *Desertification and its Control* (ed) P.L. Jaiswal. ICAR, New Delhi, pp. 6-9.

Vittal, K.P.R. 1986. Soil fertility management. In: *Technological Advances in Dryland Agriculture*, CRIDA, Hyderabad, pp. 83-99.

Vittal, K.P.R., Vijayalakshmi, K. and Rao, U.M.B. 1990 a. *Experimental Agriculture* 26: 429-439.

Vittal, K.P.R., Vijayalakshmi, K. and Rao, U.M.B. 1990 b. *Soil Science* 150: 476-481.

Vittal, K.P.R., Vijayalakshmi, K. and Rao, U.M.B. 1990 c. In: *International Symposium on Water erosion, sedimentation and resources conservation*, October 9-13, Dehradun, pp. 215-221.

Vittal, K.P.R., Vijayalakshmi, K. and Rao, U.M.B. 1991. *Soil Science* 152: 3-6.

Yadav, J.S.P. 1984. *Journal of Indian Society of Soil Science* 32: 1-8.

Yadav, J.S.P. 1986. *Journal of Indian Society of Soil Science* 34: 660-674.

6

SOIL SALINITY IN DRYLANDS OF THE INDIAN SUBCONTINENT

N.T. SINGH

INTRODUCTION

Soil salinity is invariably associated with dry lands as the two most important modulating factors responsible for salt accumulation are best expressed under dry environments. While incomplete weathering of constituent soil minerals ensures a steady source of salts, the inadequate rainfall favours their deposition in the surface layers. Little wonder that dryland areas of the Indian subcontinent, are impregnated with salts of geologic origin and products of slow weathering. Secondary salinisation in irrigated areas only accelerates their accumulation in easily discernible proportions. Primary or ancient salinity has been always a native of dry land-scapes in latent as well as perceptive forms. This is borne out by ancient scriptures and records. For example, Indians in the period 2500 B.C. to 600 A.D., which covers Harappan and Vedic periods, distinguished soils into 'Urvara' (fertile) and 'Anurvara' (sterile). The latter set was further divided into 'Usara' (salty) and 'Maru' (desert) soils. As population pressures during those days were low, such lands were conveniently left out of cultivation and served as grazing lands, hunting grounds and abodes of nomadic tribes. Population pressures in arid and semi-arid regions of the subcontinent have now increased enormously (Table 1). Now that each parcel of land is needed to augment our agricultural resources, there is a wide concern for rehabilita-

Table 1. Distribution of irrigated area and population density in arid and semi-arid regions of the subcontinent.

State	Geographical area	Semi-arid	Arid	Irrigated	Population density
	m ha	%		m ha	km^{-2}
		India			
Rajasthan	34.22	36.77	57.42	3.42	100
Gujarat	19.60	47.50	33.72	2.32	174
Punjab	5.36	60.18	26.84	3.72	336
Haryana	4.42	59.71	29.32	2.35	294
Maharashtra	30.77	61.17	0.42	1.88	204
Andhra Pradesh	27.51	44.66	7.18	3.55	195
Karnataka	19.12	72.60	4.27	1.82	193
		Pakistan			
Punjab	20.62	28.94	57.86	9.30	229
Sind	14.10	4.27	95.73	5.60	135
Baluchistan	34.72	56.95	43.05	0.32	12
N.W.F.P	10.20	22.73	8.32	0.37	146

tion of salt lands and prevention of secondary salinisation of productive soils.

HISTORY OF SOIL SALINITY IN THE SUBCONTINENT

The incidence of salinity in the soils of the subcontinent and its adverse effects on productivity have been known to the Indians since long but it was only in the middle of the last century that soil salinity was recognised as a potential threat to agriculture. In 1865, an executive engineer of the Punjab irrigation department wrote to his superior, "whole of the country in Punjab lying west of Jamuna is impregnated with the elements of 'reh' (local term for salts)". He went on to describe their presence in areas upto Multan and foot of the Soolimani mountains [Whitecombe, 1971]. This report was at a time when canal irrigation was being introduced on a large scale in Punjab and Uttar Pradesh area. Thus salinity in Punjab soils existed before the introduction of canal irrigation. In fact, when settlement was done in the Rangpur and Mailsi canal commands in Punjab (now in Pakistan) the land allottees abandoned their lands as they confronted soil salinity called by them 'Chit Dain' (white devil). At the time of canal colonisation in the Punjab, sizeable patches of salt affected land were excluded from the canal command. The production of saline and non-saline soil varied from place to place but the Government reserved 10 to 20 percent of the land in each village for giving it in exchange to the allottees if they complained of getting a saline patch [Choudhri et al., 1978].

The introduction of canal irrigation accelerated the appearance of salts in those places where they had existed at some depth in the soil profile. The gestation period between canal irrigation and appearance of waterlogging and salinity was rather small. The first complaint of soil salinity came in 1855 from village Moonak in Karnal situated near Western Jamuna canal which was constructed by Feroze Shah Tuglak in 1351 along an abandoned bed of Jamuna river and later remodelled in 1939 by the British. Soil samples from the village were collected in 1865 along with water samples from the canal and sent for analysis to the Royal School of Mines in England with a forwarding note that the purpose of soil analysis was to determine whether the opinion about the origin of salts through canal water was borne by facts. This was done to dispel the fear lurking in the minds of farmers that canal water was a source of soil salinity. Canals were responsible for the damage but in some other way. The Upper Bari Doab canal was opened in 1860 after remodelling the existing Hasli canal. In 1874 loss of water through seepage from the main canal was 12.5% of the head discharge [Ahmed and Chaudhry, 1988]. Seepage losses measured by different methods ranged from 7 to 13 cusecs per million sq.ft. of the wetted perimeter for different canals [McKenzie Taylor and Mehta, 1941]. In 1880 waterlogging was reported in its command. Sirhind canal was opened in 1872 and reports of waterlogging appeared as early as 1880. In 1895 efforts were made to line the canals with oiled paper, bitumen and clay but very soon the lining was washed away.

The problem of soil salinity in the then United Provinces (now Uttar Pradesh) was not different from Punjab. The British had embarked upon big irrigation projects as a revenue earning exercise but they were confronted with this unforeseen problem. In 1863, Mr. T.E. Brown, Chemical Examiner for Punjab had analysed Punjab soils and traced the problem to presence in the soil of minerals rich in 'alkali', the decomposition of which was promoted by irrigation water. At that time reports of soil salinity had started pouring in from Faizabad, Baharaich, Lucknow, Sitapur and Hardoi districts. Since canals did not exist at that time the presence of salts was related to natural release of salts by mineral decomposition. The correspondence on 'reh' deterioration published in 1864 was the first collected documentation on the problem and its analysis [Whitecombe, 1971]. The Ganga Canal the largest canal system in the world at that time started irrigating lands in Uttar Pradesh in 1855. In 1876, Mr. Roberts, a big indigo grower petitioned the U.P. Board of Revenue that his lands in Meerut and Aligarh, had deteriorated following canal irrigation and demanded relief. This fortified earlier complaints and led to the setting up of the 'Reh' committee in 1877 to investigate the cause of 'reh' formation. The committee submitted its report next year attributing accelerated salinity development to introduction of canal irrigation.

The Government in 1866 deputed Mr. J.W. Wilson an irrigation engineer to visit California and discuss with Dr. E.W. Hilgard the irrigation, drainage and salinity problems. Earlier, experiments on leaching of salts were ordered in 1869 followed by setting up of experiments on reclaiming salt affected soils by the newly created Department of Agriculture. In 1880 practical experiments to reclaim 'usar' patches through plantations for fuel and fodder were attempted. But these were rarely successful. Between 1893 and 1896 Dr. J.W. Leather, Imperial Agricultural Chemist, reported investigations on distribution and composition of salts in affected soils of U.P. He observed that the principal salt was sodium carbonate. He calculated gypsum requirements of these soils as simply the quantity of gypsum which was equivalent to soluble carbonates present in them [Tiwari and Sharma, 1989]. The researches of Leather in earlier part of the twentieth century on water quality standards, leaching and gypsum use and bench mark sites, though missed many scientific facts not known at that time, were a precursor for later researches on salt affected soils in the Indo-Gangetic plains.

Investigations of the problems of waterlogging and soil salinity had simultaneously started in the then Punjab. The impetus came from the large appearance of soil salinity in the upper reaches of lower Chenab canal which was opened in 1903 with increased capacity to irrigate fertile lands of Rachna Doab. The Punjab Government surveyed the command area of the canal to investigate the problem at the instance of Agricultural Chemist Punjab. It was only in 1917 that the State Government took serious note of the problem of soil salinity and set up the Drainage Board to investigate the problem of waterlogging and suggest remedial measures. This Board was later split and Waterlogging Enquiry Committee was set up in 1925 which was subsequently converted to Waterlogging Board in 1928. The problem was assuming such serious dimensions that the Royal Commission of 1926 stated, "the future of the Punjab could rest within its own discretion and it could rely upon itself to see that its future was not a vast expanse of dead alkali lands" [Ahmed and Chaudhry, 1988]. The State Government had in 1922 set up an irrigation research laboratory at Lahore which in 1930 was upgraded to the Irrigation Research Institute charged with the responsibility of research on reclamation of saline soils. Meanwhile in 1927, the Waterlogging Enquiry Committee had decided to conduct salinity surveys and its successor, the Waterlogging Board, was given the task of checking waterlogging and salinity. In 1929 experimental work in the laboratory and field was started at the Chakanwali farm which emphasized that drainage could reclaim these lands. Standards were laid out for the survey of waterlogging and salinity problems in the Lower Chenab Command in 1933 and in 1938-40 pre-irrigation survey was attempted for the Thal canal command area. A Land Reclamation Board was set up in 1940 to extend land reclamation to the farmers' fields and in 1945 the Land Reclamation Directorate was established at Lahore.

The introduction of canals in the Peninsular India aggravated the salinity problem especially in the black soil region (Table 2). The Nira canal was opened in 1885. In the beginning of this

Table 2. Changes in the extent of area damaged by some Deccan canals.

Canal	Year of construction	Command area	Area damaged (ha)		
			1928-30[1]	1938-40[1]	1988-89[2]
Nira L.B.C. 1	1885-1886	60632	3840	7323	1988
Nira R.B.C.	1885-1886	77308	371	3940	1591
Godavari canals	1915-1916	88822	6939	10233	619
Pravara canals	1926	63741	5472	9160	846

Source : 1 = Talati [1941] 2 = IRD, Pune

century, the problem of salinity had become so acute that in 1903 the then Government of Bombay appointed a committee to consider measures to check the spread of damage. The committee suggested improvement of surface drainage by reshaping the existing 'nallas'. Mann and Tamhne [1910] reported that 6 to 7% area under the Nira command had already been salinised. Basu and Sirur [1938] investigated soils of Deccan Canals and reported that soil degradation in black soils under sugarcane occurred due to impeded drainage. Talati [1941] classified damaged lands in the Deccan and traced soil salinity to weathering and leaching of parent rock, presence of salts in soil profile and poor quality of ground water. Basu and Tagare [1943] while investigating the soil of Deccan canals found that alkali soils formed in lands characterised by impermeability, extreme hardiness and occasional presence of undesirable salts on the surface, all of which adversely affected plant growth.

PEDOGENESIS OF SALT LANDS IN THE SUBCONTINENT

The two factors essential for the genesis of salt affected soils are the *in situ* and foreign sources of salts and conditions favouring their accumulation in the earth's crust. About 92 chemical elements are known to exist in earth's crust and 8 of them form 98 per cent of it by weight. The important components of salt affected soils, i.e. sodium (Na), calcium (Ca) and magnesium (Mg) comprise 2.8, 3.6 and 2.1 percent by weight, respectively. These elements combine with other elements to form minerals. These minerals, which usually occur in mixtures, form the rocks of the earth which constitute the parent material for the soil. Minerals like halite, albite, gypsum, calcite and dolomite are important sources of Na, Ca and Mg and occur extensively in dry regions where limited water keeps chemical weathering to a minimum. The accumulation factors are related to climate, land scape, time and human interventions. The different source and accumulation factors make an impressive number of combinations thereby creating a variety of salt affected soils with varying degrees of complexities.

Sources of salts

Relic salts : The lesser Himalayas and Siwaliks have local concentration of salts whose origin may be traced back to geological and orographical history of these formations. The Tethy sea extended to a large part of the present Indo-Gangetic plains, Western Rajasthan upto Aravallis and on to Kutch. The Orogeny of Himalayas in the Tertiary, started the filling of the foredeep between these lofty mountains and the Peninsular plate with detritus. This alluvial action has since been continuing and passed through a number of phases in the uplift of Himalayas and climatic periods. Some of the sea salts precipitated during the process of deposition while others entrapped in the alluvium are present mainly in the more arid parts [Raheja, 1966]. The climate also played its role in segregation of salts from the solution. For example, excess presence of sulphate salts in semiarid parts of Punjab has been attributed to their preferential deposition during the Glacial period [McKenzie Taylor and Mehta, 1941].

The Tethyan sediments in the Punjab region of the subcontinent cover Himalayas between rivers Jhelum and Sutlej. Their extrusive forms are salt deposits of Pre-cambrian and Shali formation in the Himachal Pradesh and the Salt Range in Pakistan where the Cambrian is represented by 900 m of fossiliferous sandstones, shales and dolomites underlain by sea marl known as saline series. Numerous brine springs have been noticed in the Shali belt, particularly north of Mandi town. In the Megal area there are several seepages of brine, some with discharge of as much as 14550 l hr^{-1} during winter months. The thermal spring at Tatta Pani in the Sutlej valley within the area of Shali formation contains 2055 ppm Na and 3243 ppm of Cl. This alludes to the possibility of extension of salt formations, perhaps intermittently, over a wide area in the Himalayas than is believed hitherto [Srikantia and Sharma, 1972]. According to them the Shali basin had extended right up to the saline series of the salt range during the Pre-Cambrian-Cambrian period. It is also assumed that a saline belt borders Peninsular India along its northern edge and is now covered by the frontal Himalayan thrusts. Khandelwal [1975] has reported existence of halite beds (salt domes) in the Aravalli system and these are supposed to be the source of salts in saline lakes of Didwana.

The significance of the above saline formations in relation to salinity of the Indo-Gangetic plains has never been explored. The present Himalayan river courses contribute very little salts, but this may not always be the case. The tectonic upheavals have been changing river courses down the ages. The now defunct Saraswati, a mighty river of pre-vedic-vedic times as also many of its tributaries, had their origin in the region of Shali formations (Fig. 1). There are other areas of strongly saline sedimentary rocks which originated during Tertiary and Pleistocene. Therefore, the salinity in lakes and soils

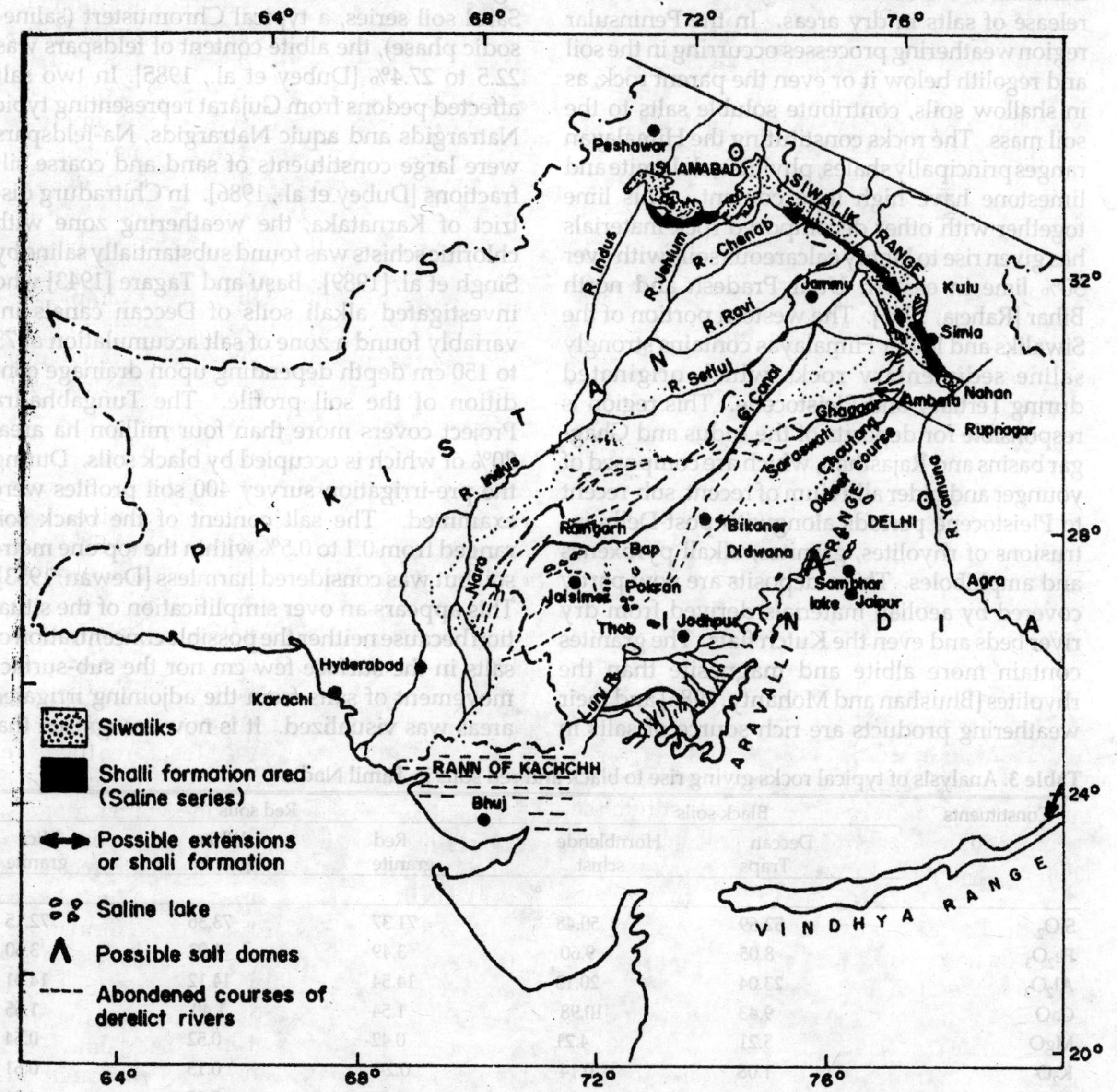

Fig. 1. Saline series as possible source of salts in arid zone of the subcontinent

in Rajasthan can also be traced to subterranean transport of salts alongwith surface and sub-surface waters originating from Mandi and other areas with saline beds. The very deep alluvium of the Indo-Gangetic trough with fairly good horizontal permeability does allow subterranean flows which may appear in basin situations and deposit their salt loads.

In-situ weathering : The weathering of soil minerals is a slow but steady source of *in-situ* release of salts in dry areas. In the Peninsular region weathering processes occurring in the soil and regolith below it or even the parent rock, as in shallow soils, contribute soluble salts to the soil mass. The rocks constituting the Himalayan ranges principally shales, phyllites, dolomite and limestone have high lime content. This lime together with other decomposed rock materials has given rise to highly calcareous soils with over 30% lime in eastern Uttar Pradesh and north Bihar [Raheja, 1966]. The western portion of the Siwaliks and lower Himalayas contains strongly saline sedimentary rocks which originated during Tertiary and Pleistocene. This region is responsible for deposits of the Indus and Ghaggar basins and Rajasthan, which are composed of younger and older alluvium of recent, sub-recent to Pleistocene periods alongwith post-Delhi intrusions of rhyolites, granites, alkali pyroxenes and amphiboles. These deposits are now partly covered by aeolian materials derived from dry river beds and even the Kutch flats. The granites contain more albite and magnesite than the rhyolites [Bhushan and Mohanty, 1986] and their weathering products are rich source of salts in dry areas.

The weathering of bed rocks is a constant source of salts in the black soil region. An analysis of typical rocks like Deccan traps and hornblende schist and various granites, that form parent materials for red and black soils, show large content of soda and lime (Table 3). Feldspars, a common constituent of coarser fraction of basaltic rocks, account for 22 to 29% of light minerals in black soils. In the silt fraction of Sarol soil series, a typical Chromustert (saline-sodic phase), the albite content of feldspars was 22.5 to 27.4% [Dubey et al., 1985]. In two salt affected pedons from Gujarat representing typic Natrargids and aquic Natrargids, Na-feldspars were large constituents of sand and coarse silt fractions [Dubey et al., 1986]. In Chitradurg district of Karnataka, the weathering zone with chloritic schists was found substantially saline by Singh et al. [1989]. Basu and Tagare [1943] who investigated alkali soils of Deccan canals invariably found a zone of salt accumulation at 75 to 150 cm depth depending upon drainage condition of the soil profile. The Tungabhadra Project covers more than four million ha area, 80% of which is occupied by black soils. During the pre-irrigation survey 400 soil profiles were examined. The salt content of the black soil ranged from 0.1 to 0.5% within the top one metre soil but was considered harmless [Dewan, 1953]. This appears an over simplification of the situation because neither the possible concentration of salts in the surface few cm nor the sub-surface movement of salts from the adjoining irrigated areas was visualized. It is now recognised that

Table 3. Analysis of typical rocks giving rise to black and red soils in Tamil Nadu.

Constituents	Black soils		Red soils		
	Deccan Traps	Hornblende schist	Red granite	White granite	Mica granite
			%		
SiO_2	52.69	50.48	71.37	73.36	72.15
Fe_2O_3	8.05	9.60	3.49	1.92	3.60
Al_2O_3	23.04	20.13	14.54	14.12	14.31
CaO	9.43	10.98	1.54	1.40	1.65
MgO	5.21	4.21	0.42	0.52	0.54
K_2O	1.08	0.14	0.24	0.13	0.61
Na_2O	2.74	2.23	3.54	3.69	4.21

the origin of soluble salts in the Peninsular region is due largely to weathering of salt bearing strata.

The alluvial and aeolian sediments in the hot arid zone of the subcontinent are rich in Na-bearing minerals. Almost 44% of light minerals in sand dunes of Rajasthan contain orthoclase feldspars which weather to release minerals like illite and montmorrolinite [Choudhary and Dhir, 1982] with soluble salts as by-products of such weathering. Alluvial deposits in the same area may show different minerology due to change in the source of sediments brought by the same river. A typical case is Mundlana in Haryana where two distinctly different soils have developed in close proximity. Feldspars make up 6.6 to 6.9% in the saline soil but their content in the neighbouring alkali soil is 16.4 to 18.2% (Table 4). The feldspar content of the black soils is more than the alluvial soils but the mechanical separates in both groups follow the sequence Na-feldspars > Ka-feldspar > Ca-feldspar. Their weathering occurs in two stages releasing alkaline earth cations and silicate and aluminate units. The silica and Na ions are leached to the regolith but under favourable conditions these may join the water table also [Goyal, 1984].

by minor channels whose salinity is due to salt releasing marine lithological formations [Kar, 1989]. Such micro-niches have low infiltration rates. Thus most of the collected water evaporates leaving behind its load of salts. Overland flows originating upstream or caused by local water concentration, are source of salt accession to soils at lower elevations both in the alluvial and peninsular regions. For example, in Karnataka, the naturally occurring salt affected soils called 'Karl' soils occur in comparatively low lying situation in the black soil region [Venkata Rao, 1967]. Similarly most alkali soils in the Indo-Gangetic plains are met at lower reliefs especially where depression situations favour concentration of run-off waters.

Rivers and streams may also carry large salt load. A water sample from Guhiya, a tributary of Luni river, contained 12.5 g kg^{-1} of Na salts [Choudhari and Sharma, 1984]. The salt load decreased with time. River Ghaggar and its tributaries carry salts as well as alkali silt and clay eroded from root hill soils/clay hillocks which are often alkali and calcareous with salts in deeper layers. The reddish clayey material is particularly salt affected. A part of the salt and

Table 4. Some characteristics of two contiguous salt-affected soils from Mundlana, Haryana, India

Depth (cm)	pH_s	ECe	Mica		Feldspars		Sand	Composition of feldspars in sand fraction		
		dSm^{-1}	Clay	Silt	Clay	Silt		Na	K	Ca
Loam saline soil										
0-30	7.7	18.7	44.3	29.7	3.9	4.9	6.8	3.2	2.0	1.6
90-120	7.3	34.6	40.5	35.1	4.1	5.1	6.6	3.1	1.9	1.6
210-240	7.2	20.8	38.7	32.7	4.2	5.0	6.9	3.3	1.8	1.8
Sandy loam alkali soil										
0-30	8.8	6.6	42.7	26.7	8.7	11.6	16.4	10.0	3.4	3.1
90-120	8.9	5.3	38.1	27.9	9.3	12.4	17.0	10.2	3.8	3.2
210-240	9.0	4.2	38.1	27.7	10.6	12.7	18.2	10.9	4.0	3.4

The difference in feldspar content may be due to different parent materials.

Overland flows: Run-off waters pick up both salts and salt-bearing sediments on their way to places of concentration. In Western Rajasthan the natural setting of lakes and saline basins is such that they have centripetal drainage which attracts overland flows from the surrounding watersheds. In Jaisalmer district alone, the saline depressions are spread over 125 km^2 and are fed

calcareous nodules met along upper reaches of the flood plain of river Ghaggar can be traced to the above source. Salt-affected soils along the defunct and active flood plain of Ghaggar in the lower reaches across Western Rajasthan are remnants of this process. In 1979, flash floods in the Luni basin deposited thick layer of sediments composed of very fine sandy to coarse silty material charged with salts over highly productive agricultural lands rendering them so saline that cultivation had to be abandoned [Joshi and Dhir, 1990].

Subterranean flux of salts : Salts dissolved from upper reaches are transported subterraneously. In Peninsular India, much of the salinity in valley lands results from dissolved salts moving laterally along the interface between the soil and the underlying 'murrum' or through the porous murrum itself. Mann and Tamhne [1910] discussed the genesis of salt lands in commands of Deccan canals. According to them these canals pass through 'murrum' cuttings in the ridges and in embankment in the valley to avoid zigzag course along contour lines. This facilitates percolation both from banks and irrigated fields. Soil water moves through the previous sub soil towards the valley to find an outfall in a 'nalla'. Owing to relative impermeability of the intervening soil, the subsoil water rises by capillarity along with salts. The latter are left on the soil as water evaporates. Therefore damage to the soil results from surplus sub soil water and is directly proportional to the difference in the quantity of discharge passing into the subsoil and that appearing in the nallah. Soil types which are coarse and medium, are not damaged to the same extent as deep black valley soils for the simple reason that they are in the upper reaches of the ridge. The coarse soils have slopes 1 in 60, the medium black 1 in 100 and deep black 1 in 300 with a fall of say 12 m or more from coarse soil to deep soil. It is not uncommon to find gradients of salt encrustation along the slope. Soil of low rainfall, undulating areas of Madhya Pradesh, Maharashtra, Karnataka, Andhra Pradesh and Tamil Nadu are prone to this process of salinisation.

In the alluvial plains, subterranean flows occur below derelict river beds as in Rajasthan. The morphogenesis of salt lakes and palyas in western Rajasthan is related to the confluence of prior drainage channels and courses of streams which, during the humid phase of climate, formed the drainage system of the region [Ghose, 1971]. Ghose and Kar [1984] have reconstructed the course of now defunct Saraswati river which supposedly originated in Siwaliks and carried waters of Sutlej, Drishtawati and Yamuna before it emptied into the Arabian sea. Rain water sinks through such beds, flows subterraneously along the buried channels carrying soluble salts collected through the course of flow. Evaporation of water from the bed concentrates salt carrying waters along the path raising their salinity to as much as 3.2 g kg^{-1} salts [Ghose, 1971].

Rain-and wind-borne salts : The rain water picks up its salts either from strong winds that sweep over oceans or from dust storms. Thus salt content of rain water changes with season and location of sampling. On July 9, 1981, the rain water collected from Pachpadra had an EC of 1.612 dSm^{-1}. But the EC of rain showers of September 2, 1981 was 0.571 dSm^{-1}. This area has high wind activity [Choudhari and Sharma, 1984] which probably accounts for the high salt load of rain water on the first date. On the same day EC of rain water in the central part of the plain with moderate vegetation was 0.299 dSm^{-1}. Narayanaswami [1939] estimated that 110 cm rainfall near Bombay in an area outside the influence of sea spray, deposited 145 kg ha^{-1} of Cl. Salt balance of the Luni basin indicates that stream water salinity in the Indian arid zone is due to salts contributed by rains and flushing of soil surface through quick overland flows.

The salt-carrying role of strong winds is not denied though their actual contribution is disputed. Holland and Christie [1909] believed that source of salts in the salt basin of Rajasthan was the Rann of Kutch from where salts were transported by wind. This belief has not found wide acceptance due to variation in salt composition of lakes and the localised deposition. This, however, does not undermine the salt-carrying role of strong dust storms. Two facts seem to have been ignored. One is the salt accruement to these basins from various sources possibly differing not only in composition but also in concentration factor both during transport and within the lake depending upon the permeability of the substratum. The second is the differential rise of salts to the surface in summer when a large part of the Rann turns into salt-flats. Salts of Ca and Mg reach their solubility limits upon concentration as soil solution moves towards the surface and precipitate in sub surface horizons while NaCl concentration increases at the surface. The wind will sweep only that salt which is dry and at the surface and this salt load will have no relationship with the composition of sea water.

Tidal floods and sea water intrusion : Coastal areas as in Andhra Pradesh, Tamil Nadu and Gujarat experience floods from sea tides and cyclones which inundate nearby areas. The back water canals and lagoons of east coast also spread salinity. Another source of salinity in the coastal areas is the sea water intrusion into ground aquifers as a consequence of excessive withdrawal of good quality water floating over saline ground waters. Over draft has rendered more than 120 km coastal strip along Saurashtra coast in Gujarat saline.

Irrigation waters : All lands in the subcontinent experience dry period as major rainfall events are limited to few months. Irrigation through canals, tanks and wells has been practised to raise crops particularly in the dry months. While canal irrigation may salinise the soil through rise in water table, well irrigation usually causes salt build up due to poor quality waters (Table 5). According to Sharma et al. [1968] while canal irrigation mainly caused salinity, both salinity and sodicity problems arose from well irrigation. Snow-fed river waters have low salt content but total salt load of rivers and tanks fed from overland flows and seepage may be large. The salt concentration also increases during periods of low discharge when rivers and tanks are charged by seepage waters. Even minor additions of salts over long periods would build up salinity in the absence of adequate drainage. When canal irrigation was introduced on a commercial scale in Northern India, farmers suspected that progressive salinisation of their lands was due to salts in the canal water. Now canals were a big source of revenue to the Government and in the Punjab 40% revenue of the province was derived directly from irrigation and 45% indirectly. Hence, to sustain the confidence of the farmers, the then British Government always made a point to seek analysis of samples of canal water alongwith those of affected soil to prove that canal water had nothing to do with the incidence of salt lands.

Agriculture in the dry regions, where canal water is scarce, is either rainfed or depends upon

Table 5 : Effect of canal and well irrigation on soils in arid Rajasthan

Depth	Texture	pH	ECe dSm-1	Na	Ca	Mg	HCO_3	SO_4	Cl	ESP
					mel-1					
				Kharda canal irrigated						
0-17	SCL	8.4	25.0	184	30	34	8.0	107	150	14.0
17-30	SL	8.1	65.0	645	27	13	7.4	282	402	12.0
30-72	LS	8.8	15.0	132	20	8	10.0	84	64	11.2
				Kharda unirrigated						
0-17	SL	7.8	1.3	10.0	4.9	1.8	6.2	5.0	3.7	12.0
17-30	SL	8.2	0.8	50	2.7	0.9	4.1	8.0	3.5	12.8
30-72	SL	8.2	1.4	10.0	4.9	1.3	3.2	6.0	5.7	10.1
				Suryata - well irrigated						
0-27	LS	7.9	25.1	134	60.7	54.1	8.2	64.0	184	10.2
27-72	LS	8.2	9.5	62	16.0	12.7	10.0	29.0	54	12.2
72-137	SiCL	8.0	20.5	115	52.5	43.0	5.6	54.2	158	15.1
137 +	CL	8.1	8.0	40	1.3	16.0	7.1	14.0	60	19.0
				Surayata unirrigated						
0-27	SCL	7.4	1.6	10.0	2.7	0.8	3.5	7.6	4.7	5.2
27-72	SCL	7.7	0.9	6.0	1.6	1.1	3.2	5.0	1.6	6.2
72-137	SL	8.1	1.4	11.0	0.9	0.2	5.2	2.0	6.4	12.1
137 +	SCL	8.3	2.4	20.0	1.4	0.2	7.5	10.0	4.4	16.9

tank or well irrigation. The occurrence of saline or highly saline well waters is a common feature of arid lands in Rajasthan, Haryana and Gujarat in India and a major part of Pakistan. Well and tank waters in the Peninsular region, which are mostly charged by centripetal seepage are saline and sodic. Ground water surveys show saline waters of varying composition in the arid and semi-arid zone of India and Pakistan (Table 6). Use of such waters is known to be a major source of soil degradation in both countries. Some scientists in Pakistan maintain that secondary salinisation through waterlogging is of limited significance but salinity introduced through low quality pumped waters is of greater concern [Choudhri et al., 1978]. Soil salinity through tank irrigation has been observed in Rajasthan and Peninsular India (Table 7). The Sardar Samund tank in Rajasthan constructed at the confluence of Sukri and Guhiya rivers has become saline as Guhiya brings in waters with more than 9.0 g kg^{-1} dissolved salts [Kar, 1988]. In Karnataka, irrigation through sodic waters of tanks has caused salinity in Challakera Taluka of Chitradurg district [Singh et al., 1989]. The waters of Vani Vilas Sagar and other local tanks are saline and are source of salinity in their irrigation commands [Venkata Rao, 1967].

Mode of salt accumulation

Climate : The two important components of climate i.e. evaporative demand and rainfall may have separate as well as complementary role in salt accumulation in drylands. While water is a reactant, the reaction rates are promoted by high temperatures. In dry areas of the subcontinent, with annual rainfall between 100 to 500 mm and a potential evaporative demand exceeding 2000 mm, inadequate leaching leaves a positive salt balance. A part of salts released through weathering of soil minerals in the surface soil, reside at depths controlled by the leaching ability of the seasonal rainfall. Such cases represent latent soil salinity, or potential saline soils. During the occasional wet years or in the event of rise in water table or with introduction of irrigation, these salts move up and appear at the surface. Sub-surface accumulation of salts also occurs when large water deficits in the atmosphere cause soil solution to rise from lower horizons of moist soil profiles. With a deep water table, a common feature of dry areas, evaporation rates often exceed capillary conductivity of the soil causing the place of phase change in water to move down-wards. Consequently rising salts may not appear at the soil surface and an apparently salt free soil surface may at times be deceptive as it lies above the salt-impregnated horizons. The dryness of the climate is also responsible for the pedogenesis of calcareous and gypsic sub-surface horizons as low solubility salts of rising soil solutions concentrate on the

Table 6 : Frequency distribution (percent) of soluble salts in ground waters in the Indian subcontinent

India	No. of water samples tested	EC_2w dSm-1			
		< 3	3-5	5-10	>10
Punjab	12500	68.0	20.0	8.0	4.0
Haryana	3637	58.5	17.5	13.2	10.18
Uttar Pradesh (Aligarh)	390	78.4	18.7	2.9	—
Rajasthan	4540	49.0	16.7	20.8	13.5
Andhra Pradesh (coastal)	1082	78.2	16.1	5.7	—
Karnataka (Bijapur)	404	87.6	8.9	3.2	0.2
Gujarat (Ahmedabad) Kheda)	505	84.0	10.0	5.7	0.2
Pakistan	Gross area	EC_{iw} mg kg^{-1}			
	(m ha)	Upto 1000	1000-1500	1500-3000	> 3000
Punjab	0.72	65.4	12.2	9.8	12.6
Sind and Baluchistan	6.13	17.2	6.6	13.8	62.4
N.W.F.P.	0.40	82.4	5.2	12.4	—

Table 7 : Mode of salinistation and properties of affected soils in Karnataka

Soil depth	pH_2	ECe dSm^{-1}	Clay	CaCO	Na	Ca	Mg	Cl	CO_3 + HCO_3	SO4	ESP
			%		Ionic composition of saturation extract mel^{-1}						
Antrhopogenic salt affected soil											
(a) Saline water irrigated											
0-5	9.5	22.3	18.8	5.1	217.4	4.5	3.8	186.0	15.0	18.7	40.4
5-20	9.8	6.8	24.3	6.1	66.6	1.4	2.5	46.9	15.6	9.2	29.9
20-60	9.9	7.2	25.6	1.8	72.8	3.4	0.8	56.4	10.3	6.4	32.6
60-100	9.7	5.1	25.9	2.6	35.9	2.8	7.5	36.3	10.6	1.6	22.9
(b) Waterlogged due to tank irrigation											
0-10	8.0	26.4	27.6	–	233.7	54.5	5.5	241.9	22.9	30.7	29.0
10-40	8.2	8.4	31.4	–	68.7	4.6	1.7	55.0	18.8	tr	28.2
40-80	8.4	7.0	30.8	3.8	62.4	4.6	1.5	61.7	8.2	2.4	27.6
Natural salt affected soil											
0-15	9.8	13.1	31.6	4.7	131.7	9.5	2.7	70.0	34.9	30.7	34.2
15-40	9.8	5.2	33.4	9.3	43.5	5.5	4.7	20.0	28.2	6.6	23.7
40-60	9.8	5.9	34.9	9.3	48.1	8.4	1.9	27.5	28.1	tr	24.2
60-80	9.6	4.8	35.0	10.3	39.1	2.3	2.9	37.5	9.4	tr	24.9
80-100	9.3	3.1	31.2	14.4	29.4	3.6	3.6	22.5	10.6	tr	23.5

Source : Singh et al. (1989)

way, exceed their solubility product, and precipitate somewhere in the profile. The genesis of such horizons also occurs in areas which receive saline overland flows. Replacement of Ca from the surface horizons through exchange reactions and precipitation in the lower horizons is a typical case of soil sodication in regions of medium rainfall. The climate has a multifold role in dry land salinity as it is involved in weathering of soil minerals, their dissolution, transport, concentration and deposition in soils.

Physiography : The landscape factor is very strongly reflected in soil salinisation. We may treat landscape as a local factor in the five broad physiographic units recognised for this purpose in the subcontinent. The Himalayan region with its east and west extensions is a source of water as well as salts but is itself largely free from soil salinity. The exceptions are the cold deserts of Lahul Spiti and Ladakh and slight salt intrusions in the piedmont plains flanking the Siwalik range. The Indo-Gangetic plains stretch parallel to the Himalayan ranges till they merge with deltic flats in the Bay of Bengal in the east and in the south-westerly direction along the courses of Indus and its tributaries merging with deltic forms in the Arabian sea. These plains have generally flat surface intercepted by natural levees and channel remnants. An almost flat relief with slope of less than 0.002% is responsible for their poor drainage and susceptibility to salinisation. The flat topography harbours local depressions which attract runoff waters and are often salt affected.

The central high lands, bordered by the Aravalli range in the west and Satpura range in the south, separate the Gangetic plains from the Peninsular Plateau. The east Rajasthan uplands with height ranging from 250 to 500 m and the Malwa Plateau have fair amount of salinity while it is rarely found in the Bundelkhand uplands and Vindhyan scarplands. The Thar desert in western Rajasthan extends across the border into Pakistan and on the north-east into Ghaggar flood plains. This area is covered by vast aeolian sand deposits amongst rock out crops. The sand covered derelict river channels, now appear as salt basins. In the far north-west occur the loess deposits which occupy a part of the Potwar Plateau and Thar area. Baluchistan plateau is mainly hilly with small dry valleys. South of Bolan pass there exist flat lands called Kacchi plains with 50-120 mm annual rainfall and fair amount of salinity in the Nushki area.

The Peninsular plateau is composed of Archaean gneisses and schists penetrated profusely by igneous rocks and pre-cambrian sediments. Another formation, the lava out flows in horizontal sheets named Deccan Traps, occupy parts of Maharashtra, Karnataka, Madhya Pradesh and Gujarat. These are extra-ordinarily uniform in composition, correspond to basalt or dolerite and give rise to the well known black soils. The rolling landscape with ridges and valleys is mainly responsible for salinisation of valley lands which happen to be quite deep. The rivers originating in the Western Ghats cut deep gorges and later from fertile valleys and deltas as they merge with the east coast carrying black soil sediments containing minerals capable of releasing large amounts of soluble salts. The salinity of the coastal region is due both to the influence of sea and the salt content of alluvial materials. The east coast is much wider than the west coast and subject to tidal influences and cyclones. Buckingham canal, which runs parallel to the east coast from Krishna to Chinglepet in Tamil Nadu and numerous saline lakes and water bodies spread salinity in the area. On the west coast influence of sea as source of salinity is more pronounced along the 1600 km coastline of Gujarat.

Based on world-wide experience from dry regions, waterlogging and salinity are often considered twin problems. This is not always true. Salt affected soils are met in the subcontinent independent of waterlogging or shallow water table conditions. Kolarkar et al. [1980] reported that in south eastern arid Rajasthan the natural saline soils or those with relict salinity comprised 44.6% of the surveyed area, while secondary salinisation due to rise in water table or saline irrigation occurred in 22.9% of area. Similarly much of waterlogged area in Sri Ram Sagar and Gandak commands is free of salinity. Secondary salinisation may follow in due course. Large areas in canal irrigation commands are already getting waterlogged and salinised (Table 8). In Gang and Indira Canal Commands annual rate of ground water rise has been 43 to 83 cm. At this rate 60% of command area will develop drainage

Table 8. Saline and Waterlogged area in selected canal commands of India

Project	State	Water logged area	Saline/alkali
		0000ha	
1. Ram Ganga	Uttar Pradesh	195.0 (9.7)	352.4 (17.6)
2. Sharda Sahayak	Uttar Pradesh	260.0 (7.0)	253.3. (6.7)
3. Gandak	Bihar	562.0 (52.9)	400.0 (38.1)
4. Sri Ram Sagar	Andhra Pradesh	60.0 (47.6)	1.0 (0.8)
5. Nagarjuna Sagar L.B. C.	Andhra Pradesh	33.1 (9.4)	26.4 (8.0)
6. Nagarjuna Sagar R.B.C.	Andhra Pradesh	114.0 (24.0)	69.2 (14.5)
7. Tungabhadra	Andhra Pradesh	30.0 (10.0)	14.8 (5.0)
8. Tungabhadra	Karnataka	37.7 (7.4)	20.0 (4.0)
9. Ukai Kakrapar	Gujarat	16.2 (4.3)	8.3 (2.2)
10. Kadana (Mahi)	Gujarat	89.0 (38.7)	60.7 (28.7)
11. Mahi R.B.C.	Gujarat	9.8 (3.7)	35.7 (11.5)
12. Tawa	Madhya Pradesh	—	6.6 (3.8)
13. Chambal	Madhya Pradesh/Rajasthan	98.7 (20.3)	40.0 (8.2)
14. Indira Gandhi Canal	Rajasthan	43.1 (8.0)	29.1 (5.4)
15. Nira L.B.C.	Maharashtra	—	2.0 (3.3)
16. Ghod R.B.C.	Maharashtra	0.7 (6.6)	0.9 (8.1)
17. Krishna	Maharashtra	3.3 (30.9)	—
18. All Commands	Gujarat	13.8 (6.1)	540.0 (23.9)

* Figures in parenthesis indicate percent of the potential irrigated area.
Sources : 1. Kandpal (1982); 2. Ambekar (1980); 3. Das Gupta (1982); 4. Hassan and Venkata Reddy (1982); 5-7. Govt. of A.P.; 8. CSSRI, Karnal; 9-14. Joshi and Agnihotri (1984) ; 15-17. Govt. of Maharashtra; 18. Deptt. of Irrigation, Govt. of Gujarat.

and salinity problems in another 50 to 100 years [Shankarnarayana, 1985]. The same is true of other canal commands. There is increasing public awareness about the possible adverse effects of canal irrigation. Population had to be shifted to other places in search for new avenues of livelihood, as the lands in several irrigation commands were damaged to such an extent that farmers were forced to abandon them.

NATURE AND CLASSIFICATION OF SALT LANDS

The nature and composition of salt lands in the subcontinent is as varied as the sources of their salts and modes of formation. In India 27 sub-groups of salt affected soils, largely based on clay minerology, presence or absence of calcic horizon, textural variations and ground water condition, have been identified [Murthy et al., 1980]. In Pakistan, salt affected soils have been grouped on the basis of their reclaimability [Rafiq, 1980]. There have been other attempts to group salt affected lands according to their saline features, location, local experience concerning management and productivity etc. For practical usage and management considerations salt affected soils have been placed in fewer classes on the basis of reaction and content and nature of salts they contain. Names like saline, alkali, saline-sodic, sodic soils are common in the subcontinent. The Central Soil Salinity Research Institute, Karnal has placed salt affected soils of India in two broad groups of saline or alkali soils [Abrol and Bhumbla, 1978; Bhumbla and Abrol, 1979].

Saline soils are recognised by an almost singular presence of neutral salts like chlorides and sulphates of Na, Ca and Mg. The pH of the saturated soil paste (pHs) does not exceed 8.2 and electrical conductivity of the extract from the saturated paste is more than 4 dSm^{-1} at 25°C. The third distinguishing characteristic is the sodium adsorption ratio, SAR, defined as $Na/(Ca + Mg/2)^{1/2}$ or the exchangeable sodium percentage, the accepted cut off limit for most soils being 15. When neutral Na salts predominate, the SAR exceeds 15 to qualify the soil as saline-sodic according to the most prevalent definition. If there are enough Ca and Mg salts the SAR will be low, and it will not exceed 15 to qualify the soil to be called sodic. Upon leaching even high SAR saline soils may attain a low SAR and a corresponding reduction in ESP. Khosla et al. [1979] leached a soil of ECe 63.8 dSm^{-1} and SAR 123 in the surface horizon, and classified as salic Calciorthid, typical of salt affected soils of dry regions of India. Leaching resulted in desodication of soil implying that the soil could very well have been called a saline soil. Those salt affected soils where pHs exceeds 8.2 or in which the ratio ($Na/Cl + SO_4$) is more than one, there is likelihood of soil dispersion upon leaching and a rise in soil pH. Such soils are alkali. Saline soils are normally met in areas with rainfall less than 500 mm or along sea coast.

In Pakistan, most salt affected soils are either saline or saline-sodic. Real sodic or alkali soils are rare [Schroo, 1967]. An analysis of 1070 soil samples from the southern Indus plains showed 49% normal soils, 12% saline, 36% saline-sodic and only 2% alkali. Schroo classified Ca + Mg concentration in saturation extracts of 24673 soil samples from different Soil Conservation and Reclamation Projects (SCARPs) in Pakistan. The frequency distribution of Ca + Mg was 70%, 17%, 5% and 8% for concentration ranges of 0-5, 6-10, 11-15 and 15 mel^{-1} indicating increase in Na content at high EC values. Soils with more than 15 mel^{-1} $(Ca + Mg)^+$ indicated presence of gypsum and possibility of their reclamation with simple leaching. In soils where divalent cations were well below 5 mel^{-1} severe alkalinity would develop upon leaching. The same phenomenon is observed in salt affected soils on the borderline of saline and alkali soils in the rainfall range of 500-600 mm.

Alkali soils in India contain large presence of carbonates and bicarbonates of Na. Almost one hundred years ago the first diagnostic report of J.W. Leather, on salt affected soils of Uttar Pradesh, mentioned that the soils contained carbonates and bicarbonates of Na and no other 'Usar' salts. Such soils are almost saturated with exchangeable Na and the pH often exceeds 10.0. For diagnostic purposes however, alkali or sodic soils will have a pH exceeding 8.2 and ESP more than 15 with no limit on salt content if the salts mainly comprise those capable of alkaline hydrolysis. Owing to the presence of smectites in black soils, their properties deteriorate even at an ESP of 8. Alkali soils of the vertisol region have relatively low pH and ESP than alkali soils of the Indo-Gangetic plains but owing to their high clay content and presence of expanding types of clays are more difficult to manage. The salient features of alkali soils of the Indo-Gangetic plains can be summed up from the observations of Tiwari et al. [1989] about alkali soils of Uttar Pradesh.

1. The soils have very high pH recording values upto 10.9 in soil water suspension and have ESP throughout the profile recording values as high as 90.
2. The soluble salt content is high in the upper part of the pedon nearly upto 70 cm recording values of EC upto 20 dSm^{-1}. Carbonates and bicarbonates of Na are the principal salts.
3. The surface horizon usually has a platy structure and soils exhibit very low infiltration rates and hydraulic conductivity.
4. The soils are highly dispersed and usually there is a zone of clay accumulation in the subsurface horizons whose bulk density is very high. Natric horizons are present in many soils usually with blocky and not columnar structure.
5. The natric horizon overlies a zone of calcium carbonate accumulation forming calcic or petro-calcic horizons indurated from a few cm to almost 1 m thickness. The zone of calcium carbonate accumulation lies within the zone of fluctuating water table.
6. The sub-surface horizons above calcic or petro-calcic horizons possess distinct yellowish-brown iron mottles alongwith moderate to abundant amounts of ferro-manganese concretions.
7. The water table may appear nearly below the surface during monsoon but subsequently recedes to about 5 m depth. The ground water is usually of low salt content comprising mainly Na and HCO_3 ions. In some pockets high residual sodium carbonate (RSC) waters are also met. The salt affected soils in the western arid region differ from the alluvial soils of the Indo-Gangetic plains in nature of salts, profile characteristics and modes of formation. This is evident from the analysis of profiles representing some typical salt affected soils (Table 9). The salinity in the western zone including Rajasthan and Gujarat and Punjab and Sind provinces of Pakistan has developed due to its dry climate, *in-situ* weathering, subterranean and surface accretion of salts and irrigation using saline waters. The salt lands of Kutch, southern Gujarat and Saurashtra region are black with silty clay or clay texture. The pH ranges from 7.8 to 8.4 and salt distribution in the profile does not show any particular pattern with depth. The quality of ground waters in this region being poor, saline irrigation is a major cause of soil deterioration.

The salt affected soils of vertisol region exhibit typical swelling-shrinking characteristics and deterioration in physical condition even at low ESP. As ESP exceeds 20, the typical appearance of deep cracks is subdued and may even disappear. Saline soils show predominance of NaCl while alkali soils may have Na_2CO_3 and $NaHCO_3$ and some times only the latter group of salt. Quoting the earlier work of Mann and Tamhne [1910], the damaged lands of Deccan were described by Talati [1941]. In deep black soils soluble salts consisted of mixed salts of Ca, Mg and Na. The percentage of Na increased with depth. In medium black, total salts were chiefly Na while Ca and Mg were almost negligible. The pH was more than in the deep black soils. He grouped salt affected soils of the region on the basis of pH and nature of salts into the following 3 types :

1.*Mixed saline* : Profiles with mixed salts of Ca and Na which upon leaching gave characteristic low pH values like normal soils with fairly high percentage of divalent bases.

Table 9. Salient characteristics of typical salt affected soils of the subcontinent

Depth	pHs	ECe	$CaCO_3$	Organic carbon	Clay	Exchangeable Na
		dSm^{-1}		%		
Natraqualf, Ghabdan (Sangrur), Punjab						
0-6	10.1	32.4	7.6	0.22	20.9	82.9
6-24	9.7	10.5	9.4	0.17	23.7	75.4
24-48	9.4	8.0	6.2	0.14	28.0	75.0
48-73	9.3	5.6	4.4	0.14	28.0	76.0
73-97	9.2	3.7	4.1	0.10	31.4	68.1
97-124	9.2	4.1	9.1	0.10	24.6	63.2
124-145	9.1	4.6	27.3	0.09	23.0	66.0
Chromustert, saline-sodic phase, Kurnool, Andhra Pradesh						
0-21	8.4	9.5	6.6	0.48	39.7	28.7
21-74	8.1	13.8	7.3	0.48	34.9	21.4
74-104	8.2	12.7	6.0	0.47	37.8	24.9
104-125	8.3	12.7	5.2	0.39	42.7	24.6
Typic Salorthid, Dandi, Gujarat						
0-2	7.6	160	1.4	0.65	41.0	73
2-17	8.0	75	1.4	0.54	51.2	57
17-53	8.0	80	2.9	0.49	47.0	72
53-75	8.2	75	5.3	0.47	49.2	61
75-96	8.3	75	6.2	0.49	46.9	57
96-155	8.7	50	21.1	0.14	43.2	40
Typic Natrargid, Lakhpat, Gujarat						
0-10	8.7	0.2	1.3	0.29	41.6	13
10-45	8.6	1.7	2.3	0.18	51.8	22
45-83	8.4	4.5	3.7	0.20	48.8	26
83-117	8.2	6.5	1.8	0.18	46.8	29
117-150	8.3	5.5	2.1	0.10	42.0	35
Natrustalf, Maharaj Gunj, Uttar Pradesh						
0-10	10.0	36.3	3.5	0.42	11.7	94.6
10-26	10.0	15.5	0.8	0.32	22.5	97.1
26-84	10.0	14.8	2.8	0.19	24.5	95.7
84-127	9.8	5.8	2.2	0.19	21.9	91.0
127-148	9.3	2.3	10.8	0.12	11.2	75.0

2.*Saline* : Profiles consist mostly of Na salts only with pH values higher than mixed saline soils. These represent accumulation of salts under high water table and are not leached either naturally or artificially.

3.*Alkali and strong alkali* : Profiles in which leaching conditions are established. Total salts are thus low at the top. These are still soils with high degree of alkalization from 20 to as high as 60%. The pH values are over 9.0 and sometimes upto 10.0 with carbonate salts in great predominance.

The saline soils along the sea coast are met under varied conditions of soil, climate and drainage and hence exhibit different characteristics. The soils of saucer shaped 'Bhal' region in Gujarat are fine-textured, calcareous and poorly drained. The Kutch soils are silt loam to clayey salt flats while Banni soils are coarser in texture. Both groups show mottling in lower horizons and presence of gypsum crystals throughout the profile but without any regular pattern. The salt affected soils of coastal Andhra Pradesh, with water less than 50 cm deep during cropping season, have restricted drainage, are mostly saline and difficult to reclaim.

Classification : The placement of saline, alkali or sodic soils in Soil Taxonomy [USDA, 1975] is largely based on salinity criteria at different levels and at 15 and 40% ESP. In aridisols, most soils have been classified either as Natrargids or Salorthids but the occurrence of a natric or a salic horizon within few cm of the surface in a large number of soils in the Indian subcontinent needs verification. The salic horizon requirements are too high for a satisfactory placement of the types of soils met in the Uttar Pradesh, Punjab and Haryana [Srinivasan, 1976]. Salic horizons are met in Rann of Kutch soil profiles to a depth of a metre or more where salt content is more than 2%. Most entisols in the subcontinent key out as aquents and fluvents and may be saline but not always meet EC requirements of the system. Alkali soils among inceptisols key out as aquepts but may lack a typical natric B horizon. The ESP often exceeds 15 or more in the top 50 cm soil but decreases at lower depths. Based on experience from alkali soils of the Indo-gangetic plains a redefinition of natric horizon has been proposed to accommodate blocky structure which is the most prevalent soil structural unit in these soils. Natric, salic, episalic horizons at sub-group level within the orders aridisol and alfisol have been proposed to accommodate most salt lands of India in soil taxonomy [Sehgal et al., 1975; Srinivasan, 1976]. Inclusion of sub-groups salic, halic and natric in vertisols and vertic sub-groups in salorthids has also been suggested [Srinivasan, 1976].

The scope of introducing new sub-groups or intergrades in soil taxonomy was discussed by Murthy et al. [1980]. They emphasized the need for a large volume of supporting data to justify any redefinition or modification of the existing criteria or to introduce new surface horizons. Murthy et al. [1980] have classified and named salt lands of India and mapped saline and sodic pedons into 12 great group associations. According to them further classification upto phase separation with textural variation, slope, EC, ESP etc. is necessary for farm level planning. The classification of salt affected soils at the great group level in Table 10 is for the sake of brevity and convenience and does not reflect the range of variation in their characteristics.

NATURAL VEGETATION OF SALINE HABITATS

The natural vegetation of salt-affected soils in the subcontinent consists of trees, hardy shrubs, herbs and grasses. The vegetation complex of saline basins in Rajasthan can be named as *Tamarix-Salvadora-Suaeda* type with dominant species being *Tamarix dioica, Suaeda fruitcosa, Zygophyllum simplex, Cressa cretica, Salsola foetida, Haloxylon recurvum, Aeluropus lagopoides, Sporobolus helvolus,* etc. [Sen, 1984]. *Salvadora oleoides, Capparis decidua, Zizyphus nummularia, Acacia nilotica, Butea monosperma, Tecomella undulata, Calotropis procera, Leptochloa fusca, Desmostachya bipinnata, Suaeda fruticosa, Salsola baryosma, Portulaca quadrifida, Argemone maxicana, Haloxylon salicornicum,* etc. are also met either in the alkali and saline habitats or on their fringes. Some species which are met in *'Dahars'* flat, interdunal lands in Cholistan are not met in saline habitats of Rajasthan or Punjab and vice versa (Table 11). Saline habitats in Rajasthan and Cholistan are met in interdunal areas. The sands encroaching upon saline flats allow establishment of non halophytes or facultative halophytes. Thus, there is often a mixture of species of all kinds in a saline environment and pure stands of halophytes or facultative halophytes may not be encountered. There are however econiches where halophytes survive and establish pure stands [Rao et al., 1989]. In a black saline soil a pure stand of *Cressa cretica* was observed between ECe of 15 to 20 dSm^{-1}. Some *'Dahars'* in Cholistan are too inhospitable for any vegetation, some only grow *H. recurvum* and some only *Salsola foetida* but there are habitats where these species may occur in mixture with non halophytes. Similar mixing is observed in Pachpadra saline basin in Rajasthan [Saxena and Gupta, 1973]. Saline habitats in India have undergone a spectacular change with the introduction of *Prosopis juliflora* which is at times the only major shrub in an otherwise desolate countryside.

Table 10 : Extent, nature and classification of salt affected soils in the subcontinent

	Indo-Gangetic alluvial plain (India)	Sandy arid region (India)	Inland black and red soil region	Deltic alluvial and coastal region	Indus basin north and Peshawar Valley	Indus basin south and deltic region
Great group Associations	Natraqualf-Natrustalf-Halaquept saline/sodic phases of Camborthids Calciorthids Ustochrepts Saline phase of Haplaquept, Fluvaquent	Salorthid-Natrargid Gypsiorthid Camborthid Saline/sodic phases of Ustochrept	Saline/sodic phases of Pellustert Chromustert Vertic ustochrept, Ustifluvent Salorthids Natrargids, Haplaquents, Halaquept Natrudalf	Halaquept saline phase of Haplaquept Fluvaquent Vertic Ustochrepts	Natraqualf Natrudalf Halaquept saline/sodic phases of Calciorthids Camborthids Ustochrepts saline phase of Halaquept Fluvaquent Natrargids,	Halaquept-Natrustalf-saline/sodic phases of Calciorthids Camorthids Salorthids Natrustalf
Nature of surface soil	Alluvial sandy to clayey	Loamy to coarse sand	Shallow medium and deep black or coarse to loamy red	Sandy along coast, clayey in deltas	Medium to deep alluvium medium to fine textured	Alluvial medium to fine textured, aeolian deposits at places.
Nature of sub soil	Calcic or petrocalcic, argillic	Old alluvium, often calcic, petrocalcic or gypsiferous	Permeable 'murrum' layer to hard bed rock	No layering except pyritic horizon in some soils	Often calcic to petro-calcic deposits	Calcic or gypsiferous deposits
Ground water quality	Generally good with scattered sodic water	Generally saline or sodic, few good aquifers	Good to marginal, sodic water also met in tanks and streams	Saline, fresh water lenses floating over saline aquifer	Good to marginal	Generally poor except in riverine tracts
Approximate salt affected area (m ha)	3.08	0.50	2.27	2.71	3.28	2.68

EXTENT AND DISTRIBUTION OF SALT LANDS

The Indo-gangetic plains are a single unit with about 8 m ha of such lands with the Indus Basin in Pakistan alone sharing 5.97 m ha. Annual survey of all commands in Pakistan indicated that on an average 15% of land was visibly affected by salts and another 0.5% by waterlogging. The actual incidence of salinity and alkalinity may be worse [Schroo, 1967]. Agencies reporting on the extent of salt lands in Pakistan have employed different survey methods and classification criteria and come out with different sets of data on extent and nature of the problem. But the data reported by Soil Survey of Pakistan appears more authoritative [Choudhri et al., 1978] The exercise to categorise, delineate and estimate salt land area in the subcontinent is beset with a number of uncertainties as area covered by detailed soil survey is negligible. The sources of information for the data (Table 12) were reports of state departments and research papers [Singh, 1989]. The saline area has been shown separately. For soils in the arid zone, unless irrigated with sodic water, the demarcation between saline and conventional saline-sodic soils has been based on the presence of soluble carbonates and in some cases could be arbitrary and subject to sizeable shifts. For Pakistan, where salt lands with both high soluble salt content and exchangeable Na constitute about

Table 11. Natural vegetation of some salt-affected habitats

Species	Cholistan desert (dahars), Pakistan	Pachhpadra salt basin, Rajasthan India	Derelict Chaggar bed Rajasthan, India	Saraswati forest range, Haryana India	Sampla-Gohand belt, Haryana, India
TREES					
Acacia nilotica	–	O	F	A	F
Salvadora oleoides	O	F	O	F	F
S. persica	–	F	–	–	–
Prosopis cineraria	F	F	O	F	O
P. juliflora	–	A	F	A	A
Tamarix aphylla	F	R	O	O	O
T. dioica	–	–	R	R	–
T. undulata	–	–	R	R	–
SHRUBS					
Capparis decidua	F	O	F	O	F
C. sepiaria	–	–	–	F	F
Zizyphus nummularia	R	F	F	F	F
Calotropis procera	F	F	O	O	R
Leptadenia pyrotechnica	F	R	–	–	–
Calligonum polygonoides	–	F	–	–	–
Salsola baryosoma	A	F	A	A	F
Suaeda fruticosa	–	A	–	A	F
Haloxylon salicornicum	A	A	A	–	–
H. recurvum	A	R	–	–	–
UNDER SHRUBS					
Suaeda maritima	–	–	–	F	A
Fagonia arabica	F	R	–	–	–
Aerva persica	F	F	–	–	R
Crotalaria burhia	F	F	R	–	–
ANNUAL HERBS					
Heliotropium supinum	–	–	–	O	A
Portuleca quadrifida	–	–	–	A	F
P. oleracea	–	O	–	R	–
Trianthema portulacastrum	–	F	–	F	F
Convolvulus microphyllus	–	F	–	F	O
Cressa cretica	–	F	–	–	–
Tribulus terrestris	–	R	–	R	R
Argemone mexicana	–	O	–	O	R
Alysicarpus spp.	–	O	–	F	R
Corchorus spp.	–	F	–	O	R
Launaea spp.	–	O	O	O	R
Euphorbia thymifolia/ E. prostrata	–	F	O	F	F

Table Contd.....

GRASSES & SEDGES					
Sporobolus marginatus	A	A	A	A	O
Desmostachya bipinta	–	O	A	A	F
Aeluropus lagopoides	A	F	A	–	–
Cenchrus spp.	F	A	O	R	–
Dactyloctneum spp.	–	O	O	F	O
Cynodon dactylon	O	O	F	F	F
Cymbopogon jaworanchusa	A	–	–	–	–
Lasiurus sindicus	A	–	–	–	–
Saccharum spontaneum	–	–	F	R	F
Diplachne fusca	–	–	F	O	R
Elusine compressa	–	F	F	O	R
Cyperus spp.	O	F	F	A	O

A denotes Abundant or prominent F denotes Frequent
O denotes Occasional R denotes Rare and absent or not reported from the region.

Table 12 : Water logged and salt affected area in the dry zone of the subcontinent (000 ha)

State/ province	Water logged area			Salt affected area			
	Canal command	Unclassified	Total	Canal command	Outside canal command	Coastal	Total
INDIA							
Andhra Pradesh	207.1	131.9	339.0	493.9	36.1	283.3	813.3
Gujarat	13.8	470.2	484.0	300.0	612.1	302.3	1214.3
Haryana	NA	275.2	275.2	455.0	NA	NIL	455.0
Karnataka	36.0	NA	36.0	204.0	114.0	86.0	404.0
Madhya Pradesh	57.0	NA	57.0	220.0	22.0	NIL	242.0
Maharashtra & Goa	NA	111.0	111.0	446.0	NA	88.0	534.0
Punjab	NA	189.3	189.3	392.6	126.9	NIL	519.5
Rajasthan	92.8	255.5	348.3	138.2	983.8	NIL	1122.0
Tamil Nadu	NA	127.9	127.9	256.5	NA	83.5	340.9
Uttar Pradesh	455.0	1525.0	1980.0	1060.7	234.3	NIL	1295.0
Total (i)	861.7	3086.0	3947.1	3986.9	2129.1	825.1	6939.1
PAKISTAN							
N.W.F.P.	NA	60.0	60.0	14.0	506.0	NIL	520
Punjab	NA	1190.0	1190.0	1628.0	1138.0	NIL	2766.0
Sind	NA	3470.0	3470.0	1544.0	1072.0	NA	2616.0
Baluchistan	NA	90.0	90.0	110.0	NA	NIL	110.0
Total (ii)	NA	4180.1	4810.1	3296.0	2716.0	NIL	6012.0
Grand total (i) + (ii)	861.7	7895.5	8757.2	72809	4845.1	825.1	12951.1

Note : NA refers to data not available. The area from unclassified and other category can spill to this category. The area classification is therefore tentative.

3.17 m ha, no such delineation has been attempted. Hence for the limited purpose in Fig. 2 all saline-sodic soils have been shown as alkali. In Punjab, the extent of such soils is more than saline soils [Choudhri et al., 1978]. Both categories of soils are evenly balanced in Sind even though there could be saline soils treated as saline-sodic.

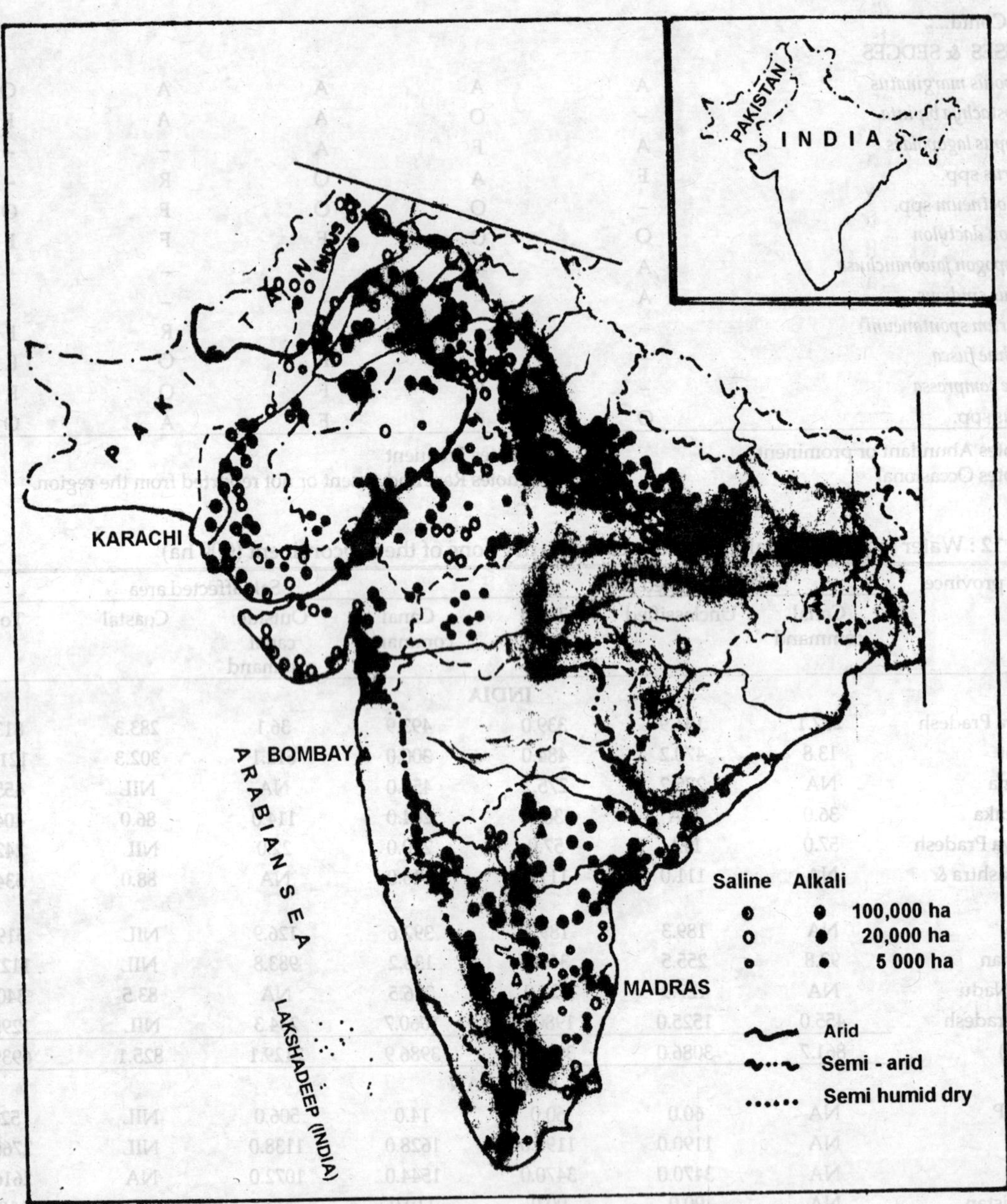

Fig. 2. Saline and alkali soils in drylands of the sub continent
(Please refer to text for details)

The area under alkali soils in the subcontinent is concentrated in the states of Punjab, Haryana, Uttar Pradesh and Bihar. In Uttar Pradesh some districts are severely affected. The other area is the semi-arid and arid west India with largely saline soils. The salt lands of Peninsular India are saline as well as alkali but both pH and salt content are not as high as in the Indo-Gangetic plains. Salt lands in the coastal and deltic regions of the subcontinent are saline and in high rainfall regions the pH may be in the acid range. A statewise brief of salt affected lands follows.

Punjab (Pakistan) : Salt affected soils are met both in and outside canal commands in an area of 2.76 m ha. Maximum affected area lies in Rachna and Bari 'Doabs'. Almost 50% area in these 'Doabs' is slightly saline with 0.2-0.5% soluble salts [Ahmed and Chaudhary, 1988]. A summary of overall chemical status of soil profiles conducted in 1977-79 showed that 27% of them were salt affected and of these 19% were either sodic or saline-sodic [Sheikh, 1989].

Sind : The salt affected area in the province is about 2.57 m ha. Out of 5.39 m ha canal command area, the cultivated area is 4.41 m ha, of which 0.56 m ha are saline. The semi-arid salt lands are either uncultivated or lie outside canal commands. The Indus delta is highly saline. Concentration of saline soils is largely confined to areas where water table is less than 1.5 metre deep and ground waters are highly saline.

Baluchistan and N.W.F.P. : Salt lands in these provinces are confined to Peshwar Vale and Sulaiman piedmont in shallow water table area and in Nushki area Kachhi plains in Baluchistan.

Punjab (India) : Salt affected lands are met both in canal and well irrigated areas. The nature of salts varies with rainfall from carbonate-chloride-sulphate types in the sub-humid areas to chloride-sulphate carbonate types in the arid zone [Singh et al., 1980]. Majority of alkali lands in the good quality water zone have since been reclaimed and grow excellent crops of rice and wheat. On the other hand, incidence of salinity in the high water table area of south west Punjab largely in districts of Faridkot and Ferozepur, where ground waters are saline, is increasing (Table 13).

Haryana : Like Punjab, large area under alkali soils in Kurukshetra and Karnal districts with good quality aquifers has been reclaimed. The saline area in the saline ground water zone is intact. Salinity in Hissar, Rohtak, Jind and Sirsa districts, where ground water has been rising, is increasing.

Uttar Pradesh : The main salt affected belt lies in the central and north-west Ganga-Yamuna Doab and trans-Ganges area. The whole of

Table 13. Influence of water table rise/fall on acreage under salt lands

Zone/State	Year of land-sat imagery	Ground water depth	Ground water quality	Affected area			
				Severely affected	Partially affected	Saline & water logged	Total
				000 ha			
South West Punjab	1972	Shallow	Marginal to poor	41	83	–	124
	1984	Partially	Marginal to poor	37	126	96	259
Rest of Punjab	1972	Shallow	Good to marginal	176	399	Not determined	575
	1984	Deep	Good to marginal	80	126	23	229
				Sodicity level (from tone)			
				Low	Moderate	High	Total
					(ha)		
Khambat Taluka, Gujarat	1975	Shallow to deep	Marginal to poor	1927	2762	2248	6937
	1986	Partially water logged	Marginal to poor	17728	6423	12130	36281
Matar Taluka	1977	Shallow to deep	Marginal to poor	2287	976	335	3598
	1983	Partially water logged	Marginal to poor	2646	5627	1721	9994

Source : Sidhu et al., 1990 : Bhagwat, 1985.

central zone from Mainpuri to Allahabad, is suffering from waterlogging. The salt affected area in Ganga canal command is 0.35 m ha. In Sharda Sahayak canal command, 0.34 lakh ha have dangerously high water table and 0.25 m ha have been salinised [Ambekar, 1986].

Bihar : Of the 1.05 m ha Gandak command area 0.4 m ha in the calcareous belt in Muzaffarpur, Vaishali, Samastipur, Saran and Gopalganj districts is salt affected [Jha, 1986]. Such soils are concentrated in basin areas which receive runoff from uplands.

Rajasthan : This largely semi-arid to arid state is virtually sitting over saline substrata in the form of salt impregnated sub surface horizons or saline ground waters. Given suitable conditions salts appear at the soil surface. Shallow water table areas of the Indira Gandhi Canal Command and of Bhilwara, Jaipur, Bharatpur and Ajmer districts have impeded drainage and are saline. The Ghaggar flood plains have vast tracts of salt affected soils with very sparse vegetation. In the Chambal command, black soils of Kota, Bundi and Jhalawar are alkali. Irrigated soils of Pali, Barmer and Bikaner districts are saline as well as alkali depending upon the quality of well waters.

Madhya Pradesh : The salinity is observed in 23 districts in the black soil region derived from Deccan Traps with 0.24 m ha area and in the alluvial zone of Chambal river in districts of Bhind, Morena and Gwalior. Waterlogging and salinity is also prevalent in Tawa Canal Command in Hoshangabad and Barna command in Raisen and Sehore districts.

Gujarat : Gujarat claims second largest area of salt lands in India with 0.5 m ha in the interior and 0.7 m ha along the sea coast. There are 0.3 m ha in a narrow strip along the coast where salinity is caused by ingress of sea water. Another important salt affected zone is the flat, fine textured Bhal tract. In Ukai-Kakarapar and Mahi-Kadana canal commands almost 25 to 30 percent area is salt affected. In Shatrunji and Fatehwadi canal commands about 66,000 ha area is saline which include coastal and inland salinity. Kutch district has large area under low level salt flats with Banni area mostly salt affected.

Maharashtra : The four coastal districts of the state which fall in a very high rainfall area have about 64,000 ha salt lands. In the interior dry region, the canal irrigated areas of Sholapur, Ahmednagar and parts of Dhulia, Jalgaon, Pune, Sitara, Nasik, Bhind and Sangli districts, with deep black low lands, are salt affected.

Karnataka : Besides the two high rainfall coastal districts with 35,000 ha saline area, salt affected lands are spread over 15 districts under different canal commands. Districts of Bellary, Bijapur, Chitradurg, Mandya, Gulbarga and Raichur contribute more than 20,000 ha each [Prasad and Malhotra, 1984]. Tungabhadra canal commands is most affected. Salinity is fast developing in Upper Krishna command. Salinity and alkali problems are also acute in tank irrigated areas.

Andhra Pradesh : Coastal districts of Krishna, Guntur, Prakasam and Nellore have large salt affected tracts, mostly saline but also with alkali patches. At present 114,000 ha are affected by waterlogging and salinity in Guntur and Prakasam districts under Nagarjunasagar right bank canal command. More than 60,000 ha are alkali in districts of Anantapur, Kurnool, Medak, Nalgonda and Mehboobnagar.

Tamil Nadu : About 0.2 m ha saline or alkali soils are met in the coastal districts of Chinglepet, South Arcot, Thanjavur, Puddukoti, Ramanathapuram and Thirunelveli. Inland salinity is present usually in canal commands in Coimbatore, North Arcot and Madurai districts. In Tiruchirapally salt affected lands are met along north bank of Cauvery. Some low lying lands have ancient salinity.

RECLAMATION AND MANAGEMENT OF SALT AFFECTED SOILS

Reclamation of salt affected soils involves rehabilitation of their original physico-chemical and productive status. Earlier attempts to reclaim alkali or saline sodic soils often met with limited success either owing to lack of knowledge about cation exchange processes at that time or due to misplaced efforts to enforce only an engineering solution like simple leaching or drainage for their reclamation. The first effort to reclaim 16 ha alkali soil at Amarau near Kanpur in Uttar Pradesh involved deep plowing and manuring with 20-25 t/ha of cattle dung [Tiwari and Sharma, 1989]. After 25 years of manure treatment, Leather [1914] analysed the soil. The physical condition had improved in the top 30 cm but no significant decrease in salts was observed. Leather was the first scientist to use gypsum for reclaiming alkali soils in India but he made incorrect assumptions to assess the quantity of amendment needed for soil amelioration. The Irrigation Department in Punjab tried to accomplish reclamation of saline-sodic soils with leaching alone. For this purpose, additional canal water was supplied through reclamation shoots and outlets. The efforts remained unsuccessful as the soils needed simultaneous application of Ca amendments to replace exchangeable Na. In Pakistan, where 1.2 m ha salt affected area is under irrigated agriculture, the salinity occurs in patches which cover more than 20% area and are mostly sodic. These patches can be reclaimed economically through application of gypsum [Choudhri et al., 1978].

Removal of excess salts from saline soils is accomplished by simple leaching provided good quality water is available for the purpose and water table is either reasonably deep or there is provision of subsurface drainage and disposal of drainage effluents. In lands with shallow but good quality ground aquifers, vertical drainage with shallow irrigation pumps has been successful in Indian part of Punjab. In Pakistan also, installation of tubewells to supplement subsurface drainage is an important component of SCARPs. According to 1988 appraisal more than 6 m ha in Pakistan have water tables within 1.5 m of surface during any part of the year [Sheikh, 1989]. Much of the area is in Sind where soils are saline or saline-sodic and require subsurface drainage to lower water table. So far about 2.7 m ha of affected land has been covered by 36 SCARP schemes. Construction of 11473 irrigation tubewells, 1723 drainage tubewells, more than 200 m cubic metres of surface drains and 5076 ha of sub surface drains was completed by 1988 [Sheikh, 1989]. In India saline waterlogged soils have been successfully reclaimed in Gujarat and Haryana through sub surface drainage. Basic information about spacing, depth and nature of drainage materials has been obtained but the disposal of drainage effluents remains a problem in the Indo-Gangetic plains where land surface is very flat. In Pakistan, Left Bank out Fall Drainage project has been planned. This system essentially consists of tile drains with sumps to pump water into open drains 1 km apart. From here the effluent is discharged into a 30 m wide spinal drain extending 270 km down through the lower Indus region out to the sea. India will face the outfall problem to discharge drainage effluents, once sub surface drainage is installed to lower water tables in parts of Haryana and Rajasthan. On experimental scale the Central Soil Salinity Research Institute, Karnal, has been recycling drainage effluents with EC of 25 dSm^{-1} or more for irrigation of wheat and mustard in conjunction with canal water. The effluents are high in sulphates and the experience is limited to few years only. Long-term effects of such effluents specially those rich in Cl could be harmful to the soil. The ultimate aim has to be to remove excess salts from the immediate soil environment.

In drylands with saline ground water or where salinity is associated with waterlogging, immediate reclamation is not possible and the solution narrows down to living with salinity. The best alternative is to plant such lands with salt tolerant plant species like *Atriplex*, *Leptochloa fusca*, etc. Both species have been grown in Pakistan. A 50% yield reduction was observed in *Atriplex amnicola* at salinity levels equivalent to ECe 33.0 dSm^{-1} and in *Leptochloa fusca*, salt tolerant grass at ECe 22 dSm^{-1}. The Nuclear Institute for Agriculture and Biology, Faisalabad has, in an undated publication, emphasized the role of salt tolerant plants for economic utilization of saline lands. *Atriplex* may have some scope in saline grazing lands but individual farmers with small holdings may prefer a more familiar plant or crop. *Leptochloa fusca* has been planted on a large acreage in Pakistan [Malik et al., 1986]. This grass adapts well in salt affected soils and is the first colonizer owing to its high tolerance to salinity and sodicity. In India it has proved much more alkali tolerant than Rhodes grass and Bermuda grass [Kumar and Abrol, 1986]. The mechanism of salt tolerance is excretion of salts through leaves as washed leaves contain less salt than unwashed leaves. It establishes rapidly in alkali soils and improves soil condition to an extent when other grasses like *Cynodon dactylon* can come up as secondary colonizers. The other plant which is not native to saline/alkali soils but has firmly established in such habitats is *Prosopis juliflora*. Field studies of the CSSRI, Karnal show that, provided proper planting and training. *P.juliflora* can produce fuel and small timber from highly alkali soils. It can grow at a pH of 10 but the growth is excellent around a pH of 9.5. In highly degraded soils, this plant is often the sole source of fuel for the poor landless villagers in India.

Earlier efforts to reclaim such lands in Punjab and Uttar Pradesh employed rice cultivation with ample quantities of water [Leather, 1893; Mehta, 1940]. Since salt-affected soils in the Indo-Gangetic plains are calcareous and some soils in Pakistan gypsiferous, such efforts had a mixed bag of success. The reclamation through mobilisation of soil Ca was slow. *Sesbania* green manure when combined with rice cultivation was shown to accelerate reclamation of alkali soils [Uppal, 1955]. Meanwhile researchers continued to employ gypsum and other calcium salts, farm manure and drainage to ameliorate alkali soils [Nasir, 1923; Singh and Nijhawan, 1932]. Amendment use as a major mode of alkali soil reclamation had to wait till next 30 years for reasons of supply, cost effectiveness and a nagging response of the Government for such a course of action. The reclamation of alkali and saline sodic soils picked up both in India and Pakistan in the sixties. The promoting factors have been high yielding rice and wheat varieties, establishment of Land Reclamation Corporations and Government support for various inputs. These factors have led to reclamation of 0.5 m ha alkali lands in Punjab, Haryana and Uttar Pradesh which now annually produce more than 2 m t food grains. There has been perceptible improvement in properties of surface soil following reclamation and cropping (Table 14). These soils will however need good management as they always run the risk of reversion to the previous conditions. This is particularly true of soils with poor internal drainage.

Table 14 : Effect of reclamation measures on properties of two alkali soils from Punjab

Depth (cm)	Clay %	pH_s 1973	pH_s 1987	ECe 1973	ECe 1987	ES 1973	ES 1987
			Ghabdan loam (Natraqualf)				
0-6	20.9	10.1	9.8	32.4	5.2	82.9	54.4
6.24	23.7	9.7	8.9	10.5	4.6	75.4	74.4
24-48	28.0	9.4	8.8	8.0	4.3	75.0	46.4
48-73	28.2	9.3	8.8	5.6	4.6	76.0	85.1
73-97	31.4	9.2	8.9	3.7	4.6	68.1	85.1
97-124	24.6	9.2	9.1	4.1	4.7	63.2	77.9
124-145	23.0	9.1	9.2	4.6	4.0	66.6	76.2
			Laungowal loam (Natric Camborthid)				
0-14	11.5	9.5	8.9	3.9	2.1	43.8	29.8
14-41	17.0	9.3	8.8	8.5	1.1	42.8	27.2
41-65	18.4	9.4	8.7	12.0	0.9	48.0	39.4
65-84	16.6	9.6	8.7	4.2	0.8	50.7	44.0
94-117	23.2	9.5	8.7	5.0	0.8	62.0	44.0
117 +	25.4	9.5	9.0	8.2	0.3	65.2	42.6

In Pakistan, Water and Power Development Authority (WAPDA) was established in 1958 and given the responsibility of tackling the problem.

Since then a large number of SCARPs have been launched and completed by WAPDA. In the latest action plan the emphasis is on soil reclamation through lowering of water table by private tubewells. About 200,000 tubewells have already been installed. It is claimed that the extent of salinity has gone down from 40% to 28% [Sheikh, 1989]. The claims of WAPDA have been contested by Choudhri et al. [1978] on the basis of a comparison of aerial photographs taken 23 years apart and subsequent confirmation of results through LANDSAT imagery. They acknowledge that few wet areas have now shrunk but maintain that large tracts of salt-affected lands are still out of cultivation as they were in 1953-54. They refer to other studies in SCARP-1 to emphasize that crop yields in SCARP area are hardly better than the adjoining non-SCARP areas and good lands have been damaged by sodicity through use of moderate to high Na hazard irrigation waters from tubewells meant for lowering the water table. A combined approach employing leaching, soil amendments and sub-surface drainage, coupled with suitable agronomic measures may resolve apparent contradiction in reclamation strategies advocated by various agencies engaged in this task in Pakistan.

The reclamation technology for alkali lands has been refined by the Central Soil Salinity Research Institute, Karnal which is the nodal research agency for this task in India. The main components of this technology are: grading and bunding of fields, application of amendments, judicious fertilizer use and growing of rice and wheat in rotation. Installation of shallow irrigation pumps lower the water table as ground aquifers in most alkali areas are of good quality. The method for determining gypsum requirement has been refined and 50% of the actual dose is mixed in the surface 10-15 cm soil. It has been established that only small amounts of additional irrigation water can complete exchange reaction and leach down reaction products below the root zone of rice [Abrol et al., 1979; Hira and Singh, 1980]. Both gypsum and pyrites are used as soil amendments in India but on equivalents basis, the former is more effective in improving soil and crop yields. About 25% additional N and 20 to 50 kg/ha zinc sulphate are recommended in the first year of reclamation. Alkali soils in the Indo-Gangetic plains are well supplied with available P [Singh and Nijhawan, 1943; Chhabra et al., 1981] making its application unnecessary unless the soils test low in available P. Similarly, higher rates of N are recommended for wheat than in the normal soils due to lowered efficiency of N in alkali soils. The clay minerals in soils of the Indo-Gangetic plains are principally illite. Hence the soils do not respond to application of K. The K release also occurs in Pakistan soils which are mostly micaceous even though major presence of clay minerals like kaolinite and, in some cases, of smectites is also reported [Niazi and Hussain, 1991].

Another aspect related to soil salinity is the salinity of ground waters in the subcontinent. Well waters are often the only source of irrigation water available but in the process they become an important cause of soil degradation [Dhir et al., 1978; Choudhri et al., 1978; Singh, 1989; Singh et al., 1989]. Irrigation water quality aspects have been widely researched in the subcontinent. The WAPDA, after studies spread over a number of years in various SCARPs, found the following safe limits for ground water use [Haque, 1989].

Nature of soil	Other conditions	Safe limits for use as irrigation		
		TDS ($mg\ kg^{-1}$)	SAR ($m.mol\ l^{-1})^{1/2}$	RSC ($me l^{-1}$)
Coarse textured	Level good, water management	1500	12	5.0
Medium textured	Fairly level lands, adequate water supply	800	10	2.5
Fine textured	Level lands, poor water supply, low level of management	500	5	1.25

Table 15 : Guidelines for using poor quality irrigation waters

Soil texture (% Clay)	Crop tolerance	Upper limits of ECiw ($ds\text{-}m^{-1}$) in rainfall regions		
		< 350 mm	350-550 mm	550-750 mm
(a) Saline waters (RSC > 2.5 mel^{-1})				
Fine	Sensitive	1.0	1.0	1.5
(> 30)	Semi-tolerant	1.5	2.0	3.0
	Tolerant	2.0	3.0	4.5
Moderately	Sensitive	1.5	2.0	2.5
fine	Semi-tolerant	2.0	3.0	4.5
(20-30)	Tolerant	4.0	6.0	8.0
Moderately	Sensitive	2.0	2.5	3.0
coarse	Semi-tolerant	4.0	6.0	8.0
(10-20)	Tolerant	6.0	8.0	10.0
Coarse	Sensitive	–	3.0	3.0
(– 10)	Semi-tolerant	6.0	7.5	9.0
	Tolerant	8.0	10.0	12.5

(b) Sodic water (RSC > 2.5 c mol kg^{-1} ECiw -4.0 dSm^{-1})

Soil texture (% clay)	Upper limits of SAR $(mmole/kg)^{1/2}$	Upper limits of RSC(mel^{-1})	Remarks
Fine (> 30)	10	2.5 - 3.5	[1] Limits pertain to kharif-fallow rabi crop rotation when annual rainfall is 350-550 mm
Moderately fine (20-30)	10	3.5 - 5.0	[2] When the waters have Na < 75 % (Ca+Mg > 25%) or rainfall is 550 mm the upper limits of the RSC range becomes safe.
Moderately Coarse (10-20)	15	5.0 - 7.5	[3] For double cropping, RSC neutralisation with gypsum is essential based on quantity of water used during winter. Grow low water requiring crops during summer. Avoid growing rice
Coarse (– 10)	20	7.5 - 10.0	

1. Texture criteria should be applicable for all soil layers down to atleast 1.5 m depth.
2. In areas where gound water table reaches within 1.5 m at any time of the year or a hard subsoil layer is present in the rootzone, the limits of the next finer textural class should be used.
3. Flourine is at times a problem and limits should be worked out.

It was assumed that under average level of water management water of TDS upto 1000 mg kg^{-1}. SAR upto 10 and RSC upto 2.5 mel^{-1} could be used safely.

The Central Soil Salinity Research Institute, Karnal [Singh, 1991], on the basis of its own experience and that of the All India Coordinated Research Project on Saline Soils and Use of Saline Waters has come up with a set of recommendations for safe use of saline and sodic waters (Table 15). Such recommendations are made to ensure sustained productivity of soils where saline irrigation is practised. However, prolonged input of salts will eventually build up salinity in sub surface horizons. Occasional wet period or rise in ground water level can cause these salts to move upwards and salinize surface soil. Therefore, saline irrigation has to be restricted to low water requirement crops on well drained soils.

REFERENCES

Abrol, I.P. and Bhumbla, D.R. 1978. Some comments on terminology relating to salt affected soils. In: *Proc. Meeting for the Sub-Commission on Dryland Saline Seep Control*. 11th Intern. Soc. Soil Sci. Cong., Edmonton, Alta, June 1976, pp. 19-25.

Abrol, I.P., Gupta, R.K. and Singh, S.B. 1979. Note on the solubility of gypsum and sodic soil reclamation. *J. Indian Soc. Soil Sci.* 27: 481-483.

Ahmed, N. and Chaudhry, G.R. 1988. Irrigated Agriculture of Pakistan. Mirajudin Printing Press,

Urdu Bazar, Lahore, 1988.

Ambekar, V.W. 1986. Status of waterlogging and soil salinity in Sharda Sahyak command area development project. In: *Sem. Waterlogging and Soil Salinity in Alluvial Soils*, WALMI, Okhla, New Delhi, pp. 10.1-10.3.

Basu, J.K. and Sirur, S.S. 1938. Soils of Deccan canal I. Genetics, soil survey and soil classification. Nira right bank and Pravara canals. *Indian J. agric. Sci.* 8: 637-697.

Basu, J.K. and Tagare, V.D. 1943. Soils of the Deccan canal IV. The alkali soils, their nature and management. *Indian J. agric. Sci.* 13: 157-181.

Bhagwat, K.A. 1985. Monitoring of crop production in saline and sodic soils of Gujarat with the help of remote sensing. Memo. M.S. Univ., Baroda, Gujarat.

Bhumbla, D.R. and Abrol, I.P. 1979. Saline and sodic soils. In: *Symp. Soils and Rice*. IRRI, Los Banos, Philippines, pp. 719-738.

Bhushan, S.K. and Mohanty, M. 1986. Mechanics of intrusion and geochemistry of alkaline granites from Siwana, Barmer district. *Indian J. Earth Sci.* 16: 103-115.

Chhabra, R., Abrol, I.P. and Singh, M.V. 1981. Dynamics of phosphorus during reclamation of sodic soils. *Soil Science* 132: 319-324.

Choudhri, B., Mian, M.A. and Rafiq, M. 1978. Nature and magnitude of salinity and drainage problems in relation to agricultural development in Pakistan. Pakistan Soils Bull. 8, Govt. of Pakistan, Lahore.

Choudhari, J.S. and Dhir, R.P. 1982. Clay minerology of dune and associated sandy plain soils of western Rajasthan. *J. Indian Soc. Soil Sci.* 30: 342-347.

Choudhari, J.S. and Sharma, K.D. 1984. Stream salinity in Indian arid zone. *J. Hydrol.* 71: 149-163.

Das Gupta, A. 1982. Problems of water management and soil salinisation in Gandak command area. *Proc. 9th Combined Workshop on Water Management and Management of Salt Affected Soils*. ICAR, New Delhi, pp. 4.1-4.12.

Dhir, R.P., Sharma, V.C. and Kolarkar, A.S. 1978. Soil resource data and mapping activity in the northwest zone of India. In: *Arid Zone Research and Development*, (ed) H.S. Mann. CAZRI, Jodhpur.

Dewan, M.L. 1953. Pre-irrigation soil survey of Tungabhadra Project. *J. Soil and Water Cons.* 2: 4-12.

Dubey, D.D., Sharma, O.P. and Bangar, K.S. 1985. Minerological studies on sodic vertisols. *Clay Res.* 4: 96-99.

Dubey, D.D., Gupta, S. and Sharma, O.P. 1986. Evaluation of weathering in Natrargids. *J. Indian Soc. Soil Sci.* 34: 861-869.

Ghose, B. 1971. Geomorphological aspects of the formation of salt basins in western Rajasthan. *Symp. Problems of Indian Arid Zone (1964)*, CAZRI, Jodhpur, pp. 79-83.

Ghose, B. and Kar, A. 1984. The former stream courses in the Indian desert and their effects on natural resources. *Proc. Indo-U.S. Workshop on Arid Zone Research*, CAZRI, Jodhpur, pp. 140-149.

Goyal, R.S. 1984. Sieving of ground water silica by soils. *J. Indian Soc. Soil Sci.* 32: 353-357.

Haque, A. 1989. Water quality criteria based on evaluation. Evaluation Report on Waterlogging and Salinity Research in Pakistan. Intern. Waterlogging and Salinity Research Institute, Lahore, pp. 10.1-10.11.

Hassan, S.T. and Venkat Reddy, P. 1982. Specific aspect of water management experienced in Sri Ram Sagar project command. *9th Combined Workshop on Water Management of Salt Affected Soils*, ICAR, New Delhi, pp. 8.1-8.10.

Hira, G. and Singh, N.T. 1980. Irrigation water requirement for dissolution of gypsum in sodic soils. *Soil Sci. Soc. Am. J.* 44: 353-358.

Holland, T.H. and Christie, W.A. 1909. The origin of the salt deposits in Rajasthan. *Rec. Geol. Surv. Ind.* 38: 154-186.

Jha, S.N. 1986. Waterlogging and salinity development in Bihar. *Sem. Problems of Waterlogging and Salinity in Alluvial Soils*. WALMI, Okhla, New Delhi, pp. 4.1-4.9.

Joshi, P.K. and Agnihotri, A.K. 1984. An assessment of adverse effects of canal irrigation in India. *Indian J. Agric. Econ.* 38: 528-536.

Joshi, D.C. and Dhir, R.P. 1990. Reclamation and saline sediments deposited during a flash flood on agricultural lands in the Indian arid zone. *Arid Soil Research and Rehabilitation* 4: 141-148.

Kandpal, S. 1982. Water management and soil salinity problems in Ram Ganga command project. *9th Combined Workshop on Water Management and Management of Salt Affected Soils*. ICAR, New Delhi, pp. 5.1-5.13.

Kar, A. 1988. Evaluation of present day geomorphic processes in the Indian desert for assessing land degradation and natural hazards. In: *Geomorphology and Environment* (eds) C.S. Singh and R.C. Tiwari. Allahabad Geographical Society, Allahabad.

Kar, A. 1989. Terrain characteristics of Jaisalmer district. *Geog. Rev. India* 51: 48-59.

Khandelwal, N.M. 1975. On occurrence of halite in Didwana salt lake area, Rajasthan. *Curr. Sci.* 44: 13-14.

Khosla, B.K., Gupta, R.K. and Abrol, I.P. 1979. Salt leaching and effect of gypsum application in a saline sodic soil. *Agric. Water Management* 2: 193-202.

Kolarkar, A.S., Dhir, R.P. and Singh, N. 1980. Characteristics and morphogenesis of salt affected soils in south-eastern arid Rajasthan. *Journal Ind. Soc. Photo-Int. and Remote Sensing* 1: 31-41.

Kumar, A. and Abrol, I.P. 1986. Grasses in alkali soils. Bull. No. 11, CSSRI, Karnal.

Leather, J.W. 1914. Investigation on 'usar' land in the United Provinces. Govt. Press, Allahabad.

Leather, J.W. 1893 and 1896. Reclamation of reh or usar lands. *Agric. Ledger* 7: 9-13, 37.

Malik, K.A., Aslam, Z. and Naqvi, M. 1986. Kallar grass plant for saline land. NIAB, Faisalabad.

Mann, H.H. and Tamhane, V.A. 1910. In: *Saline and Alkali Soils of India* (eds) R.R. Aggarwal. ICAR, New Delhi, 1979.

McKenzie Taylor, E. and Mehta, M.L. 1941. Some irrigation problems in the Punjab. *Indian J. agric. Sci.* 11: 137-169.

Mehta, M.L. 1940. The formation and reclamation of Thur lands in Punjab. Pb. Irri. Res. Inst. Res. Publ. (4).

Murthy, R.S., Hirekurpur, L.R. and Bhattacharjee, J.C. 1980. The taxonomy of salt affected soils of the Indian subcontinent. In: *Intern. Symp. Salt Affected Soils*, CSSRI, Karnal, India, pp. 67-76.

Narayanaswami, R. 1939. *Proc. Ind. Acd. Sci. 9*. pp. 518-525 quoted by Ghose 1971.

Nasir, S.M. 1923. Some observations on barren soils of Lower Bari Doab Colony in Punjab. Imp. Agr. Res., Pusa Bull. 145.

Niazi, M.H.K. and Hussain, M. 1991. Information about salt affected soils of Pakistan. Personnel communication.

Prasad, R. and Malhotra, K. 1984. Land and water resources in Karnataka. In: *Karnataka-State Environment Report* (ed) C. Saldanha. Santon Press, Bangalore, pp. 52-102.

Rafiq, M. 1980. Classification of salt affected soils of Indus plain with respect to reclamability. In: *Intern. Symp. Salt Affected Soils*, CSSRI, Karnal, India, pp. 77-82.

Raheja, P.C. 1966. Salinity and aridity - a survey of soils and land use. In: *Salinity and Aridity - New Approaches Old Problems* (ed) H. Boyko. Dr. W. Junk Publishers, The Hague.

Rao, A.R., Arshad, M. and Shafiq, M. 1989 (eds) Perennial grass germplasm of Cholistan desert and its phytosociology. Cholistan Institute of Desert Studies. Islamia University, Bahawal Pur, Pakistan, p. 84.

Saxena, S.K. and Gupta, R.K. 1973. Vegetation of Pachpadra salt basin in western Rajasthan. *J. Bombay nat. Hist. Soc.* 70: 104-127.

Schroo, H. 1967. Notes on reclamation of salt affected soils in the Indus plain of West Pakistan. *Neth. J. Agric. Sci.* 15: 207-220.

Sehgal, J.L., Hall, G.F. and Bhargava, G.P. 1975. An appraisal of the problems in classifying saline-sodic soils of the Indo-Gangetic plain in N.W. India. *Geoderma* 14: 75-91.

Sen, D.N. 1984. Plant life in Indian desert land. *Proc. Indo-U.S. Workshop on Arid Zone Research*, CAZRI, Jodhpur, pp. 277-282.

Shankarnarayan, K.A. 1985. Arid land irrigation and combating desertification. In: *Course on Irrigated Agriculture in Arid Areas* (ed) S.D. Singh. CAZRI, Jodhpur, pp. 180-189.

Sharma, D.L., Mehta, V.B. and Mathur, C.M. 1968. Salinity and alkalinity problem and fertility status of soils of Pali district (Rajasthan). *J. Indian Soc. Soil Sci.* 16: 263-269.

Sheikh, I.A. 1989. Country report on problem of waterlogging and salinity in Pakistan. *Information Seminar on Waterlogging and Salinity Research in Some Major Problem Countries*, WAPDA, Lahore, Pakistan.

Sidhu, P.S., Sharma, P.K. and Bajwa, M.S. 1990. Characteristics, distribution and genesis of salt affected soils in Punjab. *J. Indian Soc. Remote Sensing.*

Singh, D. and Nijhawan, S.D. 1932. A study of the physico-chemical changes accompanying the process of reclamation in alkali soils. *Indian J. agric. Sci.* 2: 1-18.

Singh, D. and Nijhawan, S.D. 1943. Availability of phosphates in alkaline and calcareous soils. *Indian J. agric. Sci.* 13: 131-141.

Singh, N., Sharma, B.K. and Dhir, R.P. 1989. Soil and ground water salinity in Challakere Taluka of Chitradurg district, Karnataka. *Ann. Arid Zone* 28: 235-243.

Singh, N.T. 1989. Salt affected soils of India. Discussion meet on State of environment and issues and problems on land and soil. NRSA, Hyderabad.

Singh, N.T. 1989. Water quality for agricultural use. *Proc. Regional Meeting on Desertification Control in the Asia and Pacific Region*, CAZRI, Jodhpur.

Singh, N.T. 1991. Quality criteria for saline and sodic waters. *Ind. Fmg.* 41: 7-10.

Singh, N.T., Brar, G.S. and Singh, B. 1980. The composition of soluble salts in salt affected soils of the Punjab state. *Intern. Symp. Salt Affected Soils*, CSSRI, Karnal, India, pp. 6-19.

Srikantia, S.V. and Sharma, R.P. 1972. The precambrian salt deposits of the Himachal Pradesh Himalaya and its occurrence, tectonics and correlations. In: *Himalayan Geology* (eds) A.G. Jhingaran et al. Vol.2, Wadia Institute of Himalayan Geology, Delhi, pp. 223-238.

Srinivasan, T.R. 1976. Taxonomy of saline alkali and saline-alkali soils. *J. Indian Soc. Soil Sci.* 24: 12-19.

Talati, R.P. 1941. Damaged lands in the Deccan and their classification. *Indian J. agric. Sci.* 11: 959-977.

Tiwari, K.N. and Sharma, D.N. 1989. Soil salinity research. All India Coordinated Project on Management of Salt Affected Soils. C.S. Azad University of Agric. & Tech., Kanpur.

Tiwari, K.N., Sharma, D.N. and Tripathi, S.K. 1989. Salt affected soils of Uttar Pradesh, their reclamation and management. C.S. Azad Univ. Agric. and Technol., Kanpur, p. 34.

United States Department Agric. 1975. Soil Taxonomy - A basic system of soil classification for making and interpreting soil surveys. Soil Cons. Service, U.S.D.A. Agric. Handbook 436, Washington D.C.

Uppal, H.L. 1955. Green manuring with special reference to *Sesbania aculeata* for treatment of alkali soils. *Indian J. agric. Sci.* 25: 211-235.

Venkata Rao, B.V. 1967. Nature of salt affliction in some typical soils of Mysore. *Mysore J. Agric. Sci.* 1: 44-52.

Whitcombe, E. 1971. Agrarian conditions in northern India. The United Provinces under British Rule 1860-1900. Vol.I, Thomson Press (India) Limited, New Delhi, 1971.

7

SOIL AND WATER CONSERVATION IN THE INDIAN SAT : PRINCIPLES AND IMPROVED PRACTICES

PRABHAKAR PATHAK and K.B. LARYEA

INTRODUCTION

By definition, the dominant characteristic of drylands is an insufficient water supply to support stable agriculture. Not only is there insufficient rainfall but its occurrence is also highly erratic between years, during any one year, and spatially, during any single rainfall event [Virmani et al., 1978]. In general, the variability of the rainfall pattern increases as the annual mean decreases.

Much of the rain in dryland areas falls in high intensities, causing runoff and erosion and thus, a significant part of the already-inadequate amount of water does not enter the soil. This happens even though the water storage capacity of the soil is far from full. It is not generally realized that runoff losses from a field can amount to 20-40% of storm rainfall. While an increase of agricultural production depends on numerous factors, the development of soil and water conservation technology that will enable farmers to make more efficient use of the available water is therefore an essential pre-requisite.

Alfisols, vertisols and associated soils are the two major soil orders that are extensively used for dryland agriculture in India. Soil-related constraints limit the potential productivity of these soils. Out of a total area of about 100-105 million ha under dryland agriculture in the country, nearly 30% is covered by alfisols and associated soils, 35 per cent by vertisols and associated soils. The production of most rainfed crops averages between 400 and 700 kg ha^{-1}. This chapter discusses the results from recent studies on the physical constraints and the developments that have been made in soil and water management to increase productivity of alfisols and vertisols under dryland conditions.

ALFISOLS

The Problem of Alfisols as a Resource for Crop Production

The following are some of the main problems relating to crop production in dryland alfisols.

1. Poor crop stand due to :
   - a) Crusting
   - b) Rapid drying of surface soil
   - c) High soil temperature
2. Poor crop growth due to :
   - a) Unreliable soil moisture supply
     - i) Low soil moisture storage capacity
     - ii) Low infiltration due to crusting, sealing, and consolidation
     - iii) Drought spells during the crop growing season
   - b) Low soil fertility
     - i) Inadequate nutrient supply
     - ii) Leaching of fertilizers

c) Soil workability problems

d) Compact sub-soil layer (argillic horizon)

3. Declining land productivity resulting from high soil loss due to :

a) Poor crop canopy

b) Crusting, sealing and consolidation

Poor crop stand

On some alfisols crusting is one of the primary causes of poor crop stands because it reduces germination by creating mechanical impedance to emerging seedlings. Its effects on small seeded crops such as pearl millet, finger millet and sorghum are more drastic. On alfisols, good crop establishment is also jeopardized by a high risk of rapid surface soil drying due to very high radiation early in the rainy season. On light textured soils, temperature of this dry zone causes injury to emerging seedlings [Pathak et al., 1987].

Poor crop growth

The effects of unreliability of rainfall, especially early in the rainy season are accentuated by the low moisture holding capacity of alfisols. Reliability of moisture supply for growth is further reduced by impaired infiltration caused by crusting and sealing. Moisture deficit is probably the most predominant factor for poor crop growth in the dryland alfisols of India.

Alfisols are generally low in fertility [Kanwar, 1983; El-Swaify et al., 1985]. Nitrogen is the most limiting nutrient for cropping SAT alfifols. The sandy surface texture, low cation exchange capacity and inherently low concentrations of organic matter and rapid leaching act to keep nitrogen supply low. Phosphorus is the other limiting nutrient in some of the dryland alfisols of India.

Workability of the alfisols is yet another problem which contributes to poor crop growth. These soils generally permit easy tillage when wet, but they become hard and difficult to till when dry. Tillage when the soil is too wet may result in excessive compaction. The high level of stoniness occasionally deters effective land preparation.

Characteristically, alfisols possess an argillic horizon within the profile. Effective rooting depths of crops are many times limited by this compact argillic horizon which may restrict root penetration [Ley and Laryea, 1994]. Restricted root development on these soils prevents many crops from withstanding even moderate droughts; resulting in low yields.

Declining land productivity

For sustainable agricultural development, the most serious problem of alfisols is soil erosion. Multiple, diversified sloping topography and the unstable nature of the surface structure resulting in low depression storage, surface slaking, compaction, crusting, sealing and consolidation of soil profile causes greater runoff and soil loss from alfisols.

In summary the problems of crop production on alfisols in the Indian SAT are numerous and complex. However, most of the problems are related to the physical and chemical characteristics of the soils.

Developments in improved soil and water conservation practices on alfisols

The crop yields under the traditional management on alfisols are very low (200 to 600 kg/ha). The traditional use and management which exists on the SAT alfisols are discussed in detail by Sanghi and Rao [1982], and El-Swaify et al. [1985]. The latter expressed serious concern about the excessive runoff and soil loss which are recorded from the traditional system of farming. Hydrologic studies of the traditional system at the ICRISAT Centre have shown that, of the total rainfall potentially available, an average of about 26% is lost through runoff, 34% through deep percolation and only the balance 40% is utilized for evapotranspiration by crops (Table 1). On an average a soil loss of 3.84 t/ha was recorded from the traditional system. Experimental evidence however showed that these soils are capable of producing more food with improved technology [Vijayalakshmi, 1982; Pathak et al., 1987]. The results of research that were found to be promising for improving infiltration and reducing runoff and soil loss are discussed below:

Table 1. Estimated water balance components and soil loss observed for the traditional cultivation system* on Alfisols, ICRISAT Center (1978-83).

Year	Rainfall (mm)	Runoff (mm)	Evapo-transpiration (mm)	Deep percolation (mm)	Soil loss (t/ha)
1978	1060	391	395	274	5.19
1979	671	113	335	223	1.83
1980	765	149	345	271	1.62
1981	1130	292	415	423	5.61
1982	765	181	310	274	3.69
1983	990	276	375	339	5.07
1978-83**	100	26	40	34	4.84***

* Traditional varieties of sorghum as sole crop and sorghum/pigeonpea as intercrop were grown.
** In percentage of rainfall.
*** Average annual soil loss.

Land surface configurations

For lands having slopes less than 1.5%, the flat on grade system is found to be the most effective in reducing runoff and soil loss and in increasing crop yields (Table 2). Raised land configurations (broadbed and furrow, and narrow ridge and furrow) offered no particular advantages in terms of runoff, soil loss, and yield over the flat on grade system. They also posed problems resulting from low stability and complications in accommodating certain crop combinations. In raised land configurations, the expected benefits of high infiltration rates in the planting zone, and reduced velocity of overland flow were counteracted by the extremely low surface-depression storage and other problems resulting from turning the soil and exposure of the compact argillic horizon in the furrow zone [Pathak et al., 1987]. Thus in alfisols with moderate slopes unless furrow irrigation is required, the flat on grade configuration is probably the most effective. In alfisols where

Table 2. Effect of alternative land surface configurations on crop yield, runoff, and soil loss on Alfisols at ICRISAT Center, 1981-84.

Land treatment	Crop yield (kg ha^{-1}) Intercrop system Sorghum	Pigeonpea	Runoff (mm)	Soil loss (t ha^{-1})
(A) Land slope <1.5%				
Broadbed and furrow at 0.4% slope	2740	825	315	3.79
Ridge and furrow at 0.4% slope	2910	870	282	3.02
Flat contour cultivation	2960	875	172	2.05
Flat-on-grade at 0.4% plus ridging up later	2880	840	180	2.78
SE (±)	151	65	16	0.24
(B) Land slope >1.5%				
Contour bund*	2520	710	75	0.97
Modified contour bund with gated outlet*	3020	970	160	0.92
Flat-on-grade with field bunds*	2810	900	215	3.35
Traditional flat with field bunds**	2380	220	256	4.79
SE (±)	175	60	23	0.19

* Treatment with recommended crop management practices which implies the use of acceptable or recommended variety, cropping system, chemical fertilizer and other practices for weed, pest and insect control.

** Treatment with traditional management practices which implies the use of variety, cropping system, farmyard manure, implements and other practices.

supplemental irrigation is an essential part of the system, the narrow ridge and furrow configuration is the most appropriate one. It is because this system has been found to be the most efficient in application of the irrigation water. It is also the second best land configuration for alfisols with respect to crop yields.

For alfisols having slopes more than 1.5%, a modified contour bund system is found to be the most promising (Table 2). The contour bunding is widely practised system in the SAT, particularly on rainfed alfisols in India. Results from this system showed that well designed and maintained contour bunds on alfisols conserve soil and water effectively; for this specific purpose contour bunds are more efficient than graded and field bunds [Pathak et al., 1985]. However, the system has disadvantages that include, water stagnation and restricted drainage in sizeable areas within the fields during rainy season. In most years crop yields are reduced in areas where water is impounded. In modified contour bund system the original system is modified by installing gated outlets in the lower field sections, land smoothing, and planting on grade instead of on contour, which allow runoff water to be stored above the bund for a certain period and then released at the desired rate through the gated outlet [Pathak et al., 1989]. Most of the erosional sediments are deposited so that relatively sediment-free water drains through the outlet and releasing excess water considerably reduces water logging. This system when compared with alternative land management systems on watersheds during 1981-84, consistently produced the highest crop yields, while still providing adequate control on runoff and soil loss.

Tillage

Primary tillage : Intensive primary tillage is generally found to be necessary for creating a favourable root proliferation zone and for enhancing rainfall acceptance on alfisols (Table 3). In SAT where residue management represents only a minor part of the cropping system, minimum tillage concepts are clearly at a disadvantage in dryland cropping of alfisols. Clear benefits from deep tillage (cross chiselling) were recorded (Table 4). In addition to the higher crop yields, deep tillage was also effective in reducing runoff and soil loss. There are also indications that the effects of deep tillage could last for 2 to 5 years depending upon the soil texture and rainfall.

Table 3. Effect of subsoiling on root density (cm cm^{-3}) 89 days after emergence of maize (Deccan Hybrid 103) on an Alfisol, ICRISAT Center, rainy season 1984.

Soil depth (cm)	Root density (cm cm^{-3})		
	Subsoiling	Normal tillage	S.E. ±
0-10	0.55	0.42	0.072
10-20	0.29	0.21	0.022
20-30	0.20	0.09	0.034
30-40	0.15	0.10	0.028
40-50	0.12	0.06	0.016
50-60	0.14	0.05	0.039

Table 4. Effect of normal and deep primary tillage on sorghum yield, runoff, and soil loss on alfisols at ICRISAT Center (1983).

Tillage	Sorghum yield (kg ha^{-1})	Runoff (mm)	Soil loss (t ha^{-1})
Normal tillage (mold board plowing 12 cm deep)	2160	285	3.27
Deep tillage (cross chiseling 25 cm deep)	2720	195	2.86
SE ±	130	15.0	0.234

Off-season tillage : On alfisols off-season tillage should be done whenever feasible. It has been found to be helpful in increasing rain water infiltration and in decreasing weeds problems. It minimises water evaporation by a "mulching" effect and allows the acceleration of planting operations (which must be done at the onset of the rainy season) thereby permitting earlier sowing and extension of the growing season. Experiments have shown that in most of the years off-season tillage alone can increase the crop yields by 7-9% over the control. In some years runoff from the first few storms are also reduced.

Crust and Seal Management for Improving Infiltration

On alfisols soil crust and seal formation are serious constraints to seedling emergence and soil and water conservation. One way to reduce the problem of crusting and sealing is by reducing the content of clay and silt particles in the surface layer. This is because, there is a positive relationship between the occurrence and strength of crusts with the content of clay and silt in surface soil. This implies that for alfisols a non-turning tillage system, is better than tillage with turning plows (inversion brings soil from argillic horizon which contains much higher clay and silt).

On alfisols the problems of crusting and sealing are encountered more during the early part of the crop growing season when the crop canopy is not fully developed. During this period considerable runoff was observed even when the soils were very dry. The shallow tillage imposed as additional intercultivations has been found to be effective in breaking up the crust and improving infiltration [Pathak et al., 1991]. It was found that additional shallow tillage effectively reduced runoff and soil loss (Table 5). In some years it was also effective in reducing moisture losses through evaporation by creating a dust mulch. However, significant increase in crop yields due to additional shallow intercultivations was obtained only in normal and low rainfall years.

Crop Residue Management

In the long-term, a 'farming systems' design which assures the generation and maintenance of favourable structural conditions would be critical to the successful use of alfisols. The incorporation of beneficial organic matter can be a likely means of achieving favourable soil organic matter contents and in turn a better structure in the SAT alfisols. However when undecomposed crop residue (5 t/ha) were incorporated into the soil, it created several serious problems (nitrogen deficiencies, etc.). The crop yields were drastically reduced. This suggested that under semi-arid environment large amount of undecomposed crop residue should not be incorporated into the soil.

The use of organic matter as surface mulch and incorporation into the soil in subsequent years has been useful for achieving favourable rainfall infiltration, reduced soil erosion, and structural stability. This system is found to be very effective in reducing runoff and soil loss and increasing crop yields (Table 6). Since 1981-82 was the first year of the experiments the effects seen in that year were of surface mulching alone. Whereas in 1983-84 the effects shown in Table 6 were due to combination of surface mulching and improvement in soil structure and/or soil fertility.

Table 5. Effect of interrow cultivation (shallow tillage) in addition to normal tillage on runoff, soil loss and grain yield from an Alfisol, ICRISAT Center, 1981-1983.

Year	Rain-fall (mm)	Tillage treatment	Runoff (mm)	Soil loss ($t\ ha^{-1}$)	Grain yield ($kg\ ha^{-1}$)			Sole pearl millet
					Sole sorghum	Intercrop		
						Sorghum	Pigeonpea	
1981	1092	Normal	246	5.0	2350	—	—	—
		Additional	223	4.9	2360	—	—	—
		SE ±	10.6	0.34	50	—	—	—
1982	780	Normal	159	3.1	—	2260	925	—
		Additional	120	2.6	—	2620	920	—
		SE ±	8.0	0.33	—	25	41	—
1983	990	Normal	231	4.2	—	—	—	2620
		Additional	196	4.0	—	—	—	2970
		SE ±	12.3	0.24	—	—	—	32

Table 6. Effect of different levels of organic mulch on crop yield, runoff, and soil loss on Alfisols, ICRISAT Center (1981-84).

Mulch rate (t/ha)	Crop yield (kg/ha)		Runoff (mm)	Soil loss (t ha^{-1})
	Intercrop system			
	Sorghum	Pigeonpea		
		1981-82		
No mulch	2790	1340	391	5.93
2.5	2800	1390	295	3.44
5.0	2980	1500	208	2.40
10.0	3040	1810	101	1.19
SE ±	116	99	4.9	0.221
		1983-84		
No mulch	2320	720	307	4.2
2.5	2670	810	123	2.3
5.0	2890	980	27	0.8
10.0	3040	1070	18	0.4
SE ±	132	66	12.0	0.15

Scoops

Recently, there has been an increasing interest in surface roughness viz. scoops as a means of improving in-situ soil and water conservation in the SAT (Figure 1). The machineries which can produce small scoops or pits on the soil surface are simple, less expensive and can be easily fabricated with local materials. The main aim of producing small scoops is to provide more time for water that would otherwise be lost as runoff to infiltrate into the soil. The other aim of these scoops is to reduce soil loss by a traping the eroded sediments that would otherwise be lost as

Fig. 1. Scoops with crop on an Alfisol for *in-situ* soil and water conservation.

soil loss from the field.

Experiments were conducted under simulated rainfall conditions using rotating disc type rainfall simulator to compare the performance of scoops with flat treatment. The scoops and flat land surface treatments, were studied under mulch (60% cover) and bare surface conditions. The average runoff and soil loss recorded from the scooped and flat land treatments on an alfisol are shown in Table 7. In all conditions the runoff and soil loss from scoops were significantly lower than from the flat land surface. However, the comparative advantage of scoops over flat land in terms of reducing runoff and soil loss varied considerably over the different conditions.

Table 7. Runoff and soil loss from scoops and flat treatments from an application of 46 mm rainfall on an Alfisol at ICRISAT Center.

Treatments	Runoff (mm)			Soil loss (kg/ha)		
	Scoops	Flat		Scoops	Flat	
Bare surface			SE			SE
Rainfall intensity 28 mm/h	16	27	± 3.1	1781	2906	± 210
Rainfall intensity 65 mm/h	26	34	± 4.2	4969	8344	± 479
Surface with mulch (60% cover)						
Rainfall intensity 28 mm/h	6	14	± 2.1	750	1875	± 138
Rainfall intensity 65 mm/h	15	24	± 1.9	2063	2813	± 291

The effects of rainfall amount on the comparative performance of scoops over flat in terms of runoff is shown in Figure 2. We define the term "scoop efficiency" as :

$$\frac{\text{Runoff from flat} - \text{Runoff from scoops}}{\text{Runoff from flat}} \times 100$$

The scoop efficiency is a measure of the percentage reduction in runoff due to pitting (or scoops) compared to flat. Figure 2 clearly shows that rainfall amount has an immense effect on scoop efficiency. For example, one bare surface and low rainfall intensity conditions, the scoop efficiency remains constant at 100% until at 28 mm rainfall when it decreases exponentially to only 16% at 76 mm rainfall. Figure 2 shows that small scoops of 10 mm capacity are efficient in regulating runoff and soil loss only from storms less than 55 mm. The effectiveness of this size scoops for rainfall amounts greater than 55 mm is low.

Under natural rainfall conditions, three surface roughness treatments viz., scoops, tied-ridges and flat seedbeds were compared for their effectiveness in controlling runoff and soil loss and in increasing millet crop yield on alfisols. Some of the salient results are:

- The surface roughness treatments viz., scoops and tied-ridges significantly reduced runoff and soil loss compared to flat seedbed cultivation. Using runoff and soil loss from the flat land as a basis for comparison in 1989-90, scoops reduced seasonal runoff by 69%, and soil loss by 53% while runoff and soil loss in the tied-ridged system were reduced by 39% and 28%, respectively.
- Scoops were more effective in reducing runoff and soil loss than tied-ridges. For example, 40 mm storage capacity scoops had average seasonal runoff of 81 mm and average soil loss of 2.61 t ha^{-1} compared to 160 mm runoff and 3.9 t ha^{-1} soil loss from tied-ridges of the same storage capacity.
- Scoops were relatively more stable than tied-ridges, particularly, during high intensity rainfall and runoff conditions. During high rainfall years, there was less number of breaching in scoops than in the tied-ridges.

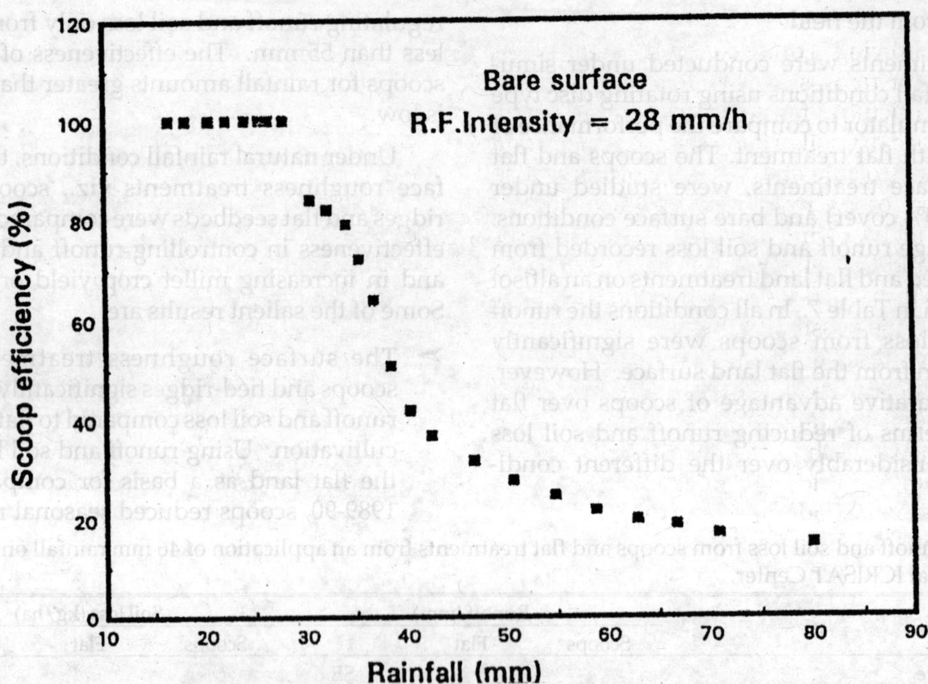

Fig. 2. Effect of rainfall amounts on the comparative performance of scoops over flat in terms of controlling runoff on Alfisol.

➢ The biggest comparative advantage of scoops over flat cultivation in terms of controlling runoff and soil loss was observed during the early part of the crop growing season. This is mainly due to the fact that surface crusting and sealing problems were rampant during that crop growth period because of sparse vegetation and poor crop canopy. Consequently, in flat seedbed cultivation, major proportion of the rains that fell during the early part of the rainy season on alfisols, was lost as runoff, while in scoops most of this runoff was stored in depressions.

➢ Throughout the rainy season, the penetration resistance and bulk density of the top 15cm soil layer were significantly lower in the scoops and tied-ridges than the flat seedbeds. Even around the harvest period of millet (65 DAS), the bulk density of 0-15 cm layer was significantly less in scoops (1.42 Mg m^{-3}) than the flat treatment (1.59 Mg m^{-3}).

➢ The stability of scoops can be greatly increased by providing a graded outlet system to the field. A graded outlet design of scoops was relatively more stable than the conventionally designed scoops during big storms. This was particularly true for small scoops of 10 and 20 mm storage capacities.

➢ There was a significant increase ($P < 0.04$) in pearl millet grain yield in the 20 mm storage capacity scoop (2.42 t ha^{-1}) over the flat seedbed (1.79 t ha^{-1}).

VERTISOLS

Physical Constraints and Their Management

The most important physical constraints to rainfed crop production on vertisols are their (i) narrow range of soil water content for tillage, (ii) high erodibility, (iii) tendency to become waterlogged and (iv) poor trafficability [Kampen and Burford, 1980]. Vertisols are hard when dry and have very plastic consistency when wet. Tillage at an inappropriate moisture content leads to compaction of the sub-soil. Traditionally, rainy season fallowing is quite common on these soils. Reasons for rainy season fallowing of vertisols are the difficulties farmers encounter in preparing the hard dry soil prior to the onset of the rainy season and/or the sticky nature of the wet soil after onset of the rainy season which does not permit timely sowing and management of crops [Krantz and Russell, 1971; Michaels, 1982]. The common threat of flooding when there are heavy rains, and the possibility that rainy season cropping may reduce soil moisture available for growing post-rainy season crops are also some of the reasons for fallowing vertisol during the rainy season. Jodha [1979] also noted that these soils are often difficult to work when dry or wet so that essential operations such as land preparation or weed control by manual labour, are difficult to perform in the rainy season.

Approach for the Management of Vertisols

The components of improved land management strategy for vertisols include [Kanwar, 1981 and Kampen, 1982]:

- a system of resource conservation, management and use based on small watersheds;
- a well designed semi-permanent or permanent broadbed and furrow system or well graded flat system;
- land smoothing and installation of drains to ensure effective surface drainage and runoff disposal from the field;
- performing the primary tillage operations for both post-rainy as well as rainy season crops immediately after harvesting the previous season's crop, when the soil has adequate moisture and is friable;
- shallow cultivation of the land (only the beds) whenever an effective rainfall (20 mm) is received;
- use of minimum tillage and stubble mulch residue management;
- application of moderate amounts of N and P fertilizer, followed by dry sowing about one week before the expected date of onset of the rainy season;
- use of high yielding varieties and other recommended agronomic practices;

Using the above techniques, vertisols have been managed successfully at ICRISAT to produce high and stable yields even under very fluctuating rainfall conditions (Table 8). Various aspects of this approach for the management of vertisols have been discussed in detail by El-Swaify et al. [1985], Kanwar [1981], Kampen [1979] and Kampen [1982].

In addition to its direct benefit to crop yield, improved soil management has been more effective than the traditional rainy season fallow system in reducing runoff and soil erosion [Pathak et al., 1985]. On an average, the improved system reduced annual runoff to one-third, soil loss to one-eleventh and peak runoff rate to one-half when compared with traditional rainy season fallow system (Table 9). As shown in Table 9, the performance of the broadbed and furrow system was consistently superior to the traditional system in reducing annual runoff, soil loss and peak runoff rate. In 1977 when rainfall was very low and moisture conservation was crucial, the broadbed and furrow system conserved most of the annual rainfall, producing an annual runoff of only 1 mm. In contrast, when removal of excess water by drainage was crucial, during the high rainfall years of 1975 and 1978, this system produced substantial runoff of 162 and 273 mm, respectively. The better control on runoff in the improved system was due to good crop cover during the rainy season, periodic land smoothing, slope adjustment, controlled runoff velocity as it flows through many furrows and disposed through grassed waterways.

The soil erosion hazard associated with a unit runoff from the different management systems is represented in Figure 3 [Pathak et al., 1985]. The sediment load per unit runoff was significantly affected by land management systems and crop cover. Relatively high values of sediment load were sustained during the season in the traditional rainy season fallow system. In

Table 8. Grain yields under improved and traditional technologies on a Vertisol at ICRISAT in 12 successive years.

Year	Rainfall during cropping period	Grain yield (t ha^{-1})			
		Improved system:double cropping		Traditional system, single crop	
		Sorghum/ Maize	Sequential chickpea intercropped pigeonpea	Sorghum or Chickpea	
1976-77	708	3.20	0.72	0.44	0.54
1977-78	616	3.08	1.22	0.38	0.87
1978-79	1089	2.15	1.26	0.56	0.53
1979-80	715	2.30	1.20	0.50	0.45
1980-81	751	3.59	0.92	0.60	0.56
1981-82	1073	3.19	1.05	0.64	1.05
1982-83	667	3.27	1.10	0.63	1.24
1983-84	1045	3.05	1.77	0.84	0.48
1984-85	546	3.36	1.01	0.69	1.23
1985-86	477	2.70	0.73	*	0.84
1986-87	585	4.45	0.38	0.37	1.27
1987-88	841	4.26	1.35	0.80	0.92
Mean	776	3.20	1.10	0.60	0.80
SD	200	0.70	0.40	0.20	0.30
CV (%)	26	21	34	27	38

* No crop sown

Table 9. Annual rainfall, runoff, soil loss, and peak runoff rate for a cropped Vertisol with broadbed and furrow system and a traditional monsoon fallow system (1975-1980).

Year	Rainfall (mm)	Broadbed and Furrow at 0.6% slope, cropped			Traditional flat, monsoon fallow		
		Runoff (mm)	Peak Runoff Rate (m^3 s^{-1} ha^{-1})	Soil loss (t ha^{-1})	Runoff (mm)	Peak Runoff Rate (m^3 s^{-1} ha^{-1})	Soil loss (t ha^{-1})
1975	1,041	162	0.06	1.39	253	0.15	5.21
1976	687	73	0.09	0.98	238	0.16	9.20
1977	585	1	0.01	0.07	53	0.06	1.68
1978	1,125	273	0.11	2.93	410	0.15	9.69
1979	690	73	0.08	0.70	202	0.15	9.47
1980	730	116	0.06	0.97	166	0.11	4.58

the both the traditional and improved management systems, the sediment load per unit runoff decreased as the crop cover increased. The broadbed and furrow system performed better in the disposal of excess water than the flat system because of lower values of sediment load in the runoff. This is highly desirable on vertisols which often have water logging conditions when disposal of runoff is essential with least soil loss.

The data presented so far clearly show that the improved management system for vertisols is superior to the traditional monsoon fallow system both in increasing productivity and in enhancing the effective use and conservation of soil and water resources.

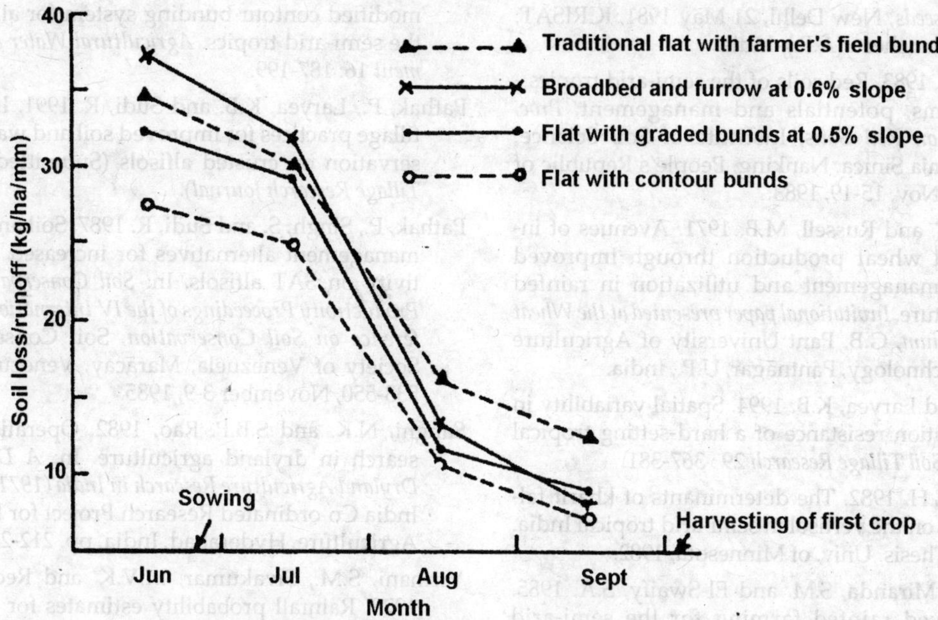

Fig. 3. Effect of land management system and crop growth on soil loss per unit of runoff from Alfisol watersheds (monthly average values from 1975-1980).

SUMMARY

Alfisols, vertisols and their associated soils constitute major soil orders in the dryland agricultural region of India. Present farming systems on these soils in the dryland areas are characterized by low and undependable yields and by an inefficient use of the rains and the soil. The risks to dependable crop production are high: farmers therefore rarely use fertilizers, high yielding crop varieties, or adopt improved soil conservation effective farming techniques. In this paper, the soil-related limitations of alfisols and vertisols for sustainable resource utilization are reviewed. The research on soil and water conservation and land management techniques is discussed. The role of using broadbed and furrow system of land management for vertisols and scoops for alfisols is highlighted. The recommended technological improvements have had immediate and positive benefits over the traditional management systems. These include highly economic increases in crop yields, amenability to small farmer's requirements, marked reduction in soil erosion and substantially low runoff losses.

REFERENCES

El-Swaify, S.A., Pathak, P., Rego, T.J. and Singh, S. 1985. Soil management for optimized productivity under rainfed conditions in the semi-arid tropics. *Advances in Soil Science* 1: 1-64.

Jodha, N.S. 1979. Intercropping in traditional farming systems. *J. Development Studies* 16: 427-442.

Kampen, J. 1979. Watershed management and technology transfer in the Semi-Arid Tropics. *Proc. Int. Symp. on Development and Transfer of Technology*, ICRISAT, Patancheru P.O., A.P., India, pp. 111-120.

Kampen, J. and Burford, J.R. 1980. Production systems soil-related constraints and potentials in the semi-arid tropics with special reference to India. *Priorities for Alleviating Soil Related Constraints to Food Production in the Tropics Symposium Proceedings*, IRRI, Los Banos, Laguna, Philippines, pp. 141-145.

Kampen, J. 1982. An approach to improved productivity on deep vertisols. ICRISAT Information Bulletin II, Patancheru P.O., A.P., India.

Kanwar, J.S. 1981. Problems and potentials of the black soils of India : some suggestions for an action plan. In: *Proceedings of the Seminar on Management of Deep Black Soils for Increased Production of Cereals, Pulses and Oilseeds*, New Delhi, 21 May 1981, ICRISAT, P.O. Patancheru (A.P.), India.

Kanwar, J.S. 1983. Red soils of the semi-arid tropics - Problems, potentials and management. *Proc. Symp. on Red Soils*, Institute of Soil Science. Academia Sinica. Nanking, People's Republic of China, Nov. 15-19, 1983.

Krantz, B.A. and Russell, M.B. 1971. Avenues of increased wheat production through improved water management and utilization in rainfed agriculture. *Invitational paper presented at the Wheat Symposium*, G.B. Pant University of Agriculture and Technology, Pantnagar, U.P., India.

Ley, G.J. and Laryea, K.B. 1994. Spatial variability in penetration resistance of a hard-setting tropical alfisol *Soil Tillage Research* 29 : 367-381.

Michaels, G.H. 1982. The determinants of kharif fallowing on the vertisol in semi-arid tropical India. Ph.D. Thesis. Univ. of Minnesota, 1982.

Pathak, P., Miranda, S.M. and El-Swaify, S.A. 1985. Improved rainfed farming for the semi-arid tropics : Implications for soil and water conservation. In: *Soil Erosion and Conservation* (eds) S.A. El-Swaify, W.C. Moldenhauer and L. Andrew. Soil Conservation Society of America, Ankeny, Iowa, pp. 338-354.

Pathak, P., Laryea, K.B. and Singh Sardar 1989. A modified contour bunding system for alfisols of the semi-arid tropics. *Agricultural Water Management* 16: 187-199.

Pathak, P., Laryea, K.B. and Sudi, R. 1991. Interrow tillage practices for improved soil and water conservation on crusted alfisols (Submitted to *Soil Tillage Research Journal*).

Pathak, P., Singh, S. and Sudi, R. 1987. Soil and water management alternatives for increased productivity on SAT alfisols. In: *Soil Conservation and Productivity Proceedings of the IV International Conference on Soil Conservation*, Soil Conservation Society of Venezuela, Maracay, Venezuela, pp. 533-550, November 3-9, 1985.

Sanghi, N.K. and S.B.P. Rao, 1982. Operational research in dryland agriculture. In: *A Decade of Dryland Agriculture Research in India (1971-80)*, All India Co-ordinated Research Project for Dryland Agriculture, Hyderabad, India, pp. 212-231.

Virmani, S.M., Sivakumar, M.V.K. and Reddy, S.J. 1978. Rainfall probability estimates for selected locations of semi-arid India. ICRISAT Res. Rept. 1, Patancheru, A.P., India.

Vijayalakshmi, K. 1982. Runoff management. In: *Dryland Agriculture Research in India : Thrust in the Eighties*. All India Co-ordinated Research Project for Dryland Agriculture, Hyderabad, India, pp. 78-87.

8

SOIL FERTILITY MANAGEMENT AND FERTILISER USE

S.K. DAS, K.L. SHARMA, NEELAM SAHARAN and K. SRINIVAS

INTRODUCTION

Most of the nutrients exist in mineral and organic matter and as such are insoluble or unavailable to plants. Nutrients become available through mineral weathering and organic matter decomposition. The nutrient availability process constitutes a sequence of reversible and irreversible reactions. Fried and Broeshart (1967) have suggested the following sequential reactions involved in migration of nutrients from soil to plants.

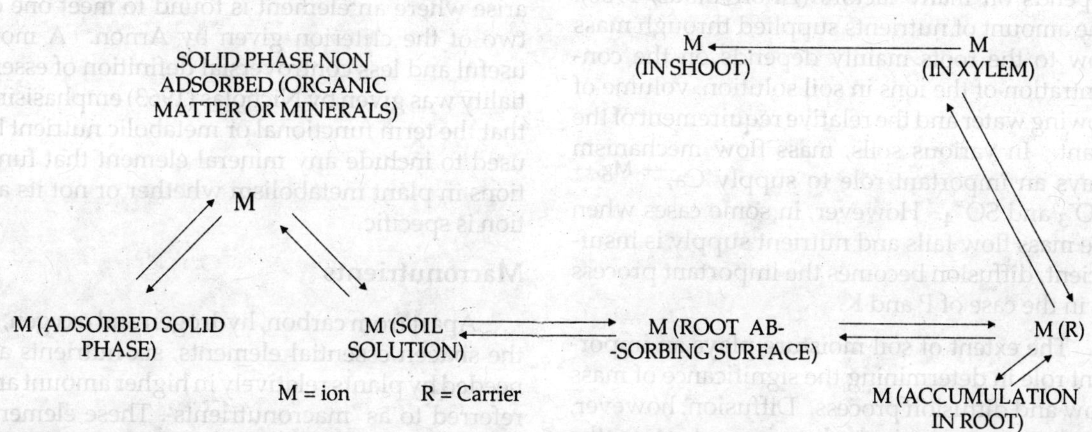

These reactions depicting soil-plant-continuum are not so simple but are definitely governed by the principles of thermodynamics and rate kinetics. Apart from these, other important factors including temperature, light, precipitation, irrigation, leaching and interactions of genotypic and phenotypic characteristics of individual plant species affect nutrient availability (Lamm, 1968). Uptake of nutrients by plants occurs in the form of ions. The major processes which transport ions to the roots are (i) root interception, (ii) mass flow, and (iii) diffusion. The process of penetration of roots through the soil and thus coming in contact with soil colloids on which nutrients are held is termed as interception. Some of the nutrients are transported to the roots with water as it is absorbed by plants. This movement is called mass flow. As the nutrients are absorbed by the plant roots, a concentration gradient occurs between the zone immediately around the root surface and the zone further away. As a result of this concentration gradient, diffusion of ions towards the root surface takes place. If the amount of nutrients transported by mass flow and interception is more than the absorption capacity of the roots, they tend to accumulate in soil zone adjacent to roots. If the nutrient supply is slower than that of root absorption, the zone near the roots gets depleted. It has been assumed that nutrients supplied through interception are almost equal to a volume of soil solution which equals the volume of the roots (Krishnamoorthy and Venkateswarlu, 1976). The concentration of nutrient ions in saturated extracts multiplied by the quantity of water absorbed by the plants gives the idea about the nutrient supplied through mass flow process. Calculation of area of root surface *in situ* is not easy, hence the contribution of diffusion process in terms of nutrient supply can be worked out by substracting the amount supplied through root interception and mass flow from the total uptake. Close to the root, mass flow and diffusion occur simultaneously. Which of these processes is significant as a rate limiting step in ion uptake depends on many factors (Anonymous, 1966). The amount of nutrients supplied through mass flow to the roots mainly depends on the concentration of the ions in soil solution, volume of flowing water and the relative requirement of the plant. In various soils, mass flow mechanism plays an important role to supply Ca^{++}, Mg^{++}, NO_3^- and $SO_4^=$. However, in some cases when the mass flow fails and nutrient supply is insufficient, diffusion becomes the important process as in the case of P and K.

The extent of soil moisture plays an important role in determining the significance of mass flow and diffusion process. Diffusion, however, becomes progressively less important as the moisture content decreases.

PLANT NUTRITION

Green plants synthesize their food material from simple constituents derived out of air and soil. As in case of other organisms, plants have their tissues made up of carbohydrates, fats, proteins and nucleoproteins. Therefore, the plant needs large quantities of carbon, oxygen, hydrogen, nitrogen, phosphorus and sulphur for building up and functioning of its tissues. It needs small quantities of iron, manganese, zinc, copper, boron and molybdenum and oftenly, cobalt for building up of its enzyme system. Similarly, it needs potassium, magnesium and calcium and sometimes sodium and chloride (Russel, 1973). To demarcate between elements which are essential from those which may be taken up by the plants but are not considered essential, Arnon (1954) has suggested the following criteria:

i) The plant must be unable to grow normally or complete its life cycle in the absence of the element,

ii) The element is specific and cannot be replaced by another, and

iii) The element plays a direct role in metabolism.

The first of the criterion is most important and is accepted without any criticism. Because of the limitations of the experimental techniques and lack of complete knowledge, many occasions arise where an element is found to meet one or two of the criterion given by Arnon. A more useful and less controversial definition of essentiality was given by Nicholas (1963) emphasising that the term functional or metabolic nutrient be used to include any mineral element that functions in plant metabolism whether or not its action is specific.

Macronutrients

Apart from carbon, hydrogen and oxygen, of the sixteen essential elements, six nutrients are needed by plants relatively in higher amount and referred to as 'macronutrients'. These elements are nitrogen, phosphorus, potassium, calcium, magnesium and sulphur. Out of the macronutrients, the first three nutrients namely N, P and K are utilised by plants in considerable quantities and mostly the soils of the world are deficient in these nutrients. The other three, viz.

calcium, magnesium and sulphur are sometime called as secondary nutrients or amendment elements. The term secondary nutrients does not indicate their less importance for plant growth but it definitely means secondary importance to the manufacturers of N, P, K fertilizers and more over these elements are added along with N, P, K fertilizers inadvertently. These elements (Ca, Mg, S) are the constituents of the amendment materials like lime and gypsum and hence often known as amendment elements. Due to intensive cultivation, these elements have gained importance in recent past.

Micronutrients

In addition to macronutrients, plants need very small quantities of certain elements for their nutrition and hence termed as micronutrients or 'trace' or 'minor elements'. These include iron, manganese, boron, zinc, copper, molybdenum and chloride.

In addition to the macro and micro nutrients, there are some elements which may or may not be essential for plant growth but have been found to increase crop growth. Under some situations and for some crops, these elements are beneficial for improving crop yield. These elements include sodium, cobalt, vanadium and silicon.

ROLE OF ESSENTIAL NUTRIENTS IN PLANT GROWTH

As has been emphasised by Arnon [1954], each of the essential nutrients has certain specific role to play in plant growth and this can not be substituted by any other element. Each essential nutrient has some critical level. If its concentration falls below that level the crop growth is hampered. Visual deficiency symptoms start appearing if deficiency is not corrected.

Nitrogen

Nitrogen is an essential part of proteins and is a constituent of physiologically important compounds like nucleotides, phosphatides, vitamins, enzymes and hormones. It promotes rapid vegetative growth and improves the quality of leafy vegetables and fodder crops. It enhances the protein content of food, fodder and fruit crops. Nitrogen is related to carbohydrate utilization. It is an integral part of the chlorophyll molecule. When plants are deficient in nitrogen, they become stunted and yellow in appearance. The yellowing or chlorosis, appears first on the lower leaves, the upper leaves remain green. In case of severe nitrogen deficiency the leaves turn brown and die. The tendency of the young upper leaves to remain green and that of lower leaves to turn yellow suggests mobility of the element in growing parts.

Phosphorus

Phosphorus is found in plants in quantities much smaller than those of nitrogen and potassium. It is a constituent of nucleic acid, phytin and phospholipid. Phosphorus plays an important role in initiating primordia for the reproductive parts of the plants. It is a major constituent of many enzymes associated with transformation of energy, in metabolism of carbohydrate and also in respiration in plants. It has important role to play in cell division and root development. Phosphorus counteracts the effects of excessive nitrogen and brings about early maturity of cereals. It improves the quality of food grains. Phosphorus has the ability to enhance activity of *Rhizobium* and increases the formation of root nodules. Phosphorus is readily mobilized in plants and when a deficiency occurs the element contained in the older tissues is transported to the active meristematic regions.

Potassium

The third so called major element required for plant growth is potassium. This element does not form part of any of the important plant constituents such as chlorophyll, proteins, fats and carbohydrates. It has following major roles to play in plant growth: (i) carbohydrate metabolism or formation and breakdown and translocation of starch, (ii) nitrogen metabolism and synthesis of proteins, (iii) control and regulation of activities of various essential mineral elements, (iv) neutralization of physiologically important organic acids, (v) activation of various enzymes, (vi) promotion of the growth of meristematic tissue, (vii) adjustment of stomatal and water relation, (viii) increased vigour and resistance to diseases and lodging.

Calcium

Calcium appears to be essential for the growth of meristems and particularly for the proper growth and functioning of root tips. It has been found necessary in the formation of middle lamella of cell wall in the form of Ca-pectate. It has been found to play an important role in neutralizing the phosphoric acid, citric acid, malic acid, oxalic acid which may become injurious to plants. It has been suggested that calcium increases the protein content of mitochondria. Calcium is also related to protein synthesis by its enhancement of the uptake of nitrate N. Calcium has been found associated with the activation of certain enzyme systems and it is directly involved in chromosome stability. In almost all plant species, the initial deficiency symptom appears as chlorosis of the young leaves followed by the obstruction of the growing points of the stem. In fruit trees, death of the growing point, followed by 'die-back' is the common symptom.

Magnesium

Magnesium is the only mineral constituent of the chlorophyll molecule and is located at its centre. It has been shown to be an essential constituent of polyribosome and chromosome. It is known to play a catalytic role as an activator of number of enzymes, majority of which are associated with carbohydrate metabolism, phosphate transfer and decarboxylation. Magnesium is related to the synthesis of oil. With sulphur, it brings about major increase in the oil content of several crops.

Magnesium is a mobile element and is readily translocated from older to younger plant parts in the event of deficiency. The deficiency of magnesium results in an interveinal chlorosis of the leaf in which only the veins remain green. If deficiency is at more advanced stage the leaf tissues become uniformly pale yellow, then brown and necrotic.

Sulphur

Sulphur is an important constituent of amino acids, viz., cystein, cystine and methionine and as such of proteins. It also forms the important part of some other biologically active components, like vitamins, lipoic acid, acetyl coenzyme A, ferredoxin and glutathione. It helps in stabilizing protein structure. It helps in the synthesis of oils and formation of chlorophyll. Disulphide linkages (-S-S-) have been associated with the structure of protoplasm and the quantity of sulfahydral groups (-SH) in plants has in some cases been related to increased cold resistance.

Sulphur is required for N fixation by leguminous plants and is a part of the nitrogenase enzyme system which is responsible for N fixing. Instances of sulphur deficiency have been increased during the recent past because of increased use of sulphur-free fertilizers, decreased use of sulphur as an insecticide and fungicide, decreased concentration of sulphur rich compounds in the atmosphere and in rainfall because of the decreased consumption of high sulphur fuels, and intensive cultivation.

The deficiency symptoms of sulphur first appear in the younger growth. The young leaves of sulphur deficient plants show chlorotic striping between the veins. In sulphur deficient plants, shoot growth is restricted, stems become stiff, woody and smaller in diameter.

Iron

Iron is necessary for the synthesis and maintenance of chlorophyll in plants. It is an essential component of many enzymes. Iron plays an important role in nucleic acid metabolism. It also has an important role to play in oxidation reduction reactions occurring in plant systems. Iron deficiency is represented by typical interveinal chlorosis, youngest leaves being first affected, tips and margins of leaves maintain their green colour longest. In case of severe deficiency, the entire leaf, veins and interveinal areas turn yellow which sometimes get bleached.

Manganese

Manganese acts as catalyst in many important enzymatic and physiological reactions in plants. It plays an important role in respiratory process of plants and is involved in oxidation of carbohydrate to CO_2 and H_2O. It activates the enzymes associated with the metabolism of N and synthesis of chlorophyll. It also governs the redox potential in plant cells during the phases of light and darkness. Manganese deficiency is denoted by chlorosis between the veins of young leaves, characterised by the appearance of chlorotic and necrotic spots in the interveinal areas. Near the base of the younger leaves a greyish area develops which slowly becomes yellowish to yellow orange. Symptoms of deficiency are popularly known in cereals as 'grey speck', in field peas 'marsh spot' and in sugarcane, 'streak disease'.

Boron

Boron is known to affect activity of certain enzymes. It enhances the permeability of the membrane and thereby facilitate carbohydrate transport. It has been found associated with lignin synthesis and cell division. It regulates Ca uptake and K/Ca ratio in plant. Boron has been found essential for protein synthesis. Its deficiency causes the death of shoot tips. The leaves have a thick texture, sometime curling and brittleness occurs. Failure of flower formation and stunted root growth occurs due to B deficiency. With B deficiency, root crops are characterised by dark spots on the thickest part of the root or splitting of centre and it is known as 'Brown heart'. Fruits such as apples develop internal and external cork symptoms.

Zinc

Zinc is involved in the biosynthesis of plant hormone indole acetic acid (IAA). It is an essential component of a variety of metalloenzymes like carbonic anhydrase, alcohol-dehydrogenase etc. It also plays role in nucleic acid and protein synthesis. It also helps in the utilization of P and N in plants. The deficiency symptoms of zinc usually appear on the second or third fully matured leaves from the top of plants. In case of maize zinc deficiency is depicted by high yellow stripping to a broad band of white or yellow tissue with reddish purple veins between the midrib and edges of the leaf, occurring mainly in the lower half of the leaf. In citrus, irregular interveinal chlorosis, terminal leaves become small and narrowed, referred to as 'little-leaf', fruit bud formation is severely reduced; twigs die back.

Molybdenum

Molybdenum is believed to be associated with nitrogen utilisation and in nitrogen fixation. It is an important constituent of nitrate reductase and nitrogenase enzyme. It is needed by *Rhizobium* in the process of N_2 fixation. In case of deficiency, chlorotic interveinal mottling of the lower leaves, followed by marginal necrosis and infolding of the leaves occurs. In case of crops like cauliflower, the leaf tissues wither leaving only the midrib and a few small pieces of leaf blade, giving the appearance of whip or a tail known as 'whiptail'. The chances of occurrence of molybdenum deficiency are more in leguminous crops.

Chlorine

Chlorine is a constituent of chlorine-containing auxin-chloroindole-3 acetic acid. It constitutes many compounds found in fungi and bacteria. It can substitute indole acetic acid in immature roots. Chlorine has been found to stimulate the activity of some enzymes and influences carbohydrate metabolism and water holding capacity of plant tissues. Wilting of leaf tips, chlorosis of leaves and finally bronzing and drying are the major symptoms of chlorine deficiency.

Copper

Copper has been found associated with cytochrome oxidase and component of many enzymes like ascorbic acid oxidase, phenolase, lactase, etc. It also promotes formation of vitamin A in plants. In case of deficiency, yellowing of the upper or youngest leaves and curling of the leaf tips occurs. Sometimes younger leaves become pale yellow; ragged edges of leaves and internodes become short in length. In case of citrus, die back of new growth, recognised as exanthema, in Brassica spp, eg. in cauliflower and cabbage, young leaves develop fine chlorotic mottling between the veins.

Cobalt

Cobalt has been found essential for higher plants. In some cases the response of crops to Co applications has been reported. It has been established that Co is needed by *Rhizobium* for the fixation of N_2.

Vanadium

It is considered as an essential nutrient for the green algae, viz., 'Scenedesmus'. However, its essentiality for the growth of higher plants has not yet been established. Increase in growth of higher plants like asparagus, rice, lettuce, barley and corn has been reported due to application of vanadium.

Sodium

It has been found that the growth of some crops is stimulated by application of Na. Sodium is reported to influence water relations in sugar-beet and to increase their resistance to drought.

Silicon

The essentiality of Si for plant growth is not known yet. It has been found that silicon is required by rice and perhaps other crops like cucumber and barley. It is believed that silicon increases the top length, number of stems and fresh and dry weight of rice plants grown in nutrient culture. In any case, rice plant is rich in silicon content.

NUTRIENTS IN DEFICIENCY FOR DRYLAND CROPS

Though much information on nutrient content in different dryland crops is not available, most of the dryland crops have been found responsive to N, P, Zn and S. Dryland regions in India are mainly represented by alluvial, black, red, laterite and lateritic, sierozemic, submontane and hill soils. Almost all rainfed soils are deficient in nitrogen. Heavy losses of N in tropical countries on one hand and low recovery on the other make nitrogen management a challenging task. Nutrient requirements and their management in drylands have been discussed by Finck and Venkateswarlu (1982), Kanwar et al. (1986) and more recently by Sharma and Das (1992). Mobile nutrients, particularly N in the non-adsorbed form in soil may be subject to excess losses in high rainfall years (Rego et al., 1982). Nutrient limitations on rainfed Vertisols are primarily due to deficiencies of N, P and Zn. As in case of Vertisols, nitrogen is the most limiting nutrient for cropping in Alfisols of semi-arid tropics. The sandy surface texture, low cation exchange capacity and low organic matter content of cropped Alfisols of Semi-Arid Tropics (SAT) require that fertilization be practised, even under rainfed conditions for effective crop production.

POTENTIAL AND SCOPE OF FERTILISER USE

India predominantly is a rainfed country. Out of 329 m ha total geographical area, 142 m ha is devoted to agriculture. The gross sown area is around 177 m ha. Rainfed agriculture occupies 97 m ha, which is about 69.3 per cent of the arable land and constitutes about 68.5 per cent of the gross sown area (Katyal et al., 1991).

Soils in dryland regions in India are, by and large, deficient in nitrogen and phosphorus. The fact that drylands are more hungry than thirsty has been amply demonstrated. In spite of this, farmers seldom apply fertilisers to crops grown under dryland conditions. The reason for not using fertilisers perhaps lies in uncertainty of return from the investment made on fertiliser use. The poor resource-base of dryland farmers is yet another reason. Available reports indicate that fertilizer use in dryland crops might vary between 5 to 40 kg/ha ($N + P_2O_5 + K_2O$) (Katyal et al., 1991).

The skewed nature of fertiliser distribution and use in different regions is another factor to be considered. It is estimated that five states namely Punjab, Haryana, Uttar Pradesh, Tamil Nadu and Andhra Pradesh which have more than 40% of the irrigated area under food crops, account for 57% of the total fertiliser consumption in the country. Another five states representing semi-arid tropics, namely, Gujarat, Rajasthan, Madhya Pradesh, Maharashtra and Karnataka, with less than 20% of irrigated area under food crops, account for only 27% of total fertiliser use in the country. Our future efforts in fertiliser promotion should be aimed at narrowing inter-regional and inter district inequalities in fertiliser use (Randhawa and Velayutham, 1989).

SOIL FERTILITY STATUS AND FERTILISER USE CONSTRAINTS

Soils in low rainfall areas vary widely in their properties. Soil formation is a function of the parent material upon which they are formed as well as the biological activity comprising of microorganisms and higher plants, the amount of weathering and the surface relief. Therefore, the climate plays a major role in determining the kind of soil formation. The sparse vegetation of low rainfall areas results in soils which are low in organic matter. Consequently, the poor fertility of soils in dryland areas is associated with near exhaustion of organic matter. Organic matter is basic source of N, S and micronutrients. About one half of the phosphorus reserves in soils have their origin in organic matter. Nitrogen in soil is generally inadequate to support plant growth. Past reports on N status of Indian soils reveal that 62.5 per cent of the districts surveyed are low in nitrogen and 32.6 per cent of them fall in medium fertility class (Ghosh and Hassan, 1980). Based on soil tests, the deficiency of P is only next in extent to N (Rao and Das, 1982; Kanwar, 1986; Tandon, 1987; Venkateswarlu, 1987). Deficiency of P is believed to be widespread in Vertisols which are developed on granite and sedimentary rocks (Swindale, 1982). An analysis of 24 districts, which accounted for 50% of the total sorghum area, showed that P fertility rating was low in 15 districts and medium in 9 districts (Tandon and Kanwar, 1984). Unlike N and P, K deficiency is not a severe problem in dryland soils (Finck and Venkateswarlu, 1982). Katyal and Rattan [1990] have made extensive review of micronutrient deficiency in Indian soils. Based on the analysis of benchmark soil samples, Katyal and Sharma (1991) have indicated the possibility of Zn deficiency in semiarid tropical regions.

MOISTURE STRESS VS NUTRIENT TRANSFORMATION

As is important in case of plant physiological functions; moisture stress plays an important role in nutrient solubilization and precipitation reactions. As the moisture content in soil decreases, the concentration of soluble salts in the soil solution rises (Power, 1990). This leads to the enhancement of the osmotic potential of the soil solution and subsequently influences plant water relations (Taylor, 1983). With the increase in the salt concentration some salts approach their solubility constants and they start precipitating. The major soluble anions are HCO_3^-, SO_4^{2-} and Cl^-. One of the salts which precipitates first is gypsum ($CaSO_4.2H_2O$). Among the cations, if sodium (Na^+) is present in the soil in significant quantity i.e. in excess of about 6% of CEC (cation exchange capacity), it results in dispersion of soil aggregates and leads to decrease in hydraulic conductivity, increase in soil resistance and crust formation. Water availability upto certain extent also affects dissolution (weathering) of soil minerals. This may keep feldspars and other potassium bearing minerals intact and these minerals serve as primary sources of plant available K. Lindsay and Vlek (1977) observed that moisture deficiency affects phosphorus availability. Drying of soil leads to precipitation of orthophosphate (PO_4) in various forms of calcium phosphate. In soils falling under temperate regions precipitated calcium phosphate exists mainly in the form of octacalcium phosphate, which is a less soluble source of plant available P. This compound remains in equilibrium not only with soluble PO_4, but also with adsorbed P and other calcium-phosphate minerals with the equilibrium concentration varying with a moisture content of soil, salt concentration, pH and other associated factors. Often, P immobilised as octacalcium phosphate can eventually become available to plants (Kissel et al., 1985). Solubility of nutrients is affected due to moisture stress by way of lowering down of root activity (Taylor, 1983). If root system is highly active, a considerable amount of CO_2 is liberated into the soil solution which lowers down the pH of the solution and thereby altering the solubility of many minerals.

BALANCED FERTILISATION AND BALANCED NUTRIENTS

Liebig's Law of Minimum envisages that nutrient that is present in least amount is the first nutrient that governs the plant growth and yield. Power (1990) has emphasised that this law is equally applicable in drylands as it holds true to other soils. However, due to the effect of water stress on plant growth in drylands the implication of the law is somewhat complicated. Whenever water availability falls, a level comes at which water deficiency hinders the plant growth to a greater extent than any nutrient deficiency. Water availability thus has an appreciable effect on the availability of N, P and other nutrients in dryland soils. While looking into the validity of the law under moisture deficit conditions, it was found that when water was very limited dry matter production by the grass was quite low, and little or no response to N fertilization was observed (Power, 1990).

Crop yields are correlated with levels of fertiliser application. The role of chemical fertilisers in increasing crop productivity has become obvious with the introduction of high yielding, non-lodging and fertiliser responsive varieties in the mid sixties. The consumption of primary nutrients increased from 0.785 million tonnes in 1965-66 to 8.211 million tonnes during 1984-85. The recovery of applied nutrients by crops is generally poor. Depending upon the soil characteristics, the recovery is less than 50 per cent in case of N, less than 15 per cent in case of P and upto 80 per cent in case of K. Major part of the applied N fertilisers is lost either through leaching or volatilisation and denitrification. In case of P and K' adsorption and fixation reactions in soil are quite common. Most of the soils are only able to supply a fraction of the nutrients required by the arable crops. Even available nutrients are rarely present in a balanced condition to meet the crop requirements. By balanced nutrition it means that the amount of nutrients in soil should be in adequate amount and in required proportions for the plants to produce maximum yield. For instance, crops must supply nutrients (N + P_2O_5 + K_2O) to the tune of 125, 45 and 165 kg for a 50 q/ha yield of wheat; 100, 55 and 150 kg for a 50 q/ha yield of rice; 130, 70 and 180 kg for a 50 q/ha yield of maize and 90, 53, 136 kg for a 40 q/ha yield of sorghum (Velayutham and Reddy, 1987). But the results of long term fertiliser experiments indicate that in general the soils in India could only supply about 78 kg/ha of N, 14 kg/ha of P_2O_5 and 76 kg/ha of K_2O when continuously cropped without external addition of plant nutrients. Therefore, it is essential to add the limiting nutrients through fertiliser in quantities that ensure balanced nutrition for a given yield of the crop.

The highest response of 14.8 (kg/kg nutrient) at 50 kg/ha of N was seen with 25 kg/ha each of P_2O_5 and K_2O. For obtaining a corresponding response (14.5) at higher level of 99 kg/ha of N application, the requirement of associated P_2O_5 and K_2O was also high, being 75 and 50 kg/ha, respectively. Thus, the data reveal that Liebig's law of minimum operates equally well for N, P or K although it is generally believed that the law is valid for N and not for P and K which follow the percentage sufficiency concept of Mitscherlich and Baule (Tisdale et al., 1985).

The concept of balanced fertilisation has much importance in dryland agriculture where moisture limitation lowers down the fertiliser use efficiency and crop yield. Apart from the major nutrients, balanced fertilisation is equally important in case of secondary and micronutrients. Some soils which are deficient in secondary and micronutrients should be supplied with these nutrients so that nutrient imbalances are corrected. Application of sulphur, zinc and boron to an alluvial sandy loam soil at Kanpur significantly increased the yield, oil and protein content of Rai (*Brassica juncea*) [Shukla et al., 1983]. However, indiscriminate use of micronutrients may lead to nutrient imbalance and may be toxic for plant growth (Venkatasubbaiah et al., 1986). Balanced fertilisation is a must for sustaining optimum yields and as has been defined by Tandon (1990) a balance means a total

$$\text{Nutrient requirement (NR) (kg/q)} = \frac{\text{Nutrient uptake (kg/ha)}}{\text{Yield (q/ha)}}$$

$$\text{Soil efficiency (SE) (\%)} = \frac{\text{Nutrient uptake (kg/ha)}}{\text{Soil test value (kg/ha)}} \text{ X } 100$$

$$\text{Fertiliser efficiency (FE) \%} = \frac{\text{Nutrent uptake} - \text{soiltestvalue X soil nutrient efficiency}}{\text{Fertiliser rate applied}} \text{ X } 100$$

$$\text{Fertiliser dose (kg/ha)} = \frac{\text{NR}}{\text{FE/100}} \text{ X Target} - \frac{\text{SE}}{\text{FE}} \text{ x soil test value}$$

plant nutrition system which is capable of taking care of all nutrient deficiencies which occur in an area, be they of macro or micronutrients. It is also seen as a dynamic approach, which represents the needs of higher productivity and emergence of any new deficiencies/disorders. Calculation of nutrient balance shows that the contributions of different sources to the P removed by crops in India is 78% from soil reserves, 14% from fertilisers and 8% from organic manures. Based on the long term experimentation, it has been observed that use of N alone can lead to accelerated depletion of soil P upto 129% over an 8-11 year period (Nambiar and Ghosh, 1984; Tandon, 1987). Similarly, application of N or N + P resulted in a quantum jump in the quantities of K removed by crops.

Balanced fertilisation can easily be practised through soil testing. The present concept of fertiliser recommendation based on yield targeting not only takes into account the relationship between the nutrients of soil and fertilisers but also amongst the nutrients themselves. The essential basic data required for formulating fertiliser recommendation for targeted yields are (i) nutrient requirement in kg/q of produce (grain or other economic part); (ii) the percent contribution from the soil available nutrients; and (iii) the percent contribution from the applied fertiliser nutrients. These parameters can be computed as follows :

Nitrogen

Because of the tropical climate, Indian soils are very low in N content. The high mean annual temperature in central and Peninsular India hinders organic matter accumulation in the soils. Indus alluvium, Gangetic alluvium, black cotton soils, red soils and laterite soils contain 0.05, 0.04, 0.05, 0.03 and 0.04 per cent total N content, respectively. The ionic inorganic forms of N in Indian soils constitute less than 2 per cent of the total N. The main inorganic forms of N occurring in soils are nitrite, nitrate and ammonium, besides the elementary N and some oxides of N which remain in gaseous forms in soil air. These inorganic forms of N are in a state of dynamic change in soil and hence their contents in soil are highly variable. A substantial amount of the total nitrogen in soil is held rather tenaciously, in the form of fixed NH_4+ (Rodrigues, 1954). Fixed or non-exchangeable ammonium may constitute 1-10% of the total nitrogen in surface soil. Exchangeable ammonium in various soils of India varies from 10 to 50 ppm.

The major part of the total N in soil is in the organically combined form associated with the organic matter. About 18-30 per cent of the total N in most surface soils is in the form of bound amino acids, 3-7 per cent as amino-sugars and 18-43 per cent as non-hydrolysable N (Puranik et al., 1978; Ramamoorthy and Velayutham, 1976; Sharma et al., 1992 a).

Among the different N fractions, NO_3- and NH_4+ which hardly constitute 5% of the total soil nitrogen, are readily available to the plants. However, organic nitrogenous compounds present in soil play a very important role in supplying nitrogen to the plants. These organic compounds are converted to mineral forms by soil micro-organisms - a process known as mineralisation. The microbes which are associated with mineralisation, need part of the N mineralised for their cell synthesis. This entrapping of N in the body cells of mineralising bacteria is referred to as immobilization. The major losses of N in soils occur through removal or uptake by plants and micro-organisms, volatilization, denitrification and leaching. Nitrogenous fertilisers are more susceptible to various losses in drylands and consequently reduce the fertiliser N use efficiency.

Nitrogen response

Historically, the advantage from applications of organic manures (FYM, compost and green manure) to rainfed crops was established by Kanitkar et al. (1960). As the result of addition of organic manures, crop yields improved by 20% to 50% over control yielding 150 to 400 kg/ha. Serious efforts on fertiliser use research in dryland regions were initiated only in the late 1960s or early 1970s. Since this period onward, a positive and significant response to fertiliser N (generally in combination with P) with rainfed crops has been reported from the semi-arid tropics of India (Kanwar et al., 1973; Mahapatra et al., 1973; Venkateswarlu and Spratt, 1977; Spratt and Chowdhary, 1978; Randhawa and Venkateswarlu, 1980; Kanwar and Rego, 1983; Tandon and Kanwar, 1984; El-Swaify et al., 1985). Rao and Das (1982) have summarised the results on response of different rainy and post rainy season crops to N and P application under varying agro-climatic conditions. Response to N application depends on (i) crop, (ii) soil type, and (iii) rainfall distribution. Sharma and Das (1992) have elaborately discussed response to N application in relation to tillage practices, moisture availability, cultivars, soil properties, crops and cropping systems.

Tillage systems have considerable effect on the level of N derived from the soil by the crop (Olson, 1984). The reduction in availability of N with minimum tillage has been attributed to lower temperature caused due to shading and insulating effects of residues (Phillips et al., 1980). Greater response to applied N with minimum tillage has been recorded under drier than normal weather conditions when crops failed with conventional tillage methods.

A close association between soil moisture and response to N application has been established. Soil depth which determines the moisture availability for the crop between rains can be as important as the mean annual precipitation (Power and Alessi, 1978). Prihar et al. (1981) quantified the response of rainfed wheat grown on deep alluvial soil to added N as a function of available water supply. Another important parameter which determines the nature of response to applied N is type of cultivar. In many instances cultivars of sorghum have been found varying in their ability to use the applied nutrients. Hybrid sorghum which has well developed root system and high vegetative vigour is more responsive to added N compared with other varieties. Similarly soil properties, viz., profile depth, available nutrient status, water retention capacity of the soil have been found influencing the response of crops to applied N. Many efforts have been made so far to improve response of crops to applied nitrogen. For post-rainy season sorghum, growing of legume as preceding crop helped to utilise the applied fertilizer economically (Umrani and Patil, 1983).

FERTILIZER N USE EFFICIENCY

The term fertilizer use efficiency is defined as the capacity of the fertilizer N to increase yield over that of control plot. The synonym to this term is 'agronomic efficiency' and is usually expressed as kilogram grain yield (more appropriately economic yield) increase per kilogram of fertilizer N (Katyal, 1989). The major concerns about the N fertiliser are to quantify the N removed by crops and lost from the crop point of use. If the 'difference method' is the criteria, the fraction of N fertiliser absorbed by crop denotes the apparent N recovery (ANR). To exclude the contribution of soil N, ^{15}N-labelled fertilizers are used. This method is more satisfactory to ascertain the fertiliser N recovery. The ability of the fertilizer N absorbed by the crop to increase yield is termed as physiological efficiency. Under rainfed situations, precipitation, water holding capacity of the soil, form of fertiliser N and method of application are the most important factors which determine N use efficiency. Results of the joint studies conducted by International Fertilizer Development Corporation (IFDC) and International Crops Research Institute for the Semi-arid Tropics (ICRISAT) revealed that in deep soil fertiliser recovery of ^{15}N-labelled urea can be as high as 55% can if half of urea is banded at seeding and remaining half is applied one month later in bands 5-7 cm deep and 5-7 cm away from the seed row and covered with soil (Katyal, 1989). Sorghum recovered upto 64 per cent of the fertiliser N when grown on a deep Alfisol. However, with shallow Alfisol the fertilizer N recovery was 46 per cent. Similar patterns in N use efficiency were observed when the study was done in Vertisols. Nitrogen loss of applied fertilizer was higher (29-30%) in shallow soils than with deep soils (< 15%). It has been established that fertiliser N recovery and loss varied with fertiliser N source, method and time of application and seasonal rainfall and its distribution. Nitrogen use efficiency in dryland crops has been extensively reviewed by Sharma and Das (1992).

Depending upon the rainfall pattern, method and time of application have much significance in determining the fertiliser use efficiency and loss occurred. In case of average rainfall (seasonal rainfall 516 mm) year in SAT, surface application of urea did not show excessive N loss when compared with its placement in bands. However, in good rainfall year (rainfall 907 mm) surface treatment of urea suffered a heavy N loss (28%), whereas the N loss from urea placed in bands continued to be low (6%). Under limited moisture situation urease activity is retarded and probability of accumulation of $NH^{+}{}_{4}$-N and its volatilization at the surface soil is also decreased [Sahrawat, 1984; Myers, 1978]. The placement of fertiliser in receding moisture condition also assumes importance in governing the N use efficiency. Compared with broadcast method, placement of fertilizer 20 cm below the seed rows produced more than a two fold increase in wheat yield (Chaudhary and Prihar, 1974). Placing the fertiliser at greater depth (60 cm) also resulted in greater extraction of water from layer below 60 cm (Prihar et al., 1977). A significant improvement in agronomic efficiency with decreasing distance of the fertiliser row from the seed row was observed (Katyal et al., 1987).

Because of erratic distribution and unpredictable nature of rainfall, time of application of fertiliser assumes considerable importance. In this context, it is not judicious to go for single application of N fertilizer in *kharif* crops. There are many evidences to support the superiority of split application of N in two or three splits over single application. Apart from enhancing the fertiliser use efficiency, split application provides the opportunity to make seasonal decisions to increase the dose according to the rainfall availability. Fertiliser N use efficiency and its losses much depends on source and form through which it is applied. If N is applied in the form of urea under air dry conditions, its hydrolysis does not take place. The nitrification of ammonium resulting from urea hydrolysis decreases when saturated (anaerobic) conditions prevail. Under those circumstances, nitrate-containing fertilisers will tend to denitrify whereas, urea may be protected. Since the deep soils exhibit high water holding capacity, denitrification is less likely a problem. Under such circumstances, it is logical to believe that the efficiency of urea and potassium nitrate (KNO_3) will not vary with rainfall when both these N sources are tested on deep soils.

PHOSPHORUS IN INDIAN SOILS

Phosphorus exists in soil in many primary and secondary compounds. Indian soils vary in phosphorus content from 0.006 to 0.5 per cent (Chakravarthy, 1964). The original source of phosphorus which constitutes about 95 per cent or more of the soil phosphorus is the apatite group of the primary minerals. The various phosphorus compounds in soils can be generally classified as fluoro-carbonate and hydroxy phosphates of Fe, Al, Ti, Mn, Ca and Mg. Out of these compounds Fe, Al and Ca phosphates are found in considerable amount (Ramamoorthy and Velayutham, 1976).

The extent of common forms of inorganic P is mainly determined by the degree and intensity of weathering and it is possible to predict it by the prevailing pH. The soils with neutral-alkaline soil reaction dominate in Ca-bound P, whereas, soils very acidic in reaction have Al-P, Fe-P, reductant soluble P and occluded P as the predominant forms. Bapat et al. (1965) have reported that in a Vertisol with pH 7.5-8.2, containing 69% clay, Al, Fe and Ca-bound P forms constitute 5, 15 and 36% of the total P, respectively. In Alfisol, the sequence of dominance of P forms can be Fe-P>Ca-P>Al-P (Goswami and Sahrawat, 1982).

The amount of organic P in Indian soils varies from 2.6 to 75 per cent of the total P (Ghani and Aleem, 1942; Basu and Kibe, 1948) depending on the amount of soil humus. Low organic matter and wide average ratio of carbon to phosphorus (100:1) in Indian soils have resulted in less focus on tapping organic P. Organic P constitutes 13 and 20 per cent of the total P in alluvial and Vindhyan soils, respectively. Among the humus fractions in Rajasthan soils, humic acid, fulvic acid, hymatomelanic acid and humus constitute 3 to 13, 10 to 25, 0.16 to 0.55 and 12 to 24 per cent of total organic P, respectively. However, non-humus fractions contain 30 to 80 per cent P (Joshi and Saxena, 1972). Saxena (1979) reported that organic phosphorus was absent in some grey-brown desert soils of river basin of Rajasthan. However, it was present in traces (0.3 ppm P) in some unidentified alluviums of Rajasthan.

Phosphorus availability in soils

Soil testing surveys conducted by Ghosh and Hassan [1979] have revealed that soils of 42% of the districts in India are low, 56% are medium and 2% are high in available soil phosphorus. Deep and medium black soils, grey brown, desert and red loam soils of the foot hills, alluvial strips of northern region and coastal alluvium which are not sandy in nature largely have a medium soil phosphorus. At the same time the vast alluvial tract of the central, eastern and southern parts of the country, latosols, red sandy mixed red and black soils, sandy coastal alluvium and red loams of the humid regions are usually low in available soil phosphorus (Roy and Kanwar, 1979). For estimating available P status, presently two methods, viz. (i) Olsen et al. (1954) for neutral, calcareous and alkaline soils, and (ii) that of Bray and Kurtz (1945) for acid soils are in practice in India.

Availability of P in soil can be explained by the quantity intensity relationship (Kamprath, 1991). The quantity factor in acid soils essentially comprises of the phosphate sorbed by aluminium hydrate oxides and some iron oxides. In neutral and calcareous soils quantity factor is represented by phosphorus held up by exchangeable Ca and P sorbed on the surface of calcium carbonate. Phosphorus is taken up by the plant from the soil solution, and its critical concentration, the intensity for plant growth is dependent on soil texture, and is higher for sandy soils than for clay soils. The concentration of P in soil solution is governed by the heterogenous equilibria in which it takes part. Larsen (1967) has represented this equilibrium as :

P adsorbed on solid phase $\leftrightarrows$ P in soil solution $\leftrightarrows$ P precipitated form

According to Gunary and Sutton (1967), the quantity factor is the amount of P required to give a certain soil solution concentration of P (intensity). The change in the intensity factor introduced by a given change in the quantity factor is referred to as the P buffering capacity. In case of highly weathered soils, the P buffering capacity is determined by the amounts of hydrated oxides of Fe and Al. Because these sesquioxides are associated with the clay fraction, the buffering capacity is correlated with the clay content of the soil. Kamprath and Watson (1980) reported that as the clay content increased, the critical soil solution P levels required for optimum growth decreased from 0.40 mg kg^{-1} for coarse textured soil to 0.06 mg kg^{-1} for clayey soils. The mechanism (diffusion) by which P moves to plant roots is most affected by concentration of soil solution P (Olsen and Watanabe, 1963).

Other factors which are also important in controlling diffusion are volumetric water content and more tortuous diffusion path than fine textured soils, high concentrations of P are required to move a given amount of P to plant roots in sandy soils than in clayey soils. Moreover sandy soils also have lower P buffering capacities and, therefore, soil solution P concentration decreases more with a given amount of P uptake. Thus, the P availability is controlled by the complex relationship between the quantity, intensity and capacity factors and their effect on the P diffusion.

P fixation and transformation

In general, the type and extent of reaction of added phosphate in soils is influenced by number of variables such as pH, type and amount of clay, redox potential, nature of phosphate fertiliser added, reaction time, soil moisture and temperature, organic matter content, $CaCO_3$ content, mineralogical make up, free sesquioxides and soil texture. Vertisols representing semi-arid tropics were studied for their adsorption/desorption reactions recently. The amount of P adsorbed by the soils representing Kasireddipalle, Barsi and Begumganj series at 0.2 ug ml^{-1} equilibrium solution P concentration was low and ranged from 34.3 to 79.5 ug g^{-1} soil. The correlation studies indicated that the DTPA-extractable Fe was the most important factor accounting for P adsorption in these soils. Clay and $CaCO_3$ content were found to be relatively less important factors affecting P adsorption in the soils (Shailaja and Sahrawat, 1990). Sharma et al. (1992b) observed that Vertisols of Akola, Solapur, Bijapur and Kovilpatti varied in their P adsorption capacity from 83 to 99 per cent. However, 57 to 66% of the adsorbed P was desorbed during two successive desorptions with 0.5 M $NaHCO_3$ indicating the non-specific nature of adsorption. As the phosphate availability to plants has been variously correlated with the soil P supply factors of quantity, intensity, and buffering capacity, the adsorption/desorption reactions of phosphate in soil have much importance. The phosphate buffering capacity of the soils has been found to decrease with the increase in solution P concentration when calculated from Langmuir and Freundlich isotherms (Shailaja and Sahrawat, in press).

Sahrawat et al. (1987) pointed out that adsorption and desorption of P are very important for P uptake by plant. It should be emphasized that the reaction of P with soil is a surface reaction, and it has been described by a number of adsorption equations. The use of these equations has been reviewed by Sample et al. (1980) and Olsen and Khasawne (1980).

Representative light textured Alfisols of Semi-arid-Tropics were studied for their P fixation and release pattern by Ravinder Singh (1991). Based on these studies it was found that 90 per cent of the added P was adsorbed at lower solute concentration of P. Olsen's reagent (0.5 M $NaHCO_3$, pH 8.5) could extract or desorb considerable quantities of adsorbed P from soils of Hayathnagar, Gunegal and Anantapur in Andhra Pradesh. The adsorbed P in soils of Ranchi and Bangalore could be desorbed with the extractant only at higher soluble P concentration thereby confirming the capacity of these soils to fix considerable amount of P into non-extractable (unavailable) form.

Response to P application

According to Khanna et al. (1982) two-thirds of the phosphatic fertilisers are applied to cereal crops, 15 per cent to sugarcane, 10 per cent to oilseeds, 7 per cent to cotton and the remaining 5% to all other crops. Several important dryland crops such as pearlmillet, small millets, pulses and others receive hardly any phosphatic fertiliser. Response of cereals, millets, pulses and oilseeds to P application under rainfed conditions has been comprehensively reviewed by Tandon (1987). Field data from different regions in India showed that P increased water use efficiency (WUE) by 15-20 per cent in wheat, 22-55 per cent in fingermillet, 41-99 per cent in chickpea, 17 per cent in linseed and up to 19 per cent in the wheat + chickpea mixed crops. P application enhanced the root growth and enabled the crop to draw water from deeper soil layers [Singh et al., 1983, 1985]. Application of 75 kg P_2O_5/ha to chickpea grown on soils having 27 kg available P_2O_5/ha increased grain yield by 78 per cent, water use by 27 per cent and water use efficiency (WUE) by 41 per cent (Singh and Bhushan, 1979). The results of a large number of field experiments summarised by Kanwar (1978) revealed that there was significant response to phosphorus which was applied in combination of 30 kg N and 40 kg P_2O_5/ha, the yield was double than that obtained with 30 kg N/ha. The response pattern of sorghum and cotton to P application in black soil was found to be variable and did not show a clear relationship with Olsen-P (Sharma et al., 1992). The important parameters which influenced the response of crops to added P were soil solution P concentration, soil test values, interaction with nitrogen, cropping systems, carriers of P, methods of P application (Murthy, 1988; Sharma and Das, 1992).

In case of intercropping system of sorghum/pigeonpea, 31% more P was removed by the crops than sorghum crop grown alone [Kanwar, 1986]. Based on six year study on sorghum/pigeonpea grown as intercrop followed by castor, it was observed that total productivity of the system was highest when 50 kg P_2O_5 ha^{-1} was applied to the main system and castor fed on residual fertility (Venkateswarlu et al., 1986).

SECONDARY AND MICRONUTRIENTS

Intensification of agriculture has led to the removal of secondary and micronutrients to a considerable extent thus causing deficiencies of these elements in soil. Earlier, the secondary and micronutrients supplied by the soil were sufficient to take care of plant needs or were supplied incidentally along with other nutrients.

The increased productivity and use of high analysis fertilisers result in rapid depletion of secondary and micronutrient resources of the soil unless replenished. Replenishment of secondary and micronutrients has unfortunately not received adequate attention in dryland areas. Consequently, the deficiencies of many secondary and micronutrients especially of sulphur and zinc have become evident.

Calcium

The CaO content of acid soils varies from less than 0.1 to about 1 per cent, though in soils with pH ranging from 5 to 6, it may not exceed 0.5 per cent (Mandal, 1954). Being neutral to moderately alkaline in reaction, alluvial soils of northern region have sufficient bases on their exchange complex and as such Ca deficiency is rarely seen in these soils. Deficiency of Ca is expected if Ca saturation is less than 25% or less than 1.5 milliequivalent exchangeable Ca/100 g soil, especially in humid region soils with rainfall over 750 mm per year, soils derived from acid igneous rocks or siliceous sand stones, acid peat soils and alkali or sodic soils where exchangeable sodium and pH are high. The crops that have responded well to liming are essentially legumes, followed by cotton, maize, sorghum, wheat, linseed, etc. Significant response to applied lime has been reported in greengram, jute and mustard in Assam, jute in West Bengal, horsegram in Orissa, rice and groundnut in Mysore, rice and cotton in Tamilnadu, wheat and maize in Himachal Pradesh and rice in Kerala. Mandal et al. (1975) have conclusively shown that NPK become highly effective in increasing crop yield in limed acid soil when compared with unlimed soil.

Apart from response of crops to Ca application the beneficial role of lime in acid soils has been attributed to the greater availability of other nutrients. The application of Ca as $CaSO_4$ increased the uptake of N and B in groundnut (Sankaran et al., 1977). The synergistic effect of Ca with P has been observed by Ramakrishnan and Jain (1965). However, the antagonistic effect with Mg and K was reported by Kansal and Sekhon (1973) and Nad (1979). Various materials like agricultural limestone (80-95% $CaSO_4$), dolomite (20-45% CaO), gypsum (40% CaO), single superphosphate (25-30% CaO) and rock phosphate (39-48% CaO) could be effective sources of Ca for different crops.

Magnesium

Crop response to Mg application has been discussed at length by Dev and Vinod Kumar (1982), Tandon (1989), Tiwari (1990) and Saharan (1992). This nutrient is of appreciable importance in acid soils. Except in case of acid soils of

Ranchi (Bihar) which contain Mg bearing minerals like hornblende, Indian acid soils are low (0.1-1%) in Mg, the average being 0.3 to 0.6 per cent. Application of Mg @ 250 and 500 kg/ha as $MgCO_3$ to groundnut in acid soils of Kerala increased the nodulation and yield but decreased oil content (Tejuddin, 1971). Application of Mg increased yield of corn in soils of Hisar (Gupta and Singh, 1985) and Varanasi (Singh and Singh, 1985) and of wheat in Delhi.

Sulphur

Sulphur is present in soil mostly as sulphides, sulphates and in organic compounds. The earth's crust contains about 0.06 per cent sulphur. The original source of sulphur in soil are plutonic rocks. During the process of weathering sulphur from sulphides is converted to sulphates and some part of it got dissolved, some precipitated and some reduced even to elemental sulphur, depending upon the situation. In humid regions, the predominant form is organic sulphur whereas in arid zones, the sulphates of Ca, Mg, Na and even K predominate. Other important source through which sulphur is added in soil is combustion of fuel which produces SO_2 which is brought down by rain.

Less attention has been paid towards the research on sulphur as compared to that on N, P, K and Zn. It is during the sixties that some efforts were initiated when deficiencies to the extent of 75% of the groundnut growing soils of Ludhiana district were noticed by Kanwar (1963). Similarly, 74% of the soils of Kangra valley of Punjab, now in Himachal Pradesh, were reported to contain less than 10 ppm of 0.15% $CaCl_2$ extractable S and were categorised as deficient in sulphur. Consequently, this led to more extensive efforts to initiate studies for assessing S status of Indian soils. Sulphur deficiencies have been reported in 72 countries recently (Morris, 1987). The several factors which have aggravated sulphur deficiencies are as follows.

1. Increased use of high-analysis fertilisers containing little or no S.
2. Multiple cropping and high yielding crops which remove greater amounts of S from the soil.
3. Decreased use of high sulphur fuels in industries.
4. Decreasing trend of S reverses in soil.
5. Use of crop residues for feed and fuel in developing countries and decreased use of pesticidal S.

The extent of sulphur deficiencies and the response of crops to S application has been comprehensively reviewed by Tandon (1986a, 1986b], Saharan (1992) and Tiwari (1990). In India S deficiencies are experienced across Indo-Gangetic alluviums, tea growing soils of Himachal Pradesh, red-yellow soils of Bundelkhand regions of Uttar Pradesh, red and lateritic soils of Karnataka, Kerala and West Bengal, alluvial and black soils of Madhya Pradesh and several black soils areas of Maharashtra, Gujarat and Southern Rajasthan.

CROP RESPONSE

The information of crop response to S application under dryland is limited and scattered. However, Tandon (1988) and Saharan (1992) have reviewed the results of studies conducted on response of crops to sulphur application under dryland conditions. Under irrigated conditions, over 30 crops have been found responsive to sulphur application (Tandon, 1986). Some of the oilseed crops have shown response to S application. Kanwar and Mudahar [1986] have reported that amounts of S absorbed per tonne of yield production varied from 3-4 kg for cereals, 5-8 kg for millets, 8 kg for pulses and 12 kg for oilseeds. In Andhra Pradesh groundnut crop has significantly responded to S when applied either as gypsum or in elemental form (Raman and Subba Rao, 1979). Groundnut grown on coarse textured, well drained soil having neutral to alkaline soil pH responded to S applied @ 40 kg ha^{-1} through gypsum under rainfed conditions (Pasricha and Rana, 1985). A significant response to direct application of 20 kg S ha^{-1} in groundnut and to residual sulphur in wheat that followed groundnut in a sequence has also been reported (Pasricha et al., 1987). Combined use of sulphur with boron increased groundnut yield by 1.5 times and oil yield by 1.8 times in medium black soils (Karle and Babula, 1985).

An increase of 17-28% in yield of rainfed mustard has been reported by Ankineedu et al. (1983) and Aulakh and Pasricha (1988). Application of sulphur led to an increase in water use efficiency by 60 per cent. Application of S @ 60 kg ha^{-1} under green house conditions using soil representing semi-arid Alfisols has shown significant increase in seed yield of groundnut and sunflower (Saharan et al., 1989). Increase in seed yield of Taramira (*Eruca sativa*) with the application of sulphur has been reported in Haryana.

Much data are not available on the efficieacy of different sources of S under drylands. It is the gypsum or single superphosphate which is frequently used in groundnut. Ammonium sulphate and ammonium phosphate sulphate are applied for non legumes (Tandon, 1989). As has been suggested by Takkar (1987), in case of S deficiency, the application of 20-40 kg S ha^{-1} is highly economical and essential for boosting crop production particularly of oilseeds and pulses.

Rainfed groundnut grown on Alfisol of Bangalore performed well with the application of 10 kg S ha^{-1}, whereas, rapeseed and mustard require, 20 kg S ha^{-1} if soils are deficient in sulphur (DOR, 1984). Sulphur application @ 60 kg/ha increased the seed yield of soybean and pigeonpea in an intercropping system by 18 and 23%, respectively over the control yield (soybean = 12.5 q/ha, pigeonpca = 15.6 q/ha) in black soils of Indore. Similarly, groundnut and sesamum yields were increased by S application in black soils of Arjia, gypsum being the better source of S compared to elemental sulphur and pyrite.

Micronutrients

Among the micronutrients, only Zn and Fe have received attention so far in drylands. Soil tests have indicated the possibility of wide occurrence of zinc deficiency in a number of dryland locations [Katyal et al., 1987]. Zinc deficiencies have been found to be widespread and are more frequent in arid and semi-arid soils than in the humid or sub-humid zones (Katyal and Vlek, 1985). Based on the analyses of 93,000 soil samples under ICAR Micronutrient Project, it was found that about 47% of the soil samples contained less than 0.6 ppm DTPA extractable Zn and were grouped as deficient in zinc. The percent samples found deficient in Andhra Pradesh, Gujarat, Haryana, Karnataka, Madhya Pradesh, Punjab, Tamil Nadu and Uttar Pradesh (Semi-Arid Tropical areas) were 51, 26, 77, 24, 63, 53, 36 and 69%, respectively. Soil reaction, organic matter and clay content influenced the zinc content. The extent of zinc deficiency increased with increase in soil pH and decreased with decrease in organic matter and clay content (Takkar and Nayyar, 1986). Among the climatic factors, temperature and moisture affect micronutrient transformations most. Under limited moisture situations, uptake of micronutrients is reduced [Kanwar and Youngdahl, 1985].

CROP RESPONSE

Zinc

Although much information is available on crop responses to zinc (Katyal, 1985; Takkar and Nayyar, 1986; Randhawa and Nayyar, 1982) under irrigated conditions, the information on response of crops to zinc application under drylands is scanty. Application of 5 kg Zn ha^{-1} increased the grain yield of rainfed sorghum by 80-410 kg ha^{-1} (Subba Rao, 1975). Zinc application @ 10 kg $ZnSO_4$ ha^{-1} increased the yield of dryland maize grown in alfisol of Bangalore by 510 kg ha^{-1}. There is probability to enhance the groundnut yield with the application of micronutrients (Joshi et al., 1987). Differential responses of genotypes to zinc application have been observed by Kaur (1987), Subba Rao (1975) and Sakal et al. (1985). Zinc sulphate ($ZnSo_4$, $7H_2O$) containing about 21% Zn still remains the important source of zinc. To correct the simultaneous deficiency of P and Zn, the desirability of incorporating Zn into suitable P carriers has been advocated by Singh (1983).

Iron

In general, iron deficiency may be one fourth

as extensive as that of Zn (Tandon, 1989). The possibility of iron deficiency in highly alkaline soil, under drought conditions and long dry spells in Andhra Pradesh has been indicated by Subba Rao (1975). Groundnut variety TMV-2, which is extensively grown in the state, is particularly sensitive to iron deficiency under dryland conditions even in red soils (Venkat Raju et al., 1987). Instances of iron chlorosis have been reported on several crops in large areas of Maharashtra (Daftardar, 1987). Still more efforts are needed in this direction.

INTEGRATED NUTRIENT SUPPLY SYSTEM

Application of high input technologies like inorganic fertilisers, pesticides, herbicides and energy intensive farm power machinery no doubt has increased production but deteriorated soil productivity and environmental quality. Soil fertility is a pre-requisite to land productivity. From the stand point of crop yields, nutrient supply both from the soil and fertiliser sources are important. It is thus of paramount importance to substitute the inorganic fertilisers with nutrients from other organic sources. The integrated nutrient supply system envisages the use of the organic manures, crop wastes, biological nitrogen fixation, phosphate solubilisers in conjunction with inorganic fertilisers. The scope and potential of integrated nutrient management in drylands have been comprehensively reviewed by Sharma and Das (1992). Based on this review following strategies have emerged :

(i) There has been considerable improvement in soil fertility status and in soil physical properties like soil structure, stability of aggregates and hydraulic conductivity with the incorporation of farm based crop residues in semi-arid Alfisols.

(ii) The incorporation of residue enhanced the yield of rice at Ranchi by 41 percent over a base yield of 4.4 q/ha. Similarly, at Rakh Dhiansar, 50% recommended rate of fertilizer together with organic residue to meet 50 per cent N, gave the highest yield of maize (37.8 q/ha) which was more than the full fertiliser dose. The release of N from sorghum residue can be computed from the following models (Das et al., 1993) :

$N = 47.55 + 0.1685\ D$ at 50% field capacity
$N = 51.24 + 0.1552\ D$ at field capacity

where, N = Predicted mineralised N (g/kg soil)
D = nth day of incubation

(iii) Incorporation of loppings (leaves + twigs) of *Leucaena* in semi-arid Alfisols has been proved effective in substituting fertiliser nitrogen for crop use. The N release models are as follows (Das et al., 1993) :

$N = 61.67 + 0.1000\ D$ at 50% field capacity
$N = 66.17 + 0.0923\ D$ at field capacity

where, N = Predicted mineralised N (g/kg soil)
D = nth day of incubation

(iv) Growing of legumes like cowpea and greengram in the inter row spaces of long duration and wide spaced crops and to incorporate cowpea in soil after 40-45 days of sowing is a viable practice as green manuring. This practice, can contribute, on an average, 30 kg N/ha on fertiliser basis.

(v) Legumes like cowpea, clusterbean, mothbean and soybean were found to enrich the soil fertility. In alluvial soils mothbean was found to produce the highest residual effect. Residual effect of groundnut was found equivalent to 15 kg N/ha.

(vi) Free living N fixer like *Azospirillum* has been found effective in fixing nitrogen when inoculated with seeds of pearlmillet and sorghum under rainfed situations. Contribution of about 20 kg N/ha on fertiliser equivalent basis has been reported.

(vii) Grain legumes are estimated to fix about 60-80 kg N/ha. Inoculation of *Rhizobium* has been tested for crops like greengram, blackgram, cowpea, clusterbean, pigeonpea, chickpea under diversified agro-climatic conditions. Under rainfed situations, clusterbean has responded to *Rhizobium* inoculation. Since water availability is a major limiting factor, the performance of strains suitable for stress situations and methods of inoculation need more attention.

(viii) Response of dryland crops to P application in Alfisol and Vertisol have been found non-persistent. To mobilise the insoluble phosphate in soils, V-A mycorrhizae (VAM) have been found to play crucial role. Artificial inoculation of soybean and cowpea with *Rhizobium* and VAM showed improvement in root nodulation and phosphate uptake. Lack of crop response to P application in some cases could possibly be explained due to the role of VAM in enabling the crop to make a more efficient use of soil P.

Thus, keeping in view the above discussion, it is of interest to comment that still there is scope to give more emphasis on integrated nutrient supply systems.

CONCLUSIONS

Review of past research findings and their critical appraisal made by the scientists can help in devising strategies for optimum fertiliser use and management for dryland crops grown under different agro-climatic situations. In order to sustain productivity of dryland crops, it is crucial to offset nutrient depletions through appropriate fertilisation practices. Addition of organic materials in light textured low rainfall areas should receive priority in managing soil fertility problems. Crop residue incorporation, tree based nutrient cycling, green leaf manuring and use of FYM are the potential avenues to reverse nutrient removal induced depletions (Katyal et al., 1991). Legumes effect needs to be capitalised wherever legume finds a place in cropping system. Turning over legumes after first picking will help to add organic matter in soil as well. Conjunctive use of fertilizers and on-farm residues is a sound management strategy in risk prone dryland areas. In comparison, profitability of fertiliser use is expected to be higher and assured in medium to high rainfall areas. However, even in such environments, the importance of harmonising the integrated nutrient supply and management is not undermined.

REFERENCES

Ankineedu, G., Rao, J.V. and Reddy, B.N. 1983. Advances in fertiliser management for rainfed oilseeds. *Fert. News* 28(9): 76-90, 105.

Anonymous 1966. Tech. Rep. Series 65. IAEA, Vienna, p. 145.

Arnon, D.I. 1954. In: *Trace Elements in Plant Physiology* (ed) T. Wallace. Chronica Botanica Waltham, Massachusetts, pp. 31-39.

Aulakh, M.S. and Pasricha N.S., 1988. Sulphur fertilisation of oilseeds for yield and quality. In: *Sulphur in Indian Agriculture,* Proc. TSI-FAI Symp., SII/3/1-14.

Bapat, M.V. *et al.* 1965. Forms of phosphorus in Vidarbha soils. *J. Indian Soc. Soil Sci.* 13: 31-36.

Basu, J.K. and Kibe, M.M. 1948. *J. Univ. Bombay* 14(3A): 29.

Bray, R.H. and Kurtz, L.T. 1945. Determination of total, organic and available forms of P in soils. *Soil Science* 59: 39-45.

Chakravarthy, S.N. 1964. *Sci. Cult.* 30: 475.

Chaudhary, M.R. and Prihar, S.S. 1974. Comparison of banded and broadcast fertiliser applications in relation to compaction and irrigation in maize and wheat. *Agron. J.* 66: 560-564.

Daftardar, S.Y. 1987. Recent studies on problems of iron chlorosis of upland rice and sugarcane in Maharashtra. In: *Natn. Symp. Micronutrient Stress in Crop Plants : Physiological and Genetical Approaches to Control them,* Rahuri, pp. 158-177.

Das, S.K., Subba Reddy, G., Sharma, K.L., Vittal, K.P.R., Venkateswarlu, B., Narayana Reddy, M. and Reddy, Y.V.R. 1993. Prediction of nitrogen availability in soil after crop residue incroporation. Fert. Res. 34

Dev, G. and Vinod Kumar 1982. Secondary nutrients. In: *Review of Soil Research In India. 12th International Congress of Soil Science,* 8-16 Feb. 1982, New Delhi.

DOR 1984. Rapeseed, Mustard, Sunflower, Linseed. Extn. Bull. No.1. Directorate of Oilseeds Research, Rajendranagar, Hyderabad.

El-Swaify, S.A., Pathak, P., Rego, T.J. and Singh, S. 1985. Soil management for optimized productivity under rainfed conditions in the semi-arid tropics. *Adv. Soil Sci.* 1: 1-64.

Finck, A. and Venkateswarlu, J. 1982. Chemical properties and fertility management of vertisols. Vertisols and rice soils of the tropics. *12th Int.*

Cong. Soil Sci., New Delhi, pp. 61-79.

Fried, Maurice and Broeshart, Hans 1967. The Soil-Plant System in Relation to Inorganic Nutrition. Academic Press, New York.

Ghani, M.O. and Aleem, S.A. 1942. *Indian J. agric. Sci.* 13: 29.

Ghosh, A.B. and Hassan, R. 1979. Phosphorus fertility status of soils of India. Bull. 12, Indian Society of Soil Science.

Ghosh, A.B. and Hassan, R. 1980. *Fert. News* 25(11): 19-24.

Goswami, N.N and Sahrawat, K.L. 1982. Nutrient transformations in soil - Macronutrients. In: *Review of Soil Research in India*. 12th Int. Soil Science Congress, New Delhi, 123-145.

Gunary, D.C. and Sutton, C.D. 1967. *J. Soil Sci.* 18: 167-173.

Gupta, V.K. and Singh, B.J. 1985. *J. Indian Soc. Soil Sci.* 33: 204-207.

Joshi, D. and Saxena, S.N. 1972. *J. Indian Soc. Soil Sci.* 20: 117.

Joshi, Y.C. *et al.* 1987. Use of Micronutrients in groundnut : Technologies for better crops. No.31, ICAR, New Delhi.

Kamprath, E.J. 1991. Appropriate measurements of phosphorus availability in soils of the semi arid tropics. In: *The Role of Phosphorus in Agriculture* (eds) F.E. Khasawneh, E.C. Sample and E.J. Kamprath. American Society of Agronomy Madison, WI, USA.

Kanitkar, N.V., Sirur, S.S. and Gokhale, D.H. 1960. Dry Farming in India. Indian Council of Agricultural Research, New Delhi, India.

Kansal, B.D. and Sekhon, G.S. 1973. *Agrochemica* 18: 94.

Kanwar, J.S. 1963. *Indian J. agric. Sci.* 33(3): 196-98.

Kanwar, J.S. 1978. Trends in Consumption and Production. *Proc. FAI-IFDC Fertiliser Seminar*. FAI, New Delhi, AGR II/3, 1-16.

Kanwar, J.S. 1986. Crop production techniques and fertiliser management with special reference to P fertiliser in rainfed areas. *Proc. IMPHOS-FAI Seminar.*

Kanwar, J.S., Das. M.N., Sardana, M.G. and Bapat, S.R. 1973. Are fertiliser applications to jowar, maize, and bajra economical? *Fert. News* 18(7): 19-28.

Kanwar, J.S. and Mudahar, S.S. 1986. Fertilizer Sulfur and Food production. Martinus Nijhoff/Dr. W. Junk Publishers, Dordrecht, The Netherlands.

Kanwar, J.S., Rego, T.J. 1983. *Fert. News* 28: 33-43.

Kanwar, J.S. and Youngdahl, L.J. 1985. Micronutrient needs of tropical food crops. *Fert. Res.* 7: 43-67.

Karle, B.G. and Babula, A.V. 1985. Effect of boron and sulphur on yield attributes and quality of groundnut. In: *Proc. National Seminar on Sulphur in Indian Agriculture*, Coimbatore, pp. 158-168.

Katyal, J.C. 1985. Research achievements of the All India Coordinated Scheme on Micronutrients in Soils and Plants. *Fert. News* 10(4): 67-81.

Katyal, J.C. 1989. Nitrogen fertilizers - their use and management in the Indian semi-arid tropics. *Proc. Colloquium, Soil Fertility and Fertilizer Management in the Semi-Arid Tropical India*, held at ICRISAT Centre, Patancheru, A.P.,India, October 10-11, 1988, pp. 61-70.

Katyal, J.C, Das, S.K., Sharma, K.L. and Neelam Saharan 1991. Effective nutrient management in improving productivity of dryland crops. Paper presented at the *Annual Seminar of FAI* held at New Delhi from Dec. 5-7, 1991, p. SV1/1-1.

Katyal, J.C., Hong, C.W. and Vlek, P.L.G. 1987. Fertiliser management in vertisols. In: *Management of Vertisols in Semi-Arid Conditions*. IBSRAM Proc. No.6, International Board for Soil Research and Management, Bangkok, Thailand, pp. 247-266.

Katyal, J.C. and Rattan, R.K. 1990. Micronutrient use in the 90's. In: *Soil Fertility and Fertilizer Use - Nutrient Management and Supply Systems for Sustaining Agriculture in the 1990's*. Vol.IV, IFFCO, New Delhi, pp. 119-135.

Katyal, J.C. and Sharma, B.D. 1991. *Geoderma* 49: 165-179.

Katyal, J.C. and Vlek, P.L.G. 1985. Micronutrient problems in tropical Asia. *Fert. Res.* 7: 69-94.

Kaur, N.P. 1987. The concept of fitting plants to soil. Identification of crop cultivars efficient in micronutrient stress tolerance. In: *Natn. Symp. Micronutrient Stresses in Crop Plants : Physiological and Genetical Approaches to Control them*, Rahuri, pp. 214-236.

Khanna, S.S., Pathak, A.N. and Saxena, S.N. 1982. Phosphorus. In: *Review of Soil Research in India, Part I.* 12th Int. Congress of Soil Sci., New Delhi, India, 8-16 February 1982, pp. 323-330.

Kissel, D.E., Sander, D.H. and Ellis, R. Jr. 1985. Fertilizer - plant interactions in alkaline soils. In: *Fertilizer Technology and Use* (ed) O.P. Engelstad. Soil Sci. Soc. Am., Madison, Wisconsin, USA, pp. 153-196.

Krishnamoorthy, Ch. and Venkateswarlu, J. 1976. Nutrient availability concepts and relationships. In: *Soil Fertility Theory and Practice* (ed) J.S. Kanwar. Indian Council of Agricultural Research, New Delhi, pp. 65-90.

Lamm, C.G. 1968. *J. Indian Soc. Soil Sci.* 16: 9.

Larsen, S. 1967. Soil Phosphorus. *Adv. Agron.* 19: 151-210.

Lindsay, W.L. and Vlek, P.L.G. 1977. Phosphate minerals. In: *Minerals in Soil Environments* (eds) J.B. Dixon and S.B. Weed. Soil Sci. Soc. Am. Madison, Wisconsin, USA, pp. 639-672.

Mahapatra, I.C., Bapat, S.R., Sardana, M.G. and Bhendia, M.L. 1973. Fertiliser response of wheat, gram and horsegram under rainfed conditions. *Fert. News* 18(4): 57-63.

Mandal, S.C. 1954. *Proc. Indian Sci. Congr. 3.*

Mandal, S.C., Sinha, M.K. and Sinha, H. 1975. Indian Council of Agricultural Research. Tech. Bull. 51.

Morris, R.J. 1987. In: *Proceedings of the Symposium on Fertiliser sulphur requirements and sources in developing countries of Asia and the Pacific.* Bangkok, January 26-30, 1987.

Murthy, A.S.P. 1988. *Adv. Soil Sci.* 8: 151-214.

Myers, R.J.K. 1978. Nitrogen and phosphorus nutrition of dryland grain sorghum at Katherine, Northern territory - 3. Effect of nitrogen carrier, time and placement. Aust. *J. Exp. Agric. Anim. Husb.* 18: 834-843.

Nad, B.K. 1979. Ph.D. Thesis. Indian Agricultural Research Institute, New Delhi.

Nambiar, K.K.M. and Ghosh, A.B. 1984. LTFE Bulletin No.1. Indian Agricultural Research Institute, New Delhi.

Nicholas, D.J. 1963. In: *Plant Physiology - A Treatise* (ed) F.C. Steward. Academic Press, New York, 3: 363.

Olsen, S.R., Cole, C.V., Watanabe, F.S. and Dean, L.A. 1954. Estimation of available phosphorus in soils by extraction with sodium bicarbonate. US Department of Agriculture Cir. No. 939.

Olsen, S.R. and Khasawneh, F.E. 1980. Use and limitations of physical criteria for assessing the status of phosphorus in soils. In: *The Role of Phosphorus in Agriculture* (eds) Khanawneh et al. ASA, CSSA, SSSA, Madison, Wisconsin, USA, pp. 361-410.

Olsen, S.R. and Watanabe, F.S. 1963. Diffusion of phosphorus as related to soil texture and plant uptake. *Soil Sci. Soc. Am. Proc.* 27: 648-653.

Oison, R.A. 1984. Nitrogen use in dryland farming under semi-arid conditions. In: *Nitrogen in Crop Production* R. D. Hauck (ed) South Segoe Road, Madison, WI, USA, pp. 335-347.

Pasricha, N.S., Aulakh, M.S., Bahl, G.S. and Baddesha, H.S. 1987. Nutritional requirements of oilseed and pulse crops in Punjab. Tech. Bull., PAU, Ludhiana.

Pasricha, N.S. and Rana, D.S. 1985. Fertiliser use in arid lands for increasing the production of oilseeds. *Proc. FAI (NRC) Seminar on Available Technology and Strategy for Promoting Fertiliser Use in Dryland Farming.* pp. 135-146.

Phillips, R.E., Blevins, R.L., Thomas, G.W., Frye, W.W. and Phillips, S.H. 1980. *Science* 208: 1108-1113.

Power, J.F. 1990. Fertility management and nutrient cycling. *Advances in Soil Science* 13: 89-149.

Power, J.F. and Alessi, J. 1978. *J. Agric. Sci.* 90: 97-108.

Prihar, S.S., Kheza, K.L. and Bajwa, M.S. 1977. Growth water use, and nutrient uptake by dryland wheat as affected by placement of nitrogen and phosphorus. *Indian J. Econ.* 4: 23-31.

Prihar, S.S., Sandhu, K.S., Singh, Y. and Singh, R. 1981. Effect of N rates on dryland wheat in relation to mulching previous crop or fallow. *Fert. Res.* 2(3): 211-219.

Puranik, R.B., Ballal, D.K. and Barde, N.K. 1978. *J. Indian Soc. Soil Sci.* 26: 169.

Ramakrishnan, P.S. and Jain, R.S. 1965. *Proc. Natn. Int. Sci. India*, 31: 219.

Ramamoorthy, B., Narasimham, R.L. and Dinesh, R.S. 1967. *Indian Farming* 17(5): 43-45.

Ramamoorthy, B. and Velayutham, M. 1976. Nitrogen, phosphorus and potassium in soil chemistry, forms and availability. In: *Soil Fertility Theory and Practice* (ed) J.S. Kanwar. ICAR, New Delhi, pp. 156-198.

Raman, K.V. and Subba Rao, I.V. 1979. Soil plant nutrient relationships. *Proc. Symp. Fifty Years of Research in the Service of Andhra Pradesh Farmers.* Andhra Pradesh Agricultural University, Hyderabad, pp. 29-67.

Randhawa, N.S. and Nayyar, V.K. 1982. Crop responses to applied micronutrients. In: *Review of Soil Research in India, Part I.* 12th International Congress of Soil Science held at New Delhi.

Randhawa, N.S. and Velayutham, M. 1989. Fertiliser and soil fertility research needs in the nineties with special reference to semi-arid tropical India. *Proceedings of Colloquium, Soil Fertility and Fertilizer Management in Semi-Arid Tropical India* held at

ICRISAT Centre, Patancheru, India. October 10-11, 1988, IFDC, Muscle Shoals, Alabama, USA, pp. 129-135.

Randhawa, N.S. and Venkateswarlu, J. 1980. Indian experience in semi-arid tropics : Prospect and retrospect. *Proc. of the International Symposium on Development and Transfer of Technology for Rainfed Agriculture and the SAT Farmer*. ICRISAT, Patancheru, India, pp. 207-220.

Rao, A.C.S. and Das, S.K. 1982. Soil fertility management and fertilizer use in dryland crops. In: *A Decade of Dryland Agricultural Research In India (1971-80)*. AICRPDA, Hyderabad, pp. 120-139.

Ravinder Singh 1991. Phosphorus fixation and release patterns in dryland red soils. M.Sc. (Ag) Thesis, APAU, Hyderabad.

Rego, T.J., Moraghan, J.T. and Sardar Singh 1982. Some aspects of soil nitrogen relating to double cropping of a deep vertisol in the SAT. Abstract 486 p. 12th International Congress of Soil Science, Feb. 1982, New Delhi, India: Indian Soc. of Soil Science, 116 p.

Rodrigues, G. 1954. Fixed ammonium in tropical soils. *J. Soil Sci.* 5: 264-274.

Roy, R.N. and Kanwar, J.S. 1979. Importance of P in balanced fertilisation. *Phosphorus in Agriculture* 76: 97-120.

Russell, E.W. 1973. Soil conditions and plant growth (ed). The English Language Book Society and Longman.

Saharan, N. 1992. Management of secondary and micronutrients in dryland cropping systems. In: *Dryland Agriculture in India - State of Art Research in India* (eds) L.L. Somani, K.P.R. Vittal and B. Venkateswarlu. Scientific Publishers, Jodhpur, pp. 352-373.

Saharan, N., Sharma, K.L., Das, S.K. and Srinivasa Rao, P. 1989. Importance of sulphur fertilization in sunflower in dryland alfisols. Paper presented at *National Seminar on Sulphur in Agriculture* held at UAS, Bangalore, 7-8 September.

Sahrawat, K.L. 1984. Effects of temperature and moisture on urease activity in semi-arid tropical soils. *Plant Soil* 78: 401-408.

Sahrawat, K.L., Mohanty, S.K. and Burford, J.R. 1987. Paper presented at the *National Symposium on Micronutrients in Soils and Crops* held at Punjab Agricultural University, Ludhiana, December 21-23, 1987.

Sakal, R. et al. 1985. Screening of fingermillet varieties for their relative response to zinc in calcareous soil. *J. Indian Soc. Soil Sci.* 33: 440-442.

Sample, E.C., Soper, R.J. and Racz, G.J. 1980. Reactions of phosphate fertilisers in soils. In: *The Role of Phosphorus in Agriculture* (eds) F.E. Khaswaneh, E.C. Sample and E.J. Kamprath. ASA, Madison, Wisconsin, USA, pp. 263-310.

Sankaran, N., Sennian, P. and Morachan, Y.P. 1977. *Madras Agric. J.* 64: 253.

Saxena, S.N. 1979. *Bull. Indian Soc. Soil Sci.* 12: 353.

Shailaja, S. and Sahrawat, K.L. 1990. Adsorption and desorption of phosphate in some semi arid tropical Indian vertisols. *Fer. Res.* 23: 87-96.

Shailaja, S. and Sahrawat, K.L. Phosphate buffering capacity and supply parameters affecting phosphorus availability in vertisols (in press).

Shailaja, K.L., Bajaj, J.C., Das, S.K., Rao, U.M.B. and Ramalingaswami, K. 1992 a. Nutrient transformations in soil due to addition of organic manure and growing crops. I. Nitrogen. *Fer.* 32 : 303-311 *Res.*

Sharma, K.L. and Das, S.K. 1992. Nitrogen and phosphorus management in dryland crops and cropping systems. In: *Dryland Agriculture in India - State of Art Research in India* (eds) L.L. Somani, K.P.R. Vittal and B. Venkateswarlu. Scientific Publishers, Jodhpur, pp. 305-350.

Sharma, K.L., Das, S.K., Neelam Saharan and Narayana Reddy, M. 1992 b. Soil testing to evaluate phosphorus needs of dryland crops raised on vertisols. *Rainfed Agriculture Research Newsletter* (1), CRIDA, Hyderabad, India, pp. 9-10.

Shukla, M.P., Shankar, H. and Pathak, R.K. 1983. *J. Indian Soc. Soil Sci.* 31: 517-520.

Singh, G. and Bhushan, L.S. 1979. Water use efficiency and yield of dryland chickpea as influenced by P fertilisation, stored soil water and crop season rainfall. *Agric. Water Management* 2: 299-305.

Singh, Mahendra 1983. Innovations in Zinc Fertilisers - Retrospect and Prospect. *Fert. News* 28(2): 17-26.

Singh, Nathu *et al.* 1985. Yield and water use efficiency of rainfed wheat + chickpea as affected by N and P application in clay loam, sandy loam and loamy sand soil. *Indian J. agric. Sci.* 55: 13-17

Singh, Ranjodh *et al.* 1983. Dryland cropping in submontane Punjab, 1971-81. Punjab Agricultural University, Ludhiana.

Singh, V.K. and Singh, R.P. 1985. *Plant and Soil* 87: 439-440.

Spratt, E.D. and Chowdhury, S.L. 1978. Improved

cropping systems for rainfed agriculture in India. *Field Crop Res.* 1: 103-126.

Subba Rao, I.V. 1975. Nutritional disorders of crops in Andhra Pradesh. Tech. Bull. 1, APAU, Hyderabad.

Swindale, L.D. 1982. Distribution and use of arable soils in the semi-arid tropics. Transactions, *12th Int. Cong. of Soil Sci.* Plenary Session Papers, New Delhi, pp. 67-100.

Takkar, P.N. 1987. *Proc. Symp. Fertiliser requirements and sources in developing countries of Asia and Pacific.* Bangkok, Jan. 26-30.

Takkar, P.N. and Nayyar, V.K. 1986. Integrated approach to combat micronutrient deficiency. *Proc. FAI Seminar on Growth and Modernisation of Fertiliser Industry*, PS III/2/1-16.

Tandon, H.L.S. 1986. Sulphur research and agricultural production in India. Fertiliser Development and Consultation Organisation, New Delhi.

Tandon, H.L.S. 1986 a. Sulphur research and agricultural production in India. 2nd Ed. FDCO, New Delhi, p. 76.

Tandon, H.L.S. 1986 b. Sulphur research and development in Indian agriculture. *Fert. News* 31(9): 9-16.

Tandon, H.L.S. 1987. Phosphorus in dryland agriculture. In: Phosphorus Research and Agricultural Production in India. Fertiliser Development and Consultation Organisation, C-110, Greater Kailash, New Delhi, pp. 73-84.

Tandon, H.L.S. 1988. Sulphur fertiliser for tropics and sub-tropics. *Proc. TSI-FAI Symp. Sulphur in Indian Agriculture*, New Delhi, S.IV/2/1-12.

Tandon, H.L.S. 1988. Available information on micronutrients in India - Some Observations. *FAI Meeting on Status of Micronutrients in Indian Agriculture* (unpublished).

Tandon, H.L.S. 1989. Management of fertiliser nutrients other than nitrogen in the semi-arid tropics of India. *Proc. of Colloquium, Soil Fertility and Fertilizer Management in Semi-Arid Tropical India*, held at ICRISAT, Patancheru, India, October 10-11, 1988, pp. 71-88.

Tandon, H.L.S. 1990. Fertilisers and sustainable agriculture. In: *Soil Fertility and Fertiliser Use.* Vol.IV, IFFCO, New Delhi, pp. 15-27.

Tandon, H.L.S. and Kanwar, J.S. 1984. A review of fertiliser use research on sorghum in India. Res. Bull. No. 8. International Crops Research Institute for the Semi-Arid Tropics, Patancheru, India.

Taylor, H.M. 1983. Managing root systems for efficient water use - An overview. In: *Limitations to Efficient Water Use in Crop Production* (eds) H.M. Taylor, W.R. Jordan and T.R. Sinclair. Am. Soc. Agron., Madison, Wisconsin, USA, pp. 87-114.

Tejuddin, E. 1971. *Agric. Res. J. Kerala* 8: 89-92.

Tisdale, S.L., Nelson, W.L. and Beaton, J.D. 1985. Soil Fertility and Fertilizers. Macmillan Publishing Company, New York.

Tiwari, K.N. 1990. Secondary Nutrients - Scope in 90's. In: *Soil Fertility and Fertiliser Use*, Vol.IV. IFFCO, New Delhi, pp. 95-117.

Umrani, N.K. and Patil, N.D. 1983. *Fert. News* 28(9): 57-61.

Velayutham, M. and Reddy, K.C.K. 1987. Balanced fertilisation for increasing fertiliser use efficiency. In: *Proc. IFIA, Regional Agricultural Meeting of the International Fertiliser Industry Association* held at New Delhi from Dec. 8-12, 1986.

Venkatasubbaiah, V., Venkateswarlu, J., Subba Rao, I.V. and Prabhu, P. 1986. *J. Indian Soc. Soil Sci.* 14: 222-224.

Venkateswarlu, J. 1987. *Adv. Soil Science* 7: 165-221.

Venkateswarlu, J., Das, S.K. and Bhaskara Rao, U.M. 1986. Phosphorus management for castor - sorghum + redgram intercropping rotation in semi-arid alfisol. *J. Indian Soc. Soil Sci.* 34: 799-802.

Venkateswarlu, J. and Spratt, E.D. 1977. Some suggestions for increasing fertiliser efficiency in dryland farming. *Fert. News* 22(12): 39-43.

Venkat Raju *et al.* 1987. Iron chlorosis of groundnut on alfisol. In: *Natn. Symp. Micronutrient Stresses in Crop Plants : Physiological and Genetical Approaches to Control them*, Rahuri, p. 109.

9

AGRONOMIC ASPECTS OF PLANT NUTRIENT MANAGEMENT IN RAIN DEPENDENT FOOD CROP PRODUCTION SYSTEMS IN INDIA

R.P. SINGH and S.K. DAS

INTRODUCTION

India is predominantly a rainfed country. Of the total geographical area (329 m ha), about 143 m ha is sown to crops annually. The rainfed area is 95 m ha which constitutes 68% of the net sown area (Table 1).

Table 1. India - current land use, 1989-90

1. Geographical area	:	328.73 m ha
2. Net sown area	:	139.52 m ha
3. Net irrigated area	:	45.14 m ha
4. Rainfed area	:	94.40 m ha
5. % rainfed area to net sown area	:	67.70

Source : Agricultural Statistics at a Glance, Directorate of Economics and Statistics, G.O.I., 1993.

According to the projections of the National Commission on Agriculture, India is likely to generate an irrigation potential of 113 m ha by the year 2025 AD. Despite this, about 47% of the total cultivated area will continue to be rainfed. This implies that the country will continue to grapple with the problems of rain-dependent agriculture for decades to come.

In 1989-90, the total area under foodgrains in India was 126.77 m ha, of which 65.8 % was rainfed. The per cent rainfed area of the total cropped area under fine cereals (rice, wheat), coarse cereals (sorghum, pearl millet, etc.), pulse crops (chickpea, pigeonpea, etc.) was 65.8, 90.7 and 90.5, respectively. About 94% area of the total area sown to sorghum and pearl millet - the two of the more important staple food crops of the dryland farmers, was rainfed (Table 2). The per cent coverage under all major rainfed oilseed crops put together was 80.2 - the rainfed groundnut and rapeseed-mustard occupying 85.5 and 44.9%, respectively (Table 3).

Table 2. Per cent rainfed area of the net cropped area under foodgrains in India, 1989-90

1. Area (m ha) under foodgrains	:	126.77
2. % coverage under irrigation	:	34.2
3. % coverage under rainfed	:	65.8
O Rice	:	54.8
O Wheat	:	21.6
4. % rainfed area under coarse grains	:	90.7
O Sorghum	:	94.6
O Pearlmillet	:	93.8
5. % rainfed area under pulse crops	:	90.5
O Chickpea	:	80.2
O Pigeonpea	:	95.6

Source : Agricultural Statistics at a Glance, Directorate of Economics and Statistics, G.O.I., 1993

Table 3. Per cent rainfed area of the net cropped area under oilseed crops in India, 1989-90

1. Area (m ha) under oilseed crops	:	22.80
2. % coverage under irrigation	:	19.8
3. % coverage under rainfed	:	80.2
O Groundnut	:	85.5
O Rapeseed-mustard	:	44.9

Source : Agricultural Statistics at a Glance, Directorate of Economics and Statistics, G.O.I, 1993

CLIMATIC CONDITIONS

According to rainfall amount received per annum, the country has three broad zones, viz., dry (0-750 mm), medium (750-1150 mm), and assured (more than 1150 mm), occupying among themselves 30.0, 42.0 and 28.0 per cent of the total area (Table 4). South-West monsoon contributes the highest (73.7%) to the annual rainfall, followed by post-monsoon (13.3%), pre-monsoon (10.4%) and north-east monsoon (2.6%) (Table 5).

Table 4. Distribution of area according to annual rainfall in India

Rainfall amount		Approximate percentage of area receiving rainfall
Dry	0-750 mm	30.0
Medium	750-1150 mm	42.0
Assured	1150-2000 mm	20.0
	Above 2000mm	8.0
	Total	100.0

Source : Fertilizer Statistics, 1992-93, Fertiliser Association of India, New Delhi.

Table 5. Distribution of annual rainfall according to seasons in India

Rainfall	Duration	Approx. % of annual rainfall
S.W. monsoon	June-Sept.	73.7
Post-monsoon	Oct.-Dec.	13.3
Winter or NE monsoon	Jan.-Feb.	2.6
Pre-monsoon	Mar.-May	10.4
Total		100.0

Source : Fertiliser Statistics, 1992-93. Fertiliser Association of India, New Delhi.

The distribution of net sown area (143 m ha) according to rainfall amount received during the year is given in Table 6. About 51 m ha of the net sown area falls in the rainfall range of less than 350 mm to 750 mm. About the same acreage (51 m ha) experiences an average annual rainfall of 750 mm to 1125 mm. The net sown area having more assured rainfall (more than 1125 mm) is 41 m ha.

Table 6. Distribution of net sown area according to rainfall amount received during the year - India

Rainfall (mm)	m ha
< 350	19
350 - 750	32
750 - 1125	51
> 1125	41
Total	143

Source : Fertiliser Statistics, 1985-86. Fertiliser Association of India, New Delhi

AGRO-ECOLOGICAL FEATURES AND CROP PRODUCTIVITY

In areas receiving less than 400 mm average annual rainfall, sierozemic soils (Aridisols) predominate. Among crops, grasses, pearl millet, short duration pulse crops like green gram, black gram, cowpea, mothbean, maturing within 75 days are the important rainfed crops grown. Mean annual temperatures range between 20° and 30°C (Table 7). Sierozemic (Aridisols), red (Alfisols and related soils) and black soils (Vertisols and vertic inceptisols) are commonly encountered in regions receiving between 400 and 1000 mm average annual rainfall. Pearl millet, sorghum, maize, oilseed and pulse crops, and cotton are the most common rainfed crops grown. The length of the growing season ranges between 80 and 200 days. The mean annual temperature ranges between 20° to 30°C. In more assured rainfall regions (rainfall between 1000 to 1800 mm), alluvial (Entisols), black (Vertisols), and laterite soils (Alfisols and related soils) predominate. With equitable temperatures (20-30°C), the growing season is longer (200-300 days), and wide variety of crops are grown, including maize, paddy, wheat, barley, mustard, soybean, etc. In the sub-montane (Inceptisols)

Table 7. Agro-ecological conditions in different rainfall zones of India

Mean annual rainfall in mm	Predominant soil type(s)	Mean annual temperature	Length of the growing season under rainfed conditions days	Predominant rainfed crops
< 400 mm	Sandy soils (Aridisols)	20-30°C 0-10 °C	30-80 30-90	Grasses Pearlmillet Short duration pulses
400-1000 mm	Sandy soils (Aridisols) Red soils (Alfisols and related soils) Black soils (Vertisols and vertic Inceptisols)	20-30°C	80-200	Pearlmillet Sorghum Maize Oilseeds Pulses Cotton
1000-1800 mm	Alluvial soils (Entisols) Laterite soils (Alfisols and related soils) Black soils (Vertic and vertic Inceptisols)	20-30°C	200-300	Maize Paddy Wheat Barley Mustard Soybean
> 1800 mm	Sub-montane soils (Inceptisols)	15-20°C	More than 300	Paddy Plantation crops

regions experiencing an average annual rainfall of more than 1800 mm, having more than 300 days of growing season, paddy and plantation crops predominate.

The average productivity of pearl millet ranges between 0.2 to 0.8 t/ha in arid to 0.3 to 1.1 t/ha in semi-arid zones. Sorghum yields are 0.4 to 0.9 t/ha in the semi-arid zones. The average productivity of maize ranges between 1.0 and 2.5 t/ha in the semi-wet (1000 to 1800 mm) rainfall zone. The average productivity of paddy ranges between 0.9 and 2.2 t/ha in semi-wet to 1.0 to 2.0 t/ha in wet rainfall areas (Table 8).

Table 8. Crop productivity in different rainfall zones of India.

Mean annual rainfall (mm)	Crop(s) grown under rainfed conditions	Average productivity (t/ha)
< 400 mm	Pearl millet	0.2 to 0.8
400-1000 mm	Pearl millet	0.3 to 1.1
	Sorghum	0.4 to 0.9
	Groundnut	0.6 to 1.2
1000-1800 mm	Maize	1.0 to 2.5
	Paddy	0.9 to 2.2
	Wheat	0.9 to 2.5
> 1800 mm	Paddy	1.0 to 2.0

There are intra-state and intra-region differences as well. The irrigated crops yield two to three times more per unit area than the rainfed crops in the same region/state. In Gujarat, the groundnut bowl of the country, the average productivity is 0.86 t kernels/ha under unirrigated conditions as compared to 1.12 t kernels/ha under irrigated conditions (Table 9). More clearer picture of productivity of some of the important food grain crops grown in rainfed and irrigated areas emerges from Table 10. On an average, irrigated areas given three times more productivity of food grain crops than the unirrigated areas.

Table 9. Productivity of kharif groundnut under irrigated and unirrigated conditions in India, 1981-82

State	(t kernels/ha)	
	Irrigated	Unirrigated
East (Orissa)	0.82*	0.44*
North (Punjab)	1.00	0.89
South (Karnataka)	1.75	1.51
West (Gujarat)	1.12	0.86

Note : *Figures relate to the year 1979-80

Source : Fertiliser Statistics, 1985-86, Fertiliser Association of India, New Delhi

Table 10. Productivity of some foodgrain crops in rainfed and irrigated areas - 1980-81

Crop	Rainfed (t/ha)	Irrigated (t/ha)
Rice	0.50	2.42
Wheat	0.74	2.02
Coarse grains	0.67	1.05
Pulses	0.43	0.80
Total foodgrains	0.60	2.00

Source : Sampath, T.V. 1988

POPULATION, POPULATION DENSITY AND FOOD INTAKE

There has been phenomenal growth in India's population over the years. The population which stood at 397.3 million in 1956 rose to 690.1 million in 1981 and 781.4 m in 1987 (Fig. 1). The population was close to 800 m at the end of 1988. The net availability of foodgrains was 52.42 mt in 1956, 104.85 mt in 1981, and about 122.47 mt in 1987, inclusive of net production and imports. The per capita net availability per day of cereals and pulses is given in Table 11. The per capita availability of cereals has increased from 360.5 g/day in 1956 to 434.2 g/day in 1986. The availability of pulses has been steadily decreasing, being 42 g/capita/day in 1986 as against 70 g/capita/day in 1956. This indicates that production of pulses has not matched the population growth.

Table 11. Per capita net availability of cereals and pulses.

Year	Population	Per capita availability/day (g)		
		Cereals	Pulses	Total
1956	397.3	360.5	70.4	430.9
1966	493.2	360.0	48.2	408.2
1976	617.2	373.8	50.5	424.3
1986	766.1	434.2	41.9	476.1

Source : Fertiliser Statistics, 1987-88, Fertiliser Association of India, 1988.

About 68% of India's total population (800 m) lives in SAT (Semi-arid Tropics) states. The population density per cultivated sq km in SAT states is about half (335/km) than that of non-SAT states (675/km). Among the SAT states, the highest population density per cultivated sq km is in Tamil Nadu (651) and the lowest is in Rajasthan (173).

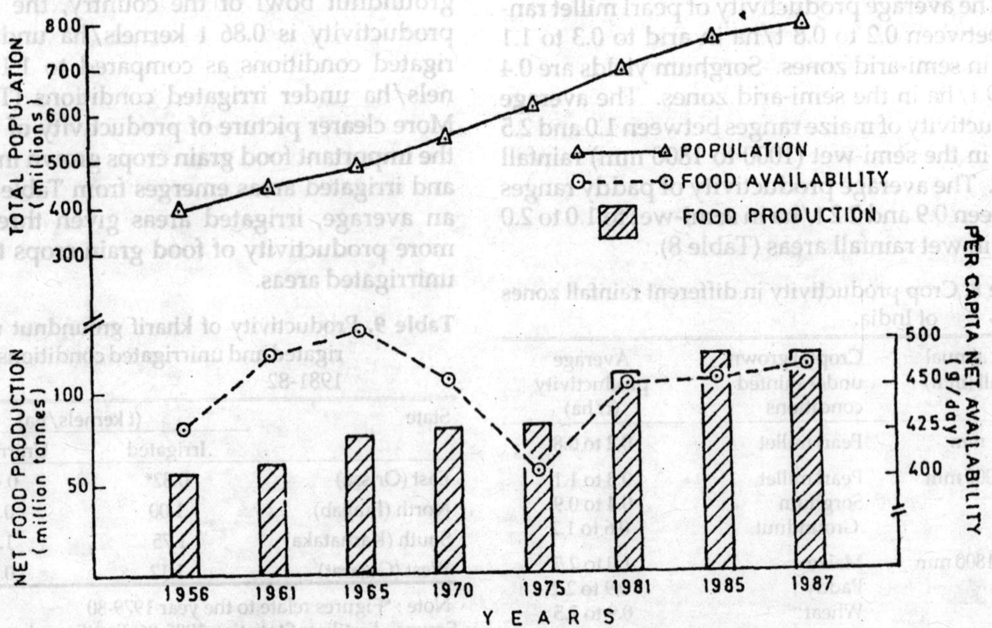

Growth in population and food supply in India (1956–87)

Growth in population and food supply in India (1956–87)

The consumption (g/day) of various kinds of foods is given in Table 12, which shows that except cereals, the daily per capita food consumption is much below the recommended levels. The discrepancy is more glaring in case of pulses, green leafy vegetables, fruits, milk, fats/oils, meat and fish etc. The data on per capita availability of foodgrains, food and calories intake per person per day for people living in unirrigated and irrigated areas, are not separately available. However, the average intake of nutrients per capita per day in different states (Table 13) revealed that as against the suggested allowance of 2400 calories per capita per day, the intake of nutrients was 2040 calories in Andhra Pradesh and 2779 calories in Madhya Pradesh. Among the predominantly rain-dependant agricultural states in India, Gujarat and Tamil Nada were having low average intake of 1612 calories and 1498 calories per capita per day, respectively. In India, the calories intake per capita per day is less than two-thirds than that of more advanced countries like UK, USA, and USSR (Table 14).

Table 12. Daily per capita food consumption and recommended levels

Food materials	Consumption (g)	Recommended level (g)	
		ICMR*	Task force on nutrition
Cereals	434	412	376
Pulses	34	60	64
Green leafy vegetables	21	125	116
Other vegetables	71	50	61
Roots and tubers	-	87	69
Fruits	10	30	40
Milk	69	100	189
Fats/oils	12	40	39
Meat and fish (and egg)	14	30	23
Eggs	-	30	15
Sugar/jaggery	19	35	38

*Indian Council of Medical Research

Source : Gopalan et al., 1971; Gopalan and Narasinga Rao, 1971; NCST, 1972

Table 13. Average intake of nutrients in different states (1968-69)* (per capita per day)

State	Calori-es	Protein (g)		Total fat (g)
		Total	Animal	
Andhra Pradesh	2,040	53	5	21
Bihar	1,865	56	3	13
Gujarat	1,612	54	2	13
Jammu & Kashmir	2,265	63	1	25
Kerala	1,842	47	11	25
Madhya Pradesh	2,779	98	7	45
Maharashtra	2,281	68	6	30
Mysore (Karnataka)	2,220	66	3	18
Punjab	2,832	84	16	52
Rajasthan	2,044	77	5	34
Tamil Nadu	1,498	36	4	12
Uttar Pradesh	2,307	66	5	32
West Bengal	1,927	48	7	24
All India	1,985	55	6	24
Suggested allowance	2,400	44	–	–

*Reports of nutrition work done in States (1968-69), National Institute of Nutrition, Indian Council of Medical Research, Hyderabad; Recommended daily allowances of nutrients and balanced diets recommended by the Nutrition Expert Group ICMR (1968) : Nutrition activities carried out in the States and Union Territories compiled by Nutrition Cell of the Directorate General of Health Services, New Delhi

Source : Report of the National Commission on Agriculture 1976, Part II - Policy and Strategy, p. 132

FOODGRAIN REQUIREMENT, GAP AND IMPORT

India's foodgrain requirement was 150 to 160 m t in 1985-86 with a population of about 750 million. As against this requirement, the production of foodgrains was 148-149.5 million tonnes - a figure quite close to requirement. Despite this, Government of India imported ten thousand tonnes of foodgrains with a view to meeting the gap, bringing stability in market price, and maintaining a buffer stock (Table 15). The estimated food requirement of India in 2000 AD has been worked out at 225 million tonnes. With the current trend of production, it is expected that the production of foodgrains during the same period would be around 230 million tonnes (Table 16) and even more.

Table 14. Calories per caput per day

Country	Total				By major commodities		
	1961/63	1969/71	1979/81	1983/85	Cereals	Roots & plantains	Livestock & other
World	2316	2450	2599	2665	1407	155	1103
Africa	2045	2097	2150	2051	945	507	599
China	1713	1976	2289	2564	1885	158	521
India	2037	2022	2104	2161	1382	43	736
Pakistan	1704	2027	2222	2185	1333	9	843
Latin America	2383	2517	2678	2703	1137	160	1416
Thailand	2011	2258	2405	2440	1639	63	738
USSR	3172	3349	3385	3403	1326	210	1867
UK	3355	3337	3174	3130	784	210	2136
USA	3263	3467	3605	3652	781	116	2755

Source : FAO, 1988

Table 15. Food requirement, production, gaps and imports in India during 1985-86

Commodity	Food requirement (m t)	Food production (m t)	Gap	Imports (000' t)
Foodgrains (cereals+ pulses)	150-160	148-149.5	–	10

Source : Agricultural Situation in India, 1986

Table 16. Supply and demand balance in foodgrains in 2000 AD

Commodity	Estimated food requirement (m t)	Expected production (m t)	Gap	Imports (m t)
Foodgrains (cereals+ pulses)	225	230	–	–

Source : Indian Agriculture, H.R. Arakeri. (1982)

CROPS AND CROPPING SYSTEMS IN DRYLANDS

Crops and varieties that are efficient utilizers of rainfall and stored soil moisture have been identified for different regions. The yield potential of newly identified efficient crops, as compared to traditional crops, is set out in Table 17.

Rain-dependent lands are generally monocropped. Crops are grown either during the monsoon season (June-September) or on conserved soil moisture in the post-rainy season (October-March). The productivity of monsoon season crops is unstable, while that of post-rainy season crops is low. Efficient utilisation of resources and stability in yield are best achieved through intercropping systems. Double cropping is possible where rainfall is more than 650 mm per annum and the moisture storage capacity of the soils is about 30 cm per metre of profile depth.

Table 17. Relative yields of traditional and efficient crops grown on different soils

Soil	Traditional crop	Yield (t/ha)	Efficient crop	Yield (t/ha)
Vertisols	Cotton	0.20	Sorghum	2.67
Entisols	Wheat	0.86	Chickpea	2.85
Alfisols	Upland rice	2.88	Maize	3.36
Vertisols	Greengram	1.18	Soybean	3.33
	Wheat	1.12	Safflower	2.42
Entisols	Wheat	1.03	Mustard	2.04
Aridisols	Wheat	0.32	Eruca sativa	1.61
Vertic Inceptisols	Maize	1.80	Hybrid sorghum	2.90

Source : Singh, S.P. (1987)

The efficient food crop-based intercropping and double cropping systems identified for different rain-dependent regions of India are presented in Tables 18 and 18 (a), respectively. The water availability period varies from less than 110 days to more than 250 days in different parts of the country. In drier regions having light textured soils (Aridisols), pearlmillet and clusterbeans are the predominant crops grown, which are intercropped in 2:1 row ratio with either cowpea or greengram or blackgram. The LER of such a system ranges between 1.00 to 1.39. On heavier soils, groundnut is intercropped with pigeonpea. The LER of such a system ranges between 1.51 to 1.87 (Table 18). Pearlmillet or sorghum or clusterbean go very well with greengram, mothbean or pigeonpea in regions having moisture availability periods of 110 to 175 days, the LER being 1.40 or so. In regions where moisture availability period is 175 days or more, there is a wider choice of principal crops available (pearlmillet, groundnut, sorghum, maize, pigeonpea, soybean, rice, etc.) for intercropping with pigeonpea as the first choice and blackgram, sesamum, as the second choice, the LER ranging between 1.16 to 1.93.

Table 18. Efficient food crop-based intercropping systems grown during rainy season for different rain- dependent regions of India (based on water availability periods)

Water availability period (day)	Base crop	Inter crop	LER
Upto 110	Pearlmillet/ clusterbean	Cowpea/gre engram/ blackgram	1.00-1.39
	Groundnut/ pearlmillet	Pigeonpea	1.51-1.87
110-150	Pearlmillet/ sorghum/ clusterbean	Greengram/ pigeonpea	1.17-1.71
150-175	Pearlmillet	Pigeonpeal mothbean	1.33-1.55
175-200	Pearlmillet/ groundnut/ sorghum	Pigeonpea	1.16-1.53
	Maize/ pigeonpea	Blackgram	1.59-1.70
	Pigeonpea	Sesamum	1.29
200-250	Sorghum/ pearlmillet/ maize	Pigeonpea	1.42-1.93
	Soybean	Pigeonpea	1.61
>250	Rice/finger-millet/soybe an/maize	Pigeonpea	1.40-1.81

Table 18(a). Efficient food crop-based intercropping systems grown during post-rainy season for different rain-dependent regions of India (based on water availability periods)

Water availability period (days)	Base crops	Intercrops	LER
150-175	Chickpea	Barley/ linseed/ mustard/ safflower	1.01-1.32
175-200	Chickpea	Safflower	1.23
	Barley	Mustard	1.17
200-250	Chickpea	Mustard	1.43
> 250	Wheat/ barley	Chickpea/ mustard	1.16-1.25
	Chickpea	Mustard	1.22

In the post-rainy season, chickpea, wheat, barley are the principal crops grown on conserved soil moisture. With chickpea, crops like barley, linseed, mustard, safflower are intercropped in varying row ratios, depending on soil moisture storage and rainfall received during the growing season. The LER of these systems is around 1.20.

Double crop systems are more common in regions having moisture availability period of 175-200 days and more. A wide variety of crops are grown, including short season (70 days) pulse crops (greengram, blackgram, cowpea), cereal crops (pearlmillet, maize, rice) maturing in 95-100 days, and sesame (110 days). Short duration pulse crops are followed by *rabi* sorghum, safflower, barley, mustard, etc. while cereal/oilseed crops are followed by chickpea, mustard or wheat. Sorghum/safflower double cropped system is more common in better water availability-endowed regions (200-250 days). Soybean in rainy season followed by safflower or wheat in post-rainy season is common. In wetter areas (moisture availability of 250 days), rice is more common, although groundnut, soybean, also find a place in the double crop systems (Table 19).

Table 19. Major efficient food crop-based double cropping systems for different rain-dependent regions of India (based on water availability periods)

Water availability Period (day)	Rainy season crop	Post rainy season crop
110-150	Cowpea/ blackgram	Safflower/ chickpea
	Soybean	Mustard/ safflower
	Greengram	Mustard
	Pearlmillet	Chickpea/ barley
150-175	Greengram	Sorghum
175-200	Cowpea	Sorghum
	Greengram	Safflower
	Blackgram	Barley/mustard
	Pearlmillet	Chickpea
	Maize	Wheat/wheat + chickpea/ mustard
	Rice	Chickpea/ wheat
	Sesame	Chickpea
200-250	Sorghum/ groundnut/ maize	Safflower
	Soybean	Wheat
	Soybean/maize	Safflower/ chickpea
> 250	Rice/maize/ fingermillet/ groundnut	Wheat/ chickpea/linseed /lentil/ horsegram/ barley
	Soybean	Fingermillet
	Pearlmillet	Wheat

UNTAPPED YIELD POTENTIAL

The potentiality of increasing productivity of rainfed crops with improved management is great. The state average, ORP average, and JNKVV Indore Research Farm average in respect of the important rainfed crops grown in Madhya Pradesh is presented as an example in Table 20. The potential is one and one-half times in case of post-rainy season crops and two to three times in case of rainy season rainfed crops. Productivity yield-gaps in case of cereals grown under irrigated conditions are anywhere between 1.15 t/ha to 2.80 t/ha in national average and national demonstrations (Table 21). Whereas the yield gaps in case of rainfed crops are of the order of 0.62 to 0.81 t/ha, in case of fine cereals, 0.31 t/ha in case of coarse cereals, 0.40 t/ha in case of pulse crops, and 0.26 t/ha in case of edible oilseed crops (Table 22).

Table 20. Average yield of crops (t/ha)

Crop	Madhya Pradesh (State average)	Operational Research Project	JNKVV Indore Research Farm
Sorghum	0.74	2.30	3.00
Maize	1.11	2.75	3.50
Soybean	0.50	1.22	2.00
Groundnut	0.66	1.20	1.50
Wheat	0.71	0.96	1.25
Chickpea	0.60	0.97	1.50

Table 21. Yield gaps in different crops under irrigated conditions

Crop	Average yield (t/ha)		Yield gap (t/ha)	Range of gaps in different states
	National demonstration	National average		
Paddy	5.27	2.47	2.80	1.24-3.59
Wheat	3.63	1.62	2.01	1.46-4.15
Pearlmillet	2.62	1.11	1.51	1.72-2.80
Sorghum	3.30	1.39	1.41	1.07-3.95
Maize	3.91	1.50	2.41	1.44-5.23

Source : Directorate of Economics and Statistics, 1973

Table 22. Yield gaps in dryland agriculture

Crop	Average yield (t/ha)		Yield gap (t/ha)
	National demonstration	National average	
Cereals			
Kharif	1.40	0.78	0.62
Rabi	1.73	0.92	0.81
Millets	0.92	0.61	0.31
Pulses	0.75	0.35	0.40
Oilseeds			
Edible oilseeds	0.60	0.34	0.26
Castor	0.52	0.23	0.29

Source : ICRISAT, 1980

AGRONOMIC ASPECTS OF PLANT NUTRIENT MANAGEMENT

It has to be recognised that fertilizer management is one of the important components of the several crop and soil management practices required for obtaining high and sustained productivity in rainfed lands. Some of the pre-requisites for judicious fertilizer management in rainfed lands are available nutrients and moisture status of the soil at sowing, responsive genotypes, timely sowing, optimum plant stand per unit area, appropriate plant protection and efficient and timely weed control. For example, response of pearlmillet to nitrogen application at 25 kg/ha increased from 8.6 kg to 18.5 kg grain/kg N, when the traditional method of pearlmillet cultivation was substituted by the improved technology (Table 23). Low native soil fertility is the characteristic feature of dryland soils. They are low to medium in phosphorus and generally adequate in potassium at their present level of productivity. Some humid Oxisols are responsive to lime particularly for the food legumes and peanuts. With continuous high crop management, zinc deficiency has been prominent. Sulphur deficiency has surfaced in many soil types.

Table 23. Response of pearlmillet to nitrogen under two management systems on a Vertic Inceptisol

Kg N/ha	Traditional method		Improved technology	
	Grain yield (t/ha)	Response (kg grain /kg N)	Grain yield (t/ha)	Response (kg grain /kg N)
0	0.67	-	1.07	-
25	0.88	8.6	1.53	18.5
50	1.37	14.1	1.98	18.1
Mean	0.98	-	1.53	-

1. Local variety, narrow spacing, no control of population, traditional weed control
2. BJ-104, 45 cm spacing, 15 plants/m^2, clean plots for 30 days

Source : Randhawa, N.S. and Singh, R.P. (1983)

Soil depth in case of Vertisols and related soils and soil texture in case of Alfisols, Entisols, Inceptisols, determine, by and large, the moisture storage capacity of the soils. The organic matter content and soil mineral constituents determine the moisture release characteristics of soils besides other factors. The moisture storage capacity of the broad soil groups found in rainfed areas of India are set out in Table 24 (Randhawa and Singh, 1983).

Table 24. Broad soil groups and their moisture storage capacities in rainfed areas of India.

Broad soil group	Sub-group (based on soil depth)	Moisture storage capacity (mm)
Vertisols and related soils	Shallow to medium (upto 45 cm)	135-145/45 cm
	Medium to deep (45–90 cm)	145-270/90 cm
	Deep (> 90 cm)	300/m
Alfisols and related soils	Shallow to medium	40-70/45 cm (sandy loam)
	Deep	70-100/45 cm (loam)
		180-200/90 cm
Aridisols	Medium to deep	80-90/90 cm
Inceptisols	Deep	90-100/m (loamy sand)
		110-140/m (sandy loam)
		140-180/m (sandy loam)
Entisols	Deep	110-140/m (sandy loam)
		140-180/m (loam)

Source : Randhawa, N.S. and Singh, R.P. (1983)

Different management practices have been developed for increasing *in situ* moisture conservation for dryland soils of India (Singh and Das, 1988). Higher moisture status in the profile provides a favourable environment for crop growth and ultimately leads to increased response to applied nutrients.

Mechanical structures like contour/graded bunds and bench terraces are the pre-requisites in soil and water conservation in sloppy agricultural lands. However, to have uniform and higher recharge of moisture in the profile, interterrace land treatments have been found to be effective. Such treatments are tillage, mulching, dead furrows, flat-on-grade and ridging later, broad bed and furrows, graded border strips, inter-plot and inter-row water harvesting. The main features of these are described below :

On an average, 10-11% of the total rains are received during the pre-monsoon season. Adoption of proper tillage practices helps the rain water to enter into the soil profile. Besides this, off-season tillage is needed for weed control. Shallow preparatory tillage was found to be beneficial in case of Alfisols and related soils. There was an increase in the yield of sorghum by 43% over the base yield of 1.43 t/ha where direct seeding was restored to. The effect was, however, more pronounced in sub-normal rainfall years. However, excessive tillage could be detrimental to Aridisols due to soil loss in wind erosion. Hard pan below the plough layer and textural profiles having high clay content (due to illuviation) prevent moisture infiltration. Under such situations, deep ploughing (> 22 cm) helps to increase crop yield, the effect being perceptible during the next two years. Deep tillage is also effective in controlling perennial weeds.

Mulching provides physical barrier in movement of moisture from soil into atmosphere. Experiments conducted at several locations have revealed that use of organic mulch in *rabi* (post-rainy season) crops increase crop yield to the tune of 30%. Sorghum vertical mulch (placing sorghum stubbles in trenches of 40 cm deep, 15 cm wide and protruding 10 cm above the ground level) enhanced available soil moisture by 4-5 cm and increased grain yield of sorghum by 40-50% in Vertisols of Bellary (Karnataka).

Making of 'idle furrows', generally referred to as dead furrows, at 3.6 m interval in groundnut helps in increasing grain yield by about 10% in arid Alfisols of Anantapur, A.P.

Sorghum and pigeonpea grown in semi-arid Alfisols of Hyderabad gave higher yield by 19 and 24%, respectively over control when the crops were sown as intercrops on a grade and ridging was done later.

Broad bed and furrows with a bed of 100 cm and 50 cm of shallow furrow which has a depth of 15 cm is useful in improving moisture regime and draining the excess water under situations where water stagnation is a problem. This situation arises in high rainfall Vertisol areas. Sorghum yield increased by 26% at Akola (Maharashtra) under this system over flat sowing.

Graded border strips of 130 x 11 m with 0.1% grade were found to be effective in deep Alfisols with medium rainfall. Fingermillet yield was increased by 42% due to this treatment at Bangalore (Karnataka).

This practice consists of growing maize and rice in ridge-furrow system in Inceptisols of sub-humid areas. Maize is taken on the ridges and water requirement of rice is met by the water harvested in the furrows. Maize and rice yield was increased by 17 and 21%, respectively over control.

In arid areas having Aridisols, pearlmillet yield was increased by 50% by inter-plot water harvesting of 1:1 of cropped to uncropped land (the slope of uncropped land being on both sides), and inter-row water harvesting with a treatment of ridges with runoff inducing substances like tank silt or polyethylene.

The recommended inter-terrace land treatments are summarized in Table 25.

FERTILIZER AND MOISTURE RELATIONSHIP

The utility of inorganic fertilizers in increasing the productivity of dryland crops has been well established through large number of experiments conducted by All India Coordinated Research Project for Dryland Agriculture. The responses of different rainfed crops to nitrogen and phosphorus application are set out in Tables 26 and 27. It is evident that both rainy and post-rainy season crops responded to application of nutrients (N and P), the level of response being higher in good rainfall years.

Table 25. Interterrace land treatments for different agro- climatic regions

Rainfall pattern	Suggested land configuration options
Arid (< 400 mm)	- Inter-plot water harvesting of 1:1 of cropped to uncropped land (slope of micro-catchment on both sides)
	- Inter-row water harvesting
	- Dead furrows at 3.6 m intervals (suitable for Alfisols)
Semi-arid (400-1000 mm)	- Contour farming (cultivation and sowing)
	- Sowing across the slope and ridging the land between rows
	- Graded bunds on grade of 0.2 to 0.3% with provision of drainage
	- Graded border strips (130 x 11 m) with 0.1% grade (for deep Alfisols)
	- Compartmental bunds for raising crops on conserved soil moisture (Vertisols and Vertic Inceptisols)
Semi-wet (1000-1800 mm)	- Broad beds and furrows (Vertisols having problem of drainage)
	- Ridges and furrows
	- Graded furrows at a grade of 0.2 to 0.3%
Wet (about 2000 mm and above)	- Raised and sunken beds
	- Inter-plot water harvesting (growing high water requiring crops in beds)

Source : Singh, R.P. (1988)

Table 26. Response of rainy and post-rainy season crops to nitrogen under dryland conditions

Rainy season		Post-rainy season	
Crop	Range of response (kg grain/kg N)	Crop	Range of response (kg grain/kg N)
Sorghum	3.4-43.4	Wheat	4.9-52.3
Pearlmillet	2.1-24.8	Barley	4.0-31.6
Fingermillet	5.0-42.4	Oats	8.3-20.0
Maize	4.1-67.4	Chickpea	1.9-23.5
Rice	4.5-33.9		
Setaria	5.9-17.9		

Source : Rao, A.C.S. and Das, S.K. (1982)

Table 27. Response of rainy and post-rainy season crops to phosphorus under dryland conditions

Rainy season		Post-rainy season	
Crop	Range of response (kg grain/kg P)	Crop	Range of response (kg grain/kg P)
Sorghum	2.4-59.0	Wheat	2.6-41.3
Pearlmillet	1.7-14.3	Barley	1.9- 9.2
Fingermillet	6.4-38.9	Chickpea	4.5-38.2
Clusterbean	6.7-10.7	Peas	3.7-10.9
Rice	5.1-31.8	Lentil	3.2- 6.3
Maize	6.8-80.0		
Greengram	1.5-11.6		
Blackgram	1.8- 6.7		
Pigeonpea	3.1- 8.3		

Source : Rao, A.C.S. and Das, S.K. (1983)

Soil moisture and fertilizer interaction was found to be quite prominent in dryland agriculture. Higher yield of post-monsoon sorghum was obtained under deep soils as compared to shallow soils (Table 28). The deep soil had a moisture storage capacity of 110 mm. While it was 50 cm in case of shallow soils. Sorghum responded upto 25 kg N/ha in shallow soil and upto 50 kg N/ha in deep soil. Similarly, higher responses to applied nitrogen were obtained in sandy loam soils as compared to loamy sand soils using wheat and maize as test crops (Table 29).

Table 28. Response of *rabi* sorghum to nitrogen at different soil moisture availability on a Vertic Inceptisol)

Soil depth (cm)	Available moisture storage (mm)	Grain yield (t/ha) N (kg/ha)				SE	CD
		0	25	50	75		
0-30	50	0.66	1.23	1.30	1.30	0.111	0.327
0-60	110	1.66	1.78	2.13	2.33	0.120	0.333

Source : Randhawa, N.S. and Singh, R.P. (1983)

Table 29. Effect of N application on yield of wheat and maize in different soil types

Crop	N rate (kg/ha)	Grain yield (t/ha)	
		Loamy sand	Sandy loam
Wheat	0	1.27	2.11
	40	1.79	3.93
	80	1.98	3.45
	120	2.05	-
Maize	0	1.26	2.07
	40	2.17	2.89
	80	2.00	3.41
	120	1.97	3.12

Source : Randhawa, N.S. and Singh, R.P. (1983)

The response to applied nitrogen varied from 6.5 kg grain/kg nitrogen at Bellary (500 mm rainfall), 9.7 kg at Bijapur, 19.0 kg at Solapur and 27.7 kg at Kovilpatti in case of post-rainy season sorghum, thereby establishing a distinct relationship between rainfall received and responses to N application obtained (Table 30). Seed yield of chickpea increased with increase in available moisture and fertilizer application (Table 31). Water use by wheat was found to increase by fertilizer application and application of FYM to previous crop of maize as shown in Table 32. Mulching is useful for moisture conservation for post-rainy season crops. It has also been found to increase the response of applied fertilizers (Table 33). This is evidently due to favourable interaction between nitrogen and moisture.

Table 30. Response of *rabi* (post-rainy season) sorghum to application of nitrogen

Centre	Average amount of rainfall (mm)	N dose (kg/ha)	Grain yield (t/ha)	Response kg grain/kg N
Bellary	500	40	1.58	6.5
Bijapur	680	30	1.56	9.7
Solapur	722	30	1.45	19.0
Kovilpatti	730	30	2.32	27.7

Source : Randhawa, N.S. and Singh, R.P. (1983)

Table 31. Response of chickpea to fertilizer application under varying soil moisture conditions

Available moisture	Grain yield (t/ha)			Mean
	Control	50% of recommended fertilizer dose	Full recommended fertilizer dose	
25%	1.04	1.32	1.77	1.38
50%	1.27	1.57	2.33	1.73
75%	1.43	2.46	2.56	2.15
Mean	1.25	1.78	2.22	

Note : Recommended fertilizer dose : 20 kg N and 26 kg P/ha
CD at 5% main plot (moisture) 0.352
Sub-plot (fertilizer level) 0.384

Source : Singh, R.P. and Das, S.K. (1983)

Table 32. Water use of dryland wheat as affected by N-rates applied to wheat and FYM applied to previous maize

Soil	FYM to maize @ 15 t/ha	Water use (mm) by dryland wheat at N rates				
		0	40	60	80	Mean
		(kg/ha)				
Sandy Loam	No	256	281	292	321	287
	Yes	309	339	340	380	342
Loam	No	265	315	337	359	319
	Yes	332	367	381	398	370

Source : Kumar, Dinesh. Ph.D. Thesis, Department of Soils, Punjab Agricultural University, Ludhiana [1982], cited by Prihar et al. (1983)

Table 33. Yield and water use of dryland wheat following maize as affected by mulching the previous crop and N-rates

Soil	Average over years	No mulch			Mulch		
		0	40	80	0	40	80 (kg N/ha
		Yield (t/ha)					
Loamy sand	4	1.90	2.57	2.96	2.60	3.18	3.48
Sandy loam	3	2.52	3.56	3.99	3.69	4.49	4.72
		Water use (mm)*					
Loamy sand	4	221	246	259	247	270	272
Sandy loam	3	239	254	262	262	275	283

*Sum of water depletion from 180 cm profile plus seasonal rainfall

Source : Prihar et al. (1983)

Table 34. Variations of crop response to applied fertilizer with rainfall

Soil Order	Crop	Rainfall distribution		Yield without fertilizer (t/ha)	Kg grain/kg nutrient At 20-40 kg/ha	At 60-80 kg/ha
NITROGEN						
Aridisols	Pearl millet	Above normal	(3)	1.09	20.8	-
		Normal	(3)	0.75	13.7	-
		Below normal	(1)	0.25	6.3	-
Alfisols	Rice	Above normal	(2)	1.35	18.1	16.0
		Below normal	(3)	1.57	22.1	18.3
	Barley	Above normal	(1)	0.62	6.3	5.3
		Below normal	(2)	0.44	9.2	6.9
Inceptisols	Pearl millet	Above normal	(2)	2.31	7.3	-
		Below normal	(2)	0.77	6.4	4.9
Entisols	Barley	Normal	(1)	2.50	10.0	11.8
		Below normal	(2)	1.29	14.8	13.7
	Wheat	Above normal	(1)	0.87	15.3	11.9
		Normal	(1)	0.72	3.2	0.5
Vertic	Sorghum	Above normal	(1)	1.84	36.1	29.6
Inceptisols		Below normal	(1)	0.28	13.5	7.4
Entisols	Barley	Above normal	(1)	3.15	14.8	14.6
		Below normal	(1)	1.38	29.0	27.0
PHOSPHORUS						
Aridisols	Greengram	Above normal	(1)	1.32	2.7	1.8
		Normal	(1)	0.30	1.0	1.3
		Below normal	(1)	0.12	2.7	1.8
Vertic	Sorghum	Above normal	(1)	2.87	13.0	-
Inceptisols		Below normal	(2)	1.33	4.5	-
Inceptisols	Maize	Above normal	(1)	2.90	40.5	11.3
		Normal	(1)	1.93	7.2	0.1

(Figures in parentheses indicate number of seasons)
Source : Vittal, K.P.R. (1987)

The variation in crop responses to fertilizer application, according to amount of rainfall is presented in Table 34. It is evident that higher response was obtained in situations where rainfall was above normal. The response was poor in below normal rainfall years.

FERTILIZER USE EFFICIENCY

Fertilizer use efficiency in sorghum was increased by placement of nutrients at 5-10 cm below seed. For post-rainy season crops, deep placement of fertilizer has been found to be useful for increasing the fertilizer use efficiency (Singh and Das, 1984).

For monsoon season crops, fertilizer nitrogen has to be applied in splits. Under high rainfall situations split application of nitrogen increased the use-efficiency (Tables 35 and 36) of applied nitrogen. For low rainfall situations, fertilizer can be saved by split application in case the weather turns aberrant. Deep tillage increased fertilizer use efficiency in pearl millet by providing favourable crop growth environment (Table 37) (Singh and Vittal, 1989).

Table 35. Split application of N in maize grown on an Inceptisol

Treatments (kg N/ha) applied at Sowing	Knee-high	Tasseling	Grain yield (t/ha) 1976	1977
100	0	0	4.35	5.12
75	25	0	4.98	6.15
33	33	33	4.99	6.25
50	25	25	4.58	5.63
CD 5%			NS	0.826

Source : Randhawa, N.S. and Singh, R.P. (1983)

Table 36. Effect of time of nitrogen application on grain yield sorghum grown on Vertic Inceptisols

Treatments (kg N/ha) applied at			Grain yield (t/ha)			
Sowing	25 DAS	45 DAS	1972	1974	1975	Mean
80	0	0	1.26	0.77	2.50	1.51
60	0	20	1.66	1.09	2.84	1.86
40	20	20	1.57	1.09	2.89	1.85
Control (without N)			0.70	0.19	0.81	0.58
CD 5%			0.30	0.15	0.21	-

Source : Randhawa, N.S. and Singh, R.P. (1983)

Table 37. Effect of profile modification and nitrogen application on grain yield of pearl millet grown on Aridisol

N level (kg/ha)	Grain yield (t/ha)	
	Shallow (~12 cm)	Deep (>20cm)
0	0.90	1.38
20	1.13	1.68
40	1.31	1.87
80	1.46	2.00

(Average of three below normal seasons)

(< 320 mm rainfall)

$Y = 8.98 + 0.1339\,x - 0.00079x^2$ 84 kg

$Y = 13.82 + 0.1690 - 0.0011477x^2$ 73 kg

Source : Singh, R.P. and Vittal, K.P.R. (1989)

CONJUNCTIVE USE OF ORGANIC AND INORGANIC FERTILIZER

Use of sheep manure, farm yard manure, leaves and twigs of *Leucaena leucocephala* were found to be useful as supplement to fertilizer N for dryland crops (Singh et al., 1981; Narkhede and Ghugare, 1987). By using these materials 20-25 kg fertilizer N could be saved. Taking a legume crop in rotation/sequence, helps to save fertilizer N to the tune of about 20 kg N/ha. The efficient legumes are pigeonpea, blackgram, cowpea and peanut. Better results with blackgram were obtained when the crop was green manured after the first picking (Singh and Das, 1983). In such a case, blackgram stovers served as a source of organic nitrogen. Similarly, cowpea intercropped with wide spaced castor and turned into the soil as green manure after 45 days of growth led to a saving of 30 kg fertilizer N in castor (Venkateswarlu, 1984). Continuous use of on-farm residues helped to improve physico-chemical properties of Alfisols (Venkateswarlu, 1984). This is considered to be one of the useful methods not only for supplying the plant nutrients but also for improving the physical properties of the soils. Higher yield of maize was obtained by on-farm residue incorporation (Table 38) (Hadimani et al., 1982).

Table 38. Effect of soil incorporation of maize residue on the yield of subsequent crop

Year	Yield of maize grain (t/ha)	
	Residue incorporated @ 4 t/ha	No residue
1975	3.57	2.95
1976	1.14	0.75
1977	3.54	2.80
1978	2.52	2.23
1979	0.99	0.69
Mean	2.35	1.88

Source : Hadimani et al. (1982)

RESPONSE TO FERTILIZER APPLICATION IN INTERCROPPING SYSTEMS

The response of crops to nitrogen under sole and intercropping systems were studied. By and large, sole crop of sorghum, pearlmillet, maize and rice produced higher response to applied nitrogen than when these crops were intercropped. It was because of competition for moisture between component crops (Table 39). Similarly, response of pigeonpea to phosphorus application was drastically reduced under intercropping system (Table 40).

NUTRIENT MANAGEMENT IN DOUBLE CROPPING SYSTEMS

Application of FYM @ 15 t ha^{-1} increased the maize yield with and without nitrogen (Table 41). The increase in yield was to the tune of 12 per cent. Distinct advantage of residual effect of application of FYM to maize was observed in the grain yield of wheat in maize-wheat rotation

Table 39 Response of crops to nitrogen in sole and intercrop systems

Crop (intercrop)	Variety	Soil type	Sole crop system			Intercrop system		
			N dose (kg/ha)	Yield (t/ha)	Response (kg grain/ kg nutrient)	N dose (kg/ha)	Yield (t/ha)	Response (kg grain /kg nutrient)
Sorghum (Pigeonpea NPWR-15)	CSH-5	Vertic	15 60	1.52 2.04	11.7	15 60	1.58 2.03	10.2
Sorghum (Pigeonpea UPAS)	CSH-5	Vertic Inceptisols	30 75	1.85 2.90	23.2	30 75	2.20 2.66	10.2
Sorghum (Pigeonpea AS 71-37)	CSH-6	Vertisols	30 120	2.36 3.09	8.1	30 120	1.66 2.31	7.2
Sorghum (Pigeonpea C-11)	CSH-1	Vertisols	50 100	2.52 3.15	11.0	50 100	2.38 2.62	4.7
Pearl miller (Greengram)	WCC-75	Entisols	30 60	1.28 1.69	13.7	30 60	1.43 1.76	10.8
Maize (Blackgram T-9)	Ganga-5	Vertic Inceptisols	20 40	0.75 0.93	9.0	20 50	0.56 0.82	8.7
Mize (Pigeonpea BR-65)	Suwan-1	Alfisols	50 100	3.17 4.34	23.4	50 100	2.57 3.92	27.0
Maize (Pigeonpea BR-65)	GS-2	Alfisols	50 125	4.29 5.41	14.9	50 100	3.80 5.02	24.4
Rice (Pigeonpea R-60)	DR-92	Alfisols	30 60	1.53 1.63	3.1	30 60	0.89 0.98	3.1

Source : Bhaskar Rao, U.M. [1989] - Personal communication

Table 40. Response of crops to phosphorus in sole and intercrop systems.

Crop	Soil type	Sole crop system			Intercrop system		
		P dose (kg/ha)	Yield (t/ha)	Response (kg grain/ kg nutrient)	P dose (kg/ha)	Yield (t/ha)	Response (kg grain/ kg nutrient)
Pigeonpea NPWR 159	Vertic Inceptisol	0	0.67	—	0	0.25	—
Sorghum (CSH-5)		32	1.33	8.8	32	0.38	1.7
Pigeonpea (C-11) +	Vertisols	0	0.44	—	0	0.36	—
Greengram (Kopergaon)		35	0.64	2.5	35	0.56	2.5

taken on an Inceptisol (Table 42). Wheat yield was increased by 18% under the unfertilized conditions due to FYM application (@ 10 t ha^{-1}) to previous crop of maize. Nitrogen application increased wheat yield almost linearly. On an average, 4.5% increase in wheat yield was recorded as residual effect of FYM. Similarly, higher yield of safflower was obtained in a Vertisol due to FYM application either directly to safflower or applied to the previous crop of soybean.

Table 41. Response of maize to N application with and without FYM in maize-wheat sequence taken on an Inceptisols.

N level applied to maize (kg/ha)	Maize yield (t/ha)		
	FYM (t/ha)		
	0	15	Average
0	1.65	1.99	1.82
40	2.28	2.50	2.39
60	2.53	2.86	2.69
80	2.93	3.16	3.05
Average	2.35	2.63	—

Table 42. Response of wheat to N application and residual effect of FYM in maize-wheat sequence grown on an Inceptisols (average of 4 years)

N level (kg/ha) to wheat	Wheat yield (t/ha) FYM (t/ha) to maize		
	0	10	Average
0	2.38	2.61	2.49
40	3.10	3.23	3.16
60	3.36	3.41	3.39
80	3.51	3.67	3.59
Average	3.09	3.23	—

RECOMMENDED FERTILIZER MANAGEMENT PRACTICES

The fertilizer recommendations for major rainfed rainy season and post-rainy season crops are given in Tables 43 and 44. These recommendations have been summarized according to average annual rainfall, moisture storage capacity and native fertility status of the soil. In case of rainy-season cereal crops (sorghum, pearlmillet, maize, finger millet, upland rice), the recommended levels of N and P range from 40-80

Table 43 : Fertilizer recommendations for principal rainfed rainy season crops for different rain-dependent regions of India (based on average annual rainfall, moisture storage, and native fertility status.

Crop	Average annual rainfall	Moisture storage (mm)	Native fertility status *		Fertilizer recommendations (kg/ha)			Management practices
			N	P	N	P	K	
Pearl millet	300-400	80-90/90 cm	Low	Medium	40	8	0	
	600-700	80-90/90 cm	Low	Medium	75	11	0	
		135-145/45 cm	Low	Medium	50	13	20	Nitrogen to be applied in two splits, half at basal and half at heading. P and K all as basal.
Sorghum	600-800	40-70/45 cm	Low	Low	40	13	0	
		135-145/45 cm	Low	Medium	90	13	0	
	800-1000	140-270/90 cm	Low	Medium	80	17	0	
Finger millet	800-1000	180-200/90 cm	Low	Low	50	22	25	Nitrogen to be applied in two splits, of which half at 3 weeks after sowing. P and K as basal
	1400-1600	180-200/90 cm	Low	Medium	40	8	20	
Maize	800-1000	180-200/90cm	Low	Low	75	22	25	Nitrogen to be applied in two splits, half at basal and half at tillering
	1000-1200	90-100/m	Low	Medium	40	9	10	
		110-140/m	Low	Medium	80	18	20	
Upland rice	1000-1200	145-270/90 cm	Low	Medium	40	18	0	N in two equal splits, at sowing and at panicle initiation stage
	1400-1600	180-200/90 cm	Low	Medium	60	18	0	N in three splits ¼ at sowing/transplanting, 1/2 three weeks therafter and 1/4 at panicle initiation stage.
Clusterbean/ mothbean	300-400	80-90/90 cm	Low	Medium	0	13	0	All as basal
Horsegram/ blackgram/ cowpea pigeonpea	400-600	145-270/90 cm	Low	Low to medium	20	13	0	All as basal

* Low N = < 280 kg/ha ; P = < 10 kg/ha ; Medium N = 280-560 kg/ha ; P = 10-20 kg/ha

Table 44. Fertilizer recommendations for principal rainfed post-rainy season crops.

Crop	Average annual rainfall	Moisture storage (mm)	Native fertility status*		Fertilizer recommendations (kg/ha)			Management practices
			N	P	N	P	K	
Rabi sorghum	150-200	135-145/45 cm	Low	Medium	30-50	13	0	All to be drilled below seed at sowing
Wheat	40-80	110-140/m	Low	Medium	30-40	6-13	0	"
	100-150	90-100/m	Low	Medium	40	9	0	"
	150-200	110-180/m	Low	Medium	60-80	18	0	"
	200-250	140-180/m	Low	Medium	60	13	0	"
Barley	50-100	110-140/m	Low	Medium	60	13	0	"
	100-250	140-180/m	Low	Medium	40-80	9	10	"
Chickpea	60-70	110-140/m	Low	Medium	10	11	0	"
	100-150	145-270/90 cm	Low	Medium	20	17	0	"

kg and 8-22 kg/ha. For maize and finger millet 10-25 kg K/ha is also recommended. For pulse crops, 20 kg N and 13 kg P/ha are recommended.

For wheat, barley, *rabi* sorghum grown on conserved/residual soil moisture, 30-80 kg N and 6-18 kg P/ha are considered adequate to obtain fairly good productivity. In case of chickpea, 10-20 kg N and 11-17 kg P/ha are recommended. However, the exact amount of nutrients to be added will depend on rainfall, moisture storage and soil test values.

Based on the work carried out at the cooperating centres of the All India Coordinated Research Project for Dryland Agriculture (AICRPDA), fertilizer management practices have been developed for different food crop-based production systems operating in different parts of India (Randhawa and Singh, 1983). Important points that emerge from these recommendations are :

- The recommended doses of fertilizers are higher in areas where incident rainfall and moisture storage capacity of the soils are higher.
- By and large, crops respond to application of N and P, the response for K being sporadic in Alfisols and Inceptisols.
- A starter dose of N @ 15-20 kg ha^{-1} is recommended for leguminous pulse crops.
- N is recommended to be applied in splits, whereas P and K are to be applied as basal.
- Fertilizers are recommended to be placed below seeds all at sowing for post-rainy season crops.
- For acid soils, application of lime is recommended as amendment for crops like maize, blackgram, greengram and fingermillet. Paper mill sludge can, however, substitute lime as amendment.
- Inoculation with appropriate *Rhizobium* culture is recommended for pulses.

In dryland research, no serious attempt has been made to study the effect of fertilizer sources. Generally, easily mineralizable and water soluble materials like urea, single super phosphate and triple superphosphate have been used in the experiments.

SOCIAL ASPECTS (INCLUDING LABOUR REQUIREMENT)

Many socio-economic factors are known to govern the choice of crops, varieties, cropping systems and above all, a farming system in a given area, besides physical factors, like climate (including rainfall) and soil type. Socio-economic studies have shown that risk attitudes differ little between small and large farmers in semi-arid regions; nearly all of them are moderately risk averse (Binswanger, 1978). This suggests that not much can be gained by developing technologies with differential risk/return characteristics for small and large farmers. Emphasis has to be laid on evolving relatively profitable and stable technologies for all farmers. Support to small farmers via assured access to credit and input is yet another focus.

In studies on economics of recommended dryland technology vis-a-vis traditional technology it was revealed that improved technology provided additional employment to the extent of 25 per cent in case of pearlmillet, 25 to 50 per cent in case of sorghum, 31 per cent in case of fingermillet, and 21 per cent in case of barley (Reddy and Rastogi, 1985). It is suggested that additional labour was required in adopting improved technologies in dryland agriculture.

SMALL SCALE FARMERS - INPUT SUPPLY AND OUTPUT

In India, the small and marginal farmers who constitute over 72 per cent of the land holders operate barely 24 per cent of the agricultural land. The average size of operational holding in India has shrunk from 2.5 ha in 1970-71 to 2.0 ha in 1980-81 and further to 1.8 ha at present. It is primarily due to steadily growing population.

In a study it was revealed that there was no relationship between farm size and productivity in dryland farming regions and in some cases, an inverse relationship existed between these two variables (Rastogi and Reddy, 1982). Many a studies have revealed that irrigation as an input trebled the production of rice and wheat. Among the improved dryland technology crop production components, improved seed contributed 25 per cent to the productivity, the contribution of other factors being : fertilizer 18 per cent, deep ploughing 13 per cent, weeding 11 per cent, placement of seed and fertilizer 12 per cent, optimum plant population 6 per cent and other residual factors 12 per cent (Reddy et al., 1985). A direct relationship exists between inputs use and output (yield) irrespective of category of farmers (Reddy and Rastogi, 1982).

MARKETING, EXTENSION AND CREDIT

Food crops in dryland farming regions are grown mainly for subsistence. Of late, some dryland farmers in certain regions are growing crops for marketing. Availability of improved seeds has enabled this to happen. The 'marketed surplus' and 'marketable surplus' of dryland crops varied from farmer to farmer and season to season. Imperfect market conditions and channels existing in dry regions are responsible for marginal 'effective income'. Barbara Harriss (1982) reported that small and marginal farmers sell a greater proportion of their produce than middle and rich peasants, a proportion on par with that sold by rich, but again, marginal and small peasants purchase. 'Marketable surplus' is generally more in case of rich farmers. It is also observed that prices of food crops crash to very low/unremunerative levels when normal/favourable conditions prevail for higher production in any year. Risk due to failure of monsoon and crashing of prices in high production years is perceived by farmers as an impediment to adoption of dryland technology.

Human resource development is a pre-requisite and often a necessary condition for the success of any development programme. Hence people's participation in a programme is a must for success. If skills are imparted to farmers through extension education, including training, the farmers gain knowledge and skills, leading to more confidence in adopting technology. Though different State Govts./Govt. of India have been making concerted efforts to improve the knowledge/skills of farmers for higher productivity, the adoption rate is abysmally slow and low. Hence, it is imperative to gear up the machinery of transfer of technology from research stations to farmers fields, more effectively and efficiently.

The credit is channelised through Cooperative Banks, Commercial Banks, Grameena Banks, Cooperative Credit Societies, etc. National Bank for Agriculture and Rural Development (NABARD) has been set up to refinance agricultural enterprise. Govt. of India/Banks have set out Rs 25,000 crores towards agricultural loans. Thus, much emphasis is given to finance for agricultural development in India. There are short term, medium term and long term loans covering from crop loans to purchase of tractors/digging wells, etc. However, timely credit is still a problem in majority of cases, in addition to complicated procedures followed in sanctioning loans. Hence simplification of procedures in sanctioning loans is required.

FERTILIZER CONSUMPTION IN RAIN-DEPENDENT REGIONS

No systematic survey data are available to draw definite conclusions on the consumption of fertilizer in case of dryland crops. However, it was revealed that of the total SAT districts (151) surveyed by Bapna et al. (1979), the total consumption of fertilizer (N + P + K) per hectare of cropped area was less than 5 kg/ha in 29 non-irrigated SAT districts. In other non-irrigated districts, the fertilizer consumption was 5 to 10 kg/ha, 10 to 20 kg/ha and 20 to 40 kg/ha in 22, 27 and 22 districts, respectively. In only two districts, the consumption of fertilizer was above 40 kg/ha. In case of irrigated SAT districts, the fertilizer consumption ranged, on an average, between 10 to 40 kg/ha in 39 districts (Table 45).

Table 45. Consumption of fertilizer (N + P + K) per hectare of cropped area in SAT districts

Fertilizer level (kg/ha)	No. of districts	
	Non-irrigated SAT	Irrigated SAT
Less than 5	29	-
5-10	22	2
10-20	27	14
20-40	22	25
Above 40	2	8

Source : Bapna et al. (1979)

In the study carried out by Desai (1982) it was revealed that fertilizer use spread to about 18% of total unirrigated area by 1976-77. Interestingly, crops more commonly fertilized under unirrigated conditions were the same as those more commonly fertilized under irrigated conditions.

After 1970-71, there was an acceleration in diffusion of fertilizer use in cotton, peanuts and sorghum under unirrigated conditions. The average fertilizer consumption per hectare of cropped area in 112 unirrigated districts (less than 25 per cent irrigated) was found to be 18.5 kg ($N + P_2O_5 + K_2O$) as against 57.5 kg/ha in 78 irrigated SAT districts (Jha and Sarin, 1984).

There is dominant influence of certain crops, irrigation and fertilizer responsive varieties on the pace and pattern of growth of fertilizer consumption. The slow but steady growth in fertilizer use under unirrigated conditions, even in case of traditional varieties, indicates farmers willingness to tap the cost-effective potential of fertilizer use under such conditions.

SCOPE OF IMPROVING FERTILIZER USE IN DRYLANDS

Studies conducted by Singh et al. (1988) have clearly brought out that fertilizer use is now practised in dryland crops grown in Alfisols and Vertisols. However, the amount applied and the ratio of nitrogen and phosphorus utilized are the major issues. Top dressing is not in vogue for various reasons. High cost of fertilizer and lack of finance are the major constraints under both the soil conditions. Availability of the fertilizer with subsidy and loan would help boost fertilizer use in the Vertisol villages. Provision of subsidy and loan can increase fertiliser use in Alfisols. Although the response of the farmers under both the soil conditions were similar, the fertilizer use per unit area was higher in Vertisols as compared to Alfisols, because of better moisture regime obtainable in Vertisols. Involvement of the farmers in the training and demonstration as done in KVK (Krishi Vigyan Kendras - Farm Science Centres) and watershed villages helped in increasing fertilizer use considerably.

Application of FYM is a general practice in both Alfisols and Vertisols. Farmers rely more on FYM than chemical fertilizers, although the quantity used is meagre. Commercial crops like chillies receive FYM dressing every year.

A review of rural credit supply system revealed that non-institutional agencies (money lenders etc.) have been an important source of borrowings. The role of Cooperatives, Commercial Banks and Government has been insignificant. Of late, the flow of institutional credit for agriculture in India has increased from Rs 885.16 crores in 1970-71 to Rs 5555.99 crores in 1984-85 (Table 46), the major share being of Cooperatives in 1970-71. In 1984-85, the wide credit gap between Cooperatives and Commercial Banks considerably narrowed down, the Cooperatives sharing a little more than half of the total institutional credit for agriculture.

Table 46. Flow of institutional credit for agriculture in India (Rupees in crores)

Source	1970-71	1984-85
Cooperatives (short, medium and long term)	678.79	2995.99
Commercial banks/RRB (short term and term loan)	206.37	2560.00
Total	885.16	5555.99

Source : Agricultural Statistics at a Glance, Directorate of Economics and Statistics, G.O.I., 1987

FUTURE OUTLOOK IN FERTILIZER USE IN DRYLANDS

As stated earlier, the fertilizer use rate in dryland crops is very low and in some cases, even nil. Whatever amount is applied goes to cash crops. Cereals, millets and pulse crops do not receive fertilizer dressings. Surveys conducted by Singh et al. (1988) have revealed that high cost of fertilizer and lack of finance are the major constraints for fertilizer use in dryland crops grown on Alfisols and Vertisols.

Lack of response to applied fertilizers in low rainfall areas is yet another problem to reckon with. Strategically, fertilizer use can find favour in Vertisols with high water retention capacity and in areas where rainfall is greater than 750 mm per annum. Farmers also lack conviction in fertilizer use in dryland crops. Surveys conducted by Singh et al. (1988) have shown that involvement of the farmers in the training and demonstrations helped in increasing fertilizer use considerably. Further, the availability of the fertilizer with subsidy and loan would help boosting fertilizer use. This fact has been substantiated by the experience gained in model watersheds launched by the Indian Council of Agricultural Research in 1983-84 in 47 locations spread in different agro-climatic zones of the country. Farmers training in improved crop management practices was an essential part of the watershed management programme. Besides, soil and water conservation practices adopted in the watershed villages led to generation of water resources in the dryland areas. Exposure to improved management practices and increased water resources in the watersheds have helped in increasing fertilizer use. This fact has been illustrated in Table 47. Within 3 years, fertilizer use has become a practice in dryland crops. In areas where fertilizer use was nil, the fertilizer ($N + P_2O_5$) application has now gone upto 36.8 to 43.0 kg ha^{-1}. Government of India have now launched 4440 watersheds in the country covering an area of 4.2 million hectares. In future, development of dryland agriculture on watershed basis would receive priority in India. In this process, fertilizer use is expected to rise considerably in the watersheds. Exposure of the farmers to improved management practices would also result in increased fertilizer use in drylands, particularly in more assured rainfall areas.

Table 47. Fertilizer use pattern in some model watersheds.

Watershed (soil type)	Mean annual rainfall (mm)	Fertilizer use (kg $N+P_2O_5$ ha^{-1}) Base year (1983-84)	1986-87
Yernal (Vertic Inceptisols)	622	0.0	43.0
Mittemari (Alfisols and related soils)	731	0.0	37.5
Chevella (Vertisols and Vertic Inceptisols)	871	0.0	36.8
Tejpura (Vertic Inceptisols and Alfisols)	938	0.0	62.2

PROJECTED PRODUCTION INCREASE THROUGH FERTILIZER USE

An exercise has been made to project the increase in yield of major rainfed cereals/millets, viz., sorghum, pearlmillet, fingermillet and pulses through fertilizer use. Based on the data available during 1984-85, rainfed sorghum, pearlmillet, fingermillet and pulses produced 9629, 3725, 1969 and 10498 thousand tonnes of grains in the country. For projection purpose a moderate dose of fertilizer use @ 40 kg N and 20 kg P_2O_5 per hectare has been considered for sorghum, pearlmillet and fingermillet. The same for pulses has been taken at 30 kg P_2O_5 per hectare. The response of cereal/millets has been taken to be 10 kg grain/kg N applied. The response of pulses to applied P_2O_5 has been considered at 6 kg grain/kg P_2O_5. Based on this assumption, it has been computed that the production of sorghum, pearlmillet, fingermillet and pulses would be raised to 15742, 7751, 2698 and 14278 thousand tonnes, respectively due to moderate fertilizer application. This will amount to an increased production of sorghum by 6113, pearlmillet by 4026, fingermillet by 729 and pulses by 3780 thousand tonnes from the rainfed areas alone.

REFERENCES

Agricultural Situation in India, 1986. Commodity Reviews (Annual), 376-396.

Agricultural Statistics at a Glance, 1987. Directorate of Economics and Statistics. Ministry of Food and Agriculture, Government of India, New Delhi.

Arakeri, H.R. 1982. In: *Indian Agriculture*. Oxford & IBH Publishing Co., New Delhi, Bombay, Calcutta, p. 289.

Bapna, S.L., Jha, D. and Jodha, N.S. 1979. Agro-economic features of semi-arid tropical India. *Workshop on Socio-economic Constraints to Development of Semi-arid Tropical Agriculture*. International Crops Research Institute for Semi-Arid Tropics (ICRISAT), Patancheru, India, 19-23 February.

Barbara Harriss, 1982. Exchange relations and poverty in dryland agriculture. Report, Department of Human Nutrition. London School of Hygiene and Tropical Medicine, U.K.

Binswanger, H.P. 1978. Risk attitudes of rural households in semi-arid tropical India. *Economics and Political Weekly* (Review of Agriculture) 13: 25.

Desai, Gunvant, 1982. Sustaining Rapid Growth in India's Fertilizer Consumption : A Perspective Based on Consumption of Use. Research Report 31. International Food Policy Research Institute (IFPRI), Washington, p. 71.

European Nitrogen Service Programme, 1980. Fertilizer Use in Dryland Agriculture. Fertilizer Information Bulletin No. 13, p. 19.

Fertiliser Statistics, 1987-88. Fertiliser Association of India, New Delhi.

Gopalan, C. and Narasinga Rao, B.S. 1971. Dietary allowances for Indians. National Institute of Nutrition, Hyderabad, pp. 97-98.

Hadimani, A.S., Hegde, B.R. and Satyanarayana, T. 1982. Management of red soils. In: *Review of Soil Research in India*. Part II. Transactions 12th International Congress of Soil Science, New Delhi, 8-16 February, pp. 689-700.

Jha, D. and Sarin, R. 1984. Fertilizer Use in Semi-Arid Tropical India. Research Bulletin No. 9, ICRISAT, Patancheru, India, p. 76.

Narkhede, P.L. and Ghugare, R.V. 1987. Organic recycling in drylands - II. Effect of organic recycling of *subabul* on yield, nutrient uptake and moisture utilization by winter wheat. *J. Indian Soc. Soil Sci.* 35: 417-420.

Prihar, S.S., Sandhu, K.S., Singh, Y. and Singh, R. 1981. Effect of N rates on dryland wheat in relation to mulching previous crop or fallow. *Fertilizer Research* 2: 211-219.

Randhawa, N.S. and Singh, R.P. 1983. Fertiliser management in rainfed areas - available technologies and future needs. *Fertiliser News* 28(9): 17-32.

Rao, A.C.S. and Das, S.K. 1982. Soil fertility management and fertilizer use in dryland crops. In: *A Decade of Dryland Agricultural Research in India 1971-80*. AICRPDA, Hyderabad, pp. 120-139.

Rastogi, B.K. and Reddy, Y.V.R. 1982. A Study on Farm Structure in Dryfarming Areas. Report, AICRPDA (ICAR), Hyderabad.

Reddy, Y.V.R. and Rastogi, B.K. 1985. Economics of Recommended Technology in Dryland Agriculture. Res. Bull. No. 2, CRIDA, Hyderabad.

Reddy, Y.V.R., Rastogi, B.K., Rao, K.P.C. and Korwar, G.R. 1985. Package Versus Graded Technology. Res. Bull. No. 3, CRIDA, Hyderabad.

Sampath, T.V. 1988. Integrated watershed development in dryland agriculture - the key to developing agriculture (ed) J.S. Kanwar. Agricole Publishing Academy, New Delhi, pp. 515-522.

Singh, R.P. and Das, S.K. 1983. Dryland agriculture - a glimpse of recent research in India. *Adv. Indian Agric.* 1: 147-173.

Singh, R.P. and Das, S.K. 1984. Nitrogen management in cropping systems with particular reference to rainfed lands of India. In: *Nutrient Management in Drylands with Special Reference to Cropping Systems and Semi-arid Red Soils.* Project Bull. No. 8, AICRPDA, Hyderabad, p. 87.

Singh, R.P. and Das, S.K. 1988. Strategies in water conservation and planning for dryland agriculture. *Second IWRS Symposium on Water Conservation for National Development*, 11-12 December, Bhopal, pp. 1-9.

Singh, R.P. and Raje, S.R. 1984. Optimising production in rainfed Vertisols - A case study of an Operational Research Project. *Fert. News* 29(4).

Singh, R.P. and Vittal, K.P.R. 1989. Managing rainfed sandy soils for higher productivity. Abstract. *Int. Symp. on Managing Sandy Soils*, 6-10 February, CAZRI, Jodhpur, pp. 622-628.

Singh, R.P., Das, S.K. and Reddy, Y.V.R. 1988. Fertilizer use in SAT soils - economic and social factors. IBSNAT/ICRISAT/IFDC Colloquium on Soil Fertility and Fertilizer Management in Semi-Arid Tropical India. ICRISAT, Patancheru, A.P., India, 10-11 October.

Singh, R.P., Singh, H.P., Daulay, H.S. and Singh, K.C. 1981. Effect of periodical application of nitrogen in organic and inorganic form on the yield of rainfed pearlmillet. *Indian J. agric. Sci.* 51(6): 409-416.

Venkateswarlu, J. 1984. Nutrient management in semi-arid red soils. In: *Nutrient Management in Drylands with Special Reference to Cropping Systems and Semi-Arid Red Soils*. Project Bulletin No. 8, AICRPDA, Hyderabad, p. 56.

Vittal, K.P.R. 1987. Soil fertility management. In: *Technological Advances in Dryland Agriculture*. CRIDA, Hyderabad, pp. 83-99.

10

INTEGRATED NUTRIENT MANAGEMENT IN DRYLANDS

R.K. AGGARWAL and PRAVEEN KUMAR

INTRODUCTION

Food grain production has increased from 52 Mt in 1950-51 to 176.2 Mt in 1990-91 and the country will need to produce 235 Mt foodgrains, in addition about 26 Mt of oil seeds to feed the population of a billion people by the turn of century. In the past increases in food grain production were achieved both through expansion of cultivated area and through increased productivity of irrigated areas. At present nearly 70% of the country's gross cropped area is under rainfed conditions. Rainfed agriculture accounts for more than 40% of total food grain production, nearly 75% of oil seed, 90% of pulses and 70% of cotton. Even if the irrigation at the current pace continues the gross irrigated area is likely to increase to 75 M ha by the turn of century but still we will have nearly 55% of gross cropped area under rainfed condition.

Most of the drylands are typified by highly fragile natural resource base, the rainfall is low, soils are often coarse textured sandy and inherently low in fertility, organic matter, water holding capacity and are easily susceptible to wind and water erosion. In this review we would like to relate the nutrient management practices in these soils to their increased availability in soil either from their direct addition in form of fertilizer, organic and other on farm residues or by alteration of soil climatic conditions resulting in increased availability.

CLIMATE AND RAINFALL

In agronomic sense, the year in dry climate can be divided into four phases :

1. Prerainy season, which may extend from 1 to 2 months, and is also some times called period of "short rains".
2. The rainy season which is the period during which monsoon is active. It is also called the period of "long rains".
3. The post rainy season which is generally cool and dry. The crops during this season thrive on moisture stored in soil profile.
4. A hot dry season, generally precedes the rainy season. During this period the temperature and the atmospheric demand for water is quite high but the extent of it varies with the location. It is particularly high at the beginning of rainy season.

High rates of evaporation coupled with low and variable rainfall in dry climate often lead to periods of water deficit and have serious implication for stability of crop production in this ecological zone. It is only during rainy period of the year when the rainfall exceeds PE that the soil moisture reserves are charged. This period varies from 2 to 4.5 months in this climate.

Drylands generally receive less than 750 mm of rainfall annually and can be divided into arid (<500 mm) and semi arid (500-750 mm). However, an absolute value can not be stated because of seasonal fluctuations in rainfall distribution, temperature and crop water requirement. The drylands of India are characterised by short and intense rainy season and may have one peak (unimodal) or two peaks (bimodal) of rainfall. The pattern of rainfall in the dryland lying between 5 and 120 N latitude is distinctly bimodal and in areas located above 20^0 N latitude it is unimodal.

The annual rainfall variability for five locations (Fig. 1) showed that the mean annual rainfall based on the data from 1941-1970 ranged between 382 mm (Jodhpur) and 1001 mm (Indore) and the coefficient of variation ranged between 20% (Hyderabad) and 42% (Jodhpur). In the drylands of India, the droughts occur frequently and in the past 30 years they occurred 8, 6, 2, 4 and 6 years in Jodhpur, Anantpur, Hyderabad, Dharwad and in Indore, respectively.

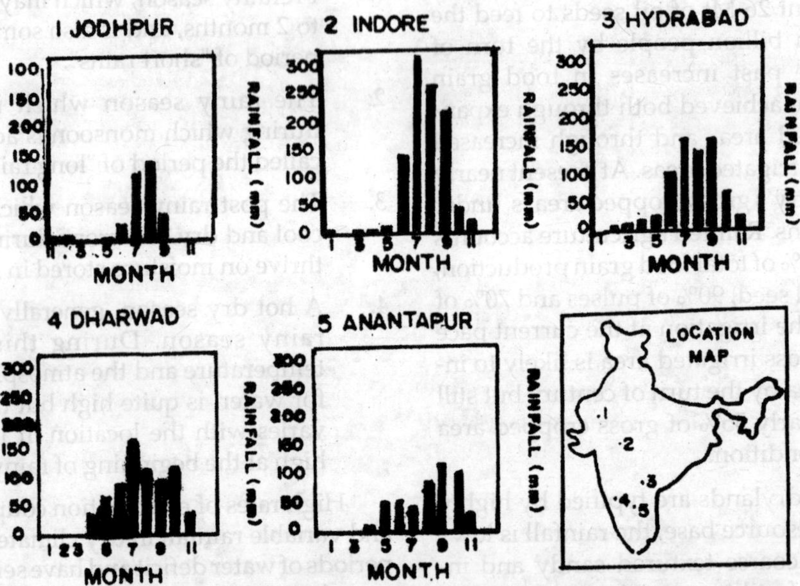

Fig. 1. Mean monthly rainfall at five representative SAT stations in India. Source Kanemasu *et al.* 1990

SOILS OF DRYLANDS IN INDIA

Sierozems (Aridisols)

Camborthids and Solorthids are the major groups in Aridisols . These soils are extremely light, sandy, loamy sand, or sandy loams. Calcium carbonate occurs at various depths influencing the effective depth of the soil. Crusting is a serious problem in these soils. Soil erosion by wind is even more serious. The thickness of the solum varies from 70 to 100 cm.

Because the soils are sandy, they hold only low amount of water and nutrients. They are universally poor in nitrogen and sometimes inadequate even in phosphorus. Subsoil salinity is not uncommon due to extreme aridity. The predominant clay minerals are smectite, mica, and kaolinite.

The texture largely determines the soil water storage capacity.In deep sandy loams, either a rainy or postrainy crop can be produced. But in loamy sands or sand, only rainy season cropping is possible.

Submontane Soils (Mollisols, Inceptisols, and Entisols)

Hapludolls (Mollisols), Haplaquepts (Inceptisols), and Udorthents (Entisols) are more common in the region occurring on level to gently sloping lands, lower elements of slopes, and steep slopes, respectively. These soils are generally deep, the texture being silty loam to loam, and hold as much as 300 mm of water per meter of profile. Land slides and soil erosion are common.

They respond well to nitrogen. Phosphorus becomes limiting only at high production levels. In lower reaches (Haplaquepts), water stagnation is possible. The predominant clay minerals are illite, smectite, and chlorite/smectite.

Alluviums (Inceptisols, Alfisols, and Entisols)

The more common groups are Ustocherpts (Inceptisols), Ochraqualfs (Alfisols), and Ustifulvents (Entisols). These soils are generally loamy sands or sandy loams, and depth being a constraint. With low clay content, they hold less water (0.1-0.2 kg of water per kg of dry soil). The nutrient-holding capacity is also less. They are neutral to alkaline in reaction. Water intake is high. Crusting could be serious problem due to either a beating rain or high temperatures. This might accelerate the problem of runoff with subsequent rains. The slopes are mostly incipient.

The productivity of these soils is high, but they require liberal applications of nitrogen for high yields. Phosphorus is needed only for very high yields. Potassium is adequate. The predominant clay minerals are illite, smectite, kaolinite and chlorite.

The lower reaches of the catena develop alkalinity and salinity. As the soils become heavier in texture, the water holding capacity increases, leading to the scope for taking two crops in a year.

Red Soils (Alfisols)

The most prominent groups are Haplustalfs, Rhodustalfs, and Palcustalfs. They are usually encountered on mild slopes of interfluves and higher terraces. These are more weathered soils. They are generally low in clay content and the predominant clays are kaolinite and illite. Most of them are sandy loams with an alluvial of higher clay below the plow layer. Due to illuviation, the subsoil is more compact, interspersed with saprolite. These soils tend to crust after a beating rain or with high temperatures. Crusting leads to higher runoff.

These soils are universally deficient in nitrogen and often in phosphorus as well. Potassium may be limiting with high production systems. Sometimes, calcium can be deficient, more so when the soils are monocropped with high calcium-demanding crops like peanuts (*Arachis hypogaea* L.). The profile water storage is about 0.15 m^3m^{-3}. As the clay content increases, it might move up to 0.20 m^3m^{-3}.

Black soils (Vertisols)

The important groups are Chromusterts and Pellusterts. Vertisols are more productive than Alfisols. As the soil organic matter is low in Indian Vertisols, the productivity is largely dependent on the quantity of clay. These soils vary both in depth (15-240 cm) and clay content (30-70%). They are rich in bases. The predominant clay minerals are smectite, montmorillonite, nontronite, and beidellite. They hold more water (0.30-0.35 m^3m^{-3}) than other predominant soils and are alkaline in reaction. With semi arid conditions, subsoil salinity is common. Boron toxicity can also be a problem.

Due to the high clay content, these soils tend to crack on drying. In low rainfall areas the cracking sometimes occur in late December when the conventional long duration crops are at flowering stage. This leads to increased loss of soil moisture. Consequently, the long duration crops suffer.

These soils are prone to sheet erosion. The runoff varies from 10% to 40% increasing with increase in rainfall. Many deep soils are calcareous. Heavy structure and water logging make it difficult for early management of these soils. To cope with hard pans and/or increased weed infestation, these soils are commonly deep plowed (25 cm) once in 3 years.

FERTILIZERS AND MANURES

Since the soils of the dry regions are often inherently poor in fertility, therefore, the crops generally respond to the fertilizer application more so in the years of good rainfall (Table 1).

Table 1. Response of different crops to fertilizer application (kg grain/kg nutrient).

Nutrient/crop	Response
Nitrogen	
Pearlmillet	7.5-18.0
Sesame	4.0-14.7
Phosphorus	
Clusterbean	4.2-5.3
Greengram	1.4-4.1
Mothbean	0.0-0.9
Pearlmillet	1.3-17.1
Sesame	1.0-2.7

Source : Aggarwal and Venkateswarlu [1989].

Singh et al. [1979] have reported that the yield of pearl millet doubled with the application of 40 kg N ha^{-1}. Similar trend was also observed by Singh et al. [1981]. But the efficiency of inorganic N fertilizers in general and that of urea in particular, in drylands, is often very low and results in undesirable economic burden to the user. Various studies in Jodhpur have revealed that the mixing of elemental S with urea [Aggarwal et al., 1987] or application of small quantity of ammonium sulphate before the application of urea [Praveen-Kumar and Aggarwal, 1988] increases its efficiency. But in view of high cost of chemical fertilizers and the characteristic uncertain yield levels in this climate, Aggarwal and Venkateswarlu [1989] suggested the supplementation of chemical fertilizers, with bulky organic manures. Singh et al. [1981] observed that under arid conditions of Jodhpur continuous application of sheep manure in general gave substantially higher yields than the application of urea alone (Table 2).

Table 2. Grain yield of pearl millet as affected by FYM and inorganic nitrogen (average 1975-79).

Treatment	Grain yield (Mg/ha)
Control	0.88
20 kg N/ha (Inorganic) every year	1.39
40 kg N/ha (Inorganic) every year	1.60
40 kg N/ha (Inorganic) once in two years	1.41
10 tons FYM + 10 kg N/ha (Inorganic) every year	1.54
20 tons FYM once in two years	1.54
40 tons FYM once in two years	1.78

Source : Singh et al. [1981].

Rao and singh [1993] showed that substitution of 50% of fertilizer N requirement by FYM resulted in yield levels nearly similar to those obtained with complete fertilization (Table 3). Aggarwal and Praveen-Kumar [1993] on the basis of a six years long study showed not only a beneficial effect of FYM application alone but also a synergistic effect of simultaneous application of FYM and inorganic fertilizers on crop yield (Table 4). They have also shown that application of FYM not only increases the N-use efficiency of urea (Fig. 2) but also improves the fertility status of soil (Table 5).

Table 3. Effect of farm yard manure vis-a-vis fertilizers on crop yields at different locations (Mg/ha).

Treatment	Pearlmillet Agra	Pearlmillet Hissar	Soybean Indore	Safflower Indore	Mustard Agra
Control	0.97	1.34	1.09	0.73	0.78
Full recommended fertilizer	1.71	2.01	1.58	1.11	1.53
Half recommended fertilizer	1.35	1.78	1.46	0.98	1.25
FYM to meet 50% N	1.30	1.94	1.75	1.43	1.18
Treatment 3+ Treatment 4	1.62	1.90	1.84	1.56	1.39
Recommended fertilizer	60-40-0	40-0-0	20-40-0	40-40-0	60-30-0
Number of year	4	4	–	–	–

Source : Rao and Singh [1993].

CROP RESIDUES

In less affluent countries grain is directly used for human consumption and crop residues are main source of fodder for animals and therefore very seldom left in field. However, the situation in the developed countries is just opposite as a result a lot of work has already been done in these countries with regards to the effect of crop residues on soil properties, and on crop yield while in India the work on this aspect started late in 80s.

In our country the estimated yield of crop

Table 4. Effect of Farm Yard Manure on yield of pearlmillet (Mg/ha).

Treatment	Grain			Straw			
	1983	1984	1985	1983	1984	1985	1986
$F_oN_oP_o$	1.18	0.93	0.03	3.14	2.53	1.23	0.83
$F_oN_oP_{40}$	1.52	1.06	0.09	3.65	2.84	1.30	0.89
$F_oN_{40}P_o$	-	1.39	0.05	-	3.39	1.42	1.15
$F_oN_{40}P_{40}$	-	1.66	0.05	-	4.38	1.81	1.35
$F_{10}N_oP_o$	1.60	1.26	0.08	3.78	3.24	1.59	1.00
$F_{10}N_oP_{40}$	2.11	1.50	0.15	4.68	4.11	1.84	1.22
$F_{10}N_{40}P_o$	-	1.87	0.12	-	4.71	1.97	1.35
$F_{10}N_{40}P_{40}$	-	2.07	0.13	-	5.16	2.35	1.64

Source : Aggarwal and Praveen-Kumar [1993].
N.B. Grain yield was not obtained in 1986 due to terminal drought.

residues is 185.3 Mt per annum for important arable crops [Bharadwaj, 1981]. The content of different nutrients in these residues varies depending on crop (Fig. 3, 4 & 5) but it is estimated that they contribute about 1.24 Mt of nitrogen, 0.16 Mt of phosphorus and 2.0 Mt of potassium. It is estimated that 1/3 of the total quantity of residues is available for incorporation as residues in arable lands. These residues can be left on soil as mulch cum manure or can be incorporated in soil, or can even be burnt in field and the ashes may then be incorporated. The incorporation of crop residues in soil may alter the physical and chemical properties of the soil and as a result can influence the crop yield.

Table 5. Effect of six years of continuous application of FYM (10 ton ha^{-1}) to pearl millet on the fertility status of soil.

Properties	Initial value	Level of FYM addition (ton ha^{-1} $year^{-1}$)	
		0	10
Organic carbon %	0.27	0.25	0.33
Available P	6.31	5.68	8.00
Available N (Kg ha^{-1})	140.00	138.60	144.30
Available Mn (ppm)	5.54	5.60	5.86
Available Fe (ppm)	2.00	2.09	2.18
Available Cu (ppm)	0.16	0.16	0.19
Available Zn (ppm)	0.31	0.37	0.45

Source : Aggarwal and Praveen-Kumar [1993].

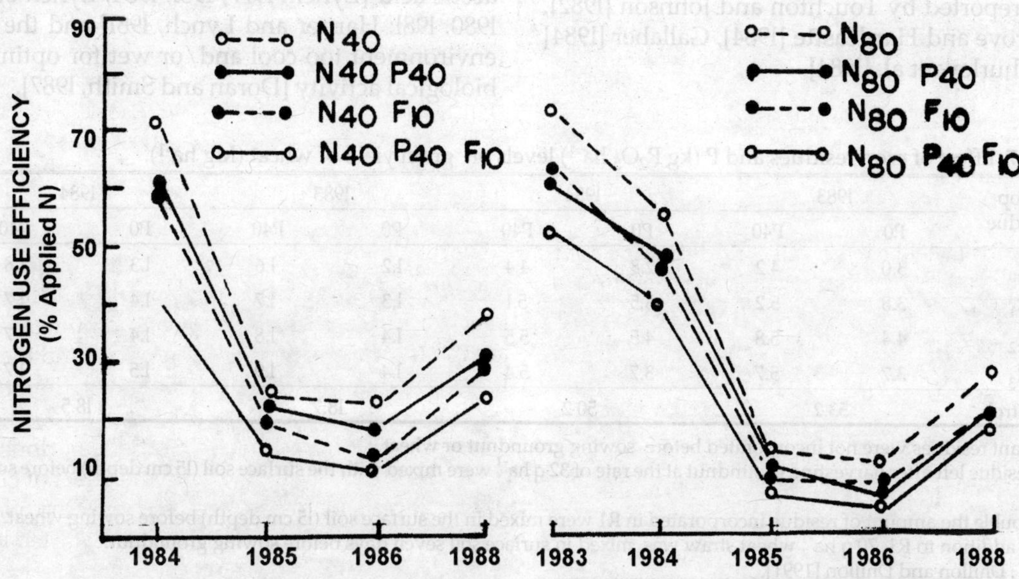

Fig. 2. Effect of FYM and phosphorus on the nitrogen use efficiency from urea.

Grain yield

Leaving the crop residues in soil may have positive, negative or no effect on grain yield depending on the condition of soil and climate. For example, Kitur et al. [1984], Al-Darby et al. [1986], Christian and Millar [1986] and Maurya [1986] reported no significant difference in yield of crops as a result of different residue management practices. However, Hadimani et al. [1982], Hegde et al. [1982], Venkateswarlu [1984], Dhillon and Dhillon [1991], Aggarwal et al. [1992] and Hegde et al. [1993] have reported higher crop yields in the plots where residues were applied (Tables 6, 7, 8, and Fig. 6).

Table 6. Effect of different residues on the yield of some crops.

Type of residue	No. of years	Yield (Mg ha^{-1}) With residue	Yield (Mg ha^{-1}) Without residue	Reference
		Corn		
Corn	5	2.36	1.88	Hegde et al., 1982
		Pearl millet		
Cowpea	8	1.93	1.75	Venkateshwarlu, 1984

In contrast to these results lower crop yields were reported by Touchton and Johnson [1982], Hargrove and Hardcastte [1984], Gallaher [1984] and Thurlow et al. [1984].

Table 8. Effect of maize residue incorporation on crop yields in maize-ragi and maize-groundnut rotation.

Levels of maize stover (Mg/ha)	Maize-Ragi rotation: Direct effect on ragi	Maize-Ragi rotation: Residual effect on maize	Maize groundnut rotation: Direct effect on groundnut	Maize groundnut rotation: Residual effect on maize
0	3.89	3.63	1.70	4.08
1	3.97	3.18	1.75	4.08
2	4.00	3.73	1.76	4.23
3	3.87	4.13	1.76	4.31
4	3.92	4.10	1.86	4.52

Source : Hegde et al. [1993].

These differences in the response to the residue application suggest that the response to residue management varies under different conditions and a number of factors regulate the plant growth and crop yield response to residue management system. Factors identified as possible causes of yield reduction are nitrogen immobilization, problem with seed placement, variable plant population, increased number of pests i.e. smugs in humid climate [Oliphant, 1982], incidence of plant pathogens [Freebairan et al., 1986], production of phytotoxins such as acetic acid [Lynch, 1977, 1981, 1984; Lynch et al., 1980, 1981; Harper and Lynch, 1981] and the soil environment too cool and/or wet for optimum biological activity [Doran and Smith, 1987].

Table 7. Effect of crop residues and P (kg P_2O_5 ha^{-1}) levels on grain yield of wheat (Mg ha^{-1}).

Crop residue	1983 P0	1983 P40	1984 P0	1984 P40	1983 P0	1983 P40	1984 P0	1984 P40
R_0	3.0	4.2	2.8	4.4	1.2	1.6	1.3	1.8
R_1	3.8	5.2	3.5	5.1	1.3	1.7	1.4	1.7
R_2	4.4	5.8	4.5	5.5	1.4	1.8	1.4	1.7
R_3	3.7	5.7	3.7	5.4	1.4	1.9	1.5	1.7
Control	53.2		50.2		18.7		18.5	

R_0 = Plant residues were not incorporated before sowing groundnut or wheat.
R_1 = Residue left after harvesting groundnut at the rate of 32 q ha^{-1} were mixed with the surface soil (15 cm depth) before sowing wheat.
R_2 = Double the amount of residue incorporated in R1 were mixed in the surface soil (15 cm depth) before sowing wheat.
R_3 = In addition to R1, 70 q ha^{-1} wheat straw was mixed in surface soil seven days before sowing groundnut.
Source : Dhillon and Dhillon [1991].

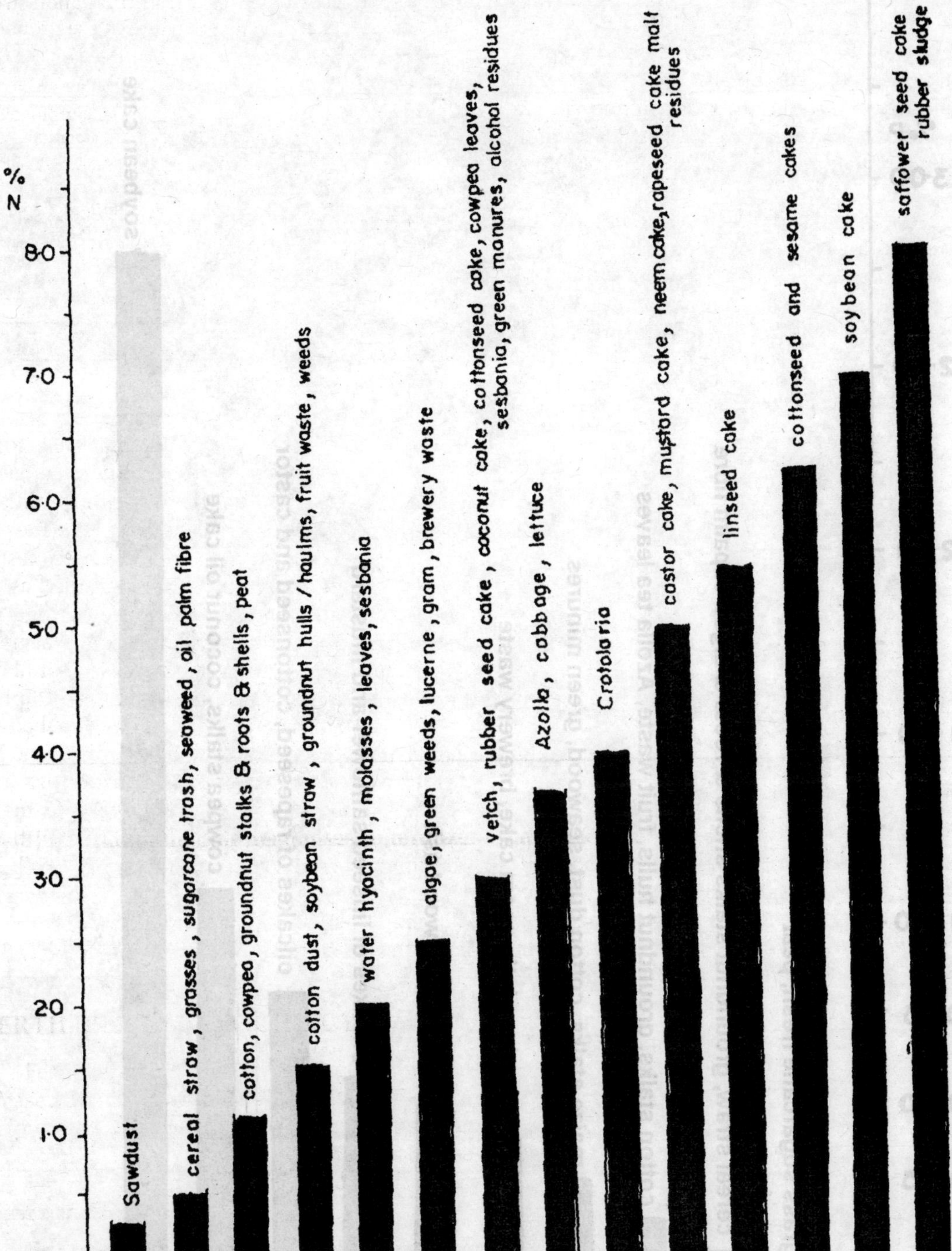

Fig. 3. Average nitrogen contents and manurial vegetable materials (oven-dry basis).
Source : Misra and Hesse [1975].

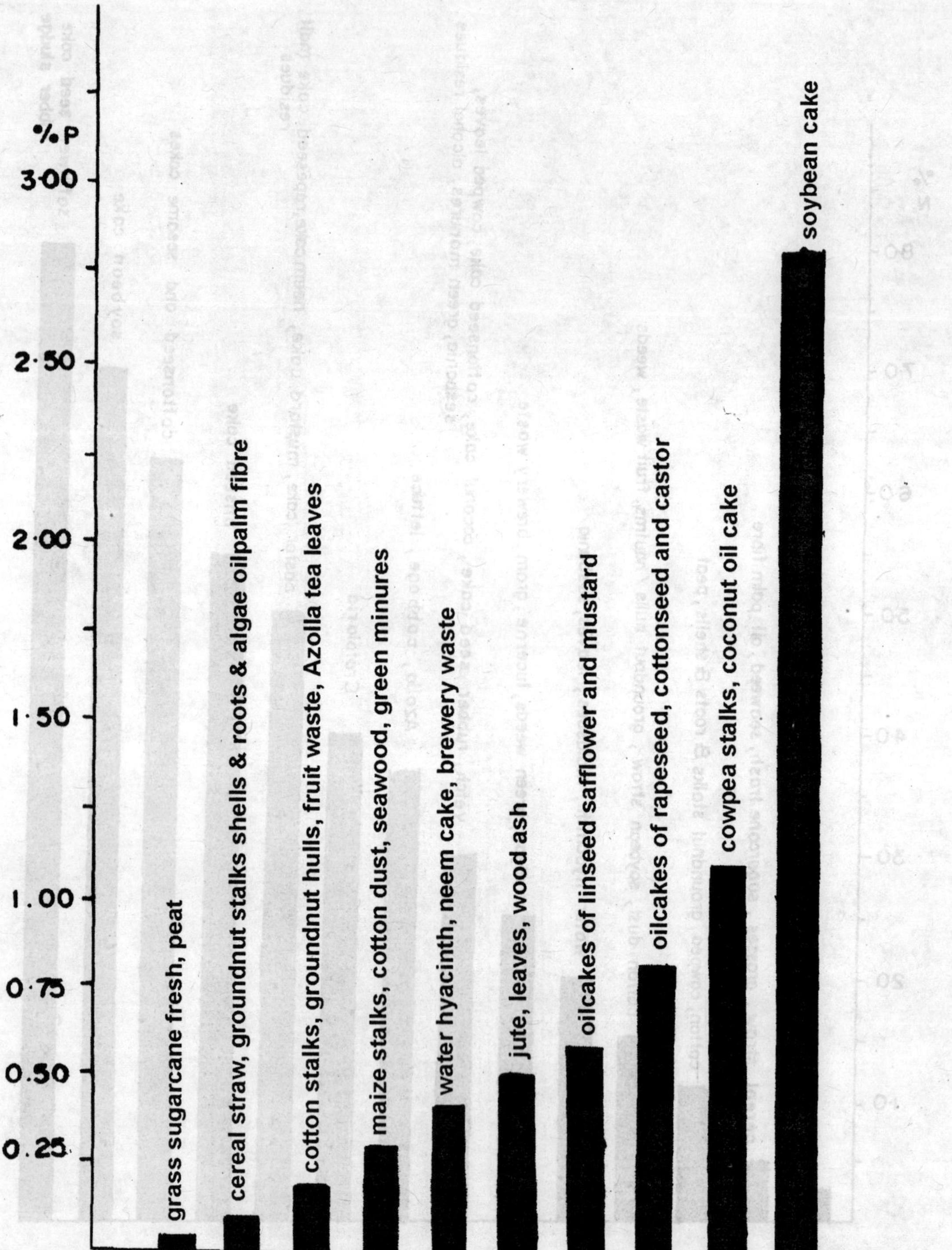

Fig. 4. Average phosphorus contents of manurial vegetable materials (oven-dry basis).
Source : Misra and Hesse [1975].

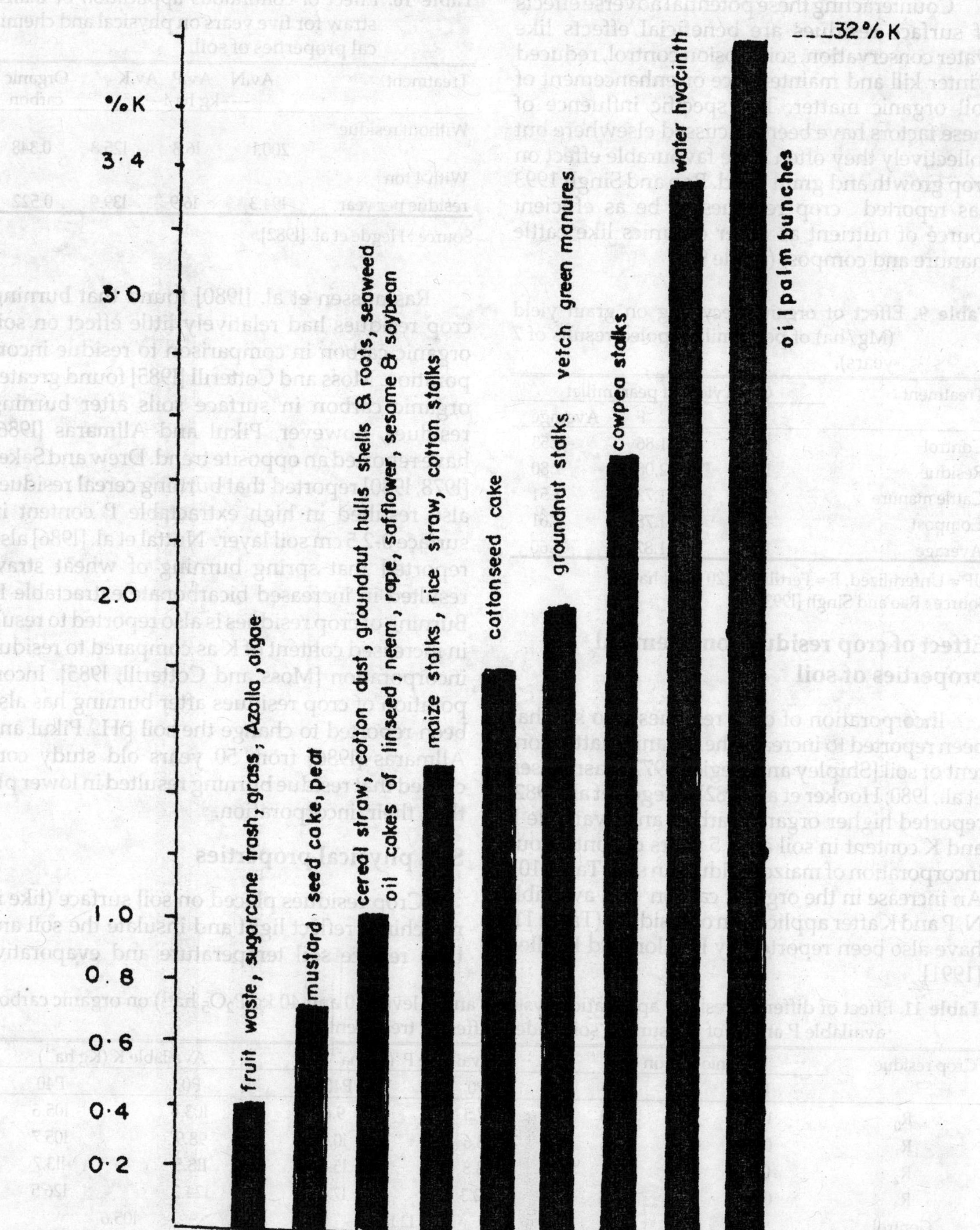

Fig. 5. Average potassium contents of manurial vegetable materials (oven-dry basis).

Source : Misra and Hesse [1975].

Counteracting these potential adverse effects of surface residues are beneficial effects like water conservation, soil erosion control, reduced winter kill and maintenance or enhancement of soil organic matter. The specific influence of these factors have been discussed elsewhere but collectively they often have favourable effect on crop growth and grain yield. Rao and Singh 1993 has reported crop residues to be as efficient source of nutrient as other organics like cattle manure and compost (Table 9).

Table 9. Effect of organic recycling on grain yield (Mg/ha) of pearl millet (poled results of 7 years).

Treatment	Grain yield of pearl millet		
	UF	F	Average
Control	1.18	1.86	1.53
Residue	1.53	2.08	1.80
Cattle manure	1.29	1.73	1.51
Compost	1.42	1.78	1.61
Average	1.36	1.87	1.60

UF = Unfertilized, F = Fertilized (20 kg N ha-1).
Source : Rao and Singh [1993].

Effect of crop residues on chemical properties of soil

Incorporation of crop residues into soil has been reported to increase the organic matter content of soil [Shipley and Regier, 1977; Rasmussen et al., 1980; Hooker et al., 1982]. Hegde et al. [1982] reported higher organic carbon and available P and K content in soil after 5 years of continuous incorporation of maize residues in soil (Table 10). An increase in the organic carbon and available N, P and K after application of residues (Table 11) have also been reported by Dhillon and Dhillon [1991].

Table 10. Effect of continuous application of maize straw for five years on physical and chemical properties of soil.

Treatment	Av.N	Av.P	Av.K	Organic carbon
	-----kg ha-l---			
Without residue	200.1	16.8	125.8	0.348
With 4 ton residue per year	194.3	16.9	139.9	0.522

Source : Hegde et al. [1982].

Rasmussen et al. [1980] found that burning crop residues had relatively little effect on soil organic carbon in comparison to residue incorporation. Moss and Cotterill [1985] found greater organic carbon in surface soils after burning residues, however, Pikul and Allmaras [1986] have reported an opposite trend. Drew and Saker [1978, 1980] reported that burning cereal residues also resulted in high extractable P content in surface 0-2.5 cm soil layer. Nuttal et al. [1986] also reported that spring burning of wheat straw resulted in increased bicarbonate extractable P. Burning of crop residues is also reported to result in increased content of K as compared to residue incorporation [Moss and Cotterill, 1985]. Incorporation of crop residues after burning has also been reported to change the soil pH. Pikul and Allmaras [1986] from 50 years old study concluded that residue burning resulted in lower pH than their incorporation.

Soil physical properties

Crop residues placed on soil surface (like in mulching) reflect light and insulate the soil and thus reduce soil temperature and evaporative

Table 11. Effect of different residue application systems and P levels (0 and 40 kg P_2O_5 ha^{-1}) on organic carbon, available P and K of the surface soil under different treatments.

Crop residue	Organic carbon (%)		Available P (Kg ha^{-1})		Available K (Kg ha^{-1})	
	PO	P40	PO	P40	PO	P40
R_0	0.19	0.20	7.5	9.6	103.1	105.6
R_1	0.25	0.26	8.6	10.0	98.9	105.7
R_2	0.31	0.31	8.8	15.5	118.5	113.7
R_3	0.35	0.35	12.3	17.4	124.2	126.5
Control	0.27		12.1		105.6	
Initial	0.24		9.2		110.4	

Details of R0, R1, R2 and R3 have been indicated in Table 9.
Source : Dhillon and Dhillon [1991].

Table 12. Effect of mulching on physical and water characteristics of soil (Values reported in table are final values after 3 years).

Treatment	Moisture retention characteristics (% moisture w/w)			Bulk density (g/cm^3)	Soil strength (kg cm^{-2})	Soil infiltration rate (cm h^{-1})
	0.01 MPa	1.5	AWC			
No tillage, no mulch	11.05	2.83	8.22	1.58	9.90	15.60
No tillage, mulch	13.23	3.86	9.37	1.51	6.70	18.0
Reduced tillage, no mulch	10.62	3.05	7.57	1.52	5.90	10.8
Reduced tillage, mulch	11.21	3.68	7.53	1.46	4.65	13.2
Excessive tillage, no mulch	10.05	3.40	7.74	1.51	6.30	9.6
Excessive tillage, mulch	10.68	3.88	6.80	1.46	4.85	10.8

Source : Gupta [1986].

losses of water [Bond and Willis, 1969]. Cannell et al. [1980], Tanaka [1985] and Gupta and Gupta [1986] reported that under arid condition mulching with local weeds increased the moisture content from 3-7% to 4-9% (Fig. 7) and decreased temperature (Fig. 8) but it was more effective under no tillage system (Table 12). Crop residue management practices also improve soil aggregation [Venkateswarlu, 1987, Table 13). This is mainly attributed to increased microbial activity during decomposition [Elliot and Lynch, 1984; Elliot and Papendick, 1986], adhesive action of decomposition products [Elliot and Lynch, 1984] or increased activity of earth worms. Blevin et al. [1977], Bauder et al. [1981], Mielke et al. [1984] and Skidmore et al. [1986] reported no significant differences between different residue management practices on soil bulk density. However, Venkateswarlu [1984] and Gupta [1986] have reported a decrease in bulk density and hydraulic conductivity of soil with the practices of residue management. They also reported an increase in hydraulic conductivity of soil. Prasad and Power [1991] after reviewing a series of results concluded that leaving organic residues on soil surface is likely to increase the hydraulic conductivity and infiltration rate in soil.

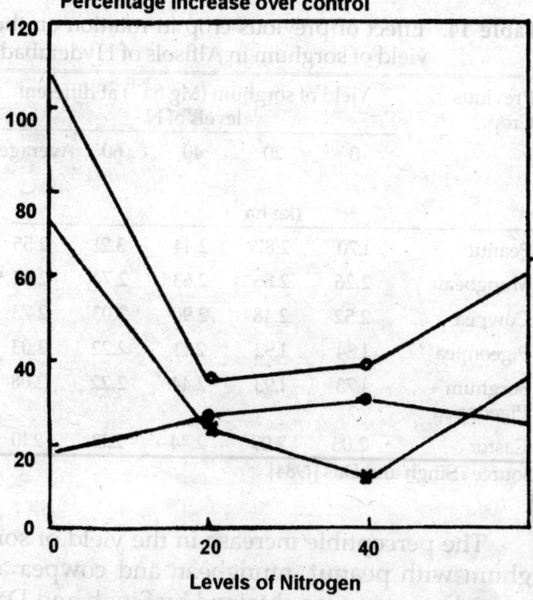

Fig. 6. Effect of crop residues and levels of N on the yield of pearlmillet (grain).
Source : Aggarwal et al. [1992].

Table 13. Soil characteristics prior to and after crop residue management for five years in an Alfisol of Hyderabad.

Soil Characteristics	Loamy sand		Sandy loam	
	Initial	After 5 years	Initial	After 5 years
Percentage aggregate (>0.25 m)	62	84	80	87
Mean weight diameter (mm)	3.21	4.19	2.63	3.04
Hydraulic conductivity (cm/hr)	2.84	8.00	4.19	10.53
Bulk density (g/cm^3)	1.78	1.65	1.81	1.69
Soil crusting (Penetrometer reading using 1/4" diameter piston needle)	40	11	35	18

Source : Venkateswarlu [1987].

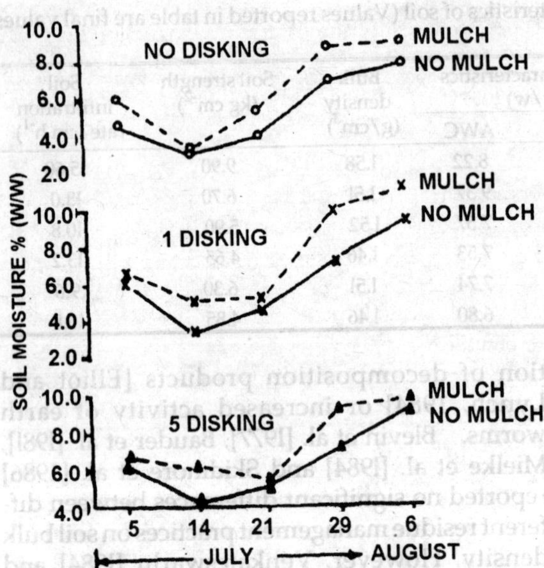

Fig. 7. Effect of tillage and mulching on the soil moisture variations during the growth period of cowpea.

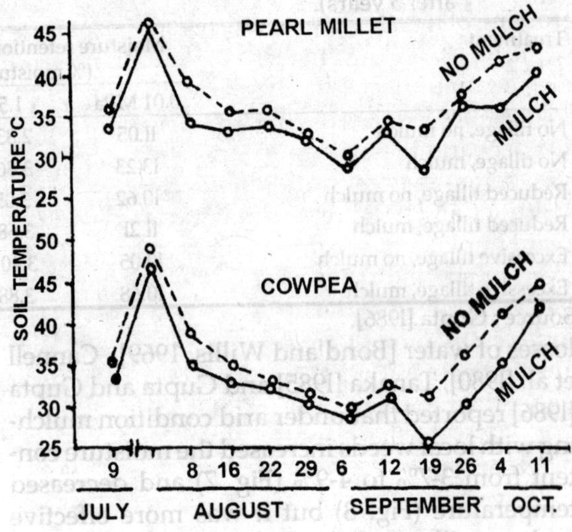

Fig. 8. Soil temperature variation during crop growth periods.

CROPPING SYSTEM

Crop Rotation

Growing one crop on same piece of land may have adverse effect even under good fertility management conditions. Mann and Singh [1977] observed 62% reduction in pearlmillet yield in pearlmillet-pearlmillet rotation in contrast to greengram-pearl millet. Singh [1980] reported that among single crop systems, pearlmillet fallow rotation proved to be most remunerative from both yield and monetary return point of view. Among double cropping systems pearlmillet-clusterbean gave highest returns per unit area [Mishra, 1971]. On the basis of the results of a long term study, Singh et al. [1985] also reported that in pearlmillet-clusterbean rotation pearl millet yield was 11% higher in comparison to continuous growing of pearlmillet. Similar results (Fig. 9) were also obtained by Oswal et al. [1989]. The beneficial effects of legume cultivation may be attributed to the improved soil fertility. Singh et al. [1985] in a long term study found an increase in soil organic carbon by 12% and available soil phosphorus by 25%. Das and Rao [1986] and Oswal et al. [1989] reported an increase in NO_3-N and available N,P,K (Fig. 10) due to inclusion of legume in crop rotation.

Table 14. Effect of previous crop in rotation on the yield of sorghum in Alfisols of Hyderabad.

Previous Crop	Yield of sorghum (Mg ha^{-1}) at different levels of N				
	0	20	40	60	Average
	(kg ha^{-1})				
Peanut	1.70	2.87	2.44	3.21	2.55
Mungbean	2.26	2.86	2.63	2.74	2.62
Cowpea	2.52	2.48	2.90	3.03	2.73
Pigeonpea	1.84	1.92	2.13	2.22	2.03
Sorghum + Pigeonpea	1.73	1.93	2.44	2.22	2.08
Castor	2.03	2.09	2.44	2.13	2.10

Source : Singh and Das [1984].

The perceptible increase in the yield of sorghum with peanut, mungbean and cowpea as preceding crop was obtained by Singh and Das [1984]. Singh and Singh [1977] on the basis of a long term study reported that cultivation of greengram in rotation with pearlmillet supplied with 20 kg N ha^{-1} gave similar yield as with direct application of 40 kg N ha^{-1}. In other words, growing of legume had an effect equivalent to 20 kg N ha^{-1}. However, there are differences on such legume effect with different grain legumes. For instance Singh et al. [1985] observed that rotation of pearlmillet with greengram or clusterbean was better than its rotation with moth bean. Reddy et al. [1993] reported that the yield and total N uptake of sorghum was maximum after green gram cultivation (Table 15). Kathju et al. [1987] reported that the yield levels of pearlmillet under cultivation of greengram, moth bean and cluster bean for past three years were much higher than could be obtained even with 80 kg N ha^{-1}. They further reported that P application increased the N fixing capacity of legumes, the hydrolysable NH_4-N rather than total N (Table 16).

Table 15. Effect of previous crop in rotation on grain and stover yield and N uptake of Sorghum.

Previous crop	Average yield Mg ha^{-1}		Total N uptake
	Grain	Stover	kg ha^{-1}
Groundnut	2.55	4.47	69.6
Greengram	2.62	4.09	79.5
Cowpea	2.73	4.63	75.4
Pigeonpea	2.03	3.95	64.6
Sorghum + Pigeonpea	2.08	3.77	58.6
Castor	2.10	3.74	61.9

Source : Reddy et al. [1993].

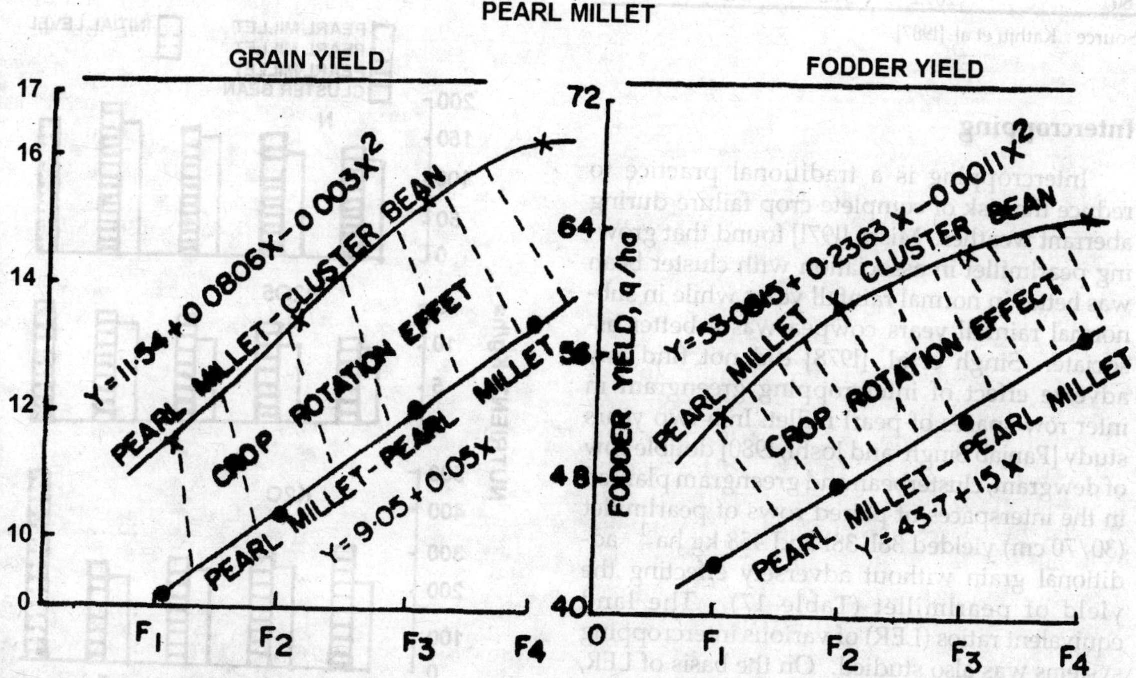

Fig. 9. Effect of legumes on grain and straw yield.
Source : Oswal et al. [1985].

Table 16. Changes in N status of soil after cultivation of legumes for three years at three levels of P application.

P_2O_5 ($Kg\ ha^{-1}$)	N2-fixed ($Kg\ ha^{-1}$)	Increase in total N over initial status(%)	NO_3-N $+NH_4$-N (ppm)	Hydro-lysable NH_4-N ($Kg\ ha^{-1}$)
Mungbean				
0	131.6	9.5	32.0	39.2
40	273.0	34.0	35.7	85.8
60	299.4	40.4	35.8	35.3
80	215.3	24.1	30.1	47.0
Mothbean				
0	84.6	4.7	35.8	43.1
40	247.6	34.0	44.8	95.8
60	264.5	36.3	47.0	107.6
80	231.3	31.3	42.5	74.6
Clusterbean				
0	75.6	8.3	47.0	81.4
40	145.7	19.4	48.2	105.3
60	195.7	27.8	38.0	117.4
80	209.2	34.0	45.9	125.4

Source : Kathju et al. [1987].

Intercropping

Intercropping is a traditional practice to reduce the risk of complete crop failure during aberrant weather. Misra [1971] found that growing pearlmillet in association with cluster bean was better in normal rainfall years while in subnormal rainfall years cowpea was a better associate. Singh et al. [1978] did not find any adverse effect of intercropping greengram in inter row spaces of pearl millet. In a two years study [Panjab Singh and Joshi, 1980] double row of dewgram, clusterbean and greengram planted in the interspaces of paired rows of pearlmillet (30/70 cm) yielded 381, 381 and 458 kg ha^{-1} additional grain without adversely effecting the yield of pearlmillet (Table 17). The land equivalent ratios (LER) of various intercropping systems was also studied. On the basis of LER, growing dew gram in double grows gave maximum advantage of 54 owing to least competition offered by this crop. Growing two rows of green gram, though resulted in comparatively lower LER (1.19) and only 19% better land utilization, gave maximum gross return, from the system. Singh et al. [1978] observed that pearlmillet could also be taken as an intercrop in greengram-pearlmillet system. They reported higher returns with intercropping than with sole crop of greengram. Studies on planting systems of pearl millet-greengram intercropping revealed that paired planting (30/70 cm) of base crop was more efficient giving 7-24% higher yield than uniform planting (50 cm) in subnormal rainfall years and was as good as uniform planting in normal rainfall years [CAZRI, 1984, 1985]. Daulay et al. [1984, 1986] have studied various intercropping systems in Jodhpur but found higher total productivity in case of intercropping of cowpea or mungbean with pearlmillet in the years of good rainfall only (Table 18). Dhingra et al. [1991] have also reported higher yield of maize in intercropping systems with mungbean. They reported that the total productivity of different intercropping systems was 3 to 5.7 q/ha higher than the sole maize.

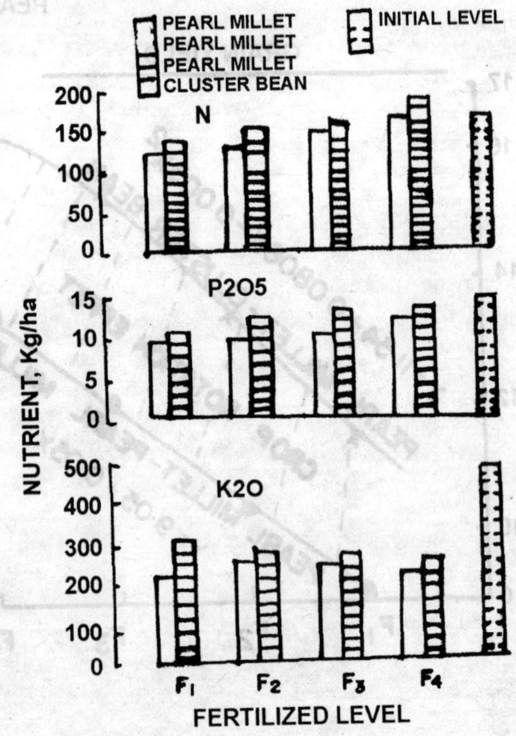

Fig.10. Soil fertility status before and after the experiment.
Source : Oswal et al. [1985].

Table 17. Yield and land equivalent ratios (LER) of pearlmillet intercropping systems.

Treatment	Mean grain yield Mg/ha Pearl millet	Inter crop	Mean total (LER)
Uniform rows(T1)	1.447	–	–
Paired rows(T2)	1.385	–	0.95
T2+Dewgram(SR)	1.249	0.165	1.30
T2+Dewgram(DR)	1.274	0.381	1.54
T2+Greengram(SR)	1.197	0.291	1.09
T2+Greengram(DR)	1.157	0.458	1.19
T2+Clusterbean(SR)	1.327	0.268	1.14
T2+Clusterbean(DR)	1.249	0.381	1.21
T2+Cowpea(SR)	1.665	0.142	0.94
T2+Cowpea(DR)	0.944	0.273	0.91
T2+Sesame(SR)	1.049	0.161	1.00
T2+Sesame(DR)	1.195	0.291	1.31

Source : Punjab Singh and Joshi [1980].

Table 18. Yield levels in various intercropping systems (Mg/ha).

	1984		1986	
	Pearl millet	Companion crop	Pearl millet	Companion crop
Pearlmillet intercropped with cowpea (2:2)	0.94	0.753	0.162	0.234
Cowpea	–	1.161	–	0.43
Pearlmillet	1.251		0.29	
Pearlmillet intercropped with sorghum (2:2)	0.913	0.838	0.183	0.185
Mung pure	–	1.245	–	0.378
Pearlmillet pure	1.164		0.285	–
Pearlmillet intercropped with Guar (2:2)	0.805	0.262	0.170	0.149
Guar pure		0.547	–	0.360
Pearlmillet pure	1.179	–	0.304	–
Pearlmillet intercropped with mothbean (2:2)	0.929	0.138	0.208	0.101
Moth pure	–	0.472	–	0.387
Pearlmillet pure	1.235	–	0.353	–

Source : Daulay et al. [1984, 1986].

There are limited number of studies on nutrient management in drylands. However, the results of these studies clearly indicate a positive effect of organics, legume based crop rotation and crop residues incorporation in maintaining soil fertility for a sustainable crop production in rainfed agriculture. They also suggest a positive synergism with blend of organics and chemical fertilizer; thus assuring higher return from applied nutrients in this drought prone ecosystem.

REFERENCES

Aggarwal, R.K. and Praveen-Kumar. 1993. Sustainable productivity through organic manures in arid land ecosystem. *Proc. of International Symposium on Environmental Degradation*, CAZRI, Jodhpur.

Aggarwal,R.K., Raina, P. and Praveen-Kumar. 1987. Ammonia volatilization losses from urea and their possible management for increasing nitrogen use efficiency in an arid region. *J. Arid Environ.* 13: 163-168.

Aggarwal, R.K., Rao, A.V., Sharma, B.M., Singh, Y.V., Praveen-Kumar and Tarafdar, J.C. 1992. Mid term project report of Indo-US Project, Enhancing fertilizer use efficiency in conjunction with residue management in dryland crops and cropping system in low rainfall conditions. CAZRI, Jodhpur.

Aggarwal, R.K. and Venkateswarlu, J. 1989. Long term effect of manures and fertilizers on important cropping systems of arid region. *Fert. News* 34: 67-70.

Al-Darby, A.M., Lowery, B. and Daniels, T.C. 1986. Evaluation of corn (*Zea mays* L.) growth and productivity with three conservation tillage systems. *Agron. J.* 78: 901-907.

Bharadwaj, K.K.R. 1981. Potential and problems of recycling of farm and city wastes on land. In: *Recycling residues of agriculture and industry* (ed) M.S. Kalra. Punjab Agricultural University, Ludhiana, pp. 57-76.

Blevin, R.L., Thomas, G.W. and Cornelins, P.L. 1977. Influence of no tillage and nitrogen fertilization on certain soil properties after five years of continuous corn. *Agron. J.* 69: 383-386.

Bond, J.J. and Willis, W.O. 1969. Soil water evaporation in surface residue rate and placement effects. *Soil Sci. Am. Proc.* 33: 445-448.

Cannell, R.O., Ellis, F.B., Christian, D.G., Graham, J.P. and Douglas, J.J. 1980. The growth and yield of winter cereal after direct drilling shallow cultivation and ploughing on non-calcareous clay soil (1974-1978). *J. Agric. Sci.* 94: 345-359.

CAZRI 1984. Annual Report. Central Arid Zone Research Institute, Jodhpur, p. 217.

CAZRI 1985. Annual Report. Central Arid Zone Research Institute, Jodhpur, p. 113.

Christian, D.G. and Miller, D.P. 1986. Straw incorporation by different tillage systems and the effect on growth and yield of winter oats. *Soil Tillage Res.* 8: 239-252.

Das, S.K. and Rao, A.C.S. 1986. Effect of preceding crops on nitrogen management for rainfed sorghum in black soil. *J. Indian Soc. Soil Sci.* 34: 510-513.

Daulay, H.S., Henry, A. and Bhati, T.K. 1984. Studies on intercropping systems. AICRPDA, Jodhpur, pp. 6-7.

Daulay, H.S., Henry, A. and Bhati, T.K. 1986. Studies on intercropping systems. AICRPDA, Jodhpur, pp. 5-7.

Dhillon, K.S. and Dhillon, S.K. 1991. Effect of crop residues and phosphorus levels on yield of groundnut and wheat grown in rotation. *J. Indian Soc. Soil Sci.* 39: 104-108.

Dhingra, K.K., Dhillon, M.S., Grewal, D.S. and Sharma, K. 1991. Performance of maize and mungbean intercropping in different planting patterns and row orientations. *Indian J. Agron.* 36: 207-212.

Doran, J.W. and Smith, M.S. 1978. Organic matter management and utilization of soil and fertilizer nutrients (eds) J.J. Mortvedt, D.R. Buxton., S.H. Mickelson, Soil Fertility and Organic matter as Critical Component of Production System. *Soil Sci. Soc. Am. Spec. Pulbe.* 19: 53-72.

Drew, M.C. and Saker, L.R. 1978. Effects of drilling and ploughing on root distribution in spring barley and on concentration of extractable phosphates and potassium in upper horizons of a clay soil. *J. Sci. Food Agric.* 29: 201-206.

Drew, M.C. and Saker, L.R. 1980. Direct drilling and ploughing : Their effects on distribution of extractable phosphorus and potassium and of roots in the upper horizon of a clay soil. *J. Sci. Food Agric.* 99: 411-423.

Elliot, L.F. and Lynch, J.M. 1984. The effect of available carbon and nitrogen in straw on soil and ash aggregation and acetic acid production. *Plant and Soil* 78: 335-343.

Elliot, L.P. and Papendick, R.I. 1986. Crop residue management for improved soil productivity. *Biol. Agric. Hort.* 3: 131-142.

Freebairan, D.M., Ward, L.D., Clarke, A.L. and Smith, G.D. 1986. Research and development of reduced tillage systems for vertisols in Queensland, Australia. *Soil Tillage Res.* 8: 211-229.

Gallaher, R.N. 1984. Soybean roots as affected by tillage in old tillage studies. In: *Proc. Southeast No Tillage systems 7th Conf.* (eds) J.T. Touchton and R.E. Stevenson. Auburn University, Auburn AL, pp. 102-105.

Gupta, J.P. 1986. Effect of tillage and mulch on soil and the growth and yield of cowpea grown in arid tropics. *Arid Soil Res. Rehabil.* pp. 161-172.

Gupta, J.P. and Gupta, G.N. 1986. Effect of tillage and mulching on soil environment and cowpea seedling growth under arid conditions. *Soil Tillage Res.* 7: 233-240.

Hadimani, A.S., Hegde, B.R. and Satyanarayana, T. 1982. Management of red soil. *Review of soil research in India, Pt.II.* 12th International Congress on Soil Science, February 8-16, 1982, New Delhi, India, pp. 689-702.

Hargrove, W.L. and Hardcastte, W.S. 1984. Conservation tillage practice for winter wheat production in the Appalachian Piedmont. *J. Soil Water Conserv.*

Harper, S.H.T. and Lynch, J.M. 1981. The kinetics of straw decomposition in relation to its potential to produce phytotoxin acetic acid. *J. Soil Sci.* 32: 627-637.

Hegde, B.R., Havanagi, G.V., Reddy, M., Venugopal, N., Viswanath, A.P. and Satyanarayana, T. 1982. Studies on incorporation of maize residue on soil properties and yield of maize under rainfed conditions. *Indian J. Agron.* 27: 254-258.

Hegde, B.R., Krishnegowda, K.T. and Paravathappa, H.C. 1993. Organic residue management in red soils under dryland conditions. In: *Recent Advances in Dryland Agriculture.* Scientific Publishers, Jodhpur, pp. 277-285.

Hooker, M.L., Herron, G.M. and Penas, P. 1982. Effect of residue burning removal and incorporation on irrigated cereal crop yields and soil chemical properties. *Soil Sci. Soc. Am. J.* 46: 122-146.

Kanemasu, E.T., Stewart, J.I., Dank, S.J. van and Virmani S.M. 1990. Agroclimatic approaches for improving agricultural productivity in semi arid tropics. Adv. Soil. Sci. 13 : 273-309.

Kathju, S., Aggarwal, R.K. and Lahiri, A.N. 1987. Evaluation of diverse effects of phosphate application on legumes of arid area. *Trop. Agric.* 64: 91-96.

Kitur, B.K., Smith, M.S., Blevins, R.L. and W.W. Frye 1984. 15_N depleted ammonium nitrate applied to no tillage and conventional tillage corn. *Agron. J.* 76: 240-243.

Lynch, J.M. 1977. Phytotoxicity of acetic acid produced in anaerobic decomposition of wheat straw. *J. App. Bacteriology* 42: 81-87.

Lynch, J.M. 1981. Promotion and inhibition of soil aggregate stabilization by micro organisms. *J. Gen. Microbiol.* 126: 371-375.

Lynch, J.M. 1984. Interactions between biological processes, cultivation and soil structure. *Plant and Soil* 76: 307-310.

Lynch, J.M., Ellis, F.B., Harper, S.H.T. and Christian, D.G. 1981. The effect of straw on establishment and growth of winter cereals. *Agric. Environ.* 5: 321-328.

Lynch, J.M., Gunn, K.B., and Panting, L.N. 1980. On concentration of acetic acid in straw and soil. *Plant and Soil* 56: 93-98.

Mann, H.S. and Singh, R.P. 1977. Crop production in Indian Arid Zone. In: *Desertification and its Control.* ICAR, New Delhi, pp. 215-234.

Maurya, P.R. 1986. Effect of tillage and residue management on maize and wheat yield and on physical properties of an irrigated sandy loam soil in northern Nigeria. *Soil Tillage Res.* 8: 161-170.

Misra, D.K. 1971. Agronomic investigations in arid zone. *Proc. Symp. Problems of Indian Arid Zone*, pp. 165-169.

Misra, R.V. and Hesse, P.R. 1975. Improving soil fertility through organic recycling. FAO Project Field Document No. 24.

Moss, S.R. and Cotterill, E.G. 1985. The influence of straw ash on some soil properties that can effect herbicide performance. *Soil Tillage Res.* 5: 361-370.

Nuttal, W.F., Bowren, K.E. and Campbell, C.A. 1986. Crop residue management practices and nitrogen and phosphorus fertilizers effect on crop response and some physical and chemical properties of a black chernozem over 25 years in a continuous wheat rotation. *Can. J. Soil Sci.* 66: 159-171.

Oliphant, J.M. 1982. The effect of stubble on yield of winter wheat after cultivation or direct drilling. *Exp. Husb.* 38: 60-68.

Oswal, M.C., Bakshi, R.K., Kumar, V. and Kumar, S. 1989. Response of dryland pearl millet to fertilizer under two cropping sequences. *J. Indian Soc. Soil Sci.* 37: 337-342.

Pikul, J.L. Jr. and Allmaras, R.R. 1986. Physical and chemical properties of a haploxeroll after 50 years of residue management. *Soil Sci. Soc. Am. J.* 50: 214-219.

Prasad, R. and Power, J.F. 1991. Crop residue management. *Adv. Soil Sci.* 15: 205-251.

Praveen-Kumar and Aggarwal, R.K. 1988. Reducing ammonia volatilization from urea by rapid nitrification. *Arid Soil Res. Rehabil.* 2: 131-138.

Panjab Singh and Joshi, N.L. 1980. Intercropping of pearlmillet in arid areas. *Indian J. agric. Sci.* 50: 338-341.

Rao, U.M.B. and Singh, S.P. 1993. Crop responses to organic sources of nutrients in dryland conditions. In: *Recent advances in Dryland Agriculture* (ed) L.L. Somani. Scientific Publishers, Jodhpur, pp. 287-304.

Rasmussen, P.F., Allmaras, R.R., Rohde, C.R. and Roager, N.C.J. 1980. Crop residue influence on soil carbon and nitrogen on wheat fallow system. *Soil Sci. Soc. Am. J.* 44: 496-500.

Reddy, G.S., Das, S.K. and Singh, R.P. 1993. Prospects of green leaf manuring as an alternative to fertilizer nitrogen in dryland farming. In: *Recent Advances in Dryland Agriculture* (ed) L.L. Somani. Scientific Publishers, Jodhpur.

Shipley, J.L. and Regier, C. 1977. Water response in production of irrigated grain sorghum-High plains of Texas. Texas Agric. Exp. Stn. Rep. MP-l348C.

Singh, K.C. and Singh, R.P. 1977. Inter cropping of annual grain legumes with sunflowers. *Indian J. agric. Sci.* 47: 563-567.

Singh, R.P. 1980. Cropping systems for drylands of Indian Arid Zone. *Ann. Arid Zone* 19: 443-447.

Singh, R.P. and Das, S.K. 1984. Nitrogen management in cropping systems with particular reference to rainfed lands of India. Part A of Project Bulletin No. 8, AICRPDA, Hyderabad.

Singh, R.P., Singh, H.P., Daulay, H.S. and Singh, K.C. 1979. All India Coordinated Research Project on Dryland Agriculture. Annual Report, Jodhpur Centre, pp. 22-27.

Singh, R.P., Singh, H.P., Daulay, H.S. and Singh, K.C. 1981. Effect of periodical application of nitrogen in organic and inorganic form on the yield of rainfed pearlmillet. *Indian J. Agric. Sci.* 51: 409-416.

Singh, R.P., Singh, K.C. and Ramakrishna, Y.S. 1978. Effect of systems of planting of pearlmillet on yield, total productivity, moisture use and monetary returns. *Indian J. Agric. Sci.* 48: 138-142.

Singh, S.D., Bhandari, R.C. and Aggarwal, R.K. 1985. Long term effects of phosphate fertilizers on soil fertility and yield of pearlmillet grown in rotation with grain legume. *Indian J. agric. Sci.* 55: 274-278.

Skidmore, E.L., Layton, I.B., Armbrust, D.V. and Hooke, M.L. 1986. Soil physical properties as influenced by cropping and residue management. *Soil Sci. Soc. Am.J.* 50: 415-419.

Tanaka, D.L. 1985. Chemical and stubble mulch fallow influences on seasonal water content. *Soil Sci. Soc. Am. J.* 49: 728-733.

Thurlow, D.L., Elkins, C.B. and Hillhold, A.E. 1984. Effect of in row chisel at planting on yield and growth of full season soybean. In: *Proc. Southeast No-tillage Systems 7th Conf.* (eds) J.T. Touchlon and R.E. Stevenson. Auburn University, Auburn/AL, pp. 124-126.

Touchton, J.T. and Johnson, J.W. 1982. Soybean tillage and planting method effects on yield of double cropped wheat and soybean. *Agron. J.* 74: 57-59.

Venkateswarlu, J. 1984. Nutrient management in semi-arid red soils. Part II, Project Bulletin No.8, AICRPDA, Hyderabad.

Venkateswarlu, J. 1987. Efficient resource management for drylands of India. *Adv. Soil Sci.* 7: 165-221.

11

MOISTURE RETENTION BY SOILS AND ITS AVAILABILITY : PRINCIPLES AND PRACTICES

RAM K. GUPTA

INTRODUCTION

The phenomenon of water retention and its subsequent release by soil has long been recognised as one of the primary functions of the soil in sustaining plant growth under conditions of intermittent water supply. We are, therefore, most concerned with soil properties that affect the (a) movement of water into and within the soil, (b) the moisture retention and release characteristics of the soil, and (c) the availability of soil moisture to higher plants.

When water coming from rains or irrigation enters into the soil, air is displaced and the soil pores become occupied with water. The wetting front moves downward displacing air. Eventually the upper part of the soil becomes saturated with respect to water as it does not retain any more water. Rather it simply conducts it to the lower layers. The soil is said to be at its maximum retentive capacity. If we now cut off the water supply to the soil surface - say it stops raining or we shut off the irrigation water the water from the upper layers will move downward for sometime at progressively reduced rate till the rate becomes negligibly small. The soil at this stage is said to be at field capacity. At this stage one will observe that the water from so called macropores moved out whereas micropores remained filled with water. If a crop is allowed to grow on a soil at field capacity, the loss of water due to transpiration from crop canopy and evaporation from exposed soil surface occurs. The soil water progressively depletes at a rate dependent on prevailing soil moisture suction and weather conditions. Stage is set in when the growing plants are unable to maintain their turgidity because of very low water availability in comparison to the evapotranspirational demand. The crop exhibits the signs of wilting at this stage when there is still a substantial amount of water left in the soil. The moisture content at which the plant exhibits signs of irreversible wilting is termed as critical moisture content and it is not a constant value characteristics of a soil. It rather depends on crop species, prevailing evaporation rate, plant characteristics etc. [Denmead and Shaw, 1962].

MECHANISM OF WATER RETENTION

In order to clearly comprehend the mechanism of water retention by soil let us first understand the phenomenon of capillarity. When a capillary is placed on the surface of a free water body, the water is attracted by the capillary to a certain height depending on the radius of the capillary. This rise is because of (a) the attraction of the solid surface (glass) for water molecules (adhesion) and (b) the attraction of water molecules for each other (cohesion). By adhesion, the water molecules are held rigidly at the solid surface. These water molecules, in turn, hold water molecules further away from the solid surface, by cohesive force. The height of rise of water column in a capillary of radius r is given by

$$h = \frac{2\gamma \cos \theta}{\rho g r}$$

where h= height of rise in a capillary tube (cm)

r = radius of capillary tube (cm)

p = density of water (g cm^{-3})

g = gravitational force per unit (dyne g^{-1})

γ = surface tension of water (dyner cm^{-1})

θ = contact angle between glass surface and liquid water and is 0.

The rise of water, h, is inversely proportion to the radius of a capillary tube.

Soil, in essence, contains numerous capillaries of various shape and size. A soil column when brought in contact with a free water surface, therefore attracts water and the water rises into the column through the interconnected pores (visualised as capillaries) till the capillary force is counter balanced by the weight of rising water column or gravitational force. Thus a soil with greater proportion of smaller pores will retain more water at a given externally applied suction. Numerous forces act upon water in a soil. They include earth's gravitational force, matric force due to surface tension, osmotic force due to presence of solutes and electric double layer at micellar surface, force due to weight of water and also due to weight of soil particles where soil matrix lacks rigidity and adsorptive forces. All these forces influence energy status of soil water, retention and movement in soil. It is beyond the scope of this chapter and the book to consider the individual forces and associated energy (potential) components. Matric forces however, are the most important from water retention and release point of view in normal soils. A detailed description of potential energy per unit quantity of water associated with matric force as it relates to water retention is called for.

MATRIC POTENTIAL VS WATER RETENTION

Matric potential is associated with the attraction of soil particle surface for water as well as the attraction of water molecules for each other. Since an unsaturated soil has a tendency to attract water in response to existing matric forces, the pressure in the soil water is less than that in the soil air, which is usually atmospheric, all air-water interforces must be curved for there to be equilibrium. This pressure deficit also called negative pressure, or soil moisture suction, is a measure of the tenacity with which water is held by the soil and therefore, of the energy status of soil water. It is negative of matric potential expressed per unit mass basis. A saturated soil has zero matric potential and the water contained is said to be under no suction/tension. As the soil drains under certain force or dries out, the soil moisture suction increases and the water content associated with that particular suction decreases. Thus there exists a relationship between soil moisture suction (matric potential on weight basis) and soil water content. The relationship between soil water content and its energy status in soil is not unique and depends on whether the soil is wetting (sorption) or drying (desorption). This difference in soil water relationship has been referred to as "hysteresis". The soil moisture suction is usually expressed in terms of bar, atmosphere or cm of water.

1 bar = 10^6 dyne cm^{-2}

=1022 cm of water column

=10^3 millibar (mb)

=0.99 atm

1 atm = 1.013 x 10^6 dyne cm^{-2}

= 1035 cm of water column.

Soil moisture content is expressed on volume basis, weight basis or depth basis.

$$\theta_v = \frac{Vw}{Vs} \quad \ldots (2)$$

$$\theta_w = \frac{Mw}{Ws} \quad \ldots (3)$$

$$\theta_v = \frac{\theta w \times B.D.}{Pw} \quad \ldots (4)$$

$$\theta_D = \frac{\theta w \times B.D.}{Pw} \times Ds \quad . . (5)$$

where θv, θ_w, θ_D refer to soil water content on volume, weight and depth basis, respectively.

Vw is volume of water in soil,

Vs is bulk volume, Mw is the weight of water contained in a soil sample weighing Ws.B.D. is soil bulk density and Ds is the soil depth, containing θ_D depth of water. The density of water, Pw, has been assumed to be 1 g cm^{-3} and all units are inegs system.

MOISTURE RETENTION CHARACTERISTICS

Moisture retention characteristics of a soil provide basic information on ability of a soil to retain water at different suctions as also the amount of water a soil undergoing water depletion will release per unit increase in suction.

Moisture retention data particularly in the 15-bar suction range are required for working out the so called water availability periods for dryland crops and irrigation schedule and also for proper land use planning. The water retention curve of a soil represents its pore-size distribution. Therefore, any change in the pore size distribution due to swelling or shrinkage, soil compaction and structural changes of soil will modify the soil water relationship [Gupta and Gupta, 1978]. Laboratory estimated water retention curves have been observed to differ significantly from in situ determined curves (Arya et al., 1975; Tomar and Ghildyal, 1978; Gupta et al., 1981]. It is therefore important that the water retention characteristics are estimated in-situ. It is not uncommon for a fully developed and differentiated soil profile to exhibit vertical heterogeneity in soil water retention characteristics. Kauraw and Gupta [1985] estimated soil moisture content thermo-gravimetrically and corresponding moisture suction using thermocouple psychrometers in a black soil profile under receding moisture condition. The moisture content at comparable suction was higher in deeper soil layers.

The moisture retention data generally conform to the following relationship [Kauraw and Gupta, 1985; Ghosh, 1976].

$$h = he\ (\theta / \theta_e)^{-b}$$

where h is soil moisture suction in mb, he is air entry value in mb and θ and θ_e are corresponding volumetric soil water content values, and b is a constant characteristic of the soil.

The value of b can be estimated empirically from a small number of moisture content suction data points which can be conveniently obtained or are already available for different soil series occurring in the dryland regions of the country.

MOISTURE RELEASE CHARACTERISTICS

Soils differ not only in their capacity to retain water but also to release it in response to the changes in suction. The amount of water released per unit increase in suction by a unit volume of wet soil is defined as the specific water capacity. The water capacity of a soil is the slope of moisture retention curve, θ vs h.

Water release characteristics of a soil are as important as moisture retention characteristics as they determine the quantity of water that a soil is potentially capable of releasing and making it available to a crop. Soils with coarse texture release most of the retained water at a low suction whereas soils with high clay content release water slowly but steadily over an extended suction range. Thus the ability of a heavy textured soil to retain relatively large amount of water and to release it over a wide suction range makes such soils more suitable for crop production than soils with low clay content.

Amount of water retained by black soils, classified as vertisol, (fine textured montmorillonitic family of typic Chromusterts) at different suctions as estimated in-situ is depicted in Tables 1 and 2. A comparison of the moisture retention and release estimated in-situ with that based on laboratory estimated values (Table 3) reveals that water retention is under-estimated and the water release between 0.3 and 1.5 MPa suctions grossly over estimated if determinations are made on the disturbed soil samples in the laboratory. In spite of these well established facts, the laboratory estimated values of water retention and release are still most commonly used for estimation of water requirement of crops.

Table 1. Water retention characteristics of a black soil (Kheri series) determined in-situ [Kauraw and Gupta, 1985].

Soil depth (cm)	Water retention, v/v percent			
	Suction (MPa)			
	0.03	1.5	2.0	3.0
0-20	35.5	26.5	26.0	24.4
20-40	35.5	27.8	26.8	25.5
40-80	35.5	29.8	26.8	25.0
80-120	36.5	30.3	29.7	28.8
120-160	36.0	30.0	29.5	28.9
160-180	36.0	30.0	29.5	28.8

Table 2. Water retention in a vertisol profile (Kheri series) at different suctions.

Profile depth (cm)	Water retention (cm)			
	Soil moisture suction (MPa)			
	0.03	1.5	2.0	3.0
100	35.7	28.8	27.2	25.7
150	53.8	43.8	42.0	40.2
180	64.6	52.9	50.8	48.8

Table 3. Water retention and release characteristics of some black soils as determined in laboratory using disturbed soil samples.

Location	Suction (MPa)	Water content v/v per cent	Release (cm per meter)
Indore (Vertisol)	0.03	37.3	18.5
	1.5	18.8	
Jabalpur	0.03	38.0	16.0
	1.5	22.0	-
Hyderabad	0.03	37.9	20.3
	1.5	17.6	

Source : Gupta and Varade, 1988.

Table 4 presents the soil profile moisture content at seeding and harvesting of a dryland wheat crop grown on vertisols at three different locations. While the profile water content at seeding is comparable to the field capacity values, the water content at the harvest of wheat is much lower than the water content values at 1.5 MPa (15 bar) suction. The data suggest that a dryland wheat crop grown on deep black soil extracts soil water in the suction range extending much beyond 1.5 MPa. Thus there is a need to have water retention characteristics of fine textured soils in the suction range extending 3.0 MPa or more.

Table 4. Moisture storage in some vertisols at sowing and harvesting of dryland wheat crop.

Soil profile depth		Moisture content (cm)			
		Sowing	Harvesting	Depletion	Source
Jabalpur	100	35.0	20.7	14.3	Jaggi et al., 1976
	150	55.0	32.0	23.0	
	180	65.0	40.4	24.6	
Indore	100	40.9	27.0	13.9	Sharma, 1976
	150	60.0	41.2	18.8	
	180	72.0	49.6	22.4	
Hyderabad	100	–	–	12.5	Russell, 1980
	150	–	–	20.0	
	180	–	–	22.6	

FACTORS AFFECTING WATER RETENTION IN SOILS

Water retention in soils is mainly a function of pore-size distribution which is influenced by soil texture, arrangement of soil particles, soil bulk density, organic carbon content and soil depth under field conditions. Soils with high content of fine particles, and organic matter content retain more water. Moderate compaction of soils, structural degradation due to sodification and high content of swelling type clay mineral increase water retention of a soil particularly at soil moisture tensions encountered under dryland agriculture. Deep soils store proportionately greater depth of water and are therefore put under crops or cropping systems requiring longer water availability periods.

AVAILABILITY OF WATER TO PLANTS

The fact that water deficit so common in dryland regions interferes with plant growth, and if severe, causes irreversible injury, damage or even the death of plants, is one of the most evident observations which can be made. As discussed earlier, the soils differ not only in their ability to retain water but the rate at which water is released at progressively increasing soil moisture tension also differs. As the soil water is depleted by evaporation or extraction by roots, the soil moisture tension progressively increases, less water is released by the soil, the amount of work required from the plant to remove water from the soil increases and the mobility of liquid water drastically decreases. The overall effect is reduced water availability to a crop growing on receding soil moisture. As is evident from Table 4, a large quantity of soil water is left unutilised for these reasons.

MECHANISM OF WATER UPTAKE

It is generally agreed that the forces bringing about the absorption of water by transpiring plants originate in the shoots rather than in the roots and the rate of water uptake is controlled by the rate of water loss from the aerial parts rather than by the activity of the roots themselves.

As transpiration occurs, the reduction of water potential in the leaf cells causes water to move into them from the xylem of the leaf veins. This causes reduction in the water potential of xylem. This reduction is transmitted through the continuous water column of the xylem elements down to the root system. Reduction in the water potential of root cells, produces a gradient in water potential across soil root interface. Water from soil moves into roots in response to this gradient.

Water moving in soil-plant atmosphere continuum encounters a series of resistances such as soil resistance, root cortex and xylem resistance, cuticular and stomatal resistance. Among these, the soil, stomatal, and boundary layer resistances to water flow vary depending on soil, plant and atmospheric conditions, respectively.

With the development of our understanding of the state and movement of water in the soil, plant and atmosphere and that of interrelationships of soil water content, potential, hydraulic conductivity and water flux, it is amply clear that in a dynamic system like soil-plant-atmosphere continuum (SPAC) the static concepts involving soil water constants (field capacity, permanent wilting point, critical moisture, capillary water, moisture equivalent etc.) are often physically meaningless. The classical concept of water availability which enjoyed wide acceptance for many years has been rendered obsolete and invalid. There is an overwhelming experimental evidence [Kauraw, 1982] to suggest that the lower limit of plant available water does not necessarily correspond to 1.5 MPa soil moisture suction. This is particularly so for dryland crops which have been observed to thrive at soil water suctions much beyond 2.5 MPa. Upper suction limit of water extraction by wheat has been observed to be 4.0 MPa (40 bars) or higher [Papendick et al., 1971; Kauraw, 1982].

FACTORS AFFECTING WATER UPTAKE

Transport of water in the soil plants-atmosphere system can be simplified by defining two functions of many variables. First is the supply function which depends on the soil water content together with the hydraulic characteristics of both the soil and the plant, which is a measure of the "ability" of the soil plant system to supply liquid water to the sites of vaporization within the plant, mainly the stomatal cavities of leaves. Second is the potential evaporation rate function which depends primarily on environmental factors.

Soil factors : The relationship between available soil moisture (ASM) and evapotranspiration rate (ET) is important in working out the field water balance for dryland crops. In general, ET rate necessarily decreases with decreasing soil water content on account of slowing down of capillary conduction to plant roots, due to stomatal closure resulting from mounting soil moisture suction and consequent decrease in leaf water potential. Under these circumstances ET rate falls short of potential evaporation rate, Eo [Lawbr, 1972; Hansen, 1974a, 1974b] : the decline in the ET rate from its maximum commences at low soil moisture stress (high soil water potential) under conditions of high atmospheric evaporativity and is more gradual in a fine textured soil than on coarse textured soil. This is because hydraulic conductivity of the soil is a major controlling factor of water availability at limiting values of soil moisture content. Yang and De Jong [1972] observed that permanent wilting in wheat occurred at soil water potential of -20 to -25 bars on a loam soil and at -45 to -50 bars on a clay soil. At these potentials the hydraulic conductivity values for both the soils were about equal.

Plant factors : During early stages of crop development, the soil resistance to water flow near the root surface is the dominant factor controlling evapotranspiration. Later in the growing season as plant begins to intercept most of the incoming radiation, plant factors begin to have an increasing influence on the evapotranspiration as long as the water supply to plant roots is non-limiting [Gates and Hanks, 1967; Ritchie, 1971]. Under the constraint of soil moisture availability, however, soil water potential becomes a dominant factor. The point at which water movement through soil becomes the major factor limiting transpiration rate depends on root density and distribution as well as water retention, release and transmission characteristics.

Ultimate control of water movement through the plant depends on stomatal conductance of the leaves which is affected by radiation flux density, leaf-air-vapour pressure deficit, leaf temperature, intercellular CO_2 concentration and leaf water status. There exists a critical leaf water potential at which pronounced stomatal closure would occur [Miller and Denmead, 1976] although critical leaf water potential is not unique value as plants make some adjustment in osmotic component of leaf water potential as drying progresses [Seaton et al., 1977].

Environmental factors : Meteorological conditions significantly influence water uptake by crops. The relationship between transpiration and soil matric potential is drastically altered by changes in atmospheric evaporative potential. Denmead and Shaw [1962], Yang and De Jong [1972] and Rajput and Gupta [1978] observed that if the evaporative demands were high, the relative transpiration rate decreased rapidly even though the soil matric potential was fairly high. However, if evaporative demands were low, the relative transpiration rate did not decrease until soil water potential was considerably lowered. The decline in the transpiration rate from its maximum commences at higher soil water potential under conditions of high evaporative demand and is more gradual on a clay than on a loam soil presumably due to higher capillary conductivity of the former under unsaturated conditions. The water uptake or relative transpiration of a crop, therefore, is a result of interaction of soil, plant and environmental factors.

A study conducted by Rajput and Gupta [1978] involving a clay soil, three plant species viz. wheat (*Triticum aestivum* L.), niger (*Guizatia abyssinica* Cass) and lentil (*Lens esculentus* Moench), and three atmospheric evaporative potentials clearly demonstrated the relative importance of these factors and their interacting effect on water uptake. As depicted in Fig. 1 the transpiration rate of three plant species was not influenced upto -10 bar soil water potential beyond which the rate sharply declined under an evaporative demand of 3.2 mm/day, whereas corresponding figures of soil water potential under 5.5 and 9 mm/day evaporative demands were -2 and -0.3 bar. The transpiration rate was the highest in case of lentil followed by niger and wheat under comparable soil and atmospheric conditions. Atmospheric evaporativity had no effect on transpiration rate at soil moisture potential of -15 bars or lower suggesting that at these soil water potential, water availability was a limiting factor and that the soil factors had domenated the effect. For a coarse textured soil the critical soil water potential would be much higher. At very high evaporative potential which may cause leaf dehydration, the transpiration rate though initially high, reduces substantially due to closure of stomata.

Fig. 2 [also taken from Rajput and Gupta, 1978] depicts the interacting effect of soil water potential and meteorological factors on relative transpiration rate (ratio of transpiration rate to potential transpiration rate) of the three plant

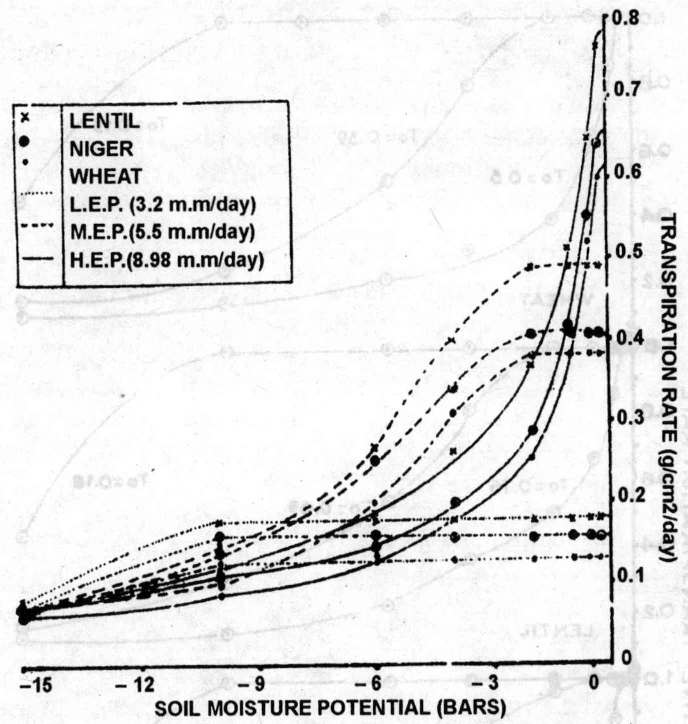

Fig. 1. Transpiration rate vs. soil moisture potential at different evaporative potentials (each point is an average of 21 values).

species. it is revealed that the critical soil water potential is not a constant value but it depends on prevailing atmospheric conditions and plant characteristics including root growth.

ESTIMATION OF CROP WATER USE

The information on water use by crops and cropping systems grown on different soils under varying agro-climatic conditions is required for ensuring efficient use of available water resources. Crop water use studies often involve the determination of all the components of the water conservation or so called field water balance equation as below.

$$P + I - S = \Delta D + \Delta M + \int_{ts}^{th} ETdt + \int_{ts}^{th} qdt \quad ..(7)$$

where P = Precipitation received in the area for which the balance is being assessed (cm)

I = Irrigation (cm)

S = Surface runoff (cm)

ΔD = Increase in surface detention (cm)

ΔM = Increase in soil water storage of the root zone (cm)

ET = Evapotranspiration rate (cm day^{-1})

q = Water flux across the lower boundary of the root zone (cm day^{-1})

t = stands for time (days) whereas t_s and t_h refer to the seeding and crop harvest time, respectively.

Precipitation, P, and irrigation, I, can be conveniently estimated using rain gauge and water meters or other devices, the net run-off can be monitored using water stage level recorders mounted on a calibrated flume. Thus the net input defined as P + I - S is determined over the cropping period. For rainfed situations, I is taken to be zero, whereas P + I - S will be zero when a crop is being grown on residual soil moisture.

The term ΔD is zero or negligible for upland crops which are planted and harvested under the

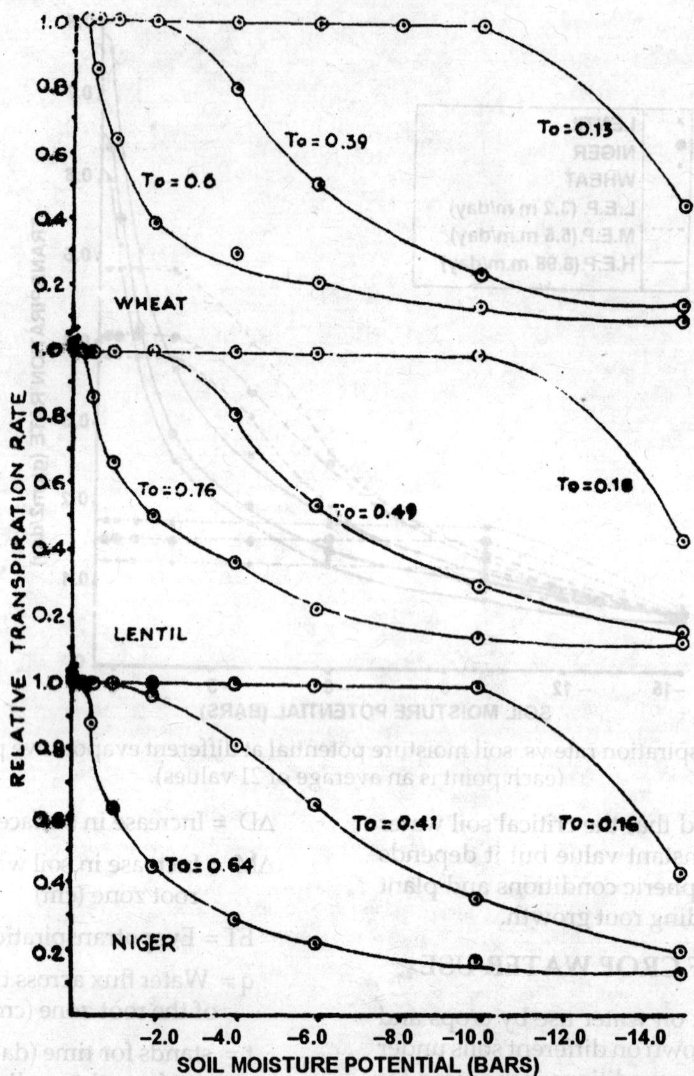

Fig. 2. Relative transpiration rate vs. soil moisture potential at different evaporativity for wheat, lentil and niger. T_o is the transpiration rate g/cm/day at -0.3 bar (each point is an average of 21 values).

condition of no surface detention of water. The determination of term ΔM requires estimation of soil moisture profiles within the root zone at planting and crop harvesting. Gravimetric moisture content values can be converted to equivalent depth of water for different soil layers, sum of which will give depth of water stored in the soil profile. Although several methods of determination of soil moisture profiles are available, thermogravimetric method is still the most reliable method. Besides, unlike other methods, no sophisticated instrumentation is required.

The term $\int_{ts}^{th} q \, dt$ is net seasonal water flux.

Estimation of this term requires regular monitoring of hydraulic head gradient and soil moisture content at the lower boundary of the soil volume under consideration. Also the hydraulic conductivity water content function at that depth must be known. Several procedures are now available to determine hydraulic conductivity at different soil moisture contents [Hillel et al., 1972; Klute, 1972; Gardner, 1970]. Even some predictive methods for computing hydraulic conductivity from water content suction relationship are available [Jackson, 1972; Campbell, 1974]. The information thus gathered is used to compute water flux by employing Darcy's equation. Water flux integrated over the crop growth period yields net seasonal water loss through deep percolation. While deep percolation losses during rainy season cropping may be significant, upward water movement from deep soil layers into the root zone of unirrigated post-rainy season crops accounts for a major fraction of crop water requirement.

The term $\int_{ts}^{th}$ ET dt which represents seasonal evapotranspiration by a crop can be evaluated by seeking solution of the water conservation equation. If desired the period between seeding to harvesting can be divided into a few time intervals of desired length. All terms of the equation can be evaluated over each period of time during the growth period.

Neglect of water movement across the root zone of a crop can lead to too high estimates of evapotranspiration of a rainy season or well irrigated crop whereas water use by post-rainy season crops may be grossly under estimated [Jaggi et al., 1977; Gupta and Jaggi, 1978; Gupta and Agrawal, 1977]. It is, therefore, advisable to estimate net seasonal water flux across the root zone as it must be accounted for while computing crop water use through field water balance.

The practices that improve water retention capacity of a soil without adversely altering water release characteristics depend on soil type and depth. With the exception of the sandy types, most soils have sufficient water holding capacity. However, a thin compacted layer at the immediate surface layer reduces the amount of infiltration, or intake of rainfall and enhances surface runoff. Low infiltration capacity of clay soils, presence of a hard pan at shallow depth, presence of exchangeable sodium in amounts exceeding 10-15% of cation exchange capacity and soil compaction also contribute to high runoff and less retention of plant available water. The problem is compounded by the fact that soils of dryland regions are low in organic matter and therefore prone to structural degradation. Practices that improve soil aggregation such as incorporation of organic manures and crop residues at least once every year, tillage practices that loosen hard compacted surface soil and maintain high porosity at the surface cut down runoff by permitting the rapid intake of water at the beginning of the monsoon and contribute to water retention and its availability to crops.

All other factors being equal, deep soils possess greater plant available moisture holding capacities than do shallow soils. For dryland crops, this is of practical significance, as crop yields in dryland regions are directly related to water storage capacity of soils. Soil stratification and presence of hard pans or impervious layers, slow down drastically the rate of water movement and also influence unfavourably the growth and penetration of plant roots. This effectively reduces the soil depth from which moisture is drawn. Sub-soiling or chiselling is recommended to partially break hard pans which exist at shallow depths. Vertical mulching may be useful for soils with hardpans at greater depths.

Soils with high content of exchangeable sodium and water soluble salts are encountered even in dryland areas where natural drainage has been disturbed. Though the water retention capacity of such soils is relatively high but water release becomes difficult due to structural degradation. Amelioration of these soils is essential in order to improve root zone aeration, root growth and water availability Use of chemical amendments such as gypsum, iron pyrites etc. is resorted to partially reclaim sodified soils.

Evaporation of water from soil profile particularly during post rainy season enhances depletion of plant available water from the root zone. In swell-shrink soils, shrinkage cracks furnish additional surface area for evaporation. Blade harrowing in between cropped rows prior to development of shrinkage cracks has been observed to be a very effective practice of moisture conservation in black soil region of Central India [Sharma et al., 1982]. The practice is commonly followed in row-planted rainfed crops such as soybean, sorghum, maize and pigeonpea in western Madhya Pradesh and adjoining areas of Maharashtra. Use of stubble straw mulch between cropped rows has led to enhanced rain water storage, improved water availability, water use efficiency and crop yields of chickpea, wheat, linseed and safflower under unirrigated conditions [Sharma et al., 1982, 1985].

Dryland grain sorghum grown after fallow yielded 1780 and 3990 kg/ha with 0 and 12 mt/ha wheat straw mulch treatments, respectively on a clay loam at Bushland, U.S.A. [Unger, 1978]. Mulching maize with green twigs of basooti (*Premna mucronata*) a few weeks before harvest increased its yield by 16%. Water storage particularly in the upper soil layers and yield of wheat grown on residual moisture after maize harvest and on fallow were also increased significantly [Prihar et al., 1979].

In conclusion, soil water retention and transmission characteristics alongwith environmental conditions interact to regulate water availability to growing crops. Water retention and its availability to plant can, however, be improved through tillage, mulching and other practices that enhance profile storage and conservation of rain water.

REFERENCES

Arya, L.M., Farrell, P.A. and Blake, G.R. 1975. A field study of soil water depletion patterns in presence of growing soybean roots. I. Determination of hydraulic properties of soil. *Soil Sci. Soc. Am. Proc.* 39: 424-30.

Campbell, G.S. 1974. A simple method for determining hydraulic conductivity from moisture retention data. *Soil Sci.* 117: 311-14.

Denmead, O.T. and Shaw, R.H. 1962. Availability of soil water to plants as affected by soil moisture and meteorological conditions. *Agron. J.* 54: 385.

Gardner, W.R. 1970. Field measurement of soil water diffusivity. *Soil Sci. Soc. Am. Proc.* 34: 832-33.

Gates, D.M. and Hanks, R.J. 1967. Plant factors affecting evapotranspiration. In: Irrigation of Agricultural Lands (eds) Hagan and Edminster. ASA Monograph No.11.

Ghosh, R.K. 1976. Model of the soil characteristics. *J. Indian Soc.Soil Sci.* 24: 353-55.

Gupta, Ram K. and Agrawal, G.G. 1977. Consumptive use of water by gram and linseed. *Indian J. agric. Sci.* 47: 22-26.

Gupta, Ram K. and Gupta, U.S. 1978. Volume change in a swelling clay and its influence on moisture relations. *J. Indian Soc. Soil Sci.* 26: 325-29.

Gupta, Ram K. and Jaggi, I.K. 1978. Seasonal water flux balance in the root zone of wheat as influenced by frequency of irrigation. *Indian J. agric. Sci.* 48: 34-37.

Gupta, Ram K., Jaggi, I.K. and Chaurecy, R.G. 1981. A comparison of field and laboratory estimated water relations of a vertisol. *J. Indian Soc. Soil Sci.* 29: 297-300.

Gupta, Ram K. and Varade, S.B. 1988. Hydrophysical properties and management of vertisols. WALMI, Aurangabad Pub. No. 32.

Hansen, G.R. 1974 a. Resistance to water transport in soil and young wheat plants. *Acta Agric. Scand.* 24: 37-48.

Hansen, G.K. 1974b. Resistance to water transport in soil and plants, plant water status, stomatal resistance and transpiration of Italian Rye grass as influenced by transpiration demand and soil water depletion. *Acta Agric. Scand.* 24 : 83-92.

Hillel, D., Krentes, V.P. and Stylianon, Y. 1972. Procedure and test of an internal drainage method for measuring soil hydraulic characteristics *in-situ*. *Soil Sci.* 114 : 395-400.

Jackson, R.D. 1972. On the calculation of hydraulic conductivity. *Soil Sci. Soc. Am. Proc.* 36: 380-83.

Jaggi, I.K., Gupta, Ram K. and Russel, M.B. 1976. Seasonal change in soil water content and water potential profiles in fallow land and in fields under irrigated and rainfed wheat. *Indian J. agric. Sci.* 46: 493-502.

Jaggi, I.K., Gupta, Ram K. and Russel, M.B. 1977. Field water balance in a vertisol under fallow and cropped conditions. *Indian J. agric. Sci.* 47: 617-21.

Kauraw, D.L. 1982. In-situ hydraulic properties of a vertisol and water uptake by plant roots. Ph.D. Thesis. Submitted to Jawaharlal Nehru Krishi

Vishwa Vidyalaya, Jabalpur, M.P., India.

Kauraw, D.L. and Gupta, Ram K. 1985. Prediction of *in-situ* measured hydraulic conductivity of a swelling clay soil. *Indian J. Soil Sci.* 13: 32-38.

Klute, A. 1972. The determination of hydraulic diffusivity of unsaturated soil. *Soil Sci.* 113: 264-76.

Lawler, D.W. 1972. Growth and water use of *Lolium perenne*. I. Water Transport. *J. Appl. Ecol.* 9: 79-98.

Miller, B.D. and Denmead, O.T. 1976. Water relations of wheat leaves in the field. *Agron.J.* 68: 303-7.

Papendick, R.J., Cochran, V.L. and Woody, W.M. 1971. Soil water potential and water content profiles with wheat under low spring and summer rainfall. *Agron.J.* 63: 731-34.

Prihar, S.S., Singh Ranjodh, Singh Nathu, and Sandhu, K.S. 1979. Effect of mulching previous crops or fallow on dryland maize and wheat. *Expt. Agric.* 15: 129-134.

Rajput, R.P. and Gupta, Ram K. 1978. Effect of soil water potential on transpiration rate and leaf water potential of wheat, niger and lentil. *J. Agron. & Crop Sci.* 147: 60-67.

Ritchie, J.T. 1971. Dryland evaporative flux in a subhumid climate. I. Micrometeorological influence. *Agron. J.* 63: 51-55.

Russel, M.B. 1980. Profile moisture dynamics of soil in Vertisols and Alfisols. *Proc. Int. Workshop on Agroclimatological Research Needs of the Semi Arid Tropics*, ICRISAT, Patancheru, A.P., India, pp. 75-87.

Seaton, K.A., Lands Beg, J.J. and Sedgley, R.H. 1977. Transpiration and leaf water potential of wheat in relation to changing soil water potential. *Aust. J. Agric. Res.* 28: 355-67.

Sharma, R.A. 1976. Studies on moisture losses through evaporation from vertisols under unirrigated conditions. Ph.D. Thesis, JNKVV, Jabalpur, M.P., India.

Sharma, R.A., Verma, G.P., Gupta, R.K. and Katre, R.K. 1982. Moisture depletion pattern and use by unirrigated wheat and safflower grown on vertisol as influenced by cultural practices. Z. *Ackher-und-Pflanzenbau* 151: 267-274.

Sharma, R.A., Upadhyay, M.S. and Tomar, R.S. 1985. Water use efficiency of some rainfed crops grown on a vertisol as influenced by soil and straw mulching. *J. Indian Soc. Soil Sci.* 33: 387-391.

Sharma, R.A. and Gupta, R.K. 1984. Soil water evaporation from clay and sandy loam soils as influenced by straw and dust mulches. *Indian J. Soil Consv.* 12: 49-56.

Tomar, V.S. and Ghildyal, B.P. 1978. Estimation of field soil water retention characteristics. *J. Indian Soc. Soil Sci.* 26: 93-97.

Unger, P.W. 1978. *Soil Sci. Soc. Am. J.* 42: 486-491.

Yang, S.J. and De Jong E. 1972. Effect of aerial environment and soil water potential on the transpiration and energy status of water in winter plants. *Agron. J.* 64: 574-578.

Vishwa Vidyalaya, Jabalpur, M.P., India.

Katraw, D.L. and Gupta, Ram K. 1985. Prediction of *in-situ* measured hydraulic conductivity of a swelling clay soil. *Indian J. Soil Sci.* 13: 32-38.

Klute, A. 1972. The determination of hydraulic diffusivity of unsaturated soil. *Soil Sci.* 113: 264-76.

Lawler, D.W. 1972. Growth and water use of *Lolium perenne*. I. Water Transport. *J. Appl. Ecol.* 9: 79-98.

Miller, B.E. and Denmead, O.T. 1976. Water relations of wheat leaves in the field. *Agron. J.* 68: 303-7.

Papendick, R.I., Cochran, V.L. and Woody, W.M. 1971. Soil water potential and water content profiles with wheat under low spring and summer rainfall. *Agron. J.* 63: 731-34.

Prihar, S.S., Singh Ranjodh, Singh Nathu and Sandhu, K.S. 1979. Effect of mulching previous crops or fallow on dryland maize and wheat. *Expl. Agric.* 15: 129-134.

Rajput, R.P. and Gupta, Ram K. 1978. Effect of soil water potential on transpiration rate and leaf water potential of wheat, niger and lentil. *J. Agron. & Crop Sci.* 147: 60-67.

Ritchie, J.T. 1971. Dryland evaporative flux in a subhumid climate. I. Micrometeorological influence. *Agron. J.* 63: 51-55.

Russel, M.B. 1980. Profile moisture dynamics of soil in Vertisols and Alfisols. Proc. Int. Workshop on Agroclimatological Research Needs of the Semi-Arid Tropics. ICRISAT, Patancheru, A.P., India, pp. 75-87.

Seaton, K.A., Landsberg, J.J. and Sedgley, R.H. 1977. Transpiration and leaf water potential of wheat in relation to changing soil water potential. *Aust. J. Agric. Res.* 28: 355-67.

Sharma, R.A. 1976. Studies on moisture losses through evaporation from vertisols under unirrigated conditions. Ph.D. Thesis, JNKVV, Jabalpur, M.P., India.

Sharma, R.A., Verma, G.P., Gupta, R.K. and Kaure, R.K. 1982. Moisture depletion pattern and use by unirrigated wheat and safflower grown on vertisol as influenced by cultural practices. *Z. Acker- und Pflanzenbau* 151: 267-274.

Sharma, R.A., Upadhyay, M.S. and Tomar, R.S. 1985. Water use efficiency of some rainfed crops grown on a vertisol as influenced by soil and straw mulching. *J. Indian Soc. Soil Sci.* 33: 387-391.

Sharma, R.A. and Gupta, R.K. 1984. Soil water evaporation from clay and sandy loam soils as influenced by straw and dust mulches. *Indian J. Soil Consv.* 12: 49-56.

Tomar, V.S. and Ghildyal, B.P. 1978. Estimation of field soil water retention characteristics. *J. Indian Soc. Soil Sci.* 26: 93-97.

Unger, P.W. 1978. *Soil Sci. Soc. Am. J.* 42: 486-491.

Yang, S.J. and De Jong E. 1972. Effect of aerial environment and soil water potential on the transpiration and energy status of water in winter plants. *Agron. J.* 64: 574-578.

12

WATER LOSSES AND THEIR CONTROL IN RAINFED AGRICULTURE

J.P. GUPTA

INTRODUCTION

Dryland agriculture is practised in most of the arid and semi arid areas. The main problem of these regions is low rainfall and meagre availability of irrigation water. A major portion of rainfall is lost in unproductive losses like evaporation, deep drainage, and seepage, thereby leaving a very small amount of water for crop use. The crop-water use can be enhanced and water use efficiency (WUE) increased by controlling unproductive losses of water. Water use efficiency is the yield of marketable crop produced per unit of water used in evapotranspiration. It can be expressed as WUE = Y/ET, where Y = yield of marketable crop and ET = evapotranspiration or seasonal water use. The term evapotranspiration (ET) and water use (WU) are generally considered identical. These are, however, different in the sense that the former term refers to loss of water in both evaporation from the bare soil and transpiration from the crop, while the latter is the beneficial utilization of water for producing dry matter and seed. Therefore, for achieving higher water use efficiency the evapotranspiration losses need to be minimized and the water efficiently utilized for producing higher yields. This is particularly important in dry areas where water is both a scarce and precious commodity. This warrants a discussion on water losses and their control for achieving higher water use efficiency.

EVAPORATION DEFINED

Evaporation is a process of losing water from a moist or wet surface to the atmosphere due to vapour pressure differences. From the moist soil surface the water is lost at about the same rate as from a free water surface having the same exposure and temperature. Harrold et al. [1959] reported that during the growing period of an annual crop about half the evaporated water comes from the soil. It, however, depends upon the type of soil, climatic conditions, the crop and its stage of growth. In arid regions, the loss of water due to evaporation from the soil is relatively low, the main component of loss being plant transpiration. It is because the soil dries quickly in such areas and acts as a mulch.

EVAPORATION PROCESS

Three conditions are necessary for evaporation to take place. First, there must be a continuous supply of heat to meet the latent heat requirement (590 cal/g of water evaporated at 15°C). This can come from the evaporating body or from outside as radiated or advective energy. Second, the vapour pressure in the atmosphere over the evaporating body must remain lower than the vapour pressure gradient between the body and the atmosphere and the vapour pressure must be transported away by diffusion or convection or both. These two conditions are influenced by meteorological factors such as temperature, humidity, wind velocity and radiations etc. The third condition relates to the supply of water through the body to the evaporating surface. This condition depends upon the amount, energy and the conducting properties of the evaporating body. Therefore, the evaporation from a soil surface or drying of soil depends upon climatic factors as well as the conducting properties of soil from lower moist layers to the evaporating site. The evaporation or drying of the soil takes place in three stages. The stage one takes place when the soil is wet and is controlled by external meteorological factors and soil surface conditions. Stage second is governed by the rate of supply of water to the evaporating surface. In stage third soil surface becomes extremely dry and the movement of water is in vapour phase. This stage persists for a longer period. The rate of loss of water in this stage is extremely low. In arid areas the first stage persists for a very short time but the second stage prolongs depending upon the soil moisture conditions and its transmission properties.

EVAPORATION CONTROL

Evaporation from the soil surface can be controlled by (i) reducing external evaporativity, (ii) reducing energy supply to the evaporating site, (iii) decreasing the conductivity/diffusivity of the soil, and (iv) reducing the potential or the force deriving water upwards through the profile. Based on these principles various measures have been devised and are discussed as under.

Shelterbelts/Shelter barriers : In arid and semi arid areas wind is one of the most important factors responsible for increased evaporation and loss of water from soil. Hot dry winds quickly dry the soil surface and create vapour pressure gradient for continuous vaporization. Shelterbelts or shelter barriers are one or multi rows of trees, shrubs and other plants planted in the field or field boundaries across the wind direction for reducing external evaporativity. The importance and utility or shelterbelts in controlling wind erosion, reducing evaporative demand and protecting crops from hot desiccating winds particularly in arid and semi arid areas has been widely recognised by Stockler [1962], Bagley [1964], Rosenberg [1967] and many others all over the world. In India, Bhimaya et al. [1958, 1984], Ganguli and Kaul [1969], Raheja [1963], Gupta et al. [1983, 1984] identified different types of shelterbelts for farmers' fields and roadside plantations in the arid areas of western Rajasthan and also discussed their effectiveness in reducing wind speed and controlling wind erosion. It has been observed that vegetative barriers as shelterbelts placed in the path of wind reduces its velocity near the ground by exerting a drag on the wind and deflecting the wind stream. The effectiveness of shelterbelts in reducing wind velocity depends on many factors such as wind velocity itself, direction, shape, width, tree height and density etc. A series of experiments were conducted at Central Arid Zone Research Institute, Jodhpur to evaluate the effectiveness of different types of shelterbelts raised in the arid zone [Gupta et al., 1983, 1984]. The results revealed that though the effectiveness of shelterbelts varies with type, age, canopy growth, design, geometry etc. the three row shelterbelt of *Cassia siamea - Albizia lebbek* and *Cassia siamea* was found most effective (36% and 46%) in reducing wind speed at 2H distance in leeward side of the shelterbelts both during summer and monsoon season respectively [Gupta and Ramakrishna, 1988]. At larger distances the effectiveness ranged from 2 to 24% only (Table 1).

Table 1. Percentage reduction of mean wind speed at different distances in the leeward side of different shelterbelts (1977-79).

Shelterbelt type	Summer season Distance from shelterbelt			Monsoon season Distance from shelterbelt		
	2H	5H	10H	2H	5H	10H
Prosopis juliflora	33	17	12	38	26	21
Cassia siamea	36	17	13	46	36	24
Acacia tortilis	36	25	13	46	36	20

Table 2. Influence of *Cassia siamea* shelterbelt on pan evaporation values (1978).

Shelterbelt side	Pan evaporation (mm/day)			
	April	May	June	July
Windward side	11.0	14.5	13.4	6.0
Leeward side	9.5	13.4	12.4	5.7
	(14)*	(8)	(8)	(5)

*Data in parentheses indicate per cent decrease in pan evaporation values.

Shelterbelts have been reported [Gupta and Ramakrishna, 1988] to reduce evaporation from the sheltered area (Table 2). Though the effectiveness has been observed upto 10H distance, it was higher (8 to 12%) at 2H distance. Konstantinov and Struzer [1969] reported a reduction in evaporation by 15% from sheltered field compared to an open field. Warren [1941], however, reported reduction of a higher order (20-30%).

Due to reduction in evaporation the moisture status of sheltered field was generally 2-4% higher than unsheltered field [Gupta and Ramakrishna, 1988]. Gorschenin [1941], however, reported an increase in moisture status by 25-30%.

Surface mulching :Mulching is a practice of covering the surface of soil with plastics, organic materials like straw, grass, and stones, etc. to reduce evaporation, to keep down weeds and also to moderate wide fluctuations in diurnal soil temperatures. It controls external evaporativity and also reduces energy supply to the evaporating site by cutting off solar radiation falling on the ground. Its main function is limited to controlling first stage of drying which helps in improved moisture status, reduced soil temperature, besides checking seedling mortality and improving crop stand (Table 3). Besides the above, mulching helps in increasing downward movement of water. Its storage deep in the profile escapes evaporation due to reduction in thermal gradients and exchange of vapours. The effectiveness of mulches in conserving moisture has generally been found to be higher under more frequency of rainfall, drought conditions and also during early periods of plant growth when canopy cover remains scanty.

Table 3. Soil moisture status, temperature and cowpea seedling stand as affected by mulching.

Treatments	Mean moisture content(%)		Mean max. Temp. (^oC)		Post emergence seedling mortality(%)	
	1982	1983	1982	1983	1982	1983
No mulch	6.5	5.2	41.3	42.7	80	50
Mulch	7.8	6.3	37.2	38.0	45	22

Recent studies under arid conditions have shown [Gupta and Gupta, 1983, 1985] that organic mulches like grass, weeds and crop residues applied to the crop at the rate of 6 t/ha reduce the maximum temperature of soil at 10 cm depth by 1 to 7^oC during monsoon season (July to September) and by 4 to 10^oC during summer season (April to June). The magnitude of reduction of soil temperature was found to be higher during clear periods of intense radiations. White transparent plastic mulch raised the soil temperature by 2 to 3^oC. Mulching was found to warm the soil by raising its temperature in winter. This was found to reverse the thermal gradients and control the upward movement of water vapours. Soil warming in winter was specially found useful in minimizing the root damages due to surface soil freezing.

Mulching suppresses the growth of weeds and their population, thereby reducing the crop-weed competition for water. Field trials carried out at CAZRI have shown (Table 4) marked reduction in weed production and significant increase in yield of grain legumes like green gram, mothbean and cluster bean grown under rainfed conditions of 1980 and 1981. Straw mulching was also found highly useful for vegetable crops grown with limited irrigations during summer months (March to May 1981 and 1982) under conditions of intense radiations (Table 5). The increase in yield with mulch under both cases could be attributed to improved soil environment by way of increased moisture availability, reduction in soil temperature to optimum levels and thus higher water use and water use efficiency. The increased water availability and water use efficiency by mulching has also been reported by various workers [Lal, 1974; Prihar and Arora, 1980; Gupta and Gupta, 1983, 1986; Gupta, 1987, 1989].

Though mulches have been found quite useful in mitigating the moisture stress effects on plants and for increasing water use efficiency, the main constraint in their use lies in their availability and cost. However, depending upon the availability, organic mulches can be used for cash crops like vegetable crops and pulses. Organic materials like weeds, waste grass, etc.

which are normally burnt before kharif sowing could be collected and used for mulching purposes.

Table 4. Effect of grass mulching on weed production and yield of grain legumes.

Treatments	Weed production (q/ha)			Mean grain yield (q/ha)		
	Green gram	Moth bean	Cluster bean	Green gram	Moth bean	Cluster bean
Control	11.3	4.4	3.6	1.4	2.1	3.8
Mulching	4.8	3.3	3.0	3.9	4.0	6.5

Table 5. Effect of straw mulch on water use efficiency and yield of tomato and okra crops.

Treatments	Yield (kg/ha)		
	Tomato		Okra
	1981	1982	1981
No mulch	4830 (9.31)*	1140 (2.79)	400 (0.58)
Mulch	10,000 (18.02)	2200 (4.87)	1950 (2.82)

*Figures in parentheses indicate water use efficiencies (kg/ha - mm).

Tillage: Tillage is a practice of loosening soil, killing weeds and increasing porosity with the help of some tool or implement. These attributes, in turn, affect soil-water-relationships, aeration status, thermal characteristics and the mechanical impedance to the root penetration. Though tillage increases moisture loss from the soil surface by increasing exposed area and the free exchange of vapours, the dry layer thus created acts as a mulch and restricts the upward movement of water to the evaporating site by reducing diffusivity gradients. Though the water loss from the surface is higher, the cumulative loss from soil profile is reduced with this practice. This is particularly observed in medium sandy loam type of soils. In aridisols the beneficial effects of tillage are specially derived from increased water recharge, reduced crop-weed competition, better root growth and proliferation, improved soil thermal conditions, leading to increased water use and water use efficiency. The results of field trials at CAZRI revealed that soil preparation by providing reduced tillage (once discing) before kharif sowing is important and was found useful in increasing the yield of pearlmillet (Table 6). However, tillage effects were found variable depending upon the type of crop and the weather conditions. Reduced or minimum tillage system was generally found better than other systems.

Table 6. Effect of different tillage systems on the yield of pearlmillet.

Treatment	Pearlmillet yield (q/ha)				Mean
	1982	1983	1984	1985	
No tillage	9.05	24.75	16.75	4.09	13.53
Reduced tillage	13.47	23.10	21.25	4.22	15.51
Excessive tillage	15.11	30.66	20.70	3.65	17.53
LSD(0.05)	2.17	2.30	1.41	NS	

Post-emergence cultivation of loamy sand soil was also found useful even under weed-free situations. This is seen from increased yield of pearlmillet obtained under arid conditions of Jodhpur (Table 7). The favourable effect could be due to improved soil-water-air-temperature scenario.

Table 7. Effect of post emergence cultivation on growth, water use efficiency and yield of pearlmillet.

Treatment	1977			1978		
	Grain yield (q/ha)	Straw yield (q/ha)	WUE kg/ha -mm	Grain yield (q/ha)	Straw yield (q/ha)	WUE kg/ha -mm
No cultivation	10.9	17.8	4.41	8.2	32.7	3.21
One cultivation after 20-25 days of seedling emergence	12.3	20.4	4.51	11.4	42.0	3.90
LSD (0.05)	1.1	1.5	–	2.8	6.9	–

TRANSPIRATION AND ITS CONTROL

Transpiration is a process of water loss from the plant surface. About 99% of water absorbed by the roots is lost to the atmosphere as transpiration. The loss is mainly from the stomatal pores present on the plant leaves. As the stomata are the focal points on the plant leaf surface where the vapour pressure gradient between the leaf surface and the atmosphere is the steepest, therefore, these are important sites for the application of anti-transpirants. Antitranspirants can be defined as the materials which decrease water loss from the plant leaves by reducing the size or number of stomatal openings, decreasing thereby, the rate of diffusion of water vapours [Fuehring, 1973]. Two most important points in the use of antitranspirants are : (i) the application of antitranspirants should restrict the water loss from the leaf surface without reducing photosynthesis, as carbondioxide diffuses through stomata and is necessary for photosynthesis, (ii) transpiration causes cooling of the leaf surface and the use of antitranspirants should not completely stop transpiration and thus raise the leaf temperature. Williamson [1963] reported that antitranspirants resulted in rise in leaf temperature of tobacco by as much as 9°C over control.

Antitranspirants can be effective in two ways [Davenport et al., 1972], viz., (i) through films that coat the leaf surface, and (ii) chemicals that close the stomata. Synthetic films that are sprayed on the leaf surface should ideally be resistant to water and completely permeable to carbondioxide. This is, however, very difficult to achieve. Antitranspirants such as phenylmercuric acetate and certain alkenyl succinic acids act by inhibiting stomatal opening [Zelitch, 1965, 1969). Film forming antitranspirants (waxy or plastic emulsions) produce an external physical barrier to retard the escape of water vapour from plants [Gale, 1961]. Silicones penetrate the leaf and may act directly on the wet cell walls and white reflecting materials (white wash or kaolinite) lower leaf temperature and reduce the vapour pressure gradient from leaf to atmosphere [Abou-Khaled et al., 1970].

Role of Antitranspirants

The main role of antitranspirants is to conserve water which otherwise is lost through transpiration. Since efficient crop production depends upon timely availability of water, any water conservation practice which enhances this availability should be useful. This is particularly important in arid areas where droughts are frequent and crop growth suffers due to short availability of water. In irrigated farming the interval between two irrigations can be increased by using antitranspirants and thus water can be saved.

In many cases the primary purpose of using antitranspirants is not water conservation *per se* but improvement of plant performance by increasing plant water potential. Antitranspirants can be used to control temporary shocks to plants during transplanting. Fuehring [1973] reported the use of stomata inhibiting or film forming antitranspirants for sorghum grown with limited irrigation with 5-17% increase in yield. Under hot arid environment conditions, use of antitranspirants has been reported to check shrinkage and improve the size and quality of peach and olive fruits [Davenport et al., 1972 d].

Although antitranspirants have been found to reduce both transpiration and photosynthesis, they have been found quite useful in conserving water for making it available during stress periods for sustained growth and production especially in arid and semi arid areas.

DEEP PERCOLATION AND ITS CONTROL

Sandy soils occupy an important place in many arid and semi arid areas. These soils are single grained, have poor moisture retention and storage capacity, allow heavy leaching and percolation losses and are thus infertile and unproductive. Soils of arid areas of western Rajasthan have been found to contain 58 to 71% fine sand and 24-28% coarse sand. These soils retain about 6 to 9% moisture at 0.1 bar tension which comes to about 90 to 140 mm of water per metre of soil profile. During periods of high rainfall or high rainfall intensity about 40-50% rainfall is lost as deep percolation retaining thereby, low amounts of water in the root zone for plant use [Gupta, 1986]. Though the deep percolation water in the strict sense is not a loss as this goes to recharge the ground water and is available to perennial vegetation, yet it is not available to the field crops or plants with shallow root system. Therefore, research efforts were made to intercept this water for utilization of the crops/orchard plants.

Sub-surface moisture barrier

Sub surface moisture barrier is an artificial layer of a material with less permeability placed at a suitable depth in profile for intercepting the free flow of water. The water thus retained over the surface of barrier is low tension water and is easily available to the plants during the periods of drought. The barrier concept originated from the observation that sandy soils with clay sub soil are world's most productive soils. Erickson et al. [1968] used bentonite clay and plastic films placed at 60-70 cm depth as sub surface barriers. Later, Saxena et al. [1971] and many other workers throughout the world used asphalt film as sub surface barrier for improving the moisture retention of sandy soils and productivity of vegetable crops. Gupta and Aggarwal [1980], however, tried bentonite clay, asphalt and pond sediments as sub surface barriers and reported the use of 2 mm thick asphalt film at 60 cm depth to be most effective as sub surface barrier. Its use doubled the moisture retension capacity and reduced the percolation losses from loamy sand soil by about five times.

Though sub surface barriers were found quite useful, their large scale use could not be achieved because of non availability of suitable machinery for placement, large scale soil disturbance and high cost involved. However, sub surface barrier of bentonite clay [Singh, 1982] and pond sediments [Gupta and Muthana, 1985] were effectively used for orchard crops and establishment of trees in the arid areas. An integrated use of circular water harvesting technique with sub surface moisture barrier of pond sediments was found highly effective for the growth and yield of *Zizyphus* (Ber) (Table 8) and establishment of *Acacia tortilis* (Babul) and *Prosopis cineraria* (Khejri), plants and reducing their mortality under rainfed arid conditions of western Rajasthan.

Table 8. Effect of integrated water harvesting on the yield (kg/ha) of *Zizyphus* (ber) plants.

Treatments /Rainfall	1986 (248)	1987 (111)	1988 (270)	1989 (230)	1990 (844)	Mean
Control	195	288	971	4319	6190	2392
Integrated water harvesting	640	591	2029	8445	12077	4756

Figures in parentheses indicate the amount of rainfall received during the season.

Soil compaction

Compaction is a process of bringing soil particles closer to attain higher microporosity and thereby increased water retention in soil. Though higher levels of compaction have been reported to produce high mechanical impedance to root penetration and thus adversely affect plant growth, low levels create favourable conditions by way of increased moisture and nutrient retention. Studies conducted at Jaipur [Somani et al., 1989] revealed the usefulness of soil compaction for increasing the water and nutrient use efficiency in wheat under irrigated conditions. However, its benefits in terms of improved plant performance could not be achieved under hot dry rainfed conditions of western Rajasthan.

Agroforestry system

Agroforestry is yet another technique to control deep percolation. Different trees like *Prosopis cineraria, Acacia senegal, Acacia nilotica, Zizyphus rotundifolia, Zizyphus mauritiana* if planted in agricultural fields of the arid areas, utilize the deep percolation water efficiently, because of their deep root system. The system thus helps in meeting grain, fuelwood and fodder requirements beside protecting and preserving the environment.

REFERENCES

Abou Khaled, A., Hagan, R.M. and Davenport, D.C. 1970. Effect of kalonite as a reflective antitranspirant on leaf temperature, transpiration, photosynthesis and water use efficiency. *Water Resources Res.* 6: 280-289.

Bagley, W.T. 1964. Response of tomatoes and beans to wind shelterbelts. *J. Soil and Water Cons.* 19: 71-73.

Bhimaya, C.P., Mishra, D.K. and Das, R.B. 1958. Importance of shelterbelts in arid zone farming. In: *Proc. Farm Forestry Symp.* I.C.A.R., New Delhi, pp. 65-72.

Davenport, D.C., Fisher, M.A. and Hagan, R.M. 1972. Some counteractive effects of antitranspirants. *Pl. Physiol.* 49: 722-724.

Erickson, A.E., Hansen, C.M. and Smucker, A.J.M. 1968. The influence of sub surface asphalt moisture barrier on the water properties and productivity of sandy soils. *Trans. 9th Int. Congr. Soil Sci.*, Adelaide, pp. 331-337.

Fuehring, H.D. 1973. Effect of antitranspirants on yield of grain sorghum under limited irrigation. *Agron. J.* 65: 348-351.

Gale, J. 1961. Studies on plant antitranspirants. *Pl. Physiol.* 14: 777-786.

Ganguli, J.K. and Kaul, R.N. 1969. Wind erosion control. I.C.A.R. Tech. Bull. (Agric.) 20, New Delhi.

Gorschenin, N.M. 1941. Agricultural improvement through forestry. Moscow : Government Publishers Kelkoz Sovkhoz (in Russian).

Gupta, J.P. 1986. Moisture and thermal regimes of the desert soils of Rajasthan, India and their management for higher plant production. *Hydrol. Sci. J.* 31: 347-359.

Gupta, J.P. 1987. Effect of tillage and mulch on soil and the growth and yield of cowpea grown in arid tropics. *Arid Soil Research and Rehab.* 1: 161-172.

Gupta, J.P. 1989. Integrated effect of water harvesting, manuring and mulching on soil properties, growth and yield of crops in pearlmillet mung bean rotation. *Trop. Agric.* 66(3): 233-239.

Gupta, J.P. and Aggarwal, R.K. 1980. Use of an asphalt sub surface moisture barrier for improving the productivity of desert sandy soils. *J. Arid Environ.* 3: 215-222.

Gupta, J.P. and Gupta, G.K. 1986. Effect of tillage and mulching on soil environment and cowpea seedling growth under arid conditions. *Soil and Tillage Research* 7: 233-240.

Gupta, J.P. and Gupta, G.N. 1983. Effect of grass mulching on growth and yield of legumes. *Agric. Water Man.* 6: 375-383.

Gupta, J.P. and Gupta, G.N. 1985. Response of tomato and Okra crops to irrigation and mulch in an arid region of India. *Agrochimica.*

Gupta, J.P. and Muthana, K.D. 1985. Effect of integrated moisture conservation technology on the early growth and establishment of *Acacia tortilis* in the Indian desert. *Indian For.* III(7): 477-485.

Gupta, J.P. and Ramakrishna, Y.S. 1988. Role of shelterbelts in checking wind erosion and increasing crop production in sandy wastelands of western Rajasthan. In: *Wastelands and Their Utilization.* Scientific Publishers, Jodhpur.

Gupta, J.P., Rao, G.G.S.N., Rao, B.V.R. and Gupta, G.N. 1983. Soil drying and wind erosion control as affected by different types of shelterbelts planted in the desertic regions of western Rajasthan, India. *J. Arid Environ.* 6: 53-58.

Gupta, J.P., Rao, G.G.S.N., Ramakrishna, Y.S. and Ramana Rao, B.V. 1984. Role of shelterbelts in arid zone. *Indian Farming* XXXIV: 29-30.

Harrold, L.L., Petres, D.B., Dreibelbis, F.R. and Guiners, M.C. 1959. Transpiration evaluation of corn growth in plastic covered Lysimeters. *Soil Sci. Soc. Amer. Proc.* 23: 174-178.

Konstantinov, A.R. and Struzer, L.R. 1969. Shelterbelts and crop yields (Translated from Russian by A. Bavouch, USDA, Washington, D.C.).

Lal, R. 1974. Soil temperature, soil moisture and maize yield from mulched and unmulched tropical soils. *Plant and Soil* 40: 129-143.

Prihar, S.S. and Arora, V.K. 1980. Crop response to mulching with crop residues. Bull. Dept. of Soils, Punjab Agril. University, Ludhiana, India.

Raheja, P.C. 1963. Shelterbelt in arid climate and special techniques for tree planting. *Ann. Arid Zone* 2:

Rosenberg, N.J. 1967. The influence and implication of wind breaks on agriculture in dry regions. In: *Ground Level Climatology* (ed) R.A. Shah. Am. Assoc. Advan. Sciences, Washington, D.C., pp. 327-349.

Saxena, G.K., Hammond, L.C. and Lundy, H.W. 1971. Effect of an asphalt barrier on soil water and on yields and water use by tomato and cabbage. *J. Amer. Soc. Hort. Sci.* 96: 218-222.

Singh, H.P. 1982. Management of desertic soils. Review of soil research in India. *12th Int. Congr. of Soil Sci.*, New Delhi, Vol.II, pp. 676-688.

Somani, L.L., Gulati, I.G. and Kumawat, B.L. 1989. Effect of compaction and claying on physical properties of loamy sand soil, nitrogen use efficiency and crop yields. *Int. Symp. Managing Sandy Soils.* CAZRI, Jodhpur, India, pp. 220-224.

Stockler, J.H. 1962. Shelterbelt influence on great plains field environment and crops - a guide for determining design and orientation. *USDA Forest Serv. Produ. Res.* 62, p. 25.

Warren, G.H. 1941. Shelterbelts. *Afforestation Journal* 7(1): 8-19.

Williamson, R.E. 1963. The effect of a transpiration suppressant on tobacco leaf temperature. *Soil Sci. Soc. Amer. Proc.* 27: 106.

Zelitch, I. 1965. Environmental and biochemical control of stomatal movement in leaves. *Biol. Rev.* 40: 463-482.

Zelitch, I. 1969. Stomatal control. *Ann. Rev. Pl. Physiol.* 20: 329-350.

13

IMPROVING WATER USE EFFICIENCY UNDER DRYLAND CONDITIONS

B.R. HEGDE

DEFINITION AND SCOPE

Water Use Efficiency (WUE) has been defined in several forms depending upon its intended use. The most commonly accepted definition given by Viets [1962] is :

$$\text{Water use Efficiency} = \frac{\text{Dry weight produced (Y)}}{\text{Evapotranspiration (ET)}}$$

Quite often the yield (Y) or dry matter produced is referred mainly to grain. However, it should be remembered that in many dry areas straw is as important as grain itself. Dry matter accumulated in roots (as much as 0.1 to 0.2 of the total biomass) is not usually taken into consideration. The units of WUE are, mass produced per volume of water used. Water used is restricted to that portion removed from the soil by way of evaporation and transpiration and excludes non-productive losses occurring by deep percolation or surface runoff. Thus WUE can be expressed in various ways, such as kg or tonnes of dry weight produced per cm of water evapotranspired or as a simple ratio of kilogram of dry weight per kilogram of evapotranspiration. Water Use Efficiency in the latter dimensionless form is similar to the reciprocal of the older term, transpiration ratio except that water use efficiency also includes the water evaporated from the soil surface concurrent with the water transpired by the plant.

The hydrological definition of WUE given by Stanhill [1986] appears to be more meaningful under dryland conditions. According to this definition 'Water use efficiency is the ratio between volume of water used productively (ET) to the volume of water potentially available for the purpose'. Potential availability naturally will include the whole of rain water and the water already stored in the soil before sowing of the crop.

Onken et al. [1988] have discussed the advantages of each of the methods of expression of water use efficiency taking into consideration the precipitation (WUE_P), evapotranspiration (WUE_{ET}) or transpiration (WUE_T). When rainfall data are used in connection with field crops, WUE includes 'non-productive' water which involves all types of water loss/remains unused in the soil. This concept is useful in long term agronomic studies where evaluation of conservation and cultivation practices regarding their effectiveness in making the water resource available to crops. This form of expression is also termed as 'rainfall use efficiency' to distinguish it from the conventional use of the term WUE.

Eventhough the most common way of expression of WUE is in terms of grain yield or total dry matter produced per unit of water used, in some Indian literature we do come across the expression Rs/ha/mm. For example, Dhoble et al. [1987] have expressed the WUE of kharif crops at varying dates of sowing in Maharashtra on the basis of net return per unit of water expended in ET. This appears to be a convenient way of comparing different crops in terms of economic value when their biological productivity is vastly different and cannot be compared. However, it makes the matters more complicated in view of the highly fluctuating market prices of different commodities.

If yield were completely independent of ET, any factor which caused an increase of yield or decrease of ET, would have a favourable effect on water use efficiency. Actually, the numerator and the denominator of the formula Y/ET are not independent of each other. From purely mathematical point of view, increasing the denominator (ET) will decrease the value of WUE, but from the physiological point of view it is essential to increase ET (upto a certain level) in order to achieve an increase in the numerator Y - which is our basic objective [Arnon, 1975]. However, WUE can be improved by either increasing the numerator in the equation or by decreasing the denominator. Though both are possible, increasing the numerator seems to be more feasible. The practical conclusion is to make every effort to increase the amounts of water available to the crop for production and reduce to the minimum the losses of water due to evaporation and transpiration.

Gregory [1988] has reviewed the previous literature and come to the conclusion that about 33 to 61 per cent of the total water used is due to transpiration in wheat, barley and range soils and there is a great potential to reduce E which leads to improved WUE. In addition, increased water storage in the root zone can be achieved by control of runoff, improved infiltration and by water harvesting methods. Crop management practices such as tillage and land layout are also effective in this effort and can incidentally improve the yields.

Following agronomic practices are listed by Gregory [1988] to improve the WUE :

1. Practices for reducing evaporation from soil surface:

Mulching, modifying plant population and spacing, selecting varieties with rapid early growth, early sowing, application of fertilizers.

2. Practices for increasing water supply to plants:

Rain water harvesting, supplementary irrigation, cultivation to improve infiltration and reduce runoff, weed control, application of fertilizers, multiple/relay cropping and selecting varieties with deeper roots.

The literature available on the subject of water use efficiency is vast and some important references are listed at the end of this chapter. The scope of this chapter is to list out practices that help in reducing the evaporation (E) and transpiration (T) while aiming at improving the yield component (Y). Passing references will, however, be made to the methods available to conserve more water in the profile though it is a topic for discussion in another chapter.

IMPROVING WATER AVAILABILITY

In dry regions, water availability is the major limiting factor in crop production. Even when the rainfall is fairly high it is often lost as surface runoff when the soil surface is poorly structured and lost as deep percolation in highly porous soils. Therefore, the major focus in dryland farming system should be to improve the surface infiltration and retention of moisture within the root zone in the soil profile. As a second step, the stored water should be protected against evaporation so that the maximum quantity of water is made available for plant absorption and subsequent transpiration. In the third step, a proper selection of crops and their management should lead to complete use of stored water for biomass production by the crops. Even with all the efforts of moisture conservation *in situ* it is inevitable that high intensity rains cause runoff. Such runoff could be stored in depressions and dug out ponds and used for irrigations during periods of moisture stress. This is a package for achieving a high rainfall use efficiency (RUE) but may not necessarily mean a high water use efficiency.

It is possible that a set of practices leading to high WUE (Y/ET) may not always result in complete use of water from the root zone and may not fully make use of the available rain water. A short duration crop grown on a fully saturated profile will certainly result in high WUE but may waste lot of residual moisture unutilized which cannot be carried over to the next season.

Tillage practices

Cultivation of soil with various implements provides more pore space for storage and increases the amount of water stored. Another, perhaps more important objective is to keep the surface soil loose so that a higher rate of infiltration is maintained and runoff reduced. This statement is more relevant to dry farming areas where the rate of infiltration is low and the intensity of rainfall is high.

Musgrave and Free [1936] found that cultivation of the surface greatly enhanced the water intake of soils. For example, the infiltration rates of water into the field soil (a) in its natural packing, (b) cultivated 4 inches deep and cultivated 6 inches deep were 0.85 inch, 1.77 inches and 1.87 inches per 15 minutes, respectively. The first hour rates for the same conditions were 1.49", 2.31" and 2.93", respectively. After one hour the curves were practically parallel. This shows that the cultivation reduced the runoff in the beginning of the storm considerably. In the absence of tillage, the highly crusting red soils produce as much or even more runoff than the low permeability vertisols under similar rainfall situations. At ICRISAT, Patancheru, India, it was observed that with a total rainfall of 499 mm the runoff in cropped alfisol was 140.6 mm as against 72.0 mm in cropped vertisols [Krantz et al., 1978].

Larson [1962] has pointed out that pulling a tillage implement through soil results in the total porosity and thickness of the tilled area being greatly increased. For example, a soil layer of 7" thick before ploughing may store 3.3" of water whereas after ploughing the same layer may detain or temporarily store 6.1" of water and the total porosity increased from 47 to 62 per cent. The extra storage volume is temporary, however, because the loosened layer generally settles when water is added.

From their critical review of the available soil management practices for efficient water use, Unger and Stewart [1983] have concluded that surface roughness and microdepressions play a great role in higher retention of water in the soil profile. Profile modification by very deep ploughing (40-60 cm depth) with mould board ploughs in soils with sandy surface but clayey subsurface resulted in better storage at the surface layers.

Tillage refers to different mechanical manipulations of the soil used to provide the necessary soil conditions favourable to the growth of crops. Tilth is the physical condition of the soil in its relation to plant growth. In dry lands, tillage is required (i) for moisture conservation, (ii) to break the hard pan and help root penetration, (iii) to control weeds, (iv) provide better aeration, and v) to provide a good seed bed. One of the good examples of tillage for moisture conservation is the work carried out at Bangalore under red soil conditions (Table 1).

Table 1. Penetration of moisture as influenced by ploughing before the onset of rains.

Depth of soil (cm)	Per cent moisture after a total of 81mm rain in May	
	Ploughed area	Unploughed area
0-15	10.74	3.59
15-30	13.22	7.13
30-60	13.27	8.59
60-90	13.44	Dry

Source : Dryland Agriculture Project, Bangalore [Anon., 1983].

Tillage operations like ploughing and chiselling open up the hard layers of soil and improve infiltration. Depth of penetration increases and thus the runoff reduces.

One of the core projects under All India Coordinated Research Project for Dryland Agriculture initiated in 1971 was to study the beneficial effects of off-season or early tillage. The results were quite revealing and benefits were striking both in terms of water stored and the crop yields obtained almost in all types of soils [Sriram et al., 1981].

While surface tillage helps in improving the

infiltration of rain water into the soil, contour cultivation further helps increasing the temporary storage capacity in furrows opened across the slope. It is estimated that shallow furrows of 8-10 cm depth opened 20-25 cm apart, can hold about 15 mm of water in the furrow depressions. By this, the surface water will get greater opportunity time for entry into the soil profile.

increased considerably the yield of maize as well as of the following wheat. In an year of more favourable rainfall viz., 1973-74 tillage treatment did not affect the yield. During the high rainfall year of 1975, deep ploughing and chiselling slightly reduced the yields (Table 2).

Cultivation on the contour helps in reducing the runoff and *in situ* moisture conservation.

Table 2. Effect of tillage treatments on crop yields (kg/ha) and residual moisture (mm, 180 cm depth) [Ranjodh Singh et al., 1983].

Treatments	Maize yield			Residual moisture			Wheat yield		
	1973	1974	1975	1973	1974	1975	1973	1974	1975
Shallow ploughing	1960	1950	2740	375	259	325	2450	630	4060
Deep ploughing (25 cm)	2250	2430	2210	373	257	349	2470	1300	3870
Chiselling (1 m apart)	1970	2550	–	392	286	–	2410	2050	–
Chiselling + organic residue	2170	3390	2390	400	302	344	2610	1820	4010
LSD (0.05)	NS	450	330	–	–	–	NS	890	NS
Rainfall(mm)	782	577	822	–	–	–	77	152	227

Deep tillage has been extremely beneficial in the shallow soils of Anantpur, Andhra Pradesh, India. In this area, the soils are as shallow as 6-30 cm with compact murrum underneath. Opening up of this compact layer helped in increasing the infiltration from 3.5 cm/hr in shallow tillage to 6.0 cm/hr in deep tillage. Depth of root penetration increased by 15-20 cm and the crop yields improved. Eventhough, deep tillage with tractor drawn implements is highly useful it may at times be difficult for resource poor farmers. Under such situations, deep tillage with pedda madaka (big wooden plough), using two pairs of bullocks was found to be as effective as that of tractor drawn plough. The residual effect of such an operation was observed in the subsequent three years and about Rs 60/- per ha was the extra expenditure only in the first year [Chary et al., 1979].

In the sandy loam soils at Ludhiana, Punjab, the effect of deep tillage on crop yield was strongly linked with the amount and distribution of rainfall [Ranjodh Singh et al., 1983]. During 1974 when the monsoon withdrew quite early causing moisture stress, deep ploughing and chiselling had large favourable effect on soil moisture and

Studies carried out on the farmers' fields in the Operational Research Project, Kabbalanala Watershed in Karnataka revealed the effectiveness of contour cultivation in prolonged availability of moisture after cessation of rains [Nagaraju et al., 1990]. Other crop management practices like opening ridges and furrows have also shown beneficial effects. Crops like maize are either planted in a ridge and furrow system across the slope or earthed up after 45 days of sowing to create ridges. The ridges thus formed have been found to reduce the runoff. Observations with maize crop during 1980 on the Dryland Project Farm at Bangalore have shown that the runoff with the ridges and furrow for each crop row on a gradient of 0.3 to 0.4 per cent was 17.77 per cent of the total rainfall as compared to 31.21 per cent in bedding system (furrows across the slope after 3 m). Further observations at the same place have also shown that runoff from maize plots after ridging up was almost nil even with rains that caused 17-22 per cent runoff from fallow land.

Inter-terrace management practices like vertical mulching, tied ridging, scooping and compartment bunding provide more opportunity time for infiltration of rain water by its surface impounding and make more water available for crop growth [Patil et al., 1981]. Even with such conservation practices, there will be inevitable runoff to an extent of 20-40 per cent in the dry farming areas with 750-850 mm rainfall. El-Swaify et al. [1985] have reviewed the management practices both under alfisol and vertisol situations of the SAT and have very clearly shown that at ICRISAT Center, Patancheru only about 40 per cent of the total rainfall was used for evapotranspiration during the growing season and more than 25 per cent was lost as runoff. Deep percolation losses, however, were low in vertisols (9.5%) compared to 33 per cent in alfisols. Their observation in alfisol watersheds has also indicated that high runoff produced during the early part of the season provides a dependable surface water source for supplemental irrigation for much of the crop season. Even in low rainfall years, with least runoff of the nine year period (1974-82), they were able to store 35 mm of water (on a catchment area basis) during periods with high drought probability. It is of great interest to note here that the potentialities of runoff water harvesting are bright throughout the SAT areas where high intensity rains are common. A very good review on efficient resource management systems for drylands of India is brought out by Venkateswarlu [1987].

Fallowing for moisture conservation

Traditional dryland farming systems of deep vertisols involve leaving the land fallow during rainy season and raise crops only during the post rainy season on profile stored soil moisture. As the main intention of fallowing is to provide sufficient moisture for the main crop, the land during the fallow period is kept weed free either by frequent cultivation (cultivated fallow) or by chemicals. In areas with low and relatively undependable rainfall, the system assures at least one crop during the post rainy season and also helps in taking a decision on whether the moisture stored is sufficient to raise a successful crop or not.

In many sub-tropical and temperate regions fallowing has been used to increase the supply of nutrients or water in a soil for production in a subsequent crop. In nutrient - poor soils, fallow has been used primarily to maintain the productivity of the soil over time [Steiner et al., 1988]. Many of the earlier studies in North America have been able to get a very good correlation between the wheat yield and the amount of rain received during the fallow period. The type of relation obtained by Cole [1938] is found to be applicable to most of the dryland situations where crops are raised on conserved moisture.

$$\hat{Y} = (\text{Precipitation} - 8.02)\ 2.19 \text{ bushels/acre}$$

[Cole, 1938]

Based on the equation, 8.02 inches of precipitation are needed for zero yield of spring wheat in the northern plains thereafter 2.19 bushels of wheat per acre can be expected per inch of additional precipitation.

Fallow is efficient in conserving moisture only if the distribution of rainfall during fallow is such that water penetrates at least upto 20 cm depth, because moisture in the first 20 cm can be easily lost by evaporation. Therefore, the conditions for infiltration must be satisfactory.

$$\text{Efficiency of fallowing (E)} = \frac{M_2}{M_1 + R \text{ or } SC} \times 100$$

whichever is less

where, E = Efficiency of moisture conservation

M_1= Available moisture at the beginning (cm)

M_2= Available moisture at the end of the period (cm)

R = Rainfall received (cm)

SC= Storage capacity of the profile.

However, in most of the recent references, fallowing efficiency is simply taken as the proportion of the total rainfall conserved in the root profile to the total quantity of rainfall received during the fallow period expressed on percentage basis.

$$\text{Fallow Efficiency} = \frac{\text{Water stored in root profile}}{\text{Fallow period precipitation}} \times 100$$

[Steiner, 1988]

For fallow system to be more effective it should be combined with conservation tillage systems that leave the crop residue on the soil surface. In the conventional fallowing systems with 7-10 tillage operations the fallowing efficiency was seldom more than 18 to 23 per cent whereas with limited tillage, efficiencies as high as 45-55 per cent have been achieved [Greb, 1979].

Fallowing is likely to be useful where,

(i) rainfall during the fallow period is sufficient to wet the surface layer of soil upto 30 cm depth.

(ii) soil is heavy enough to store water in the first 100-120 cm depth. But it should neither be a coarse sand nor a heavy clay.

(iii) the soil is weed free.

(iv) the soil at the surface is loose.

In the classical experiments reported by Kanitkar et al. [1968] growing sorghum in alternate years at Bijapur, India, gave 67 per cent higher grain yield and 64 per cent increased straw yields over continuous growing of sorghum, on an average of six years. The reduced expenditure on cultivation was found to compensate for the loss of yield in years of fallow (Table 3).

Table 3. Yield of post-rainy season sorghum in black soils with and without fallowing (kg/ha).

Treatments	Grain	Straw
Fallow alternate years	490	803
Sorghum every year	292	436

The experimental results at Raichur and Hagari in Karnataka with rabi sorghum and cotton have also given increased yields though not of the same order.

Favourable effects of summer fallowing in the Weibei Loess Plateau region of China have been reported by Li et al. [1988]. Planting two crops per year resulted in lower yields than did planting only one crop per year (Table 4).

Table 4. Crop yields (t/ha) for different cropping systems in Weibei Loess Plateau of China.

Cropping system	First year		Second year		Third year		Total in 3 years
	wheat	corn	wheat	corn	wheat	corn	
One crop a year	3.00	–	3.20	–	2.05	–	8.25
Two crops a year	1.45	1.45	1.10	1.45	0.66	0.93	7.03

Source : Li. *et al*, 1988.

Papastylianou and Michael [1988] analysed the conditions in the Mediterranean region and have concluded that the extent and value of conservation by fallowing depended on soil factors (texture and depth), management factors (type of fallow and time of tillage) and the rainfall itself. In particular, the soil depth is critical for water storage. During the long, hot Mediterranean summer, any plant available water in the top 30-45 cm is usually lost by evaporation, only water stored below that depth could be useful to the next crop. Shallow soils therefore have little value for water storage. Similarly, if the fallow year rainfall is less than 200 mm, its penetration is only shallow and no water is stored for the subsequent year, regardless of soil depth.

Considered from several angles, fallowing is not an efficient system for moisture conservation. Summer fallow increased wind and water erosion and resulted in decline of organic matter levels [Steiner et al., 1988]. With increasing world demand for food and the realization of the need for keeping the land covered with vegetation to reduce the risk of soil erosion, general use of fallow is being questioned even in countries where plenty of land is available for agriculture.

The consequences of traditional rainy season fallowing in the dependable and high rainfall areas are further analysed by El-Swaify et al., [1985] in terms of directly reduced land productivity and the frequent occurrence of excessive water losses as runoff and soil loss by erosion. If a crop is not grown in the rainy season the potential utilization of received rainfall (by a post-rainy season crop) is limited to the moisture storage capacity of the soil profile plus whatever little rainfall may occur in that season. Hydrologic studies of this traditional system have shown that of the total rainfall potentially available, about 28 per cent was lost through runoff, 24 per cent lost through evaporation from the bare fallow soil 9 per cent through deep percolation and only 39 per cent is actually utilized for evapotranspiration by a post-rainy season crop. A Broadbed and Furrow (BBF) system of land configuration developed at ICRISAT centre, India meets the requirements of utilizing vertisols for kharif cropping where fairly high amounts of (750-1250 mm) assured rainfall is received.

CROP MANAGEMENT AND WATER USE

Efforts made to improve the crop yields by adjusting population, row width, intercropping and fertility management have a tremendous influence on the total quantity of water used and hence have a role in influencing the water use efficiency. Earlier reviews on these aspects by Viets [1962], Pendleton [1966], Loomis [1983] and Stewart and Steiner [1990] have thrown sufficient light on the subject and are worth reading.

Plant Population : It has been a general belief that a lower population is always good for dryland crops as this will result in lesser demand on the conserved water. Earlier studies by Kanitkar et al. [1968] have also supported this idea where the cropping was mainly under conserved moisture situations. Contrary to this belief, many crops have done well under higher densities. For example, the studies carried out under the All India Coordinated Research Project for Dryland Agriculture have shown that rabi sorghum at Solapur, Maharashtra performed better at 1,25,000 plants/ha than at the prevailing recommendation of 75,000 plants/ha. At Bellary, Karnataka, when the population was increased from 34,000 to 1,28,000 plants per hectare the yield of sorghum variety R-16 increased from 3370 kg/ha to 5970 kg/ha. The optimum plant population for sorghum at Bijapur, Karnataka was 1,80,000 plants/ha as compared to the traditional practice of 1,00,000 plants/ha. With higher populations there was marginal increase in grain but substantial increase in fodder.

Selecting an optimum plant population and row width continues to be a challenging decision for the dryland farmer. Because, too high or too low plant densities can reduce grain yields as high densities use too much of the available soil water early in the season and low densities do not fully exploit the available soil water.

Early and fast cover of land by crop plants to a large extent helps in reducing the E component of ET. Plant density is one of the major determinants of evapotranspiration by its leaf area coverage. It does not really matter whether this coverage is by better fertilization of the lesser number of plants or by higher seed rate. Viets [1966] observed that usually, the increased yield in a wet year ET remains almost constant though yield may increase. However, another difficulty in a dry year with limited stored water can be that most of the stored water may be exhausted during vegetative growth leaving little moisture at grain filling/maturity stage which again calls for striking a balance between the plant population and the available water.

Planting geometry

Planting geometry refers to the spacial arrangement of plants or colonies in the field. Normal geometry can be square, rectangular, elongated rectangle or irregular. These arrangements for a given population can be adjusted either by changing the inter-row or intra-row distances. Such adjustments have special effects on aspects like utilization of net radiation, root growth nutrient uptake and water use. For example, Steiner [1987] observed that the net radiation was 5 per cent higher over wide compared to narrow rows (0.76 and 0.38 m, respectively) when averaged over three population levels from 33 to 110 days after sowing.

The need for suitable geometry under dryland condition is mainly for better utilization of available water. Rectangular planting is found to be more advantageous than square planting for crops raised under conserved moisture situations. When plants are spaced closely within a row and increasing the distance between rows, the soil moisture supply is not exhausted as rapidly as in narrow rows. Brown and Shrader [1959] and Karchi and Rudich [1966] observed that when soil moisture was not severely limited, highest grain yields of sorghum were obtained with narrow row spacings (50 cm or less) and high populations, whereas in dry climates wider row spacings were desirable.

Water extraction in the initial stages was usually higher with closer row spacings leaving smaller amounts for future use [Hegde et al., 1976]. In wide rows, usually the inter-row contained more moisture where the roots penetrated and derived their requirements. Though there is a feeling that wider rows may encourage more evaporation, in reality the rate of evaporation comes down due to surface drying in a short time. Experiments conducted by Bhan and Misra [1970] showed that in a poor rainfall season with a total precipitation of 186.8 mm (plus an equivalent of 75.1 mm stored in the soil upto a depth of 1 m before sowing) the highest water use efficiency was obtained from a planting distance of 60 x 25 cm. In the following year with more favourable moisture conditions due to rainfall of 357.3 mm (plus 67.2 mm stored water) the highest WUE resulted from 45 x 25 cm spacing. In both the years the close spacing of 30 x 25 cm resulted in lowest WUE.

Arnon [1972] postulated that when soil moisture was limited, there would be an advantage by increasing the inter-row spacings and decreasing the intra-row spacings. Intra-row competition would result in smaller plants and inter-row reservoir of soil moisture would be maintained over a longer period of the growing season. The results of research under rabi situations at Solapur, Bijapur and Bellary in India, support this view.

It is difficult to predict the season especially under kharif rainfall situations. Therefore, it may be advisable to start the season with permissible optimum spacing and optimum population and later on thin out the alternate or third row to provide for wider spacing or by removing the unhealthy seedlings within a row itself if the season is not favourable. The experience at Bellary indicated that in subnormal years every third row of sorghum should be removed to obtain higher yields [Venkateswarlu, 1983].

Table 5. Effect of row spacing on the yield of rabi sorghum.

Variety	Spacing (cm)	Grain yield (kg/ha) (Average of 3 years)
CSV 7R/8K	45	2620
	90	2860
M–35–1	45	2340
	90	2450

Source : Annual Report, Bellary Soil Conservation Centre

Intercropping and Mixed Cropping Systems

Most of the common intercropping systems adopt two or more crops with different root systems. It is a general observation that when a deep rooted crop is intercropped with shallow rooted cereals the extraction of moisture from the entire root profile is complete. Because of this, higher water use efficiencies are also observed. In an intercropping study involving barley and other crops viz., wheat, chickpea, sarson and linseed at IARI, New Delhi it was observed that the consumptive use of water in barley + sarson combination was 174 mm compared to 149 mm in sole crop of barley. In addition, the extraction of moisture from deeper layers (90-120 cm depth) was also higher in intercropping system [Hegde, 1964]. In other studies also higher water use efficiency and higher water use was observed in intercropping than in sole crop system. Singh [1985] observed that WUE in the intercropping systems under rainfed conditions was 58.8-89.5 kg grain/ha/cm compared to 51.3-84.3 kg/ha/cm in pure stands of pearl millet.

Agroforestry systems which occupy the land for the whole year can make use of the rain water more effectively than annual crops which may occupy the land for 4-6 months only. In a study by Ong et al. [1991] at ICRISAT Centre, India involving a two year hedge-row intercropping experiment with single hedgerows of perennial pigeon pea and alleys under four rows of groundnut, it was found that the total annual transpiration in pigeon pea was 887 millimetres corresponding to 84 per cent of the annual rainfall compared to 313 mm in groundnut during 1988 and 297 mm in 1989. Thus, the total transpiration was considerably higher under hedgerow intercropping with perennial pigeon pea and groundnut than under groundnut planted in pure stands and the system enhanced the utilization of rainfall leading to overall increased productivity. However, the trees tended to compete with crops for moisture which depressed crop production.

Fertilizer Use

The total amount of water used being dependent on the energy input from the solar radiation, is fairly conservative especially under adequate supply of water. While the ET remained constant, the total yield (biological or economical) can tremendously be increased by improving the nutrient status of soils. In one of the earlier studies, Zubriski and Norum [1955] from the analysis of 12 North Dakota trials showed that there was only 7-8 inches of total water, not available, remaining in the soil profile to a depth of 5 feet whether the wheat without fertilizer yielded 13.1 bushels/acre or 17.9 bushels with fertilizers.

Fertilization does sometimes permit deeper penetration of the soil by roots and thus the amount of water available for extraction is increased. Under dryland conditions, the deeper subsoil frequently contains no available water for exploitation but under more humid conditions subsoil moisture is a resource that can be tapped. Large number of field experiments under Indian conditions have shown that yield increased without substantial increase in the total quantity of water used (Table 6).

Table 6. Water use efficiency as influenced by fertilizer use in finger millet.

Fertilizer level (kg/ha)	Grain yield (kg/ha)	Water used (cm)	WUE (kg/ha/cm)
Control	2318	33.60	68.99
$N_{25}P_{25}$	2656	33.89	78.37
$N_{50}P_{50}$	3076	35.92	85.64

Source : Hegde *et al.*, 1978.

In general, the ET values will increase with availability of water in the soil, but not much influenced due to fertility especially under humid and sub-humid conditions.

In a study carried out on wheat by Ramig and Rhoades [1963] in western Nebraska, application of N upto 80 lb/acre hardly made a difference of 1 to 2 inches in the total water used but water use efficiency improved substantially due to fertilizer application.

Deeper soils normally store more available water even under the same rainfall situations and account for higher ET. The studies carried out at Solapur to identify suitable crops for different depths of soil and amounts of water stored have led to the conclusion that for good production of rabi crops, soils used should not be less than 60 cm in depth [Patil et al., 1981].

There are examples under sub-humid conditions where application of nitrogenous fertilizer substantially increased yields of winter wheat and also increased total water use [Brown, 1971]. Without N, soil water extraction was largely limited to the upper 91 cm while with N wheat plants extracted water to double this depth. Placement of fertilizers in the soil has also a great influence on the root proliferation. The classical example is that of Gliemeroth [1955] quoted from Black [1966] (Table 7).

Table 7. Distribution of barley roots in a loam soil with NPK fertilizer placed at different depths.

Depth of soil layer (cm)	Percentage of total weight of roots		
	Fertilizer in 0-18 cm layer	Fertilizer in 18-36 cm layer	Fertilizer in 36-54 cm layer
0-18	50	35	27
18-36	30	45	33
36-54	19	20	40

Though the total amount of water used under fertilized as well as unfertilized conditions remains almost the same, the pattern of extraction changes with fertilizer application. Linscott et al. [1962] studied the moisture uptake pattern in fertilized and unfertilized corn in Nebraska. In their study they observed that water was used more rapidly from fertilized than from unfertilized soil while the water supply was ample, and more slowly after the available water was exhausted. The total evapotranspiration during the period of measurement was 22.2 and 22.4 cm on the unfertilized and fertilized plots and the corresponding yields of corn grain were 1.8 and 2.5 metric tonnes/hectare (Fig. 1).

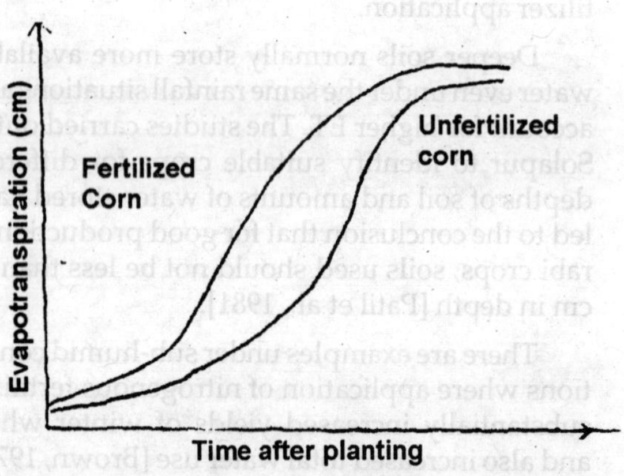

Fig. 1. General trend of water used under fertilized and unfertilized conditions [Linscott et al., 1962].

Based on the type of curve, it may be concluded that all the fertilizers may not be applied as basal and some amount could be saved for top dressing at appropriate time of growth. In case the season is not favourable, the amount that would otherwise have been spent on fertilizers can be saved. Split application of N, especially under kharif cropping situations has usually resulted in higher yields [Singh and Das, 1984].

As Loomis [1983] has observed in his review on 'Crop manipulations for efficient use of water', improved nutrition alters rooting habit thus changing access to stored moisture. In drylands, timing of fertilizer application has great significance. Early application results in better expansion of leaves, use of stored water during vegetative stage and serious moisture deficiency can develop late in the season.

Organic manures like Farm Yard Manure (FYM) can cause similar effects as fertilizers. Earlier observations at Hagari in Karnataka have clearly shown that during low rainfall years, addition of FYM at 10,000 lb/acre resulted in higher straw but lower quantity of grain [Kanitkar et al., 1968]. Even during the recent years, the studies of Eck [1988] in Bushland, Texas, have shown that excessive application of beef feed lot manure increased wheat growth but reduced grain yield. The reduction in wheat yield is attributed to increased growth expanding available water before the grain filling period. However, correction of nutritional deficiency by addition of manures and fertilizers is one of the most important options for higher yield and thus help improve the water use efficiency. What is more is the early maturity possible under dryland conditions by correction of nutritional deficiency which helps in maturing the crop when there is still some moisture in the soil.

EVAPORATION CONTROL

Since most of the water loss from agricultural land occurs through surface runoff or evaporation from soil and plant surfaces, it is logical that the most successful water conservation procedures relate to runoff and evaporation control.

Lemon [1956] describes three stages of evaporation from soil surface. The first stage is characterised by a rapid and steady loss of water with the rate dependent on the net effects of water transmission rate through the soil and above ground boundary conditions such as wind speed, temperature, relative humidity and radiant energy. The second stage is one of rapid decline in the rate of loss as the soil reservoir is depleted. Here, the above ground conditions are no longer as important and the intrinsic soil factors assume a dominant role in governing the rate of moisture flow to the surface. The third stage is extremely slow and is governed by adsorptive forces of molecular distances at the liquid solid interface. Lemon [1956] points out that the greatest potential for suppressing evaporation from soil lie within the first and second stage. Potentials include methods for (i) decreasing turbulent transfer of water vapour to the atmosphere, (ii) decreasing the capillary continuity, and (iii) decreasing capillary flow and moisture holding capacity of the surface layers.

Air turbulence near the soil surface can be reduced by plants, wind breaks, straw mulch, stone mulch and rough soil surfaces. Good review on various mulching practices are brought out by Jacks et al. [1955] and Unger [1975].

A mulch is any material at the soil surface used for covering with a primary objective of reducing the evaporation loss. Numerous soil properties and conditions are influenced either directly or indirectly. Among those favourably influenced are soil water through runoff control, increased infiltration, decreased evaporation and weed control. Gardner [1959] has pointed out that surface mulch will be useful only when subsurface soil has plenty of water. In the laboratory, when dust mulch ground corn cob and gravel were tested, only gravel was found to be very effective and dust much was least effective. These results are consistent with the concept that a perfect mulch must have an artificial capillary break or poor capillary conductivity. Bertrand [1966] observed that artificial mulches are not more effective partly because of the 'self-mulching action' of the moist soils. As soil dried 'a dry mulch' or 'a dust mulch' usually forms. Since most of the difference in moisture content between a mulched and a bare soil is found near the surface, they are very useful in seedling establishment although the total difference in water content between the mulched and bare soil may be very small. As Lemon [1956] has indicated, stubble mulch can decrease evaporation significantly as long as the surface remains wet. After the soil near the surface dries, the effect of mulch decreases in importance and cumulative evaporation from a mulched or bare soil is nearly equal after prolonged drying.

Crop residue mulches

Crop residues and soil mulch are of importance in the Indian context. Stalks of various crops like maize and bajra, groundnut haulm and shells as well as paddy husk and coir pith dust have been used with success. Soil mulch is practical both in kharif as well as rabi situations. In kharif it helps in reducing the weed population and increasing infiltration whereas in rabi its main function will be to reduce evaporation.

Prihar and Arora [1980] have summarised the work on mulching with crop residues in Punjab. As observed by them, the most important effect of mulching was its impact on erosion control and runoff reduction. A four year study on the alfisols at ICRISAT, Patancheru, India [1986] has conclusively proved the beneficial effects of groundnut shell mulch on crop yield and soil loss reduction.

Table 8. Effect of groundnut shell mulch on crop yield, runoff and soil loss of Alfisol at ICRISAT Center, Patancheru (1981-85).

Rate of groundnut shell mulch t/ha	Crop yield (t/ha) (Intercropping)		Runoff (mm)	Soil loss t ha^{-1}
	Sorghum	Pigeonpea		
No mulch	2.07	0.84	282	3.92
2.5	2.27	0.97	147	1.95
5.0	2.42	1.11	88	1.21
10.0	2.63	1.34	26	0.35
SE ±	0.17	0.09	13	0.23

Source : ICRISAT Annual Report, 1986.

The surface mulch was incorporated into the soil in the subsequent years and the addition of organic matter had its own advantages on soil physical properties and crop yields.

At the Dryland Agriculture Project, Bangalore, maize residue incorporation at the rate of 4 t/ha per annum continuously for three years revealed the effects of mulching in a dry year with only 652 mm annual rainfall. The moisture content at the sowing time in residue incorporated plots was 11.24 per cent and 14.03 per cent in the surface at a depth of 0-15 cm and 15-30 cm when compared to 8.50 and 13.39 per cent, respectively at the same depths without residue, grain yield of finger millet with residue incorporation was 3528 kg/ha compared to 2521 kg/ha without residue [Anon., 1983]. Giri et al. [1983] have found that in North-West India WUE increased from 5.32 to 9.31 kg/ha/mm as a result of adding straw mulch because it increased the amount of stored soil water available during the flowering and grain filling growth stages.

Reviewing the extensive work carried out at the centres of the AICRP for Dryland Agriculture, Randhawa and Venkateswarlu [1980] concluded that mulching has very little place in crop production during rainy season, but where soil mulch is created as part of weed control practice, it may serve a purpose. However, post rainy season experimental data have invariably proved it to be a beneficial practice (Table 9).

Table 9. Effect of mulches on yields of post-rainy season crops.

Center		Crop	Yield of grain (t/ha)	
			Control	Mulches
Hoshiarpur	(Inceptisol)	Wheat	2.86	3.51
Dehradun	(Inceptisol)	Wheat	2.33	2.93
Anand	(Entisol)	Tobacco	1.33	1.84
Kovilpatti	(Vertisol)	Sorghum	0.53	0.94
Varanasi	(Entisol)	Barley	1.75	1.91
Solapur	(Vertisol)	Sorghum	0.98	1.64

Source : Randhawa and Venkateswarlu [1980]

Polyethylene mulches

One of the most promising practices for reducing evaporation is to place vapour barriers like polyethelene films. However, complete cover of the soil surface with such films may not be desirable as it prevents the entry of rain water into the root zone. In the study carried out by Willis et al. [1963] in North Dakota 90 per cent cover gave very high yield advantages, where soil moisture between August and May was about 4" under a 90 per cent film cover compared to 1" in bare plots.

Table 10. Influence of soil surface covering on corn yields and moisture conservation.

Treatment	Grain yield (lb/acre)	Water used (inches)	WUE lb/ac/inch
1960 (wet year)			
Check	2860	15.9	180
Ridged, 90% cover	3870	14.9	260
1961 (dry year)			
Check	1450	6.3	230
Ridged, 90% cover	3510	7.8	450

Source : Willis et al. [1963]

Plastic Film Mulch Farming (PFMF) has become popular in several dry farming areas of the world. Ma Shijun [1988] has described two methods of PFMF in North China.

(i) Sowing before mulching and perforating the film at the early seedling stage

(ii) Mulching and then perforating the film and then sowing.

The first method is adaptable to all kinds of sowing, especially to machine work. It is necessary that within row or in hill seeding, the rows or hills be 3 to 4 cm under the mulch to prevent young seedlings from touching the plastic film directly thus avoiding burn damage. There is also a green house effect under the film for about a week before perforation. The second method is good for soil-water conservation because the fields are covered with plastic film in early spring when soil is still frozen and a large amount of the thaw water can be kept in the soil. This method, however, requires more manual labour at seeding.

The good performance of covered plots was due to increased soil temperature, better utilization of light and suppression of evaporation losses. Ridges formed also help by diverting small amounts of water into the ditches which otherwise may be lost by runoff and evaporation. For large scale adoption of plastics, economics is the major hurdle.

Gravel Mulch Farming

Large pores between gravel permit rapid infiltration of water to the soil but retard evaporation. Water can return to the atmosphere only in vapour phase. Thus gravel tends to act as a one way water valve for the soil. Gravel sand mulch farming system as practised in China is described by Ma Shijun [1988] as follows.

Before mulching, the field should be ploughed deeply, manured with large amounts of farm yard manure, harrowed and levelled in the summer fallow time as for preparing a seed bed. Gravels and sands of mixed mulch pebbles, gravels and sand in different sizes taken from river alluviums and glacier deposits are spread to a thickness of 10-16 cm to the frozen soil in winter. During the growing season, the soil or mulched fields are not tilled. There are special tools for sowing and manuring to prevent the mulch layer from being mixed with the soil.

The porosity of the mulch can be maintained for about 30 years. When the gravel and sand mulch is mixed with soil to an extent of about 25 per cent by weight, its functioning will diminish gradually. Then the old mulch should be removed.

Series of experiments and observations on the gravel-sand mulched fields indicated that water content (March-September) increased by 2-7 per cent in the surface 20 cm depth, evaporation reduced by 30 per cent and temperature below the mulch increased by 1.5 to 2.3°C.

In Gansu and Qinghai provinces, soil water content under sand mulch was higher at any time compared to no mulch (Table 11).

The chief difficulty of putting gravel-sand mulch farming into practice is that it requires much strenuous manual labour, that is why the practice is limited to a small area. The practice appears to be suitable for tree and fruit crops where small area of the basin can be manually maintained.

Table 11. Water content (g/kg) in sand covered land.

Treatment	Mar.	Apr.	May	June	Jul.	Aug.	Sep.
Sand covered	134	123	133	157	126	138	189
General area (not covered)	68	62	71	79	88	85	102

Source : Li *et al*,. 1988.

Cultivation

As a result of the classical experiments of King [1907] on the effect of dust mulch (loose surface soil) of 5-8 cm deep, on the evaporation of water from soils, the generally accepted reason for cultivation has been to conserve soil water. Following King's experiment and conclusions, a soil mulch for moisture conservation became standard procedure and dicta such as 'two cultivations can replace one irrigation' became very popular [Arnon, 1975]. The Chinese proverb 'hoes have water' gives an idea about the farmers' view on this aspect. Subsequent studies, however, have disproved the theory of dust mulch except in soils with high water table. The too frequent cultivations impaired soil structure and increased the soil loss by erosion. Surface cultivation in hard soils provides better aeration and in cracking vertisols helps in covering up the cracks to avoid direct evaporation from deeper layers.

Surface cultivation as a measure of evaporation control was tried by Bhumbla and Prihar [1968, unpublished] in Punjab and the practice was found to be useful in reducing the loss on sandy loam as well as silty clay loam soils in winter but not in summer (Table 12).

Table 12. Average rate of moisture loss from mulched and unmulched plots for 16 days after treatments.

Treatments	Water loss (mm/day)			
	Winter		Summer	
	Sandy loam	Silty clay loam	Sandy loam	Silty clay loam
No mulch	1.94	2.06	2.64	2.56
Soil mulch	1.19	1.03	3.16	1.92

Source : Bhumbla and Prihar, 1968

In the light of the results, it can be concluded that intercultivation for moisture conservation is advantageous. Most of the dry region crops are grown in kharif and rabi, but not in summer and in winter season the soil mulch has a definite advantage.

TRANSPIRATION CONTROL

A large portion of water lost in cropped fields is due to transpiration. Gregory [1988] has concluded that about 33 to 61 per cent of water lost in wheat, barley and range soils is due to transpiration. The practical approaches to reduce transpiration by making use of anti-transpirants are reviewed by Davenport and Hagan [1975].

An anti-transpirant is any material applied to transpiring plant surfaces with the aim of reducing water loss from the plant. Materials used for reducing transpiration have been grouped into following categories:

(i) Metabolic inhibitors that induce partial stomatal closure
(ii) Film forming materials
(iii) Radiation reflecting materials
(iv) Growth retarding substances which induce plant stress

Anti-transpirants may be useful under dryland conditions in two ways :

(i) Conservation of water supplies particularly where water is scarce and costly
(ii) Improved plant performance via an increase in plant water potential

The effectiveness of an anti-transpirant depends on the extent to which the control plants are subjected to and also on the stage of spraying. They are most effective when water losses from the leaves are not restricted by natural stomatal closure in response to leaf water deficits or darkness. Thus an anti-transpirant will be of little benefit to an already wilted plant but delays wilting onset if applied when the plant is transpiring fully. While anti-transpirants of reflecting type reduce leaf temperature, the film forming and stomata closing types tend to increase leaf temperatures. Stanhill [1986] suggests that reflecting type of anti-transpirants which are selective especially for the wave lengths of light which are not much effective in photosynthesis is a practical way. Even non-selective reflectants like kaolin have been found to be effective on crops like barley and sorghum in several studies conducted in India [Agarwal and De, 1979; De et al., 1983].

Antitranspirants, to be practical may have to be used as a regular spray because one cannot predict as to when the dry spell is likely to occur. Once the dry spell hits, the plants lose turgidity and most of the chemicals will be ineffective. Cost need not be an alarming factor as the materials are needed in very small quantities. However, identification of ideal anti-transpirants for dryland situations is still a far off reality.

CONCLUSIONS

Water is a limiting factor and a scarce resource under dryland conditions. Rainfall being the main source of water should be made use of by effectively conserving and utilizing the same for production of maximum possible biomass per unit of potentially available water. Water use efficiency in this sense has a wider scope and can perhaps be termed as Rainfall Use Efficiency.

Achieving a higher value of water use efficiency is not the ultimate aim in efficient utilization of available water. Effective utilization of all sources of water for crop production should be the goal in dryland farming. Highest WUE may not be feasible due to economic reasons. However, a constant effort in this direction should be the endeavour of dryland farmer where the resource constraint is the biggest challenge to make the practices economically viable.

REFERENCES

Agarwal, S.K. and De, R. 1979. Effect of nitrogen rates, mulching and antitranspirants on water use and water use efficiency of barley (*Hordeum vulgare* L.) varieties grown under dryland conditions. *J. agric. Sci. Camb.*, 92: 263-268.

Anonymous 1983. Highlights of dryland research in

red soils of Karnataka 1971-83. AICRP on Dryland Agriculture, Bangalore Centre, India (mimeographed). pp. 98.

Arnon, I. 1972. Crop production in dry regions. Barnes and Noble Books, New York, pp. 650.

Arnon, I. 1975. Physiological principles of dryland crop production. In: Physiological Aspects of Dryland Farming, (ed) Gupta, U.S. Oxford and IBH Publishing Co., 3-145.

Bhan, S. and Misra, D.K. 1970. Water utilization by groundnut as influenced due to variety, plant population and soil fertility level under arid zone conditions. *Indian J. Agron.*, 15: 258-263.

Bertrand, A.R. 1966. Water conservation through improved practices. In: *Plant Environment and Efficient Water Use*, (eds) Pierre, W.H. et al., Amer. Soc. Agron., Madison, Wisc., 207-235.

Black, C.A. 1966. Crop yields in relation to water supply and soil fertility. In: *Plant Environment and Efficient Water Use*, (eds) Pierre, W.H. et al., Am. Soc. Agron., Madison, Wisc., 177-206.

Brown, P.L. 1971. Water use and soil water depletion by dryland wheat as affected by nitrogen fertilization. *Agron. J.*, 63: 43-46.

Brown, P.L. and Shrader, W.D. 1959. Grain yields, evapotranspiration and water use efficiency of grain sorghum under different cultural practices. *Agron. J.*, 51: 339-343.

Chary, G.V.N., Rao, G.P.P. and Padmaraju 1979. Technical Bull. No.1, Dryland Centre, Anantapur.

Cole, J.S. 1938. Correlations between annual precipitation and the yield of spring wheat in great plains. US Dept. of Agri. Tech. Bull. 636, p. 40.

Davenport, D.C. and Hagan, R.M. 1975. Role of antitranspirants in arid agriculture. In: *Physiological Aspects of Dryland Farming*, (ed) Gupta, U.S., Oxford & IBH, New Delhi, 315-326.

De, R., Bheemaiah, G., Ramsheshaiah, K. and Yogeswara Rao, Y. 1983. Effect of mulches and antitranspirants on the grain yield of sorghum grown under limited irrigations on a deep vertisol. *J. agric. Sci.* Camb., 100: 159-162.

Dhoble, M.V., Khating, E.A. and Thete, M.R. 1987. Productivity and water use efficiency of kharif crops at varying dates of sowing. *J. Maharashtra agric. Univ.*, 12: 310-313.

Eck, H.V. 1988. Effect of manure on dry farmed soil - 12 years after application. In: *Proc. Int. Conf. on Dryland Farming*, Amarillo, Bushland, pp. 449-451.

El-Swaify, S.A., Pathak, P., Rego, T.J. and Singh, S. 1985. Soil management for optimized productivity under rainfed conditions in the semi-arid tropics. *Advances in Soil Sci.*, 1: 1-64.

Gardner, W.R. 1959. Solutions of the flow equation for the drying of soils and other porous media. *Soil Sci. Soc. Amer. Proc.*, 23: 183-187.

Giri, G., Singh, R.R. and De, R. 1983. Grain yield and moisture use efficiency of dryland wheat as affected by mulch and transpiration suppressants. *Indian J. agric. Sci.*, 53: 899-904.

Greb, B.W. 1979. Technology and wheat yields in the Central great plains : Commercial advances. *J. Soil Water Conserv.*, 34: 269-273.

Gregory, P.J. 1988. Plant and management factors affecting the water use efficiency of dryland crops. In: *Proc. Int. Conf. on Dryland Farming*, Amarillo, Bushland, pp. 171-175.

Hegde, B.R. 1964. Studies on the associated growth of barley with a cereal, a legume and oilseed crops and the response of pure and mixed stands to fertilizer application under barani conditions. M.Sc. thesis, IARI, New Delhi.

Hegde, B.R., Major, D.J., Wilson, D.B. and Krogman, K.K. 1976. Effects of row spacing and population density on grain sorghum production in southern Alberta. *Can. J. Plant Sci.*, 56: 31-37.

Hegde, B.R., Havanagi, G.V. and Reddy, N.M. 1978. Fertilizer need and water use efficiency of dryland ragi in a cowpea-ragi sequence. *Mysore J. agric. Sci.*, 12: 433-436.

Jacks, C.V., Brind, W.D. and Smith, R. 1955. Mulching. Tech. Comm. No. 49. *Commonwealth Bur. Soil Sci.*

Kanitkar, N.V., Sirur, S.S. and Gokhale, D.H. 1968. Dry Farming in India. ICAR Monograph, New Delhi, pp. 470.

Karchi, Z. and Rudich, Y. 1966. Effects of row widths and seedling spacing on yield and its components in grain sorghum grown under dryland conditions. *Agron. J.*, 58: 602-604.

King, F.H. 1907. Text Book of the Physics of Agriculture. Madison Wisc., pp. 161-203.

Krantz, B.A., Kampen, J. and Virmani, S.M. 1978. Soil and water conservation and utilization for increased food production in the semi-arid tropics. Paper presented at the 11th Congress of International Society Soil Science 19-27 June, 1978, Edmonton, Canada.

Larson, W.E. 1962. Tillage requirements for corn. *J. Soil Water Conserv.*, 17: 3-7.

Lemon, E.R. 1956. Potentialities of decreasing soil moisture evaporation loss. *Soil Sci. Soc. Amer. Proc.*, 20: 120-125.

Linscott, D.L., Fox, R.L. and Lipps, R.C. 1962. Corn root distribution and moisture extraction in relation to nitrogen fertilization and soil properties. *Agron.J.*, 54: 185-189.

Li Shengxiu, Xiao Junzhang and Cheng Suyun 1988. Soil water management in drylands in China. In: *Proc. Int. Conf. on Dryland Farming*, Amarillo, Bushland, pp. 201-204.

Loomis, R.S. 1983. Crop manipulations for efficient water use : An overview. In: *Limitations to efficient water use in crop production* (eds) Taylor, H.M. et al., Am. Soc.Agron., Madison, pp. 345-380.

Ma Shijun 1988. Advances in mulch farming in China. In: *Proc. Int. Conf. on Dryland Farming*, Amarillo, Bushland, pp. 510-513.

Musgrave, G.W. and Free, G.R. 1936. Some factors which modify the rate and total amount of infiltration of field soils.. *J. Am. Soc. Agron.*, 28: 727-739.

Nagaraju, Arunkumar, Y.S., Ibrahim, S. and Krishnappa, A.M. 1990. Contour cultivation for improved moisture retention in rainfed red soils. *Curr. Res.*, 19: 149-150.

Ong, C.K., Odongo, J.C.W., Marshall, F. and Black, C.R. 1991. Water use by trees and crops : Five hypothesis. *Agroforestry Today*, 3(2): 7-10.

Onken, A.B., Wendt, C.W. and Halvorson, 1988. Soil fertility and water use efficiency. In: Proc. Int. *Conf. on Dryland Farming*, Amarillo, Bushland pp. 441-444.

Papastylianou and Michael, J. 1988. Replacement of fallow in the rainfed areas of the Mediterranean region. Proc. *Int. Conf. on Dryland Farming*, Amarillo, Bushland, pp. 822-825.

Patil, N.D., Umrani, N.K., Shende, S.A. et al. 1981. Improved crop production technology for drought prone areas of Maharashtra. Tech. Bull. of MPKV, Solapur pp. 96.

Pendleton, J.W. 1966. Increasing water use efficiency by crop management. In: *Plant Environment and Efficient Water Use* (eds) Pierre, W.H. et al. Am. Soc. Agron., Madison, Wisc., pp. 236-258.

Prihar, S.S. and Arora, V.K. 1980. Crop response to mulching with crop residues. Bull. of the Deptt. of Soils, Punjab Agril. University, Ludhiana, pp. 33.

Ramig, R.E. and Rhoades, H.F. 1963. Inter-relationships of soil moisture level at planting time and nitrogen fertilization on winter wheat production. *Agron. J.*, 55: 123-127.

Randhawa, N.S. and Venkateswarlu, J. 1980. Indian experience in the semi-arid tropics - Prospects and Retrospects. Proc. *Int. Symp. on Development and Transfer of Technology for Rainfed Agriculture and SAT Farmers.* ICRISAT, Patancheru (A.P.), pp. 207-220.

Ranjodh Singh, Singh, N., Singh, M. et al. 1983. Dryland cropping in Submontane Punjab. Punjab Agricultural University, Ludhiana, pp. 47.

Singh, B.P. 1985. Effect of intercropping with pearl millet on productivity, profitability and water use on aridisols. *Indian J. Agron.*, 30: 408-413.

Sriram, C., Yadav, G.C., Gupta, P.D. and Atwal, J.S. 1981. Tillage and seeding practices relevant to drylands. In: *A Decade of Dryland Agriculture Research in India* (1971-80). AICRP for Dryland Agriculture, Hyderabad, India, pp. 140-152.

Singh, R.P. and Das, S.K. 1984. Nitrogen management in cropping systems with particular reference to rainfed lands of India. Project Bull. No.8, AICRP for Dryland Agriculure, Hyderabad.

Stanhill, G. 1986. Water use efficiency. *Advances in Agron.*, 39: 53-85.

Steiner, J.L. 1987. Dryland grain sorghum water use, light interception and growth responses to planting geometry. *Agron. J.*, 79: 259-265.

Steiner, J.L., Day, J.C., Papendick, R.I., Meyer, R.E. and Bertrand, A.R. 1988. Improving and sustaining productivity in dryland regions of developing countries. *Advances in Soil Sci.* 8: 75-122.

Stewart, B.A. and Steiner, J.L. 1990. Water use efficiency. *Advances in Soil Sci.*, 13: 151-173.

Unger, P.W. 1975. Role of mulches in dryland agriculture. In: *Physiological Aspects of Dryland Farming*, (ed) Gupta, U.S., Oxford & IBH, New Delhi, pp. 237-258.

Unger, P.W. and Stewart, B.A. 1983. Soil management for efficient water use : An overview. In: *Limitations to Efficient Water Use in Crop Production*, (eds) Taylor et al., Madison Wisc., Am. Soc. Agron., pp. 419-460.

Venkateswarlu, J. 1983. Contingent crop production strategy in rainfed areas under different weather conditions. Project Bulletin No.5, AICRP for Dryland Agriculture, Hyderabad, pp.79.

Venkateswarlu, J. 1987. Efficient resource management systems for dryland of India. *Advances in Soil Sci.*, 7: 165-221.

Viets, F.G. 1962. Fertilizers and the efficient use of water. *Advances in Agron.*, 14: 223-264.

Viets, F.G. 1966. Increasing water use efficiency by soil management. In: *Plant Environment and Efficient Water Use*, (eds) Pierre, W.H. et al. Am. Soc. Agron., pp. 259-274.

Willis, W.O., Hans, H.J. and Robins, J.S. 1963. Moisture conservation by surface and sub-surface barriers, and soil configuration under semi-arid conditions. *Soil Sci. Soc. Amer. Proc.*, 27: 577-580.

Zubriski, J.C. and Norum, E.B. 1955. N Dakota Agri. Expt. Sta. Bimonthly Bull., 17: 126-127.

14

WATER HARVESTING PROCEDURES FOR DRYLAND AREAS

G.D. RADDER, M.I. BELGAUMI and C.J. ITNAL

INTRODUCTION

In ancient days, the idea of utilising rain water collected in depressions, valleys and low lying areas for meeting the water demands of human beings, livestock and crop production was known to man. The techniques of rain water harvesting and utilising the same at the time of need were then followed. The ponds prevailing in the Southern parts of India give ample evidence about the knowledge of man regarding harvesting and utilization of run off water in low rainfall areas.

The soil topographic features and climatic factors prevailing in most dryland areas are highly conducive to generation of runoff. Further intense and heavy showers (characteristics of tropical areas) are a rule rather than an exception in these areas. The combination of these factors results in excessive runoff from the fields with the soil profile generally remaining unsaturated. Research conducted in several areas of drylands in the country and elsewhere has revealed the feasibility of harvesting and storing excess runoff in small dugout ponds and reutilising the same for providing supplemental irrigation to crops during prolonged periods of drought with success and high payoff.

FARM PONDS - TYPES.

Farm ponds are small storage structures used for collecting and storing runoff water. As per the method of construction and their suitability for different topographic conditions farm ponds could be classified into three categories, viz., excavated farm ponds suited for flat topography, embankment ponds for hilly and rugged terrains with frequent wide and deep water courses, and excavated-cum- embankment type ponds. For soils having mild to moderate sloping topography, excavated farm ponds are more suitable.

Excavated farm ponds are of three types according to their shape. They are : 1. Square, 2. Rectangular, and 3. Circular ponds. Among these, the circular ponds have the geometrical advantage that they have the highest storage capacity and have least circumferential length for a given surface area and side slopes. However, their curved shape is disadvantageous in as much as a substantial area is normally lost for agricultural operations. Hence, either square or rectangular ponds are suggested (Fig-1).

Table-1. Surface area, perimeter and storage capacity of different shapes of ponds.

Shape of pond.	Side slope	Surface area (m^2)	Perimeter (m)	Storage capacity (m^3)
Circular	1.5:1	900	106	1499
Square	1.5:1	900	120	1464
Rectangular	1.5:1	900	122	1458

SELECTING A CATCHMENT FOR A FARM POND

Selection of catchment for locating a farm pond is extremely important for several reasons.

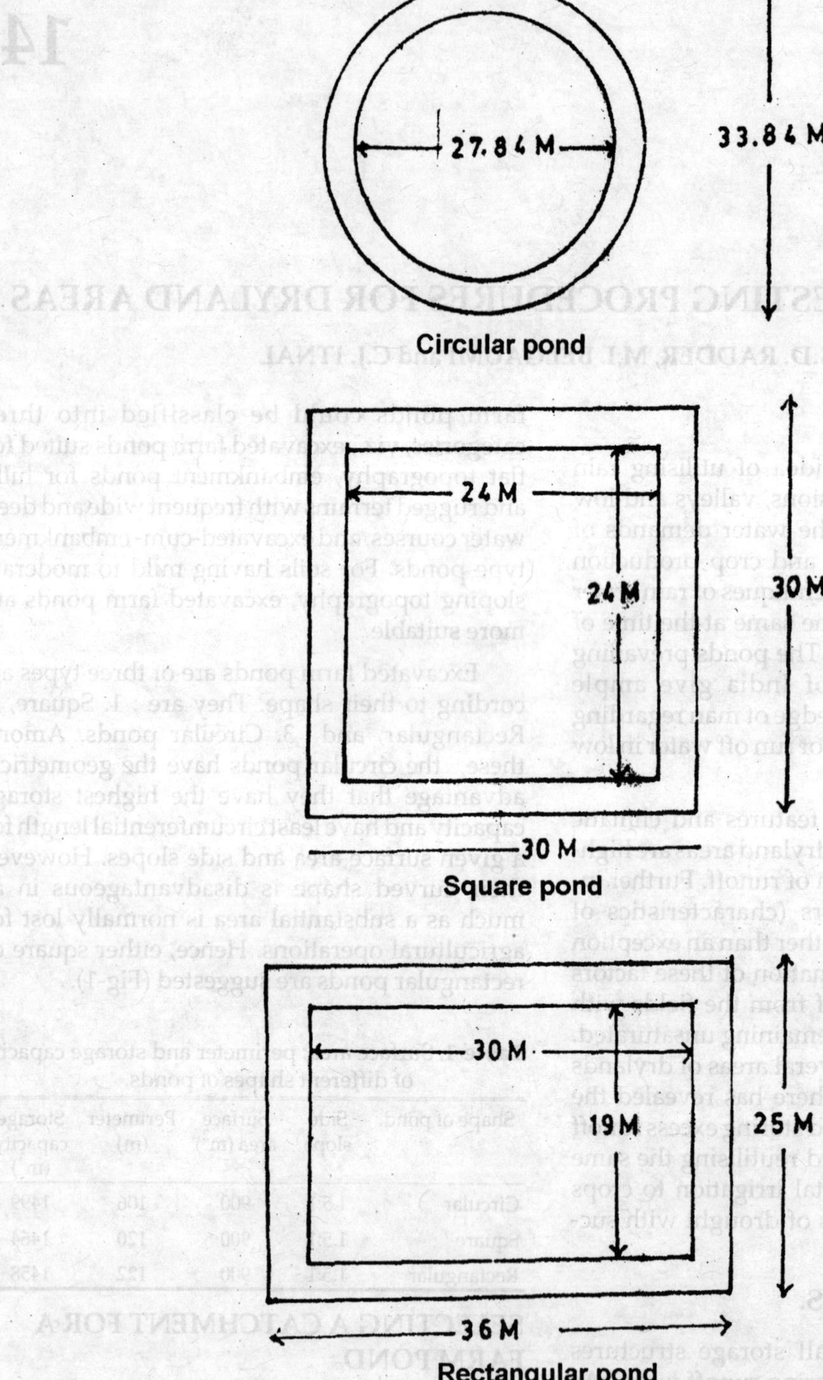

Fig. 1. Geometric features of ponds of different shapes.

It is selected on the basis of its potentiality for yielding sizeable quantity of runoff. Too big a catchment results in rapid silting, while too small a catchment may not bring in enough water into the pond. The runoff amount from different catchment will depend on several factors tabulated below.

Table-2. Factors governing amount of runoff.

Major factor	Associated factors.
1. Rainfall	1. Rainfall intensity 2. Rainfall duration 3. Rainfall distribution 4. Events of rainfall causing runoff.
2. Land topography	1. Degree of land slope 2. Length of run 3. Size and shape of the catchment 4. Extent of depressions and undulations on the catchment.
3. Soil type	1. Soil infiltration rate 2. Antecedent soil moisture 3. Soil texture 4. Soil structure 5. Soil erodibility characteristics
4. Land use Pattern	1. Cultivated or uncultivated or partially cultivated 2. Under pasture or forests 3. Bare fallow or with vegetation 4. Soil and moisture conservation measures adopted or not 5. Crop cultural practices that are adopted

The expected water yields in different soil types for several locations in India are summarised by Sastry *et al.*, (1985). The data are presented in Table-3.

RUNOFF INDUCEMENT

Significant increases in the yield of runoff can some times be achieved by simple treatment of the soil surface. Sometimes mere shaping of the top surface could improve runoff yield (Frasier, 1974). Proper shaping of rocky surface areas and compacting the top layers of clayey soils would yield higher runoff. Smoothing out of the land reduces surface depression storage, and evaporation and infiltration losses are reduced by running the water quickly into the farm ponds. Infiltration losses may further be decreased by compacting the soil surface by rolling, or by spreading a clay blanket before rolling in catchment with sandy soils if clay is available. The principle of roaded catchments, by reshaping the land into roads for conducting runoff was investigated by Hollick (1974), for collection of rain water in Australia. Contoured catchment consisting of contour drains are used to collect and convey runoff intervening areas discharging into a pond through grassed water way. Water harvesting studies conducted at Jodhpur (Chatterji, 1978) from a catchment of gravelly soils having 3 to 6 per cent slope indicated that runoff could be substantially increased by constructing diversion bunds and increasing the slope.

Table-3. Expected water yield into farm ponds in different regions of India.

Region	Soil type	Annual rainfall (mm)	Land slope (%)	Expected water yield (%).
Dehradun, U.P.	Silty loam	1600	2-4	16.5
Bellary, Karnataka	Deep black	508	1-2	20.0
Shiwalik region	Sandy loam	1150	10-15	55.0
Anantapur, A.P.	Shallow red	540	3-5	20.0
Hyderabad, A.P.	Red sandy loam	770	2-5	10.0
Bangalore, Karnataka	Redsoil	830	2-3	20.0
Karnal, Haryana	Alkali	650	0-2	40.0
Chitradurga , Karnataka	Redloam	612	5-10	10.0
Bunga, Haryana	Silty clay loam	1116	10-12	50.0

Source : Sastry *et al* (1985).

Application of sodium salts to the soil surface was found to have increased the runoff from treated catchment in clay loam and sandy loam soils (Hillel, 1967; Murthy et al., 1980 and Tiwari and Sharma, 1987).

Sodium compounds disperse and swell the clay particles present in the soil as they come in contact with water thus seal the soil pores and decrease the soil permeability. The sodium salts used for inducement of runoff affect water quality and the same was found acceptable for irrigation (Tiwari et al., 1987). Various hydrophobic compounds viz., sodium methyl silanolate, sodium silanolate, sodium rosinate, alkoxy-silane, di-ethyl ammonium salts, fatty aminoacetate and metallic soap were suggested for use as a surface coat and increasing runoff (Myers, 1967). The runoff water harvesting efficiency with silicane and paraffin water repellent varied between 50 to 80 per cent and 60 to 90 per cent, respectively (Frasier, 1974).

The surface treatment of soil with bitumen and asphalt was found to be effective for inducement of runoff. The quality of runoff water collected from asphalt and bitumen compounds treated catchments was good and could be used for irrigation and animal consumption (Tiwari, 1991). An alternative way of reducing infiltration and inducing runoff is by adding bentonite, a very fine colloidal clay used for sealing the floor of leaking reservoirs or mixing cement into the soil surface to form a soil-cement stable layer.

Hundred per cent run off harvesting efficiency could be achieved by covering the land surface by galvanised sheet coverings, Butyl rubber cover (Frasier, 1974) and low density polyethylene sheet (Tiwari, *et al.*, 1987). Concrete membranes have a reasonably long life but they are very expensive and the runoff is surprisingly low, from 56 to 80 per cent. Corrugated iron sheets have also been used and do give higher runoff but the cost limits their application. A cheaper material is roofing felt which can be laid in strips with the edges sealed together (Hollick, 1982).

Most of the land treatments for inducing runoff are location specific depending upon the method and cost of application, the lasting qualities of treatment, maintenance to ensure that the treatment remains effective during the expected life of the treatment etc., Hence the choice of practice depends on analysis of the costs and the value of the stored collected runoff water. The performance of different land treatments on inducement of runoff are presented in Table-4.

Table-4. Water harvesting efficiency of different surface treatment for inducing runoff.

Surface treatment of the catchment	Water harvesting efficiency (%).
1. Compacted earth surface	30-60
2. Removing the vegetation	7-21
3. Roaded catchments	24-41
4. Coating of bitumen on compacted earth.	65-89
5. Sodium chloride	10-64
6. Sodium carbonate	35-71
7. Mixture of Bentonite and sodium chloride	48-61
8. Bentonite clay	19-56
9. Asphalt	60-90
10. Asphalt fibre glass sheet	85-95
11. Asphalt roofing	52
12. Bitumen with kerosene oil	77
13. Concrete membranes	56-80
14. Low density polyethylene sheet	60-85
15. Silicane and paraffin	50-80

LOCATING A FARM POND

The selection of a suitable site for the location of a farm pond should begin with a few preliminary studies. Good command area with level land requiring little or no land shaping should be available near the pond. It is advisable to locate the pond at elevation differences between two fields or in a valley portion or depression which favours storage excavation ratio and facilities for gravity flow irrigation. The catchment of the pond should be well protected for arresting rapid siltation. This would mean that the catchment should have well laidout bunds and all other agronomic measures of soil conservation like contour cultivation, contour strip cropping, crop rotation etc. The pond should be so selected that the catchment size is optimum. Consideration should be given to the impounding effect of soil conservation measures while deciding the capacity of the farm pond. The construction of runoff harvesting ponds involves consideration of yield of catchment, storage available at site, water requirement for project need and ground water conditions. Any type of pond, irrespective of its purpose, should be located on oneside of the water course and not in the water course itself to avoid rapid siltation. When soils of different permeabilities occur is succession, it is then feasible to locate it in soils with low permeability to minimise seepage losses. Ponds in shales, basalt or on shattered rocks are likely to loose more water. It is therefore advantageous to know the nature of sub-strata. Trial pits may indicate the nature of sub-strata.

SURVEY OF CATCHMENT AND POND SITE

Survey of the catchment gives a knowledge of the extent of the contributing catchment and other catchment characteristics that are vital for successful planning of the pond size. In undulated areas and areas with defined topography, deleniation of the catchment boundary is not a difficult task. However, when the topography is relatively flat as in case of black soils, contouring may become necessary to demarcate the contributing catchment. For this purpose grids of either 15 or 30 m may be used. The grids are to be extended well beyond the tentative boundary from a reconnaissance survey of the area. A counter map with a contour interval of either 0.3 of 0.5m is prepared for demarcation of catchment boundary. The area can be measured with a planimeter or by means of graphical methods. In case of a bounded catchment, the boundaries are artificially fixed by the main and side bunds. Hence in such cases the reconnaissance would clearly indicate the catchment contributing to the proposed pond site. This can be easily accomplished by moving in the area and the area contributing can thereafter be decided by running a closed traverse with a prismatic compass around the boundary. Survey of the pond site involves the determination of the exact location of the pond with reference to the water course, the position of the inlet and outlet to the pond, the location of silt trap and demarcation of the area around the pond where the excavated spoil obtained from the excavation will be stacked. Spot levels on the four corners of the proposed pond and a few additional points, using a dumpy level will be necessary to determine the earth work involved in the digging of the first segment of the pond by the stepping method of construction.

STORAGE CAPACITY OF POND

The capacity of the pond depends on the purpose for which the pond is created, the annual water yield, storage losses and the siltation rates. The pond should be of sufficient capacity to meet fully the purpose for which it is constructed. If the farm pond is for supplemental irrigation, the extent of area to be irrigated, the estimated soil moisture deficiency which is required to be replenished, the storage losses in pond, the efficiencies of conveyance and application systems have to be considered. The capacity of the pond also depends upon the catchment size and factors affecting its water yield. The water yield from the catchment is a product of the interaction effects of the rainfall factors and the physiographic features of the catchment. Some times it may not be possible for the technical personnel involved to collect all the aforesaid information. Alternatively, a reasonable estimate of the water yield can be arrived at by taking the information on annual runoff per cent available from a representative research station. On a conservative estimate, a value of 10 to 15 per cent of the seasonal rainfall can be expected to go as runoff in medium and deep black soils.

CONSTRUCTION OF FARM POND :

Construction of an excavated farm pond involves the following works.

Site clearing :

The area where the pond is to be dugout should be cleared to an extent of about 10 to 15 m from all sides of the top dimensions of the pond. All bushes, shrubs, stumps and other unwanted materials like roots etc., should be removed. It may also be necessary to plough the area and harrow it to get a more or less even topography.

Demarcation of pond area :

The farm pond site is demarcated by driving pegs to indicate the four corners and if necessary the ropes can be extended beyond the actual site of the pond.

Establishing reference level :

Spot levels at the four corners and at the mid point of the demarcated farm pond site are taken with reference to a nearby temporary bench mark. The average of the level is transferred on to a permanent/semi-permanent object at an approximate distance of about 15 to 20 m from the pond site (Fig.-2).

Stepping method of construction

Since it is not possible to have the cutting/excavation exactly to the trapezoidal shape with uniform side slopes straight away either by human labour or by mechanical means, a segment wise construction known as stepping method is adopted during the time of actual excavation (Fig.-3). These steps like formation not only help in calculation of earth work and payment of wages for the days work but also become easy for lifting the spoil as the pond gets deeper. The steps formed while constructing can be subsequently eased out to get the required shape and designed side slopes for the farm pond.

Excavated spoil

Since considerable quantity of spoil is obtained from dugout farm ponds, it is essential to dispose the same systematically and in a proper manner. The top 30 cm depth of soil could be utilised for filling depressions and gullies and land levelling near the pond. Some of the spoil can be used for bunding. About 50% of the spoil is required for constructing an embankment along the sides leaving a berm of about two metres (Fig.-4).

Shoulder bund and toe drain :

The rain water falling on the spoil bank and the berm is likely to enter into the pond and create rills alround. To prevent such rilling, shoulder bund with a small toe drain which would run all along the spoil bank may be provided. The rain water collected on the berm is allowed to flow out through earthen ware pipe (Fig.-4).

Silt trap :

A silt trap of suitable dimensions should be formed in water course just near the entrance to inlet inorder to check the bed load entering the pond along with runoff water. The length of such silt trap should be slightly greater than the width of the water course and depth may be about 0.5 to 1 m with side slopes of 1.5:1.

STORAGE LOSS OF WATER IN FARM PONDS

Because of high evaporation under dry climatic situations in arid and semi arid tropics and the seepage and percolation losses occuring in a particular soil type, the storage losses could be considerable in some locations and the water available for supplemental irrigation during the prolonged drought or under period of soil moisture stress could be limited. Ponds located in soils with high storage losses turnout to be uneconomical. Hence it is essential that farm pond need to be located in soils having high clay content at the sub-surface or the farm ponds should be lined so that the storage losses could be minimised. The other alternative to overcome storage loss is to use the water immediately as and when the pond is full. Under unavoidable circumstances, the problem of water losses through seepage and evaporation could be mitigated by use of suitable lining material and evaporation retardants.

Lining of farm ponds :

To avoid excessive seepage/percolation, lining of farm ponds becomes necessary. The selection of a lining material depends upon the cost and availability of materials, durability and the maintenance costs. An ideal lining material should he impermeable to water, reasonably durable, flexible over a wide range of temperatures, resistant to mechanical damage, reasonably low investment and maintenance costs, and easily/locally available.

There are a variety of materials viz., natural clays, bentonite, bitumen, soil-cement mixture, stones or bricks in cement mortar, cement concrete, asphalt compounds, rubber and

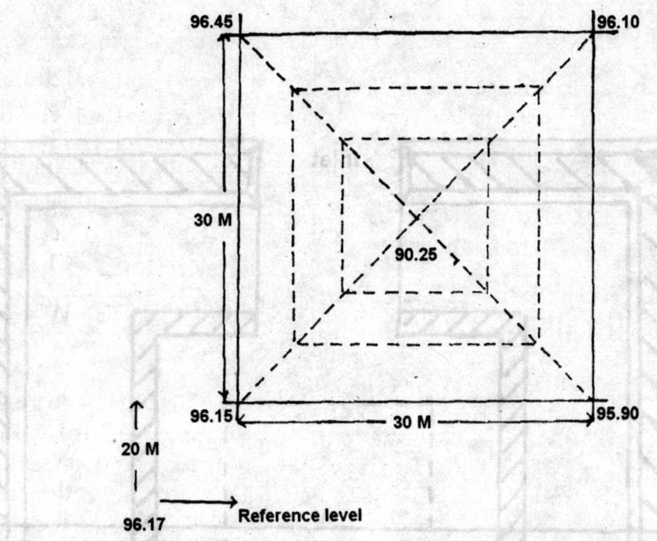

Fig. 2. Demarcation of pond area and establishing reference level

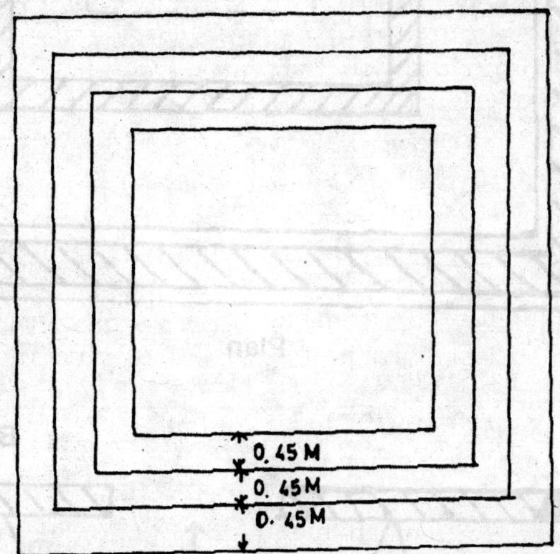

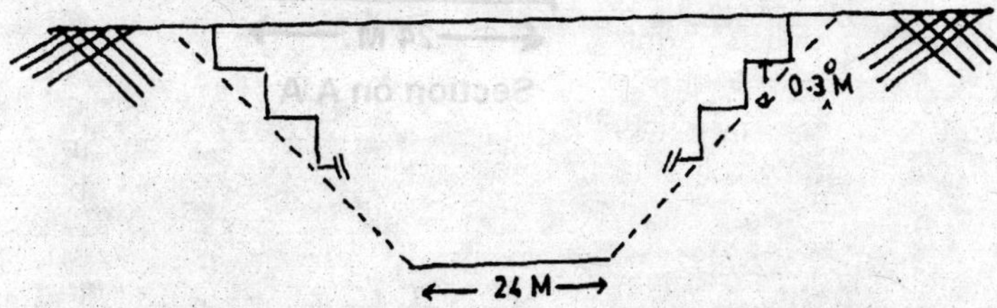

Fig. 3. Stepping method of construction of farm pond

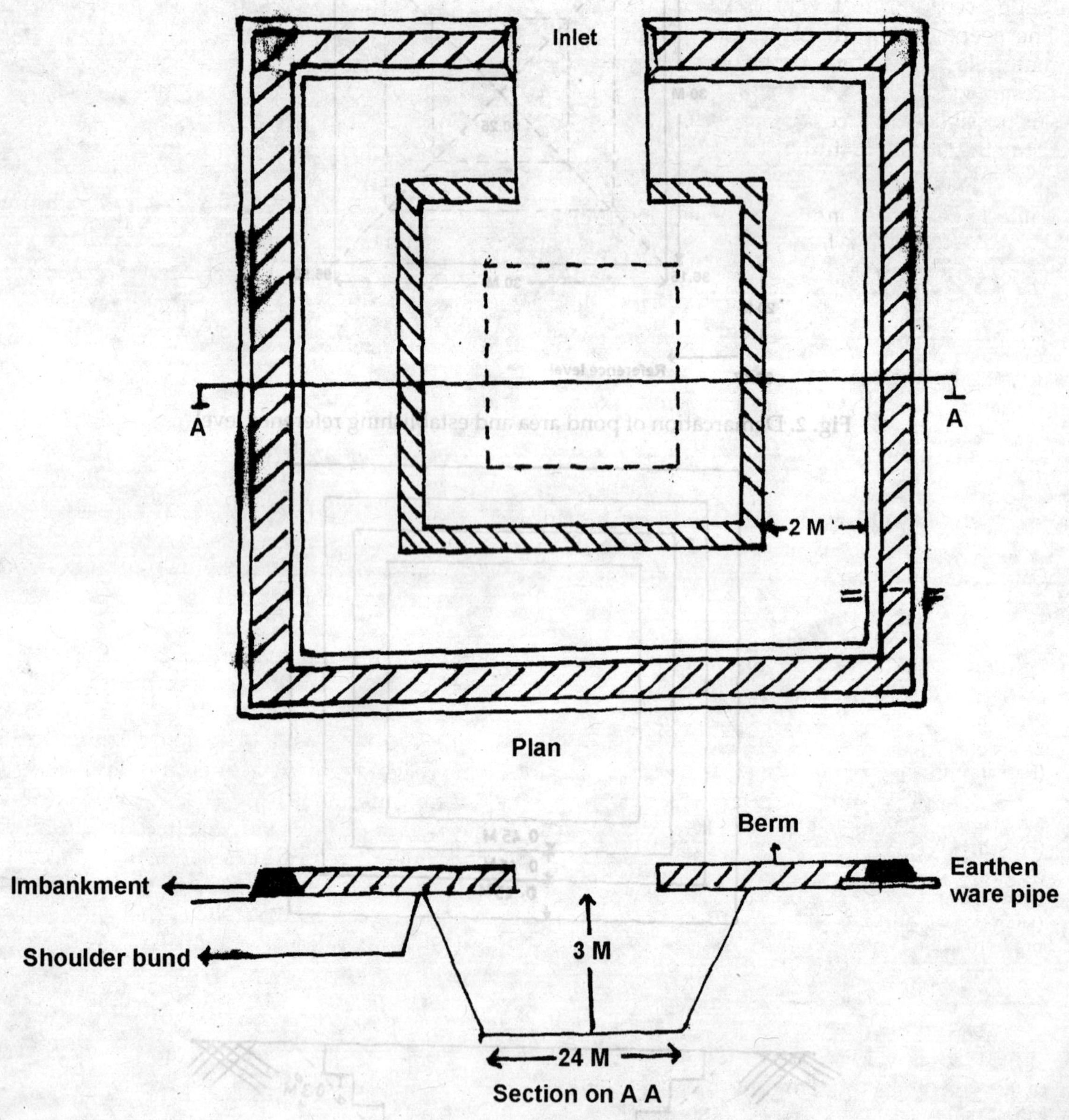

Fig. 4. Complete view of farm pond

polyethylene sheets, chemical additives etc., that can be used for lining of farm ponds. Since all the materials are not suitable for all situations due to one reason or the other, the selection has to be made according to the local specific situation. The seepage as influenced by various lining materials in different parts of the country, is presented in Table-5. The data clearly reveal that it is possible to reduce the losses considerably by using the sealant materials.

Table-5 : Seepage as influenced by different sealant materials in farm ponds.

Region	Treatment	Soil type	Seepage	
			Control	Treated
Kota (Rajasthan)	High density polyethylene sheet	Medium black	45 cm /day	1.3 cm /day
Vasad (Gujarat)	Soil + cement (8:1)	Alluvial sandy loam to loamy sand.	668.9 (lit/m^2 /day)	231.9
Ooty (Tamilnadu)	Bitumen	Laterite silty clay loam	72.3 (lit/m^2 /day)	2.3
Hyderabad (Andhra Pradesh)	Plastic sheet	Red sandy loam	422.2 (lit/m^2 /day)	3.2
Bangalore (Karnataka)	Soil + cement (8:1)	Redsandy loam	8.9 (cm /day)	3.4
Hoshiarpur (Punjab)	Polyethylene sheet	Sandy loam	520.0- (lit/m^2 /day)	7.0
Dehra Dun (Uttar pradesh)	Cement + bricks	Silty loam to silty clay loam.	10.0 (cm /day).	2.0

At Solapur it was noticed that the total loss of stored water from farm pond in medium deep black soil was about 72 per cent, out of which 80 per cent was due to seepage and 20 per cent was through evaporation. Among the various lining materials tried, saline sodic soil gave the least losses per day (362 lit/day) and plastering with soil + cowdung + straw recorded a loss of 818 lit/day.

The use of lining materials to reduce seepage is found to be escalating the cost of investment and hence not economical. As an alternative to the use of lining materials, compaction is the least expensive of the different sealants and is suitable to well graded soils and smaller depths of impounding. Natural clay soil linings are generally the most economical in controlling the seepage losses, especially in fairly large sized ponds and are preferred when low initial cost is desired. The minimum thickness of clay soil lining should be 30 cm upto 3 m pond depth and 5 cm extra for each additional 30 cm pond depth. This method of lining is best adopted where bands of clay deposits are available near the farm pond site.

EVAPORATION RETARDANCE IN FARM PONDS

The evaporation losses are of great concern in farm ponds located in arid regions which are characterised by high temperature, low relative humidity and high wind speeds. Several anti-evaporants viz., plant residues (Straw, wood bark and saw dust), oil emulsions, fatty alcohols, gum mixtures, polyethylene oxides and cationic, anionic and non-ionic chemicals etc., could be used in prevention of evaporation losses. Wax is an unusual, recently tested evaporation suppressant. Floating blocks of wax are added to the water, in sunlight they soften and flow to form a flexible continuous film. Even if the film cracks and breaks during cold weather, the sun's heat subsequently reforms it. Floating material covering the water surface reduces the area where evporation can occur. Rubber and plastic floats can cut evaporation by about 80 per cent (Singh, 1983).

An economical, efficient, and durable evaporation retardant for farm ponds is yet to be found by research under varying agro-climatic regions and soil types. Until such time, to over come the storage loss of water in farm ponds, it is advisable to use the water immediately by irrigating and storing the water in the soil profile for subsequent utilization by deep rooted perennial crops.

CARE AND MAINTENANCE OF FARM PONDS

The farm ponds can get silted up fast. Hence periodic desilting to restore the original storage capacity is required. Breaches and rill formations in the spoil banks and shoulder bunds should be attended to and plugged promptly. The silt accumulated in the silt trap should be removed periodically and preferably as and when it gets filled up after the showers causing runoff. In order to know the depth of water collected in the farm pond and thereby the volume of water stored, depth gauges should be installed. The farm ponds should be properly protected through barbed wire fencing in order to prevent human beings and animals from slipping and falling into the pond.

SUMMARY

The climatic factors in most of the dryland areas are highly conductive to excessive generation of runoff. Research conducted in several dryland areas in the country has revealed the feasibility of harvesting and storing this excess runoff in small dugout farm ponds and re-utilising the same for providing supplemental irrigation to crops during prolonged periods of drought. Among different shapes of farm ponds, the circular ponds have the geometrical advantage of maximum storage capacity and least perimeter for a given surface area and side slope. But their curved shape is a disadvantage as substantial area is normally lost. Therefore square or rectangular ponds are more feasible for harvesting the runoff water.

The ideal location of the pond is in the middle of the cultivated field so that a sizeable amount of runoff could be collected and used for irrigation through gravity flow. A farm pond of 150 m^3 capacity, with side slopes of 1.5:1 is sufficient for each hectare of catchment area in black soils with provision of emptying the pond after its fillup to accommodate subsequent events of runoff. In areas where generation of runoff water is low, significant increase in the yield of runoff can be achieved through mere shaping of the soil surface or through surface treatment of soils with sodium salts, benotonite clay, hydrophobic compounds, asphalt, bitumen or water proofing membranes.

While excavating a farm pond the stepping method of construction will be most convenient. For preventing silting of ponds, a silt trap with suitable dimensions should be located near the inlet of the farm pond. The percolation losses are generally low in ponds located in deep black soils. For preventing percolation losses in ponds located in coarse textured soils, materials such as natural clays, bentonite, bitumen, soil + cement mixture, stones or bricks in cement mortar, asphalt compounds, polyethylene /rubber sheets and chemical additives could be used. To reduce the evaporation losses from the free water surface in the ponds, antievaporants such as plant residues, oil emulsions, fatty alcohols, gum mixture, polyethylene oxides and non toxic chemicals may be used.

REFERENCES :

Chatterji, P.C. 1978. Aspects of conservation of Water resources in Rajasthan for water supply. Current trends in arid zone hydrology. *Proc. Symp. Study and management of water resources in arid and semi-arid regions.* Ahemedabad, April 5-8, 1978 : 225-242.

Frasier, G.W. 1974. Water harvesting for live-stock, wild life, and domestic use. *Proc. water Harv. Symp. Phoenix, Arizona,* March 26-28, 1974 : 40-49.

Hillel, D. 1967. *Runoff inducement in arid lands.* Final technical report submitted to USDA by Volcani Instt. of Agril. Research and Herbrew University of Jersusalem, Rehovat Israel, Project - Alo - SWC-36, 142 pp.

Hollick, M. 1974. The design of roaded catchments for maximum runoff. *Proc. Water Harv. Symp.* Phoenix, Arizona, March 26-28, 1974:201-220.

Hollick, M. 1982 Water harvesting in the arid lands. *Scientific Review on Arid Zone Research,* Vol.-1 : 173-247. Scientific publishers, Jodhpur, India.

Murthy, K.N.K., Sharma, K.D. and Vangani, N.S. 1980 Efficient rain water havesting from arid zone catchments. *National. Symp. on Soil and Water Management* in 1980's Dehradun, March 12-14 1980.

Myers, L.E. 1967. New water supplies from preecipitation harvesting. *Int. Conf. Water for peace.* Wahington, D.C. Vol.2:631-639.

Singh, R.P. 1983. Farm ponds. Project Bulletin No.6 AICRP for Dryland Agriculture, Hyderabad, India.

Tiwari, K.N. and Sharma, P.B.S. 1987. Water harvesting by run off inducement. *J. Institution of Engineers* (India) 68 (AG-1) : 52- 55.

Tiwari, K.N. 1991. Rain water harvesting technology–A reivew. *Ind. J. Soil Conservation* 19(3) : 37-44.

15

PROTECTIVE IRRIGATION (LIFE SAVING) : PRINCIPLES AND PRACTICES

G.D. RADDER, C.J. ITNAL and M.I. BELGAUMI

INTRODUCTION

Rain water is the most important factor in India's agricultural production. Rainfed/unirrigated areas are spread over an area of 75 per cent of the total cropped area. The major cereal production, cotton, oilseeds and millets is obtained from dryland areas. However, the productivity levels of these crops are low mainly because the crops are subjected to vagaries and vicissitudes of the monsoon. Since water is the major limiting factor for production of rainfed crops and precipitation is the only source of water in most of the arid and semi-arid regions, effective rain water management assumes greater significance in maintaining and stabilising crop production.

It is estimated that the country receives an average annual rainfall of 370 million hectare metres; of which 210 million hectare metres is lost as river flow and deep percolation. About 160 million hectare metres gets evaporated through soil surface and crop canopy. In dryland areas, water stored in the soil profile and rainfall received during the growing season constitute the potential water supply available to the crop. When rainfall exceeds water storage capacity of the soil and/or rainfall intensity increases the infiltrability of the soil, there occurs run-off. The inevitable run off may be collected and stored in suitable reservoirs and used for crop life saving irrigations with high pay off. The concept of water harvesting and re- cycling is not new to India. There are evidences in several parts of the country of collection of run off in excavated ponds and utilising such water during scarcity years to save the withering crops (VonOppen and Subba Rao, 1980).

Rain water harvesting is a technology of run off farming which is most feasible in dryland areas. The objectives and technology of rain water harvesting are highly location specific. In very low rainfall areas there may arise a need to induce run off by treating the uncultivated catchments with the objective of its collection in the cropped microwatersheds(Prihar and Gajri, 1970). In areas receiving medium rainfall (500 to 700 mm), the run off is bound to occur and such water could be collected in excavated ponds.

Water collected in the farm ponds or similar structures is of immense importance and needs to be utilised most judiciously. Every care needs to be exercised to cut down seepage and evaporation losses so as to get the maximum benefit out of the collected runoff water. The different aspects that need to be considered for efficient utilization of stored water are: (i) conveyance of the water from the pond to the field, (ii) time of irrigation, (iii) methods of irrigation, (iv) quantity of irrigation and (v) selection of crops and cropping systems, including agro-forestry and agro-horti systems that respond most to limited water supply.

RESPONSE OF CROPS TO SUPPLEMENTAL IRRIGATION

The results of application of supplemental irrgation were summarised by Singh, 1983 (Table 1). It is evident that application of 5 cm of water at a critical stage of crop growth increases crop yields considerably.

Table-1 : Effect of supplemental irrigation (5 cm) on crop yields at different locations

Research Centre	Crop	Yield (q/ha)		Per cent increase over control
		Control	Irrigated	
Agra	Wheat	21.9	27.4	25
Bellary	Sorghum	4.3	13.5	213
Chandigarh	Wheat	6.7	20.0	198
Dehra Dun	Wheat	21.4	35.5	66
Hyderabad	Sorghum	13.9	19.9	43
Jodhpur	Pearlmillet	19.8	36.0	82
Vasad	Beedi tobacco	14.9	17.5	17

Gajri *et al.* (1981) summarised the response of different crops in different locations of the country to supplemental irrigation. The data clearly indicate that response to one supplemental irrigation at critical stages of the crop would enhance the yield considerably (Table 2).

The results of supplemental irrigation in medium deep black soils at Bijapur over a period of 4 to 5 years indicated that crop yields with one life saving irrigation could be increased by 33 to 92 per cent (Table 3).

The studies conducted in medium deep black soils of Solapur, have shown that the yield of wheat could be increased by 60 per cent with two supplemental irrigations over control (Anon., 1988). Supplemental irrigations recorded significantly higher grain and fodder yield of *bajra* (pearl millet) at Kovilpatti (Anon., 1990). Due to better moisture availability through supplemental irrigations, the crops may even respond to higher levels of nutrient application. The grain yield of *rabi* sorghum at Bijapur over a period of 2 years was increased with an increase in the level of nitrogen application under supplemental irrigations (Table 4).

Table 2. Response of different crops to supplemental irrigation

Research Centre	Crop	No. of Years	Average response (kg/ha)	Per cent increase over control
Lundhiana	Wheat	10 (16)	1101	54
	Barley	3 (7)	870	38
	Gram	2	260	14
	Yellow sarson		550	35
	Brown sarson	1	420	29
	Raya	1	730	54
Dehra Dun	Wheat	5 (6)	1473	81
	Barley	3	934	32
Ranchi	Wheat	2	238	30
	Barley	4	561	55
	Linseed	3	307	51
	Safflower	4	288	40
	Rapessed	4	203	141
Garhshankar	Wheat	7	777	31
Bellary	Sorghum	6	1263	83
	Cotton	2	219	58
	Safflower	2	291	123
Solapur	Sorghum	4	801	83

Note : Figures in parenthesis denote the number of experiments when it is different than number of years.

Table 3 : Response of crops to supplemental irrigation in medium deep black soils (Bijapur)

Crop	Yield (kg/ha)		Per cent increase over control
	Irrigated	Unirrigated	
Rabi sorghum	2574	1845	40
Safflower	1702	1196	42
Bengalgram	887	569	33
Sunflower	1429	873	64
Pigeonpea	381	198	92
Hybrid cotton	1514	1094	38

Introduction of high value crops under protective irrigation further helps in enhancing the income of dryland farmer. Growing of hybrid cotton or following sequence cropping are some of the examples through which the farm income could be increased, besides increasing the cropping intensity (Table 5).

Table 4 : Grain yield of rabi sorghum (kg/ha) as influenced by nitrogen levels under supplemental irrigation

Nitrogen levels	No irrigation			One irrigation			Two irrigations			Mean		
	1984-85	85-86	Mean	1984-85	85-86	Mean	1984-85	85-86	Mean	1984-85	85-86	Mean
50 Kg N/ha	2263	2495	2379	2289	2966	2628	2598	3068	2833	2383	2843	2613
75 kg N/ha	2271	2657	2464	2572	2993	2783	3086	3286	3186	2643	2979	2811
100 kg N/ha	2770	3007	2889	3039	3093	3066	2770	3360	3065	2859	3153	3006
Mean	2435	2720	2577	2633	3018	2826	2818	3238	3028	–	–	–

	1984-85		1985-86	
	S.EM. ±	C.D.(0.05)	S.Em. ±	C.D. (0.05)
N Levels	98	288	48	145
Irrigation levels	98	288	48	145
N x I	167	N.S.	83	249

NS - Non significant.

Table 5. Evaluation of cropping system under farm pond water utilisation

Croopping system	Gross income (Rupees) with irrigations			
	Control	1	2	3
1. Hybrid cotton (DCH-32)	7348	10968	14904	19228
2. Bengalgram + Safflower intercrooping (4 : 2)	4227	5321	6117	6646
3. Rabi sorghum + Bengalgram intercropping (2 : 1)	6315	6413	6920	7129
4. Sunflower - Bengalram sequence cropping	7664	8222	10205	11182

In another experiment on the influence of supplemental irrigations in medium deep black soils at Bijapur with different horticultural crops, it was observed that guava, *ber* (*Zizyphus mauritiana*) and fig responded in a range of 41.7 to 122.6 per cent with irrigations over control. The highest response to supplemental irrigations was recorded in guava (122.6%) and the lowest response was found in fig (41.7%). However, the income from *ber* was the highest (Rs.26, 180/ha), followed by fig (Rs.8,046/ha) and guava (Rs.5,544/ha) after third year of their planting (Table 6).

Table 6 : Influence of supplemental irrigations in different horticultural crops (Bijapur)

Crop	Fruit yield (kg/ha)		Gross income (Rs./ha)	
	Irrigated	Control	Irrigated	Control
Ber	5236	3265	26,180	16325
Fig	1341	946	8,046	5676
Guava	2772	1245	5,544	2490

Time of Irrigation

Time of irrigation is extremely important so as to derive the maximum benefit of the water collected. In the *kharif* regions the supplemental irrigation may be provided to save the crop from cyclic stress that will generally occur in July and August. In potentially double cropped areas, the supplemental irrigation may be used to start of the second crop as the surface soil is dry enough to adversely affect the crop stand after *kharif* harvest. The results obtained from Dehra Dun (Table 7) illustrated the beneficial effects of supplemental irrigation under such conditions.

Under *rabi* conditions, the stress is continuous as the crops are grown on stored soil moisture. Supplemental irrigation is found to give maximum returns under these conditions as is evident from the experiments conducted at Bijapur, Bellary and Solapur (Table 8).

In red lateritic soils of Bangalore, one irrigation given during the critical stage of crops increased the crop yield significantly. It is evident that the extent of response by different crops in different years largely depends upon the nature

Table 7. Yield of wheat (kg/ha) as influenced by supplemental irrigations.

Treatment	1975-76	76-77	77-78	78-79	Average	Per cent increase over control
1. Control	1249	1110	2458	2528	1931	–
2. Irrigation at crown root initiation (5 cm)	1999	1339	4437	3837	3328	82
3. Pre-sowing irrigation+ irrigation at CRI (10 cm)	2955	4978	4785	4938	4189	128
4. Pre-sowing irrigation (5 cm)	2083	2981	4600	4142	3452	85
5. S.Em. ±	163	192	249	139	–	–
C.D. (0.5)	564	863	861	471	–	–

Source : Gurmel Singh et al, 1981

and intensity of dry spell. The late sown crops like *ragi* (finger millet) or long duration crops like chilly or cowpea sown early in May are likely to be more benefited by protective irrigation. The same studies further reveal that a single irrigation of 5 cm is more beneficicial than two irrigations of 2.5 cm each (Table 9).

Table 8 : Effect of minimal irrigation on sorghum grain yield (kg/ha) in black soils

Treatment	Bijapur	Bellary	Solapur
Control	1664	541	966
One irrigation (4-6 cm)	2639	2109	1797

Source : Ramamohan Rao and Ranga Rao, 1981.

Table 9 : Yield of different crops as influenced by protective irrigation

Crop	Yield (q/ha) (average of 2 years) Irrigated	Unirrigated	Per Cent increase over control
May sown crops			
1. Cowpea + Maize (Fodder)	96.8	54.8	75.3
2. Cowpea (Grain)	6.34	2.35	167.0
September sown crops			
1. Chilly (dry)	4.22	2.37	81.4
2. Tomato	32.83	21.34	29.7
3. Ragi (grain)	21.11	11.24	122.5
4. Ragi (straw)	31.12	14.93	106.7

Source : Hegde *et al.*, 1981

Field investigations by Prihar and Gajri, 1980, suggest that in areas which do not receive rain during early growth of *rabi* crops, a small early irrigation with harvested water helped in improving and stabilising wheat and barley yields in light textured soils. Gajri *et al.* (1981) worked out the magnitude of response of wheat to one irrigation of 5 to 7 cm applied at seeding or early in the growing season (Fig. 1), which shows the response of wheat is a function of control yield.

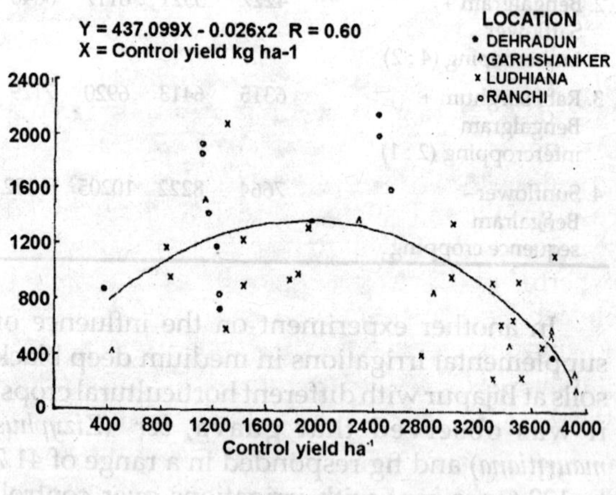

Fig. 1. Wheat response to one irrigation as a function of control yield.

Initially the response increased with increase in the control yield and after a threshold it decreased with increase in the control yield. This relation has been developed irrespective of soil type, moisture content at seeding, nutrient supply conditions and amount and distribution of growing season rain. Low initial response may be due to low moisture content at seeding and/or low rains during crop growth. Decrease in response with increase in control yield after the threshold may be due to adequate and timely rains during the crop growth period. To get maximum advantage of applied water, it is necesary to remove other factors that limit crop production. This is to say that to derive maximum benefit from small doses of applied water, there is need to use responsive cultivars and follow the recommended package of practices.

Investigations at Bhopal in vertisols revealed that irrigation to rice at transplanting and grain filling and to soybean at grain filling stage accounted for an increased yield of 90 and 40 per cent respectively as compared to control. For *rabi* crops, pre-sowing irrigation was found most essential. Increase in yield by 40,64,37 and 17 per cent was observed in case of wheat, chickpea, safflower and linseed respectively due to supplemental irrigation over control (Nimje and Bhandarkar, 1990). Water may be applied as pre sowing irrigation, if seed zone is dry or as an irrigation to remove water stress and promote root growth. Results of research with limited water supply showed that on an average of 12 experiments over 7 years, one irrigation of 5 to 6 cm depth at pre sowing or after 30 to 40 days after sowing increased wheat yield by 6.6 q/ha and stabilised wheat yields over years (Verma *et al.*, 1984).

Depth of Irrigation

The depth of irrigation (supplemental) has got direct bearing on the yield of crops depending on the time of irrigation, available soil moisture in the root zone and the critical stage of crop growth. The data obtained at Hyderabad clearly bring out that whereas one to two centimetres of water application may still increase the yield of shallow rooted crops, at least 3 to 5 cm of wetting is essential for deep rooted crops (Table 10). This only means that the rooting depth has to be wetted to achieve increases in yield (Vijayalakshmi and Vittal, 1981).

Table 10 : Response of long duration kharif crops to supplemental irrigation

Crop	Year	Level of irrigation (cm)	Yield (q/ha)		Per cent increase
			Irrigated	Unirrigated	
Castor	1977	1	9.8	7.4	32
	1978	2	9.6	5.3	81
	1979	5	13.2	10.1	31
Redgram	1980	3	1.7	0.5	240
(Sole)	1980	5	3.5	0.5	600

Source : Vijayalakshmi and Vittal, 1981.

The water-use-efficiency of collected run-off in ponds could be increased by irrigating the cash crops/vegetables. A small quantity of water can boost the crop yields significantly and add to the income of the farmer (Table 11).

Table 11 : Response of different vegetable crops to supplemental irrigation

Crop	Year	Level of irrigation (cm)	Yield (q/ha)		Per cent increase over unirrigated
			Irrigated	Unirrigated	
Brinjal	1977	0.5	15.8	11.8	34
		1.0	26.7	17.7	51
Chillies	1978	0.5	16.8	12.7	28
		1.0	17.1	12.7	35
Beans	1977	1.0	14.3	1.1	1200

Source : Vijayalakshmi and Vittal, 1981.

From the above presentations, it is evident that the response for extensive irrigations (with limited supply) is more compared to intensive irrigations. Extensive irrigation provide opportunity to irrigate more area with the available water. Besides extensive irrigation, another important consideration which needs to be kept in view is the method of irrigation.

WATER APPLICATION METHODS

Water application methods need be benefit of the stored water. Application methods to increase the efficiency of water and cut down on unproductive losses were tried over years and it has been clearly shown that for shallow rooted narrow spaced crops like sorghum, combined middle alternate furrow has an advantage over furrow irrigation (Anon., 1988). At Bijapur, it has been observed that drip method of irrigation saved 50 per cent of pond water compared to modified sprinkler and surface methods of application (Anon., 1991). The response for a given depth of irrigation in different fruit crops with various methods of irrigation was maximum in drip method of water application (Table 12).

Table 12 : Response of fruit crops to different methods of water application in black soils at Bijapur

Method of irrigation	Yield increase over control (%)		
	Ber	Fig	Pomogranate
1. Drip	252	228	133
2. Modified sprinkler (Jet method)	189	128	52
3. Surface	149	82	60

At Solapur, it was observed that sprinkler method of irrigation increased the grain yield of rabi sorghum (12.5 q/ha) compared to contour furrow irrigation (19.95 q/ha) as against a yield of 13.46 q/ha obtained in control (Anon., 1988). The subsoil application of water is found advantageous as it cuts down evaporation. However, it may be difficult to implement (Vijayalakshmi and Vittal, 1981).

Table 13 : Yield and gross income of different farming systems as influenced by supplemental irrigation

Farming system	Yield (kg/ha) (average of 2 years)		Gross income (Rs./ha)	
	Irrigated	Control	Irrigated	Control
1. Ber (Umran)	7452	5901	37,260	29,505
2. Ber (Kadaka)	5592	4731	27,962	23,657
3. Curry leaf	8888	–	26,664	–
4. Sericulture (Mulberry)	7216	6145	16,870	14,385
5. Pastures	22774 +7141	19023 +5294	5,107	4,121
6. Hybrid cotton	1884	1501	18,840	15,010
7. Bengalgram + Safflower (4 : 2)	611 + 115	440 + 73	4,937	3,488

FARMING SYSTEMS UNDER SUPPLEMENTAL IRRIGATION

A farming systems approach was developed by utilising farm pond water for realising higher returns. The data obtained at Bijapur on different farming systems revealed that horticultural component with *ber* plantation (cv. Umran and Kadaka) gave gross income ranging between Rs. 27,962 to Rs. 37,260/ha with two to three supplemental irrigations as against Rs. 23,657 to Rs. 29,505/ha in control from *ber* plantation. The next highest income was realised from the cultivation of curry leaf (Rs. 26,664/ha) with two irrigations, followed by hybrid cotton (Rs.18,840/ha). Intercropping of bengalgram + safflower in row proportions of 4:2 gave only Rs.4,937/ha with supplemental irrigations, whereas the increase from the arable cropping systems was Rs.3,488/ha under control. Cultivation of mulberry pastures (grasses + legumes) under protective irrigations gave Rs. 16,870/- and Rs.5,107/ha respectively, against corresponding income of Rs.14,385/- and Rs.4,121/ha under control (Table 13).

EFFICIENT USE OF RUNOFF WATER THROUGH FERTILIZER MANAGEMENT

Judicious fertilizer use may go a long way in optimising the use of limited runoff water. In an experiment on different levels of supplemental irrigation (no irrigation, 5 cm, 10 cm and irrigation at 60% depletion of available moisture), in conjunction with four levels of fertility (N_oP_o,

$N_{20}Po$, $N_{40}Po$ and $N_{40}P_{30}$) carried out at the Central Arid Zone Research Institute, Jodhpur during Kharif 1978, it was found that one supplemental irrigation (5cm) given to unfertilized pearl millet (N_oP_o) increased the grain yield of pearl millet by 35 per cent and a 10cm irrigation increased the same by 48.5 per cent over unirrigated and unfertilised control (11 q/ha). A supplemental irrigation (7.5 cm) given at 60 days after sowing, coinciding with 60% depletion of available moisture resulted in about 100 per cent increase in grain yield over control. An application of a modest dose of fertilizer ($N_{20}P_o$) increased the grain yield of pearl millet by 24.5, 67.8, 89.3 and 135.4 per cent with no irrigation, 5 cm, 10 cm and 7.5 cm supplemental irrigation to 60% depletion of available soil moisture over unirrigated and unfertilized control, respectively. A further increase in the respective grain yields of the order of 67.6,113.5,158.5 and 192.8 per cent was obtained with an application of $N_{40}P_{30}$ fertilizer dose (Singh, 1983).

In another experiment at Regional Research Station, Bijapur, *rabi* sorghum crop responded to higher levels of nutrient application due to better moisture availability through supplemental irrigations. The grain yield of *rabi* sorghum over a period of two years was increased with an increase in the level of nitrogen application under supplemental irrigations (Table 14).

Table 14 : Grain yield (kg/ha) of rabi sorghum as influenced by nitrogen levels under supplemental irrigation..

N. Levels	No irrigation			One irrigation			Two irrigations			Mean		
	1984-85	85-86	Mean	1984-85	85-86	Mean	1984-85	85-86	Mean	1984-85	85-86	Mean
50 Kg N/ha	2263	2495	2379	2289	2966	2628	2529	3068	2833	2383	2843	2613
75 kg N/ha	2271	2657	2464	2572	2993	2783	3086	3286	3186	2643	2979	2811
100 kg N/ha	2770	3007	2889	3039	3093	3066	2770	3360	3065	2859	3153	3006
Mean	2435	2720	2577	2633	3018	2826	2818	3238	3028	–	–	–

	1984-85		1985-86	
	S.EM±	C.D.(0.05)	S.Em±	C.D. (0.05)
N Levels	98	288	48	145
Irrigation levels	98	288	48	145
N x I	167	N.S.	83	249

ECONOMICS OF FARM PONDS

Several studies in the country have been conducted to find out the feasibility of farm ponds in dryland areas. All these investigations reveal that farm ponds are economical and assure sustained crop production. An evaluation of farm ponds at Dehradun costing Rs. 32,919 at 1978 price and assuming a life span of 50 years and Rs. 50,765 at assumed active life of 30 years at 10 per cent rate of interest revealed a benefit-cost ratio of 1.96 for 50 years project life and 1.85 for 30 years of life. The studies clearly indicated that the farm ponds are viable proposition (Anon, 1980). Rambabu *et al* (1985) reported that the farm ponds excavated in Fakot watershed project in Tehri Garhwal of Uttar Pradesh gave a benefit cost ratio of 2.71:1 considering 25 years project life and at 12 per cent discount rate. When analysis was done expecting 75 per cent yield only, the benefit cost rato was 2.03:1. These studies demonstrate the economic feasibility of such projects in hill areas.

A dug out farm pond at Bellary was constructed in 1956-60, with a total storage capacity of 0.28 ha m, having catchment area of 16.2 ha. At 1980 price level, the construction cost of the farm pond was worked out to be Rs. 10,650. At 12 per cent interest, a benefit cost ratio of 3.4 was obtained when 5 cm water was applied to *rabi* sorghum as protective irrigation. The benefit cost ratio came down to 2.5 with 10 cm of irrigation. (Table 15).

Table 15 : Benefit-cost evaluation of supplemental irrigation to rabi sorghum in vertisols at Bellary.

Particulars	Av. annual equivalent cost (Rs.)	Net amount benefits (Rs/ha)	Irrigated area (ha)	Total net annual benefits (Rs)	B.C. ratio.
Irrigation depth					
5 cm	2169	1418	5.2	2374	3.4
7.5 cm	2169	1707	3.6	5975	2.8
10.0 cm	2169	2086	2.6	5424	2.5

Narayan *et al* (1987) analysed the feasibility of farm ponds in deep black soils at Bijapur (Karnataka) and reported that an additional returns ranging between Rs. 1300 to Rs. 1800/ha from *rabi* sorghum or safflower with two supplemental irrigations could be realized.

SUMMARY

Soil and water are the two most important natural resources which need to be conserved, developed and used more efficiently for crop production. This is particularly important in semi arid and arid areas in view of the shortage of soil moisture. In most of these areas there is possibility of one or two events of runoff which could be collected in excavated farm ponds and the water thus collected could be made use of for life saving irrigation during periods of moisture stress. Considerable information has been generated in the past decade and a half both in the country and elsewhere. It has been established that supplemental irrigation once or twice, can boost the crop yields by two to three times over control.

Since the harvested water in the farm ponds is a costly commodity, there cannot be any room for its extravagant use. Therefore all possible efforts need to be taken to use this water more judiciously. This could be achieved by (i) extensive irrigation and (ii) growing of high value, commercial and orchard crops. The data on the effect of one or two supplemental irrigations to orchard crops like *ber*, guava, fig and Sapota clearly indicate more advantages in their favour compared to seasonal/annual crops with regard to returns per unit area.

REFERENCES

Anonymous, 1980, *Twenty five years of Research on soil and water conservation in semi-arid deep black soils.* Monograph No. 1 of Central Soil and water Conservation Research and Training Institute, Dehradun.

Anonymous, 1988, *Annual progress report, 1987-88, A.I.C.R.P.*,D.L.A., Solapur.

Anonymous, 1990, *Annual progress report, 1989-90,* A.I.C.R.P.,D.L.A., Kovilpatti.

Anonymous, 1991, *Annual progress Report,* 1990-91, A.I.C.R.P.,D.L.A., Bijapur.

Gajri, P.R. Verma, H.N. and Prihar, S.S., 1981. Rain water harvesting and its recycling for maximisation of crop production, *Ind. J. Soil. Cons.*, 9(2&3): 116-123.

Gurmel Singh., Koranne, K.D. and Bhushan, L.S., 1981, *Advances in Rainfed Farming, Bulletin No.* 10/D-8, C.S.W.C.R., and T.I., Dehradun.

Hedge, B.R. Channappa, T.C. and Ananda Ram, B.K., 1981, water harvesting and recycling in the red soils of Karnataka. *Ind. J. Soil. Cons;* 9 (2&3):107-115.

Narayan, H.C., Itnal, C.J., Gopalkrishna Hebbar, G and Patil, V.S. (1987). An evaluation of supplemental irrigation through farm ponds on drylands. *Ind. J. Soil Cons.* 15(1):1-6.

Nimje, P.M. and Bhandarkar, D.M., 1990, Effect of supplementary irrigation through farm pond on crop productivity in vertisols. *Ind. J. Soil. Cons;* 18 (1 &2):9-14

Prihar, S.S., and Gajri, P.R., 1970, Water harvesting-a management practice in dryland farming. *Indian Farming* 20 (8):22.

Prihar, S.S. and Gajri, P.R. 1980. Magnitude and mode of response of dryland crops to limited irrigation. Presented at the *Symposium on "Rain water and Dryland Agriculture"* Sponsored by Indian Nata. Sci. Acac, New Delhi, Oct. 3,1980.

Ramamohan Rao, M.S. and Ranga Rao, V., (1981), Pushing up yields in rabi black soils. *Intensive Agriculture*, April, May, 1981.

Ram Babu, Gupta, R.K., Gurmel Singh, and Katiyar, V.S., (1985), Ecomomic evaluation of water resources development in hills- A case study at ORP, Fakot (Tehri-Garhwal) *Ind. J. Soil. Cons.* 13 (2&3): 71-76.

Singh, R.P. 1983, *Farm ponds. Project Bulletin No.6*, All India Co-ordinated Research Project for Dryland Agriculture, Hyderabad.

Verma, H.N., Prihar, S.S. and Singh, R., 1984. Feasibility of storage of runoff in dugout ponds and its use for supplemental irrigation in submontone Punjab, *Ind. J. Soil. Cons.*, 12(1), : 31-36.

Vijayalakshmi, K. and Vittal, K.P.R., 1981, Rain water Harvesting and Re-cycling in Semi-Arid Red and Sierozem soils, *Ind. J. Soil. Cons.*, 9 (2&3) : 87-99.

VonOppen, M. and Subba Rao, K.V. 1980. History and ecomomics of tank irrigation in semi-arid India. *Presented at the symposium on "Rain water and dryland agriculture"*, sponsored by Indian Nata. Sci. Acad., New Delhi, Oct. 3,1980.

16

TILLAGE IN DRYLANDS — PRINCIPLES AND PRACTICES

V.M. MAYANDE

INTRODUCTION

Tillage may be described as the practice of modifying the state of the soil in order to provide conditions favourable to crop growth, (Cuplin, 1986). The main objectives of tillage are to (a) create suitable tilth or soil structure, (b) conserve soil moisture, (c) control weeds, and (d) incorporate crop residues, manures, fertilisers and agrochemicals into the soil. Agricultural soils are heterogenous collection of mineral particles existing either singly or as small crumbs. The spaces between soil particles are filled with air or water. The amount of air or water in a given soil can vary widely according to cultivation and traffic over land. Soil in compacted conditions does not provide a suitable environment for growing crops and one of the chief uses of tillage implements is to restore a tilth favourable to crop development.

Tillage has been also defined as those mechanical soil stirring actions carried on for the purpose of nurturing crops. The goal of proper tillage in drylands is to provide a suitable environment for seed germination, root growth, weed control, soil erosion control, and moisture control avoiding moisture excesses and reducing moisture stress. Any tillage practice in dryland which does not return more than its cost by increasing yield and improving soil conditions should be eliminated. Soil needs to be worked only enough to assure optimum crop production and weed control and any tillage activity beyond that is of questionable value.

The tillage operations in drylands should be evaluated on the basis of following goals : (a) moisture management (b) erosion control (c) weed control (d) management of crop residues (e) Improvement of soil tilth (f) soil aeration (g) provide good seed-soil contact (h) prepare fine surface for seeding operation (i) Incorporation of manure and agro chemicals into the soil (j) Insect control and (k) temperature control, for seed germination.

The tillage research in Indian drylands has been by and large on agronomical aspects to quantify the effects of intensity of tillage operations on crop yield. Very little or no work has been done on engineering aspects of tillage with reference to quantification of effects of each of the factors mentioned above. In this chapter, an effort is made to review and compile the tillage research conducted in Indian drylands. Theoretical aspects of tillage practices and principles involved are discussed. The contents of this chapter are targeted to benefit the researchers and post graduate students involved in tillage research in drylands.

TILLAGE OBJECTIVES

Tillage may also be defined as manipulation of soil for any purpose. Although one of the major objectives of tillage is to provide optimum environmental conditions for crop growth, the desired soil conditions cannot be quantitatively identified or specified. The forces applied to tillage tool to produce a given effect can be accurately measured but we cannot reliably predict the effects of changes in tool design (Kepner et al 1982). Therefore it is not surprising that the design of tillage equipment is still largely an art rather than a science. Some of the objectives of tillage in drylands are to :

(1) Develop desired soil structure for a seed bed which allows rapid infiltration and good retention of rainfall, provides adequate air capacity and exchange within the soil, and minimises the resistance to root penetration. A good seedbed implies finer particles and firmness in the vicinity of the seeds.

(2) Minimise soil erosion by following practices as contour tillage, tillage across the slope etc.

(3) Control weeds and remove unwanted crop plants.

(4) Manage crop residues, through mixing of trash is desirable for achieving good tilth and decomposition of residues. However, the retention of trash on top layers is also useful in reducing erosion. On the other hand, complete coverage of residues sometimes necessiates control of insects or to prevent interference with precision planting operations.

(5) Obtain specific soil configurations for *in-situ* moisture conservation, drainage, planting etc.

(6) Incorporate and mix manures, fertilisers, pesticides or soil amendments into the soil.

(7) Accomplish segregation by moving soil from one layer to another, removal of rocks or root harvesting.

TILLAGE MECHANICS

The reactions of soil to forces applied by tillage tool is affected by the resistance of the soil to compression, shear, adhesion (attractive forces between the soil and some other material), and friction. Practically all tillage tools consist of devices for applying pressure to the soil, often by means of inclined planes or wedges. As the tool advances, the soil in its path is subjected to compressive stresses which in a friable soil results in a shearing action. The shearing of the soil may extend to a considerable distance on either side of the shear plane because of internal friction and the cohesive action of moisture films.

Some commonly used terms in tillage are explained below:

Force : It is any action that changes the state of rest or motion of body. It is completely specified by its magnitude and direction and the position of its line of action.

Pull : It is the total force exerted upon the implement by a power unit. With tillage implement it is generally with some angle above horizontal, and it may or may not be in a vertical plane parallel to the line of motion.

Draft : It is horizontal component of pull, parallel to the line of motion.

Side Draft : It is horizontal component of pull, perpendicular to the line of motion.

Specific Draft : The draft per unit area of tilled cross section. The unit in SI system is N/cm^2.

Power : It is rate of doing work. The unit in SI system is KW.

Kilowatt-hour : It is the quantity of work performed when one KW is used for one hour.

Cohesion : It may be defined as the force that holds two particles of the same kind together. Two soil particles are held together due to cohesion.

Adhesion : The attractive force between the soil and some other material is called adhesion.

Friction coefficient of soil : It is the ratio of frictional force and the normal force. The magnitude of the frictional forces (T) arising when a soil slides along a metal depends on the magnitude of the force (N) normal to the contact surface, the mechanical composition and moisture of the soil, the roughness of the metallic surface, the velocity of sliding, and several other factors. It can be explained as :

$$F = \frac{T}{N}$$

Where F= Friction coefficient of soil

T= Frictional force

N= Normal force

The forces during working of straight and inclined wedge and its effect on soil deformation and displacement were given by Sinekov (1977).

SOIL DEFORMATION AND DISPLACEMENT DURING WORKING OF STRAIGHT WEDGE

The straight wedges are used in a limited number of cases. The nature of soil deformations under the action of a straight wedge is determined by the angle of cutting (β) and the soil properties. In the case of slightly cohesive sandy soils, when the angle of cutting is medium (β=20-60^0), the deformations of the soil in front of the wedge consists of two periodically recurrent phases as shown in Figure 1. (a) compaction by the wedge of the soil and penetration of its particles into the still underformed soil mass, when the wedge blade moves from the point 0 to the point 01, (b) instantaneous appearance of the sheer plane 01A1, making an angle θ with the furrow bottom, in the soil in front of the wedge blade. The ultimate result of the above described process is the formation of a prismatic lump with a trapezoidal cross section 01 A1 AB. Furtheron without suffering fresh deformations, this lump starts moving along the surface of the wedge with a velocity Vr and simultaneously slides along the shear plane 01 A1 with a velocity Va.

We find from the triangle of velocities that;

$$Va = V \frac{\sin \beta}{\sin (\text{symbolb} + \phi)} \quad \ldots (1)$$

$$Vr = V \frac{\sin \phi}{\sin (\text{symbolb} + \phi)} \quad \ldots (2)$$

Where V is translatory velocity of the wedge. Thus, during the deformation of a soil by shear, the relative velocity of the furrow slice along the surface of the wedge Vr < V and its thickness a1 more than the depth of penetration of wedge. The ratio a:a1 characterising the shrinkage of the slice is :

$$\frac{a}{a_1} = \frac{\sin \beta}{\sin (\beta + \phi)} \quad \ldots (3)$$

The above relationship shows that with the increase in the angle of cutting B, the relative displacement of lumps increases, i.e. the srinkage of furrow slice increases.

In case of loamy soils the wedge fractures individual lumps of irregular shape. In this case, the thickness of the furrow slice is equal to the depth of penetration of the wedge. Hence there is no srinkage of the chip and the relative velocity of the furrow slice Vr is equal to the translatory

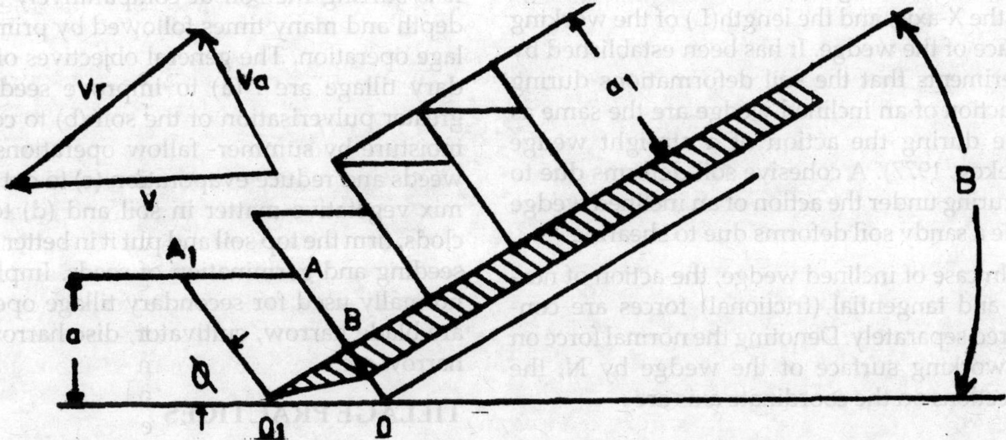

Fig. 1. Diagram of deformation of a sandy soil during the working of a straight wedge.

velocity V of the wedge.

Soil resistance : The soil resistance to be overcome by a straight wedge may be subdivided into four types : (1) the soil resistance to the penetration of the blade, (2) the soil resistance to the deformation, (3) the weight of the furrow slice, and (4) the internal force of the furrow slice. The compressive stress in the soil in front of the wedge increases with increase in the length of the wedge, the volumetric weight of the soil, the angle of friction between soil and steel, the angle of cutting of the wedge, and the velocity of the wedge, but does not depend on the cross section of the furrow slice.

Efficiency of Straight Wedge

The frictional forces on the working surface of a wedge and the resistance due to the blunting of its blade may be regarded as the parasite resistance during the working of a wedge. The efficiency of a straight wedge with a sharp blade may be taken as the ratio of the tractive resistance of a wedge with an antifriction surface to the tractive resistance of an ordinary wedge.

SOIL DEFORMATIONS AND SOIL RESISTANCE DURING THE WORKING OF INCLINED WEDGE

The geometrical and dimentional characteristic of an inclined wedge should include the angle of cutting(β) made by working surface with XOY plane, the angle (r) between the blade edge and the X-axis, and the length(L) of the working surface of the wedge. It has been established by experiments that the soil deformations during the action of an inclined wedge are the same as those during the action of a straight wedge (Sinekov, 1977). A cohesive soil deforms due to fracturing under the action of an inclined wedge while a sandy soil deforms due to shearing.

In case of inclined wedge, the action of normal and tangential (frictional) forces are considered separately. Denoting the normal force on the working surface of the wedge by N, the projections on the coordinate axis are :

$$Nx = N \sin \beta \sin r \quad \ldots (4)$$

$$Ny = N \sin \beta \cos r \quad \ldots (5)$$

$$Nz = N \cos \beta \quad \ldots (6)$$

To determine the projections of the frictional force (T) on the coordinate axis, it is resolved into two components : T1-Parallel to the share blade, and T2- Perpendicular to it. The projections on the coordinate axes are :

$$Tx=T(\cos^2 r + \sin^2 y \cos \beta) \quad \ldots (7)$$

$$Ty=-T \sin r \cos r (1-\cos \beta) \quad \ldots (8)$$

$$Tz=-T \sin r \sin \beta \quad \ldots (9)$$

TILLAGE METHODS

Tillage operations for seedbed preparation are often classified as primary and secondary.

Primary tillage : It is an initial major soil working operation and normally designed to reduce soil strength, cover plant materials and rearrange aggregates. It cuts and shatters soil and may bury trash by inversion, mix it into the tilled layer or leave it basically undisturbed. Primary tillage layer is more aggressive, relatively deeper operation and usually leaves the surface rough. Implements normally used for primary tillage are : country plough, M.B.Plough, disc plough, chisel plough, sub soilers etc.

Secondary tillage : It is intended to create refined soil conditions following primary tillage. It is stirring the soil at comparatively shallow depth and many times followed by primary tillage operation. The general objectives of secondary tillage are : (a) to improve seedbed by greater pulverisation of the soil (b) to conserve moisture by summer- fallow operations to kill weeds and reduce evaporation (c) to cut up and mix vegetative matter in soil and (d) to break clods, firm the top soil and put it in better tilth for seeding and germination of seeds. Implements normally used for secondary tillage operations are blade harrow, cultivator, disc harrow, tyne harrow etc.

TILLAGE PRACTICES

A tillage intensity to create a desirable

seedbed conditions is determined depending upon the power available, soil type and moisture available in the soil. In the countries dependent on human and animal power for tillage, the tilling of soil limits only to available time and energy. With the introduction of the tractors the tillage became easier and excessive tillage practiced resulting in excessive breakdown of soil particle size, avoidable erosion, unnecessary compaction from wheel traffic and waste of time and fuel. The tillage practices are often classified depending upon the objective it served, intensity of tillage etc. The common tillage practices referred are given below :

* minimum/optimum/reduced tillage
* conservation/mulch tillage
* zero tillage

Minimum/optimum/reduced tillage systems involve principles rather than well defined practices. The objectives of these systems include (a) reducing energy input and labour requirement for crop production (b) conserving soil moisture and reducing erosion (c) providing optimum seedbed rather than homogenising the entire soil surface, and (d) keeping field compaction to minimum.

The objectives of conservation/mulch tillage are to achieve soil and water conservation and energy conservation through reduced tillage operations. Both systems usually leave crop residue on the surface and each operation is planned to maintain continuous soil coverage by residue or growing plants. The conservation tillage practices may advance some of the goals of alternative farming such as increasing organic matter in soil and reducing soil erosion, but some conservation tillage practices may increase the need for pesticides. Conservation tillage changes soil properties in ways that affect plant growth, and reduce water runoff from fields. The mulched soil is cooler and soil surface under the residue is moist, as a result many conservation tillage systems have been successful (Annonymous 1989). In zero tillage or no-till system the crop residue is usually shredded and planting is done without preliminary tillage. No till planting has problem of adequate weed control.

PERFORMANCE EVALUATION OF TILLAGE PRACTICES, TOOLS AND IMPLEMENTS

The purpose of tillage tool is to manipulate a soil as required to achieve desired soil condition. Three abstract design factors are (1) initial soil condition (2) tool shape and (3) manner of soil movement. These can also be termed as input factors. The final soil condition and forces required to manipulate soil are output factors.

The forces and changes in soil condition are the two basic aspects of tillage-tool performance. The tool should accomplish the necessary soil manipulation with minimum energy input and the final desired soil condition. The degree of soil breakup, segregation of clod sizes in relation to depth, and uniformity of mixing throughout the tilled depth determines the final soil condition. The performance evaluation involves the comparison of actual final soil condition with the desired soil condition. The desired soil condition is entirely determined by the intended use of the tilled soil. Any tillage practice which does not return more than its cost by increasing yield and improving soil conditions should be eliminated or changed. The soil needs to be worked only enough to assure optimum crop production and weed control.

In Indian drylands most of the tillage research has been dominated by studies related to the tillage effect on increasing crop yields. Erosion control, in-situ moisture conservation, moisture management, weed control and residue management have been the main aspects of the tillage study undertaken by various dryland research centers. A review of tillage research is presented below :

EFFECT OF TILLAGE ON YIELD

Tillage operations consume a large proportion of the total energy expenditure in all field operations. The experiments conducted in alfisols at ICRISAT center indicated that effect of tillage treatments on crop yield was not consistent (Table 1) and thus there is no conclusive evidence to suggest that any one treatment was better (Klaij, 1983). The split-strip plowing gave significantly more yield of sorghum (3080 kg ha^{-1}) in a year of low rainfall. The crop responded well to the increased tillage because of enhanced infiltration. Sorghum grain yield varied considerably from year to year in a soil where clay content was greater.

Table 1 : Effect of primary tillage on sorghum grain yield (kg ha^{-1}) in alfisols, ICRISAT center.

Treatment	Year		
	1979	1980	1981
Split-strip ploughing	2970	1800	3080
Strip ploughing	3080	1750	3010
Chiseling	2870	1450	2680
Shallow cultivation	3040	1910	2870
Mean	2990	1740	2910
SE ±	111	120	260

Source : Klaij 1983

The deep plowing with inversion plough increased crop yields (Vijayalaxmi, 1987). The data from Hyderabad alfisols showed that the deep ploughing is advantageous to sorghum in normal rainfall years, and to castor in normal and above normal years. Therefore in sorghum- castor rotation, it is better to deep plough for castor. The effect of ploughing is however transitory. At Anantapur and Bangalore also the deep- ploughing has shown an increase in yield (Table 2)

Table 2 : Effect of tillage on crop yields.

Center	Crop	Yield (t ha^{-1})	
		Shallow tillage	Deep tillage
Anantapur	Castor	1.01	1.30
	Pigeonpea	0.71	1.04
	Pearl Millet	0.68	1.00
Bangalore	Groundnut	0.86	1.17
	Maize	3.20	4.51
	Pigeonpea	1.23	1.90

Source : AICRPDA. 1982

Hegde *et al.*(1987) reported that one additional ploughing with a mould board plough to a depth of 15 cm increased the finger millet yield by 44 percent (Table 3).

Table 3 : Effect of mold board ploughing on the yield of finger millet.

System	Yield (kg ha^{-1})	
	1981	1982
Local practice (2-3 times wooden plough	1220	1040
One additional mold board ploughing in July	1670	1570

Source : Hegde *et al* 1987.

Yadav (1988) showed that there was no significant difference in wheat yield due to 3,4,5, and 6 times tillage operation. It is also reported by AICRPDA center, Rajkot (Gujarat), that 4-5 times shallow ploughing with local plough and 30 cm deep ploughing (tractor) + 2-3 discing did not give significant difference in the yield of pearl millet, sorghum, groundnut and sunflower (Annonymous, 1981).

The tillage with M.B. Plough + twice disc harrowing gave highest yield of sorghum (4.77 t ha^{-1}) in black soil of Bhopal, compared to 4 times blade harrow, twice cultivator tillage and no tillage. The tillage with M.B.Ploughing + twice disc harrowing also gave lowest bulk density (1.06 g m^{-3}) after tillage operations, lowest penetration resistance at 15 cm depth, and lowest mean dry weight of weeds. (RautRay, 1983).

The highest yield of castor (1.23 t ha^{-1}), pearl millet (2.75 t ha^{-1}) and sorghum (2.71 t ha^{-1}) in alfisol was obtained with tillage using 6 tyne cultivator (twice) + sowing by M.F. Seed drill (Sriram and Vishnumurthy, 1987).

The studies conducted at Ranchi, recommended the deeper tillage by moldboard plough every year which increased yield by 2.4 times compared to farmers practice (Annonymous, 1985). The effect of summer tillage prior to seeding showed increase in sorghum yield when compared to plough seed technique (Yadav *et. al.* 1984).

ENERGY REQUIREMENTS FOR TILLAGE OPERATIONS

Tillage consumes about 23 - 31 percent of total energy for field operations (Table 4). The magnitude of energy requirement of tillage varied greatly from location to location in the range of 284 - 816 MJ ha^{-1}. The lowest energy required for tillage was reported from Raichur. It is predominently kharif fallow- Rabi Jowar area where the tillage by blade harrow is repeated during *kharif*. The Varanasi and Bangalore centres reported highest energy for tillage. These variations are mainly due to the variations in tillage intensity depending upon soil type, cropping patterns, weed control to get desired tilth.

Table 4 : Energy requirement for tillage operations in Indian drylands.

Location	Energy for tillage operation	
	(MJ ha^{-1})	(Per cent of total energy for field operation)
Hyderabad	450	31
Varanasi	816	23
Raichur	284	24
Bangalore	744	29
Rewa	562	26

Source : Annonymous (1991).

Table 5 : Effect of tillage methods on soil moisture on an alfisol during no-rain periods at ICRISAT center

No. of days	Tillage Methods					
	Split strip ploughing	Strip ploughing	Chiseling	Shallow cultivation	Mean	SE ±
	Soil moisture (mm) in top 20 cm layer					
0	45	43	44	43	44	0.6
2	31	29	34	35	32	0.7
4	28	29	30	33	30	0.5
9	22	25	28	32	27	0.5

Source : Bansal et. al 1987.

The energy expenditure in primary tillage was estimated by measuring the draft and the travel distance required to cover 1 hectare (Awadhwal and Thierstein, 1984). The split strip plowing consumed a much higher level of energy (28.2 MJ ha^{-1}) as compared to strip plowing (11 MJ ha^{-1}).

EFFECT OF TILLAGE ON MOISTURE CONSERVATION

The experiments were conducted at ICRISAT centre on broadbed and furrow configuration where the traffic zone is clearly separated from the cropping zone. Four primary tillage treatments evaluated in alfisols were (a) split-strip ploughing and (b) strip ploughing (using left and right hand M.B.Plough), (c) chiseling to 120 mm depth where crop rows were to follow, and (d) shallow cultivation by duck foot sweeps (Bansal et. al. 1987). Table 5 shows trends of soil moisture depletion in the top 20 cm layer during the early growth of sorghum crop. Split-strip tillage holds marginally more of water, but moisture depletion during stress is faster. Consequently, at the end of dry spells moisture in intensive tillage is less than in shallow tillage treatments. Two hypothesis to explain differences in soil moisture in tillage plots could be, first, the intensive tillage provided better environment for crop growth and root development resulting in higher rates of transpiration that led to faster depletion of moisture. Secondly, in shallow tillage the soil is stratified in two layers : a top tilled layer and sub surface untilled layer. The stratification causes increased resistance to moisture movement.

Tillage operations before the onset of the rainy seasons are essential for increasing infiltration and for effective weed control. The effect of ploughing before onset of mansoon is shown in Table 6.

Table 6 : Soil moisture as influenced by ploughing before the onset of the rainy season.

Depth of Soil (cm)	Moisture percentage after total of 81 mm rainfall in May	
	Ploughed area	Unploughed area
0 - 15	10.7	3.6
15 - 30	13.2	7.1
30 - 60	13.3	8.7
60 - 90	13.4	Dry

Souce : AICRPDA 1977-82.

The deep tillage using M.B.Plough and chisel plough is beneficial for rain water conservation (Table 7).

Table 7 : Effect of tillage implement on soil moisture conservation.

Treatment	Average soil moisture (mm) in 100 cm soil profile
M.B. Ploughing + discing	58.3
Chisel ploughing + discing	60.7
Tractor cultivation	47.4

Source : Mayande, 1992

EFFECT OF ORGANIC RESIDUE INCORPORATION ON SOIL TILLAGE PROPERTIES

The organic residue incorporated through either pearl millet or cowpea improved all physical parameters (Venkateswarlu, 1987). The physical soil properties as affected by organic residue incorporation given by Mohd. Shakeel Hameed (1978) are presented in Table 8.

EFFECT OF TILLAGE ON INFILTRATION RATE AND RUNOFF

The tillage operations such as ploughing or ridging will loosen the top 10-15 cm and therefore the infiltration rate will be high. To increase the total volume of water infiltrated under rainfall, two options are available : an increase of the surface storage, or an increase in the infiltration rate. The runoff for two surface storage values and under till and no-till condition for three years were reported by Hoogmed (1987) and results are presented in Table 9. It is clear that the effect of a permanet storage of 10 mm, and a tilling at the beginning of the season, are most effective for runoff control.

EFFECT OF TILLAGE TOOLS ON PHYSICAL CHARACTERISTICS OF ALFISOLS

The experiment conducted in alfisol at ICRISAT center (Table 10) revealed that the tillage by shovel resulted in more water in top 20 cm layer as compared to other tillage methods (Annonymous, 1985). The bulk densities under various tillage methods were not different. The cone index values in the top and middle layers in sweep tillage were significantly higher than in other treatments. There was no significant dif-

Table 8 : Physical soil properties as affected by organic residue incorporation :

Soil constituent	Initial value		After 5 years	
	Red sandy loam	Red loamy sand	Red sandy loam	Red loamy sand
Mean weight diameter (mm)	2.63	3.21	3.04	4.19
Hydraulic conductivity (cm h^{-1})	4.19	2.34	10.50	8.00
Bulk density (g cm^{-3})	1.81	1.73	1.69	1.65
Soil crusting (Penetrometer reading)	35.00	40.00	18.00	11.00
Moisture content at field capacity (g/100 g^{-1})	15.70	12.40	19.00	12.90

Source : Mohd. Shakeel Hameed, 1978.

Table 9 : Run-off for different tillage practices and surface storage values (ss) for three growing seasons.

Year	Rain (mm)	Tillage treatment	Run off (mm)			
			SS = 1.0 mm		SS = 10.0 mm	
			MM	(%)	MM	(%)
1977	368	Tillage	76	18	11	3
		No tillage	155	41	48	13
1978	271	Tillage	49	18	19	7
		No tillage	104	38	33	12
1979	361	Tillage	80	22	38	11
		No tillage	141	39	70	19

Source : Hoogmed, 1987.

Table 10 : Effect of tillage tools on physical characteristics of Alfisols.

Tillage Treatment	Mean water held (mm) in 20 cm surface layer	Bulk densities (Mg M^{-1})			Cone index (kpa) 3 weeks after tillage operation			Crop yield kg ha^{-1}
		(0-7 cm)	(7-14 cm)	(14-20 cm)	(0-7 cm)	(7-14 cm)	(14-20 cm)	
Shovel	21.5	1.42	1.48	1.57	134	212	445	950
Chisel	18.8	1.36	1.42	1.49	124	186	282	940
Sweep	18.2	1.38	1.48	1.52	194	347	469	980
Split-strip ploughing	19.2	1.36	1.49	1.51	131	180	422	990
Mean	19.4	1.38	1.47	1.56	146	231	404	970
SE ±	0.93	0.017	0.017	0.017	14.9	14.9	14.9	107

Source : Report of work, Farm Power and Equipment, FSRP, ICRISAT, 1985.

ference in the crop yield due to tillage treatments.

Goyal and Leland (1979) studied the comparative performance of mould board ploughing and chisel ploughing. Mould board plough pulverises, aerates and loosens the soil. By inversion of soil, the new and unused plant nutrients are provided near the surface for the new plant seedlings. The analysis of draft requirement of M.B.Plough shows that 18% of draft is required for pulling, 48% for cutting furrow slice and 34% for turning over the slice. An increase of soil moisture by 1% decreased the draft by 10%. The depth of operation of M.B.Plough can be increased upto 12.5 cm in most soil conditions without any appreciable increase in specific draft. Mould board plough helps in burying the trash in the soil and covers plant material because of its inversion action.

Chisel plough stirs the soil more or less in place. Soil is broken but not inverted and pulverised to the extent of M.B.Plough crushes the soil. Chisel plough breaks the hard pan. It leaves trash on surface. It would mean more chances of insect-infestation and disease spread, but surface residue helps to slow down soil erosion.

Thakur (1988) outlined the factors which are of concern to the designers of soil working tools. Simplicity of design and operations is the most desirable character in any farm tool. The important design considerations of the blade for tillage are the penetrations, soil disturbance, scouring and draft requirement which are dependent upon tool geometry and the soil physical properties. The draft of tynes working at a constant depth increased with increase in rake angle. The rate of increase of draft becomes greater with more than 50^0 rake angle. The blade with approximately 20^0 rake provide maximum upward force and soil disturbance with minimum draft.

The line of action of soil resultant force acting on the blade has a big influence on the ease of penetration and is influenced by the rake angle. The closer the approach angle of the tool to the direction of travel, the better the chance of scouring. The depth/width ratio of tool termed as 'aspect ratio' varies with respect to tyne rake angle and could be helpful in deciding the width of the tool and its working depth for lower magnitude of draft.

For weed cutting the blade rake angle must be such that it disturbs the weeds sufficiently to bring them to the surface without buldozing them along and mixing them with the soil. The blades are usually swept back at angle of 20^0 to 50^0. This permits self cleaning of the blade.

The performance of the mouldboard plough, disc plough and cultivator was studied by Bukhari et al (1988). The soil volume disturbed by mouldboard plough was about 29% greater than disk plough and cultivator. The soil inversion by M.B.Plough, disk plough and cultivator was 95%, 50% and 5% respectively. The study recommended the cultivator for light soils, M.B. Plough where erosion is not a problem and in weed infested fields. The disk plough is recommended for problem soils where other tillage implements cannot operate satisfactorily.

The performance of the five types of mold

board plough designs developed in Brazil, France, Mexico and India was evaluated in vertisols (Mayande et al. 1990). These ploughs vary in their size and shape of the mould board, share type and material used for construction. Although the clods formed by these ploughs and surface roughness index did not differ significantly, it appears that there were differences in the way these ploughs acted to cut the furrow slice. The study concluded that despite the common use by farmers, the kirlosker type plough has a relatively high pull requirement even in shallow working depths in dry vertisol. This suggested that the kirlosker plough should be replaced by a better designs like ICRISAT plough, that would save energy and time.

REFERENCES

AICRPDA (All India Co-ordinated Research Porject for Dryland Agriculture) 1982. *A Decade of Dryland Agricultural Research in India 1971-80*, Hyderabad, Andhra Pradesh, India : AICRPDA.

AICRPDA (All India Co-ordinated Research Project for Dryland Agriculture) 1977-82. *Annual Reports of Bangalore Center.* Bangalore, Karnataka, India : University of Agricultural Sciences.

Annonymous, 1981. Report of main Dryland Research Station, Rajkot (1971-1980) : Gujarat Agricultural University.

Annonymous, 1985. Annual Progress Reports AICRPDA main center, Ranchi, Bhihar : Birsa Agricultural University.

Annonymous 1985. Report of work, 1985. Farm Power and Equipment, Farming Systems Research Programme : ICRISAT Patancheru, P.O. 502 324, A.P.

Annonymous, 1989. *Alternative Agriculture,* National Research Council, National Acadamy Press, Washington, D.C. 1989.

Annonymous, 1991. Report of work on Research on Mechanisation of Dryland Agriculture, Hyderabad Centre, Nov. 1991 : Central Research Institute for Dryland Agriculture, Hyderabad.

Awadhwal, N.K., and Thierstein, G.E. 1984. Evaluation of tillage methods suitable for light-textured alfisols of the semi-arid tropics. *Journal of Agricultural Engineering* 21 (1-2) : 44-54.

Bansal, R.K., Awadhwal, N.K. and Mayande, V.M., 1987. Implement development for SAT Alfisols. In *Proceedings of the Consultants Workshop on the State of the Art and Management Alternatives for Optimising the Productivity of SAT Alfisols and Related Soils,* ICRISAT Center, Patancheru, India : pp 97-107.

Bukhari Sheruddin, Bhutto, M.A. Baloch, J.M., Bhutto, A.B., and Ali Nawaz Mirani, 1988. Performance of selected tillage implements. *Agricultural Mechanisation in Asia, Africa and Latin America.* Vol. 19 No. 4, 1908 (9-14).

Cuplin Claude, 1986. *Farm Machinery*, Eleventh edition, English Language Book Society, Collins. pp 55-60.

Goyal Megh R., and Leland O. Drew. 1979. Chisel ploughing verses Mold board ploughing. *Agricultural mechanisation in Asia*, 1979 : pp 51-56.

Hegde, B.R., Chennappa, T.C. and Anand Ram, B.K., 1987. Management alternatives for increased productivity of red soils : Experience in Karnataka. In *Proceedings of the Consultants Workshop on the State of Art and Management Alternatives for Optimising the Productivity of SAT Alfisols and Related soils.* ICRISAT Center, Patancheru, India : ps 137-143.

Hoogmed, Willem, B., 1987. Some aspects of crust formation of soils in Semi-Arid Regions. In *Proceedings of the Consultants Workshop on the State of the Art and Management Alternatives for Optimising the Productivity of SAT Alfisols and Related Soils,* ICRISAT, Patancheru, India : pp 127-135.

Kepner, R.A., Roy Bainer, and E.L.Barger, 1982. *Principles of Farm Machinery.* Third edition, AVI Publishing Company, INC, West Port, Connecticut, USA.

Klaij, M.C., 1983. Analysis and evaluation of tillage on an Alfisol in a Semi-arid tropical region of India. Ph.D Thesis, Soil and Tillage Laboratory, Wageningen, The Netherlands.

Mayande, V.M., 1992. Performance of implements under rainfed management. A chapter in a book on *'Dryland Agriculture- State of Art of Research in India,* eds L.L. Somani, K.P.R. Vittal and B.Venkateshwarlu (1992), Scientific Publishers, Jodhpur (437-458).

Mayande, V.M., Bansal, R.K., and Smith, G.D. 1990. Comparison of Mold board ploughs for dry tillage of vertisols. *Soil & Tillage Research,* 15 : 349-358.

Mohd. Shakeel Hameed, 1978. Physico-chemical changes with different systems of organic recycling in the semi-arid red soil region. M.Sc. Thesis, Andhra Pradesh Agricultural University, Hyderabad, Andhra Pradesh, India.

RautRay, S.K., 1983. Tillage systems for production of sorghum in black soils. *Journal of Agricultural Engineering*, Vol. XX, No. 3&4 : (27-38).

Sinekov, G.N., 1977. *Design of soil tilling machines*. Published by Indian National Scientific Documentation Center, Hillside Road, New Delhi.

Sriram, C., and Vishnumurthy, T. 1987. Minimum tillage requirements for castor, sorghum / pearl millet rotation. *Journal of Agricultural Engineering*, Vol XXIV, No. 3(239-244).

Thakur, T.C., 1988. Design aspects of soil engaging hand tools. *Agricultural Engineering Today*, Vol. 12 No. 4, July- August, 1988 : pp 14-21.

Venkateswarlu, J, 1987. Soil Fertility Management of Red Soils, In *Proceedings of the Consultants Workshop on the State of the Art and Management Alternatives for Optimising the Productivity of SAT Alfisols and Related Soils*, ICRISAT Center, Patancheru, India : PP : 115-122.

Vijayalaxmi, K., 1987. Soil Management for increasing productivity in semi-arid Red soils : Physical aspects. In *Proceedings of the Consultants Workshop on the State of the Art and Management Alternatives for Optimizing the Productivity of SAT Alfisols and Related Soils*, ICRISAT Center, Patancheru, India : pp 109-113.

Yadava, G.C., Anderson, D.T. and Rajput, D.S., 1984. Farm implements and practices for small farmers of drylands. Paper presented at *XVII Annual Convention of Indian Society of Agricultural Engineers, held at IARI, New Delhi.*

Yadav, S.S., 1988. Effect of varying levels of presowing tillage on the yield of wheat. Paper presented at *the XXIV Annual Convention of Indian Society of Agricultural Engineers*, held at PKV, Akola.

17

IMPROVED AGRICULTURAL IMPLEMENTS FOR DRYLANDS

V.M. MAYANDE

INTRODUCTION

Sixty eight percent of total cultivated area in India is rainfed and it contributes 44% food and supports 40% of the population. The productivity of these farms depends upon the behaviour of monsoon which often causes dry spells and moisture stresses during the cropping season. The field operations performed in drylands aim primarily at conservation of moisture for crop growth. Therefore the implements used for performing various operations in drylands should have features which meet the objectives of dryland cultivation practices.

The major objectives of cultivation practices in drylands are (i) to produce a rough cloddy surface to increase moisture absorption and reduce runoff (ii) the control and destruction of weeds, and (iii) rapid tilling and timely seeding at low cost.

TIMELINESS AND PRECISION

The timeliness of operations and precision in seeding pays a lot of dividend in dryland farming. The major power sources (human and animal) available in drylands operate at slow pace. Using hand labour or draft animals often requires that the farmers wait until there is sufficient moisture in the soil to facilitate primary tillage. Timeliness is critical in a low rainfall region where conservation of moisture is essential to achive maximum production potential (Stout and Downing, 1974). Some crops are very sensitive to planting dates and each days delay in planting reduces the yield. Clayton (1972) reported that cotton yields in Uganda increased 25 per cent due to the fact that tractor mechanisation facilitated timely seed bed preparation in May rather than in June as was the practice in traditional agriculture.

The use of tractors for seedbed preparation is also expanding in India. Binswanger (1978) reported that delaying sowing from period 1 to period 4 (each period = 2 weeks) implies an yield loss of about 20 per cent. Jodha (1974) attributed the very rapid spread of tractors in Rajastan to the fact that on these sandy to sandy loam soils with very scarce rainfall, a safe sowing period is often 5 to 6 days which puts a much higher premium on timeliness than is the case in heavy soils. We thus can conclude that timeliness of operation should be most important in dry areas with scanty rainfall and shallow red and sandy soils. Umakesan (1971) reported that the tractor farms complete field preparation and sowing in exactly half the time required by bullock farm.

In a typical double cropping system timeliness assumes a special importance, as the period between harvest of first crop and sowing of second crop is very limited. The timeliness factor also varies depending upon soil types, rainfall and cropping patterns. Binswanger (1978) reported that in heavy soils bullock farms do better than tractor farms.

The land holdings owned by small and marginal farmers in dryland regions of Bangalore, Varanasi and Hyderabad are 54-68 per cent, whereas in Raichur and Rewa region it is 14-24 per cent (Annonymous 1991). This group of farmers is fully dependent on human and draught animal power. These categories of farmers, in all regions are in deficit of human and animal power for land preparation, sowing, interculture and harvesting operations, if they have to achieve timeliness. The deficit of human and animal power for achievement of timeliness is presented in Table 1. The highest deficit of power in Varanasi followed by Bangalore indicates the need to mechanise the field operations either by improving the efficiencies of present implements or to shift to high power equipments to complete operations faster.

The implements for tillage commonly used by the farmers are country plough, M.B.Plough, blade harrows etc. The tillage research has been mainly centered around the intensity of tillage by using single or combination of the available implements. The major emphasis of development of implements at CRIDA and AICRPDA in the past has been on seeding machines as seeding is a crucial operation in drylands. Half of the battle is won if the seed is placed in moist soil at appropriate time. Therefore the development of seeding machines has been a priority of the drylands.

Table 1. Power deficit in achieving timeliness on small and marginal farms in drylands.

S. No.	Region	Small and marginal farms		Deficit of power for achieving timeliness	
		Holdings	Population	Human	Animal
		(%)		(Kw)	
1.	Bangalore	54	81	64	398
2.	Raichur	14	30	0	69
3.	Rewa	24	35	48	11
4.	Varanasi	68	77	268	733

The second important aspect in the drylands is the precision in placement of seed and fertilizer in the soil. The most common method of sowing in drylands is line seeding using furrowing implements and mannual dropping of seeds. Good crop establishment depends on the quality of seeds sown as well as good seedbed and weather conditions. Under field conditions the factors which affect the seed emergence include the seed and soil characteristics as well as methods/machines used for seeding. The precision in seed metering, and placement in moist zone at appropriate depth is essential for good crop establishment in drylands.

During the last two decades several implements were developed at Central Research Institute for Dryland Agriculture (CRIDA) and All India Coordinated Research Project on Dryland Agriculture (AICRPDA) to achieve the timeliness and precision. As the available power sources in drylands are predominently human and animal labour, the emphasis was on development of implements to suit the available power.

APPROACHES AND CRITERIA OF DEVELOPMENT

The approach adopted for development of machines can be categarised into two groups (a) the improvement of local methods by modifying traditional implements to improve the efficiency (b) the development of new implements for achieving precision and timeliness. Apart from this the labour saving, low cost, simple construction and ease of operation are also the important criteria adopted in designing of implements for drylands.

REQUIREMENTS OF DRYLANDS

The tillage implements in drylands are used for mechanical manipulation of soil to develop desirable soil structure, to allow rapid infiltration and good retention of rainfall to provide adequate air capacity and exchange within the soil to produce finer soil aggregates desirable for good seedbed, and to control weeds. The field should be weed free for first 30-40 days after sowing. The weeds compete with crop for nutrients and moisture and result in 50% or more yield loss, if unchecked.

Seeding operations involved placing seed in the soil at a predetermined depth. A seed planter performs mainly four functions - (i) open the seed furrow to the proper depth, (ii) meter the seed, (iii) deposit the seed in the furrow in an acceptable pattern, (iv) cover the seed and compact soil around seed to proper degree. The seed should be placed in moist soil at even depth to promote rapid and uniform germination and emergence. However, some additional requirements in dry farming are:

(1) The soil surface after seeding should be in such condition as to (a) prevent rapid evaporation from seed zone. (b) allow for rapid absorption of subsequent rainfall (c) resist erosion.

(2) Seedrates and row spacings must be closely controlled in order to provide a plant population the moisture requirement of which does not greatly exceed the available supply.

(3) Seed should generally be placed in shallow furrows within the moist zone.

Timely harvesting is also an important aspect in drylands. The delayed harvesting causes lodging and shattering losses. The harvesting in dryland areas is presently done by human labour using local sickle. It requires 150 man hours/ha. The serrated sickle (125 man hours/ha) increased the efficiency by about 20 percent. The mannual harvesting is tedious. The shortage of labour at peak demand time causes delays.

PRESENT STATUS OF IMPROVED IMPLEMENTS

Seedbed Preparation Equipments

The tillage implements used for seedbed preparation are country plough, chisel plough, mould board plough, blade harrow, tyne harrow, disc harrow etc. The research on seedbed preparation in drylands has been concentrated by and large on intensity of tillage using single or combined operations of available implements to get desirable seed bed conditions.

The chisel ploughing (once) + disc harrowing (once) conserved highest moisture (63.7%) in 100 cm profile for rabi seeding at Varanasi. The introduction of blade harrow for tillage in Rewa region reduced weed intensity by 90 percent. The year round tillage with country plough + interculturing increased the yield of paddy by three times at Ranchi. The use of mould board plough + one blade harrowing gave highest yield of kharif sorghum at Hyderabad (Rajput et al. 1980).

The design of Navin bakhar blade developed at Central Institute of Agricultural Engineering, Bhopal has been recommended for black soils (Annonymous 1982). The 'V' shape blade is about 1 metre wide with 120 ^{o} angle. This is self cleaning. The straight blade requires cleaning once in a 50-100 meter run depending upon weed intensity. The blade gives 0.5 to 0.6 ha/day, coverage.

Mayande et al. (1990) reported the performance of moldboard ploughs for dry season tillage of vertisols. The commonly used moldboard plough by farmers required very high draft power. The design of cutting edge and shape of moldboard in coventional ploughs is made of cast iron. The cutting edge gets blunt after use and increases the power requirement. The convex shape of moldboard does not turn the soil properly. The ICRISAT design of M.B. plough which has self sharpening type share and moldboard with desired curvature required almost half of the draft power.

Equipments for seeding and fertilizer application

The traditional country plough is used very commonly by dryland farmers for seeding. In most of the small and marginal dryland farms in India, the seeding and fertilising is done manually by dropping the seed and fertiliser in a furrow opened by country plough. This requires two labourers, one for dropping the seeds and other for fertilizer. Farmer then makes a dumy pass with his plough to cover the seed and fertilizer. His attempt to provide soil cover pushes uneven soil mass onto the seed. The result is unequal depth of seeding and unusually deep seed placement and poor germination resulting in overall low yields. This practice has high labour requirements and slow pace. Seven man-days and one bullock pair are engaged to complete seeding of one hectare in two and half days. Also irregular crop stands are promoted by random seed and fertiliser depth by inexperienced farm hands. The drylands which influences over 100 million hectares, the span of sowing period is very restricted. The seeding operation must be accomplished to cover large areas. The equipments for seeding and planting developed for drylands are discussed below.

Fespo Plough

A single row, bullock drawn seeding implement used for seeding and band placement of fertiliser, has been developed at Central Research Institute for Dryland Agriculture (CRIDA) Hyderabad. Its name is derived from the words fertiliser-seed-pora-plough (Fig. 1). This implement is a modified version of country plough with attachments for carrying out sowing and fertilizing. Two funnels are mounted on the plough with a flat iron frame. Two plastic pipes connect the funnels to the pipe fitted behind the share of the plough. One is about 5 cm below the other, which facilities the dropping of seeds deeper than the fertilizer. An U-shaped blade frame covers the soil on the seed and fertiliser. One pair of bullocks with an operator and two persons for manually metering the seed and fertiliser are required for operation. The coverage is 0.15 to 0.20 ha/day depending on the row spacing of the crop.

Crida Drill-Plough

Appreciating the farmers attachment to country plough, an efficient and inexpensive seed cum fertiliser placement device, as an attachment to the coutry plough has been developed at CRIDA (Fig. 2). With this attachment the country plough has been upgraded with expanding its versatility. The basic character of country plough remained undisturbed.

The hopper box of unit drill is made of 20 gauge M.S. Sheet and stores seed and fertilizer in two separate compartments. Seed compartment holds 2.5 kg seed; while fertiliser compartment is 5 kg capacity. Each compartment is fixed with tube to guide seed and fertilizer in furrow. Hopper box is provided with see-through plastic sheet to keep an eye on the quantity of seed and fertiliser. The steel shaft (16 mm diameter) with rubber agitator is fixed to the bottom. The drive wheel (30 cm dia) with flat pegs is attached to the left side. Metering plates are made up of 6 mm thick flat with standardised orifice centered exactly below the rubber agitator. The diameter of the orifice is standardised for different sizes of crop seeds. The orifice is shaped like funnel top and bottom to minimise clogging of oblong seeds. A "U" shape floating iron frame is bolted to hopper bottom sides covers the seed and fertiliser. It consists of an arrangement for vertical movement of the unit. The unit can be detached by unbolting the mounting frame. The drill unit is simple and can be constructed in a village workshop. It covers 0.4 to 0.6 ha/day. It required olny ploughnman to operate. If promoted faster emergence and uniform crro establishment.

Ridger Seeder

This is a two row bullock drawn seeder developed at AICRPDA Research Centre, Hissar (FIg. 3). The salient features of this seeder are that it works well under dry and receding moisture conditions. Sowing by this seeder is done on the sloped side of ridges which avoids soil crust formation in kharif season, whereas in rabi season it can seed in furrows after scrapping dry soil to the sides. Like the FESPO Plough the bullock drawn seeder has two funnels attached by plastic tubes to the furrow openers. Fertiliser is placed in a band along with the seed. Ridger seeders are giving the best performance during dry years. The yield increase of pearlmillet in the range of 15-18% over local method has been reported over the years.

Crida Seed-cum-Fertiliser Drill

This is a three row seed cum fertiliser drill developed at CRIDA. This implement has a 100 cm long tool-bar to which a wooden beam is fitted through an adjustable iron plate (Fig. 4).

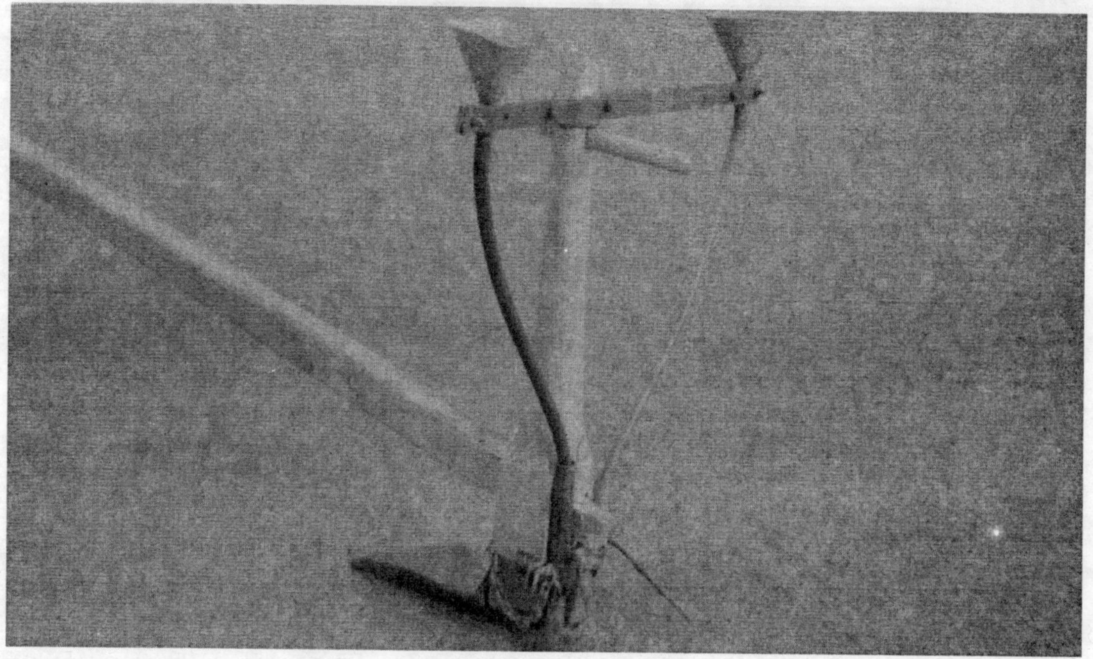

Fig. 1. Fespo Plough

Fig. 2. Crida Dril-Plough

Fig. 3. Riger Seeder

Fig. 4. Crida Seed Cum Ferti Drill

Three pipes each of 50 cm long are clamped vertically on the tool bar. The shovels are bolted at the bottom of the pipe to open the furrows. The hoppers are fitted on the top of the toolbar one each for seed and fertiliser. Metering is done manually. Plastic pipes guide the seed and fertiliser to the openers. This does not have seed covering arrangement.

One pair of bullocks with an operator and two persons for metering the seed and fertiliser are needed. The field coverage is 0.28 ha/hr. It requires 56 kg draft. This implement can be used for sowing intercrops (2:1 sorghum + pigeonpea) with additional funnel and tube. The tool frame can also be used for blade harrowing by clamping the blade in place of furrow openers.

Rayala Gorru (Seed-Drill)

This is made of all steel structure developed at AICRPDA Centre, Andhra Pradesh Agricultural University, Anantapur (Fig. 5). The body of the seed drill is constructed with an angle iron and three steel furrow openers similar to local drills are bolted to the body. A small pipe with shovel at bottom is bolted to main opener for fertiliser. Two funnel bowls are mounted on the frame and transparent tubes connects the bowls to furrow openers. A covering blade is bolted to the body of the drill. The advantage of this ferti-seed drill is that it does covering in addition to placement of seed and fertiliser at about 5 cm lateral and vertical spacing. The 47% increase in the yield of *bajra* (pearl millet) crop compared to local practice has been reported.

Ragi Seeder

This is a seed cum fertiliser drill developed by AICRPDA Center, Bangalore specifically for *Ragi* (Finger millet) (Fig. 6). The seed cum fertiliser drill consists of five coulters or furrow openers at 30 cm row spacing. The coulters are made of 16 gauge M.S. pipe of 2 cm diameter. The fertiliser coulters of 30 cm length are fixed in a light wooden head piece of 1.6 m long. The seed coulters of 10 cm length are clamped at 15 cm behind and 2 cm to the side of the fertiliser coulters so that the seeds are dropped after the fertiliser is covered with some soil. The depth of seed placement can be controlled by adjusting the height of the seed coulters with the help of clamps.

A cone type wooden seed bowl of 14 cm diameter and 12 cm height is used for hand dropping of seeds. A double walled funnel hopper of 14 cm diameter and 14 cm height which is made of mild steel sheet is used for fertiliser distribution. The inner wall of the funnel is having a single opening and the outer wall is having five openings. A tapering bolt at the centre of the funnel helps in regulating the uniform distribution of fertilisers. The seed and fertiliser hoppers are kept at about 60 cm apart on a stand so that three persons can walk easily behind the implement.

Transparent plastic tubes of 1.6 cm diameter and 0.2 cm thickness are used to connect the seed and fertilizer hoppers to the respective coulters. The markers are fixed at two ends which act as floats and help in controlling the depth of sowing.

It can cover 1.4 to 1.6 ha/day with 5 rows at a time. This ferti-seed drill was compared with farmers method and has shown 25 percent yield increase.

Malviya Seed cum Fertiliser Drill

This animal drawn seed cum fertiliser drill was developed by AICRPDA Centre at Banaras Hindu University, Varanasi (Fig. 7). It is a two row machine with adjustable row spacing and sowing depth. The seed drill mechanism is flutted roller. The equipment has common furrow opener for seed and fertiliser. Fertiliser is applied 3 cm below the seed. A vertical rotor over adjustable orifice is used for fertiliser metering. Power transmission to metering device is through sprocket and chain arrangement from ground wheel. The draft requirement is 65 kg and field capacity 0.25 ha/hr. The machine has a tool bar carrier on which matching tools for seed bed preparation and intercuturing can be attached.

Shivaji Multi Purpose Farming Machine

This was designed and developed under the Indo-Canadian Collaborative programme at

AIRPDA Centre, Sholapur. The machine can be used for land preparation, seed and fertiliser placement and interculturing. Row spacing is adjustable from 25 to 90 cm. The depth of seed and of fertiliser placement can be adjusted from 2.5 to 12.5 cm.

This machine is equipped with two seed boxes, one of which has small fluted rollers and is used for small seed metering. A larger seed box with larger flutted rollers is suitable for metering medium and large- size seeds. Seeding rate adjustment on both the boxes is accomplished by sliding the fluted rollers to provide contact with a greater width of seed. The fertiliser box works on an agitator and sliding gate principle. A hinged flap closes the openings at the time of transport. On the outside of these boxes, some graduations help to control the dropping of seed and fertiliser. In double cropping systems, after the harvest of the first crop usually the field is chiseled, harrowed and the next crop is sown. However, in this machine, a provision is made to sow the crop at the time of harrowing, taking advantage of the available moisture in the soil.

CRIDA Groundnut Planter

The groundnut is grown on 8.3 million hectares in India, of which 2.3 millon hectares is in Andhra Pradesh alone. The data collected from farmers field in Andhra Pradesh indicated that uniform crop establishment is one of the major problem as conventional seeding devices meter the seed with hands. This leads to crowding and gaps within the row.

The four row mechanical planter based on inclined plate mechanism for seeding and agitator and orifice mechanism for granular fertiliser has been developed at CRIDA (Fig. 8). The hopper is mounted on 1.7 m long tool-bar with clamps. The pipe beam is attached to tool-bar with arrangement for angle adjustment. The drive to mechanism is taken from ground drive wheel with chains and sprockets. Four furrow openers are clamped to toolbar at 30 cm row spacing. The blade is attached for seed covering. The toolbar is mounted on two rubberised wheels at 1.5 m apart. The depth can be adjusted by raising or lowering the wheel shank on tool bar. The unit is drawn by one pair of bullock and one person can operate it. The planter was used by farmers on 20 ha area in Andhra Pradesh during 1992-93. The performance of planter on farmers field is presented in Table 2. The results show that the planting operation with planter can be completed with double speed as compared to local drill. The local drill required higher seed rate due to human hand used for metering. The saving on labour and seed cost was Rs. 970.00 per hectare.

Table 2. Performance of CRIDA groundnut planter as compared to local method on farmers field.

S.No.	Treatment	Field capacity (ha/hr)	Seed rate (kg/ha)	Plant stand per ha '000	Yield (kg/ha)	Human & bullock cost (Rs./ha)	Seed cost (Rs/ha)
1.	Local drill	0.21	150	268	1562	718	2700
2.	CRIDA groundnut planter	0.40	120	278	1875	288	2160
Net gain (Rs.)					2504*	430	540

* @ Rs 8.00 per kg

The higher yield contributed Rs 2500 higher income per hectare by use of planter. This machine was operated by farmers on 13 ha area during rabi season of 1992-93 and is finding favour with farmers.

Interculture equipments

The khurpi is very common tool used for mannual weeding. The efforts were made at various centers to improve efficiency of mannual weeding by improving tools. The dryland weeder with long handle developed at TNAU, Coimbtore, gives about three times higher output compared to khurpi (Rajput et al 1980). Four shapes of weeding tools viz: straight blade, curved blade, triangular blade and sweep were evaluated in black soils (Biswas and Ingle, 1993).

Fig. 5. Rayala Gorru

Fig. 6. Ragi Seed Drill

Fig. 7. Malviya Machine

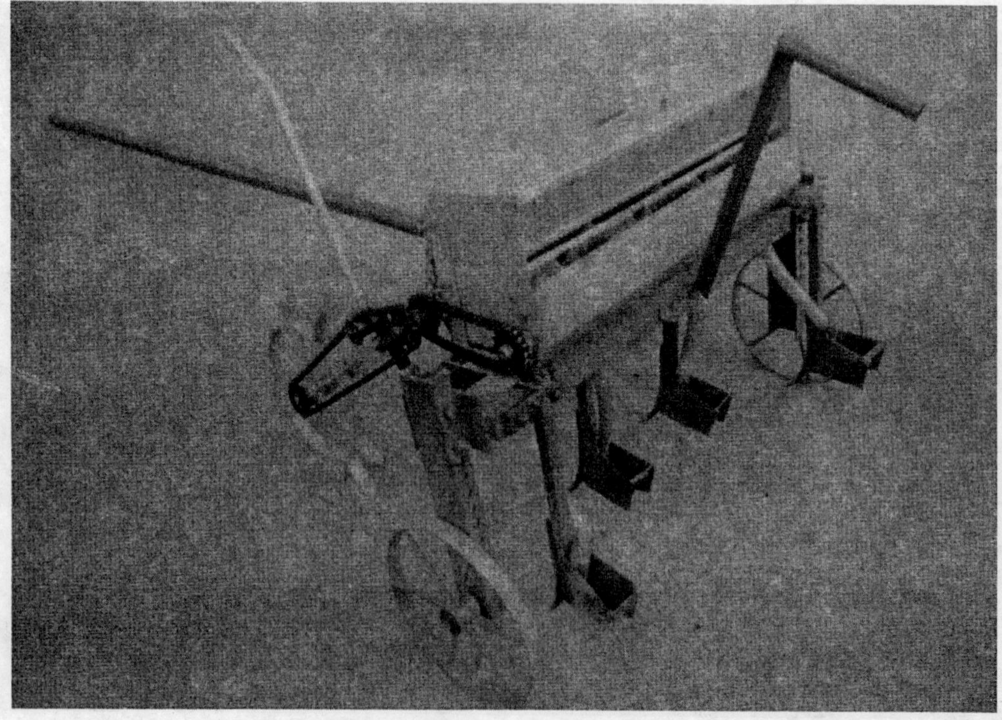

Fig. 8. Crida Groundnut Planter

The field performance index was highest for sweep followed by triangular blade, curved blade, and straight blade. The sweep which has shown better performance can be adopted for mannual and animal drawn weeding machines. Gite et al (1993) evaluated six wheel type mannual weeders to measure physiological cost and postural discomfort of workers. It is reported that oxygen was 35% less than their maximum aerobic capacity and postural discomforts were very high. The weeding efficiency of these weeders ranged between 41-57 percent. These weeders require 100 man hours/ha.

The power weeder developed by Rangaswamy et al (1993) at Tamilnadu Agricultural University, Coimbatore was compared with mannual weeding with hand hoe and dryland weeder. The field capacity of power weeder is 0.04 ha/hr with 93 percent weeding efficiency. The cost of operation of power weeder is only half of the dryland weeder and one third of the mannual weeding with hand hoe.

Harvesting equipments

The traditional harvesting tool (sickle) is not efficient and need constant sharpening and induce human drudgery during operation. Although the improved serrated sickle increased efficiency by 20%, the operation remains labour dependent. The power operated combine harvesters and vertical conveyer reapers widely used in wheat-paddy belt have the limitations under dryland conditions mainly because of small and fragmented holdings. The power tiller mounted vertical conveyer reaper has been evaluated under dryland condition for harvesting of ragi at Bangalore and found satisfactory (Annonymous 1992).

The single row rotary reaper windrower developed at IARI, New Delhi looks promising under dryland conditions (Solanki and Singhal, 1993). It is portable machine mounted on a small trolley with ground wheel. The total weight of machine is 30 kg. It uses serrated disc cutter at 18 m/s speed, and the field efficiency at 12% moisture content was 83 percent. The total labour requirement is only 30 man- hours/ha for cutting and collection of crops.

Datt and Prasad (1993) reported the development of 1 m self propelled vertical conveyer reaper for safflower crop. The safflower is an important rainfed oilseed crop. It is difficult to harvest due to spiny vegetative material, hard stem and bulky nature of plant. The harvesting losses were reported to be 0.65 -1.5 percent by use of machine. This is a promising machine for difficult to harvest crops like safflower.

Eighty five percent of the groundnut area in India is rainfed. The harvesting is highly labour dependent operation. The mannually pulling the groundnut vines is a common practice, however in some areas bullock drawn blades are also used. The groundnut digger developed at Central Institute of Agricultural Engineering, Bhopal, gave 98% digging efficiency with only 2% pod loss in black soils (Rajput, 1993). In alfisols, the soil surface is hard at harvesting. The peg type 'V' shape groundnut digger developed at ICRISAT has shown promising results.

SUMMARY

The implements discussed in this chapter have definitely shown an increase in the field coverage, operational efficiency and enhanced crop yields compared to the conventional methods. The increasing scarcity of the human and animal labour resources during peak demand period suggests that there is a need for labour saving, efficient, and precision implements in drylands. The gradual shifting to mechanically powered implements by way of selective mechanisation and improvements in animal powered equipments for increased coverage to achieve timeliness; and adopting/developing precision mechanisms needs emphasis in future thrust areas. Although the gains from the development are visible, the mechanisation of Dryland Agriculture has a long way to go.

REFERENCES

Annonymous, 1982. Research highlights 1976-1981, Central Institute of Agricultural Engineering, (CIAE), Nabibagh, Berasia Road, Bhopal, India.

Annonymous, 1991. Annual Progress Report, Indo-US Project on 'Research on Mechanisation of Dryland

Agriculture', Central Research Institute for Dryland Agriculture, Hyderabad, India.

Annonymous, 1992. Annual Progress Report, Indo-US Project on 'Research on Mechanisation of Dryland Agriculture', University of Agricultural Sciences, Bangalore, India.

Binswanger Hans P. 1978. The economics of tractors in South Asia. An analytical review, published by Agricultural Development Council, New York and International Crops Research Institute for the Semi Arid Tropics, Hyderabad, India.

Biswas,. H.S., and Ingle, G.S. (1993) Field performance of weeding tools in black soils. Paper presented in 28th *Annual Convention of Indian Society of Agricultural Engineers*, (ISAE) held at CIAE, Bhopal, from 2-4 March 1993.

Clayton, K.C.C. 1972. Mechanisation and employment in East African agriculture. *International Labour Review* vol. 105, No.4.

Datt, P and Prasad, J. 1993. Development and evaluation of self propelled vertical conveyer reaper for safflower crop. Paper presented at 28th Annual Convention of ISAE, held at CIAE, Bhopal from 2-4 March, 1993.

Gite, L.P. Bhandarkar, D.M. and Bohra C.P. 1993. Ergonomic evaluation of wheel type mannual weeders. Paper presented in *28th Annual convention of ISAE*, held at CIAE, Bhopal from 2-4 March 1993.

Jodha N.S. 1974. A case of process of tractorisation. *Economics and political weekly*, Vol IX,. No 52 Review of Agriculture, December 28, 1974.

Mayande, V.M., Bansal, R.K. and Smith, G.D. 1990. A comparision of Moldboard plows for dry season tillage of vertisols. *Soil and Tillage Research*, 15 (1990) 349-358. Elsevier Science Publishers, B.V. Amsterdam, Netherlands.

Rajput, D.S., Anderson, D.T. and Yadava G.C. 1980, Farm implements and machinery research in drylands. Paper presented in the *9th Annual Workshop for the Coordinated Scheme for Research and Development of Farm Implements* held at IGFRI, Jhansi, from January 31-February 2, 1980.

Rajput, D.S., 1993. Performance of improved CIAE animal drawn groundnut digger. Paper presented in *28th Annual Convention of ISAE*, held at CIAE, Bhopal from 2-4 March 1993.

Rangaswamy, K., Balasubramanian, M and Swaminatham, K.R., 1993. Power Weeder. Paper presented in *28th Annual Convention of ISAE*, held at CIAE, Bhopal, from 2-4 March 1993.

Solanki, R.C. and Singhal, O.P., 1993. Design development and evaluation of a single row rotary reaper windrower. Paper presented in *28th Annual Convention of ISAE*, held at CIAE.

18

CROP MANAGEMENT TECHNOLOGY FOR DRYLANDS

N.K. UMRANI

INTRODUCTION

Drylands constitute about 75 per cent of arable lands in the country. They contribute about 42 per cent of food grains, almost all the coarse grains and about 75 per cent of pulses and oilseeds. More than 90% of sorghum, pearlmillet, groundnut and pulses are grown in arid and semi-arid areas. The share of the drylands to national food basket should increase in future to cope up with the increased demands. Due to natural limitation to the irrigation and enormous cost to develop the irrigation system more reliance on drylands is the only way. Dryland agriculture, however, suffers from twin problem of low productivity and high instability. The situation must change. The available technological information indicate that increasing productivity of dryland is a reality. Sustained efforts are required to achieve this objective. While doing so appropriate modification in the technology may be required to adapt to the location.

Low crop production in the dryland agriculture can be attributed to various factors. Among others, lack of moisture conservation practices, low rate of fertilizer consumption, lack of timeliness in farm operations and lack of appreciation of soil water and plant relations are important. Weather being unpredictable, timeliness assume greater importance. Economic factors as constraints to adoption of technology emerge more important than biological and technological factors (Rastogi, 1983).

DEVELOPMENT OF DRYLAND TECHNOLOGY

Dryland research to develop management practices for low and unassured rainfall areas is in progress since 1933 in India first at Solapur and immediately followed at Bijapur, Hagari, Raichur and Rohtak (Kanitkar, *et al.*, 1968). As a result of continuous working for a decade the well known crop management system known as 'Bombay Dry Farming Method' was evolved as package for *rabi* sorghum. (*Sorghum bicolor*). The increase in the crop production was about 20% after adopting this package over already very poor base production of 2-5 q/ha. The then technology lacked responsive biological material to management and input for full expression of technology. As such it did not enthuse the farmers to adopt the same. With availability of new responsive plant material there had been considerable change in the level of production. Randhawa and Venkateswarlu (1979) compiled the crop production levels for those centres where dry farming research was initiated during 1930's. (Table 1). It would be seen that with almost the same rainfall, the level of crop production has gone high by almost 2 to 5 times during 1970's. Thus, the present knowledge of technology is quite adequate to take the challenge of improving the crop production in dryland agriculture.

Table 1 : Relative yields of crops with earlier and new technology of crop production in drylands.

Station	Normal rainfall (mm)	Crop	Period	Mean yield (q/ha)
Hagari	515	Sorghum	1935-40	5.5
Bellary*	518	Sorghum	1971-77	21.7
Bijapur	601	Sorghum	1934-41	5.1
Bijapur	606	Sorghum	1971-77	11.1
Sholapur	616	Sorghum	1934-41	2.3
Sholapur	680	Sorghum	1971-77	16.6
Rohtak	446	Gram	1935-38	6.3
Hissar*	413	Gram	1971-77	13.8

*Since Hagari and Rohtak ceased to function by 1970 the nearest stations have been used.

Source : Randhawa and Venkateswarlu (1979)

UNTAPPED PRODUCTION POTENTIALS :

Dryland agriculture has vast potentials when managed scientifically. The gap between yields of crops at research station and those with farmers' fields is very wide. This gap needs to bridged to create full impact of dryland technology. Results in dryland agriculture are not quick and spectacular as in irrigated agriculture due to poor adoption of packages. Joshi *et al* (1979) observed that salvation for rainfed agriculture can not come from technology alone, and a concentrated multidisciplinary, multiagency approach is necessary. Venkateswarlu (1981) has compiled the data on untapped production potential in dryland agriculture. (Table 2). It would be seen that in general there is a scope to push the production of dryland crops by nearly 3 to 4 times the farmer's level. Considerable gap exists for dry crops like bajra (*Pennisetum glaucum*), sorghum, finger millet (*Eleusine coracana*) and maize (*Zea mays*).

MANAGEMENT OF NATURAL RESOURCES

In dryland agriculture, soil and rain water are the two naturally available resources with farmers. Under given ecological situation it is the rainfall variation that causes fluctuations in productivity from year to year. Available stored moisture needs to be used effectively. Crop plants are found to respond differently to the situations. Some of them are efficient to use soil moisture than others. With the commencement of Dryland Agricultural Research Project in 1970,

Table 2 : Untapped production potentials in drylands of India.

Region	Soil type	Crop	% Area	Yield (t/ha) Farmers' practice	Yield (t/ha) Improved practice
Hissar	Aridisol	Pearl millet	78	0.3	3.4
Jodhpur	--do--	Pearl millet	51	0.3	1.8
Anantpur	Alfisol	Gr. nut	27	0.5	1.5
Hyderabad	Alfisol	Sorghum	35	0.4	3.4
		Castor	55	0.6	1.5
Bangalore	Alfisol	Finger millet	40	1.0	3.7
Ranchi	Alfisol	Upland rice	52	0.8	3.0
Bhubaneswar	Alfisol	Upland rice	14	1.1	2.1
Varanasi	Entisol/ Inceptisol	Upland rice	35	0.7	2.8
Dehra Dun	Inceptisol	Upland rice	–	1.2	4.4
Indore	Vertisol	Maize	–	1.2	4.3
		Soyabean	–	1.0	2.7
Akola	Vertisol	Sorghum	32	0.5	1.8
Solapur	Vertisol	Sorghum	62	0.3	2.1
Bijapur	Vertisol	Sorghum	36	0.5	2.1
Bellary	Vertisol	Sorghum	50	0.3	3.5

Source : Venkateswarlu (1981)

efforts were made to identify the most efficient crops and their varieties at all the locations.

SELECTION OF EFFICIENT CROPS AND VARIETIES

For the purpose of working out moisture use efficiency (MUE) for each crop and its varieties consumptive use of moisture (CU) was estimated. For *kharif* areas it was bit difficut due to alternate wetting and drying. Many centres however, collected such information which was used for deciding the crop pattern. The data generated at Solapur are reported in Table 3. For *rabi* cropping situation these data are very useful. This helped to identify the crops and their varieties which need to be fitted in the cropping system. In areas like Western Maharashtra, North Karnataka and Telangana region of Andhra Pradesh, *rabi* sorghum is the most efficient crop from moisture use point of view. Since it satisfies both grain and fodder requirements, its dominance in the area would continue. Hybrids of *rabi* sorghum were found efficient when only grain portion is considered. Because of low fodder yields, they have not spread in this region in spite of their higher MUE. In the process dry wheat was the least efficient and need replacement in region like Bellary. Safflower was found to be the most efficient crop (Rao and Rao, 1983).

Table 3 : Yield and moisture use efficiency of different crops at Solapur.

Crop	Variety	Yield (q/ha)	Cu (mm)	MUE (kg/mm)
Bajra	Local	7.4	268	2.8
	Local Improved	10.5	264	4.0
	Hybrid (HB3)	18.3	261	7.0
Sunflower	EC 68414	10.9	239	4.5
	Morden	9.6	241	4.0
	EC 69874	11.7	238	4.9
Groundnut	SBXI	10.5	295	3.8
	TMV-10	10.2	274	3.7
	K4-11	10.3	316	3.3
Pigeonpea	Prabhat	7.5	295	2.5
	S-5	9.2	324	2.8
	No 148	12.2	325	3.7
Sorghum	M35-1	17.5	247	7.0
	CSH8R	23.2	220	10.5
Chickpea	N59	13.8	199	7.8
	Chafa	13.6	212	6.6
Safflower	N62-8	14.4	212	6.7
	7-13-3	15.4	227	7.3

Source : Umrani *et al* (1981)

Under given quantity of soil moisture, hybrids and high yielding varieties are found efficient. Hybrids of *bajra* have yield potential 2 to 3 times higher than local varieties. Thus, it shows that just by replacing traditional varieties in *kharif* areas by high yielding varieties, crop yields can be increased (Singh, 1983)

Crop Planning as per Length of Cropping Season

One of the exercises done by the dryland research project to impart stability to crop production is to select the crop of proper duration to match the length of growing season. There was less appreciation of crop duration with length of effective cropping season in the past. Climatological analysis, however, helped to identify the regions suitable for short duration crops, sequence cropping and those where intercropping would be successful. In the region where cropping season is less than 20 weeks duration a single crop either during *kharif* season or *rabi* season is recommended. Most of the hard core areas fall in this category (Table 4). Those regions with more than 30 weeks and above length of growing season have no problem to adopt two crops. These areas have better rainfall and as such efforts are needed to increase cropping intensity by sequence cropping. Potential crop sequences are indicated in Table 5. Usually a short duration *kharif* crop follows 100-120 days duration *rabi* season crop or vice-a-versa. In areas with 20-30 weeks crop growing season it is usually risky to take two crops since the rainfall aberration may shorten the length of growing season. Such areas are ideally suited to intercropping. More detailed discussion on intercropping is presented in a separate section.

Table 4 : Length of effective cropping season at Dryland Research Centres.

Category	Regional centres and effective cropping seasons			Cropping intensity
A. Less than 20 weeks	Bellary (8) *	Jodhpur (11)	Anantpur (13)	Sole cropping
	Hissar (17)	Rajkot (17)	Bijapur (17)	
B. 20-30 weeks	Jhansi (21)	Kovilpatti (21)	Hydera-bad (22)	Intercrop ping
	Udaipur (22)	Solapur (23)	Agra (24)	
	Anand (25)	Akola (27)		
C. More than 30 weeks	Bhubneswar, (31)	Varanasi (32)		Sequence cropping
	Bangalore, (32)	Hoshiarpur, (35)	Indore, (36)	
	Rewa, Samba, (36) (44)	Ranchi, Dehra Dun (45) (44)		

* Figures in parentheses indicate the duration of cropping season in weeks.
Soruce : Randhawa & Venkateswaslu, 1980.

Improved varieties suitable for drylands

Breeding of varieties suitable for drylands was less attempted in the past. It was only with the inception of Dryland Research Project the screening and breeding of new crop varieties were attempted. Earlier, only available varieties were recommended without testing their performance. Some of the old varieties were continued for 30 to 40 years. Variety M35-1 of *rabi* sorghum is the best example.

Varieties suited to dryland conditions should have faster rate of growth in initial period. They should be short in duration to escape the late season drought. In drylands crops are grown during both *kharif* and *rabi* seasons. Problems of evolving varieties suitable for *kharif* situation is more complex due to midseason drought, creating moisture stress at critical growth stages. Ability, therefore, to survive under these conditions assume significance (Asana and Sinha, 1970). Usually high grain number per ear is a desirable attribute. Strong and penetrating root systems assume importance, since proliferation of roots into the soil mass can tap water and nutrients and can make crop stand in an adverse situation. Hybrids of *bajra* and sorghum are found to filfill these requirements.

Table 5 : Important two crop sequences for drylands

Location	Crop sequence
1. Vertisols and related black soil zone	
Akola	Green gram-Safflower
	Sorghum-Safflower
Indore	Soyabean-Safflower/Chickpea.
	Maize-Chickpea/Safflower
Rewa	Rice-Chickpea/Lentil.
	Rice-Wheat
Udaipur	Green gram-Safflower.
2. Alfisol and related soil zone	
Bangalore	Cowpea-Finger millet
	Soybean-Finger millet
Bhubaneswar	Rice-Horse gram
	Finger millet-Horse gram
	Groundnut-Horse gram
	Cowpea-Maize
Ranchi	Rice-Chickpea
	Finger millet-Chickpea
	Rice- Linseed
3. Inceptisols (Submontane Soils) Zone	
Dehra Dun	Maize-Chickpea
	Soyabean-Wheat
Hoshiarpur	Maize-Wheat + Mustard
	Maize-Wheat + Chickpea
Rakh Dhiansar	Maize-Wheat/Mustard.
	Blackgram-Wheat
4. Entisols (Alluvial soil) Zone	
Agra	Greengram-Mustard
	Pearlmillet-Chickpea/barley
Varanasi	Rice-Chickpea
	Rice-Wheat
	Blackgram-Mustard

Source : Singh (1987)

Pre-1970 era lacked such promising biological material like early maturing groundnut variety JL 24 (Phule Pragati) which made significant impact in groundnut production in Gujrat and Maharashtra and spread all over the country. Lack of efforts in breeding new varieties

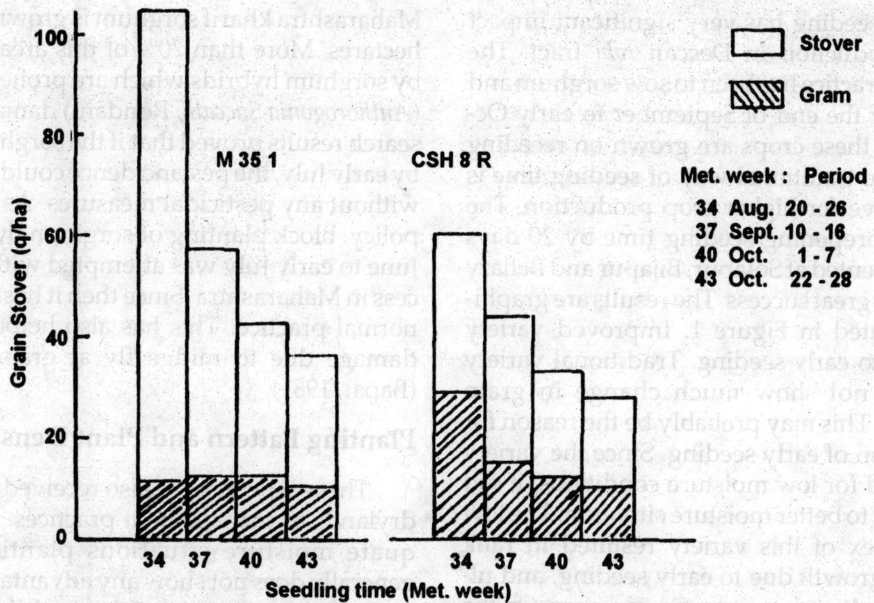

Fig. 1. Rabi sorghum production in relation to seeding dates and cultivars.

suited to crops grown under drylands can be attributed to the then low priority crops like pulses and oil seeds. With change in the priorities, greater attention was paid to evolve new varieties. 'Bhima' variety of safflower specifically bred for drylands is an another example (Umrani *et al* 1984). This variety is doing exceedingly well in Deccan *rabi* region. Breeding of low priority crops like horse gram (*Dolichos biflorus*) was taken up at Solapur resulting in releasing two high yielding varieties like Sina and Man doing exceedingly better on shallow and unretentive soils (Nimbalkar *et al.*, 1986). Other crop improvement projects have also directed their programme to evolve varieties suitable for dry tracts.

SEEDING TIME FOR DRYLAND CROPS

Much of the success in crop production in dryland depends upon proper time of seeding. Since length of growing season is likely to be shortened by late seeding, time of seeding assumes importance. The very first shower should be utilised for crop growth. It calls for preparing land prior to probable rainfall date. Climatological analysis has been done for most of the stations indicating the probabilities for potential seeding. Patil and Kale (1988) attempted such probabilities for most of the tahsils in drought prone areas of Maharashtra. Seeding prior to potential week helped to increase yield of crops. For example, seeding in 24th meteorological week at Solapur in dry soil proved beneficial in 5 out of 7 years. In remaining years, it was at par with seeding at break of monsoon (Annu.Rep.1984). Traditional practice of seeding cotton in dry soil in Vidarbha region of Maharashtra and paddy in south Maharashtra are, therefore, sound.

The concept of off season tillage helps to shorten time lapse between first rain and actual seeding time. It helps to control the weeds as well as conserves moisture from the early showers. Vijayalakshmi (1987) showed that the concept of off season tillage helped to increase the yield of sorghum at Hyderabad by 39% over the traditional system.

Table 6 : Effect of off season tillage on yield (q/ha) of sorghum and barley.

Practice	Sorghum at Hyderabad	Barley at Varanasi
Traditional, no off season tillage	18.7	13.7
Off season tillage	26.0	15.7

Source : Vijayalakshmi (1987).

Timely seeding has very significant impact on crop production in Deccan *rabi* tract. The traditional practice has been to sow sorghum and safflower by the end of September to early October. Since these crops are grown on receding soil moisture traditional way of seeding time is not conducive for higher crop production. The concept of preponing seeding time by 20 days was implemented at Solapur, Bijapur and Bellary centres with great success. The results are graphically presented in Figure 1. Improved variety responded to early seeding. Traditional variety M35-1 did not show much change in grain production. This may probably be the reason for non adoption of early seeding. Since the variety was evolved for low moisture conditions, it did not respond to better moisture situation. The low harvest index of this variety resulted in rank vegetative growth due to early seeding, and ultimately resulting in poor grain recovery. On the contrary, cultivar like CSH 8R produced very good proportion of grain with early seeding. There is also better utilization of nutrients under early seeding (Umrani and Patil, 1985 b.).

Early seeding of safflower also helped to increase the yield in black soils in Deccan (Table 7). The yield continuously declined after the optimum date. The success of early seeding of safflower, however, depends on the disposal of rain water from the September showers which are intense in the region. Since safflower suffers from *Rhyzoctonia* and *Alternaria* due to water stagnation it is very essential to pay attention to the disposal of rain water. Results of All India Coordinated Research Project on Oilseeds (Safflower) also confirm these findings (More, 1987).

Timely seeding for pest avoidence

Timely sowing has also another dimension to avoid pest damage in *kharif* sorghum. In Maharashtra kharif sorghum is grown on 30 lakh hectares. More than 70% of this area is covered by sorghum hybrids which are prone to shootfly (*Anthorogonia Soccata*, Rondani) damage. The research results proved that if the sorghum is sown by early July, the pest incidence could be avoided without any pesticidal measures. As a matter of policy, block planting of sorghum by the end of June to early July was attempted with great success in Maharashtra. Since then it has become the normal practice. This has also helped to avoid damage due to midge fly at grain formation (Bapat, 1981).

Planting Pattern and Plant Densities

These aspects have also received attention in drylands as management practices. Under adequate moisture situations planting pattern generally does not show any advantage. It is only for intercropping system the modifications were made to accommodate additional crop rows. Paired planting and wider pair (skip row) planting have helped to adopt intercropping. Under limited soil moisture, planting pattern also helps to improve the crop yields by improving efficiency of individual plants. Deeper and more proliferous root growth has been obtained by wider planting than normal (Umrani and Bhoi, 1977).

As contingent approach to face the moisture deficient situation reduction in plant density either by removing the alternate row or every third row in sorghum areas is usually recommended (Venkateswarlu, 1983). Though the practice has scientific base; removal of plant rows may not help to conserve the moisture which is already exhausted by the plants. As alternative to it changing planting pattern have yielded good results in the case of post rainy(*rabi*) sorghum. In addition to creating border effect root

Table 7 : Effect of sowing date on yield of safflower at Solapur. (Grain yield : q/ha)

Sowing time		Met. week	1973-74	1974-75	1975-76	Mean
Aug	27-2 Sept.	35	10.27	16.02	8.01	11.43
Sept	10-16	37	8.26	15.99	13.87	12.71
	24-30	39	5.45	5.62	16.37	9.15
Oct	8-14	41	–	5.19	8.97	7.08
	22-28	43	5.48	2.93	5.44	4.62

Source : Patil *et al.* (1981).

training and convenience in interculture operations add to the efficiency of individual plant to produce more dry matter and finally the yield. Data (Table, 8) show that for *rabi* sorghum under medium or deep soils paired planting pattern of 30-30-60 or 45-45-90 cm helped to increase grain yields. Rao and Rao (1980) also noticed advantage of wider planting for *rabi* sorghum at Bellary under subnormal situation of rainfall.

Table 8 : Effect of planting pattern on yield of *rabi* sorghum in vertisol.

Planting pattern	Grain yield q/ha			
	1973-74*	1974-75*	1979-80**	1980-81**
Uniform planting at 45 cm distance	8.5	9.6	32.3	29.5
Paired planting at 30-30-60 cm.	11.3	12.2	34.5	31.8
Paired planting at 45-45-90 cm	–	–	35.9	33.2

*Soil depth 60 cm ** Soil depth 150 cm.
Source : Umrani (1985).

Systematic research on plant density levels in relation to soil moisture storage was done at Solapur. These studies revealed that as in the case of fertilizer so also in the case of plant density availability of stored moisture needs to be considered (Fig. 2). For low moisture availability wider planting helps. Studies conducted at Hissar on chickpea showed that as the moisture situation improves the row spacing could be reduced (Table 9). In the similar way for low availability of soil moisture plant density could be as low as 5 plants/m^2 for post rainy sorghum. Whereas, under better moisture situation it could be between 5 to 10 plants/m^2. The trend of putting maximum population pressure does not help in drylands. Therefore, situation has to be studied before making recommendations. In tillering crop like *bajra* plant density of more than 10 plants/m^2 does not show any advantage. On the contrary in years of low rainfall the high plant density (22.5 plants/m^2) showed lower yields (Umrani *et al.* 1983). In branching crops like safflower, plants show considerable adjustment to

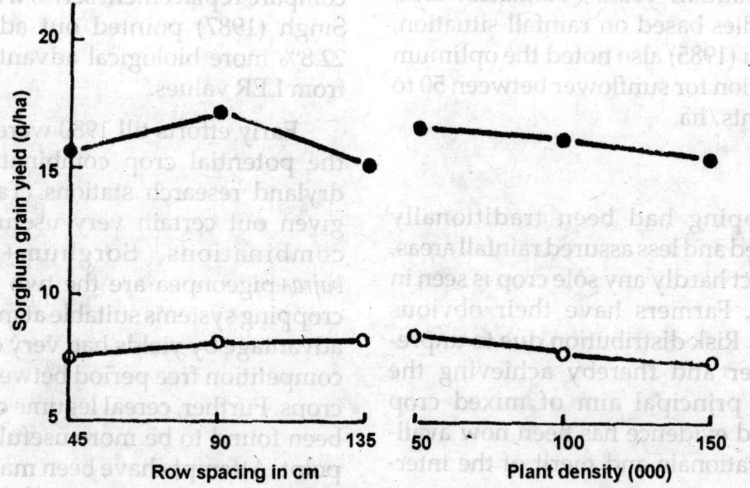

Fig. 2. Relation between moisture storage row spacing, density and production of rabi sorghum.
Source : Umrami et al (1985)

plant density and spacing by regulating branching. At low density, plants produce higher number of secondary and tertiary branches compensating the low plant density. On the contrary, high plant density produces less secondary branching. Thus, over a wide range of populations in the case of safflower yields remain at the same level (Umrani and Bhoi, 1984).

Table 9 : Interrelation of row spacing and profile moisture on chickpea yields at Hissar.

Row spacing (cm)	Grain yield (q/ha)		
	Profile moisture (mm)		
	110	150	190
30	14.7	19.1	29.4
60	25.1	22.8	27.4
90	13.4	23.5	23.9
CD (5%) interaction 5.0			

Source : Balasubramanian and Venkateswarlu (1985)

Sunflower being a new introduction in the country and being photo and thormo-insensitive is planted at any time whenever soil moisture is available. Studies conducted at Hyderabad revealed yield plateau between 56 thousand to 100 thousand plants/ha. Further, it was noticed that plant density in the range of 60-80 thousand plants/ha would be adequate to have full expression in normal years and minimise the crop failure in low rainfall years (Venkateswarlu, 1985). From studies based on rainfall situation, Umrani and Bhoi (1985) also noted the optimum range of population for sunflower between 50 to 70 thousand plants/ha.

Intercropping

Mixed cropping had been traditionally adopted in rainfed and less assured rainfall areas. As a matter of fact hardly any sole crop is seen in these drylands. Farmers have their obvious reasons to do so. Risk distribution due to unpredictable weather and thereby achieving the stability is the principal aim of mixed crop stands. Increased evidence has been now available about the rationale and merit of the intercropping system.

There had been very little research effort in the past to evaluate the merits of intercropping system. It received less attention from the research workers and was considered as a sign of primitive agriculture. Mixed cropping has been improved to the status of intercropping where parameters like planting patterns, row proportions, population combinations, nutritional needs have been studied and specified. In its early stage of development several dimension were discussed. Willey (1979) opined that intercropping can provide substantial yield advantages to sole crop. He further opined that intercropping may give greater stability of yields over different seasons. In his studies, Jodha (1979) pointed out that intercropping is more prevalent with small farmers who have neither enough capacity to risk nor enough land to diversify to put land to different sole crops. Rao and Willey (1979) from the extensive studies to verify the performance of intercropping observed that intercropping system is more widely adoptable. Failure of intercropping to obtain specified income level under constant or randomly varied prices was less frequent than the sole cropping. Earlier intercropping consisted of replacing specified rows of base crop by intercrop of pulse or oilseed. This was not so much advantageous. Later development in so called intercropping was additional population of intercrop to full population of base crop (Fig. 3). This gave rise to compare replacement series with additive series. Singh (1987) pointed out additive series had 22.8% more biological advantage as evidenced from LER values.

Early efforts till 1980 were made to identify the potential crop combinations all over the dryland research stations (Table 10). This has given out certain very useful and productive combinations. Sorghum+pigeonpea and *bajra*+pigeonpea are the two very useful intercropping systems suitable at many locations. The advantage by yields had very close relation with competition free period between the component crops. Further, cereal legume combinations have been found to be more useful from soil fertility point. Attempts have been made to work out the nutritional requirements of different crop combinations. Nutrient use efficiency has been in-

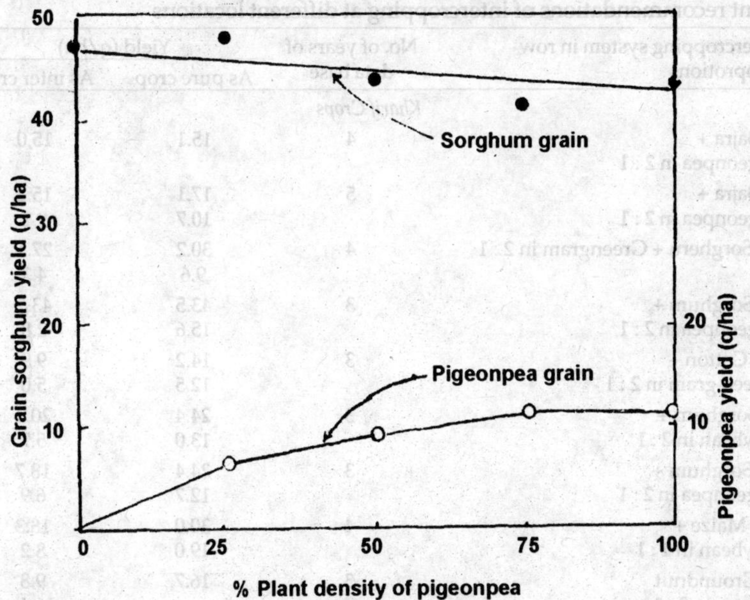

Fig. 3. Interrelation of plant density of pigeonpea with sorghum in sorghum + pigeonpea intercropping at Akola.

creased due to intercrop system indicating that the system is efficient in using natural resources.

Comparison of intercropping during *kharif* season and those from rabi season revealed that it was advantageous to adopt intercropping during *kharif* both from biological and monetary point of view whereas, during *rabi* advantage due to intercropping in general was very low. In fact, in most of the cases biological advantage is less that 5%. Chetty (1983) from his studies opined that intercropping during post rainy season is recommendable beyond north of Vindhya range. Monetary advantage in *rabi* season intercropping was mostly due to high value crops like oil seeds. Even in the past Kanitkar *et al* (1968) recommended mixed cropping of *rabi* sorghum with safflower purely from monetary gain point of view. It appears that competition free period between component crops is very narrow during rabi season since there is very limited scope to select crops in *rabi* season. Further, *rabi* crops are grown on stored soil moisture, selection of crops of shorter duration is very important. As such intercropping is not an important recommendation for *rabi* season cropping. Only at a few places like Jhansi (wheat+chickpea) and Agra (barley+mustard) intercropping in *rabi* season was found advantageous. There is less risk in *rabi* crops since one knows the moisture accumulated to decide to adopt recommended practice. Production varies only due to rains as an additional input. As such emphasis on intercropping in *rabi* is much less.

To increase biological advantage from intercropping different genotypes were screened with different plant architectures. However, these studies did not lead to any concrete findings. Length of competition free period between component crops assumes the prime consideration. Intercropping also was looked from weed management view point. Fast growing and canopy developing crops (e.g. cow peas) put up vigorous growth and competition to the weeds. Thus, smothering effects on weed growth were noticed.

FERTILIZER USE IN DRYLAND AGRICULTURE

Fertilizer use in dryland agriculture is probably the most important, next to moisture.

Table 10 : Important recommendations of intercropping at different locations

Region	Intercropping system in row proprotions	No. of years of data base	Yield (q/ha)		LER
			As pure crop	As inter crop	
		Kharif Crops			
Bijapur	i) Bajra +	4	15.1	15,0	1.52
	Pigeonpea in 2 : 1				
Solapur	i) Bajra +	5	17.1	15.2	1.74
	Pigeonpea in 2 : 1		10.7	9.1	
Akola	(i) Sorgheru + Greengram in 2 : 1	4	30.2	27.1	1.35
			9.6	4.3	
	ii) Sorghum +	3	43.5	43.2	1.17
	Pigeonpea in 2 : 1		15.6	2.8	
	iii) Cotton +	3	14.2	9.9	1.10
	Green gram in 2 : 1		12.5	5.0	
Indore	i) Sorghum +	3	24.4	20.2	1.27
	Soybean in 2 : 1		13.0	5.7	
	ii) Sorghum +	3	24.4	18.7	1.31
	Pigeonpea in 2 : 1		12.7	6.9	
	iii) Maize +	4	30.0	18.3	1.04
	Soybean in 1 : 1		19.0	8.2	
Rajkot	i) Groundnut	3	16.7	9.8	–
	Castor in 3 : 1		–	6.4	–
	ii) Groundnut +	3	16.7	8.5	–
	Pigeonpea in 2 : 1		–	9.1	
Hyderabad	i) Sorghum + 3 : 1	3	27.0	20.4	1.44
	Pigeonpea in 2 : 1		13.6	9.3	
	ii) Sorghum +	2	30.0	29.3	1.38
	Green gram in 2 : 1		7.8	3.1	
Bangalore	i) Finger millet +	3	24.2	19.1	–
	Soybean		–	3.5	
	ii) Finger millet +	3	23.4	11.8	1.13
	Chillies		9.3	5.9	
Jodhpur	i) Sunflower +	2	8.0	6.3	1.64
	Green gram		13.5	11.5	
	ii) Sunflower +	2	8.0	6.3	1.66
	Cowpeas		12.7	11.0	
	iii) Pearlmillet +	2	–	17.8	–
	Green gram		10.6	7.7	
Hissar	i) Pearlmillet +	4	23.6	16.9	0.97
	Green gram		13.3	3.3	
	ii) Cluster bean +	5	10.4	8.6	1.13
	Black gram		10.7	3.2	
Bhubanes war.	i) Castor + Sweet potato	2	7.6	7.3	1.54
			95.0	55.0	
	ii) Paddy +	2	11.3	8.1	1.09
	Black gram		5.7	2.1	
Ranchi	i) Pigeonpea +	3	10.0	6.0	1.54
	Maize		30.6	28.9	
	ii) Pigeonpea +	3	8.3	7.6	1.57
	Paddy		23.6	15.4	
Varanasi	i) Pigeonpea +	2	13.6	13.4	1.71
	Black gram		8.4	6.0	

Table Contd...

Region	Intercropping system in row proportions	No. of years of data base	Yield (q/ha) As pure crop	Yield (q/ha) As Inter crop	LER
		Rabi Crops			
Agra	i) Barely + Mustard	2	14.3 18.1	12.6 6.8	1.26
Jhansi	i) Barley + Chickpea in 2 : 1	1	20.8 11.0	17.1 3.8	1.17
Kovilpatti	i) Sorghum + Pigeonpea in 2 : 1	4	24.7 4.5	22.3 0.6	1.03
-do-	ii) Sorghum + Cowpeas	4	24.7 14.1	24.1 1.7	1.12

Source : Chetty (1993)

Kanwar (1981) described fertilizer use in drylands as kingpin for increased production. Soils being deficient in nitrogen; application of nitrogen fertilizer to the crops brings about considerable increase in the productivity. The fertilizer application in general during pre-1970 was very negligible and restricted to irrigated conditions and under assured rainfall area. Umrani and Narkhede (1983) reported from their survey conducted in the drought prone areas of Maharashtra that practically there was no fertilizer use prior to 1970. Further, the experience reveals from works related to fertilizer use in drylands that soils in dry region are more hungry than thirsty. As much as 81% increase in *rabi* sorghum yields and about 65% increase in *bajra* yield can be brought about by use of fertilizer alone (Umrani and Patil, 1985 a and b). As such fertilizer use has become an important component of dryland technology.

Vittal (1987) while reviewing the soil fertility status of important locations of dryland and agricultural research projects observed that these soils are low in organic matter, low in available phosphorous and medium in potassium. The work on fertilizer use between 1970-1980 was related to establishing the response of crops/varieties to applied nitrogen and phosphorus, under varied conditions of rainfall situations. While analysing, the response situations like *kharif* (rainy season) and *rabi* (post rainy season) are required to be considered. In the former situation alternate wetting and drying occurs due to rainfall. Prolonged breaks at critical stages also cause lot of variation in expression of crop growth and production whereas, in latter situation crops are grown on stored soil water and possibility of rainfall during the crop growth is low (Fig. 5). Thus, under both the situation improving soil water storage becomes the most important consideration to get reasonably good production from applied fertilizer. Review of the work done on fertilizer requirements under drylands conditions proves that the fertilizer application is a tool to optimise the soil moisture availability. There is less risk involved than it is made out to be. Improving management techniques like responsive varietial adoption, optimum plant density, timely weed control and controlling pest and diseases are the prerequisites (Umrani, 1979).

Responses of crops to nitrogen application

Summary of the work on nitrogen fertilizer application reveal some important observations.

(i) Cereals have responded very well compared to oilseed and pulse crops. Early efforts were diverted to improve productivity of cereals (Table 11).

(ii) Response to applied nitrogen are comparable to the irrigated situation as in Dehra Dun (2314 mm precipitation). Response upto 200 kg/ha was noticed for both rice and maize.

(iii) Under moderate rainfall situations like Varanasi, Ranchi or Rewa response of rice to nitrogen application was noticed upto 120 kg N/ha.

In all these situations there is practically no

Table 11 : Response of crops to nitrogen application under different situations.

Crop	Season	Region	Annual Rainfall (mm)	Number of years data	Nitrogen (kg/ha)					Reference
					N_0	N_2	N_3	N_4	N_5	
Rice	Kharif	Ranchi	1462	6	13.8	17.0	21.6	24.0	25.8	Singh (1979).
Rice	Kharif	DehraDun	2314	3	2802	37.44	41.01	43.24	56.23	Singh *et al* (1979)
Maize	Kharif	Dehra Dun	2314	3	26.20	34.34	42.28	45.70	50.47	-do-
Wheat	Rabi	-do-	-do-	4	20.61	27.03	29.88	29.87	28.75	-do-
Rice	Kharif	Varanasi	1114	4	12.8	20.9	27.0	31.8	–	Sharma *et al* (1979)
Maize	Kharif	-do-	-do-	3	14.3	19.4	21.5	–	–	-do-
Sorghum	Kharif	Indore	1054	2	–	18.3	23.7	28.1	–	-do-
Wheat	Rabi	-do-	-do-	2	14.3	16.7	16.1	–	–	-do-
Rice	Kharif	Rewa	1168	2	15.4	22.9	29.0	33.0	–	-do-
Wheat	Rabi	-do-	1168	2	7.0	10.3	12.8	13.8	11.4	-do-
Sorghum	Kharif	Jhansi	1000	2	5.7	12.0	15.9	17.7	–	-do-
Wheat	Rabi	-do-	1000	2	6.5	8.1	10.5	12.8	–	-do-
Pearl millet	Kharif	Agra	766	2	10.5	16.1	18.2	15.8	–	-do-
Ragi	Kharif	Bangalore	924	3	19.3	24.9	26.6	30.6	–	Havangi (1979)
Maize	Kharif	-do-	-do-	2	19.5	41.7	44.6	–	–	-do-
Sorghum	Kharif	Parbhani	895	3	20.7	28.7	40.7	40.6	39.2	Dhobale *et al* (1987)
Sorghum	Rabi	Bijapur	573	2	15.98	20.88	23.73	23.15	–	Havangi (1979)
Sorghum	Rabi	Bellary	519	4	7.0	10.0	9.6	–	–	-do-
Pearl millet	Kharif	Solapur	724	3	10.8	16.0	20.6	23.5	25.7	Umrani and Bhoi (1981)
Setaria	-do-	-do-	-do-	2	4.9	8.2	11.5	13.6	–	Umrani and Bhoi (1982)
Sorghum	Rabi	-do-	-do-	3	8.5	13.6	17.7	19.0	–	Umrani and Bhoi (1981)
Safflower	Rabi	-do-	-do-	3	6.0	10.0	11.0	14.0	14.1	Umrani *et al* (1981)
Gr. nut	Kharif	Rajkot		3	8.6	9.4	9.5	–	–	Parmar, (1979)
Castor	Kharif	Anand		2	9.7	11.3	12.2	11.9	–	-do-

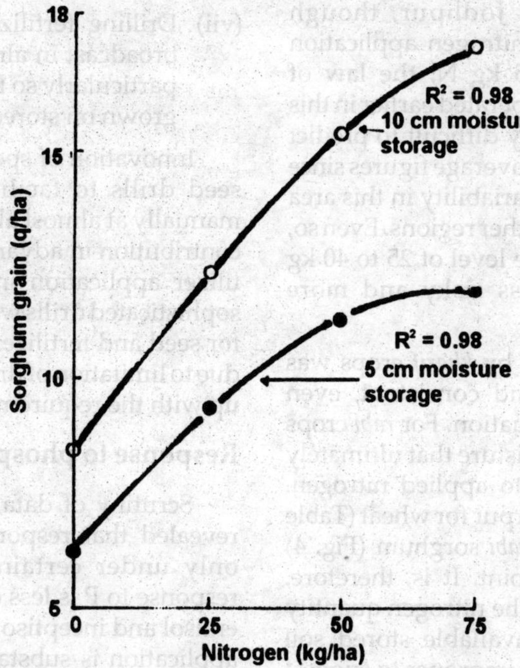

Fig. 4. Interrelation between sorghum grain yield and soil moisture storage capacity

risk to apply fertilizer. Improving fertilizer use efficiency through split application of nitrogen either in two splits or three splits is a desirable practice (Table 12).

Table 12 : Split application of fertilizer nitrogen.

Centre	Crop	Method of Application	Yield (q/ha)
Bangalore	Finger millet	All at planting	25.1
		Three splits	27.8
Dehra-Dun	Maize	All at planting	34.7
		Three splits	38.1
Ranchi	Upland rice	All at planting	24.8
		Two splits	33.9
Rewa	Upland rice	All at planting	27.0
		Three splits	41.0
Udaipur	Sorghum	All at planting	9.3
		Two splits	13.0
Agra	Pearlmillet	All at planting	8.5
		Three splits	13.8

Source : Spratt and Choudhary (1978)

(iv) In areas receiving rainfall between 800-1000 mm response to nitrogen application to crops like sorghum, *ragi* was noticed upto 70 to 80 kg N/ha. Even in this zone there was practically no risk in application of fertilizer. Top dressing part of nitrogen was found to be very useful practice (Table 13).

Table 13 : Effect of split application of nitrogen on crop production on sandy loam soils.

Centre	Crop	Mode of application	Yield (q/ha)
Bangalore	Ragi	All basal application	24.3
		2 splits[1]	26.5
		3 splits[2]	28.0
	Maize	All basal application	35.7
		2 splits[1]	39.0
		3 splits[2]	41.93

1. 50 per cent of the quantity applied as basal.
2. 50 per cent to 33 per cent of the quantity applied as basal.

Source : Havangi (1979).

(v) In low rainfall areas like Solapur, Rajkot,

Anand, Hissar and Jodhpur, though average response to nitrogen application was noticed upto 75 kg N, the law of diminishing returns operated earlier in this area. Further, it is very difficult to predict the response from the average figures since year to year rainfall variability in this area is more compared to other regions. Even so, operating at moderate level of 25 to 40 kg N appears to be less risky and more rewarding.

(vi) Response to nitrogen by *kharif* crops was more pronounced and consistent, even under low rainfall situation. For *rabi* crops it is the stored soil moisture that ultimately affects the response to applied nitrogen. The data from Hoshiarpur for wheat (Table 14) and Solapur for *rabi* sorghum (Fig. 4) amply proved the point. It is, therefore, essential to workout the nitrogen quantity after assessing the available stored soil moisture. It would be improper to suggest a blanket quantity of nitrogen for different soil moisture situations. The split application of N as basal followed by foliar feeding has no advantage at any of the locations. (Table 15).

Table 14 : Effect of soil type storing differential moisture on yield of wheat under dryland conditions.

N rate (kg/ha)	Soil type	
	Loamy sand	Sandy Loam
0	12.7	21.1
40	17.9	30.3
80	19.8	34.5
120	20.5	34.5

Source : Singh and Prihar (1979).

(vii) Drilling fertilizer nitrogen is superior to broadcast in almost all the situations. It is particularly so for *rabi* crops when they are grown on stored soil moisture (Table 15).

Innovation of specially fabricated two bowl seed drills to facilitate fertilizer application manually at almost all the centres is a substantial contribution in advancing and popularising fertilizer application in drylands (Plate I). More sophisticated drills with mechanical devices both for seed and fertilizer have not become popular due to limitation of draft of bullock power to cope up with the requirements by these drills.

Response to phosphorus

Scrutiny of data on phosphate fertilization revealed that response to phosphorus is seen only under certain situations. Further, the response to P is less conspicuous. Under alfisol, entisol and inceptisol conditions the response to application is substantial to recommend in the package (Table 16). Response to P in vertisol is very marginal in some cases and almost absent in others. Even cash crops like cotton and groundnut did not show substantial response. Further, in all the cases response to phosphate was limited to medium level (25-40 kg P_2O_5/ha) of fertilizer application. Absence or low response to P under vertisol is attributed to high phosphate fixation due to calcium content. Venkateswarlu (1987) opined that deep rooted crops and the crops with higher exchange capacity of roots are found to feed more effectively on native phosphorus. Deep placement of fertilizer particularly P is very useful. Further, for *rabi* season conditions where crops are grown on receding soil

Table 15 : Method of application of fertilizer nitrogen on yield

Centre	Crop	Method of Application			Reference
		Broad cast :	Drilled :	1/2 Drill + 1/2 Foliar	
		(Grain yield in q/ha)			
Hoshiarpur	Wheat	36.5	41.2	36.5	Singh and Prihar, 1979
Rewa	Rice	31.4	30.1	29.1	Sharma *et al.* 1979
Dehra Dun	Wheat	26.2	28.2	28.0	Singh, 1979
Solapur	Rabi sorghum	–	8.91	7.29	Umrani et al 1978
Bijapur	Rabi sorghum	–	21.50	20.63	Havangi, 1979.

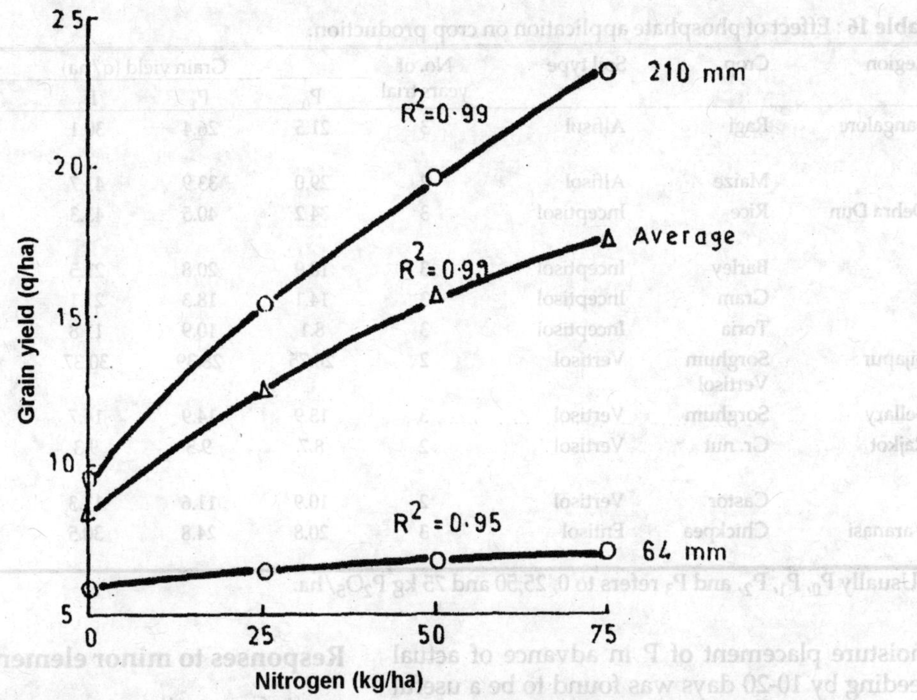

Fig. 5. Relation between yield of rabi sorghum with port-seeding rainfall.

Plate I. Two bowl seed drill : useful for proper fertilzer application

Table 16 : Effect of phosphate application on crop production.

Region	Crop	Soil type	No. of years trial	Grain yield (q/ha)				Reference
				P_0	P_1	P_2	P_3	
Bangalore	Ragi	Alfisol	3	21.5	26.4	30.1	30.6	Havangi, (1979)
	Maize	Alfisol	2	29.0	33.9	41.7	42.4	-do-
Dehra Dun	Rice	Inceptisol	3	34.2	40.5	43.3	–	Singh *et al.* (1979)
	Barley	Inceptisol	3	18.9	20.8	21.5	–	-do-
	Gram	Inceptisol	3	14.1	18.3	21.1	25.3	-do-
	Toria	Inceptisol	3	8.1	10.9	11.8	14.5	-do-
Bijapur	Sorghum Vertisol	Vertisol	2	26.75	28.39	30.37	28.82	Havangi (1979)
Bellary	Sorghum	Vertisol	3	13.9	14.9	14.7	15.4	-do-
Rajkot	Gr. nut	Vertisol	2	8.7	9.5	9.3	–	Parmar (1979)
	Castor	Vertisol	2	10.9	11.6	11.3	11.3	-do-
Varanasi	Chickpea	Entisol	3	20.8	24.8	30.5	27.9	Sharma *et. al.* (1979.)

* Usually P_0, P_1, P_2, and P_3 refers to 0, 25,50 and 75 kg P_2O_5/ha.

moisture placement of P in advance of actual seeding by 10-20 days was found to be a useful practice. A detailed study on advanced application of P to rabi sorghum at Solapur resulted not only in higher production of both grain and fodder but also increased uptake of P due to advanced placement (Table 17). A detailed analysis of situation revealed that post application rains affect the response. Else early application without any rainfall may be as good as application of P at seeding. However, it is felt that P application may not prove economical unless soil test shows deficiency of P.

Responses to potassium

Soil analysis at various locations revealed that K status is medium to rich (Vittal, 1987). The responses to K_2O were not as pronounced as N and P. Hence K may not be considered at this stage. In some of the permanent experiments in the All India Coordinated Project on Cropping System Research no limitation of K was noticed even after 14 crops in sorghum-wheat sequences. (Patil *et al.*, 1989). After 14 crops in succession K status did not change significantly. As such K may not be critical in dryland agriculture.

Responses to minor elements

A few studies are reported in the dryland research project, since attention so far was focused only on N and P, the major nutrients. Reports of response of groundnut to lime and sulphur application are reported from red soil. Maize also reported to respond to zinc in red soil (Table 18). In Maharashtra under light textured soils boron application is reported to be useful for groundnut. (Shinde and Kale, 1985). Application of borax at 5 kg/ha in alternate years is recommended. (Daftardar, 1989).

FERTILIZER MANAGEMENT IN CROPPING SYSTEM

Intercropping and sequence cropping under better rainfall situations leave enough scope to manage fertilizer application. In intercropping system involving cereal+legume combination, there is evidence to economise the fertilizer application. Application of higher levels of N(100-120 kg N/ha) in cereal + legume intercropping seems to affect legume component adversely resulting in low recovery of legume. (Rego, 1979). Further, nodulation in intercropping legume appears to be very poor resulting in low fixation of nitrogen (Shivashankar, 1979). Thus, it leads to

Table 17 : Effect of advanced application of P on *rabi* sorghum in vertisol at Solapur.

Time of application	Yield (q/ha) Grain	: Stover	P uptake kg/ha	P concentration in index leaf %
Without P	12.0	31.6	11.44	0.24
P applied at seeding	14.4	36.7	15.41	0.28
P applied 20 days in advance of seeding.	15.7	40.5	19.23	0.31
S.E.	0.39	1.36	-	-
C.D. 5%	1.10	3.76		

Source : Annu. Rep. Chief Scientist, AICRPDA, Solapur, 1989.

Table 18 : Response of dryland crops to micronutrients..

Location	Crop	Nutrient	Yield (q/ha) Without	With	Level of nutrient /ha
Bangalore	Maize	Zn	29.4	33.2	5 kg
Bangalore	Gr.nut	Lime	16.4	25.0	2.5 t.
	Gr.nut	Sulphur	16.4	20.7	10 kg

Source : Venkateswarlu and Singh (1982)

conclude that in cereal + legume intercropping, there is no need to apply additional fertilizer to legume over and above the cereal component(Umrani et al., 1984).

In areas with better rainfall like Dehra Dun, Varanasi and Hoshiarpur two crops under dryland conditions are possible. Even in regions like Solapur and Bijapur in some seasons two crops are possible. Under these situations a short duration legume like green gram/black gram preceding cereal or oil seed are found to benefit the succeeding crops. There had been conclusive evidence to show that fertilizer cost could be minimised to cereal or oil seed if it precedes legume. The estimated benefit may be about 25 to 30 kg N/ha. (Table 19). It is further recommended that legumes after picking pods can be incorporated in soil to the benefit of the succeeding crops in a better way.

Table 19 : Nitrogen management in cereal and legume based cropping sequences at Varanasi.

Nitrogen Levels (kg/ha)	Grain yield of Mustard (kg/ha) After blackgram	After maize	Nitrogen levels (kg/ha)	Grain yield of barley (q/ha) After blackgram	After maize
0	5.93	5.14	0	16.01	12.06
20	8.76	6.89	30	26.63	21.87
40	10.20	8.85	60	28.75	23.90
60	9.62	9.50	90	33.34	27.65

Source : Singh (1983).

WEED MANAGEMENT

It is estimated that weeds cause about 37 to 79% loss to the crop production (Table 20) in dryland agriculture, where moisture is the most limiting factor. As such removal of weeds for moisture conservation and thereby good yields assumes important consideration. Not that farmers do not realise its importance, but timeliness of weed control becomes a crucial factor. In one of the surveys conducted in SAT region about the extent of weed control measures taken by farmers in Akola, Solapur and Mahbubnagar districts revealed that most of the farmers carry out weeding operations 26 or more days from sowing where rainfall pattern is less assured (ICRISAT, 1976-77). Farmers give preference to irrigated crops and cash crops for weed control. Hence it is important to emphasize the timeliness. Earlier work in the Dryland Research revealed critical period to weed and crop competition. Usually in crops with 80 to 120 days duration first 20-30 days period is more sensitive to competition from weeds. Checking weed growth during this period helps to minimise the loss in crop production. Results of such crop weed competition in pearl millet are shown graphically (Fig. 6). It can be noted that there is sharp decline in production due to weeds. Fall in production is relatively less in later period in-

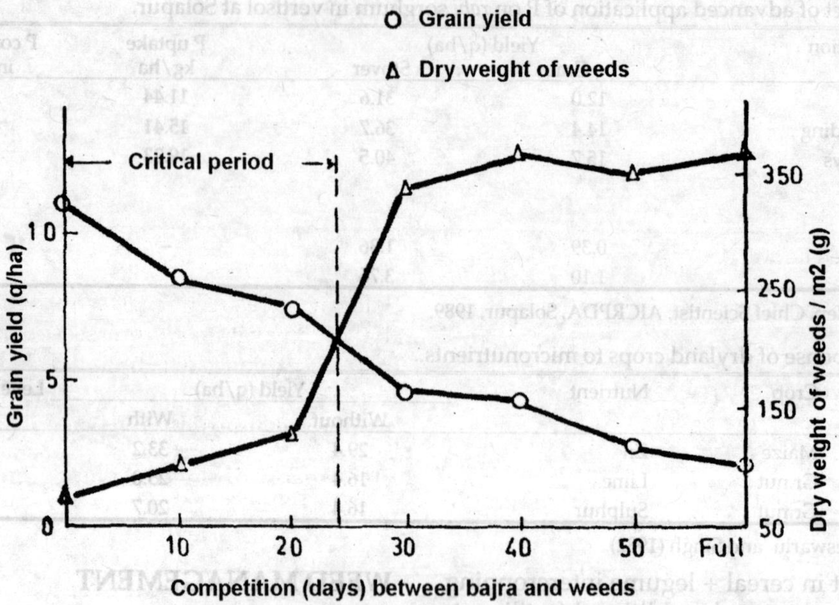

Fig. 6. Bajra yield and dry weight of weeds affected by competition at Solapur
(Source : Umrani et al 1980)

Table 20 : Crop losses caused by weeds under different environmental situations in drylands (1972-81).

Crop	Centre	Crop yield (q/ha)		
		Weeds controlled	Weeds not controlled	% Yield Loss
Upland Rice	Dehra Dun	33.7	7.1	79
	Varanasi	27.0	17.0	37
	Rewa	27.0	16.6	39
Maize	Dehra Dun	46.0	17.6	62
Ragi	Bangalore	27.2	15.0	45
Pearl millet	Agra	15.2	6.8	55
	Solapur	11.9	3.6	70
Gr.nut	Bangalore	19.1	4.2	78
Soyabean	Dehra Dun	18.4	9.2	50

Source : Sanghi and Korwar (1987).

dicating thereby importance of weed competition in early period. Since crops do not develop enough canopy weeds compete very severely with crops. Weed growth also shows steep rise between 20 and 30 days and thereafter remaining at the same level. This type of relation would be similar in crops with 80-100 days growth cycle.

Weed control through chemicals

Chemicals are found to be as effective as mechanical measures in controlling weeds as in dryland agriculture. Use of atrazine @ 0.5 to 1.0 ai/ha as pre-emergence for number of cereal crops (sorghum, maize, *bajra*) is effective (Umrani and Bhoi, 1982). However, due to cheap availability of labour the cost involved in chemical weed control is yet prohibitive. Times, however, are changing fast. Certain weeds like striga can be more effectively controlled by 2, 4-D than mechanical measures (Table 21). Several chemi-

Table 21 : Effect of chemical treatments on incidence of striga and yield of sorghum, Hyderabad.

Treatments	Time of application	Av.number of striga plants/m^2	Average yield (kg/ha)
Weedy control		22.3	979
Hand weeded	–	23.5	1046
2, 4-D (Na salt)	4 weeks after sowing	8.1	2825
Atrazine 1.0 kg per ha.	-do-	16.8	1436

Source : Sanghi and Korwar (1987).

cals can be recommended looking to their effectiveness.

Integrated weed control

The concept of integrated weed control is more suitable for dryland agriculture to save the cost and get efficient control. A combination of mechanical and chemical weed control methods has been effectively used in important crops. In maize crop at Dehra Dun simazine was found to reduce the weed flora by 23% when combined with tillage (Sanghi and Korwar, 1987). Such a combination was found to increase the yield of maize by 30% over weeding alone and reduced soil nutrient loss by 30 to 40 per cent.

Normal weeding or interculture operation required to control weeds are not possible under certain situations. Further, initial weed population which is very small in growth, is some times ignored for some time to make enough growth to carry field operations. Such very early growth can be effectively controlled by pre-emergence application of herbicides in cereals as well as in legumes. However, a combination of herbicide with interculture/weeding greatly enchances the utility of weed control. Studies carried out at Solapur in respect of *kharif* groundnut showed that on an average of four seasons, combination of pre-emergence of propanil @ 3 kg/ha + weeding at 30 days growth increased yield of groundnut almost by 100 percent (Table 22).

Off season tillage for weed control

Tillage operations immediately after harvest of the crop are more effective in controlling weed population. This point is probably overlooked due to hurry in threshing. The off season tillage needs to be done. This would help in two ways. In one way, it would help to sow the crop with the first shower of rain. In drylands due to uncertainty of continued rains the first phase of accumulated moisture should be utilized for crop growth. In another way, off season tillage would control the weeds reaching maturity after the harvest of first crop and checking multiplication. Studies carried at Hyderabad have proved off season tillage to be a very good practice (Table 23). Efficicacy of off season tillage is probably more rewarding during the years of low rainfall. Thus, this is a very useful practice in dryland agriculture.

Table 22 : Effect of integrated weed control on yields of groundnut at Solapur.

Weed control treatment	Yield (q/ha)				
	1978-79	1980-81	1981-82	1982-83	Average
Weedy control	1.45	5.00	3.89	5.51	3.96
Only hand weeding	4.70	5.42	6.83	9.33	6.57
Only hoeing	5.56	5.32	6.30	9.56	6.69
Only propanol pre-emergenc @ 3.0 kg/ha.	4.19	5.88	4.79	6.80	5.43
Propanol + weeding	7.59	6.20	6.51	10.88	7.80
Propanol + hoeing	6.23	4.94	6.46	10.43	7.02
CD (0.05)	2.63	NS	1.42	1.29	–

Source : Annu.Rep. AICRPDLA, Solapur for 1978 to 80.

USE OF MINIMAL IRRIGATION

Moisture stress due to low rainfall and limited soil moisture due to soil depth are the situations normally experienced in dryland agriculture. Mid seasonal droughts and long dry spells during crop growing periods are also not uncommon. One has to prepare to face such events whenever they occur. Severity of mid season droughts and deficiency of soil moisture could be mitigated by adopting protective irrigation approach. The irrigation could be life saving and not productive. The objective would be to save crop from disaster and impart stability. The water required could be from harvested water in a farm pond or could be available from the seasonally functioning wells.

In areas receiving more than 800 mm annual rainfall water could be stored in a farm pond at the end of watershed or the farm. In *kharif* cropping regions water thus stored can be used as life saving irrigation to the crops. In *rabi* cropped areas dominated by cropping during post rainy seasons, storage of water in a farm pond for supplemental irrigation has less chances. Normally under such situation dug out wells which usually serve as seasonal source of irrigation water can be used for supplemental irrigation.

Number of Dryland Research Centres generated useful data on advantages of life saving/protective irrigation. (Table 24). It is seen that substantial increase in yield is noticed at all the locations. The advantage of critical irrigation is more pronounced in drier regions like Bellary, Solapur and Bijapur where the crops are grown on receding soil moisture. Even in Rewa region for paddy critical irrigation has helped to relieve the mid season drought. Studies conducted at Banglore to give supplemental irrigation from harvested water from May rainfall to fodder cow pea the increase in yields have been substantial. (Hedge, *et al*, 1981). Attempts have also been made to grow vegetables like chilli, tomato on the supplemental water. (Vijayalakshmi and Vittal, 1981). There could be a number of avenues once the principle is understood. The approach should be the maximization per unit of water rather than optimization.

Table 23 : Effect of off season tillage on yield of sorghum in red soils of Hyderabad.

System	Yield (q/ha)			
	1977	1978	1979	Mean
No off season tillage	19.50	9.34	10.52	13.12
Off season tillage	24.30	13.36	19.65	19.10
Percent increase due to off season tillage	24	43	31	46
Rainfall in growing season (mm)	595	391	508	-

Source : Sanghi and Korwar (1987).

Table 24 : Effect of critical irrigation on yield of crops

Region	Crop	Seasons averaged	Yield (q/ha)	
			Without irrigation	With irrigation
Dehra Dun	Wheat	4	21.4	35.5
Varanasi	Barley	2	26.0	33.6
Ludhiana	Wheat	4	19.2	41.1
Agra	Wheat	2	21.9	27.4
Bijapur	Sorghum	5	16.5	23.6
Bellary	Sorghum	4	4.3	13.7
Solapur	Sorghum	5	9.8	18.2
Rewa	Upland rice	4	16.2	27.8
	Wheat	4	5.7	18.8

Source : Randhawa and Venkateswarlu (1979).

CROP PLANNING

Crop planning based on various scientific considerations has received very less attention in general and in dryland agriculture in particular. Even to-day not much scientific information on crop planning is available. The prime consideration for crop planning at farmers' level had been the home requirements and the low risk. Thus, the crops were selected on the consideration of low risk and home consumption rather than their suitability to the given situation. This approach is not conducive for higher production.

In general, review of available information revealed that pre-1970 period was characterized by non availability of responsive plant material to high input management technology. In fact crops were so adopted that they would require low moisture and nutrients. Even today nutrient stress still continues in the low rainfall areas. To alleviate the situation and making drylands to contribute increased share in crop production cropping pattern has to be reoriented.

The type experiments during 1970-78 generated good information on efficient crops and varieties. Soil moisture was monitored alongwith yield to workout moisture use efficiency (MUE). This also gave information on the quantum of moisture needs to mature the crop. This, ultimately helped to select crops to given available soil moisture. For example at Hissar depending on conserved moisture in the soil profile, efficient crops were suggested (Table 25). When available moisture is more than 30 cm. wheat, chickpea and field peas are recommended. At medium availability of moisture storage crops like *sarson* (*Brassica campestris* var. brown sarson) and chickpea were found suitable whereas, under the situation of very low moisture storage (less than 7.5 cm) it is only *taramira* (*Eurica sativa*) that can stand to the situation. Such type of information is available in all regions. The information on crop pattern based on soil moisture availability for black soil region of Indore is presented in Table 26. The availability of moisture would be affected by either rainfall or the soil depth. In Deccan *rabi* area variable soil depth is noticed due to undulating topographical situation. Solum depth may very from a few centimeters to some meters in the valley portion. Naturally the crops will have to be selected according to the soil depth (Umrani, 1979). In Solapur, Bijapur and Akola regions soils less than 15 cm are found suitable for perennial grasses and tree plantation (Table 28). As a matter of fact these shallow soils need to be put under permanent vegetation to halt land degradation due to soil erosion.

Table 25 : Suggested crop strategy based on conserved soil moisture storage in the profile at Hissar.

Conserved soil moisture (mm)	Suggested crops
More than 300	Wheat, field pea, chick pea.
200-300	Wheat, barley, lentil, chickpea.
150-200	Chickpea, barley, raya (*Brassica Juncea*)
120-150	Sarson (*B. Campestris* Var. brown *Sarson*) Chickpea.
75-120	Raya, Chickpea-in better catchment area.
50-75	Taramira (*Eruca sativa*)

Source : Singh (1987).

Table 26 : Efficient crops in relation to soil depth in Indore region.

Depth	Available moisture (cm)	Crops/crop pattern
Shallow	10	Sorghum (CSH 5, CSH 6)
		Maize (Ganga-5, Chandan 3).
Medium	15	Soybean (T-49)
		Sorghum + Pigeonpea (intercrop)
Deep	20	Double crops like Maize-Safflower Soybean- gram Sorghum-gram

Source : Randhawa and Venkateswarlu (1979).

Table 27 : Soil depth v/s yields of crops in vertisol at Solapur.

Depth (cm)	Stored water (cm)	Yield of crops (q/ha)		
		Sorghum	Safflower	Chickpea
0-30	9	12.9	3.5	6.2
0-60	18	21.3	4.9	7.1
0-90	27	–	10.5	11.9

Source : Umrani (1979).

Table 28 : Comparative yield performance of crops and grasses on shallow (0-20 cm) soil at Solapur.

System	Yield[1] (q/ha)	Money value1 (Rs/ha)	Mean Money[2] value (Rs/ha)
Bajra	8.30	1141	1000
Mothbean	3.16	926	790
Dicanthium annulatum	118.0*	2433	2024

* Green fodder yield.
1 Mean of 3 locations.
2 Mean of 3 locations of 3 years.
Source : Umrani and Shinde (1987).

Selecting efficient crops and varieties : Systematic efforts were made at Solapur to study the moisture use pattern for both *kharif* and *rabi* crops to delineate crops and suitable varieties (Table 3). For *kharif* season *bajra* was found to be more efficient. Further, between groundnut and safflower, the latter was found efficient in moisture use. As a result, large scale sunflower is grown in this region. Maharashtra alone contributes to about 60% area under sunflower which is concentrated in the drought prone area. Further, medium duration pigeonpea varieties are more suitable than early types. Amongst the rabi crops, the choice is very limited. All the traditional crops are equally efficient in the region. It is necessary to improve potential of these crops.

Varietal choice is also possible looking to the available resources. In adequate moisture regions like Varanasi, Dehra Dun, Ranchi simply change of variety from the traditional to improved type brings about substantial increase in production. (Singh, 1983). In this region short duration improved varieties not only increase the yield but also allow to adopt multiple cropping under dryland conditions.

Planning for efficient resource utilization : In region where both *kharif* and *rabi* season cropping is adopted and where soil depth variation affects the choice like areas in Deccan *rabi*, there is considerable scope to exercise scientific crop planning. Usually shallow to medium deep soils (upto 45 cm deep) are amenable to both *kharif* and *rabi* cropping depending on the situation. Due to unpredictable rains farmers have practice to cultivate transitional soils (depth 30-40 cm) with *rabi* crops. The production is obviously poor due to limited soil moisture for the crop growth as the chances are very less to replenish the depleted soil moisture. Converting these transitional soils to *kharif* crop was attempted at Solapur with considerable success (Table 29). It can be seen that these soils are more productive during *kharif* and the risk could be minimised by inter cropping. The data revealed that intercropping besides being productive have better rainfall useful efficiency, a concept need to be considered seriously for drylands. In addition, intercropping of *bajra* + pigeonpea improved organic carbon and soil structure compared to pure cereal cropping or keeping the land fallow. Thus, ul-

Table 29 : Yield, Moisture use, MUE, Rainfall use efficiency and soil properties in different cropping systems at Solapur (1978-1983).

System	Yield (q/ha)	CUM (mm)	MUE (kg/mm)	RUE* %	Organic carbon% (0-15 cm)	Structural stability index (MWD in mm)
Bajra	24.6	190	13.0	30.0	0.35	0.341
Groundnut	7.6	240	3.2	38.2	0.44	0.457
Pigeonpea	6.71	325	1.9	51.2	0.45	0.444
Bajra +	20.5	340	17.2	53.5	0.42	0.387
Pigeonpea	3.9					
Sorghum (M35-1)	10.9	172	6.28	26.8	0.35	0.331
Safflower	5.9	180	3.3	28.0	0.39	0.387
Chickpea	4.7	154	4.1	18.9	0.35	0.353
Fallow	–	–	–	–	0.33	0.329

* RUE : Rainfall use efficiency
Source : Annu.Rep. Chief Scientist, AICRPDA, Solapur, pp 26-29 (1983).

timate concept of *kharif-rabi* continuum needs to be the primary objective instead of compartmentalising *kharif* and *rabi* season cropping. This situation prevails in the hard core of Deccan *rabi* dry farming tract.

RELATIVE CONTRIBUTION OF PRODUCTION PARAMETERS

For drylands various factors like seed, fertilizer, management are important which contribute to the total yield. Though full package has been found to increase the productivity by nearly 100% or more at farmers' level simultaneous adoption of these inputs may not be feasible for the reasons beyond his control. In such a situation it is important to advise the farmers to adopt certain critical inputs which affect the productivity considerably. Looking to these needs dryland research was directed to study the contribution of the critical inputs to the total productivity. These efforts generated very good information for important crops. A perusal of the data at various locations and for different crops revealed that the contribution of fertilizer application is substantial. Fertilizer application increased the yield from 43 to 81% for cereals. (Table 30). The management like plant density, spacing also contribute significantly in areas with low rainfall and when crops are grown on stored moisture. Thus, it shows that in moisture scarce condition fertilizer use and moisture conservation become very important parameters. Hence, for drylands unless the management is improved there is little hope to increase productivity only by varietal change.

Synergestic effects of these critical inputs was also studied at different locations. One such study conducted in drought prone area of Maharashtra on farmers' fields is very interesting (Table 31). It is seen that change of variety for dry land *rabi* sorghum on farmers field increased yield by 28.6 and 21.5 per cent grain and stover respectively. Management contributed to another 18% in respect of grain but not in respect of stover, probably due to regulated plant density the stover production was lowered. Simultaneous inclusion of variety, management and fertilizer use increased the grain yield by 100% and stover yield by 63.9%. These type of studies on farmers' fields have proved substantially the merit of different technological inovations. More or less similar type of information would be available at all the locations.

ALLEY CROPPING FOR FERTILITY MAINTENANCE

Fertilizer application has been identified as the important input for dryland technology. The dryland farmers are known to be resource poor and the technology needs to suit their sources. Chemical fertilizers are expensive and farmers may not afford to use them. As an alternative, alley cropping was considered to be the best suited under certain situations. In Deccan *rabi* region *rabi* sorghum is the most important crop which is taken after fallowing and thereby conserving moisture during *kharif* season. Green leaves can be incorporated to increase organic carbon and thereby production. Narkhede *et al.* (1984) have shown for the first time that established rows of *subabul (Leucaena leucocephala)* can be pruned to obtain green matter for incorporation to increase the yield. They reported that sorghum grain yield can be increased by 73% by leaf incorporation. Increase in N uptake was noticed due to leaf incorporation.

These studies were further persued at Solapur. Patil (1989) reported detailed observations on soil properties and crop production (Table 32). It is reported that on average about 12.5 t/ha green matter would be available for incorporation providing about 96 kg N/ha. The alleys were established at 3-0.9-3 m distance and were pruned at 60 cm height. About 4-5 loppings would be available. Available nitrogen at the time of planting *rabi* sorghum was in the range of 176 to 180 kg/ha against 145 kg N/ha in plots where organic recycling was not done. Thus, about 30-35 Kg N/ha was available due to organic recycling. The advantage in terms of grain production was almost equivalent to the quantity of nitrogen added through chemical fertilizer.

What is important is the improvement in physical conditions of soil. Organic carbon, soil structure, infiltration rate and bulk density were improved due to organic recycling. (Table 33). Thus, a sustained system for dryland agriculture

Table 30 : Relative contribution of production factors(%) in dryland farming.

Crop	Location	Factors under study			Author
		Management	Fertilizer :	Seed	
Bajra	Rajkot	5	43	36	Singh (1985)
	Solapur	11	65	53	Umrani and Patil (1985)
Finger millet	Bangalore	9	51	–	Hegde and Seetharam (1985)
R.Sorghum	Solapur	21	81	5	Umrani and Patil (1985)
Chickpea	Hissar	–	16	13	Balasubramanian and Venkateswarlu (1985)

Table 31 : Effect of graded technology on production of *rabi* sorghum in Maharashtra.

Level of technology			Grain (q/ha)	% increase	Stover (q/ha)	% increase
Seed :	Management :	Fertilizer				
L*	L	L	7.3	–	19.8	–
Imp**	L	L	9.4	28.6	24.0	21.5
Imp	Imp	L	10.6	46.3	23.2	17.3
Imp	Imp	Imp	14.6	100.4	32.4	63.9

N.B. : Based on 70 trials conducted on farmers' field during 1985-86.
* L : Local ** Imp : Improved.
Source : Umrani and Kulkarni (1987).

Table 32 : Yield of rabi sorghum as affected by organic recycling in alley cropping at Solapur.

Method of recycling	Sorghum grain (q/ha) Nitrogen Kg/ha			Mean	CD 5%
	0	25	50		
No alley, no recycling	7.29	11.81	14.86	11.32	0.68
Only alleys no recycling	7.36	11.08	13.78	10.74	
Only green leaf recycling without alleys.	11.47	15.45	16.39	14.44	
Alley with recycling	11.82	13.91	16.66	14.32	
Mean	9.48	13.06	15.48		
CD 5%	0.59				

Interaction CD 5% 1.18

Table 33 : Effect of organic recycling in alley cropping at Solapur.

Method of recycling :	Organic : carbon %	MWD : (mm)	Infilteration : (cm/ha)	Hydraulic : conductivity (cm/hr)	Bulk density (g/cm^3)
Control	0.364	0.357	0.45	1.27	1.10
Only alleys, no incorporation	0.402	0.369	0.50	1.35	1.09
No allyes, incorporation of green leaves.	0.441	0.501	0.61	1.43	1.04
Alleys + incorporation	0.454	0.489	0.62	1.42	1.05

is a reality and needs to be worked out for all regions.

Besides this, crop residue can be incorporated which has no fodder value. At Bangalore maize residue was incorporated. The results of five years showed that maize yielded 23.5 q/ha when maize residue was incorporated against 18.89 q/ha maize grain when residue was not incorporated. Thus, the technique was found to yield about 25% more (Venkateswarlu, 1983). More and more evidences are made available from different centres.

CONTINGENT CROP PLANNING

Since drylands totally depend on monsoon rains and the trend of monsoon is still unpredictable, the dryland technology has to be equipped with contingent crop plans to avoid total failure of crops. It is expected that under such weather aberration crop yields would not be normal, but the strategy helps to get something to sustain. Under such situation, especially, fodder for cattle becomes a problem. Large scale transport of bulky material form long distance is very difficult. The contingent crop plans, therefore, should give serious consideration for fodder production. Large scale scarcity of 1972 in Maharashtra served as a warning that fodder production is the most important consideration in contingent crop planning. Weather aberrations are normal features of dryland and one has to keep prepared to face these situations. These are inseparable part of life.

There are atleast four important aberrations in the rainfall behaviour (Venkateswarlu, 1983) :

(i) Commencement of rains may be quite early or considerably delayed.

(ii) There may be prolonged 'breaks' during the cropping season.

(iii) There may be spatial and/or temporal aberrations.

(iv) Rains may terminate early or continue for long periods.

Commencement of rains delayed : Commencement of rain quite early is rare as it happened in Madhya Maharashtra in 1925, when rains commenced as early as 28th May as against normal commencement of 10th June. Experience revealed that the normal crops can be planted as early as 15 days without any adverse effect. What causes concern is the delayed onset of monsoon. This type of aberration is very frequent in Madhya Maharashtra comprising the drought prone area. Rains may be delayed as late as by 35-40 days as in 1974-75 (Kulkarni, 1986). This causes large scale dislocation in production programmes. Usual crops and their varieties sometimes need replacement due to two situations. viz. due to late arrival of monsoon the length of growing season is considerably shortened requiring crops/varieties of shorter duration and secondly some crops get effected due to pests and disease complex. For example in Maharashtra *bajra* is likely to be affected by ergot (*Claviceps microcephala*) disease. All the dryland research centres are equipped with alternate crop strategy as contingent crop planning. One such crop strategy for Solapur regions is reported in Table 34.

'Breaks' in monsoon rains

Breaks in the monsoon rains can be of different durations whereas, breaks of shorter duration like 5-7 days may not be serious concern, breaks of more than 15 days duration create situation of plant water stress leading to reduction of agricultural production. The stage of the crop growth at which break occurs decide the severity of damage. Breaks, usually during grain formation cause considerable reduction in crop production. However, crops interact differently to the situations. Crop like pigeonpea and castor modify their growth pattern. Several flushes are noticed in flowering. Therefore, indeterminate habit of crop growth is desired under dryland conditions. Further, drought may be more pronounced on soils with limited water holding capacity (usually shallow to medium deep soils) than with more water holding capacity (deep soils). A break of 15 days duration would affect the same crop differently on shallow and deep soils.

Early withdrawal of monsoon

If monsoon terminates earlier than usual it

Table 34 : Contingent crop planning strategy for late onset of monsoon in Solapur region.

Rainfall situation	Suggested crops	Remarks
Rains during 1st Fortnight of July (Normal onset of monsoon)	Bajra, Setaria, Sorghum (CSH 6), Groundnut (M-13), astor (Aruna), Red-gram (No 148), horse gram (K 42), Black gram (T 9), Sunflower.	Adopt intercropping of bajra + red gram in 2:1 proportion.
Rains during 2nd fortnight of July. (Rains delayed by 15 days)	Setaria (Arjuna), Sunflower (EC 69874), Castor (Aruna) Red gram (No. 148), Horse gram (D 40-1)	Intercropping Red gram + Setaria in 2:2 proportion.
Rains during 2nd Fortnight of August. (Rains delayed by 40 days)	Castor (Aruna), Sunflower (EC 69874) Redgram.	Intercropping Sunflower + Redgram in 2:1 proportion.

Source : Improved Agronomic Practices for Dryland Crops in India (1983).

cuts the length of growing season. Normally crops in their maturity and those growing on shallow soils suffer heavily. Sowing of *rabi* crops may also get affected. It is, therefore, advisable to use short duration and early maturing varieties to face the situation.

Uneven distribution of monsoon rains in space and time : Such situations are encountered almost every year in one or the other part of the country during south-west monsoon leading to drought and flood situations. This situation affects the agricultural production giving instability to it. A combination of perennial crop component in agriculture can absorb shocks because of this situation. As such approach of Agroforestry is important to impart stability.

Practices For Mid-Season Corrections

If the drought occurs in the initial stages of crop growth it usually results in poor crops stand. If it is not more than 10 days delay it is advisable to resow the crop. On the other hand, if drought occurs during pre flowering stage reduction of plant stand may be useful. Further, ratooning of drought affected crop may be useful (Venkateswarlu, 1983). Use of organic mulches, though a good practice to overcome the drought situations, its operational difficulties do not make these as extensive methods for adoption. Usually weed control and dust mulching are useful.

Crop Life Saving Measures

Mid season drought creates moisture stress which can be relieved by life saving irrigation. In command areas it is possible to apply one irrigation on contingent basis. Such practices result in substantial increase in crop production. In the case of areas where water harvesting is already done the same can be recycled to remove the waterstress. Opening dead furrows at regular interval in standing crop to conserve more water in expectation of rains is also a useful practice.

Important decisions have to be taken to face drought by establishing seed banks to distribute seeds of alternate crops. As a matter of fact in dryland agriculture it is equally important to prepare for aberrant weather as that of normal weather.

FUTURE RESEARCH NEEDS

Dryland agricultural research during the past one and one-half decade came out with very useful practices to be adopted by the farmers. These have also been tested on large scale through Drought Prone Area Programmes and proved their worth. The stability, however, has not come to dryland production and to the farmers. Years of extreme drought during the past decade exposed some of the weaknesses of the technology. The field crops are vulnerable under the extreme situations which are equally likely in the region. Fodder supply becomes extremely difficult and transportation of bulky material is out of consideration. There is, therefore, immediate need to work out the technology wherein tree component is inseparable. Trees are known to survive and produce under extreme situations due to their extensive root system. Dryland horticulture and agro-forestry are the ways which need immediate attention. Their performance under different agro- ecological situation is the need of the hour. Further, management of drought as a farming system approach incorporating crops, trees and animals are the next priority for the management of drylands.

REFERENCES

Anonymous 1983. Improved Agronomic Practices for Dryland crops in India. All India Co-ordinated Res. Project for Dryland Agriculture. Hyderabad.

Annual Report 1984. The Chief Scientist, All India Co-ordinated Reserach Project for Dryland Agriculture, Solapur pp. 19-21.

Annual Report 1989. Chief Scientist, AICRPDA, Solapur.

Asana, R.D. and Sinha, S.K. 1970. Tailoring new crop varieties for dry areas. In *A New Technlogy for Dryland Farming*. Indian Agril. Res. Inst. New Delhi. pp. 55-63.

Balasubramanian, V. and Venkateswarlu, J. 1985. Chickpea. In *Efficient Management of Dryland Crops*. Ed.V. Balasubramanian and J. Venkateswarlu, Central Research Institute for Dryland Agriculture, Hyderabad. pp. 209-19.

Bapat, D.R. 1981. *Sorghum in Maharashtra*. Mahatma Phule Agric. Univ. Rahuri. pp. 17.

Chetty, C.K.R. 1983. Research on Intercropping Systems in Drylands. Project Bull. 7. All India Co-ordinated Research Project for Dryland Agriculture, Hyderabad.

Chittaranjan, S. 1981. Rainwater harvesting and recycling. *Indian J. Soil Conserv.* 9(2 & 3) : 100-106.

Chowdhary, S.L. 1979. Recent Studies in Intercropping Systems on the Drylands of India-Some Thoughts,. In Proc. *Intl. Workshop on Intercropping*. 10-13 Jan.1979 held at International Crops Research Institute for Semi-Arid Tropics. Hyderabad. pp 299-305.

Daftardar, S.Y. 1989. Boron application for increased groundnut production in Maharashtra. (personal communication).

Dhobale, M.V., Pawar, K.R. and Kale, U.V. 1987. *Improved Dryland Agricultural Technology for Marathwada*. Marathwada Agric. Univ. Parbhani.

Havangi, G.V. 1979. Fertilizer use for dryland crops in Karnataka. Proc. FAI Group discussion on Fertilizer use in drylands. Fertilizer Association of India. New Delhi. pp. 39-54.

Hegde, B.R. Chennappa, T.C. and Anand Ram, B.K. 1981. Water harvesting and recycling in red soils of Karnataka. *Indian J. Soil Conserv.* 9(2&3) : 107-115.

Hedge, B.R. and Seetharam, A. 1985. Finger millets and other millets. *In Efficient Management of Dryland Crops*. Ed.V.Balasubramanian and J. Venkateswarlu. Central Research Institute for Dryland Agricultre, Hyderabad. pp. 79-100.

ICRISAT 1976-77. Annu.Rep. International Crops Research Institute for Semi-Arid Tropics, Patancheru, Hyderabad.

Jodha, N.S. 1979. Intercropping in traditional farming systems. *In Proc. Intl. Workshop on Intercropping*, 10-13 Jan., 1979 held, at International Crops Research Instituule for Semi-Arid Tropics, Hyderabad. pp.282-291.

Joshi, A.B., Patil, N.D. and Umrani, N.K. 1979. The drought prone area of Maharashtra. *Proc. International Symp. Development and Transfer of Technology for Rainfed Agricultre and the SAT Farmer.* 28 Aug-1 Sept.1979. Intl. Crops Res. Inst. Semi-Arid Tropics. Patancheru, Hyderabad. pp. 221-225.

Kanitkar, N.V., Sirur, S.S. and Gokhale, D.H. 1968. *Dry Farming in India.* Indian Council of Agricultural Research, New Delhi.

Kanwar, J.S. 1981. Dimensions of problems in rainfed agriculture, Problems and prospects of fertilizer use in development of arid and semi-arid areas. *Ferti. News.* 26(11) : 3-7.

Kulkarni, S.D. 1986. Research Management Systems for Drought prone black soils of Maharashtra. Mimeographed. All India Co-ordinated Research Project for Dryland Agriculture, Solapur.

Miranda, S.M., Pathak, P. and Srivastav K.L. 1983. *Run off management on small agricultural watersheds*: The ICRISAT Experience. In Dryland Agric. Res. India : Thrust in Eighties. AICRPDA., Hyderabad. pp. 88-97.

More, S.M. 1987. Agronomy. *In Safflower* Ed. P.S. Patil. Mahatma Phule Agricultural University, Rahuri (India) pp. : 91-124.

Narkhede, P.L., Umrani, N.K. and Ghugare, R.V. 1984. Organic recycling of subabul for winter sorghum cultivated on drylands. *Indian J. Agric. Sci.* 54(5) : 444-6.

Nimbalkar, V.S., Umrani, N.K. and Deokar, A.B. 1986. 'Sina' and 'Man' Two high yielding varieties of horse gram(*kulthi*) for Maharashtra. *J. Maharashtra Agric. Univ.* 11(2) : 148-150.

Parmar, M.T. 1979. Fertilizer use in drylands in Gujarat State. Proc. FAI Group discussion on Fertilizer use in drylands. *Fertilizer Association of India.* New Delhi. pp. 65-78.

Patil, A.J. 1989. Effect of subabul planting, lopping addition on nutritional status, moisture utilisation, yield of sorghum and physico-chemical properties of soil. Annu. Rep. Chief Scientist, AICRPADA, Solapur.

Patil, B.P., Bendgude, S.P., Joshi, A.C and Umrani, N.K. 1989. Long range effects of continuous cropping and manuring on soil fertility and yield stability. Annu. Rep. Cropping system (Irrigated). Mahatma Phule Agric. Univ. Rahuri (India) pp. 7-11.

Patil, C.B. and Kale, S.P. 1988. Weekly rainfall probabilities of selected places in scarcity zone of Mahrashtra. Mahatma Phule Agric. University, Rahuri (India) pp 6-7.

Patil, N.D., Umrani, N.K., Shinde, S.A., Manke, B.S., Kale, S.P.and Shigte, A.K. 1981. *Improved crop production technology for drought prone areas of Mahrashtra.* Mahatma Phule Agricultural University, Rahur (India).

Randhawa, N.S. and Venkateswarlu, J. 1979. Indian Experiences in Semi- Arid Tropics; Prospects and Retrospects. In Development and Transfer of Technology. *Proc. Intl. Symp. 18 Aug-1 Spt. 1979.* International Crops Research Institute for Semi-Arid Tropics, Hyderabad. pp. 207-220.

Rao, M.R. and Willey, R.W. 1979. Stability of performances of pigeonpea / sorghum intercrop system. In. *Proc. Intl. Workshop on Intercropping.* 10-13, Jan.1979, held at International Crops Research Institute for Semi-Arid Tropcis, Hyderabad. pp. 306-317.

Rao, V.R. and Rao, R.M.M.S. 1980. Rainfed Agriculture. In Research on soil and water conservation in semi-arid deep black soils. Central Soil and Water Conservation Research and Training Institute, Research Centre, Bellary(India). pp. 77-143.

Rastogi, B.K. 1983. Yield gap analysis : Constraints and approaches. *In Dryland Agric. Res. India. : Thrusts in eighties.* Ed. R.P. Singh and V. Balasubramanian. ACRPDA, Hyderabad. pp. 139-150.

Rego, T.J. 1979. Nitrogen response studies of Intercropped sorghum with pigeonpea. *In Proc. Intl. Workshop on Intercropping,* 10-13 Jan.1979 held at International Crops Research Institute for Semi-Arid Tropics, Hyderabad. pp. 210-216.

Sanghi, N.K. and Korwar, G.R. 1987. Integrated pest management. *In Technological Advances in Dryland Agricultures* Ed.S.P. Singh *et al.* Central Research Institute for Dryland Agricultre, Hyderabad. pp. 101-119.

Sharma, H.K., Singh, U.N. and Singh, R.K. 1979. Fertilizer use in Drylands : Problems, suggestions and field experience to solve them in U.P. and M.P. *Proc. FAI Group discussion on fertilizer use in dryland agricultre.* Fertil. Assoc. India. New Delhi. pp. 92-113.

Shinde, B.N. and Kale, S.P. 1985. Boron for groundnut in dryland agriculture, *Ferti. News* 30(5) : 53-54.

Shivashankar, K. 1979. Discussion *In Proc. Intl.Workshop on Inercropping,* 10-13 Jan.1979 held at International Crops Reserach Institute for Semi-Arid Tropics. Hyderabad. pp. 128-138.

Singh, G., Chand, S. and Ghosh, S. 1979. Fertilizer use in rainfed agriculture of North-Western low hill regions. *FAI Group discussion on fertilizer use in dryland agriculture.* Fertl. Assoc. India. New Delhi. pp. : 123-130.

Singh, M. 1983. Cropping Systems for Drylands. *In Dryland Agric. Res. in India : Thrust in Eighties.* Ed. R.P. Singh and V.Balasubramanian. AICRPDA, Hyderabad. pp.15-24.

Singh, R., Singh, Y., Prihar, S.S. and Singh, P. 1975. Effect of N fertilization on yield and water use efficiency of dryland winter wheat as affected by stored water and rainfall. *Agron. J.* 67:599-63.

Singh, R. and Prihar, S.S. 1979. Fertilizer use in rainfed submontane areas of Punjab. *Proc. FAI Group discussion on fertilizer use in Drylands.* Fertl. Assoc.India. New Delhi. pp. : 85-91.

Singh, R.K. 1979. Research Achievements with special references to fertilizer use in drylands - problems encountered, suggestions, field experience to solve them in plateau region of Bihar and Orrisa. *FAI Group discussion on fertilizer use in dryland agriculture.* Fertl. Assoc. India, New Delhi. pp. 114-122.

Singh, R.P. , De, R., Malik, D.S., Hegde, B.R., Umrani, N.K. and Singh, R.K. 1981. *Dryland Agriculture.*

Indian Soc. Agron. Silver Jubilee (1975-80). New Delhi : pp. 62-78.

Singh, R.P. 1985. Pearlmillet In *Efficient Management of Dryland Crops,* Ed. V. Balasubramanian and J. Venkateswarlu. Central Research Institute for Dryland Agriculture. Hyderabad. pp. 51-61.

Singh, S.P. 1987. Efficient crops and cropping systems. *In Technological Advances in Dryland Agri.* Ed. S.P. Singh *et al.* Central Research Institute for Dryland Agriculture, Hyderabad. pp. 47- 066.

Spratt, E.D. and Chowdhary, S.L. 1978. Improved cropping systems for rainfed agriculture in India. *Field Crops Res.* 1(2) : 103-126.

Umrani, N.K. 1983. *Dryland Research Project* : A Profile. All India Co-ordinated Research Project for Dryland Agriculture, Solapur, pp. 17.

Umrani, N.K. 1985. Sorghum (Post rainy season) *In Efficient Management of Dryland Crops.* Ed.V. Blasubramanian and J. Venkateswarlu. Central Research Institute of Dryland Agriculture, Hyderabad. pp. 113-28.

Umrani, N.K. and Bhoi, P.G. 1977. Root development in relation to interrow spacing of sorghum. *Sorghum News Letter* 20:49.

Umrani, N.K., Bhoi, P.G. and Patil, N.D. 1980. Effect of Weed competition on growth and yield of pearlmillet. *J. Maharashtra agric Univ.* 5(1) : 56-57.

Umrani, N.K. and Bhoi, P.G. 1981. Response of Hybrid HB 3 (*Pennisetum americanum*) to nitrogen and P_2O_5 under dryland conditions. *J. Maharashtra agric. Univ.* 6(2) : 123-26.

Umrani, N.K., Bhoi, P.G. 1981. Response of different sorghum varieties to nitrogen under dryland conditions. *Fertil. News.* 26(11) : 18-23.

Umrani, N.K. and Bhoi, P.G. 1982. Response of *setaria italica* to N and P under two rainfall situations. *Indian J. Agron.* 27(1) : 61- 63.

Umrani, N.K. and Bhoi, P.G. 1982. Studies on weed control in bajra under dryland conditions. *J. Maharashtra agric. Univ.* 7(2) : 145-7.

Umrani, N.K. and Bhoi, P.G. 1984. Effect of plant density on growth and yield of safflower under two rainfall situations. *Indian J. Agron.* 29(3) : 282-86.

Umrani, N.K. and Bhoi, P.G. 1985. Performance of sunflower as influenced by plant density under drylands. *J. Maharashtra agric. Univ.* 10(1) : 34-35.

Umrani, N.K., Daftardar, S.Y., Patil, C.B. and Walujkar, R.B. 1983. Effect of inter row spacing, plant density and nitrogen application on pearlmillet BJ 104 under scanty rainfall conditions. *Annals Arid Zone Res.* 22(2) : 173-75.

Umrani, N.K., Deokar, A.B. and Nimbalkar, V.S. 1984. 'Bhima', High yielding variety of safflower. *Indian Fmg.* 34(8) : 7.

Umrani, N.K., Kale, S.P., Diami, S. and Pharande, K.S. 1978. Response of 'M 35-1' Winter Sorghum to levels and methods of application of nitrogen. *Indian J. agric. Sci.* 48(9) : 559-60.

Umarni, N.K., Kale, S.P. and Bhoi, P.G. 1981. Studies on Crop planning for drougth prone area of Maharashtra. *J. Maharashtra agric. Univ.* 6(2) : 110-13.

Umrani, N.K., Kale, S.P. and Bhoi, P.G. 1981. Response of safflower to application of nitrogen under rainfed conditions. *Indian J. Agron.* 26(4) : 365-70.

Umrani, N.K. and Kulkarni, N.R. 1987. Performance of graded dryland technology for *rabi* sorghum in drought prone area of Mahaarashtra - a case study. *Indian J. Dryland Agric. Res. & Dev.* 2(1) : 57-62.

Umrani, N.K. and Narkhede, P.L. 1983. A reconnaissance survey of fertilizer in drought prone area of Maharashtra *Fertil. Marketing News* 14(12) : 1-4.

Umrani, N.K. and Patil, C.B. 1985 a. Improved technology in bajra production under different environmental situations. *J. Mahrashtra agric. Univ.* 10(1) : 31-33.

Umrani, N.K. and Patil, C.B. 1985 b. Studies on the contribution of production parameters for *rabi* sorghum under dryland conditions. *J. Maharashtra agric. Univ.* 10(3) : 294-7.

Umrani, N.K. and Patil, N.D. 1983. Advances in ferilizer management for rainfed sorghum. *Fertil. News* 28(9) : 57-61.

Umrani, N.K., Shinde, S.H., and Dhonde, P.W. 1984. Studies on fertilizer application in inntercropping sorghum and groundnut. *J. Maharashtra agric. Univ.* 9(1) : 51-52.

Umrani, N.K. and Shinde, S.H. 1987. Alternate land use system for scarcity zone of Maharashtra. *Proc. Seminar on Alternate Land use systems held on 21-23 Feb. 87,* Indian Society of Agronomy, New Delhi.

Venkteswarlu, J. 1981. Maximisation of crop production in drylands. *Indian J. Soil Conserv.* 9(2 & 3) : 124-140.

Venkaateswarlu, J. and Singh, R. 1982. Crop response to applied nutrients under limited water conditions. *In Review of Soil Research in India Part-I.* 12th Int. Congress Soil Sci. 8-16 Feb. 1982. New Delhi.

pp. 442-52.

Venkatoswarlu, J. 1983. Contingent crop production strategy in rainfed areas under different weather conditions. Project Bull. No. 5. All India Co-ordinated Research Project for Dryland Agriculture, Hyderabad.

Venkateswarlu, J. 1985. Sunflower. In *Efficient crop Management of Dryland Crops*. Central Research Institute for Dryland Agriculture, Hyderabad, pp. 269-277.

Venkatesarlu, J. 1987. Efficient resource management systems for drylands of India. *Advances in Soil Science*. Springer-Verlag. New York. Inc. Vol. 7 : 165-221.

Vijayalakshmi, K. 1987. Rain water management in Drylands. *In Technological Advances in Drylands Agri*. Ed.S.P.Singh *et al*. Central Research Institute for Dryland Agri. Hyderabad. pp. 19-45.

Vijayalakshmi, K. and Vittal, K.P.R. 1981. Rainwater harvesting and recycling in Semi-arid sierozem soils. *Indian J. Soil Conserv*. 9(282) : 87-99.

Vittal, K.P.R. 1987. Soil Fertilizer Management. *In Technological Advances in Dryland Agri*. Ed. S.P. Singh *et al*. Central Res. Institute for Dryland Agriculture, Hyderabad. pp. 83-99.

Willey, R.W. 1979. A scientific approach to intercropping research. *In Proceedings International Workshop on Intercropping*. 10- 13 Jan. 1979. held at International Crops Research Institute for Semi-arid Tropics, Hyderabad. pp. 4-14.

19

EFFICIENT CROPS AND CROPPING SYSTEMS FOR DRYLAND AREAS

R.P. SINGH

INTRODUCTION

Cropping system is a general term which includes crop rotations. The term 'crop rotation', simply stated, is a crop following another. Growing one crop continuously would be a cropping system if it is done according to a plan and with a definite objective. The term cropping system refers to the number and distribution of crops grown on a given piece of land in a given year. To be precise, a cropping system is a combination of crops in time and space. Sequential cropping is a combination of crops in time, while intercropping is a combination of crops in space. Mono, inter, sequential, ratoon, and relay cropping are different versions of cropping systems. Some cropping systems may provide both temporal and spatial effects. However, the ideal system is one which makes efficient use of resources for as long a period as possible.

PROBLEMS AND APPROACHES

In many dry areas in India the problems that are commonly encountered with regard to crop production are :

Unfavourable crop growth environment; limited choice of crops, more particularly in arid regions; limited choice of crop varieties for aberrant weather situations; low cropping intensity; and low and unstable productivity.

In any cropping systems research programme, identification of suitable crops and varieties, evolving systems to increase the cropping intensity, and formulation of improved agronomic practices that increase and stabilise productivity, assume great importance.

The crops and varieties that are currently being grown by the dryland farmers are not necessarily the most efficient. Many of the cultivars of sorghum, pearlmillet, pigeonpea, groundnut, castor, cotton, and other crops, currently in vogue, do not match the rainfall pattern of the area. That is, they are of longer duration than the effective cropping season. They are invariably caught up in moisture stress of varying duration at the most critical stage of their life cycle, thereby resulting in low and uneconomical crop yields.

TRADITIONAL CROPS AND CROPPING SYSTEMS

The traditional systems followed in arid and semi-arid regions are need-based. They are not necessarily the most efficient with regard to productivity, moisture use, monetary returns and labour utilization potential.

In the arid regions, single crops system involving a long fallow period (October to June) is a rule rather than an exception. Mixed cropping as a means of insurance and risk distribution is quite common.

In semi-arid regions, on the majority of Vertisols, land is left fallow during the rainy season.

A post-rainy season crop is raised on the moisture stored in the soil profile. Sorghum, chickpea, and to a less extent, safflower are commonly grown in Central India. They may be grown as sole crops or in intercropping combinations. In the North Central Plains the main crop is wheat. It is grown mostly as a sole crop, but sometimes intercropped with chickpea. In some Vertisol areas the commonest systems are based on cotton. Cotton is commonly intercropped with occasional rows of pigeonpea. The cotton systems are found on the higher, better drained parts of the toposequence, whereas at the other extreme, rice might be found in flooded bottom lands. Rainfed rice is not commonly grown on the wetter Vertisols in the rainy season.

In Alfisols, *kharif* cropping is common, except in deeper soils where double cropping is practiced in good rainfall years. A short duration pulse crop is followed by a cereal crop.

CROPS AND VARIETAL SELECTION - CONCEPTS AND PRINCIPLES :

Land Use Capability

It is an old concept but rarely used in dryland crop production in India. In subsistance farming, food crops are grown according to the household needs of the farmer, regardless of the land capability.

For purposes of dryland crop production, it is the moisture storage capacity of the soil and water availability which govern land capability besides other factors like terrain, topography, slope, erosion hazard, soil, fertility, etc. Moisture storage capacity depends on depth and texture of the soil, the higher the soil depth and finer the soil texture, the greater the moisture storage.

Soil depth governs the available moisture for crop growth. In shallow soils where the available moisture is about 10 cm, it is possible to grow a single crop (sorghum, maize or soybean) in the *kharif* season. In medium deep soils having a moisture storage of about 15 cm it is possible to have an intercropping system (sorghum + pigeonpea). In case of deep soils with a moisture storage of 20 cm, a double crop system (maize - safflower, soybean - chickpea; soybean- - safflower) are possible.

In case of *kharif* (rainy season) crops, it is the rainfall pattern and length of the effective growing season which decides the choice of crops and varieties, but in case of *rabi* (post-rainy season) crops grown on conserved soil moisture, it is the available moisture in the soil profile at sowing time which dictates the choice of crops.

Crop Duration and Effective Cropping Season

Every place, zone or region has an effective cropping season which is governed by availability of water. Growing crops and varieties of longer duration than the effective cropping season of an area will either lead to crop failure or low and uneconomic crop yields. Based on the analysis of long term rainfall data of various locations, effective cropping seasons have been worked out for different regions of the country. These will serve as a good guide in selection of efficient crops and their sowing time.

In arid region, the effective cropping season is of 11 to 17 weeks, thereby restricting the choice of crops. In Bellary (Karnataka), the effective cropping season is hardly of 8 weeks (37 - 44 standard weeks). In semi-arid regions, the effective cropping season is longer (21 to 27 weeks) and it starts earlier than the arid areas, except Kovilpatti in Tamil Nadu which receives North-East monsoon.

Water Availability Periods

The water balance for different agro- climatic regions has been calculated and water availability periods for each of the 23 dryland research centres in the country have been worked out (Fig. 1). The water availability period, which accounts for both rainfall and stored soil moisture, ranges from 74 days at Hissar to 308 days at Rakh Dhiansar.

In regions receiving 350-600 mm rainfall and having 20 weeks of effective growing season, only single cropping (100% cropping intensity) in *kharif* (red and shallow black soils), or in *rabi* (deep black soils) is possible. In deep sierozems

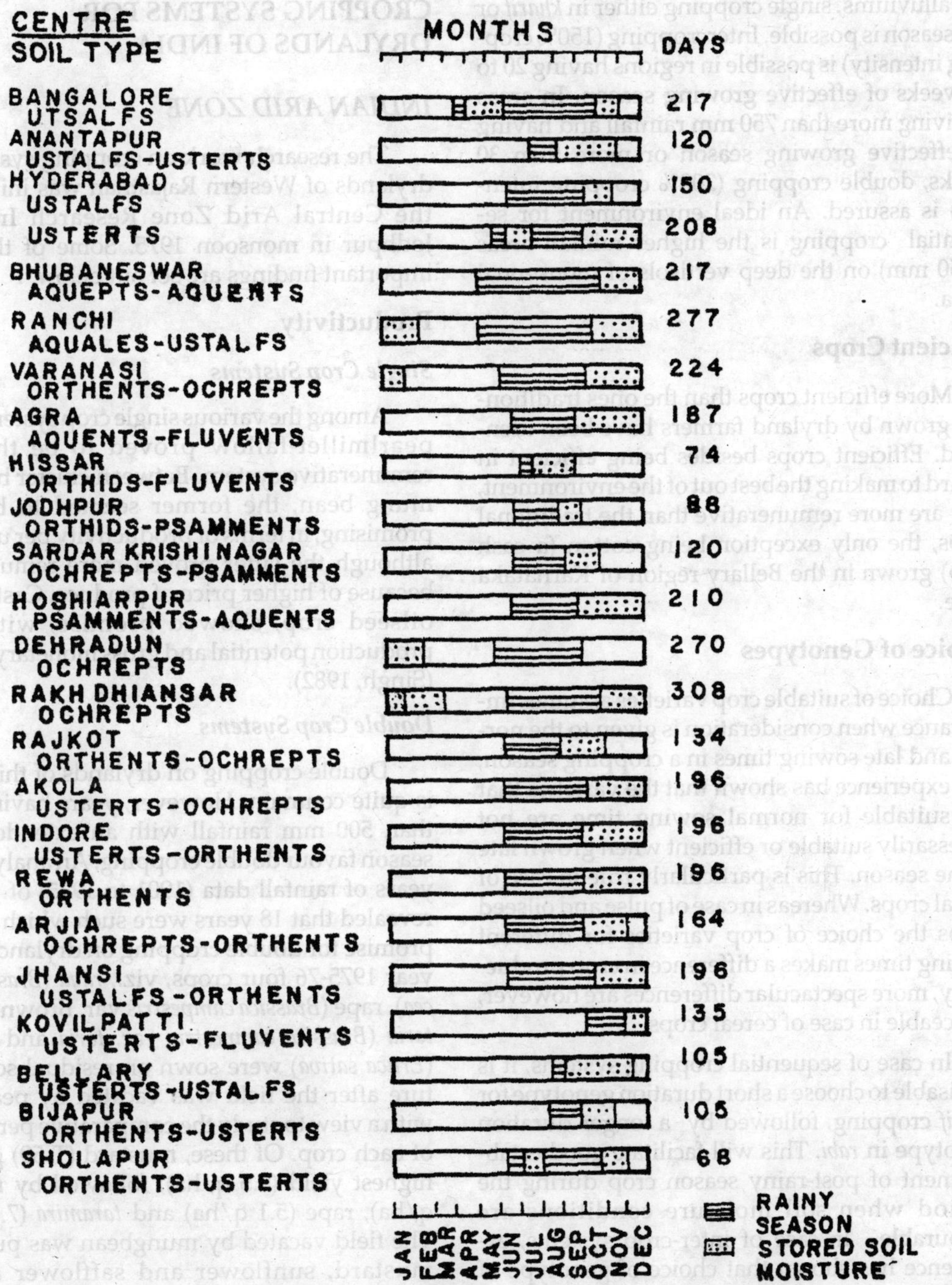

Fig. 1. Water availability periods at different dryland centres
(Source : S.M. Virmani, ICRISAT)

and alluviums, single cropping either in *kharif* or *rabi* season is possible. Intercropping (150% cropping intensity) is possible in regions having 20 to 30 weeks of effective growing season. In areas receiving more than 750 mm rainfall and having an effective growing season or more than 30 weeks, double cropping (200% cropping intensity) is assured. An ideal environment for sequential cropping is the higher rainfall areas (>750 mm) on the deep vertisols of monsoonal India.

Efficient Crops

More efficient crops than the ones traditionally grown by dryland farmers have been identified. Efficient crops besides being efficient in regard to making the best out of the environment, they are more remunerative than the traditional crops, the only exception being cotton (a cash crop) grown in the Bellary region of Karnataka State.

Choice of Genotypes

Choice of suitable crop varieties assumes importance when consideration is given to the normal and late sowing times in a cropping season. The experience has shown that the varieties that are suitable for normal sowing time are not necessarily suitable or efficient when grown late in the season. This is particularly true in case of cereal crops. Whereas in case of pulse and oilseed crops the choice of crop varieties for different sowing times makes a difference in crop productivity, more spectacular differences are however, noticeable in case of cereal crops.

In case of sequential cropping systems, it is advisable to choose a short duration genotype for *kharif* cropping, followed by a longer duration genotype in *rabi*. This will facilitate good establishment of post-rainy season crop during the period when soil moisture conditions are favourable. In case of inter-crop systems, experience has shown that choice of genotype of one component crop changes with the change in the genotype of the other crops. There is need for detailed studies on genotype compatability.

CROPPING SYSTEMS FOR DRYLANDS OF INDIA

INDIAN ARID ZONE

The research work on cropping systems for drylands of Western Rajasthan was initiated at the Central Arid Zone Research Institute, Jodhpur in monsoon 1975. Some of the more important findings are set out below :

Productivity

Single Crop Systems

Among the various single crop systems tried, pearlmillet-fallow proved to be the most remunerative system. Between cluster bean and mung bean, the former seemed to be more promising, in terms of productivity per unit area, although the latter proved more remunerative because of higher price of produce. Castor as an oilseed crop, showed promise, with good production potential and gross monetary returns (Singh, 1982).

Double Crop Systems

Double cropping on drylands of this region is quite common. However, years having more than 500 mm rainfall with an extended rainy season favour double cropping. An analysis of 75 years of rainfall data (1901 to 1975) of Jodhpur revealed that 18 years were such which showed promise for double cropping on drylands. In the year 1975-76 four crops, viz., *raya* (*Brassica juncea*), rape (*Brassica campestris* var. Brown sarson), *toria* (*Brassica campestris* var. *toria* and *taramira* (*Eruca sativa*) were sown on residual soil moisture after the field was vacated by pearlmillet with a view to study the comparative performace of each crop. Of these, mustard (T-59) gave the highest yield (9.5 q/ha), followed by *toria* (8.0 q/ha), rape (5.1 q/ha) and *taramira* (7.6 q/ha). The field vacated by mungbean was put under mustard, sunflower and safflower and the respective yields were 7.1, 3.4 and 7.3 q/ha. From the above, it is clear that pearlmillet-mustard and mung bean-mustard are the most promising double crop systems for this area. In fact, between mung bean-mustard and mungbean-safflower double crop systems, there is nothing much to choose as both these systems exhibited the same pattern of productivity and gross monetary returns (Singh, 1980).

Intercropping Systems

Among the inter-cropping systems involving *Cenchrus ciliaris* as the principal crop, intercropping of mungbean in between two rows of *Cenchrus ciliaris*, tried in 1975 and 1977 showed promise of high monetary returns. In the year 1977, three annual grain legumes, viz., mung bean, cluster bean and moth bean were intercropped with *Cenchrus ciliaris*, the yield of the grain legumes obtained was of the order of 8.2, 11.5 and 1.2 q/ha respectively; in addition to 70, 68 and 65 q/ha of green forage, respectively. In case of castor, mothbean for fodder was sown in between two rows of castor, in 1976 and 1977, the mean yields of respective crops obtained being 9.2 and 39.9 q/ha. (Singh, 1980).

An interesting feature of the intercropping systems involving *Cenchrus ciliaris* and castor as principal crops was that intercropping of the companion crop had no adverse effect on the yield of the principal crop and two together gave more remunerative returns that *C. ciliaris* and castor grown as sole crops.

The Land Equivalent Ratio (LER) in respect of each intercropping system tried was worked out. Total LER was more than unity in case of *C. ciliaris* grown in association with mung bean (1.84), cluster bean (1.73) and moth bean (1.57), suggesting thereby that better utilization of land and higher total productivity could be obtained through judicious intercropping systems.

An investigation on the intercropping systems for drylands, with sunflower as the principal crop, was carried out from 1973 to 1976. In general, the productivity of the intercropping systems was found to be 74 per cent higher than the pure cropping of sunflower. Higher monetary returns were obtained from sunflower + cowpea and sunflower + greengram intercropping systems over the other systems tried. (Singh, 1980).

In a separate study on cropping intensity it was revealed that castor-fallow was a more remunerative single crop system than fallow-mustard crop system. This suggests that efficient utilization of monsoon rains is ensured by growing a crop in the monsoon itself rather than leaving the land fallow and taking a chance with conserved moisture. Among the double crop systems, mung bean followed by mustard and pearlmillet followed by mustard systems were equally promising. Pearlmillet grown for fodder followed by mustard was as remunerative as pearlmillet grown for grains followed by mustard. Of the fodder crop sequences, the best sequence with regard to productivity and gross return was cluster bean for fodder followed by castor. Cropping systems like these seem to impart greater stability to crop production on dryland. (Singh, 1980).

Moisture Use Efficiency

Studies on moisture use indicated that total moisture use by different crops in 1977 ranged between 245 mm in the case of pearlmillet to 261 mm in the case of mung bean and cluster bean. The highest moisture use efficiency was exhibited by pearlmillet, followed by cluster bean, mung bean, and castor. A comparative study of moisture use by pearlmillet and mung bean grown in the conventional and paired row systems of planting in monsoon 1977 revealed that the paired row system of planting showed higher moisture use at the seedling stage. In the subsequent growth stages, however, the paired row system proved to be more efficient than the conventional system of planting pearlmillet and mungbean. Results of replicated trials carried out during monsoon seasons during 1975 and 1976 revealed that the paired systems of planting pearlmillet proved to be more efficient with regard to utilization of moisture from deeper layers of the soil profile than the conventional system, besides exhibiting higher moisture use efficiency. Crop canopy in the paired row system absorbed 2.0 times more radiant energy than the conventional system of planting.

Moisture use data as influenced by intercropping systems indicated that the intercropping system *per se* utilized more moisture than sole crops of *C. ciliaris* and castor. Grass + cluster bean showed higher total moisture use, followed by grass + mung bean, grass + moth bean and grass alone. Between castor alone and castor + moth bean intercropping system, the latter showed little higher total moisture use than the former. The moisture use in castor + moth bean system was particularly high during the period from 13 August to 8 September. In the studies carried out earlier on system of planting mung bean (1975 to 1976), it was revealed that the consumptive use was not materially affected by growing a companion crop of pearlmillet in the interspaces of mung bean over that of the pure crop of mung bean. The moisture use efficiency of the intercropping systems was almost double than that of the sole crop systems.

Economic Evaluation

Based on the mean productivity of 3 years (1975 to 1977), economic evaluation of different cropping systems was carried out. Of the various cropping systems tried, the highest returns (Rs. 4.53) per rupee invested were obtained from the double crop system of pearlmillet-mustard, followed by *Cenchrus ciliaris* + cluster bean (Rs. 3.67). Among the single crop systems, pearlmillet-fallow gave the highest returns (Rs. 2.76), followed by mung bean-fallow, castor-fallow and cluster bean-fallow cropping systems. *C. ciliaris* gave the lowest return per rupee invested.

The cost of production per hectare in the double crop systems involving pearlmillet and mung bean was almost twice as much as the single crop systems of the respective crops. This suggests that dryland farmers with favourable monetary resources could go in for double cropping in years of favourable rainfall distribution with extended rainy season, so as to obtain higher returns on investment. Further, double crop systems will make the most efficient use of resources.

CROPPING SYSTEM AND EMPLOYMENT POTENTIAL

Round the year gainful provision of employment is possibly one of the major consideration for evolving any cropping systems. In a typical year, it has been estimated that dryland farm families remain unemployed for almost two-thirds of the time. Change in crops and cropping patterns is one such way of generating additional employment. The employment potential per hectare was around 70 mandays in case of mung bean and cluster bean, and 84 days in the case of direct seeded pearlmillet. By replacing mung bean with castor, and the practice of direct seeding of pearlmillet with transplanting in the event of delayed onset of monsoon, additional employment potential of the order of 25 and 8 labour mandays per hectare could be created. By resorting to the double crop system of pearlmillet followed by mustard in favourable rainfall years, the employment potential per hectare of 84 man-days in the single crop system of pearlmillet could be increased to 176 mandays per hectare. By intercropping annual grain legumes (mung bean, cluster bean or moth bean) in the perennial grasses like *C. ciliaris*, additional employment potential to the extent of 20 mandays per hectare could be achieved over the sole cropping of *C. ciliaris* (59 mondays). With the adoption of improved cropping systems on dryland, more efficient utilization of machinery and equipment could be obtained, besides generating additional employment potential. The change-over from single crop to double crop system not only increases the employment potential and machinery utilization, but causes an increase in productivity. For example, the productivity of 0.8 tons/ha obtained in the mung-fallow system increased to 1.6 tons/ha in the case of mung bean-safflower cropping system. It can thus be inferred that the double and intercropping systems establish the complementarity of labour and capital productivity. Any cropping system which meets the requirement of releasing these two crucial inputs from the orbit of competitiveness is also desirable for its widespread acceptability.

INDIAN SEMI-ARID TROPICS (SAT) VERTISOLS

Vertisols and Alfisols are the two major soil groups encountered in the Indian SAT. In the cropping systems experiments the initial objective was to determine the feasibility of replacing the *kharif* fallow in Vertisols with a rainy season crop and to examine whether this jeopardised the yield of sequentially-sown traditional post rainy season crops. By comparing rainy season cropping versus fallowing over a three years period, it was established that a good rainy season crop of maize could be grown with little or no reduction in yield of post-rainy season crops of sorghum or chickpea. Operational-scale experiments have confirmed that rainy season maize can be followed by good crops of chickpea and safflower with advantage. This has dispensed with the need of keeping the Vertisols fallow during the rainy season. In ICRISAT experiments, sorghum was also tried as one of the rainy season crops, followed by post rainy season sorghum, chickpea and pigeonpea. The yield of post rainy season crops was invariably reduced, the yield reduction varying from 12 to 19 per cent. (Willey et al. 1985). The possibility of taking an early maturing (100-105 days) rainy season rice in sequential systems was tried at Rewa in Madhya Pradesh. Yield of about 2.5 tons/ha of rice has been obtained, followed by almost 2.0 tons/ha of the crucial post-rainy season wheat (Singh, 1985).

Whether based on sorghum, maize or rice, sequential systems suggest that in the wetter areas there can be a possibility of producing virtually two full crops, each of about 100 days maturity. However, such systems can be difficult to manage and they incur considerable risk. Experience in on-farm projects has revealed that farmers view sequential systems of two full crops as being very demanding and highly risk-prone. A more easily managed sequential system has been shown to be mung bean rainy season crop followed by sorghum, its major advantage being that the very early maturing mung (65-75 days) allows an early, assured establishment of the sorghum. This is a system already practiced by farmers in areas where rainfall and consequently the risk of foliar diseases and waterlogging, is not too high.

As compared to sequential systems, intercropping systems are more assured and less demanding. One of the most successful intercropping systems on the wetter Vertisols has been the cereal+pigeonpea combination. This has usually been in a 2 rows cereal : 1 row pigeonpea arrangement with the populations of each crop the same as the recommended sole crop population. Both crops are sown at the beginning of the rains, the cereal being harvested after about 100 days and the pigeonpea going on for about a further 100 days until the end of the post-rainy season.

A four year average for large watershed areas gave 2791 kg/ha maize and 1060 kg/ha pigeonpea in the intercropping system and 3197 kg/ha maize and 976 kg/ha chickpea in a sequential system (Willey *et al.*, 1985). However, a coefficient of variation of only 24 per cent for the intercropping system compared with 54% for the sequential one was found.

One very promising system in the State of Madhya Pradesh has been soybean/pigeonpea intercropping system. This has given very good yields of both crops, producing a total of 2.5-3 t/ha of high value seeds. A 3 crop system has been developed by AICRPDA (All India Coordinated Research Project for Dryland Agriculture) at its Indore centre (990 mm rainfall). In the rainy season a very early maize (cv. Satha, 65-70 days) is intercropped with soybean. Before the harvest of the maize, safflower is 'relay' sown along the sides of maize rows. The maize is harvested at physiological maturity and the safflower continues first as an intercrop between the soya rows and then as a sole crop after soya harvest.

In the drier vertisol areas with lower and less assured rainfall (i.e. < 750 mm) there is usually insufficient rainfall to support cropping during both the rainy and post rainy season. Post-rainy season crops, however, involve much less risk since they depend on moisture already safely stored in the profile and are not subject to the problems of erratic rainfall. It leaves little scope for increasing productivity by improving the system *per se*; productivity increase will mainly have to be brought about by improvements in the individual crops.

ALFISOLS

Alfisols are invariably mono-cropped in the rainy season. Double crop systems are rarely successful. Intercropping of cereals with low canopy legumes are the most common systems. Experience at ICRISAT has shown that in case of maize+groundnut intercrop system, intercrop groundnut produced 77 per cent of its sole crop while the millet produced more than twice its expected yield, i.e., 54 per cent of the sole crop. Combining yields of both crops; the intercrop has 31 per cent yield advantage over the sole crop; the returns were also 31 per cent higher than those of the sole crops grown at the same proportion as yields in intercropping (Rao *et al.*, 1982).

Intercropping with long season crops such as pigeonpea or castor provides an excellent opportunity to extend beyond the rainy season. At ICRISAT, the sorghum+pigeonpea intercrop produced good sorghum yields at 89 per cent of the sole crop but gave somewhat lower pigeonpea yield, 59 per cent of the sole crop (Rao *et al.*, 1982).

In the farming practice, groundnut is grown with pigeonpea in more varied proportions than the cereal intercrop. Quite often many rows of groundnut are planted with only occasional row of pigeonpea, as is the case of Rayalaseema region of Andhra Pradesh. In ICRISAT studies, pigeonpea was grown at 135 cm apart with five very close rows of groundnut in between (at 22.5 cm), accommodating 100 per cent population of the respective sole crops. Yields across seven locations in three years averaged 76 per cent of sole groundnut plus 89 per cent of sole pigeonpea, i.e., there was an advantage of 65 per cent over sole cropping.

SOME INTERACTIONS BETWEEN CROPS IN INTERCROPPING SYSTEMS

Legume Benefits

In the vast majority of experiments, there has been no evidence of a legume transferring any appreciable amount of N to a non-legume intercrop during the legume's active growth period. (Willey, 1988). In a series of experiments at ICRISAT, the residual N left after intercropping and sole cropping systems that incorporated different proportions of legumes was quantified with a sorghum test crop grown the following year. In comparison with a non-legumes sequential system of sorghum + safflower, the replacement of safflower by chickpea improved the residual N status by about the equivalent of a 20-kg/ha application of fertilizer N. The comparable improvement in N status after a sorghum+pigeonpea intercrop was a very worthwhile 40 kg/ha, probably because of the heavy leaf fall that occurs in pigeonpea prior to harvest. A cowpea+pigeonpea intercrop left an even larger residual effect of 60 to 70 kg/ha of N. (Willey, 1988).

Control of Weeds, Pests, and Diseases

There are quite a few instances where intercropping gave better control over weeds, pests, or diseases.

Better weed suppression is achieved when total plant populations in intercropping are higher than in sole cropping, or because of the greater overall growth and leaf area development as was observed in the millet + groundnut system.

In an intercrop system, one crop can provide a barrier to the spread of a pest or disease of another crop. In sorghum + pigeonpea, sorghum intercrop reduces the soil-borne wilt disease of pigeonpea. The most likely explanation seems to be that the sorghum produces some specific root exudate that reduces the vigor or population of the pathogen, an explanation that is supported by the fact that a maize intercrop appears to have no effect. (Willey, 1988).

INCREASING CROPPING INTENSITY THROUGH EFFICIENT CROPPING SYSTEMS

Pearlmillet-based cropping systems are more promising for arid regions, and semi-arid Deccan region of Maharashtra and Karnataka having light textured soils. In semi-arid regions, however, sorghum-based cropping systems with pigeonpea as the companion crop have been found to be more productive, stable and remunerative.

A distinct feature of these systems, is that they are all in 'additive series' rather than 'replacement series' recommended earlier. The overall mean of 59 trials showed 22.8 per cent more LER (Land Equivalent Ratio) in additive series vis-a-vis replacement series, suggesting the superiority of the former (Singh and Singh, 1985).

In these systems, it is suggested that full population of the base crop and 50-75 per cent of the companion crop be maintained, besides the use of normal recommended fertilizer levels. For instance, in the case of sorghum + pigeonpea intercropping system with 2:1 row ratio, the full seed rate of sorghum should be used, even though in the system the area occupied by sorghum would be 66 per cent. On the other hand, pigeonpea population should be atleast 50,000 plants per hectare.

Another significant consideration in designing these systems has been the introduction of pulses and oilseeds so as to produce additional quantities of these commodities required by the country.

In order to ensure complementarity between the component crops in the system, one may have to resort to staggered time of planting so as to provide competition free period to one of the component crops, preferably the slow growing one. The choice of a suitable genotype of component crops assumes added importance in this context.

YIELD STABILITY

Rao and Willey (1980) examined the 'stability' of sorghum+pigeonpea intercrop system in 51 experiments. Based on the coefficient of variation in yield approach sole pigeonpea (cv 43.6%) was rather more stable than sole sorghum (cv 48.9%), but intercropping was more stable than either (cv 39%). When regressions of yield against an environmental index were computed, it was found that sole pigeonpea will fail one year in five, sole sorghum one year in eight, but intercropping only one year in thirty-six.

Double Crops Systems

In semi-arid regions, upland rice, sorghum, maize, greengram, black gram, soybean are commonly grown in the *kharif* season. After the harvest of short duration (65-70 days) pulse crops like greengram, blackgram, cowpea, etc., it is possible to take a crop of wheat in areas where winter rains of the order of 50 to 100 mm are received. Rapeseed-mustard is another possibility in less favourable winter rainfall areas. Upland rice is usually followed by chickpea, although wheat can also be grown if soil moisture situation and winter rains are favourable. In the Malwa plateau of Madhya Pradesh, soybean and safflower are finding favour with farmers. The double crop systems found more promising for that region are : soybean- safflower; maize-safflower; sorghum-safflower; and sorghum- chickpea.

In semi-arid Bundelkhand region of Uttar Pradesh, cowpea (fodder) followed by chickpea, has been found to be promising double crop systems. Research work carried out at Dry Farming Research Centre of University of Agricultural Sciences, Bangalore has shown that in years when May rains are of the order of 100 mm or more, it is possible to follow a double crop system of cowpea grain followed by fingermillet.

Based on the decades of experimental work carried out at different locations under the ACRPDA (All India Coordinated Research Project for Dryland Agriculture) and CRIDA (Central Research Institute for Dryland Agriculture) efficient cropping systems for different dryland areas in the country have been found and the same are given in Table 1.

FERTILIZER MANAGEMENT IN SEQUENTIAL AND INTER-CROP SYSTEMS

Work carried out on alluvial soils at Indian Agricultural Research Institute, New Delhi showed that soil fertility enrichments from the legume fodders (equivalent to 40 kg N/ha or more) were highly beneficial to the succeeding barley crop. Grain legumes like groundnut or cowpea grown for the full season benefitted the

Table 1 : Efficient cropping systems for different dryland areas in India.

Soil zone and region	Water availability period (Days)	Double cropping system	Intercropping system
1. Vertisols and related Soil Zone			
Malwa plateau	210-230	Maize-safflower/chickpea	Maize+soybean (2.2)
(Madhya Pradesh)		Soybean-wheat	Soybean+pigeonpea (4 or 6:2)
	191-210	Sorghum-safflower/chickpea	Sorghum+pigeonpea (2:1)
Baghelkhand (Madhya Pradesh)	210-230	Rice-chickpea/lentil	Wheat+chickpea (2:1) Chickpea+linseed (2:1)
	190-210	Sorghum-chickpea Blackgram/greengram-wheat Groundnut-chickpea	Sorghum+pigeonpea (2:1)
Bundelkhand (Uttar Pradesh)	190-220	Sorghum-chickpea Blackgram-mustard/safflower Fodder cowpea-mustard	Pearl millet+fodder legume (1:1) Sorghum+pigeonpea (2:1)
Vidharbha (Maharashtra)	190-210	Groundnut-safflower Sorghum-safflower	Sorghum+pigeonpea (2:1) Cotton+pigeonpea (2:1 or 2) Cotton+greengram/cowpea (1:1) Pigeonpea+greengram (1:3)
	170-190	Greengram-sorghum/safflower	Pearl millet+pigeonpea (2:1) Sorghum+greengram/blackgram (2:1)
Southern (Maharashtra)	160-180	Greengram-sorghum/Safflower	Pearl millet + pigeonpea (2:1) Sunflower+pigeonpea (2:1 or 2) Groundnut+sunflower (2 or 3:1) Chickpea+safflower (3:1)
	120-130		Pearl millet+mothbean (2:1)
Southern (Rajasthan)	160-180	Greengram-safflower	Maize+pigeonpea (1 or 2:1) Sorghum+greengram (2:2) Groundnut+pigeonpea (2:2) Chickpea+mustard (4 or 7 : 1)
Northern (Central Karnataka)	130-150	Cowpea-sorghum Greengram-safflower	Pearl millet+pigeonpea (1:1) Groundnut+pigeonpea (3 or 4:1) Sunflower+pigeonpea (2:1) Chickpea+safflower (2 or 3:1) Sorghum+pigeonpea (2:1)
	100-120	—	Sorghum+coriander (2:1) Safflower+coriander (2:1)
Saurashtra (Gujarat)	130-140	—	Groundnut+castor bean/pigeonpea (3:1) Pearl millet+pigeonpea/castor bean (2 or 4:1)
Southern Tamil Nadu	120-130	—	Sorghum+blackgram/cowpea (2:1) Cotton+blackgram (2:2)
2. Inceptisols and related soil zone			
Jammu region of Jammu & Kashmir	280-320	Maize-wheat	Maize+greengram/blackgram/cowpea (1:2) Wheat+mustard (4:1) Chickpea+mustard (4:1) Barley+chickpea (2:2)

Table 1. Continued

Soil zone and region	Water availability period (Days)	Double cropping system	Intercropping system	
Sub-montane region of Uttar Pradesh	240-280	Rice-wheat/chickpea Soybean-wheat Maize-chickpea	Maize-soybean Rice+soybean Wheat+mustard Wheat+chickpea	(2:8) (1:2) (9:1) (2:2)
Sub-montane Region of Punjab	260-280	Maize-mustard/chickpea	Maize+blackgram Chickpea+mustard	(1:2) (4:1)
Himachal Pradesh	240-260	Pearl millet-chickpea	Pearl millet+blackgram Chickpea+mustard	(2:1) (4:1)
Plains of Uttar Pradesh	200-230	Rice-chickpea/lentil Maize-lentil	Maize+blackgram Pigeonpea+blackgram Pigeonpea+sesame Groundnut+pigeonpea Barley+mustard	(1:1) (1:1) (1:1) (2:1) (6:1)
	180-200	Pearl millet-chickpea Greengram/blackgram mustard/barley	Pearlmillet+greengram/blackgram Pearl millet+pigeonpea Chickpea+mustard Chickpea+barley/linseed	(2:1) (2:1) (4-5:1) (2-3:1)
3. Oxidols zone				
Sub-humid region of Orissa	200-220	Rice-horsegram Maize/finger millet-horsegram Groundnut-horsegram	Rice+greengram/blackgram Rice+pigeonpea Finger millet+pigeonpea Miaze+pigeonpea	(2-3:1) (3-4:1) (1:1) (2:1)
Sub-humid region of Southern Bihar	160-180	Maize-chickpea/safflower Groundnut-barley Finger millet/chickpea Rice-chickpea/lentil	Maize+pigeonpea Rice+pigeonpea Groundnut+pigeonpea Soybean+pigeonpea Pigeonpea+blackgram	(1:1) (2-3:1) (2-3:1) (3-1:2) (1-2:3)
4. Alfisols and related soil zone				
Southern Karnataka	190-220	Cowpea-finger millet Soybean-finger millet	Finger millet+pigeonpea Groundnut+pigeonpea Finger millet+soybean	(3:1) (4:1) (1:1)
Telangana (Andhra Pradesh)	140-160	—	Sorghum+pigeonpea Castor bean+cluster bean	(3:1) (2:2)
Rayalseema	110-130	—	Groundnut+pearl millet Groundnut+pigeonpea Groundnut+castor bean	(6-8:1-2) (6-8:1-2) (2:1)
5. Arid soil zone				
North Western Gujarat	100-120	—	Pearl millet+greengram Pearl millet+clusterbean Cluster bean+greengram	(1:3) (3:1) (2:1)
Western Rajasthan and parts of Haryana	75-90	—	Pearl millet+blackgram/greengram Pearl millet+clusterbean Clusterbean+blackgram Pigeonpea+blackgram	(2:1) (1:2) (1:1) (1:2)

N.B. : Figures in parentheses are optimum row ratios of intercropping system.

Source : S.P. Singh, 1988.

succeeding crop of pearlmillet to the extent of 60 kg N/ha (Singh, 1983).

In medium black soils, in order to obtain the full yield potential of the sequential cropping systems, judicious fertilization of both the crops in the system, has been suggested. In studies carried out at Akola Dryland Research Centre, the beneficial effect of taking a mung bean crop in the monsoon season was reflected in the seed yield of safflower following it (Singh, 1983).

In an inter-crop system, involving a non-legume and a legume, nitrogen application should be as per requirements of non-legume and phosphorus as per the requirement of the legume crop. In legume + legume intercropping systems, slightly more phosphorus should be applied.

NUTRITIVE VALUE OF CROPPING SYSTEMS

Cropping systems are now available that can meet the nutritional goals (about 10,000 calories per day for an average Indian family of five members) in dry farming areas. Among the cereal based intercropping systems, maize + soybean supplies the highest calories and other nutrients on a per hectare basis. However, the sorghum + pigeonpea (2:1 row ratio) system is more stable over the seasons and environments and, hence preferable. Among sequential cropping systems, sorghum-chickpea system holds promise of providing better nutritional standards. From the point of view of biological value and protein efficient ratio, the rice - chickpea sequence cropping system holds the greatest promise (Singh and Chetty, 1984).

EVALUATION PARAMETERS

In Table 2 are given some parameters for evaluating cropping systems. Besides productivity (yield/biomass) per unit area, nutrient value (calories, protein, oil etc.), and energy obtained from seed yield and biomass, are important considerations. For a resource-poor dryland former, the over riding consideration would he costs involved and net returns obtained from the cropping systems under consideration. Although there is not much difference in net returns

Table 2 : Evaluation parameters for some example cropping systems grown on a deep Vertisol in operational-scale at ICRISAT Center (means of 4 years)

Parameters	Cropping System			
	Intercrop	Sequential	Sequential	Sequential
1st crop	Sorghum	Mung	Maize	Sorghum
2nd crop	Pigeonpea	Sorghum	Chickpea	Safflower
Cropping period (days)	180	150	190	190
Yields (Mg/ha): 1st	2.88	0.65	2.44	3.34
2nd	0.94	2.07	1.31	0.79
Nutrient value/ha				
M.cals	11032	7956	10120	10765
kg protein	508	371	494	453
kg oil	71	48	157	271
Biomass (Mg/ha)	9510	6776	9363	9023
Energy (M. cals/ha)				
seed yields	17351	12407	17250	19464
biomass	41567	29821	41094	44287
Gross returns (Rs/ha)	6727	4770	7490	6900
Costs (Rs/ha)	1600	1750	2200	2450
Net returns (Rs/ha)	5127	3020	5290	4450
Rate of return (%)	320	173	240	182

Source : Willey, R.W. 1988.

obtainable from sorghum + pigeonpea and maize-chickpea cropping system, the former is a much attractive proposition because of lower costs involved and higher rate of return to inputs.

REFERENCES

Bhaskar Rao, U.M. and Vijayalakshmi, K. 1985. Crops and cropping systems for semi-arid red soils of Telangana region. *Research Bulletin No. 4, Central Research Institute for Dryland Agricutlure,* Hyderabad, pp. 44.

Rao, M.R. and Willey, R.W. 1980. Evaluation of yield stability in winter cropping-studies on sorghum/pigeonpea. *Exptl. Agric.* 16 : 105- 116.

Rao, M.R., Reddy, M.S. and Willey, R.W. 1982. Improved rainfed cropping systems for vertisols and alfisols of semi-arid tropical India. *Paper presented in the Seminar on 'A Decade of Dryland Agricultural Research in India and Thrust for the Eightees'*, 1982.

Reddy, M.S., and Willey, R.W. 1982, Improved cropping systems for the deep vertisols of the Indian semi-arid tropics. *Exptl. Agric.* 18 : 277-287.

Singh, R.P. 1990 Crops and cropping systems for arid and semi-arid dryland regions of India. In *Ecology and Vegetation of Indian Desert* (Ed. David N. Sen) Agro-Botanical Publication, (India), Bikaner, 257-283.

Singh, R.P. 1980. Cropping systems for drylands of the Indian arid zone. *Annals of Arid Zone* 19 : 437-443.

Singh, R.P. 1983. Fertilizer management in multiple cropping systems under dryland conditions. In : *Fertilizer use under Multiple Cropping Systems, FAO Fertilizer and Plant Nutrition Bulletin No. 5,* pp. 97-101.

Singh, R.P. 1985. Rainfed rice-based cropping systems - prospects, problems and research needs. Discussion paper presented at the *Mini- workshop on rice-based cropping systems organised at IARI, New Delhi,* 4-5 February, 1985.

Singh, R.P. and Chetty, C.K.R. 1984. Design of cropping systems to achieve nutritional goals in dry farming. In : *Interfaces between Agriculture, Nutrition and Food Science.* (ed.) K.T. Acharya. The United Nations University, Tokyo, Japan, pp. 98-109.

Singh, R.P. and Raje, S.R. 1984. Optimising production in rainfed vertisols. A case study of an operational research project. *Fertilizer News* 29 : 59-71.

Singh, S.P. and Singh, R.P. 1985. Current status of research in dryland agriculture. Paper presented at the *Annual Workshop of the All India Coordinated Agronomic Experiments Project held at APAU, Rajendranagar,* Hyderabad, November 4-7, 1985.

Singh, S.P. 1988. Efficient cropping systems for dryland areas in India. *In Challenges in Dryland Agricutlure- A global Perspective. Proc. of Int. Conf. on Dryland Agric.* Amarillo/Bushland, Texas, USA, August 15-19, p. 770-774.

Willey, R.W., 1988. Cropping Systems for Drylands. *In Challenges in Dryland Agriculture - A Global Perpective. Proc. of Int, Conf. on Dryland Agric.* Amarillo/Bushland, Texas, U.S.A. August, 15-19, pp. 703-709.

Willey, R.W., Singh, R.P. and Reddy, M.S. 1985. Cropping systems for vertisols. Paper presented at the *International Workshop on Management of Vertisols for Improved Agricultural Production* held at ICRISAT Centre, 18-22 February, 1985.

Zandstra, H.G. 1979. Cropping systems research for the Asian rice Farmer. *Symposium on Cropping Systems Research and Development for the Asian Rice Farmer. IRRI,* Philippines, 1977.

20

EFFICIENT MANAGEMENT OF DRYLAND CROPS — PRINCIPLES AND PRACTICES

R.P. SINGH

INTRODUCTION

Productivity of dryland crops is greatly influenced by soil and moisture conservation practices and crop management (including tillage). It is beyond the scope of this chapter to deal with area and production, adaptation, cropping systems, promising genotypes, pests and disease management, post harvest technology, economics of production, etc. These aspects have, however, been dealt in different chapters of this book. Some of the more promising practices are discussed in this chapter. An elaborate and free use of the material published in the book entitled, "Efficient Management of Dryland Crops", published by the Central Research Institute for Dryland Agriculture, Hyderabad in 1985, has been made to bring home the point or clarify the concept. The topic is dealt under the heads : soil and moisture conservation, tillage, seeding (rates, spacing, seeding methods and time of seeding), weed management, fertility management, relative contribution of production factors and constraints to higher productivity, in that order. In the end, some basic principles concerning efficient management of dryland crops are discussed.

SOIL AND MOISTURE CONSERVATION MEASURES

Soil and moisture conservation measures are a prerequisite for ensuring a good crop stand, growth and development of crop, and higher crop yields. These measures depend mainly on the land (topography and slope), soil type, rainfall pattern, and nature of the crop. Some specific measures are given in Table 1.

TILLAGE

Off-season tillage is necessary to get a weed-free seedbed having good tilth and better moisture conservation. Pre- monsoon showers should be utilized for deep ploughing (where necessary) and blade harrowing. This will enable sowing of large areas in limited time-a necessity in dryland areas.

Higher yield advantage with deep ploughing (22 cm) has been obtained in case of sorghum, pearl millet, maize, and rice. In case of rice, use of mould board plough to a depth of 30 cm every year increased the rice yield at Ranchi. For small-seed crops like finger millet, soil compaction below the seeding zone is absolutely necessary for getting good crop stand establishment and higher yields.

Tillage requirement for pulse crops (cowpea, mungbean, black gram) varies with the type of soil on which they are grown. For loamy sand or sandy loam soils, one or two cross cultivation using a sweep harrow, will be enough to provide a good tilth.

In medium to heavy textured soils, summer cultivation with a disc plough followed by a cross-disc harrow will be adequate for obtaining

Table 1. Soil and moisture conservation measures suggested for different dryland crops (based on research findings)

Crop	Season/Soils/Location	Suggested measures- Land treatment and others	Impact on yield, soil, and moisture conservation
Sorghum	Rainy season/Black Soils	1. Ploughing in dry season immediately after the removal of preceding crop. Smoothen the land with in individual plots.	Seedbed receptive to rainfall and moisture storage is obtained..
	Alfisols and related soils	1. Furrowing as part of interculture operation implemented on contour	15% increase in yield of high yielding varieties of sorghum, over traditional land management.
Pearl millet	Rainy season/light textured soils	1. Bunding found superior to trenching and crop strips (cowpea/pearlmillet)	
		2. Mulching; straw superior to dust mulch	Significant increase in yield, 38 to 40 per cent over control (no mulch)
Finger millet	Rainy season/light textured soils	1. Contour bunds in areas having rainfall upto 600 mm; graded bunds for higher rainfall areas. 2. Border strips (120/150 m length; 10-12 m width), levelled breadth-wise, maintaining a gradient of 0.2 to 0.4% along the length. 3. Stirring the surface soil by intercultivation for combating crusting.	
		4. Incrporation of crop residues (maize residues @ 4 t/ha)	1. Grain yield, 35.3 q/ha with incorporation, 25.7 q/ha without incorporation. 2. Soil moisture (top 15 cm - 12% as against 8.5% in control plots.
Maize	Rainy season : (a) Sub-montanne region of Punjab, Jammu & Kashmir	Land levelling	
	(b) Doon valley	Zingg terracing	
	(c) Uplands of Bhubne shwar and Ranchi	*Tati* terraces with provisions of waste weirs for each terrace	
	(d) Malwa plateau	Surface drainage is crucial	
	(e) Telengana, Bundel khand and Udaipur regions	Graded bunding or conservation drain for safe disposal of runoff as well as conservation of soil	
	(f) Red soils of Bangalore	Graded bunding or graded border strips (with a provision of grassed waterway on boundary lines)	
Upland rice	Rainy season (a) In parts of UP and M.P.	1. Levees are created around the field in order to impound rain water. Bench terraces on slopy land.	
		2. Run off farming on conservation bench terraces (cropped to collection area 3:1)	Average yield 29.6 q/ha as against 15.8 q/ha from control (1:1 system)

Crop	Season/Soils/Location	Suggested measures- Land treatment and others	Impact on yield, soil, and moisture conservation
		3. Inter row water harvesting system (planting 2 rows of maize on raised bed 1.5 m) alternated with 3 m wide strip of rice	Additional yield 4 q/ha
Pigeon pea	Rainy season (Reproductive phase commences in the post-rainy season)	Adequate recharge of profile before cessation of monsoon is important. Intercultivation highly efficaceous.	Better conservation moisture
Groundnut	Rainy season	Provision of a dead furrow in between the rows of groundnut.	Helps in deep penetration of rain water
Sesamum	Rainy season/light textured soils	Inter-plot and inter- row water harvesting	
	Heavy soils	Provision of drainage of excess water during monsoon is essential	
Sunflower	Rainy season/Black soils	Opening graded furrows across the slope at set intervals of about 10 m	
	On lands having more than 1.5% slope	Graded bunds	Prevent soil loss and conserve more water
Niger	Rainy season/On slopy lands	1. Easing the slope through graded bunds or terracing. 2. Soil mulch during and after the rainy season	
Castor	Rainy Season/Low rainfall areas	Deep tillage and ridge and furrow cultivation	Efficaceous in conservation of soil and moisture.
Cotton	Rainy season/Black soil : Soil depth : 45 to 50 cm Soil slope : 0.5 tp 1.5%	Graded bunding with grade of 0.2 to 0.4% Broad bed and furrow system	Reduces erosion, conserves both soil and moisutre. Effect of land configuration on yield of cotton (Table 1a)
Wheat	Post/rainy/alluvial	1. Minor land treatment and bunding 2. Spreading wheat or barley straw in between crop rows after crop emergence	Higher profile water storage (168 mm/90 cm) over farmers' practices resulted in one quintal extra grain yield per cm extra water conserved (Table 1b)
Barley	Post-rainy/alluvial	1. Contour bunding/field-wise proper levelling. 2. After the withdrawal of monsoon, the land should be cross-ploughed,levelled, planked and uniformly rolled to break the clods. 3. Regular hoeing and weeding.	
Sorhum	Post-rainy/black soils	1. Contour bunding or graded bunding 2. Fallowing during rainy season 3. Use of surface mulch soon after crop emergence 4. Vertical mulching-opening trenches (20 cm width; 40-50 cm deep) at 2 m intervals; trenches to be filled with organic wastes.	50% increae in grain yield.
Safflower	Post-rainy/Low to medium rainfall areas.	1. Graded bunds combined with inter-terrace cultivation across slopes 2. Compartment bunding	

* In this systems maize is taken in the upper 75% area of field which is kept slopy to provide quick drainage and paddy is taken in the remaining 25% area which is levelled to accommodate the run off from the upper portion of the plot.

Table 1a. Effect of land configuration on cotton yield at Akola

Land treatment	Seed cotton yield (q/ha)				Mean
	1979-80	1980-81	1981-82	1982-83	
1. Broad bed and furrow system	16.3	8.9	7.4	10.6	10.8
2. Flat on grade at 0.4 per cent	12.4	7.2	5.6	9.4	8.6
Rainfall (mm)	733	668	965	541	727

Table 1b. Effect of moisture conservation practices on profile moisture and yield of dryland wheat

Treatment	Profile moisture at sowing (mm/90 cm)	Wheat yield (q/ha) (Mean of two years)
Farmers' practice	147	15.0
Minor levelling and bunding	168	17.8

a good seedbed. The field preparation for pigeonpea normally consists of one cultivation with soil turning plough during summers, utilizing pre-monsoon showers, followed by a cross harrowing or ploughing with a country plough on or before onset of monsoon, and finally planking.

Groundnut requires a loose and friable seedbed for the pegs to penetrate easily and also prevent loss of nuts during harvesting. For sesamum and rape seed mustard, a well prepared seedbed is required for uniform plant stand.

An ideal seedbed for cotton is formed if the soil is loosened and pulverized to a depth of 10 to 12 cm, kept free of weeds and compacted by an inverted blade harrow before sowing. Deep ploughing is not necessary except on soils carrying weeds like *Cynadon dactylon* and *Cyperus rotundus*. In black soils, ploughing once in three or four years, keeps the weed intensity down. Good initial surface drainage and prevention of water logging is essential.

A flat seedbed was found to store more moisture, gave significantly higher yield of pearl millet at IARI (Indian Agricultural Research Institute), New Delhi, and increased the uptake of nutrients, as compared to ridge and furrow beds. In semi-arid (eg. Udaipur, Jhansi) and sub-humid regions (eg. Bhubeneswar, Ranchi, Varanasi), bed and furrow or ridge furrow system was found suitable, particularly for crops like maize, which does not tolerate wet feet for long.

Wheat in dryland areas, usually grown after fallow, needs timely tillage operations so as to get the best advantage, of limited rainwater. More importantly, tillage operations are to be carried out at optimum soil moisture conditions. One mould board ploughing plus two harrowings can achieve a clod size of 6 to 8 mm in black soil region of Madhya Pradesh, provided ploughing is done at 23 per cent surface soil moisture conditions. Barley does not need as fine a seedbed as wheat. One ploughing with soil turning plough followed by levelling and rolling would be good enough.

For *rabi* sorghum, usually grown on black soils, there is no need for deep tillage. Harrow will accomplish a good seedbed. However, intercultural operations are necessary to keep the weeds under check, close up cracks, and conserve moisture. Compaction after seeding was found advantageous, resulting in improved emergence of seedling (by more than 50% of the control), early seedling vigour, and increased root growth. The advantage of compaction was more pronounced in years of low rainfall after seeding than in the normal years.

Rape seed mustard is grown in the post rainy season on conserved soil moisture and, therefore, favourable soil moisture conditions in the seeding zone at sowing time are necessary. Towards the end of the rainy season each ploughing operation must be followed by planking so that soil be comes pulverized and favourable moisture conditions are created.

SEEDING

Sub-optimal and gappy plant stand is largely responsible for low productivity of dryland crops. Proper germination and good crop stand is the first requirement for obtaining higher yields. It is also necessary for making an efficient use of inputs-seeds, fertilizers, etc. on one hand, and labour, time, and money, on the other. Besides soil tilth which has to be good and firm with adequate moisture in the seeding zone, seeding methods and time of sowing exert significant influence on crop stand establishment. Seed viability and seed health are yet other factors to reckon with.

Seeding Method

Upland direct-sown rice is sown either in dry (friable) seedbed or in puddled soil conditions. The seed is broadcast or line sown. Pregerminated or sprouted seeds are usually used in case of puddled condition. Superiority of seeding upland rice with seed drill has been clearly established on dry seedbed. The use of seed cum fertidrill enables the placement of the seeds in moist zone and at an appropriate depth, ensuring good crop stand establishment. Almost all dryland crops are traditionally sown by a country plough. As an alternative to a seed drill, a *pora* (bamboo/aluminium/tin) funnel is tied to the handle of the country plough, and seeds are drilled through a bowl provided at the head of the funnel. This enables placement of the seeds in the moist zone. However, because of hand- metering of seeds, crop stand is not as uniform as in case of crop sown with a seed drill. In case of crops sown during the monsoon season, *kera* method of sowing is good enough. In this method, seeds are drilled behind the plough in furrows. Seeded furrows are usually covered by planking. In the *pora* method of sowing, seeded furrows are not covered.

Seeding Depth

Depth of seeding has been shown to be critical for the establishment of satisfactory stand of pearl millet, finger millet and other millets, sesamum, and rape seed mustard. Seeding depth of 2 to 3cm is about the optimum for these crops. For maize, sorghum, wheat, barley, pulse crops and most of the oilseed crops, 5 to 6 cm has been found to be the optimum seeding depth. In light textured soils, castor can be sown a little deeper. Soybean should be sown at 3 to 5 cm depth. Finger millet is often broadcast with heavy seed rate (25 kg/ha) and thinned out later by cross cultivation with fine hoes. This results in non-uniform stand and too much of population. Sesamum seed being small, it should be mixed with dry earth, sand or well sieved farm yard manure to increase the bulk and to ensure even distribution of seeds while sowing. Drill sowing is by far the best, whatever the crop.

Transplanting

Transplanting has been found to be advantageous, especially under delayed sown conditions. The transplanted crop of pearlmillet gave significantly more grain yield than the direct seeded crop. In case of finger millet, 20 to 25 days old seedlings of a medium duration variety are ideal for transplanting. As a thumb rule, the seedlings could be left in the nursery for a week for every month of its total life period.

Seeding Time

All rainy-season crops, including upland rice, should be sown after the first soaking rains when the top 22 to 30 cm soil layer is thoroughly wetted to sustain germination, emergence and early seedling growth for atleast 15 to 20 days. Such conditions normally arrive by the end of June or beginning of July, in most parts of the country. In case of sorghum, progressive delay of 7,14 and 21 days cause substantial reduction in crop productivity, the average reduction in grain yield being of the order of 319 to 370 kg/ha/day. The reduction in grain yield was attributed to decrease in number of seeds per panicle and seed weight. Late sown crop attracted more pest attack.

Time of sowing is crucial for obtaining higher yield of all dryland crops. Sowing of pearl millet at first monsoon showers was found to be the best. As a dryland crop, finger millet is traditionally sown in July in the Southern States and in May-June in the Northern States. Some of the short duration varieties can be sown late but their yields are low. Sowing long duration varieties late in the season reduced their duration as well as yield. Optimum sowing time for foxtail millet is August-September in Tamil Nadu, July in Karnataka and first fortnight of July in Andhra Pradesh. For plains of Uttar Pradesh, late May to middle of June is the optimum sowing time. Kodo and Proso millet are sown with the onset of monsoon, preferably in July.

In pigeonpea, planting date has a profound effect on productivity. Since most of the pigeonpea varieties are photo-sensitive, late planting cuts short the growth period resulting in low productivity. Early and early-medium varieties should be sown with the onset of monsoon. The medium-late and late varieties offer some flexibility in planting dates but not beyond July. The ideal time for planting post rainy season pigeonpea in Bihar region is first week of September, and around end of September in Akola region of Maharashtra. The optimum time for sowing cowpea, mungbean and black gram is at the onset of monsoon (first week of July) and for mothbean, first fortnight of July is the optimum seeding time. In Karnataka, horsegram is sown at different times depending upon the region and cropping systems. A short duration cultivar HH-2 gave maximum yield when sown in May. However, early sowing upto middle of August resulted in high yields of horsegram cultivars. For coastal Orissa, middle of September has been found to be the optimum time, while for Tamil Nadu second fortnight of October is the optimum seeding time. In Maharashtra, horsegram is sown in June or July for fodder. In post-rainy season, October-November is the usual seeding season.

The optimum seeding time of groundnut in different states is presented in Table 2.

Table 2. Optimum sowing time for planting groundnut

States	Optimum sowing time
Andhra Pradesh	
a. Telangana & Coastal region	First fortnight of June
b. Rayalaseema region	Middle of June to first week of July
Gujarat	Second fortnight of June
Karnataka	Bunch : First week of June
	Spreading : Last week of June to end of July
	Post rainy : Last week of December to end of January
Maharashtra	Last week of June to first week of July
Tamil Nadu	April in Coimbatore; First fortnight of July in the rest of the State.

Sesamum is generally taken as a rainy season crop in most of the states. However, it is grown as post rainy and summer season crop in some states. The recommended time of seeding for important sesamum growing states is given in Table 3.

Table 3. Optimum time of seeding sesamum in different States

State	Optimum time of seeding
Uttar Pradesh	Second fortnight of July
Rajasthan	Late June to early July
Madhya Pradesh	First week of July; late August or early September
Orissa	June-July; September-October, February-March
Maharashtra	Second fortnight of June to first week of July; early September
Andhra Pradesh	
a. Coastal region	First fortnight of May and second fortnight of January
b. Telangana	Second fortnight of July
Tamil Nadu	July-August; October-November; Januray-March
Gujarat	Last week of June; January-February
West Bengal	February-March (after potato and late rice); End of May to first week of June; September-October
Karnataka	
a. Northern region	June-July
b. Southern region	April-May

Probably no single cultural factor is more important to soybean production than planting date. Beginning of June to mid-July in northern hill zone of Madhya Pradesh, last week of June to mid-July in northern plain zone and mid-June to mid-July in central and southern zones are generally the optimum time of seeding. Sunflower, due to its photo insensitivity, has shown a wide flexibility in sowing time. Under Anantpur and Hyderabad conditions, best results have been obtained with late sowing, when there would be better and assured rainfall. Niger is sown either at the beginning or late in the monsoon season in different parts of the country. In Madhya Pradesh, it is sown in the third week of July; in Karnataka and Maharashtra in June- August; in Andhra Pradesh in second week of August, in Bihar and Orissa in August-September. Sowing of castor on onset of monsoon is advantageous under dryland conditions. However, in certain tracts late sowing is preferred so as to escape the attack of red hairy caterpillar and semi looper.

Timely sowing of cotton is of utmost importance. In all the cotton growing zones, the crop is sown immediately with the onset of south West monsoon. In the southern zone, sowing is done in August-September. Delayed sowing reduces the cotton yield by 0.50 to 0.80 q/ha/week.

In case of post rainy season crops grown on conserved soil moisture, moisture and temperature are important factors to be considered. Studies have shown that wheat should not be sown unless the mean daily temperature comes down to 25°C (a situation usually obtained after 15 October in most of the wheat growing areas). If the soil moisture permits, the optimum time of sowing is the last week of October to the first week of November, the latter period proving still better. When sown in the first week of November (20°C - 22°C), greater tillering and higher fertiliser and water use efficiency is obtained, as is evident from yield data given in Table 4.

Table 4. Yield of dryland wheat as affected by dates of seeding

Cultivars	Grain yield (q/ha)		
	Dates of seeding		
	15 Oct.	30 Oct.	14 Nov.
C-306	32.7	34.6	39.8
Kalyan Sona	15.5	33.8	39.8

In case of barley, crop should be sown when the day temperature falls to 23-25°C.

Seeding time is important in production technology for post rainy sorghum. Magnitude of yield improvements with advanced seeding (early September) were more marked with hybrids and improved varieties than with the traditional varieties. Locally adapted variety M-35-1, exhibited relatively low sensitivity to delayed sowing compared to high yielding hybrids. Variety R-26 and hybrids CSH-2 and CSH-8 R showed significant increase in grain yield over M 35-1 when sown late, August to early September.

Chickpea should be sown by first fortnight of October in Haryana and Jammu areas. Delay in seeding resulted in significant decrease in yields. In Rewa, seeding could be extended to even November without any reduction in yield.

Planting time is a critical operation in case of rape seed-mustard. Rate of development of oil is greatly influenced by variation in temperature and humidity. Sowing either too early or too late is harmful. Rape seed type *toria* should be sown the earliest i.e. during September, while brown *sarson* and yellow *sarson* should preferably be sown during early October. Mustard sowings may be adjusted keeping in view the soil moisture in seed zone and temperature. There is usually a high incidence of *Sclerotia* wilt at seedling stage in case maximum temperature is above 30°C at sowing time. *Eruca* can be successfully planted under late sown conditions, as it has high tolerance to aphids, frost and moisture stress.

Time of seeding determines the ultimate yield of safflower in a given region and its response to scarce and costly inputs. The recommended time of seeding for different regions in the country is mid to late September for drought prone areas of western Maharashtra and adjoining areas of Marathwada and Bijapur (Karnataka); later half of September to early October for Malwa plateau of Madhya Pradesh; early to mid October for assured moisture tracts of western Maharashtra, Marathwada and Vidarhbha (Maharashtra), north eastern parts of Karnataka except Bijapur, Andhra Pradesh and southern parts of Tamil Nadu and south eastern Rajasthan; and second week of October for all other areas viz., north west India, U.P., plateau region of Chhota Nagpur.

System of Planting

System of planting is known to exert considerable influence on yield of certain crops, particularly in normal and subnormal years. In case of post rainy season sorghum, paired row planting showed distinct advantage in grain production compared to normally recommended spacing of 45 cm (Table 5).

Table 5. Yield advantage due to crop geometry in post-rainy sorghum

Planting method	Grain yield (q/ha)	
	1973-74	1974-75
Uniform row planting at 45 cm	8.5	9.6
Paired planting at 30-60-30	11.3	12.2

Row Spacing and Plant Density

The inter-and intra-row spacing found suitable/optimum for each of the dryland crops is given in Table 6, together with optimum seed-

Table 6. Optimum inter- and intra-row spacing and optimum seed rate for dryland crops

Crop	Optimum inter and intra-row spacing (cm)	Optimum seed rate (kg/ha)	Plant population (000/ha)
Upland rice	30	–	
Sorghum	45x12	8-10	
Pearl millet	45x15	2-3	
Maize	45x20	20	
Finger millet	45x15	–	4 to 8 hundred thousand plants
Wheat	30x10	100	
Barley	25-30x5	75-100	
Post-rainy sorghum	45x15	—	90,000 to 1,30,000
Pigeon pea*	50x20 (early var.) 75 to 90x25 (late)	15-20 (early var) 12-15	1,00,000 (early) 75,000 (medium) 50,000 (late)
Cowpea, mungbean and black gram	45x10 (cowpea); 30x10	15-20 (cowpea); 12-15 (mungbean); 10-12 (blackgram)	
Horse gram	30x10	15-20	
Mothbean	45x10 25x10 (for fodder)	10-12; 20-25 (for fodder)	
Chickpea	30-60x10	50	
Groundnut**	—	—	
Sesamum	Usually sown broadcast	2.5-3	
Soybean	45-60x5 30-45x5 (central & southern zones)	80-90 for variety JS-2 70-75 for JS-72-44 20-25 for local Kalitur	
Sunflower	60x22.5		60,000-80,000
Niger	Usually sown broadcast		
Castor	45-60x30	6-8	
Rapeseed-mustard	30x10-15 (Toria) 45x10-15 (Mustard)	5	
Safflower	45-90x10	8-10	
Cotton	45-60 (Arboreum) 60-75 (Hirsutum)	80-120,000 50-80,000	

* For post-rainy season pigeonpea, a plant density of 3.0 to 3.5 hundred thousand plants per hectare was found to be optimum. This could be achieved by keeping rows at 25 to 30 cm and plants at 10 cm using a seed rate of 50 kg/ha.

ing rate/plant population. Spacing and seed rate for bunch and spreading type groundnut are presented in Table 6a.

Table 6a. Spacing and seed rate of groundnut for various situations

Situation	Spacing (cm)	Seed rate (kg/ha)
1. Spanish bunch type		
a. Asured rainfall areas	30x10	100-110
b. Low rainfall areas		
(i) Coarse textured soils	30x15	70-80
(ii) Fine textured soils	37.5x10	80-90
c. Paddy fallows and post rainy areas	30x10	100-110
d. Irrigated areas (summer crop)	22.5x10	125-130
2. For spreading type		
a. Assured rainfall areas	37.5x15	80-85
b. Low rainfall areas	45x15	75-80

Weed Management

In dryland areas, weeds compete with crops for nutrients in the early stages of crop growth, and later for moisture and light. The competition for moisture becomes critical with increasing soil moisture stress. Further, for producing equal amounts of dry matter weeds transpire more water than do most of crop plants (Tripathi *et al.*, 1992).

In general, 4-6 weeks period is more critical from competition point of view in almost all the major crops. It is important to know the critical period - a time span before and after which weed competition does not reduce the crop yield, for each crop, so that weed management could be more meaningful, efficaceous and cost effective. All crops need a weed-free period for about one fourths to one thirds of their life cycle. The critical period of important dryland crops is given in Table 7.

Based on research work carried out, the following herbicidal measures are suggested for control of weeds in different crops (Table 8).

Experiments carried out at ICRISAT indicated 50 to 75 per cent reduction in weed infestation by intercropping. In intercropping systems (sorghum/cowpea; sorghum/mungbean; sorghum/pigeonpea) one hand weeding would be adequate to get as good yields as from weed-free check.

Most herbicides are crop specific. It is difficult to find the most efficaceous herbicide for different crop combinations. Butachlor and alachlor are such herbicides which gave satisfactory control of weeds in maize/mungbean,

Table 7. Critical periods of competition and percentage yield reduction.

Crop	Critical period (DAS)*	% reduction in yield	Remarks
Upland rice	10-40	60-80	Approximate % losses caused by weeds. However the extent of reduction in yield mainly depends on the types of weeds present and other factors
Maize	15-40	40-60	
Sorghum & millets	15-35	30-40	
Cotton	30-60	45-85	
Groundnut	25-55	55-70 (erect) 20-25 (spreading)	
Sunflower & Safflower	30-60	30-50	
Castor	30-60	35-40	
Sesamum	15-30	45-50	
Soybean	20-45	25-35	
Mungbean	20-45	50-55	
Pigeonpea	25-60	40-60	
Blackgram	25-35	45-50	

*DAS = Days after sowing

Source : Tripathi *et al* (1992)

Table 8. Weed management in different dryland crops

Crop	Herbicidal control measures	Remarks
Upland rice	A pre-emergence treatment of Alachlor or Butachlor followed by post-emergence treatment with Propanil or mechanical weeding	First 20 days after emergence-most critical period
Sorghum	Pre-emergence application of Atrazine or Propazine either alone or in combination @ 0.5 kg a. i/ha	Weedicide application followed by hand weeding-most effective
Pearl millet	Pre-emergence application of Atrazine @ 1.0 kg a. i/ha	
Finger millet	Pre-emergence application of Neburon @ 1 kg a. i/ha. In transplanted crop, spraying 2, 4 D @ 0.5 kg a.i /ha between 10-15 days after sowing	
Maize	Pre-emergence application of Atrazine @ 1.0 kg a.i/ha	
Wheat	Application of 2, 4 D	Additional yield of 8 to 12 q/ha obtained through effective and timely weed control
Barley	Spraying 2, 4 D @ 0.5 kg/ha combined with 3% urea in 700- 800 litres of water 40-50 days after sowing	
Pigeonpea	1. Two hand weedings (3 weeks and 7 weeks after sowing) found adequate 2. Pre-emergence application of Ronstar (0.75 kg a.i/ha) and Basalin (1.0 kg a.i/ha) is effective	
Cowpea, mungbean, black gram	Pre-emergence application of Prometryne/Nitrofen/Trifluralin @ 1.0 kg a.i/ha	
Chickpea	Pre-emergence application of Nitrofen @ 1.0 kg a/i/ha or Prometryne @ 0.5 kg a/i/ha	
Groundnut	Pre-emergence application of Alachlor (1.5 kg a.i/ha)	
Sesamum	Fluchloralin (pre-plant) or Basalin (pre-emergence) @ 1.0 kg a. i/ha.	
Soybean	Basalin @ 1.5 to 3.0 kg/ha (pre-emergence)	
Castor	Field should be kept weed-free during the period 30-60 days after sowing	

maize/cowpea, sorghum/cowpea, sorghum/blackgram, intercrop systems. In maize/groundnut intercrop system, trifluralin was found quite effective. For sorghum/pigeonpea intercropping system, ametryn and terbutyn were found equally effective. The problem of weed management in sequential crop systems is different from that of intercropping systems. Residual toxicity of herbicides applied to the previous crop assumes importance and should be considered while suggesting weed management measures.

Off-season tillage (summer tillage) keeps the soil free of weeds, besides conserving more moisture. Crop rotations with dissimilar life cycles or cultural conditions are potential means of controlling weeds under most of the dryland situations. Inclusion of legumes, particularly fodder legumes, has proved very effective in controlling weeds.

FERTILITY MANAGEMENT

In dryland areas, soil fertility management is as important as soil and moisture conservation measures. In order to get the best out of limited moisture, it is essential that nutrient requirements of dryland crops are adequately met. Highest moisture use efficiency is obtained when the crop is sown in time, has optimum plant density, and is free from weeds, diseases and pests and above all, is nutritionally well managed. Response of rainy and post rainy season crops to nitrogen and phosphorus is given in Tables 26 and 27 of the chapter– 'Agronomic Aspects of Plant Nutrient Management in Rain Dependent Food Production Systems' in India, in this book. Variations of crop response to applied fertilizer with rainfall are set out in Table 24 of the above chapter. Response of crops to nutrients is dependent on soil moisture status and soil type, a fact well depicted by data set out in Tables 31 and 29 of the above chapter, respectively. Fertilizer recommendations, based on average annual rainfall and soil moisture storage for principal rainfed rainy and post rainy season crops are given in Tables 43 and 44 of the above chapter.

In a along term study carried out on pearl-millet at Jodhpur, it was found that the grain yields obtained with 20 kg N/ha (inorganic) every year and from 40 kg N/ha/(inorganic) once in two years were about the same, suggesting higher risk cover under the former treatment. Similarly, it makes little difference in yield whether 20 t FYM is applied once in two years or 10 t FYM/ha is supplemented with 10 kg N/ha (inorganic) every year, the mean yield of the 5 year period being the same (15.1 to 15.4 q/ha). The effect of fertility treatments on the yield of pearl millet grains is shown in Table 9. Studies were conducted on the response of finger millet to farm yard manure and green manure at different levels of fertilizers at Bangalore. It is beneficial to apply 50 per cent of the recommended N in the form of organic and the remaining 50 per cent in the form of inorganics for maximizing yield at lower cost (Hedge, 1985). In the bimodal region of Bangalore, cowpea and horsegram were sown in the month of May either for grain, fodder or green manure. The legumes had a favourable effect on the yield of late sown finger millet. A horsegram green manure crop yields 2010 kg/ha on an average, whereas a single crop of finger millet sown at the same time yielded only 1670 kg/ha (Hedge, 1985). In many situations growing of sunhemp or cowpea as green manure for incorporation *in situ*, within 45 says at tender stage has been found to be effective for conservation of soil moisture, besides addition of organic matter for improving the productivity of wheat that follows. Studies have indicated that no fertilizer or less fertilizer is required if wheat succeeds a green manure legume or a grain legume (Singh, 1985). In such cropping conditions, a dose of 30 kg P_2O_5/ha is recommended for maximization of yield of wheat. Manurial trials carried out on calcareous sandy loam soils of Pusa (Bihar) for six years showed that with the application of 8000 kg of FYM/ha, the grain yield of piegeonpea increased by 11 per cent over on manure. In dryland areas of Andhra Pradesh, Maharashtra and Gujarat, where medium-late pigeonpea is usually intercropped with sorghum/pearlmillet, 5-10 cart loads of FYM/ha is quite beneficial and meets the fertilizer needs of the crop (Ali, 1985). Applica-

Table 9. Grain yield of pearl millet as influenced by fertility treatments

Treatment	Grain yield (q/ha)					Mean of 5 years
	1975 (519.8)	1976 (548.3)	1977 (229.7)	1978 (297.7)	1979 (662.3)	
Control	20.2	6.6	10.3	4.8	2.5	8.8
20 kg N/ha inorganic every year	29.4	12.2	15.9	7.7	4.5	13.9
40 kg N/ha inorganic every year	30.1	16.7	19.3	9.0	5.0	16.0
40 kg N/ha inorganic once in 2 years	30.2	8.9	20.1	6.0	5.4	14.1
10 t FYM + 10 kg N/ha every year	23.8	16.5	19.1	10.8	6.8	15.4
20 t FYM once in 2 years	26.1	13.0	17.8	10.1	8.4	15.1
40 t FYM once in 2 years	27.0	19.8	25.0	10.8	8.6	17.8

Figures in parenthesis indicate rainfall (mm) received during the crop season.
Source : Singh. *et al.* (1981)

tion of FYM @ 5-10 t/ha is beneficial to groundnut as it lessens soil hardening, facilitates peg penetration, and increases nodulation. For seasmum, application of 5 t FYM/ha is recommended. Organic manures (FYM or compost) @ 25-30 bullock cart loads per hectare may be applied to soybean, once in 2-3 years. Studies at Nagpur for a period of four years on the effects of organic manure in combination with inorganic fertilizers in the crop rotation of cotton-sorghum brought out very clearly that application of organic manure @ 9.5 t/ha to either cotton or sorghum every year or 18 t/ha every alternate year to either of the crops increased seed cotton yield by about 3 q/ha and sorghum grain by 7 q/ha (Bhatawadekar, 1985), Similar results were also reported from Surat (Gujarat) on cotton-sorghum system.

In case of rape seed-mustard, advancing the time of application of fertilizers has shown striking advantage during two out of three years (Table 10). During the year 1980, the response due to advance application was not observed because of early withdrawal of monsoon.

Table 10. Effect of advancing the time of fertilizer application in mustard at Hissar

Time of fertilizer application	Grain yield (q/ha)			
	1979	1980	1981	Average
At sowing time	11.9	13.6	14.9	13.4
15 days before sowing	18.0	13.0	17.2	16.3

Beneficial effects of advancing the time of application of fertilizers have been obtained in case of post-rainy season sorghum and barley.

RELATIVE CONTRIBUTION OF PRODUCTION FACTORS

Relative contribution of production factors in case of different dryland crops is given in Table 11. In case of cereals (upland rice, sorghum, pearl millet, maize, finger millet) and pulse crops cow-

Table 11. Relative contribution of production factors

Crop	Relative contribution of production factors	Remarks
Upland rice	Improved genotype (26.8%), followed by fertilizer (17.6%)	Timely and proper weed control
Sorghum	Contributions of improved seed and plant protection are 7.0 and 6.2% respectively	
Pearl millet	Variety followed by improved management and fertilizer use	
Maize	Improved variety alone contributed only 1.8 q/ha of grains/ha, whereas fertiliser alone contributed 14.3 q/ha; a combination of both gave 17.7 q/ha higher yield. Efficiency of these inputs was further increased when management practices like land treatment and placement of fertilizer were followed	
Finger millet	Adoption of improved variety in conjunction with recommended level of fertilisers (50 kg N, 50 Kg P_2O_5 and 25 kg K_2O/ha) appear to be key factors for higher production	
Sorghum (post rainy season)	Early September seeding contributed about 37% increase in yield, fertiliser use (50 kg N/ha) contributed maximum (81%) and improved management contributed 21%	
Pigeonpea	Weed management was found to be most important, followed by fertiliser use	
Cowpea, mungbean and blackgram	Improved seed contributed 20 to 39%, fertiliser use 15 to 25%	
Mothbean	Improved variety gave 40% more yield	
Chickpea	Improved variety 10%, fertiliser use 10%, weed management 6%	
Niger	Maximum contribution (37.5%) comes from varieties, followed by fertilisers (33.9%) and plant protection measures (19.5%)	
Rape seed mustard	Pest control measures and fertiliser use assume great importance in increasing productivity	
Safflower	Application of fertilisers turned out to be far more critical than any other factor	
Cotton	Plant protection followed by fertiliser use	

pea, mungbean, black gram, mothbean, chickpea, and niger, maximum contribution to higher productivity was made by improved varieties, followed by fertilizer use. In case of pigeonpea, weed management was found to be most important, followed by fertilizer use. For rape seed-mustard and cotton, plant protection was of paramount importance, followed by fertilizer use. In safflower, application of fertilizers turned out to be far more critical than any other production factor. For post-rainy season sorghum, early September seeding was one single critical factor which made the maximum contribution to increased productivity. Judicious fertilizer use and improved management were necessary to make the best use of timely sowing.

CONSTRAINTS TO HIGHER PRODUCTIVITY

These are set out crop-wise in Table 12

Table 12. Crop-wise constraints to higher productivity

Crop	Constraints to higher productivity*
Upland rice	Moisture stress, weed competition, poor soil fertility, and inadequate control of pests and diseases
Sorghum	Poor germination and emergence, shoot fly (late sown), charcoal rot (moisture stress) and striga
Pearl millet	Inadequate crop stand, use of longer duration varieties, poor weed and pest management, lack of fertilizer use
Maize	Lack of high yielding hybrids/composites with shorter duration; downy mildew and leaf spots
Finger millet	Broadcasting the seeds, low level of fertilizer use, blast disease
Wheat	Delayed seeding and poor crop management
Barley	Cultivation on marginal and problematic soils
Sorghum (post rainy)	Lack of soil moisture, cropping on shallow soils, lack of fertiliser use
Pigeon pea	Moisture and nutrient stress, vulnerable to large number of pests and diseases, lack of availability of seeds of improved varieties and efficient rhizobium culture
Chickpea	Lack of moisture, cultivation pushed to low soil fertility soils, pod borer
Other pulse crops	Moisture and nutrient stress, suboptimal plant density, weeds, pests and diseases
Groundnut	Terminal drought, cultivation on less productive and marginal soils, gappy crop stands, deficiency of Ca, S and Zn; inadequate plant protection
Sesamum	Poor crop stand, moisture stress in later stage, poor crop management
Soybean	Non-availability of good quality seed of improved varieties, use of low seed rate, low fertiliser use, inadequate surface drainage
Sunflower	Poor seed setting
Niger	Grown on marginal/submarginal lands, lack of fertilizer use and plant protection measures
Castor	Inadequate availability of pure seeds of improved genotypes, root wilt.
Rapeseed-mustard	In north India, frost, aphids, and low rainfall during post-rainy season; diseases, poor crop stand
Safflower	Precarious moisture conditions and poor management
Cotton	Soil moisture limitation, cultivation extended to shallow soils of poor moisture storage and soil fertility, lack of plant protection measures

*As identified by Scientists associated with the above crops. Constraints have been listed in the order perceived by the scientists.

REFERENCES

Ali, Masood, 1985. Pigeonpea. In : *Management of Dryland Crops.* (Eds. V. Balasubramanian and J.Venkateswarlu), Central Research Institute for Dryland Agriculture, Hyderabad, A.P. 143-168.

Bhatawadekar, P.U. 1985. Cotton. In : *Management of Dryland Crops.* (Eds. V. Balasubramanian and J. Venkateswarlu). Central Research Institute for Dryland Agriculture, Hyderabad, A.P., pp. 355-372.

Hegde, B.R. and Seetharam, A. 1985. Finger Millet and other Minor Millets, In : *Management of Dryland Crops.* (Eds. V. Balasubramanian and J. Venkateswarlu), Central Research Institute for Dryland Agriculture, Hyderabad, A.P., pp. 74-100.

Singh, R.P., Singh, H.P., Daulay, H.S. and Singh, KC. 1981. Effect of periodical application of nitrogen in organic and inorganic form on the yield of rainfed pearl millet *Ind. J.* agric. *Sci. 51*, 409- 416.

Singh, R.P., 1985. Peall millet. In : *Management of Dryland Crops.*, (Eds. V. Balasubramanian and J. Venkateswarlu), Central Research Institute for Dryland Agriculture, Hyderabad, A.P., pp. 51-62.

Tripathi, B., Rao, A.N. and Ramakrishna, A., 1992. Weed Management in dryland crops and cropping systems. In : *Dryland Agriculture in India - State of Art of Research in India.* (Ed. L.L., Somani), Scientific Publishers, Jodhpur, pp. 185-210.

21

CROP MANAGEMENT OPTIONS FOR ABERRANT WEATHER SITUATIONS AND AGRICULTURAL DROUGHT

R.P. SINGH

INTRODUCTION

Aberrant weather is a common feature of rainfed agriculture. Crop production in dry areas is mainly dependent on rainwater, a major portion of which is received during the southwest monsoon. The rainfall is seasonal, erratic and highly variable with space and time. Vagaries of monsoon rains are proverbial in India, 'breaks' of varying duration being most common. Three important aberrations in the rainfall behaviour have been more commonly observed. These are : (1) the commencement of rains may be quite early or considerably delayed, (2) there may be prolonged 'breaks' during the southwest monsoon season during which most of the dryland crops are grown, and (3) rains may terminate earlier than normal cessation date or may continue beyond the normal rainy season.

Anomalies in the distribution of south-west monsoon rainfall are well known. Based on the analysis of historical rainfall records, Singh (1987), Ramana Rao (1988) have brought out the following main features of rainfall distribution in India.

1. There is large variation in the dates of commencement of the southwest monsoon rainfall from year to year in different parts of the country.
2. The monsoon rainfall is of sequential nature with dry spells interspersed with wet spells, dry spells extending from few days to several weeks.
3. There is large year to year variations in the dates of cessation of southwest monsoon rainfall in different parts of the country.
4. There is variation in the quantum of rainfall received from year to year and the coefficient of variation of rainfall increases with decrease in monsoon rainfall.
5. High intensity rainfall occurs in association with movement of cyclones/depressions resulting in sizeable loss of rain water through run-off and deep drainage.

In view of the above, there is nothing like a normal rainfall year in monsoonal India, considering the long term historical data as a guide. Even if the rainfall (quantum and distribution) happens to be near normal during the cropping season, there could be prolonged cloudy weather, thereby reducing solar radiation available to crops. In some parts of the country, temperatures may be abnormally high or go up abruptly, thereby leading to pre-mature ripening of crops. There could be more windy days than normal and stronger winds than normally encountered. Thus, highly variable rainfall coupled with wide differences in temperature, sunshine, and wind velocity present a wide spectrum of agro-ecological situations in the country.

Weather is the most variable factor in rainfed crop production. It is the one which can be least controlled. However, various approaches, often referred to as contingent plans, have been evolved over the years, for efficient weather management. The infrastructural requirements have to be developed and established by the Government and other agencies as people themselves will not be in a position to adopt contingency measures more effectively because of their poor resource base. Some of the management options that can be adopted to mellow down the adverse effects of aberrant weather are discussed in this chapter.

LATE ONSET OF MONSOON

During some of the years the onset of the southwest monsoon gets delayed so that the crops/varieties which are regularly grown in the region can not be sown in time. Delayed sowing of the crops can lead to reduced and even uneconomical crop yields. Under these circumstances, two management options are available.

Transplantation

In case of some of the crops, community nursery has to be raised at a point where water is available and transplantation has to be undertaken with the receipt of rains. This is a surer way of compensating the delay in the commencement of sowing rains, though transplantation is a labour intensive operation, Extensive studies carried out at Bangalore (Hegde and Seetharam, 1985) indicated that both long and medium duration varieties of finger millet should be transplanted for maintaining yield levels whenever sowing is delayed. Finger millet transplanted at Bangalore during the month of September yielded as high as 25.6 q/ha whereas the directly sown crop yielded 7.5 q/ha only. Under delayed conditions, transplanting of pearl millet has been found to be advantageous as the transplanted crop give significantly more grain yield than the direct seeded crop (Tomer *et al.* 1974).

Alternate Crops/Varieties

Certain crops are more efficient either due to the shorter duration or capacity to produce better yields even under relatively unfavourable moisture regimes when sown late in the season. Therefore, crops/varieties have to be chosen depending upon the date of occurrence of sowing rains. For example, it is now well established that castor (var : Aruna) produces more yield in the red soils of Telangana region compared to pigeonpea under late sown conditions.

Table 1. Choice of crop varieties for normal and late sown conditions

Centre	Crop	Choice of crop varieties for different sowing times	
		Normal	Late
Akola	Sorghum	CSH.5/CSH.9	CSH.1
Anantapur	Groundnut	Kadiri 1	TMV.2 or Kadiri.3
Bangalore	Finger millet	Indaf.1 or Indaf.3	PR.202 (Godavari)
Bellary	Sorghum	SPV-86 or CSH.7R or CSH.8R	M-35-1
Bhubaneswar	Rice	Ratna or Kumar	DR-92
Dantiwada	Sorghum	CSH.5 or CSH.6 for grain	S.1049 or Malwan for fodder
Dehra Dun	Maize	G.5 or G.2 or Vikram	Ageti
Hissar	Greengram	Varsha	S.9
Hoshiarpur	Wheat	WL.410	C.306
Jhansi	Sorghum	CSH.5	CSH.6, SPV.224
Jodhpur	Greengram	S-8	S.9
	Cowpea	FS.68	K.11
Udaipur	Maize	Ganga.5	"Bassi selected", Satha
	Sorghum	CSH.5	CSH.6
Varanasi	Greengram	Varsha or K-851	T.44

Source : Venkateswarlu *et al.* (1991).

In western Rajasthan, short duration crops like greengram and cowpea were found to perform better than pearl millet under late sown conditions. The crops and varieties that can be sown depending upon the date of sowing have been identified for dry farming tracts of India.

Choice of crop varieties suitable for normal and late sown conditions are given in Table 1.

DRY SPELL IMMEDIATELY AFTER SOWING

During some of the years, a dry spell may occur immediately after sowing the crop. This may result in poor germination due to soil crusting, withering of seedlings and poor establishment of crop stand. It is always necessary to maintain proper plant stand to ensure better yields. Therefore, if a severe dry spell occurs immediately after sowing, it is better to resow the crop than to continue with inadequate plant stand (Table 2).

Table 2. Performance of pearl millet (Var : MBH-110) in relation to occurrence of early season drought during 1986 at Solapur, Maharashtra.

Date of sowing	Plant population		Grain yield (q/ha)	Fooder yield (q/ha)
	Initial	At harvest		
28th June	1,50,000	21,990	3.7	10.9
15th Aug.	1,50,000	1,11,806	6.2	20.1

Table 2 gives a good illustration on the advantage gained by resowing pearl millet (var : MBH-110) after a severe dry spell immediately after sowing at Solapur during kharif, 1986. If the drought conditions occur at a very early stage, it is better to resow with subsequent rains rather than allowing sub-optimal plant stands to persist and yield a poor harvest.

BREAK IN MONSOON-MID OR LATE SEASON DROUGHT

Sowing of alternate crops/varieties as the monsoon progresses is now an established approach to minimise the risk of total crop failure of traditional crops/varieties. But once the crop is growing, there could be moisture stress leading to mid-season drought.

The water requirement of the crop depends upon the solar radiation intercepted by the crop canopy. If the drought conditions occur 40 to 50 days after sowing, the leaf area development will be maximum leading to fast depletion of soil moisture reserve. Therefore, reduction in leaf area index by either ratooning or thinning of the crop mitigates the ill effects of drought to a certain extent. Ratooning of a drought affected sorghum crop with a subsequent rain gave 8 q/ha of grain yield over 2 q/ha without ratooning. Similarly, pearl millet gave 25.1 q/ha with ratooning, while non-ratooned crop gave 15 q/ha. Ratooning is a high management technique and success depends on the general vigour of the drought affected crop.

For indeterminate crops like castor, pigeonpea etc., if the drought affected plants were seen to recoup, 2 per cent urea spray was found useful after the foliage was wetted with the rains.

Mulching to Extend Period of Water Availability

If the break in monsoon is very brief, soil mulching was found to be a tool in extending the period of storage of water in the soil profile due to reduced heat load on the soil and hence the evaporation losses which, in turn, leads to extended periods of water availability. Some examples on the advantages of the mulching on grain yield are shown in Table 3.

Weed Control to Improve Water Availability to the Crops

Weed control is very essential to save the crop from onslaught of drought as weeds rob soil moisture and nutrients as well. The importance of weed control in dry farming can be best illustrated from some of the examples given in Table 4.

Rain Water Harvesting and Recycling for Protective Irrigation

In situ water harvesting and/or runoff recy-

Table 3. Effect of mulches on crop productivity in drylands.

Station	Crop	Mulch adopted	Productivity (q/ha)	
			With mulch	Without mulch
Hoshiarpur	Maize	Basooti mulch	42.0	30.9
	Wheat	Basooti mulch	30.0	21.2
Rewa	Wheat	Paddy straw	17.2	11.5
Varanasi	Maize	Wheat straw	5.7	4.7
Akola	Cotton	Vertical mulch	10.3	8.0

Table 4. Effect of weed control on increase in crop productivity under rainfed conditions

Station	Crop	Weed control method	Productivity (q/ha)	
			With weed control	Without weed control
Akola	Sorghum	Two manual weedings	22.9	17.1
Rewa	Paddy	Two manual weedings	10.9	6.2
Agra	Pearl millet	Weeding after 30 DAS	6.9	4.0
Ranchi	Paddy	Manual weeding at 10, 20 and 30 DAS	20.3	11.3
	Ground-nut	Manual weeding at 15 and 25 DAS	18.4	8.9
Jodhpur	Pearl-millet	Manual weeding at 20 and 40 DAS	0.9	3.6

cling are the potent measures for crop life saving during the periods of moisture stress. There is no control over the receipt of rain water but it can be managed in a way better what is being done at present after it is received. Supplementary irrigation provided to the crop through rain water harvesting and recycling particularly during the grain formation stage will be very much rewarding as illustrated with few examples in Table 5.

Run-off recycling holds immense prospects in deep black soils where the seepage losses are very much less. High intensity rains are quite common during the rainy season contributing 75

Table 5. Effect of critical irrigation (5 cm) on yield of crops

Crop	Station	No.of trials	Yield (q/ha)		Efficiency of water use (kg/ha/mm)
			Without irrigation	With irrigation	
Wheat	Dehra Dun	4	21.4	35.5	28.2
Wheat	Ludhiana	4	19.2	41.1	43.8
Wheat	Agra	2	21.9	27.4	11.0
Wheat	Rewa	4	5.7	18.8	26.2
Barley	Varanasi	2	26.0	33.6	15.2
Sorghum	Bijapur	5	16.5	23.6	14.8
Sorghum	Bellary	4	4.3	13.7	18.2
Upland rice	Rewa	4	16.2	27.8	23.2
Sorghum	Solapur	5	9.8	18.2	16.8
Native tobacco	Anand	4	12.1	18.1	12.0

per cent of seasonal rainfall. The inevitable run-off (10 to 40 per cent of the total rainfall) could be stored in dug out ponds of appropriate sizes and recycled to the donor area, in the event of severe moisture stress.

Land Configurations for Better Conservation

Appropriate land configurations like broad beds and furrows, graded border strips, inter-row and inter-plot water harvesting systems etc. hold great promise for *in situ* moisture conservation. The productivity of sorghum under different land treatments at Kovilpatti serves as a good example for this purpose (Table 6).

Table 6. Effect of land treatments on productivity of sorghum in vertisols at Kovilpatti

Station	Land treatment	Productivity of sorghum (q/ha)
Kovilpatti	Flat bed system	6.9
	Broad-bed and furrow system	10.5
	Ridge and furrow system	12.7
	Compartmental bunding	10.0

For shallow alfisols and related soils even dead furrows serve this purpose very well as illustrated in Table 7.

Intercropping and Risk Distribution

Breaks in the monsoon rains can be of different durations. Breaks of shorter duration like 5 to 10 days may not be of serious concern. But prolonged breaks of more than two weeks can create plant water stress leading to low productivity of crops. These breaks can not be predicted in advance. The agricultural droughts caused due to prolonged breaks in the monsoon rains can be of different magnitudes and severity and affect different crops in varying degrees. Application of meteorological information in terms of the frequency and probability of these breaks can be made to select a combination of crops of different durations in such a way that there is time lag in the occurance of their growth for appropriate intercropping systems.

Systems of intercropping are age old practices and in a way cover the principle of risk distribution agronomy with particular reference to droughts. Areas with a growing season of 20 to 30 weeks are generally most suited for intercropping systems. These areas receive 625 to 800 mm rainfall with stored water being 15 to 20 cm in the root profile. Venkateswarlu (1991) indicated that the choice of the intercropping system should depend on the pattern of rainfall and indicated three possibilities as stated below :

(i) Rainfall more uncertain in the early part of the season,

(ii) Rainfall more uncertain in the later part of the season, and

(iii) Rainfall more or less uniformly distributed (in other words, the risk is uniformly distributed).

For areas with uncertain rains in the early part of the season (e.g. Bangalore region) early planting of deep rooted drought tolerant crops like pigeonpea might be useful followed by planting of other component crop. But the farmers may not accept such a proposition of sowing twice. Then the high yielding systems of

Table 7. Influence of land treatments on the yield of crops in alfisols at Mittimari Watershed

Land treatment	Productivity (q/ha) of Groundnut 1986	Groundnut 1987	Pigeonpea 1986	Pigeonpea 1987
Farmers practice	8.6	11.4	7.3	4.2
Furrows at 3.3 m interval on 0.2% grade	10.0	12.0	8.1	5.4
Contour trench at 15 m interval	11.7	13.5	8.8	6.5
Contour sowing	10.9	12.6	9.0	5.9

suitable crops need be compared to the modified intercropping system. For areas with uncertain rainfall in the later part of the season, the companion crop should be shorter in duration than the base crop (e.g. sorghum - green-gram). In areas where the rainfall is more or less uniformly distributed, it would be ideal to take suitable base crop and a companion crop of either longer or shorter duration depending on the growing season. In each of these cases it would be always better to develop the system over the best base crop of the region. On these principles, different intercropping systems have been tested and some of the suitable systems for various regions are given in Table 8.

Every soil has a capacity to hold rainfall received and supply water to plants. The water availability in the soil depends upon the water holding capacity of the soil. Deep black soils have a capacity to store as much as 300 mm of water in one meter depth while the desert soils can store only as little as 100 mm for one meter depth. Therefore, the impact of the drought will be more pronounced in the shallow soils compared to deep soils. While deciding the drought management options, the soil factor has to be adequately considered.

Fertilizer use helps the crop to guard against drought by encouraging development of a root system which will utilize soil water more efficiently.

PERMANENT DROUGHTS

Drought is a recurring feature in the arid regions with variation in its intensity only from year to year. In these regions, the average rainfall may exceed potential evapotranspiration for a period of not more than four weeks. Even the drought resistant crops grown in these regions are likely to be subjected to moisture stress even during the years with above normal rainfall. Alternate land use systems have to be introduced in these regions for sustainability of production and stability of income.

In the black soil regions of Deccan Plateau, the farmers generally sow *rabi* crops like sorghum, safflower, chickpea etc., after the cessation of the rainy season during the month of October. These soils start cracking up by the middle of December and as such the yield of *rabi* crops are likely to be adversely affected. It is now established that early sowing of *rabi* crops can improve the yield substantially in black soil regions of peninsular India. The choice of the crop has to be decided based on the available soil moisture in the soil profile. In the black soil region of Maharashtra and Karnataka wherever wheat is grown under conserved moisture conditions, crops like chickpea and safflower can bring manifold returns to the farmers. *Rabi* sorghum

Table 8. Some suitable intercrops for different dry farming regions

Region	System	Productivity of base crop (q/ha)	Productivity of intercropping system (q/ha)	
			Base crop	Companion crop
Hisar	Pearlmillet + cowpea	21.0	13.1	7.4
	Pearlmillet + greengram	21.0	16.4	5.2
Agra	Pearlmillet + greengram	9.1	8.4	3.3
Bangalore	Pigeonpea + cowpea	30.4	27.2	2.5
Dantiwada	Pearlmillet + cowpea	2.7	2.7	2.2
Hyderabad	Sorghum + pigeonpea	34.4	33.5	5.5
Akola	Sorghum + greengram	33.5	30.8	7.3
Ranchi	Pigeonpea + rice	14.9	9.3	18.7

was found to be advantageous over cotton in the black soils of Bellary region. In the event of subsequent failure of rains in these tracts, thinning of plant population can remarkably improve the crop yields as illustrated in Table 9.

(d) early weed control;
(e) application of phosphorus and nitrogen (two-thirds at sowing time); and
(f) water harvesting and recycling.

Table 9. Effect of thinning on yield of rabi sorghum (Bellary)

Thinning system	Grain yield (q/ha)			
	1974-75 (Subnormal)	1975-76 (Above normal)	1977-78 (Sub normal)	1978-79 (Above normal)
Original population	15.5	24.2	5.4	24.9
Third row removed	21.1	20.2	9.2	23.3

CROP MANAGEMENT STRATEGIES FOR LATE KHARIF AND EARLY RABI CROPS

The success of late *kharif* and early *rabi* crops will depend on :

(a) selection of suitable crops;
(b) appropriate varieties that fit into the short growing season;
(c) optimum plant stand;

The crops that can be sown in late August and early September are : Setaria, fodder crops, sunflower, toria, castor, niger, sesamum (Purva-1) in the north western states.

For obtaining optimum plant stand, it is necessary that seeds are sown precisely in the moist zone and appropriate measures for termite control are taken. Use of seed-cum-fertiliser drill and application of Aldrin for termite control are absolutely necessary.

Selection of alternate crops and varieties

(For North and North-Western India-Gujarat, Rajasthan, Haryana and Punjab states).

The following crops and varieties may by preferred for late *kharif* and early *rabi* situations :

STATE	CROPS AND VARIETIES	
Gujarat	Sesamum	— Purva 1
	Castor	— Gauch-1, GCH-2, SHB-18
	Sunflower	— BSH-1, Morden
	Clusterbean	— Malosan
	Sorghum fodder	— S 1049
	Tobacco	— Gujarat-1
	Fodder crops	— *Bajra* + cowpeas
Western Rajasthan	Sunflower	— BSH-1, Morden
	Clusterbean	— HFG-75, RGC-471
	Mothbean	— JMM-259, Jadia, Jwala
	Fodder crops	— *Bajra* + moth
	Taramira	— T-27
Haryana	Clusterbean	— HFG-75, RGC-471
	Mothbean	— JMM-259, Jadia, Jwala
	Toria	— Sangam, PT 303, ITSA
	Fodder crops	— *Bajra* + moth
Punjab	Setaria	— Local
	Toria	— Sangam, PT 303, PL-15
	Taramira	— T-27
	Fodder crops	— Maize + cowpea

The following agronomic practices have to be followed for the crops and varieties mentioned above :

(i) Use higher seed rate than normal. It would be preferable to use about 20 per cent higher seed rate.

(ii) BHC (10%) application is a must in light textured soils.

(iii) Moderate quantities of nutrients (20 kg N and 20 to 30 Kg P/ha) will have to be used at sowing time to impart seedling vigour for over coming short dry spells.

(iv) Early and effective weed control (within 20 days of sowing) will have to be done to remove competition for limited water and nutrients.

(v) A thorough surveillence for pests and diseases with timely plant protection will have to be undertaken.

(vi) In the event of further aberrant situations, the crops may have to be harvested for fodder and the land prepared and kept ready for late rabi crops.

LOCATION SPECIFIC STRATEGIES FOR COMBATING WEATHER ABERRATIONS :

Two notes one for the Bangalore region pertaining to contingent crop planning for delayed onset of monsoon, and the other for Solapur region (Maharashtra) regarding strategies for abberant weather, are presented here. The Bangalore region represents red soils (alfisols and related soils), while the Solapur region represents black soils (vertisols and vertic inceptisols). These notes were prepared by the respective chief Scientitist of the above centres and the same are reproduced below.

Contingent Crop Planning For Delayed Onset of Monsoon-Bangalore Region

The Bangalore, Kolar and parts of Tumkur districts receive an average annual rainfall of 768 mm with monthly distribution of rainfall as follows :

January	6.2 mm	May	93.0 mm
February	6.9 mm	June	63.0 mm
March	9.6 mm	July	82.3 mm
April	38.0 mm	August	110.0 mm

September	146.6 mm
October	133.9 mm
November	66.2 mm
December	12.5 mm

Eventhough the main cropping season starts in July and most of the sowings are completed by the beginning of August, pre -monsoon thunder showers received in April/May are extensively utilized for raising short duration catch crops like cowpea and sesamum. Long duration crops like pigeonpea and castor may also be sown in May.

Delayed onset of monsoon may affect the agricultural activities in the following ways :

1. Delay in land preparation,
2. Delay in sowing of crops,
3. Inadequate plant population due to insufficient moisture, and
4. Increased incidence of pests and diseases.

Though it is difficult to predict the exact nature of the deficiency and its effect on agricultural operations, some type of preparedness is possible to reduce the impact of the delayed onset of rainfall.

Delay in land preparation.

If pre-monsoon thunder showers fail in April-May, advance preparation of the land becomes difficult and early crops like cowpea, sesamum and groundnut cannot be sown. When the rains set later in June-July there will be very little time to catchup with the season. Therefore, all efforts should be made to mobilize the power available for primary tillage including the tractor power for land preparation taking the advantage of rains received in June.

(i) If the rains are not received till the end of May, no attempts should be made to sow early crops as there will be no time for taking second crop after harvesting the first one.

(ii) Sowing of pigeonpea in wider rows of 90 to 120 cm, keeping a provision for transplanting finger millet later in August can be taken up till the end of June. Paired row sowing of pigeonpea at 3.3 m interval is another option available.

Delay in sowing of crops

When normal season sowings are not possible, crop and varietal changes are the major strategies to reduce the ill effects of delayed rainfall.

(i) Sow pigeonpea (TTB-7), groundnut (JL-24) and maize (DH-103) in July. Avoid sowing of these crops if season is not favourable till the end of July.

(ii) Select downy mildew resistant maize variety Ganga-11 especially when sowing is to be takenup towards the end of July.

(iii) Sow long duration *ragi* varieties Indaf-8 and MR-1 till end of July, use PR-202, Indaf-5 or Indaf-9 for August sowing.

(iv) Staggered *ragi* nursery may be prepared in July-August, so that transplanting of seedlings could be adopted in later part of August or early September.

(v) When rains are received only in August, sow sunflower (variety Morden).

(vi) Sow cowpea (TVX) during August second fortnight and horsegram (KBH-1) during September.

(vii) Fodder crops like *Jowar* (Local), maize (South African tall) with fodder cowpea, barnyard millet and *Pennisetum pedicellatum* can be sown in September when acute shortage of fodder demands some area to be kept aside for fodder production.

(viii) Dry seed medium or short duration *ragi* if rains are not received till the beginning of August as September rains are assured in the region.

(ix) Sowing of any crop after first week of October is not feasible unless protective irrigations is possible. However, seedlings of chilli and tomato may be transplanted.

Inadequate population due to in-sufficient moisture

When rains are received in smaller intensities and amounts, there may be gaps in germination, or the seedlings that have established might die subsequently. Under such situation either gap filling or resowing may become essential.

(i) Gap filling with seedlings is better than resowing in case of *ragi*. Seelings from areas with higher population may be removed and planted in areas with lower population.

(ii) Nursery should be raised utilizing either farm pond water or any other source of available water during August so that the gaps could be filled by transplanting seedlings.

(iii) Alternate crop seedlings like chillies and tomato to be kept ready in staggered nurseries.

(iv) Gap fill the fields of groundnut and *ragi* with the seeds of cowpea, horsegram, niger or sunflower in the event the gaps are more than 25 per cent in the field.

General strategies to overcome ill effect of drought.

(i) Frequent intercultivation to keep the weeds under control and reduce evaporation losses.

(ii) Mulching with organic residues in case of transplanted crops like chillies.

(iii) Providing protective irrigation whenever the soil moisture conditions become critical.

(iv) Removing unhealthy seedlings to reduce population.

(v) Top dressing with nitrogenous fertilizer on alleviation of the drought when the rain revives.

(vi) Providing effective plant protection

measures. Aphids in several crops, leaf minor in groundnut, soybean and pigeonpea become predominant when the dry spell persists.

(vii) Harvesting of crops like maize for fodder leaving the intercrop component pigeonpea for full maturity.

There is a practice of removing top portion of maize plant (above the cobs) after milky stage is maize prevailing with farmers.

Alternate crop strategies to face the aberrant weather situations demand the timely availability of required types of seeds and inputs. Government machinery should be geared up for this purpose. Similarly, one of the important strategies to combat the drought effects is to provide protective irrigation. Therefore, developing water resources by excavation of farm ponds or construction of gully embankments become essential.

Technologies and strategies for aberrant weather - solapur region

The production of field crops suffer from instability due to aberration in weather conditions from time to time. The common weather aberrations experienced are delayed onset of monsoon, early withdrawal of monsoon and dry spells during the rainy months (monsoon). Under these situations, crop production suffers heavily. For example, delayed onset of monsoon results in non-sowing of traditional *kharif* soils which account for nearly 25 to 30 per cent of the total area under crops. Further production of pulses and oil seeds suffers to a great extent as major pulse production (redgram, greengram, blackgram, mothbean, horesgram etc.) and oil seed production (groundnut, sesamum) comes from *kharif* harvest. Further, these soils are shallow and moisture storage capacities are rather limited. Production potentials of these soils during *rabi* are also poor.

Early withdrawal of monsoon interferes with sowing of *rabi* crops which is the mainstay of crop production in this region. Similarly, breaks in monsoon also affect the crop production adversely. Appropriate technology is developed to combat all these weather aberrations.

Analysis of rainfall data for 108 years at Solapur revealed that these situations described have occurred during 1971, 1974, 1976, 1981, 1982, 1985 and 1986 in the recent years.

Delayed onset of monsoon

Even though monsoon is expected to be arrived by mid-june, such situation is experienced very rarely. In the recent past, since 1981 it was possible to sow the crops by June except during the years 1984-85 and 1986-87. Fluctuations in the arrival schedule of monsoon and further optimum conditions to take up sowing are found to vary from late July to late August. During 1974, it was possible to sow crops by early August. Under these circumstances, normal crop pattern requires to be modified as the traditional crops are found to suffer heavily from either disease-pest or very low production. In general, farmers do not take up sowing as late as end of July or early August. Experience with alternate crops is very encouraging (Table 10). Based on these experiences, suitable cropping schedule is recommended (Table 11).

In addition, recent experience gathered from the experiment conducted for contingency planning during the year 1988-89 to 1991-92 revealed that sunflower, pearlmillet and horesgram yielded better as contingent crops in scarcity areas, whereas mothbean failed completely due to excess rainfall (Table 12). Sunflower proved to be the best crop for delayed onset of monsoon followed by pearlmillet. It also sustained well under dry spell during the year 1991. Sunflower (SS-56 and Morden), pearlmillet (ICMH-88951) and (HHB-67), horesgram (Sina) and mothbean (Local) were observed to be better as contingent crops for scarcity area of Maharashtra State.

Good start of monsoon followed by dry spells

This situation is also very common. Usually one or two dry spells are experienced during *kharif* crop season. If the dry spell exceeds two weeks (Table 13) corrective measures are re-

Table 10. Average grain yields (q/ha) of various crops under mid-season correction trial sown at fortnightly intervals during 1974-1980.

Fortnight	Cereals				Oil Seeds			Pulses		
	Pearl millet	Setaria	Sorghum CSH-1	Sorghum CSH-6	Ground-nut	Sunflower	Castor	Pigeonpea	Horse gram	Green gram
June-II	17.5 (2)	19.9 (2)	17.5 (2)	*	10.4 (2)	9.3 (2)	17.0 (2)	5.4 (2)	8.8 (2)	6.0 (1)
July-I	19.1 (5)	22.7 (5)	23.6 (5)	24.9 (1)	7.1 (5)	10.3 (5)	20.1 (5)	5.6 (5)	11.9 (2)	2.2 (1)
July-II	9.6 (5)	12.2 (5)	15.1 (5)	12.3 (1)	5.9 (5)	8.9 (5)	14.8 (5)	9.4 (5)	13.0 (2)	1.8 (3)
Aug.-I	5.5 (5)	7.4 (5)	5.4 (1)	6.0 (5)	3.1 (5)	4.5 (5)	10.2 (5)	6.5 (5)	6.2 (2)	0.5 (5)
Aug.-II	4.0 (5)	4.4 (5)	2.7 (5)	*	2.5 (5)	4.2 (5)	10.8 (5)	5.4 (5)	4.0 (2)	Failed (2)
Sept.-I	0.4 (4)	5.8 (4)	3.3 (1)	*	4.6 (4)	2.7 (4)	6.0 (4)	3.2 (4)	4.4 (2)	Failed (2)

Figures in parenthesis indicate number of years.
* Not tried.

Table 11. Cropping schedule for late sowing

Rainfall situation	Suggested crops
(1) Rains during July and sowing of *kharif* crops by the end of July or early August	Setaria (Arjun), redgram (No. 148), Sunflower (EC-69874), castor (NPH-I), horsegram.
(2) Rains during August and sowing upto August end	Redgram, sunflower and castor
(3) Rains during late August and sowing late August to early September	Castor

N.B. : Varieties as above.

Table 12. Grain and stover yield (q/ha) and monetary returns (Rs/ha) as affected by different treatments.

Crop and sowing date	1988		1989		1990		1991	
	Grain	Monetary returns	Grain	Monetary returns	Grain	Monetary returns	Grain	Monetary returns
Horsegram								
15th July	7.36 (21.55)	2186	9.92 (28.92)	3125	8.29 (27.67)	3045	7.26 (24.94)	2659
25th July	5.12 (14.41	1549	12.81 (19.31)	3462	8.09 (30.50)	3140	6.70 (23.76)	3055
Mothbean								
15th July	– (16.85)	338	– (7.81)	156	1.39 (12.72)	1093	2.86 (18.26)	2803
25th July	– (19.77)	393	– (5.50)	108	2.65 (17.84)	4151	3.92 (20.57)	3724
Sunflower								
15th July	2.54 (21.00)	1742	14.96 (14.40)	9125	13.47 (15.17)	2956	9.38 (24.17)	10566
25th July	0.64 (12.75)	516	8.93 (20.85)	5579	8.22 (18.00)	6764	7.17 (19.54)	8148
Pearlmillet								
15th July	17.55 (30.00)	4690	26.29 (22.89)	6809	11.73 (19.28)	3042	13.89 (16.72)	4372
15th July	4.40 (24.00)	1344	16.88 (25.97)	4516	11.59 (20.83)	2626	13.36 (24.94)	3951

Figures in parenthesis indicate the stover yield.

Table 13. Start and ceasation of rainfall and dry spell at Solapur

Year	Date of start of monsoon	Date of cessation of monsoon	Dry spell during rainy season Period	Days
1981	1st June	5th Oct.	23rd June – 22 July	30
			6th Aug. – 21st Aug.	16
			6th Oct. – 31st Oct.	26
1982	1st June	30th Oct.	25th June – 8th July	14
			20th July – 7th Aug.	19
			9th Aug. – 10th Sept.	33
			28th Sept. – 26th Oct.	29
1983	3rd June	29th Oct.	26th June – 4th July	16
			15th Aug. – 11th Sept.	28
1984	5th June	23rd Oct.	17th June – 14th July	28
			10th Aug. – 10th Sept.	32
1985	3rd June	9th Oct.	19th June – 20th July	32
1986	2nd June	27th Sept.	29th June – 31st July	33
			14th Aug. – 7th Sept.	25
			28th Sept. – 31st Oct.	33
1987	5th June	19th Oct.	6th June – 24th June	19
1988	4th June	30th Sept.	9th June – 20th June	12
			12th July – 26th July	15
1989	6th June	2nd Oct.	15th June – 25th June	11
			25th July – 12th Aug.	19
1990	5th June	28th Oct.	7th Sept. – 23rd Sept.	16
			10th Oct. – 26th Oct.	17
1991	5th June	2nd Oct.	12th July – 5th Aug.	23
			26th Aug. – 17th Sept.	23

quired to be taken with the hope that the dry spell would be broken after about 3 to 4 weeks. Such measures are usually aim at conserving available moisture by way of controlling plant population, checking weed growth to reduce moisture loss, increasing interculturings to prevent evaporation etc. Some useful data are available at Solapur on some of these aspects.

Early withdrawal of monsoon

This type of situation is more dangerous in drought prone area. Early withdrawal creates difficulties of two types. In one case, sowing of *rabi* crops may be jeopardised in case crops are not sown in early September. In the second case, moisture conserved in the soil is required to be carefully used by appropriate practices. The suggested practices are as under :

Reduce plant density : *Rabi jowar* sown during early September at 1 to 1.35 lakh plants/ha requires 50 per cent reduction. This was practical during 1976-77 season. Plant population needs to be adjusted before plants go for their grand period of growth (30-35 days after sowing).

Use of surface mulch : Moisture can be conserved by using organic surface mulches. For this purpose, mulch material at 5 t/ha is required. During 1976-77, deficient year (rains during September and October were nearly 88 per cent deficient), by using mulch sorghum grain production was increased by 22 per cent.

Protective irrigation : If possible protective irrigation may be given. Usually protective irrigation is proposed to be given at 55 to 56 days growth. Due to early withdrawal of monsoon, the same may be applied at 35 to 40 days growth. Data collected on this aspect during 1976-77 at Solapur is worth studying.

Increase frequency of intercultivation : Early stoppage of rainfall brings about cracking early

Table 14. Efect of reduced plant density on production during dry spells of 1976-77 on sunflower

Plant density/ha	% Reduction	Yield (q/ha)	Add. grain yield (%)
74,000	(Normal)	7.12	–
56,000	24	9.87	39
49,000	34	9.12	28

Table 15. Effect of weed control on conservation of moisture during dry spells (Yield of bajra during 1976-77)

Treatment	Yield(q/ha)	Per cent reduction	Consumptive use of moisture (mm)
No control of weeds	2.83	74	176
Clean cultivation	12.00	–	175

Table 16. Effect of protective irrigation applied at 40 days stage on yield of *rabi* Sorghum (1976-77)

Treatments	Yield of grain (q/ha)	Add. grain (q/ha)	Grain (q/cm) of water
Without irrigation	8.22	–	–
2 cm irigation in alternate furrow	12.89	4.67	2.33
4 cm irrigation in all furrows	15.29	7.37	1.84

in soil. To prevent cracking and thereby loss of moisture, frequency of intercultivation may be increased. Usually three interculturings are recommended. If the same are increased to 5 or 6 it would be helpful in creating dust mulch. Same practice was adopted during 1976-77.

Stripping of leaves : It helps to control moisture loss temporarily. This was not found to be useful practice for prolonged drought.

Extended monsoon

Such aberration is rarely experienced. This is a happy situation as double cropping would be possible in medium deep soils. In other cases, extended monsoon may require postponement of sowings of *rabi* crops. In such a case, replacement of normal crops in certain areas would be essential. It is usual experience that extended monsoon results in cool spells. To combat the cool spell, sorghum may require replacement in certain cases as the crop is not found to do better under cool spell. Gram and wheat may find place under such circumstances.

DROUGHT PROOFING

There are two very potent options for drought proofing of drylands. These are : (1) alternate land use systems, and (2) management of natural resources (in particular soil and water) on watershed basis.

For imparting stability in the farming system, alternate land use systems incorporating tree component such as alley cropping, agrohorticulture, silvipasture management systems have been found to be very rewarding. These systems are discussed in details in the chapter 'Alternate land use systems for drylands' in this book.

To reduce the severity of effects of drought on a permanent basis, soil and water conservation programmes play an important role. For this approach, basic unit of development is a watershed, which is a hydrological unit. Watershed management is a holistic approach arrived at optimizing the use of land, water, and vegetation in an area, and thus, providing solution to alleviate drought, moderate floods, prevent soil erosion, improve water availability and increase fuel, fodder, and agricultural production on a sustained basis. In line with watershed approach, four important components of dry farming incorporated in the watershed development programme are :

1. Improvement of water resources,
2. *In situ* soil and moisture conservation,
3. Increase in cropping intensity, and
4. Alternate land use systems for efficient use of lands as per land capability to provide stability to production.

The recurring nature of occurrence of drought calls for finding permanent solutions to the problem of drought. Drought management requires precise information on the impact of drought in all spheres of human and livestock activity. This information is very much lacking at present. It should also be remembered that drought conditions have to be managed by judicious utilization of natural resources available within the region. The major emphasis should be on the conservation of resources for the future. Therefore, the management practices which can protect the regions from the onslaught of drought and provide much needed assets, employment and confidence of the people have to be implemented and perfected.

REFERENCES

Hegde, B.R. and Seetharam, A. 1985. Finger millet and minor millets. In : *Efficient Management of Dryland Crops* (Ed. V. Balasubramanian and J. Venkateswarlu), Central Research Institute for Dryland Agricltulture, Hyderabad, pp. 79-100.

Ramana Rao, B.V. 1988. *Operational Agricutlural Meteorology - Problems and Priorities*. Indian Society of Agronomy, New Delhi, pp. 59.

Singh, R.P. 1987. Research Efforts to Mitigate the Advers Effects of Drought (Mimeograph), Central Research Institute for Dryland Agriculture, Hyderabad, 42 p. .

Singh, R.P. and Ramana Rao, B.V. 1988. *Agricultural Drought Management in India - Priniciples and Practices.* Technical Bulletin No.1, Central Research Institute for Dryland Agriculture, Hyderabad, 45 pp.

Tomer, D.S., Singh, R.S. and Saini, S.L. 1974. Effects of direct seeding and transplanting on the growth and yield of Hybrid Bajra - 3' pearl millet grown at different nitrogen levels under late sown conditions. *Ind. J. agric. Sci.*, 44 : 317-328.

Venkateswarlu, J., Vishnumurthy, T. and Padamanabhan, M.V. 1991. *Contingent Crop Production Strategy In Rainfed Areas Under Different Weather Conditions*, Project Bulletin No. 5. Central Research Institute for Dryland Agriculture, Hyderabad, pp. 76.

22

ADVANCES IN OILSEEDS PRODUCTION UNDER DRYLAND SITUATIONS

M.V.R.PRASAD and B.N.REDDY

INTRODUCTION

The oilseed group of crops occupy an important place in India's agricultural economy. Next to food grains, they hold a sizeable share of the country's gross cropped are (=13%) and contribute around 5 per cent of its gross national product and 10 per cent of the value of all agricultural products. The country depends for its indigenous vegetable oil supplies on a wide array of oil yielding species of plant origin which include the annuals, perennials, minor oil bearing species of forest and tree origin, non-traditional sources such as rice bran, cotton seed, corn etc. Of the multitude of rich oleaginous plant genetic wealth, the country is blessed with the annual oilseeds which comprise of seven edible viz. groundnut, rapeseed-mustard, soybean, safflower, sunflower, sesame, niger and two non-edible sources namely castor and linseed form the main stay of India's vegetable oil economy; together the contribution more than 85 per cent of the country's indigenous vegetable oil supplies. Among these the first two crops namely groundnut (89-90=8.7 million hectares) and rapeseed-mustard (89-90=5.0 million hectares) alone account for more than 38 per cent and 22 per cent of the country's total acreage under annual oilseeds (88-89=23.0 million hectares) and about 48 per cent (8.1 million tonnes) and 25 per cent (=4.1 million tonnes) of its aggregate output (=16.75 million tonnes) respectively. Barring soybean (area=9.3%, production=10.2%) and sesame (area=10.3%, production=4.3%) and the contribution of other oilseed crops namely sunflower (area= 6.2%; production=3.5%), safflower (area=3.7%; production=2.9%; linseed (area=5.1%; production=2%), castor (area=3%; production=3%), niger (area=2.7%, production=1%) rarely exceeds 12 per cent of the country's total oilseed output.

The Indian oil seed scenario and the advances in its production under dryland conditions have been highlighted in this chapter

INDIA IN THE WORLD'S OILSEED MAP

India holds premier position in the world not only in terms of rich diversity of oil crops it grows but also in area as well. As much as 23 million hectares out of the current world's areas of 125 million hectares under nine annual oilseeds is from India alone (Table 1). If one adds to above list coconut and a number of other currently less/underexploited oil yielding sources such as cotton seed, rice bran, maize, corn, oilpalm the country's share in the world's oilseed area would go up by 20 per cent. Interestingly, this country ranks either first (groundnut, sesame, castor, linseed and safflower) or second (rapeseed) or third (sunflower) position in the world in terms of the area they occupy. Ironically, its contribution to world's output of annual oilseeds is less than 10 per cent (India=16.7 million tonnes; World = 181 million tonnes). One of the principal reasons for

Table 1. India in world's vegetable oilseeds scenario

Crop	Area (Lakh hectares)		Production (Lakh tonnes)		Productivity (kg/ha)	
	India	World	India	World	India	World
Soybean	26.0 (4.5)	583.0	18.5 (1.7)	1073.5	711	1841
Groundnut	81.0 (40.3)	200.9	80.0 (35.4)	225.9	988	1124
Castor Beans	6.3 (41.4)	15.2	5.0 (43.5)	11.5	800	759
Sunflower	14.0 (9.1)	153.8	6.0 (2.7)	218.7	377	1339
Rapeseed	49.0 (28.7)	170.8	44.1 (19.8)	223.0	907	1136
Sesame	22.0 (33.4)	65.8	6.0 (25.5)	23.5	273	357
Linseed	11.8 (27.6)	42.7	3.5 (16.5)	21.2	295	497
Safflower	7.8 (61.9)	12.6	4.3 (47.2)	9.1	549	721

FAO Production Year Book 1989
Figures in parenthesis indicate per cent of World total

such a paradoxical situation is the predominant cultivation of oilseeds under conditions of scanty rainfall/moisture and input starvation with poor management and their concentration in regions characterized by low and uncertain rainfall. Thus between early 50's and late 80's while the area under irrigation went up by 2 to 5 fold in crops like wheat (1950-3.4 m.ha; 1986=17.3 m.ha), rice (9.84 to 17.3 m.ha), cotton (0.46 to 2.1 m.ha), the country's much neglected group of oilseed crops continue to be exposed to the vagaries of monsoon even after more than three decades (1950=1.71 m.ha; 1986=3.1 m.ha). The result of such hostile crop growing situations and the uncertainties associated with their production and pricing on account of high speculative trade practices is extremely low per hectare yields of most oilseed crops when compared to their corresponding average yields in the world as well as other major oilseed producers in Asia namely China. Even under assured moisture conditions, the general productivity levels of the country's two premier oilseed crops namely groundnut and rapeseed-mustard seldom exceed 1.5 t/ha as against the corresponding yields of 2 to 3 t/ha or more registered elsewhere in other parts of the world. In addition, the diseases and pests take a heavy toll of the oilseed crops in India.

EDIBLE OIL NEEDS

Starting from the decade of 60's, India's position in the International Vegetable Oil Scenario steadily slided from one of the net exporter to the worlds' largest importer for vegetable oils today. Apart from the ever increasing demand of the growing population the consumption of vegetable oil is increasing not only for edible purposes but also for industrial purposes as the vegetable oil consumption is income elastic and as the standard of living of people is improving, there is a corresponding increase with regard to consumption of more fats. The survey conducted by National Nutritional Monitoring Bureau (NNMBO) of Indian Council of Medical Research indicates that a sizeable section of our society is consuming far less fat than what is needed. Based on the survey conducted in 10 states namely, Andhra Pradesh, Karnataka, Gujarat, Kerala, Madhya Pradesh, Maharashtra, Orissa, Tamil Nadu, Uttar Pradesh and West Bengal indicates the consumption level of the fats and oil in different income groups of our country is as under:

Income group	Consumption of fat in grams per day
High	43
Middle	35
Low	22
Industrial Labourers	23
Slum dwellers	10

Indian Council of Medical Research group has recommended an intake of fats and oils of 30 to 35 g/day. The above data indicates that around 80 per cent of our population which has predominantly low income category of people consumes far lower quantum of fats than recommended.

Principal Areas of Oilseeds Cultivation

While nodoubt, oilseeds are raised in almost all parts of the country, as much as 73 per cent of the current area of 23 million hectares and 70 per cent of their total output (=16 to 17 million tonnes) comes from only 6 states, namely Gujarat (area=2.8 million hectares; production = 2.5 million tonnes), Andhra Pradesh (area=2.9 million hectares; production=2.2 million tonnes). Rajasthan (area=2.9 million hectares; production = 1.8 million tonnes), Maharashtra (area=2.9 million hectares; production=1.8 million tonnes), Karnataka (area=2.3 million hectares; production=1.4 million tonnes) and Madhya Pradesh (area=3.4 million hectares; production=2.2 million tonnes). All other states together account for the remaining area (=6.2 million hectares) and production (=4.7 million tonnes).

MAJOR CONSTRAINTS IN OILSEED PRODUCTION ON DRYLANDS

Most of the oilseed crops are energy rich; but are grown under energy starved conditions. The analysis of the behaviour of the monsoon and its impact on oilseeds production has been reported by Ranga Rao (1991). Although country has experienced an equally favourable monsoon in 1983-84 and to a lesser extent in 1984-85, the output of oilseeds in neither of the above two years exceeded 12.5 to 13 million tonnes. Even when more than 60 per cent of the mateorological subdivisions in the country received subnormal rainfall during 1987-88 the total production remained at the level largely due to the intensive efforts launched to enhance the level of oilseed crop management under the Technology Missions. Obviously weather was not only the factor but other improved technologies played significant role in boosting the oilseeds production. (Ranga Rao, 1990).

Due to diversification of oilseed cultivation both in space and time, prevailing system of management of rainfed oilseeds practiced by farmers is the major constraint for increasing their production. For instance, major groundnut growing areas such as Suarashtra region of Gujarat, Rayalaseema area of Andhra Pradesh and southern Karnataka region need different set of practices for better utilization of rainfall. Most of the oilseed crops are generally raised on marginal and submarginal lands having poor soil fertility and low water holding capacity. In spite of beneficial effect of fertilization, hardly dryland farmer either applies the low dose of fertilizer or leaves the crop without fertilization. Many of the crops experience water stress at one or the other crucial development stage of their growth. Occurrence of stress conditions at critical stage will not only result in the reduction in yield but also pose many pests and disease problems. For example, leaf miner in groundnut becomes more severe under stress conditions. Similarly, bud necrosis poss problem under late sown dryland conditions, is a major biotic constraint limiting groundnut productivity. Most of the oilseed crops are sensitive to waterlogging which is not uncommon under rainfed conditions. Waterlogging conditions in sesame for three hours result in the total crop failure. Non-availability of quality seed of high yielding varieties in sufficient quantities at right time and problem encountered in large scale coverage of seeding under limited moisture due to non-availability of suitable implements are some of the production constraints in achieving higher productivity under dryland situations.

PERFORMANCE OF OILSEED SECTOR BEFORE AND AFTER THE SETTING UP OF THE TECHNOLOGY MISSION

Contrary to spectacular breakthroughs witnessed in food grains more particularly wheat and rice since late 60's the country's oilseed scenario presented until recently, a picture of virtual stagnation. Thus between late 60's and mid 80's while total food production more than doubled and the output of wheat and rice went up by four and two fold respectively, oilseeds registered a mere 5 lakh tonnes additional production which is indeed a matter of serious concern considering their sizeable share in the country's gross cropped area. Equally unimpressive are the growth rates of productivity of oilseeds when compared to those of food grains in general and wheat and rice in particular (Table 2). Added to their virtual insulation from the impact of green revolution which transformed wheat and rice, the yield levels of oilseeds suffer from wide fluctuations in space and time.

Table 2. Growth rate of oilseeds vis-a-vis food grains in india during 1967-68 to 1988-89

Crop	Per cent compound growth rates of		
	Area	Production	Productivity
Rice	0.56	2.60	2.06
Wheat	2.02	5.27	3.17
All Cereals	0.20	2.89	2.26
Pulses	0.30	0.74	0.51
All Food grains	0.22	2.68	2.10
Groundnut	0.08	1.23	1.14
Rapeseed-Mustard	1.50	4.01	2.48
Sesame	0.43	1.22	1.67
All nine Annual Oilseeds	0.17	1.91	1.42

Since 1987-88 the country's vegetable oil scenario has been however, showing distinct signs of transformation from its decade old inertia, thanks for the setting up of the Technology Mission on Oilseeds in May, 1986 by the Govt. of India and the series of synergistic and innovative farmer oriented policies and measures it has launched, backed up by better availability of inputs and services, attractive and guaranteed price support operations. Thus, the country's aggregate output of oilseeds which in the past never surpassed 13 million tonnes even under the most favourable seasonal conditions of 1984-85 with more than 85 per cent of meteorological subdivisions receiving normal/above normal rainfall, not only touched close to the earlier peak in 1987-88 in spite of one of the worst droughts that characterized the year but scaled to all time record of 18 million tonnes in 1988-89 (Target=15.6 million tonnes) and around 17 million tonnes in 1989-90 (Target=16.5 million tonnes). The above production marks 38 per cent increase over the previous best of 13 million tonnes recorded during the pre-mission era and 59 per cent over that of 1986-87. Simultaneously, the quantum imports of vegetable oils and the outgo of scarce foreign exchange reserves on account of it came down from 18 lakh tonnes and Rs. 920 crores (foreign exchange equivalent) in 1987-88 to as low as 4.0 lakh tonnes and Rs. 250 crores, respectively by 1989-90. What is more important, the export earnings from oils, oilseeds and oilseed extractions registered a phenominal spurt after 1986-87 (1986-87–Rs. 225 crores; 1989-90–Rs. 954 crores) resulting thereby in a net saving of Rs. 670 crores in imports of vegetable oils besides an additional income of Rs. 729 crores on account of enhancement in exports. These encouraging developments in the country's most sensitive and vital sector namely oilseeds have not only catapulated it from a position of one of the major sources of drain on the exchequer to a net foreign exchange earner in recent years but also created the much needed sense of optimism and confidence in the latter's potentials and capabilities to accomplish the goal of self-reliance. Notwithstanding the above positive changes in the output and the increasing resilience and strength the oilseeds front acquired in recent years, the general per hectare yields of several oilseeds still fall far short of those realised elsewhere in the world.

Table 3. Demand and supply of edible oils in india

Year	Demand (Lakh tonnes)	Indigenous production (Lakh tonnes.)	Gap (Lakh tonnes)	Per capita availability (kg/year)
1970.71	26.29	25.00	1.29	4.88
1975-76	39.10	33.40	5.70	4.30
1980-81	37.73	27.00	10.73	4.90
1985-86	51.15	38.50	12.65	6.30
1989-90(P)	61.30	54.00	7.30	6.50

P = Provisional.

THE PARADOX OF WIDENING DEMAND-SUPPLY GAP IN OILSEEDS

Despite the rich natural endowments the country possesses in terms of multiplicity of oil yielding species of plant origin both annuals and perennials, diversity of soils and crop growing situations for year round culture, vast research infrastructure and the wealth of technologies generated by it over the last two decades or so, the indigenous supplies of vegetable oils fell far short of the country's aggregate demand (Table-3). Thus, during the last 22 years the output of oilseeds hardly registered a growth rate of just around 2 per cent per annum which is in fact less than the growth of population (=2.2% /annum), whereas the domestic demand for vegetable oils and fats has been rising at about 4 per cent/annum (income elasticity of demand : 1-1.2%). As a result of the failure of the indigenous oilseed output to keep pace with the country's growing requirements for vegetable oils and fats the Govt. of India has been forced to resort to huge imports every year to bridge the yawning shortfall and ensure per capita availability of oils and fats at around 6 kg/annum as against 14 kg minimum nutritional requirements prescribed by Indian Council of Medical Research (world average = 11 kg/annum; developed countries = 22 kg/annum). From a net exporter of oilseeds, oils and oilseed extractions until early 70's, the country's import bill which was barely Rs. 15 crores (foreign exchange equivalent) in mid 70's reached a staggering level of Rs. 1300 crores in 1983-84 earning thereby the country the dubious distinction of one of the world's largest importer of vegetable oils and fats. A multitude of factors are responsible for the above distressing situation on the country's ailing oilseeds front. Principal among these are the predominant cultivation of oilseeds which are energy rich at the mercy of wayward monsoon under conditions of input starvation and poor management. This and the highly unfavourable and uncertain prices, their wide fluctuations to the disadvantage of the farmer consequent upon high speculative activity so characteristic of the commodity provided little or no incentive to the farmers to invest in the new technologies. Non-availability of quality seed of high yielding varieties of different crops in required quantities and at right time and the poor economic conditions of bulk of the farmers are various other constraints holding adoption of new technologies. Barring sporadic and isolated attempts, developmental efforts in oilseeds prior to setting up of the Technology Mission on Oilseeds were quite far and few. Neither the researchers had the full confidence on the yield worthiness of the new technologies they advocated nor were they equipped to provide the required feed back to the policy makers and extension personnel engaged in oilseeds promotion on their implementability, feasibility and viability from real farm situations to support their claims based on data from experimental farms and enthuse the latter to push up and spread the newer varieties and technologies to farmer's fields. Little surprise, the wealth of new generation varieties, hybrids, remunerative and productive cropping systems, agro-production and protection technologies generated by the extensive network of research centres over the last one and one-half decade or more, largely remained confined to the four walls of the research laboratories for want of patronage and support from developmental personnel and the end users, the farmer.

SOIL AND MOISTURE CONSERVATION PRACTICES

Among the various oilseed crops, groundnut is a erosion resistant crop because of its thick basal crop canopy which acts as a soil cover. In the low lying areas, a dead furrow is made in between the two rows of groundnut so as to provide scope for deeper penetration of moisture. In the event of excess rains the rainwater needs to be collected in the ponds for the use of crop during the dryspells between two rains. In the areas of heavy rainfall, zones of shallow water table and saline soils, throwing the field into ridges and furrows and raised beds is preferable to flat beds. In broad bed furrow system method advocated by ICRISAT after every 4 or 3 rows of groundnut spaced at 30 cm, one row is left blank and it can be converted as drainage or irrigation furrow and in between two furrows, bed is raised by about 10 to 15 cm (ICRISAT 1987). In the case of sesame, provision of excess water during monsoon is essential particularly in heavy soils. The excess water could be collected in small farm ponds and could be used for life saving irrigation during the moisture stress period. Like other oilseeds sunflower can not withstand water stagnation even for few days. So drainage should be provided to prevent waterlogging. This can be achieved in black soils by opening graded furrows across the slope at set intervals of about 10 m. In areas where the soils are slopy (more than 1.5 per cent) graded bunds may be adopted for preventing soil loss and conserving rain water (Venkateswarlu 1985). Since niger is grown on hill tops and slopy lands graded bunds or terracing is suggested. Castor should be planted in ridges and furrows across the slope for moisture conservation. The study conducted at Anantapur indicated that yield of castor could be increased through deep tillage, which benefitted the crop in subsequent years also. (Patel, 1985).

Moisture conservation is perhaps the most important aspect in management of rapeseed-mustard. Land should be perfectly levelled to prevent run off losses. Use of ridger seeder is becoming popular for conservation of moisture *in situ* in parts of Haryana (Kataria *et al.* 1982). Safflower is basically a dryland crop grown predominantly in heavy soils. It is important to check soil erosion and conserve the maximum possible rain water through compartmental bunding. In all low to medium rainfall areas, graded combined with inter terrace cultivation across slopes has been quite effective. Provision of drainage, land configurations such as broad based bunds, graded ridge furrows is important in ill drained soils to overcome the yield losses through wilts and root rot. The conventional harrowing after each rain 3 to 4 times during the monsoon has been found to be as effective as deep ploughing in controlling weeds and conserving moisture in black soils. Linseed crop too yields well with measures to conserve the moisture by arresting run-off in black soils and graded bund and terracing in light textured soil. The performance of oilseed crops is excellent in the areas covered under watershed development programme. For instance, at Mettimari watershed in Southern Karnataka the mean yield of groundnut over six years (1984-89) was 825 kg/ha under watershed area as against the bench mark yield of 300 kg/ha (Hegde 1992).

SUPERIOR VARIETIES AND HYBRIDS OF OILSEED CROPS

The crop improvement programmes launched in the project since seventies, have resulted in the identification of over 65 varieties of groundnut, 54 varieties of rapeseed-mustard, 39 varieties of sesame, 23 varieties of linseed, 14 varieties of safflower, 11 varieties of sunflower, 10 varieties of castor and 10 varieties of niger (Table 4) including four hybrids of sunflower and 2 hybrids of castor. As compared to the traditional oilseed cultivars, the new generation of improved cultivars recommended for different agro-ecological situations displayed on an average 15 to 40 per cent yield superiority. A number of these new varieties developed in the recent past possess inbuilt resistance/tolerance to one or other major biotic stresses such as bud necrosis of groundnut (ICGS 11, ICGS 44, KADIRI 3), late leaf spot ICGS (FDRS) 4, ICGS (FDRS) 10, ICGV 86590), rust (ALR 1 & ICGV

Table 4. Some high yielding varieties and hybrids released / identified after 1985

Crop	Name of variety released/identi-feed (during VII plan)	Year of release/ identi fication	Specific areas of adaptation	Increase in yields over best check	Other desirable agronomic features
Groundnut	MA-16	1986	Uttar Pradesh	43.0	Bold kernels, semi-spreading, ARS grade. Tolerant to leaf spots.
	VRI-1	1986	Tamil Nadu	R 40.01 19.0	High shelling %(74.4) and possess fresh seed dormancy for a week.
	ICGS-11	1986	Andhra Pradesh, Maharashtra and Karnataka and parts of MP in India	30.0	Released for *rabi*/summer cultivation. Less susceptible to bud necrosis disease and tolerant to end season drought and is photo-insensitive
	SG-84	1986	Punjab under irrigated conditions for spring planting	—	Well suited for cultivation after potato and late *toria*
	M-335	1986	Sandy soils of Punjab under irrigated conditions	17.0 R 6.0	Tolerant to tikka and leaf webbers Early maturing by 11 days than M-13. Resistant to aphids.
	ALR-(AFDRS-20)	1987	Tamil Nadu for rust endemic areas	230	Resistant to rust and late leaf spot diseases
	Girnar-1	1988	Peninsular India	23.0	Tolerant to Alternaria, rust, leaf spots & drought. Resistance to jassids
	TAG-24	1988	Maharashtra	R 54.01 36.0	Matures early by 15 days than UF 70103, more tolerant to bud necrosls than existing varieties.
	VRI-2	1988	Tamil Nadu	R 22.0 19.0	Tolerant to drought
	ICGS-44	1988	Gujarat, Northern Maharashtra and MP for summer cultivation	33.0	Tolerant to bud necrosis as compared to JL-24 and has good recovery from mid season drought
	ICGS-76	1989	Tamil Nadu, Karnataka, Kerala, Southern Maharashtra, AP except coastal districts.	33.0	Has good pod yield recovery from mid season drought
	ICGS-37	1989	Gujarat, North Maharashtra and MP under irrigated *Rabi*/summer situations	27.0	Tolerant to end season drought adapted to irrigated post rainy season and insensitive to photo period.
	ICGS-1	1990	Haryana, Punjab, UP and Rajasthan	17.0	Also suitable for spring season cultivation in UP and released as SG-84 in Punjab for spring season
	ICG(FDRS-10)	1990	Peninsular India (AP, Karnataka, parts of Maharashtra and Tamil Nadu)	8.0	Foliar disease resistant variety with resistance to rust and tolerance to late leaf spot, peanut mottle virus, stem and pod rots and field tolerance to bud necrosis disease

Table Contd...

Crop	Name of variety released/identi-feed (during VII plan)	Year of release/identi fication	Specific areas of adaptation	Increase in yields over best check	Other desirable agronomic features
	MH-4	1989	Haryana	11.0	Low seed requirement, less problem of weeds due to increase in height
	RH-141	1989	Udaipur and Chittorgarh areas of Rajasthan	14.9	Spanish bunch variety suitable for heavy soils for kharif seasons.
	Somnath (TGS-1)	1990	Gujarat	23.0	Non-dormant virginia runner.
	VR -I -3	1990	Tamil Nadu	4.76	Early maturing (90 days) bunch variety.
	ICGV-86590	1991	Zone-V comprising of AP (except North coastal districts). Karnataka, Tamil Nadu, Kerala and Southern Maharashtra during rainy season	13.7	High yielding variety with resistance to rust and late leaf spot.
	J(E)-2 (GG-3)	1991	Zone-III	10.0	High yielding and early maturing variety.
Rapeseed-mustard	YSB-19-7-C	1986	West Bengal	10.0	–
	PT 507 B	–	Eastern zone	10.0	Matures in 86 days
	Sarama (RW 85-59)	1986	West Bengal	24.0	Suitable for late sown conditions and responds to high doses of fertilization
	Panchali (TWC-3)	1986	West Bengal	21.00	–
	PT-30	1986	Uttar Pradesh	5.0	Tolerant to Alternaria blight, white rust and downy mildew.
	RAUTS-17	1986	Bihar for Diara area	33.0	Early maturing.
	TK-8401 (Bhawani)	1986	Irrigated areas of Uttar Pradesh	3.0	Extra early in maturity 15-20 days earlier than check variety T-9 and fits in multiple cropping system.
	Bhagirathi (RW-351)	1986	West Bengal	19.0	Suited for mixed cropping
	RH-819	1986	North western Zone	9.0	–
	RL-1359	1987	Punjab for irrigated areas	11.0	Non lodging and non-shattering and less susceptible to aphid attack compared to Kranti, Krishna and Varuna

Table Contd...

Crop	Name of variety released/identi-feed (during VII plan)	Year of release/identi fication	Specific areas of adaptation	Increase in yields over best check	Other desirable agronomic features
	TL-15	1988	High altitude areas of Punjab for irrigated summer plantings	20.0	Matures earlier by 15-20 days than ITSA
	NDR-8501 (Narendera Rai)	1988	UP and MP under irrigated conditions	14.0	Early maturing by 5 days than Krishna and resistant to Alternaria blight, white rust, aphids and leaf miner. Bold seeded, tolerant to salinity.
	TH 68	1988	Irrigated areas of Haryana	8.0	Earlier in maturity than Sangam by one week and suitable for toria wheat rotation.
	RH 8113 (Saurabh)	1988	Haryana, Jammu, HP, Punjab, Gujarat and Rajasthan under irrigated conditions	14.0	Tolerant to white rust, Alternaria and downy mildew.
	RSK-10 (GM-1)	1989	Gujarat	13.0	Matures early by 6 days than Kranti and lodging resistant
	Rh-781	1988	Assam, West Bengal, Orissa, Bihar and Haryana under irrigated conditions	14.0	Tolerant to frost and suitable for intercropping with wheat (9 : 1) in Haryana
	RW-4C-6-3-II	1986	West Bengal	25.0	Suitable for intercropping with wheat under irrigated conditions
	DIRA-367	1989	West Bengal, Bihar, Orrissa	23.0	Resistance to downy mildew, non-lodging, non-shattering and tolerant to powdery mildew
	DIR-247	1989	West Bengal, Bihar, Orrissa	17.0	Matures earlier by 4 days than Varuna
	RH-7361	Pre-release	Jammu, HP, Punjab, Haryana and Rajasthan.	–	Frost tolerant
Mustard	TM-2	1987	Assam (Brahmma putra valley).	–	Compact plant type with appressed pods
	TM-4	1987	Assam (Brahmma putra valley).	–	
Taramira	TMH-52	Identified, yet to be relased	Haryana	17.6	–
Yellow *sarson*	PYS-842	1990 (Identified)	Central zone	33.8	Moderately resistant to Alternaria blight, white rust and downy mildew. Matures early than check variety YST-151.
Sunflower	MSFH-8	1986	Maharashtra, AP, Karnataka and Tamilnadu	26.0	Moderately resistant to rust, Alternaria leaf and stem spots

Table Contd...

Crop	Name of variety released/identi-feed (during VII plan)	Year of release/identi-fication	Specific areas of adaptation	Increase in yields over best check	Other desirable agronomic features
	Co-2	1986	Tamil Nadu, AP and Maharashtra	2.0	Tolerant to rust. Early maturing (4 days earlier than EC-68414)
	MSFH-10	1987	All sunflower growing areas of the country	41.0	Resistant to rust and other foliar diseases, lodging resistant and performs well even under water stress conditions
	APSH-11	1987	Andhra Pradesh for all seasons and late sowings	8.0	Resistant to Alternaria and major insect pests
	NSFH-17	1988	All sunflower growing areas of country	29.0	Resistant to rust, leaf blight, and downy mildew
	SS-56	1987	Scarcity zone of Maharashtra	25.0	Matures 5-8 days earlier than Morden
	LDMRSH-1 (LSH -1)	1988	Downy mildew endemic areas of Maharashtra	52.0	Resistant to downy mildew. Non lodging and non shattering
	EC-68415 C (Haryana Suraj mukhi-1)	1989	Haryana	31.0	
Safflower	HUS 305	1986	UP, Bihar, West Bengal, Madhya Pradesh	12	Possess higher oil (36-38%) content than check A-1 (30%). Highly tolerant to Alternaria leaf spots, wilts and root rots. Tolerant to salinity
	Nira (NRS-209)	1987	Irrigated areas of Maharshtra	17	Fairly resistant to powdery mildew-Responds well to limited number of irrigation. Early maturing (15 days earlier than Bhima)
	JSI-7	1988	Malwa plateau of MP	6.0	A spineleas variety
	JLSF-88	1989	Maharashtra	14.1	An early maturing variety suitable for low rainfed areas of Marathwada
Sesame	TC-289	1986	Punjab and HP	3.0	–
	Krishna	1987	MP, Maharashtra, Orrissa, Bihar, West Bengal, AP, Gujarat for semi- *rabi*, UP and Bihar for *kharif*	16.0	Tolerant to Alternaria leaf spot and capsule borer
	Tapi (JLT-7)	1987	Maharashtra	31.0	Early maturing (10-12 days earlier than Phule) bold seeded, fits in double cropping system
	Rajeswari N 62-39	1988	Telangana region	54.0	Moderately tolerant to stem rot and phyllody

Table Contd...

Crop	Name of variety released/identi-feed (during VII plan)	Year of release/identi fication	Specific areas of adaptation	Increase in yields over best check	Other desirable agronomic features
	RT-46	1989	J&K, HP, Punjab, Haryana, Rajasthan and western UP	11.0	Resistant to dreaded oozing complex and Macrophomina root rot and also tolerant to stress
	OMT-11-6-3	1990	MP, eastern UP, Bihar, West Bengal, Orissa, Assam and north-eastern states	11.0	Matures earlier by 10-15 days than TC-25 and synchronous in maturity and is non-shattering
	Imp Set-5	1988	West Bengal and Eastern India	30	Moderately resistant to stem rot, tolerant to leaf spot, phyllody and powdery mildew
	JLT-26	1990	Khandesh and adjoining areas of Vidharaba (Maharashtra)	30	Matures earlier by 7-8 days than Phule Til No. 1 and suitable for double cropping
	RT-54	1991	Maharashtra, Gujarat and Telangana regionof Andhra Pradesh	11.0	Tolerant to fusarium, macrophomina, root rot and antigastra
Linseed	Jeavan(KPL-21)	1987	HP, Punjab, Haryana, Rajasthan, UP, Bihar, West Bengal for irrigated areas	2.0 Seed 64 0 Fibre	A dual purpose variety resistant to rust, wilt and powerdry mildew and also resistant to lodging, frost and shattering
	Gaurav(LCK-15 2)	1987	UP, West Bengal, Bihar and Rajasthan	40 Seed 4.2 Fibre	Dual purpose variety resistant to wilt and rust
	Kiran(RLC-6)	1988	MP, Karnataka, Orissa, Rajasthan, Maharashtra, Bun-delkhand of UP	2.0 Seed	First multiple resistant variety, besides tolerant to drought
	Janaki(KL-43)	1988	Himachal Pradesh	18.0	Resistant to rust, wilt and powdery mildew and responds to inputs
	KL-31(Nagarkot)	1988	HP, Punjab, Haryana, UP, Bihar, West Bengal and Assam	22.0 Seed 5.0 Straw	Dual purpose variety resistant to rust, wilt, powdery mildew
	ICK-8528	1989	UP(excluding Bun-delkhand), Bihar, West Bengal and Assam	33.0 Seed + 20.0	Dual purpose variety resistant to rust and tolerant to powdery mildew and wilt
Niger	RCR-317	1987	Gujarat	32.30	–
	Shiva(GA-10)	1988	Orissa, Karnataka, Bihar, Rajasthan, West Bengal	28.5	Non lodging and non-shattering type, tolerant to drought and relatively free from cuscuta infestation
	Bhabani(GA-5)	1988	Orissa, Karnataka, Rajasthan, MP, Bihar and West Bengal	24.61	Non lodging and non-shattering type, tolerant to drought and relatively free from cuscuta infestation
Castor	GCH-4(SHB-18)	1987	Gujarat and other castor growing states in India	13.62	Tolerant to wilt and root rot. Highly responsive to irrigation

86590) and leaf miner (ICGS 2271 & ICGV 865031) in groundnut; tolerant to Alternaria, white rust, aphids and leaf miner (NDR 8501 and white rust and downy mildew (PT 30) in rapeseed-mustard, wilt, powdery mildew and rust in linseed (Jawahar 23, J-552, G-1, Kiran, Gaurav, Garima, LCK 8528, KL 31); rust (APRR-3) and powdery mildew (Nira) in safflower downy mildew (LDMRSH 1 and 3) in sunflower, wilt and root rot in castor (GCH 4 and 48-1), the popularisation of these varieties possessing inherent tolerance and resistance to the specific diseases imparted considerable stability to oilseeds production. In sunflower (Morden, SS 56, Co2) *toria* (Bhavani, TL 50) and mustard (RW 4-6), the availability of short duration and early maturing varieties has facilitated restructuring of cropping patterns in which the above could be fitted in leading to remunerative returns as described later. These varieties have offered the farmers a wide spectrum of choice to suit different crop growing situations/areas as for instance salt affected areas, earlier late plantings, mixed/intercropping systems, high altitude and frost prone areas, assured and limiting moisture conditions as well as seasons.

A critical examination of the varieties of the oilseed crops developed during the period upto 1975 revealed that they offer yield advantages of the tune of 15 to 30 per cent only over their traditional and local counterparts. This is more true in respect of groundnut, sesame. However, the intensive and wide ranging genetical and breeding researches carried out in the project have resulted in the development of new genotypes which not only possess conspicuous levels of tolerance and resistance to specific biotic stresses, some of the new breeding materials available are also paving way towards breaking the yield barriers particularly in groundnut and castor. In the All India Coordinated Network breeding lines possessing new genes for total dry matter production and calcium uptake facilitating in the development of genotypes of groundnut with consistently stable and higher levels of productivity despite abiotic stresses have been developed and are being used in breeding programmes. The drought being the most common and important abiotic stress limiting the productivity of wide range of oilseed crops, the efforts are underway to develop drought tolerant genotypes which can give consistent levels of productivity. Screening techniques have been developed for identifying drought tolerant genotypes. Another important research activity which is being strengthened in the Coordinated Network Programme is the identification and development of suitable plant types of oilseed crops for intercropping with cereals and non-cereals. Already such genotypes are available in castor (Sowbhagya), groundnut (Spanish and virginia bunch varieties), rapeseed-mustard and *toria*, sunflower (Morden), sesame (Madhavi and Gouri) and linseed (Gaurav, Garima, LK-8528).

INTERCROPPING AND CROP SEQUENCES INVOLVING OILSEED CROPS

Intercropping as a measure of risk minimisation leading to higher incomes to the farmers is prevalent in most of the oilseed growing states. It is the only way to stabilise the *kharif* oilseeds production as these are more often subjected to the vagaries of monsoon. Further, the area increase per se can be achieved through intercropping. Reddy (1990) reported that groundnut can be successfully intercropped with sorghum, pigeonpea, sunflower, cotton, castor and tapioca in various agro-ecological situations. According to Agasimani *et al.* (1991) groundnut + maize, groundnut + chillies, groundnut + pulses intercropping systems are also remunerative when grown in definite row proportion. Some of the productive and profitable intercropping systems with oilseeds are suggested by Prasad (1991) based on the work done under All India Coordinated Research Project on Oilseeds (Table 5). According to Singh (1988) intercropping system is possible in areas receiving annual rainfall from 600 to 800 mm and those characterised by soils capable of storing moisture for 20 to 30 weeks only.

Since resource utilization in intercropping is of utmost importance, the work done at ICRISAT indicated that there was no transfer of fixed

Table 5. Some productive and profitable intercropping systems with oilseeds

Base crop	Region(s)	Profitable and productive intercropping systems
Groundnut	Saurashtra region of Gujarat	Groundnut + Sunflower(1:1)
	Khandesh tract of Maharashtra	Groundnut + Castor(6:2)
	Karnataka	Groundnut + Castor (3:1)
	Gujarat	Groundnut + Castor (3:1)
	Northern A.P.	Groundnut + Castor (6:1)
Castor	Gujarat	Castor + Sunflower (1:2)
	Gujarat	Castor + Soyabean (1:1)
Wheat	Deccan Rabi	Wheat + Safflower (2:1 or 3:1)
	All mustard growing areas of India	Wheat + Mustard (9:1)
Chickpea	Potential Safflower belt in Deccan *Rabi*, southeastern Rajasthan, eastern Uttar Pradesh.	Chickpea + Safflower (2 :1 or 3 : 1)
	Uttar Pradesh, Haryana & Rajasthan	Chickpea + Mustard (3 : 1)
	West Bengal	Chickpea + Linseed (1 : 1)
Coriander	Maharashtra and Karnataka	Coriander + Safflower (2:1)
Potato	Uttar Pradesh	Potato + Mustard (3:1)
	Uttar Pradesh, Haryana	Potato + Linseed (3:1)
Cotton(Hybrid)	Maharashtra, Gujarat, Karnataka	Cotton + Soyabean
Ragi	Karnataka	Ragi + Sunflower (4:2)
Maize	Northern States	Maize + Soybean (1:1)
	All parts	Sorghum + Soybean (1:1)
Linseed	Deccan Rabi	Linseed + Safflower (2:2 or 4:2)

nitrogen to maize when it was intercropped with groundnut. However, the relative yield advantage of intercrop was 44 per cent as compared to sole crop of maize at no nitrogen level. Reddy *et al.* (1985) reported that pendimethalin @ 1 kg a.i./ha controlled the weeds in groundnut intercropped with sunflower, sorghum and pigeonpea. Groundnut + pigeonpea recorded higher returns with the same herbicide. Intercropping of linseed with gram is found to reduce the incidence of wilt disease of chickpea (AICORPO 1984). Some of the remunerative intercropping systems involving safflower are safflower + chickpea, safflower + coriander and safflower + wheat and linseed + safflower.

The experiences show that dryland areas receiving more than 800 mm rain annually with moisture availability of more than two weeks only, are suitable for double cropping. Efficient crop sequences are groundnut- chickpea, groundnut-cowpea and groundnut-clusterbean (Sankar Reddy *et al.*, 1981). In Rayalaseema areas of Andhra Pradesh, Saurashtra region of Gujarat and parts of Tamil Nadu, groundnut is grown year after year. In residual moisture conditions of coastal areas of Andhra Pradesh, Karnataka and Tamil Nadu paddy-groundnut sequence was found to fetch more income. Due to photoinsensitive nature, sunflower finds a better place in legume/non-legume sequential system. Sesame as a catch crop during early *kharif* or semi *rabi* crop is popular in coastal areas of Andhra Pradesh, Orrisa and Vidharbha region of Maharashtra, respectively. A number of viable and productive crop sequences identified for different regions having safflower or mustard as one of the components have been suggested by Ranga Rao and Reddy (1987). The successful and profitable crop sequences involving oilseed crops are given in Table 6. Linseed as a 'Utera' crop under residual soil moisture conditions in parts of Uttar Pradesh, Madhya Pradesh and Bihar has become popular.

Table 6. Some productive and profitable crop sequences involving oilseed crops

Region	Sequence
Deccan rabi situation	Greengram/sesame-safflower
Malwa region of Madhya Pradesh	Soybean-safflower
South eastern Rajasthan	Maize/cowpea-safflower
North Indian mustard belt	Maize/*bajra*/green legumes Rapeseed-mustard
Uttar Pradesh and Bihar	Sesame-mustard/yellow sarson
Himanchal Pradesh and eastern India	Jute/paddy-toria-mustard/yellow *sarson*
Southern States	Paddy-groundnut/sesame/sunflower
Gujarat region	Wheat/tobacco/paddy/cotton-*rabi*/summer groundnut
Eastern region	paddy/potato/sesame
Gujarat	Paddy/cotton/*rabi*-summer groundnut
Eastern India	Paddy/potato/sesame
Indo Gangetic Belt	Paddy/cotton/pigeonpea mustard
Southern States	Toria/potato-*spring* sunflower
Eastern India	Rice-*toria*/mustard or toria-sesame
North Indian Mustard Belt	Sesame/maize/*bajra*-mustard
Northern Region, Punjab and Gujarat	Paddy-toria-sugarcane-wheat-*maize* toria-wheat/sugarcane, groundnut/*bajra*-mustard-sugarcane

CROP MANAGEMENT

Groundnut

The crop is cultivated as a rainfed *kharif* crop in dryland areas of Rayalaseema, Southern Karnataka, Saurashtra region of Gujarat and parts of Tamil Nadu and Maharashtra. It is considered sufficient to secure good surface tilth to a depth of 12.5 to 18 cm. Preparing the soil, after the harvest of previous rice crop to obtain a good seed bed for the post rainy crop is a problem. The soil does not dry up thoroughly and ploughing is done once or twice but the soil is at high moisture content leading to a very cloddy seed bed resulting in poor germination. Under such situation, rotovator can be employed.

Selection of a suitable variety for a particular area depends on soil type, rainfall distribution, season, resistance to major pests and diseases. The crop experiences early, mid season or late season drought which cannot be predicted. *Spanish* bunch type varieties like JL-24 and TMV-2 are preferred where the rainy season is delayed and is short and where soils have high clay content and harvesting is a problem. Under late drought situation also early maturing bunch varieties are preferred by virtue of their early maturity and synchronous podding. In the case of mid-season drought, early maturing *virginia* bunch varieties such as Kadiri 3, G201 and ICGS 76 may be recommended. In order to avoid the chance of total failure, interplanting of *spanish* bunch and *virginia* bunch are suggested. Planting time is crucial in dryland areas to take advantage of the early monsoon. Planting on 12th June not only increased the yield but also reduced the incidence of disease (Table 7). Among the various methods of planting, broad bed furrow method has an edge over conventional method. It is essential to treat the seed either with thirum, capton or Mancozeb @ 3g/kg of seeds against seed borne disease. Application of *rhizobium* culture has been found beneficial where the crop is grown first time or under paddy-fallow system. Although spacing depends on soil type, variety fertility and cultural practices, it is suggested to adopt 30 X 10 cm spacing for bunch types and 30 X 15 cm for *virginia* types of groundnut. Seed rate varies depending on the test weight.

The groundnut crop requires adequate quantities of nutrients for higher yield. Every tonne of unshelled nuts and two tonnes of haulm produced by a groundnut crop, remove 63 kg N, 11 kg P, 46 kg K, 27 kg Ca and 14 kg Mg. Hence, the crop needs balanced fertilization. Application of 5 to 10 t/ha of farm yard manure (FYM) not only improves the soil physical conditions

Table 7. Impact of simple low cost inputs minimising losses from disease- planting time and varietal choice in *Kharif* groundnut

Cultivar	Reaction of variety to bud necrosis	Early plantings + (12 June)		Delayed plantings + (10 July)	
		1	2	1	2
Kadiri-3	Tolerant	14.9	1162	31.8	752
TMV-2	Susceptible	25.6	835	66.3	366
JL-24	Susceptible	29.1	848	79.6	214

+ 1 = % incidence of bud necrosis, 2 = Yield kg/ha.

but has also been observed to increase the nodulation in groundnut. Agasimani and Hosmani (1989) reported that application of 7.5 t/ha has increased the pod yield by 68 per cent over control in coastal sandy soils. Lower doses of nitrogen (10-20 kg/ha) would be required before symbiotic fixation of nitrogen commences. Basal dose of nitrogen is applied at the time of sowing in addition to phosphatic and potassic fertilizers. The quantity of P_2O_5 depends upon the phosphorus status of the soil. The dose varies from 20 to 40 kg P_2O_5/ha depending upon soil test value. In scanty rainfall areas, localised application of phosphatic fertilizer in a band placement is recommended. Response to potassium can be obtained when the available K_2O is less than 150 kg/ha. Application of potash (25 kg K_2O/ha) has resulted in the increased yield of groundnut in soils of Tamil Nadu and coastal Karnataka. Calcium and sulphur are taken up from the zones of peg and pods. Gypsum is a cheap source of calcium and sulphur. Application of 500 kg/ha of gypsum at peak flowering/pegging not only enhanced the yield but also provide strength to the plant. Application of Zinc sulphate (25-50 kg/ha) in red and sandy soils of Andhra Pradesh, Karnataka and Tamil Nadu and correction of iron deficiency through foliar spray of $FeSO_4$ (0.2%) in black soil areas of Karnataka and Maharashtra proved beneficial. The recent investigations (Prasad, 1992) have shown that canopy development at peak flowering has a strong positive correlation to yield. Therefore crop management technology to achieve higher leaf area indices at preflowering and flowering stage would result in higher levels of productivity.

Weed management is crucial in groundnut to avoid competition of weed with crop for moisture. Two hand hoeings followed by interculterring will take care of the weed control. Dust mulch through repeated interculture will be helpful to save the crop under drought situation to some extent. Under the conditions of severe pests and diseases, yield loss may touch as high as 40per cent. Integrated pest management practices involving choice of varieties, scientific crop rotation, seed treatment and need-based chemicals pay rich dividends.

Sesame

Uttar Pradesh, Madhya Pradesh, Maharashtra, Andhra Pradesh, Tamil Nadu, Gujarat, West Bengal and Karnataka are the major sesame growing states of India. The crop is grown during early *kharif*, *kharif* and late *kharif* and early *rabi* seasons under dryland situations. About 90 per cent of total area is under rainfed conditions. The crop is grown in shallow to medium soils in Andhra Pradesh, Orissa, Madhya Pradesh and Tamil Nadu and deep soils in Rajasthan, Maharashtra and Gujarat.

Sesame requires well pulverised seed bed for uniform germination and stand. The crop growth in the beginning is slow and as such hand weeding becomes very difficult. Therefore, field should be thoroughly prepared with one deep ploughing and subsequent harrowing. Sharma (1984) suggested a list of varieties for various agro-ecological situations. However, few varieties like Gauri and Madhavi in Andhra Pradesh, Gujarat til in Gujarat, Tapi In Maharashtra, Vinayak in Orissa, Co.1 in Tamil Nadu, JT-7 in Madhya Pradesh, T.13 in Uttar Pradesh, RT-54 and Pratap in Rajasthan are promising varieties. Optimum time of seeding has been worked out under AICORPO network where in onset of monsoon resulted in higher yield. However, late sowing of the crop has been found to reduce the incidence of phyllody. Reddy and Sharma (1983) reviewed the agronomic research in India, where in important agronomic practices in different regions have been highlighted. Simple practices such as seed treatment, line sowing at appropriate plant geometry, split application of nitrogen recommended for each regions, interculturing to keep the field weed free have been emphasized to attain high production of sesame under dryland conditions. Pest and disease management of sesame is very important aspect of management. Leaf roller (*Antigastra catalunalis* L.) and Gall fly (*Aspondylia sesami* felt) can be controlled by three sprays of phosphomidon and endosulfaan 0.07 per cent, respectively. Cultivation of multiple resistant varieties for various diseases is the only effective control measure in disease management programme.

Sunflower

In India, sunflower as an edible oilseed crop has been introduced in early seventies. The investigations carried out under the Dryland Research Project have conclusively established the potential of sunflower for dryland cropping as well as crop for mid season correction on drylands in the event of aberrant weather. The area of sunflower has been increasing year after year due to its photo insensitive nature, short duration and high oil content. The major sunflower producing states are Karnataka, Maharashtra, Andhra Pradesh and Tamil Nadu. Recently, the area of sunflower is extending to indo-gangetic belt during *spring* season. The crop is cultivated in both black and red soils during *kharif* season under rainfed situations. Any normal system of tillage would be adequate for sunflower. The seeding should not be beyond 5 cm depth, therefore, seed bed should be well prepared. Choice of early maturing variety like Morden is preferred for dryland conditions. Hybrids can also be planted where rainy period is longer and moisture retention capacity of the soil is good as in vertisols. Sunflower has shown wide flexibility in sowing time due to photoinsensitivity. Early sowing in first week of June or late sowing in second fortnight of August is desirable in view of adequate pollination and better seed set. Presoaking of seeds will enhance germination in low moisture conditions. Sunflower has the capacity to adjust itself to variations in population densities. In scanty rainfall areas, it is preferable to go for low plant population or wider spacing. The range of sixty to one lakh population seems to be optimum depending on the genotype. Sunflower being an exhaustive crop, supply of adequate nutrients is essential. In scanty rainfall it is recommended to apply 60:40:0 N, P_2O_5, K_2O kg/ha. Split application of nitrogen, half as basal at the time of sowing and remaining half 30 to 45 days after sowing will be advantageous as it increases the fertilizer use efficiency. Hand pollination at flowering stage between 8 to 11 AM. on alternate days for 7 to 10 days has been found to improve the seed set especially during rainy season when the bee activity is expected to be low. Application of Borax @ 2 kg/ha dusting at ray floret stage proved to be effective in improving the seed set and increasing the seed yield.

Niger

Although it is a minor oilseed crop, it is of considerable importance for lateritic or sandy soils in states like Maharashtra, Madhya Pradesh, Bihar, Karnataka, Orissa and Andhra Pradesh. It is an important contingent crop as it can be planted as late as second fortnight of August without much reduction in yield. Tillage by an iron or country plough combined with blade harrow not only keeps the field weed free but also helps in maintaining better soil moisture profile. Suitable varieties for different states have been identified. Among the new varieties, N-71, IGP-76, N-35, RCR 317, GA-10 and GA-5 have adaptability combining with good yielding ability. Sowing at the onset of monsoon invariably gave higher yield in all the niger growing states. Niger crop is generally sown as broadcast. However, line sowing has distinct advantages in seed economy, uniformity of population, ease in weeding and interculturing. In cuscuta infested areas, care should be exercised to separate cuscuta seeds from that of niger. A spacing of 30 X 10 cm with a seed rate of 5 to 6 kg/ha was found to be optimum to realise higher yield.

Although farmers do not apply any fertilizer to the crop, it has been found to respond positively upto 40 kg N/ha. However, the optimum dose is around 20 kg/ha. Application of 5 to 6 t/ha of farm yard manure and split application of nitrogen and 20 kg P_2O_5/ha have been found to be beneficial in major niger growing states (Ankineedu *et al.* 1983). Niger when intercropped with sunflower has been found to increase the bee activity and improve the seed setting of sunflower in Karnataka state.

Castor

The castor is mainly cultivated under rainfed conditions in the states of Andhra Pradesh, Karnataka and Tamil Nadu and parts of Gujarat and Rajasthan. Castor requires good seed bed which can be achieved by working with country plough followed by two harrowings. The varieties like Aruna, Bhagya, Sowbhagya, RC-8, SA-2, TMV-4 are recommended in different states. Recently, GCH-4 hybrid has become popular both in major castor growing areas like Andhra Pradesh and Gujarat. Soaking of seeds before sowing improves the germination and plant stand. Use of quality seeds of improved varieties has resulted in increased yield considerably. Wherever wilt problem is severe, planting of 48-1 or GCH-4 is recommended avoiding susceptible hybrids like GAUCH1 and GCH 2. Castor as a contingency crop can be planted late in first fortnight of August with close spacing of 60 X 30 or 60 X 15 cm as against normal spacing of 90 X 20 cm. Under late sown conditions the crop escapes the attack of red hairy caterpiller and semilooper. However, sowing at the onset of monsoon helps the crop to take advantage of larger quantum of moisture and long duration under dryland conditions of Andhra Pradesh and Gujarat leading to higher yields. A seed rate 10 to 15 kg/ha is adequate for sowing behind the furrows opened by a plough or seed drill. The field should be kept weed free during the period 30 to 40 days after sowing. Application of 40N:40P_2O_5:20K_2O kg/ha results in higher yield. In the event of drought before or at flowering/primary spike development, application of 20 kg N/ha coinciding with different spike orders has been proved beneficial. Management of castor semilooper through bioagents and wilt by growing *bajra* crop in rotation go a long way in augmenting castor production.

Rapeseed Mustard

It is second important oilseed crop contributing nearly 25 per cent to the total edible oilseeds production in India. Rapeseed-mustard group includes *toria*, brown *sarson*, yellow *sarson*, mustard and *taramira*. Among the various *rabi* crops, mustard has been found to be more stable compared to wheat, barley, chickpea etc. Taramira is mostly grown in Rajasthan under late sown conditions. *Toria* being surface feeder requires adequate moisture conditions at sowing. As a catch crop in between maize-wheat rotation it has become popular in indo-gangetic belt. The choice of these crops/varieties depends largely on moisture availability at the time of sowing and length of growing season. When the crops are raised on the conserved moisture, *taramira* crop is preferred where the soil moisture is upto 175 mm per meter of soil depth, brown *sarson* and mustard when the profile moisture is more than 175 mm per meter soil depth (Surinder singh and Sanghi, 1985). Saini *et al.* (1989) reported that optimum time of sowing varies with the species and regions. Sowing may be done when the maximum temperature is around 30°C. Border method of sowing (skipping every fourth row) resulted in saving 25 per cent seed and fertilizer without reduction in mustard yield. Optimum and uniform plant population is a prerequisite for obtaining maximum yield. Thinning may be resorted to 15 to 25 days after sowing to maintain optimum population.

Balanced fertilization with macro (NPK and S) and micro nutrients (Zn and B) is essential for achieving higher yield. Weeds can be effectively controlled by the use of herbicides like Fluchlorolin or Isoproturon as pre-emergence @ 0.5-1.0 kg a.i/ha. Management of aphids through systemic insecticides like metasystox @ 0.05 per cent has been found effective in reducing the infestation and resulted in higher yield. The spray of cycocel at flower initiation has been found advantageious under rainfed conditions. (Saini *et al.* 1989).

Safflower

The area of safflower is mostly concentrated in Deccan plateau region. The crop hitherto neglected is more stable and remunerative than number of traditional crops currently cultivated in Maharashtra, Karnataka, Andhra Pradesh, Malwa plateau of Madhya Pradesh, south eastern Rajasthan and eastern Uttar Pradesh. Tillage studies carried out at various centres in black soils did not reveal any significant effect of deep tillage over usual practice of 3 to 4 harrowings during the monsoon period for effective weed control and conserving the moisture (Ranga Rao, 1985). For better germination and stand establishment, seed bed should be free from clods and have a firm subsoil with adequate moisture. Use of fertiseed drills which can place the seed and fertilizer in the moist zone can be used for planting. Crop improvement programmes launched under AICORPO (Safflower) resulted in identification of high yielding varieties like Bhima, A-1, Manjira, HUS-305, JSI-7, JLSF-88, NIRA etc. for major safflower growing states. The time of seeding is critical and vary depending on the withdrawal of southwest monsoon, onset of north east monsoon and initial soil moisture at planting. Mid to late September for drought prone areas of Maharashtra, second half of September to early October for Malwa plateau region, second week of October for Northwest India are recommended. Safflower displays considerable plasticity to plant density and geometry because of its superior branching potentials. Seed rate of 8 to 10 kg/ha has been found sufficient to achieve optimum plant stand. Wider row spacing of 45 to 60 cm is preferable to cover larger area per unit time.

Application of 20 to 30 kg N/ha in scanty rainfall areas, 40 to 50 kg N/ha in areas with assured moisture along with 20-30 kg P_2O_5 /ha gave higher yields. Placement of fertilizer 10 cm (traditional areas) or 10 to 15 cm (northern region) below the soil is essential to improve the nutrient availability of the crop. Management of aphid and scientific crop rotation to prevent build up of fungal diseases and seed treatment are simple crop management practices to realise high yields of safflower.

Linseed

Madhya Pradesh, Uttar Pradesh and Maharashtra are the important linseed growing states in India. The most distarting feature is a high degree of instability in production resulting in negative growth rate primarily because most of the linseed area is rainfed. According to Jeswani (1984) two distinct ecological types have been recognised in linseed : (1) small to medium seeded Indogangetic alluvial types with shallow and abundant root system, and (2) the bold seeded deep rooted peninsular types. About 25 per cent of the linseed area is under *Utera* cropping to utilise residual moisture in rice fields where tillage is difficult. Under *Utera* cropping linseed is broadcast in standing paddy field. Research under AICORPO network has resulted in standardization of cultivation practices for both sole as well as *Utera* cropping. For sole cropping soil should be brought to a fine tilth with 2 to 3 harrowing. It is advisable to create a soil mulch with the help of harrow after each good shower to conserve moisture. Optimum time of sowing varies widely (October to 15th November) in different states depending on the moisture availability. The row spacing varies from 25 to 30 cm with a seed rate of 20 to 30 kg/ha. Shallow sowing of 2 to 3 cm below the soil is advantageous. Seed should be treated with Bavistin @ 1.5 g/kg or Thiram @ 3 g/kg to protect the crop from seed borne diseases. Application of 40:20:15 N, P_2O_5, K_2O kg/ha is recommended. For *Utera* cropped linseed, sufficient farmyard or green manure should be applied along with phosphatic fertilizer to rice. Seed rate should be 30 to 40 kg/ha and sowing should be done at the dough stage or 10 to 15 days before rice is harvested. Crack system of sowing in which 5 cm deep cracks are allowed to develop in paddy field and watered for planting. A dose of 10 to 20 kg N/ha should be applied before linseed is sown. Field should be kept weed free by manual weeding once or twice. Timely control of rust and Alternaria blight in Northern India and of linseed bud fly and powdery mildew in Central and Peninsular India have been found to give higher yields.

IMPLEMENTABLE NEW TECHNOLOGIES

The establishment of AICORPO in late 60's gave a momentum to organised and sustained researches in oilseeds and led to a plethora of improved high yielding varieties/hybrids in different oilseed crops and development of matching crop production and protection technologies required to harness their full potentials under diverse oilseeds growing areas and situation in the country. During the last more than two decades of its existence the vast network of research centres (AICORPO) from central and state sector engaged in oilseeds improvement has developed around 240 improved cultivars of different annual oilseed crops of specific regional and multi-regional importance excluding soybean. Of these more than 145 were released during 80's alone which also include four hybrids in sunflower and two in castor. As compared to their traditional counterparts the new generation of currently recommended varieties and hybrids of various oilseed crops recommended for different regions and situations not only possess superior genetic yield potentials but are also endowed with one or the other desirable features such as mono/multiple resistance/tolerance to one or more biotic and abiotic stresses which makes them a better choice. In crops such as sunflower, *toria*, mustard, safflower, groundnut, soybean and sesame a number of improved cultivars developed by the project also complete their life cycle one to two weeks ahead of their traditional counterparts thereby brightening possibilities for successful introduction of oilseeds in proven double (drylands) and multiple cropping (irrigation) systems.

Researches undertaken both within and outside the project have also culminated in the development of a number of location and situations specific agro-production and protection practices involving simple low or no cost inputs such as right choice of crops and varieties, timely planting, use of optimum seeding rates, recommended rhizobia (groundnut and soybean) etc. and critical cash inputs, viz. nutrients (micro, macro and secondary), time, method and frequency of irrigation, integrated management of major insect pests and diseases etc.

Another important area which researches have opened in recent years is the role of oilseeds whose cultivation is, by and large, confined to rainfed areas under conditions of limited irrigation i.e. where water availability is very limited and farmers cannot afford to frequent and heavy irrigations. According to the data now available from the network, under conditions of irrigation water constraint, cultivation of oilseeds is more paying than a number of other traditional crop such as rice, wheat. In fact, in crops such as safflower, groundnut, sunflower, rapeseed-mustard ultimate yields from drylands could be pushed up on experimental farms by 40 to 50 per cent through application of one or two limited irrigation at vulnerable stages of crop growth and/or when moisture is a serious constraint.

Cropping systems research is another fertile area of research which has opened a whole range of new and hitherto unavailable a venues for augmenting production of oilseeds through both horizontal as well as vertical means thanks for the availability of a spectrum of crop and varietal choice with different growth, rhythm, maturity duration and refinements brought about in their management etc. A number of most remunerative and productive location specific/sequential /relay crops involving oilseeds either *inter se* or with other crops now available for doubling/tribling. The prevailing intensity of cropping in different regions and situations and thereby step up yields and returns per unit area, time and input as for instance sequences such as green gram/sesame/soybean-safflower in rainfed vertisols of Deccan *rabi*, toria-potato-spring sunflower in Punjab, U.P. and paddy-groundnut/sesame/sunflower in southern states under irrigation. .

The identification of a whole range of region/location specific, synergistic and compatible intercropping systems with oilseeds either as a base crop or an intercrop is yet another significant achievement which offered considerable scope for expanding the frontier of their cultivation and maximise aggregate yields and returns from rainfed and irrigated areas often coupled with stability; some examples of such synergistic, viable and feasible intercrop combinations are castor/pigeonpea/sunflower with groundnut; castor with sesame and niger; safflower with coriander, linseed, wheat, chickpea in vertisols; soybean with maize, sorghum and cotton; mustard with potato and *gobhi sarson* with *toria* (irrigated) etc.

Research data now available from several oilseed growing areas of the country also point out that the prevailing per hectare yields and returns from rainfed as well as irrigated areas could be considerably stepped up and farmers incomes stabilised through re-orientation of crop choice in favour of those which are efficient in the utilization of rainfall and/or stored moisture, viz. safflower in place of wheat, coriander, chickpea, herbaceum cottons in Deccan *rabi*; groundnut in *rabi*/summer in place of paddy in southern states; mustard in place of wheat in the Indo-gangetic belt; sunflower in place of *rabi*/summer groundnut under restricted water/moisture availability etc.

IMPACT OF IMPROVED TECHNOLOGIES ON THE PERFORMANCE OF OILSEEDS UNDER REAL FARM SITUATIONS

The All India Coordinated Research Project on Oilseeds with a net work of 64 centres and 104 functional units covering eight oilseed crops has generated the location specific, viable and feasible cropping systems, agro-production and protection technologies, improved varieties/hybrids for raising the yield and income from oilseed crops. Under the cess fund supported project large number of demonstrations each of 0.4 ha area were conducted at various locations to demonstrate the improved technologies under real farm situations. The result of rainfed demonstrations conducted for three years (1988-89 to 1990-91) are presented in Table 8. The results clearly point out beyond doubt that the existing per hectare yields and incomes from oilseed crops can be pushed up substantially even with the technology currently recommended. As compared to the local practices prevailing with the farmers, the demonstration plots involving improved technologies registered additional yield to the tune of 43 per cent in groundnut, 59 per cent in mustard, 71 per cent in sunflower, 86 per cent in safflower, 69 per cent in niger, 51 per cent in castor and 60 per cent

Table 8. Yield and economics of improved technologies in relation to prevailing farmers' practices in oilseed crops under rainfed conditions (1988-89 to 1990-91)

Crop	Number of demonstrations	Mean Yield (kg/ha) IT	Mean Yield (kg/ha) FP	% increase over FP	Mean additional cost over FP (Rs/ha)	Mean additional returns (Rs./ha)	Incremental B:C ratio
Groundnut	209	1299	910	43	1145	3118	3.48
Mustard	46	1439	596	59	1170	2505	2.60
Toria	34	746	437	71	1142	2826	2.50
Sesame	159	589	359	64	687	2272	3.83
Sunflower	9	1286	814	58	738	2964	3.61
Safflower	40	926	498	86	719	2918	5.18
Niger	147	422	249	69	715	1418	1.99
Castor	45	1295	858	51	954	2268	2.76
Linseed	166	560	349	60	742	2144	3.36

IT = Improved Technology, FP = Farmers' practice
Source : Directorate of Oilseeds Research, Frontline Demonstrations in Oilseed Crops, Annual Report 1990-91.

in linseed. In many of the crops, the additional investment incurred towards inputs and operational cost is marginal and the benefits accrued from such investment is substantial. From the data on incremental benefit-cost ratio (B:C ratio), it is conspicuous that farmers who resorted to the improved technology realised higher incremental B:C ratio to the extent of 3.48 in groundnut, 2.60 in mustard, 2.5 in *toria*, 3.83 in sesame, 3.61 in sunflower, 5.18 in safflower, 1.99 in niger, 2.76 in castor and 3.36 in linseed. The low cost practices such as choice of appropriate variety, timely planting, recommended seed rate, timely weeding, use of critical cash inputs such as fertilizer and plant protection chemicals played significant role in achieving higher productivity.

REFERENCES

Agasimani, C.A. and Hosmani, M.M. 1989. Response of groundnut crop to farm yard manure, nitrogen and phosphorus in rice fallows in coastal sandy soils. *J.Oilseeds Res.* 6(2) :360-363.

Agasimani, C.A. and Babalad, H.B. 1991. Recent advances in agronomy of groundnut. (*Arachis hypogea) J. Oilseeds Res.* 8(2):133-158.

AICORPO 1984. *Annual Progress Report.* Linseed, Directorate of Oilseeds Research, Hyderabad.

Ankineedu, G., Rao, J.V. and Reddy, B.N. 1983. Advances in fertilizer management for rainfed oilseeds. *Fert. News.* 28(9) : 76-90 & 105.

Hegde, B.R. 1992. Strategies for increasing production in drylands. *Ind. Fmg.* 42(1) : 19-25.

ICRISAT 1987. *ICRISAT Strategies For Higher Groundnut Yields in the Post Rainy Season (Rabi/ Summer).* ICRISAT, Patncheru, Hyderabad.

Jeswani, L.M. 1984. Linseed (*Linum usitatissimum* L.). *Proc. Symposium on Oilseed Production and Utilisation - Constraints and Opportunities.* New Delhi, pp. 232- 249.

Kataria, D.P., Malik, R.K. and Agarwal, M.C. 1992. Risk minimisation in mustard production in drylands through improved technology. *Abstract in Production and Processing Equipment for Oilseeds, Sixth National Convention of Agricultural Engineers.* Hyderabad, p.4.

Patel, R.R. 1985. Castor– in *Efficient management of Dryland Crops.* Central Research Institute for Dryland Agriculture, Hyderabad, pp. 293-309.

Prasad, M.V.R. 1991. *Improved Agro-economic management Practices for Annual oilseed Crops in India.* Country Report Presented at FAO Regional Expert Consultation of Asian Network on Oilseed Crops. Bangkok.

Prasad, M.V.R. 1992. *Annual Progress Report,* Groundnut. Directorate of Oilseeds Research, Hyderabad.

Ranga Rao, V. 1985. Safflower-In. *Efficient management of Dryland Crops.* Central Research Institute for Dryland Agriculture, Hyderabad, pp. 331-352.

Rango Rao, V. and Reddy, B.N. 1987. *Produciton potentials of rabi oilseed based cropping system in different zones of India.* Paper Presented at National Seminar on *'Alternate Farming System'*, New Delhi.

Ranga Rao, V. 1990. Self reliance in vegetable oils - A myth or reality. *J.Oilseeds Res.* 7(2) : 58-61.

Ranga Rao, V. 1991. What catapulated the vegetable oilseeds sector from out of its decade old intertia? two successive good monsoons or wonder technologies ? *J. Oilseeds Res.* 8(1) : 59-66.

Reddy, B.N. and Sharma, S.M. 1983. Agronomic research on sesame in India. *Oilseeds Journal.* 13 (1-4) : 47-52.

Reddy, B.N. Rao, J.V. and Muralidharudu 1985. Weed management in grondnut based intercropping system. *Indian J.agric.Sci.* 55(10) : 631-633.

Reddy, B.N. 1990. Groundnut in intercropping and groundnut based cropping sytems. *J.Oilseeds Res.* 7(2) : 31-40.

Saini, J.S.; Sahota, T.S. and Dhillon, A.S. 1989. Agronomy of Rapeseed-mustard and their place in new and emerging cropping systems. *J.Oilseeds Res.* 6(2) : 220-267.

Sankarareddy, G.H.; Soundararajan, M.S. and Venkataramanareddy, V. 1981. Sequence cropping in drylands. *Nat. Symp. on Crop Management to Meet the New Challenges.* Hissar.

Sharma, S.M. 1984. Constraints and opportunities for increasing the productivity and production of sesame in India. *Proc. Symposium on Oilseed Production and Utilisation - Constraints and Opportunities,* New Delhi. pp. 165-179.

Singh, S.P. 1988. Ways to increase oilseed production from drylands. *Proc. National Seminar on Strategies for Making India Self-reliant in Vegetable Oil,* Hyderabad, pp. 115-118.

Singh, S. and Sanghi, N.K. 1985. Rapeseed and Mustard-In *Efficient management of Dryland Crops.* Central Research Institute for Agriculture. Hyderabad, pp. 311-329.

Venkateswarlu, J. 1985. Sunflower-In *Efficient Management of Dryland Crops*. Central Research Institute for Dryland Agriculture, Hyderabad, pp. 269-277.

23

PROBLEMS PROSPECTS AND MANAGEMENT STRATEGIES OF PULSE PRODUCTION UNDER RAINFED SITUATIONS

D. KUMAR

INTRODUCTION

India has made impressive gains on food-grains, production front from 50 million tons in 1950 to 176 million tons in 1990-91. However, two disturbing features of Indian agriculture need to be highlighted. One, the essential agricultural commodities like pulses and oilseeds are currently contributing a bare minimum of 8.0 and 10.5 per cent, respectively to the national food grain basket. Second, there is a widening gap in cereal to pulse production ratio, being 5:1 in early 50's, rising to 11.5:1 in late 80's. The primary reason has been the sluggish growth rate in pulse production in reference to cereal production.

During the last four decades, there has been stagnation in pulse production (11-14 million tons), leading to short supply of protein-rich vegetarian diets. Cereal dominated Indian diets are particularly deficient in energy rich proteins, essential amino acids, vitamins and minerals.

Pulses constitute essential components of vegetarian foods by providing much needed proteins required to balance the predominated cereal diets. Pulses conserve and enrich the soils in respect of fertility gradients by way of well documented atmospheric nitrogen fixation process. Pulses are also the important components of cropping systems in marginal and sub-marginal areas of dryland farms as 92 per cent of the pulse production in India is realized from drylands or the areas depending on rains for the cultivation of crops.

The gross demand for pulses in India has been projected at 30.4 million tons by the beginning of 2000 AD. Looking at the present day growth trend of pulses, this task appears herculis.

Attempts have therefore, been made to compile relevant information on growth rate of area, production and productivity, principal areas, chief constraints in pulse production, cropping systems, components of production technology, improved cultivars, important diseases and insect pests in respect of important pulses with particular references to dryland conditions.

INDIAN PULSES SCENARIO

Over past three decades, the area (21.0 to 24.5 million hectares), production (8.3 to 13.4 million tons) and the productivity (385-589 kg/ha) of pulses have remained almost static in India (Table 1). During 1990-91 only, the corresponding values have exhibited remarkable increase to the extent of 14.3, 40.0 and 21.5 per cent, respectively over the preceding year. However, in order to have better understanding on the pulse production in India, it would be quite logic to examine production trends of two major pulses i.e. chickpea and pigeonpea which occupy 45 per cent of the total area and contribute almost 60 per cent to the total pulse production. All India compound growth rate on area, production and grain yield of chickpea, pigeonpea, other pulses and total pulses over three selected periods are presented in Table 2.

Table 1. Area, production and productivity of pulses in India during different plan periods

Periods	Area (m ha)	Production (m t)	Productivity (kg/ha)
1951-56	21.00	10.04	475
1956-61	23.71	11.77	495
1961-66	23.85	11.14	467
1966\67-1968\69	22.01	10.29	467
1969-74	22.21	10.90	491
1974-79	23.11	11.42	493
1979-80	21.75	8.37	385
1980-81	22.31	11.09	497
1981-82	23.83	11.50	483
1982-83	22.38	11.51	514
1984-85	22.80	12.65	541
1985-86	24.50	13.40	547
1986-87	23.40	11.70	506
1987-88	21.30	10.96	515
1988-89	22.36	13.50	589
1989-90	23.41	12.86	549
1990-91	24.39	14.06	576

Source : Agriculture situation in India, 1991.

It will be observed from the data that growth rate of pulse production in India was considerably higher during pre-green revolution era (1949-50 to 1964-65) compared to two latter periods, viz., 1949-50 to 1990-91 and 1967-68 to 1990-91; it however, considerably improved (1.87% per annum) during the last decade i.e. 1980-81 to 1990-91. Contrary to it, growth rate of productivity of total pulses that was negative during the pre-green revolution period (-0.50% per annum) considerably increased during the later periods, being highest (1.68% per annum) during the last decade. Situation in respect of growth rate on area of total pulses has been more disheartening. For instance, compound growth rate on area, that was maximum (1.90% per annum) in 1949-50 to 1964-65, declined to 0.28% in 1949-50 to 1990-91, slightly improved (0.33% per annum) during 1967-68 to 1990-91 and further lowered down to a minimum of 0.18% per annum during the past decade.

Table 2. All India compound growth rates of area, production and yield of pulses (% per annum)

Crop	1949-50 to 1964-65			1949-50 to 1990-91			1967-68 to 1990-91			1980-81 to 1990-91		
	Area	Production	Yield	Area	Production	Yield	Area	Production	Yield	Area	Production	Yield
Gram	1.64	2.66	1.00	-0.66	-0.09	0.58	-0.65	-0.38	0.26	-0.77	-0.10	0.68
Arhar	0.57	-1.34	-1.90	0.69	0.81	-0.09	1.58	2.02	0.43	2.32	2.28	-0.04
Other pulses	2.07	1.28	-0.77	0.90	0.86	0.16	0.61	1.66	1.05	0.18	3.30	3.11
Total	1.90	1.39	-0.50	0.28	0.46	0.19	0.33	0.93	0.59	0.18	1.87	1.68

Source : Government of India, Indian Agriculture in Brief, 1990.

The simple reason of poor production of pulses in India therefore, appears to be continued declined trends in its area over a period of last four decades. The same, in turn, may have been due to the conversion of pulses areas to assured and more remunerative crops like wheat, mustard etc. and probably due to increased irrigated area. Farmers do not like to cultivate pulses in assured irrigated and high cost inputs areas. Chickpea, the top ranking pulse in India, is a glaring example of this fact, as the growth rate of area, production and yield have gradually gone down in post-green revolution era. Mean compound growth rates on area, production and grain yield of chickpea were considerably higher (1.64, 2.66 and 1.00% per annum, respectively) during the period of 1949-50 to 1964-65 when compared to the last decade period (-0.77, -0.10 and 0.68% per annum, respectively) reflecting negative growth rate in area and production. Eventhough, rate of growth in grain yield increased during 1980-81 to 1990-91, yet it was not sufficient to compensate the declining trend in area; ultimately the production of chickpea decreased in India.

Pigeonpea the second important *kharif* and *rabi* pulse of India, reflected remarkably increased trend in growth rate of area and production during the post-green revolution era over the pre-green revolution period. To great dismay, however, growth rate in respect of grain yield gave no sign of relief as the same exhibited a negative value barring the period 1967-68 to 1990-91. The growth trends in area of pulses other than chickpea and pigeonpea was at par to total pulses which declined in post-green revolution period, being minimum (0.18% per annum) in the last decade compared to 1949-50 to 1964-65 (2.07% per annum). However, the rate in growth of grain yield was many folds higher in case of chickpea and pigeonpea and so was the case in regard to production. It may therefore, be logical that low production rate of pulses in India seems to have been due to continued declining trends in area of almost all the pulse crops and negative growth rate in productivity of pigeonpea, in particular (Table 2).

INDIA IN THE WORLD'S PULSE MAP

Statistics available indicate that world production of pulse grains during 1988 was 54.6 million metric tons produced from 68.5 million hectares (Table 3). The scenario in developed and under-developed countries was quite in contrast as the percent increase in area, production and yield of developed countries considerably shot-up to the extent of 41, 98 and 40 per cent during 1988 over the base year 1979-81. On the contrary, the corresponding figures in the developing countries were much less to the tune of 9, 14 and 5 per cent, respectively. Situation in India did not appear to be much different from the developing countries. For instance, compared to the base years 1979-81, the area and production of grain legumes during 1988 increased marginally to the extent of 3.5 and 2.9 per cent vis-a-vis 14 and 34 per cent corresponding to world pulse production figures. India, however, has a distinction of being world's largest pulse producing (25% of the world) and growing (40% of the world) country of the world. Productivity of pulses in India was however, lower (589 kg/ha) than the world productivity (797 kg/ha). Moreover, an insight into pulse productivity of different countries reflects quite discouraging productivity potential of pulses in India. For instance, productivity of pulses in China (1203 kg/ha), New Zealand (2330 kg/ha), Egypt (2699 kg/ha), France (4127 kg/ha), United Kingdom (4138 kg/ha), Austria (5670 kg/ha) and Netherlands (5751 kg/ha) were, in general, 2 to 10 times higher than the pulse productivity in India (FAO, 1989). The shortcomings and prospects of pulse production in India can be observed by examining the same figures for some of the important pulse crops individually.

Evidences gathered so far point out India as a centre of origin of pigeonpea. It is widely grown in India and occupies world's 87 per cent area and 93% production (Table 4). The south Asia, Africa and America are the regions of great significance for this crop. The pigeonpea area is also substantial in Kenya, Uganda, Malawi, Domanican Republic and Puerto Rico. Besides, these major pigeonpea countries, there are 22 countries in Asia and Oceana, 32 in Africa and 27 in America where pigeonpea are grown in limited scale (Van der Malsen, 1983, 1986; Nyabyenda, 1987; Nene *et al.* 1989). Regarding production, besides India, other important countries are: Myanmar, Uganda, Malawi, Domanican Republic, Tanzania, Nepal, Venezuela, whereas contribution of other countries like Trinidad, Guayana, Grenda, Bar-

Table 3. Pulse production in developed and developing countries in comparison to production in India in 1988.

Zone/area	Area 1988 (ooo ha)	Inerease over 1979-81 (%)	Grain Yield 1988 (kg/ha)	Inerease over 1979-81 (%)	Production 1988 (mt.)	Increase over 1979-81 (%)
Developed countries	12,811	41.0	1494	40.0	19,137	98.0
Developing countries	55,726	9.0	637	5.0	35,515	14.0
World total	68,537	14.0	797	18.0	54,652	34.0
India	23,300	3.5	589	24.3	137000	2.9

Source : F.A.O. Production Year Book for 1988 (FAO, 1989)

Table 4. World pigeonpea production of major pigeonpea growing countries.

Country	Year	Area (000 ha)	Production (000 t)	Reference
India	1979-84	2973.0	2230.0	ICRISAT 1990
Uganda	1985	63.0	22.0	Brathwaite 1981
Tanzania	1972	22.0	11.0	Sharma and Green 1975
Malawi	1972	35.0	20.0	Shrama and Green 1975
Nepal	1987-88	18.0	9.0	ICRISAT 1990
Domanican Republic	1978	13.9	14.5	Brathwaite 1981
Venezuela	1972	11.0	6.0	Sharma and Green 1975
Burma	1985-86	8.1	52.0	Wallis *et. al.* 1988
Haiti	1978	6.6	4.0	Brathwaite 1981
Puerto Rico	1972	3.0	4.0	Sharma and Green 1975
Panama	1978	2.7	3.4	Brathwaite 1981
Bangladesh	1985-86	2.4	1.7	ICRISAT 1990

bados, Australia, Antigua is marginal towards general production.

Chickpea with a total annual production of 75 million tons from an area of about 10.5 million hectares ranks among the world's three most important pulses (Summerfield *et al.* 1985). The *Desi* type chickpea with 85 per cent of world's production are mainly confined to the Indian subcontinent, Ethiopia, Mexico and Iran, whereas Kabuli types comprise only a small part in India, but account entirely in Afghanistan, South Europe, Americas and north Africa. The area and production of chickpea have fluctuated world over, including India. However, there are regions where area and production of chickpea have increased during the eighties. These are : Turkey, Syria, Iran, north central and south America and east Africa. In contrast, countries like Australia, Canada, USA and Kenya constitute marginal area, hence make little difference to the general pattern. World trends and fluctuations in chickpea production are principally dominated by the events that happen in Indian subcontinent that account for nearly 77 per cent of global production from 76 per cent of the area occupied by this crop. In India, the same increased during the sixties but have been declining till recently.

Information gathered till date point the mungbean as the wild form in India. The world mungbean production of 3.04 million tons is estimated to be harvested from 5.12 million hectares (Table 5). Mungbean therefore, contributes only 5 per cent to the total pulse production world over. The major mungbean producing countries are: Iraq, Iran, Pakistan, India, Sri lanka, Bangladesh, Burma, Thailand, Vietnam, China, Philippines and Indonesia. The major mungbean exporting countries are: Australia, Burma, China, India, Kenya and Peru. India is the

Table 5. World mungbean production of major mungbean growing countries.

Country	Year	Area (000 ha)	Yield (kg/ha)	Production (000 mt)	References
India	1986-87	3237.2	415	1883.7	Sandhu, *et. al.* 1988
China	1987	500.0	675	337.5	Ling *et al.* 1988
Thailand	1986	434.3	606	262.0	Chainuvati et al. 1988
Indonesia	1985	280.0	800	192.0	Brotonegoro et al. 1988
Burma	1984-85	240.0	667	160.0	Singh, 1988
Pakistan	1986	115.0	504	58.0	Bashir et al. 1988
Kenya	1983	39.0	307	12.0	Muthoka and Shakoor, 1988
Philippines	1986	37.5	691	25.9	Navarro, 1988
Nepal	1986	15.4	526	8.1	Navarro, 1988
Korea	1986	7.8	890	7.0	

leading country in the production of mungbean with almost 63 per cent of the world hectarage and 62 per cent of the world production. Mungbean ranks third in India among the pulses after chickpea and pigeonpea (Singh and Yadav, 1978). Compared to world productivity (602.5 kg/ha), the productivity of mungbean in India (415 kg/ha) is least (barring Kenya) in the world. The countries like Korea (890 kg/ha) and Indonesia (800 kg/ha) have more than two-times productivity than that of India.

Lentils are an old world legumes and were probably one of the first species to be domesticated. Lentils at the global scale are a minor pulse as they contribute only 3 per cent to the total pulse production of the world. Lentils are, however, major pulse in Indian subcontinent, and in the semi-arid tracts of middle east, Turkey and Northern Africa. World production of lentils during 1980 was estimated to be 1.1 million tons from 1.8 million hectares of land (Table 6). The productivity of lentils (655 kg/ha) has been static from 1961 to 1980. India ranks first in area (50.2%) and production (38.1%) of lentils in the world. Productivity (486 kg/ha) is, however, quite low compared to the remarkably higher productivity of other countries like USSR (1818 kg/ha), Turkey (1169 kg/ha), USA (1148 kg/ha), Egypt (1111 kg/ha). There are only few countries (Pakistan, Morocco, Canada) having productivity slightly lower than India. Yields of lentils have varied greatly over the years in different countries. For instance, in Italy the area declined from about 340 thousand hectares in 1927 to only 2 thousand tons in 1977 (Solh and Erskine 1981). This was probably due to heavy infection of parasitic *Orobanche* weed. In Egypt, the construction of 'Aswan dam' in 1965 was mainly responsible for declining trend in lentils production. In Turkey, which is the largest exporter of lentils, the production was enhanced due to the feasibility of replacing summer-fallow in cereal production system with annual grain legumes like lentil or chickpea. In India, stagnation in production of lentils seems to be due to lack of alternative crop to be grown in the marginal lands during the *rabi* season.

Table 6. World production of lentils during 1980

Country	Area (000 ha)	Yield (kg/ha)	Production (000 t)
A-rgentina	40	875	29
Bangladesh	85	600	51
Canada	44	450	15
Chili	53	507	26
Ethiopia	58	470	29
Egypt	06	1111	11
India	925	486	428
Iran	38	737	27
Morocco	37	432	17
Pakistan	86	422	36
Spain	63	905	54
Syria	84	952	72
Turkey	201	1169	199
USA	53	1148	59
USSA	11	1818	10

Source : FAO Production Yearbook (1980); and Nygaard and Hawtin (1981)

Peas constitute world's one of four most important grain legumes (Davies, 1976). The total area of world dry pea is estimated around 8.5 million hectares and production 8 to 10 million tons. Dry peas are more extensively grown than green peas. USA is the single largest exporter of peas and constitutes 18 per cent of the global production of grain legumes. However, China and Russia jointly produce world's 80 per cent dried peas. Area of peas have considerably increased in USSR, Australia and Czechoslovakia, it has however declined in USA, China and India in 1979-81. India constitutes only 7.7 per cent world acreage and 3.9 per cent production. The reason for poor production of dried peas in India can be explained on the basis of productivity that is least (570 kg ha^{-1}) in the world, whereas leading peas producing countries like China, USSR and USA have remarkably very high productivity to the extent of 1500, 1050 and 2290 kg/ha, respectively (Table 7).

Other grain legumes like blackgram, cowpea, horsegram, mothbean and *Lathyrus* are treated as minor pulses at the global level. Production statistics of these pulses are not well documented. The FAO publishes information on the acreage and production on total pulses and on some specific pulses which are of global significance (chickpea, dry pea, dry bean, broad bean and lentils). Moreover, mungbean and blackgram are included in dry bean. However, out of a total of 54.7 million metric tons (m mt) pulse production world over, was harvested from dry bean (15.5 m mt), broadbean (4.7 m mt), dry peas (15.5 m mt), chickpea (5.8 m mt) and lentils (2.5 m mt). Remaining 10.7 million metric tons was harvested from other legumes-cowpea, pigeonpea, blackgram, horsegram, mothbean and lupin etc. (FAO 1989). Thus, it is difficult to bring out production figures of minor pulse (major pulses in many tracts of India) at the global level.

Table 7. World production of dry peas during 1979-81

Country	Area (000 ha)	Yield (t/ha)	Production (000t)
Australia	47	1.17	55
Burundi	31	1.19	37
Canada	47	1.76	84
China	1533	1.50	2300
Colombia	56	0.59	33
Czechoslo-vakia	31	2.11	64
Ethiopia	131	0.98	128
France	26	4.13	107
Hungary	53	2.03	104
India	571	0.57	330
Iran	25	1.26	31
Morocco	49	0.48	25
Newzealand	24	2.82	67
Poland	42	1.55	65
Rwanda	60	0.78	47
USA	70	2.29	162
USSR	4181	1.05	4339
U.K.	36	2.96	107
Zaire	73	0.63	47

Source : FAO Production Yearbook (1981).

Horsegram grown on 1.8 million hectares of land in India, is however grown marginally in Australia, Malaysia, Mauritius and West Indies. Likewise, mothbean, a principal pulse crop of arid tracts, is grown in north west desert regions of Indo-Pak subcontinent, and also sporadically grown in dry habitats of Burma, Srilanka, Malaysia, South China, part of Afghanistan. In southwest USA, it is grown for fodder, feed and pasture. Official statistics on its global area are not available as in case of black gram, Lathyrus and other minor pulses.

PRINCIPAL AREAS OF PULSE PRODUCTION IN INDIA

In India over a dozen of pulses are grown on 24 million hectares of land during *rabi, kharif* and spring seasons. However, 92 per cent pulses are grown as rainfed. Chickpea, pigeonpea, mungbean and uradbean are the major pulses in India. However, chickpea and pigeonpea jointly constitute 60 per cent of the total pulse production and the species of *Vigna* group constitute nearly 25 per cent of the total production in India. Statewise statistics on area, production and productivity of pulses as a whole indicate that Madhya Pradesh (4.7 m ha), Maharashtra (3.3 m ha), Uttar Pradesh (2.94 m ha), Rajasthan (2.92 m ha), Orrisa (1.9 m ha), Kerala (1.6 m ha), Andhra Pradesh (1.5 m ha), Bihar (1.1 m ha), Tamil Nadu (0.89 m ha), Gujarat (0.77 m ha) and Haryana (0.73 m ha) were major pulse growing states of India during 1988-89 (Table 8). Pulse productivity was maximum in Haryana (926 kg/ha), followed by Uttar Pradesh (904 kg/ha), Karnataka (741 kg/ha), Bihar (688 kg/ha), West Bengal (674 kg/ha), Gujarat (635 kg/ha) whereas, it was quite low in Himachal Pradesh (171 kg/ha) and Jammu & Kashmir (114 kg/ha). Productivity during 1988-89 decreased over 1965-67 in states like Assam, Himachal Pradesh, Jammu & Kashmir and Punjab. Highest production of pulses in Uttar Pradesh was due to higher productivity; and in Madhya Pradesh due to maximum area under pulses. However, precise conclusion on contribution of states in respect of area and production can be drawn by examining the acreage and production figures for individual pulse crop over the states (Tables 9.1 to 9.3).

Pigeonpea is grown in large areas (3.2 m ha) in almost all the states of India, however, Maharashtra, Uttar Pradesh, Madhya Pradesh, Andhra Pradesh, Gujarat Karnataka, Tamil Nadu and Orissa are the principal pigeonpea growing states. Four states viz., Maharashtra, Madhya Pradesh, Karnataka and Uttar Pradesh

Table 8. State-wise area production and yield of pulses in India

State	Area (000 ha)			Production (000 t)			Yield (kg/ha)		
	1965-67	1988-89	% change	1965-67	1988-89	% change	1965-67	1988-89	% change
Andhra Pradesh	1227	1500	22.3	236	670	183.90	192	449	133.8
Assam	84	122	45.2	36	51	41.7	428	418	-2.3
Bihar	2048	1130	-44.8	1010	689	-27.72	493	688	39.6
Gujarat	421	770	82.9	153	490	220.26	363	635	74.9
Haryana	1196	730	-39.0	673	670	-0.45	562	926	64.8
Himachal-Pradesh	38	41	7.9	11	7	-36.36	289	171	-40.8
Jammu & Kashmir	48	44	-8.3	25	5	-80.0	520	114	-78.1
Karnataka	41	27	-34.1	15	20	33.3	365	741	103.0
Kerala	986	1620	64.3	278	520	87.05	281	320	13.9
Madhya Pradesh	3846	4770	36.88	1369	2680	95.76	392	562	43.4
Maharashtra	2030	3330	64.0	663	1730	160.94	326	519	59.2
Orissa	944	1930	104.4	427	1080	152.93	452	560	23.9
Rajasthan	3050	2920	-4.26	849	1610	89.63	278	553	98.9
Tamil Nadu	385	890	-131.2	89	340	282.02	231	387	67.5
Uttar Pradesh	4241	2940	-30.7	3010	2660	-11.63	709	904	27.5
West Bengal	765	310	-59.5	406	210	-48.27	530	674	27.2
Punjab	707	165	-76.7	534	101	-81.1	755	612	-18.9
Other States	20	21	5.0	13	59	353.8	650	600	-7.7
India	22,077	23,260	5.36	9797	13592	39.54	451	589	30.6

Source : Indian Agriculture in Brief Various Issues, GOI. (A.S.I. 1991)

jointly occupy nearly 67 per cent area and 73 per cent production of pigeonpea in India. However, Maharashtra is the single most important state constituting nearly 23.5 per cent area and 16 per cent production of pigeonpea in India. Due to expansion of irrigated areas and development of new high yielding varieties for northern India, the area and production of chickpea, have been declining till date. Principal chickpea growing states in India are Madhya Pradesh (2.3 m h), followed by Uttar Pradesh (1.27 m h), Rajasthan (1.20 m h), Maharashtra (0.71 m ha), Haryana (0.65 m h) and Gujarat, (0.16 m h). The major states, viz., Madhya Pradesh, Rajasthan and Uttar Pradesh jointly constitute nearly 70 per cent area and 77 per cent production in India. Madhya Pradesh is the leading chickpea producing state in India, which remarkably occupies 30 per cent area and produces 30 per cent chickpea grain in India.

Third most important pulse in India is mungbean which is planted on 3.2 million hectares of land annually. It is cultivated in almost all the states of India but largest production is realized from Andhra Pradesh, Orissa, Maharashtra, Madhya Pradesh, Rajasthan, Bihar and Gujarat (Singh *et al.*, 1988). Nearly 72 per cent of mungbean area is planted during rainy season which constitutes 61 per cent production annually (Sandhu *et al.*, 1988). Area and production of mungbean have considerably increased during eighties in India (Tickoo *et al.*, 1988). The production of mungbean in India has increased 3-folds during 1986 compared to 1965.

Table 9.1. All India area (000 ha) and production (000 t) of pigeonpea, chickpea and mungbean during 1986-87.

State	Pigeonpea		Chickpea		Mungbean	
	Area	Production	Area	Production	Area	Production
Andhra Pradesh	271.1	80.1	51.1	20.3	513.8	149.6
Assam	9.9	7.2	3.6	1.7	–	–
Bihar	82.8	105.0	186.0	149.2	190.6	92.6
Gujarat	318.7	146.1	43.6	28.1	157.6	14.0
Haryana	34.3	40.4	610.0	413.0	9.0	5.3
Himanchal Pradesh	–	–	9.3	2.0	0.2	–
Karnataka	436.7	191.7	217.2	83.2	173.9	44.6
Jammu Kashmir	–	–	0.7	0.4	3.4	4.0
Kerala	0.2	0.2	–	–	1.6	1.1
Madhya Pradesh	475.3	455.7	2093.8	1372.4	204.5	50.1
Maharashtra	759.9	372.4	482.4	129.4	520.5	158.5
Meghalaya	0.8	0.7	0.5	0.3	–	–
Orissa	144.0	93.1	41.9	28.9	692.4	384.3
Punjab	30.2	30.2	114.0	82.0	51.4	46.5
Rajasthan	16.7	3.5	1313.1	760.8	279.0	18.4
Tamil Nadu	152.9	80.5	7.3	4.8	413.9	144.9
Tripura	0.5	0.2	0.5	0.2	0.7	0.3
Uttar Pradesh	491.7	672.1	1534.7	1326.7	19.9	4.7
West Bengal	7.6	9.6	70.1	50.2	4.8	2.9
Dadra & Nagar Haveli	1.8	2.0	0.2	0.1	–	–
Delhi	0.6	0.2	0.9	0.9	–	–
All India	3,235.7	2,316.3	6780.9	4455.3	3237.2	1121.8

Nearly 0.14 m ton dried peas in India are harvested from 0.48 m ha lands annually. Madhya Pradesh, Bihar and Uttar Pradesh are the principal growing states which constitute nearly 73 per cent area. Rajasthan and Orissa are other important states. Rest of the states like, Haryana, Himachal Pradesh, Jammu & Kashmir, Maharashtra, Punjab and West Bengal occupy only marginal areas (0.2 to 5 thousand ha). Uttar Pradesh and Madhya Pradesh jointly constitute nearly 82 per cent area but produce only 2 per cent dried pea.

Lentils constitute only 7 per cent to the total pulse production in India, hence these are considered as minor pulse. Uttar Pradesh (0.45 m ha), Madhya Pradesh (0.28 m ha), and Bihar (0.16 m ha) are the principal lentil growing states which jointly occupy nearly 85 per cent area and contributes 33 per cent to the total lentil production in India. West Bengal (0.093 m ha) also constitutes considerable area. The crop has its distribution in North west, Central and Eastern zones. Southern states and the interior North eren States (Jammu & Kashmir and Himachal Pradesh) have negligible area.

Blackgram occupies 12 per cent of the total pulse area and contributes 8 per cent to the total pulse production in India. Madhya Pradesh with 27 per cent of area is the largest blackgram growing state in India. Other major blackgram growing states are Maharashtra, Tamil Nadu, Andhra Pradesh, Uttar Pradesh, Orissa, Rajasthan and Bihar constituting nearly 15, 14.3, 10.2, 7.3, 7.0, 6.4 and 3.5 per cent area of lentil, respectively in India. In the states of Madhya Pradesh, Maharashtra, Uttar Pradesh, Tamil Nadu, Orissa and Rajasthan, blackgram is cultivated as a rainy season crop whereas, in Orissa, Andhra Pradesh and West Bengal it is grown as post-rainy season crop. Andhra Pradesh produces maximum

Table 9.2. All India area (000 ha) and production (000 t) of peas, lientil and khesari.

State	Peas		Lentil		Khesari	
	Area	Production	Area	Production	Area	Production
Andhra Pradesh	–	–	–	–	–	–
Bihar	35.2	35.2	166.5	128.6	361.6	284.7
Gujarat	–	–	–	–	–	–
Haryana	6.5	4.3	23.9	14.1	–	–
Himachal Pradesh	1.8	1.8	0.5	0.5	–	–
Jammu & Kashmir	0.2	0.2	–	–	–	–
Karnataka	–	–	–	–	–	–
Kerala	6.8	5.6	–	–	–	–
Madhya Pradesh	114.4	38.1	289.7	124.9	578.8	177.7
Maharashtra	9.8	2.5	12.4	3.2	66.3	17.1
Orissa	13.7	6.9	–	–	–	–
Punjab	3.6	3.5	13.4	9.8	–	–
Rajasthan	4.5	12.1	16.6	17.1	–	–
Tamil Nadu	–	–	–	–	–	–
Tripura	0.6	0.3	0.6	0.2	–	–
Uttar Pradesh	280.0	33.1	455.6	317.5	–	–
West Bengal	5.1	2.7	93.0	50.0	99.0	63.5
All India	482.2	145.7	1072.2	1738.2	1105.7	543.0

Table 9-3. All India area (000 ha) and production (000 t) of urdbean, mothbean and Kulthi

State	Urd bean		Mothbean		Kulthi	
	Area	Production	Area	Production	Area	Production
Andhra Pradesh	294.9	212.0	–	–	131.9	55.5
Bihar	100.8	47.8	–	–	84.9	48.2
Gujarat	82.1	34.0	69.6	83.9	–	–
Haryana	2.9	1.9	0.2	0.1	–	–
Himachal Pradesh	19.5	3.1	0.3	0.1	5.4	0.9
Jammu & Kashmir	22.6	6.9	9.2	10.4	2.9	0.8
Karnataka	47.0	15.9	–	–	436.0	163,3
Kerala	2.2	1.6	–	–	22.1	15.8
Madhy Pradesh	475.1	166.1	–	–	1773.3	45.3
Maharashtra	451.9	146.6	136.6	34.7	324.5	92.2
Orissa	203.5	112.1	–	–	368.3	162.3
Punjab	11.5	6.7	1.4	0.7	–	–
Rajasthan	186.8	21.2	1154.0	42.8	–	–
Tamil Nadu	423.9	144.9	–	–	143.0	61.1
Uttar Pradesh	212.7	55.7	1.4	0.2	–	–
West Bengal	88.5	56.4	0.1	0.1	6.7	3.6
All India	2885.9	1032.9	1372.8	173.0	1700.0	649.0

Source : Agriculature Situation in India, April, 1988

(20.5%) blackgram in India due to highest productivity (700 kg/ha) which is 3-times more than the national productivity (350 kg/ha). Madhya Pradesh has largest area, however due to poor productivity (247 kg ha) produces only 166 thousand tons annually. Area and production of blackgram are lowest in Kerala, whereas productivity is lowest (105 kg/ha) in Himachal Pradesh.

Horesgram a minor pulse is grown on almost 1.8 m ha of lands in India which produces 0.64 mt of grains annually. Almost 90 per cent of the total area of horsegram is constituted by six states, viz., Karnataka, Andhra Pradesh, Madhya Pradesh, Maharashtra, Orissa and Tamil Nadu. Karnataka with 25 per cent area is the single largest horsegram growing state in India. In Karnataka, 42 per cent of the total pulse area is contributed by horsegram only, whereas the same area in Tamil Nadu is almost one-third. National productivity of horsegram is generally low (325 kg/ha).

Mothbean the principal arid pulse is grown on 1.8 m ha of land in India which produces 0.173 mt annually. It is abundantly grown on dry tracts of Rajasthan, Gujarat, Maharashtra and Haryana. Rajasthan alone occupies 70 per cent area of mothbean in India and shares only 24 per cent to the total national production, whereas Gujarat form an area of 5.0 per cent, produces 48.4 per cent of total mothbean grains in India. Differences are due to low productivity of mothbean in Rajasthan.

Lathyrus is still a minor pulse in India. Nearly 0.5 mt (3.8%) of Lathyrus is produced from 1.1 m ha (4.7%) of land in India. It is grown in 4 states viz., Madhya Pradesh, Andhra Pradesh, West Bengal and Maharashtra. Madhya Pradesh is the single largest (0.57 m h) *khesari* growing state occupying nearly 50 per cent area and sharing 32 per cent of the total *khesari* production in India.

PULSE AVAILABILITY IN INDIA

Pulses and oilseeds are the chief sources of proteins and fats for all sections of vegetarians across the different sections of income, religion and regional classes. The former constitutes an important gradient of vegetarian diets in developing and densely populated countries and provide essentially much needed vegetable proteins to balance the cereal diets. However, due to stagnation in grain legume production in concomitant with continued population pressure, the overall net availability of pulses in India has been adversely affected compared to the other commodities like cereals and oilseeds. Pulse availability that was 61 g/day/adult in 1951 was reduced to 40 g in 1987-88 and declined to 36.5 g/day/adult in 1990 as against 80 g/day/adult recommended by FAO/WHO and 70 g/day/adult by ICMR in 1968. The declining trend in availability of gram pulse was sharper that declined from 30.2 g/day/adult in 1961 to remarkably low 11 g/day/adult in 1990 (Bhatia, 1991). On the contrary, availability of oilseeds and cereals increased from 10.9 and 339.7 g/day/adult in 1961 to 17.8 and 455.0 g/day/adult in 1989, respectively (Table 10). Looking at the present growth rate of population to the tune of 2.2 per cent annum and the meagre growth output of pulses just 0.40 per cent annum, it is feared that even ICMR recommendation of bare minimum (70 g/day/adult) requirement of pulses may not be fulfilled in near future and the Indian masses may remain deprived of essentially needed food protein.

Table 10. Availability of pulses, oilseeds and cerals (g/day/copita) in India

Years	Pulses	Oilseeds	Cereals
1951	60.7	–	334.2
1961	69.0	10.9	339.7
1971	51.2	12.3	417.6
1975	39.7	10.7	365.8
1980	30.9	12.8	379.5
1985	38.1	18.0	415.6
1986	41.9	18.6	434.3
1987	35.9	17.0	436.1
1988	33.1	19.2	413.2
1989	40.4	17,8	455.0
1990	36.5	–	438.1

Source : Government of India, Economicc Survey of India, 1990.

NUTRITIVE VALUES OF PULSES

Pulses have great relevance in those regions where animal proteins are less available or where poverty or other social factors preclude the consumption of meats. Pulse proteins are important because they balance the nutritionally deficient cereals in respect of proteins, they provide abundance of food energy and also provide minerals and vitamins. However, pulse proteins as food have known shortcomings also. Pulse proteins are deficient in sulphur containing amino acids. These proteins are known to carry antinutritional component and have poor digestibility than the animal proteins.

In respect of energy provided by pulse grains, all the pulses were almost at par, however, chickpea and lens had a bit more whereas, cowpea had minimum energy per 100 g of dry edible parts (Table 11).

Protein content had a wide range from 19.4 (chickpea) to as high as 29.9 per cent (*Lathyrus*). The oil content was bare minimum (0.8 to 5.5%). Pulse legumes are rich in digestible carbohydrates. Total carbohydrates were however maximum in pigeonpea and mothbean (72.5-72.7%). Pulses also contain appreciable proportions of dietary fibre (4.3 to 15.9%). Frenchbean was richest in respect of Ca, P and Fe contents.

Table 11. Nutritional composition of pulses (100 g of dry edible parts)

Crops	Energy	Protein	Oil	Total Carbohydrate	Fiber	Ash	Ca	P	Fe
Pigeonpea	383	21.9	1.5	72.7	8.1	4.2	179	316	16.6
Chickpea	396	19.4	5.5	70.5	7.4	3.4	280	301	12.3
Cowpea	316	25.6	1.7	67.5	13.7	5.1	376	385	6.0
Mungbean	381	25.6	1.3	69.2	4.9	3.9	118	370	7.9
Urdbean	385	23.5	1.8	71.0	4.9	3.8	123	390	9.4
Frenchbean	318	22.1	1.7	69.9	15.9	6.2	381	425	12.4
Mothbean	378	20.7	1.3	72.5	5.1	3.3	120	332	–
Field peas	391	25.6	2.3	65.5	9.1	4.1	91	331	6.6
Lens	393	26.9	0.8	67.8	4.3	3.2	71	331	7.7
Lathyrus	379	29.9	1.2	65.2	8.0	3.6	–	447	10.9

Source : Chandra and Ali, 1986.

Pulses are rich sources of minerals like, calcium, iron and potassium. Mungbean is a good source of thiamin, lysine, whereas lentil is a good source of ascorbic acid B_6. Cowpea is deficient in ascorbic acid however, is a good source of thiamin. Chickpea provides rich contents of ascorbic acid, panthothemic acid and B_6 (Haytowitz and Matthews, 1986).

Proteins

In pulses, globulins are the major storage proteins which constitute 70 per cent of the total proteins. Glutelins (10- 20%) and albumins (10-20%) make the remainder proteins (Aykroyd *et al.*, 1982). Albumin and globulins which are the non-storage proteins are mainly present in enzymes, membranes and structural units. Protein contents greatly vary with production environments, production practices and also with the varieties. Grain legumes contain only small amounts of lipids as phospholipids.

Amino acids

Amino acids are important in human diets because the same are generally imbalanced in vegetable proteins as lysine, methionine, cystine, threonine and tryptophan. A general rule is that amino acids like, lysine, methionine, threonine, and tryptophan be supplied in the human nutrition in proportion of 4:2:1, respectively.

Detailed studies on pulse amino acids have been carried out (Gupta and Kapoor, 1980; Gupta, 1983; Augustin and Klein, 1989). Studies carried out by Venkat Rao *et al.* (1964) suggested that deficient cereal nutrition could be balanced by pulse proteins. Data on average composition of important pulses along with wheat, rice and egg protein amino acids are given in Table 12. Data reveal that all the pulses possessed two-folds more lysine content than rice and wheat. Pigeonpea had the maximum proportion of lysine and phenylalanine in which cereals are generally found deficient. Lentil had more cystine than the other pulses whereas, cowpea was richer in tryptophan and tyrosine. Valine was found more in blackgram. Egg protein had remarkably higher contents of all the amino acids and was almost at par to pulses in respect of valine. Pulse proteins are in general rich in lysine, isoleucine, threosine, leucine and valine but had comparatively less contents of methionine and

Table 12. Amino-acid composition of seeds of some pulses, oilseeds (soybean) and cereals (wheat and rice) and egg protein expressed as milligrams per g of nitrogen, edible portion [a b c] (Mg.)

Species	Cystine	Isoleucine	Leucine	Lysine	Methionine	Phenylalanine	Threonine	Trypto phan	Tyrosine	Valine
Mungbean	55	264	484	436	75	378	205	68	187	324
Blackgram	58	319	518	415	91	365	217	65	194	351
Cowpea	69	255	479	423	89	365	238	77	202	298
Drybean	68	276	499	429	94	338	263	74	176	327
Chickpea	84	268	445	418	82	335	232	60	155	202
Pigeonpea	72	226	446	438	70	535	221	61	153	270
Lentil	82	276	453	436	53	308	224	56	167	310
Soybean	101	303	509	416	84	326	277	91	236	312
Wheat (whole grain)	152	288	489	201	120	342	228	82	217	332
Rice (unpolished)	84	296	600	262	144	363	287	76	279	431
Egg protein	150[d]	425	488	450	213	463	325	94	–	363

a) Pulse and soybean data adopted from Haytowitz and Mathews (1986).
b) Cereal data adopted from Souci *et al.* (1986)
c) Except for cystine egg protein data adopted from Gupta (1983)
d) Gupta and Kapoor (1980)
Source : Poehlman, 1991.

cystine compared to cereals (Evans and Bandemer, 1967).

Digestibility and Biological Values

Protein quality may also be determined in terms of digestibility and biological value. Pulse protein is somewhat difficult to digest due to antinutritional compounds and flatus production in stomach and intestinal discomfort (Bressani, 1973). Comparison of digestibility coefficient values (% of ingested food nitrogen absorbed) reflected least digestibility of cowpea whereas chickpea and blackgram were easiest to digest (Table 13). Biological value (amount of nitrogen absorbed and retained in the body) is also a measure of protein quality. According to the biological values, lentil was the most digestible proteins followed by cowpea, whereas chickpea had most inferior protein. Mungbean, blackgram and pigeonpea proteins were similar but superior to chickpea proteins. The obligosaccharides (raffinose, verbacose and stachyose) are the flatus producers. Intestinal enzymes fail to digest these sugars perfectly consequently they start accumulating in the large intestine, where natural gut microflora use these sugars and produce gases (Poehlman, 1991).

MAJOR CONSTRAINTS

Information gathered clearly reveal that area, production and productivity of pulses have been fluctuating with minor changes, whereas population pressure is mounting at a faster pace in India, leading to decreased pulse availability per capita per day since 1951 till the past year. Thus, it is desired to have an idea on the principal constraints causing poor pulse production in India which are given below.

Agro-ecological Constraints

Almost 92 per cent of grain legume cropped area is rainfed, therefore, scarce availability and erratic behaviour of monsoon rains in India make the sowing of legumes difficult at the optimum time. The plant stand on such vast areas is also sub-optimal thereby, leading to population stress. The marginal and sub-marginal lands being utilized for cultivating legumes are in general poor in fertility, texture and water retention capacity, ultimately leading to reduced pulse production. Such agro-ecological reasons, therefore, contribute to poor harvest of grain

Table 13. Digestibility coefficient, biological value and nutritional value of seed protein in relation to egg-protein.

Pulse crops	Digestibility coefficient[a] (%)	Biological value[b] (%)	Nutritional value[c] (%)
Mungbean	79	72	32
Blackgram	80	72	–
Cowpea	67	65	57
Chickpea	80	79	53
Pigeon pea	73	72	39
Lentil	78	52	41
Egg protein	97	94	100

a) Adopted from Patwardhan (1962), Sevilla-Eusebio *et al.* (1968[a]). Gomzales *et al.* (1972) and Eusebio and Eusebio (1984).
b) Adopted from Patwardhan (1962) and Gonzales *et al.* (1972).
c) USAID 1971.
Source : Poehlaman 1991.

legumes in India.

Biological Constraints

Pulse legumes have remained neglected in respect of their biological behaviours ever since they were domesticated. The primitive behaviours, viz., indeterminate growth habit, high degree of flower drop, pod shattering characteristics, long life cycle etc. are some of the testimony that pulses were domesticated mainly from the survival view point and not for the high productivity behaviour. These primitive behaviours are still the parts of poor links in pulses biology, forming the chief adaptive constraints leading to their poor productivity.

ECONOMIC CONSTRAINTS

Low yield and low value productivity

It is a fact that the yielding ability of pulses is in general lower than other grain crops. Moreover, in spite of more procurement prices of pulses, the overall productivity values of grain crops remain lower. Data presented in table 14 indicate that chickpea, mungbean and pigeonpea yielded 652, 435 and 760 kg/ha compared to 2634 and 2200 kg/ha of rice and wheat, respectively. In spite of relatively much higher prices, the overall productivity values of pulses remained less than cereals and oilseeds. In terms of input and output ratios of various crops, even if being almost the same, the income from pulses would be much lower than the income from principal crops. Hence, due to this economic constraint farmers prefer to grow cereals and oilseeds vis-a-vis pulses.

Higher risk involved in pulse cultivation

The chief problem in increasing productivity of pulses is the high-risk involved with their cultivation. The yield levels greatly fluctuate over the locations and years, thereby making pulse cultivation in marginal and sub-marginal lands very risky due to poor stability in their yielding behaviour. Low productivity of pulses associated with their high risk stand in the way of expansion of pulses to more areas.

Complementary relationship with coarse cereals

Cultivation of pulses on marginal and sub-marginal lands as mix- cropping with coarse cereals is often a common practice. For instance, mungbean, uradbean, cowpea and mothbean are grown mixed with *jowar* (sorghum) pearlmillet, maize; and gram is usually mixed with wheat and barley. However, with increase in irrigation facilities, the area occupied by coarse grains has been gradually replaced by major grain crops, thereby leaving no further scope for expansion of *kharif* pulses. Decreased area was however, compensated to some extent by pigeonpea. Nevertheless over all area of pulses is not remarkably increasing in India which is a major constraint in pulse production.

Technological Constraints

(i) Eventhough, India is the largest pulse

Table 14. Grain yield and value productivity per hectare of pulses vis-a-vis major crops in India during 1990-91

Crop	Yield (q/ha)	Price * (Rs/q)	Gross value productivity (Rs)
Pigeonpea	7.60	480	3648
Chickpea	6.52	450	2934
Bengalgram	4.60	480	2208
Moongbean	4.35	480	2088
Rice	26.34	205	5400
Wheat	22.00	225	4950
Groundnut	9.50	580	5510
Repeseed-Mustard	8.50	600	5100

* Minimum support price offered by the government
Source : Bhatia, M.S., 1991.

growing and producing country in the world yet the indigenous and exotic technological back-up has not been upto the mark. Cultivars with high yield potential and tolerant to biotic and abiotic factors are still not available. National varieties like Kalyan Sona in wheat and Varuna in Indian mustard are not available in any of the pulses.

(ii) Pulses are highly susceptible to various diseases and a number of insect pests. Package of practices to control them are not as popular as in case of major crops. Even after the successful harvest of pulses, farmers have to suffer heavy storage losses due to poor grain weevil preventing measures.

(iii) Seed yield of pulses can be substantially stepped up by using efficient rhizobium culture. However, as against the requirement of $1.25x10^8$ units, the culture is produced only to an extent of, $0.01bx10^6$. The efficient culture has not been fully utilized in pulses due to lack of sufficient quantity which is also a major constraint in enhancing pulse production in India.

Sociological Constraints

Due to obvious reasons, the pulses have been occupying subsidiary status in cropping systems on the farmers' fields in India. This is possibly because: (1) cereals are the staple food, (2) pulses are not treated as cash crops, as are principally consumed by the farmers themselves, (3) pulses are not able to utilize the resource bases fully, and (4) fluctuating prices, high storage damage and other similar problems have pushed the pulses as the secondary choice of the farmers almost all over the country.

Adaptive and Extension Constraints

There has been no breakthrough in pulse technology in India. However, what is causing more concern is that whatsoever, technology developed, has not fully been disseminated and adopted by the pulse growers. One of the glaring example is the use of recommended dose of chemical fertilizers in grain legumes (Table 15). It is observed from the data that the application of the doses of the chemical fertilizers in pulses has been hardly 10-12 per cent of the recom-

Table 15. Application of chemical fertilizers in cultivation of pulses in major growing states of India (kg/ha)

State	Pigeonpea	Chickpea	Moongbean	Blackgram
Uttar Pradesh	4.36	5.27	–	6.78
Madhya Pradesh	5.65	5.35	4.36	4.10
Karnataka	4.82	–	–	–
Tamil Nadu	–	–	–	4.93
Orissa	–	–	5.53	5.32
Rajasthan	–	5.40	–	–
Haryana	–	4.95	–	–

- Not studied in these states Source : Bhatia, M.S. 1991.

mended doses. On the contrary, fertilizer application in rice and wheat ranged 40 to 60 per cent of the recommended doses even in the states of Madhya Pradesh and Uttar Pradesh. The same figures were however, 90 per cent in Punjab for major cereals. Thus, non- adoption of available technology is an important constraint in enhancement of pulse production in India.

SOIL AND MOISTURE CONSERVATION PRACTICES

Suitable soil and water conservation practices are of great significance for successful harvest of dryland crops like grain legumes which are largely grown on drylands of different regions with wide range of cropping systems over the seasons. Since the crop has to be sustained on rain water only therefore, practices leading to the preparation of fields that are receptive to rain water need to be adopted. Levelling of the soil is the simplest and the most important soil tillage operation required to effectively utilize the rain water and uniform distribution of fertilizers and seeds under scarce and erratic rainfall situations.

Off season tillage has been observed very effective in increasing rain water infiltration, reducing runoff losses, minimising weed flora, nematode population and adding to the soil organic matter. Intensive primary tillage is also necessary to create a favourable root proliferation zone and rain water intake.

For alfisols and other related soils, off season or summer ploughing should be the primary consideration (Singh, 1988). On vertisols, soil and water management that reduce runoff and erosion and ensure surface drainage, better aeration and workability of the soil are considered ideal.

For Udaipur and Bundelkhand regions, the system of graded bunding is suitable for easy and quick disposal of runoff as well as conservation of soil resources. Similarly, in Malwa region and north-eastern zones, surface drainage is essential. In red soils of Karnataka, the system of graded bunding or graded border strips has been found suitable.

Mulching the inter-row spacing with native weeds or other material is another important practice which has been observed effective in increasing organic matter from 10 to 25 per cent and moisture retention capacity of the loamy sand soils in arid areas from 6 to 20 per cent, ultimately increasing growth and dry matter production of cowpea (Gupta, 1987; Gupta, and Gupta et. al., 1986). Mulching has also been effective in increasing dry matter production of mungbean and cowpea in addition to water-use-efficiency (Gupta and Rao, 1989).

IMPROVED CULTIVARS

Improved cultivar of a concerned crop is essential for having higher and stable yield production under any agro-ecological condition. Under dryland situations also role of an improved cultivar is well documented. Important varieties of different grain legumes with their expected yield potential, days taken to maturity, adaptive zones and the specific traits have been given in Table 16. The list reveals that the UPAS-120, Pusa-84, Manak and Sagar are the most suitable cultivars for double cropping system. Pusa ageti with determinate growth habit is also suitable for double cropping system in north-west plains and central zones. Prabhat maturing in 120 days is an extra early variety specially suited to double cropping system with wheat in north-west zone. In case of chickpea, BGM 417, suited to central zone with multi resistant potential (Ascochyta blight, wilt, root rot, stunt virus and gray mould) appears most promising for rainfed situations. Pusa-209 is suited for rainfed and irrigated conditions as well. In mungbean, promising varieties like, K 851 and PDM-11 are suited for spring season, the latter is resistant to many diseases also, whereas Jawahar-45, ML-131 are suited to *kharif* season. In urdbean, UG- 128 resistant to YMV is the highest yielding. In cowpea, Co-4 promising for Tamil Nadu under rainfed situations is also tolerant to may diseases and stem fly. Early maturing Charodi and FS-68 are also promising for dryland conditions of Uttar Pradesh and Rajasthan. In mothbean, Maru Moth-1 is the latest cultivar with better yield potential than the existing ones.

Table 16. Promising cultivars of pulses

Cultivar	Adaptability	Grain yield (q/ha)	Days to maturity	Specific traits
Pigeonpea				
UPAS 120 (1976)	Northwest regions	15	140-150	Suitable for double cropping
Pant A-3 (1975)	North-plain west zone	12-15		Suited to double cropping, resistant to sterlity mosaic disease
Pusa 84 (1980)	Northwest region	15	140-150	Suited for double cropping
Manak (1983)	Northwest zone	20	130-165	Suited for double cropping
Sagar (1984)	Northwest zone	20	150-160	Suited for double cropping
Pusa Ageti (1971)	Northwest plains and central zone	15	150-160	Determinate type suited to double cropping
HY-1 (1975)	Peninsular region	10-12	160-170	Suitable for rainfed areas
Prahat (1975)	Northwest zone.	12-15	120	Extra early, suitable for double coorpping
Pusa-33 (1985)	Western UP	20	152	Tolerant to wilt, sterlity mosaic virus and pod borer
Chickpea				
Pusa 209 (1980)	Eastern UP	20-25		Irrigated and rainfed conditions, tolerant to wilt and *Bolrytis* grey mould
JG-62	Western MP, adjoining tracts of Orissa, Maharashtra and Karnataka	15-20	120	Suited to dry and short growing seasons
JG 221 (1974)	Satpura plateau	15-20		Suited to rainfed areas
JG-74	Late sown conditions in eastern Madhya Pradesh, Bihar and West Bengal after late paddy	15-20	120	Suitable for rainfed conditions
BGM 417 (1984)	Central zone	20	100-145	Suitable for rainfed areas, tolerant to *Ascochyta* blight, wilt, root rot, stunt virus and grey mould
H 208 (1977)	Haryana	20	153	Rainfed areas
GG 588 (1982)	Punjab	23		Tolerant to *Ascochyta* blight, wilt root not, nematode, root borer
ICCC-4 (1983)	Gujarat	18		Suited to rainfed areas
Mungbean				
K 851	North Central zone	10-12	60-65	Suitable for spring/summer cultivation
Pant Mung 3	North plain, West zone	10-11	75-85	Resistant to YMV, Powdery mildew, *Macrophomina* blight, and leaf curl virus
Jawahar-45	Madhya Pradesh	12-15	75-80	Suitable for *kharif* planting
PDM-11	Madhya Pradesh	10-12	60-70	Resistant, to leaf curl and powdery mildew, tolerant to YMV, cercos-pora leaf spot, suitable for summer/spring cultivations
RS-4	Udaipur and Kota regions	6-8	60-65	–
ML 131 (1983)	Peninsular zone	8	65-70	Resistant to YMV, tolerant to powdery mildew, Macrophomina, suitable for *kharif* Season
ML 5	Punjab	10	85-90	Tolerant to YMV and *cercospora* leaf spot

Table Contd...

Cultivar	Adaptability	Grain yield (q/ha)	Days to maturity	Specific traits
S-8	Northwest plains	8-10	70-75	Tolerant to *Cercospora* leaf spot
Pusa 105 (1983)	Northwest plains	10	–	Tolerant to YWV, powdery mildew *Macrophomina*, blight, aphid, jassid.
Uradbean				
UG 218	Western UP	15	75	Resistant to YMV,
Pant U-30	Central and Peninsular region	10-12	–	Resistant to YMV
TNU-1 (1979)	Tamil Nadu	10-12	70-75	–
Pant U-19 (1981)	North and eastern plains	10-12	75	Resistant to YMV,
Cowpea				
RC 19 (1984)	Northeast plain zones of Rajasthan	10-12	60-65	Moderately tolerant to salinity, synchronus maturity
Co-4 (1983)	Tamil Nadu	16	80-85	Suited to irrigated and *Barani* conditions, resistant. to stemfly, tolerant to wilt, root rot
FS-68	UP	6-7	60-65	Resistant to YMV and pod shattering
Chharodi-1 (G-2)	UP and Rajasthan	8-9	60-65	Compact plant type, suited for intercropping with *bajra*
Mothbean				
Jwala (1983)	Rajasthan, Haryana	4-5	90	Resistant to YMV
Jadia	Rajasthan	4-5	95	–
Maru Moth (1988)	Rajasthan	6-7	90	–
Moth 880 (1989)	–	3-5	80-85	–
Horsegram				
PL 77-2	Eastern U.P.	12-15	130-140	Tolerant to wilt and *Ascochyta* blight
JLS-1	Madhya Pradesh	10-15	110-120	Suitable for rainfed areas
PDM1	Andhra Pradesh	7-8	90-100	
KBR-1	Maharashtra	7-8	90-100	
Maru Kulthi (1989)	Rajasthan & Maharashtra	5-6	90-100	
Field pea				
RP-3	Rajasthan	15-20	110-120	Dry areas, susceptible to powdery mildew
Pant P-5	Uttar Pradesh	15-20	130-140	Resistant to powdery mildew
Lentil				
JLS-1	Madhya Pradesh	10-15	110-120	Suitable for rainfed situations
K 75 (1984)	NEP and Central zone	10-16	125-130	Specially suited to eastern part of U.P.
PL 77-2	North eastern plain zone	12-15	120-130	Tolerant to wilt

CROPPING SYSTEMS

An ideal cropping system in any situation must be capable of efficient use of land and water resources and provide high and sustainable return. Efficient use of basic resources in a cropping system depends on the inherent ability of the crops concerned which form the component of the system; and partly on the complementary effects between the crops (Willey *et al.*, 1981). Intercropping with legumes has been more remunerative than single cropping in a variety of ways. Efficient intercropping systems in different agroclimatic zones of India with special reference to rainfed/dryland situations have been briefed for important grain legumes.

Pigeonpea

In traditional cropping system, tall and long duration pigeonpea is very often intercropped or mixed with other crops (Osiru and Kibira, 1981). However, with the development of short duration cultivars, rich in harvest index and capable of sowing with high plant density have shown scope in sequentional cropping systems in assured irrigated areas (Willey *et al.*, 1981).

Intercropping systems

Studies at various levels have revealed the productive and efficient intercropping of pigeonpea with cereals, legumes and oilseeds.

Intercropping with cereals : Results have revealed high profitability and efficacy of pigeonpea-cereal intercropping with sorghum (Umrani *et al.*, 1984), maize (Rajaram, 1984; Srivastava and Srivastva, 1991) and pearlmillet from different parts in India. The pigeonpea-sorghum (1:2) was the most efficient system on vertisols. Medium long duration cultivars of pigeonpea (C 11, Type 7, Bahar) and early maturing sorghum cultivar as CSH-1 have offered ideal combination. Pigeonpea-pearlmillet system was productive on alfisols and entisols in Gujarat and Karnataka. Maize-pigeonpea (1:1) intercropping system has found its place in Udaipur and Ranchi regions, whereas rice-pigeonpea (4:1) was an efficient system in upland conditions of the pleateau of Bihar.

Pigeonpea-pulse intercropping : Short duration pulses like blackgram, mungbean and cowpea have been found quite compatible with pigeonpea in Punjab, Haryana, Western Uttar Pradesh, Gujarat, Andhra Pradesh, Maharashtra, Tamil Nadu and Bihar. These pulses do not depress the yield of pigeonpea, the principal crop, and give an additional yield of 400-500 kg/ha. Reports from different places have indicated that the intercropped blackgram (Saraf *et al.*, 1975), mungbean (Giri and De, 1978), and mungbean, blackgram and cowpea (Ali, 1988) with pigeonpea hold great promise.

Pigeonpea-oilseeds intercropping : In the present context of short supply of proteins and edible oils, such systems have great national relevance in dryland areas. Among oilseeds, groundnut (Rafey and Verma, 1988), sesame (Kumar and Ahlawat, 1986), soybean (Patra and Chatterjee 1986; Rafey and Verma, 1988) have proved compatible with pigeonpea. Pigeonpea-groundnut is considered one of the most popular combinations, particularly in Central and Southern India. The latter, due to its ability to attain faster growth in the early stage, is able to effectively utilize the land and water resources. On alfisols, groundnut produced 82 per cent and pigeonpea 85 per cent equivalent yield of their sole stands (Willey *et al.*, 1981). At Honnaville, pigeonpea- groundnut (1:1) with 100 and 50 per cent stands, respectively, the system offered a net return of Rs 5999 compared to Rs 1609/ha for sole pigeonpea. Planting of 4 to 6 rows of groundnut at 22.5 to 30 cm inter-row distance with 1 row of pigeonpea of 120-135 cm was considered ideal combination under dryland situations of Anantpur, Rajkot and Bangalore.

Double cropping

The short duration high yielding pigeonpea cultivars have paved the way to the introduction of pigeonpea-wheat sequential cropping system in irrigated areas of northern and central areas (Chandra and Ali, 1986). In Eastern and Peninsular India, double cropping system was found possible in the post-rainy season (Narayanan and Sheldrake, 1979). Likewise, in eastern coastal region also double cropping system has become a reality (Lanka and Satpathy, 1976).

In Punjab, Haryana, Uttar Pradesh, Madhya Pradesh, Gujarat and parts of Rajasthan, short duration cultivars (Pusa Ageti, Pusa 74, PUSA 84, UPAS 120 etc.) have been found ideal in permitting timely sowing of wheat. Results at different locations have established the profitability of pigeonpea-wheat sequential system in northern India (Table 17).

At Ludhiana, sowing of pigeonpea was found optimum in the first fortnight of June (Kaul *et al.*, 1980). Similarly, Panwar and Yadav, (1981) observed 26.0 q/ha grain yield of T 21 pigeonpea cultivar when planted on April 15 as against 17.4 q/ha planted on 30 June. Thus, in sequential system, sowing of pigeonpea upto mid of June has been observed crucial.

Table 17. Yield and monetary return in pigeonpea-wheat sequential cropping system in Northern India

Location	Pigeonpea yield (q/ha)	Wheat yield (q/ha)	Net return (Rs../ha)
Hissar	17.4	41.0	8223
Dholi	17.5	28.6	7550
Kanpur	21.1	53.4	15811

Source : Chandra and Ali, 1986.

Chickpea

Chickpea is exclusively a sole crop of post-monsoon season. In Northern India, chickpea is cultivated as a second crop in areas receiving winter rainfall of around 100 mm. In areas having comparatively less rainfall, fallow-chickpea mono-cropping system is common. In areas like Rewa either wheat or chickpea crop is cultivated. On black soils of Southern India, chickpea is grown on stored soil moisture. It has proved a very potential *rabi* crop grown on stored soil moisture. At Hisar, yield of chickpea was at par to those of mustard and *taramira* (Eruca sativa). At Varanasi, based on experimentation for 7 years, chickpea was found stable and high yielding (28 q/ha) compared to rapeseed-mustard (13.9 q/ha), lentil (14.8 q/ha), wheat (20.5 q/ha), safflower (22.6 q/ha) and barley (25.8 q/ha).

Intercropping

Chickpea is being intercropped with wheat, rapeseed-mustard and safflower in different regions. On medium to deep black soils of southern India, chickpea-safflower (3:1) intercropping has been recommended (Singh and Das, 1987), whereas in black soils of Rewa, the intercropping of wheat-chickpea (1:1) was recommended. On alluvial soils of Agra, intercropping of bengal gram with Indian mustard was found 37 per cent more productive than either crops. In south-eastern zone of Rajasthan, bengal gram and mustard (3:1) intercropping gave highest monetary return (Rs 5986/ha), which was higher than either of the sole crops and other crop combinations (Patel *et al.*, 1991). Chickpea-mustard (4:1) intercropping at Faizabad has proved most productive system giving 20.16 and 30.25 q/ha chickpea equivalent yield during 1987-88 and 1988-89, respectively, and giving a net return of Rs 6015 and 10301/ha in respective years.

Double cropping

Studies on double cropping systems have revealed that early maturing rice, sorghum or maize followed by chickpea would be quite profitable in specific regions. For instance, in Dehradun, maize-wheat, in Indore, sorghum-chickpea, in Hoshiarpur, maize-chickpea and in Rewa and Varanasi upland paddy-chickpea crop sequences have proved productive, remunerative and efficient (Table 18).

Mungbean, Urdbean and Cowpea

Kharif pulses, like mungbean, urdbean and cowpea, characterized with better compatibility and short growth period have traditionally been intercropped with cereals, legumes and other miscellaneous crops in different regions. Selection of a particular grain legume with compact growth pattern and growth habit which do not coincide with the peak growth period of the principal growth is vital for success of a system under rainfed conditions.

Table 18. Total productivity and monetary return in some sequential cropping system involving chickpea

Region	Soil type	Moisture storage (mm/m)	Winter rains (mm)	Cropping system Ist crop	2nd crop	Total productivity (Kg/ha)	Monetary return (Rs /ha)
Dehradun	Submontane	140-180	88.2	Maize	Wheat	7280	7124
				Maize	Barley	6730	5867
				Maize	Chickpea	5430	6490
Rewa	Black	270-300	36.3	Fallow	Chickpea	740	1480
	.			Upland paddy	Chickpea	1920	2517
				Upland paddy	Wheat	1790	1824
Varanasi	Alluvial	140-180	42.7	Fallow	Chickpea	3590	7180
				Upland paddy	Chickpea	5560	7949
				Upland paddy	Wheat	5180	5353

Source : Singh and Das (1987), (AICRPDA Annual Reports for respective locations.
Data Pooled : Rewa 4 years (1972-75) : Varanasi 3 years (1972-74) and Dehradun 8 years (1972-1979).

Intercropping with cereals

Intercropping of mungbean between rows of pearlmillet on light textured soils of Jodhpur has been observed more remunerative giving more water use efficiency and net return than the sole crop. The LER, in this system was found to be increased following the sowing of pearlmillet after a week of sowing of mungbean (Singh *et al.*, 1978). On light textured soils of Agra, pearlmillet-mungbean intercropping system with 100 per cent population of each component was found most productive over space and time by utilizing the resources more efficiently and giving higher LER (1.35) values (Singh and Chauhan, 1991). A 4-year consecutive study at Ludhiana, confirmed the feasibility of maize-mungbean intercropping system in 2:2 paired rows north-south planting, giving 35.1 q/ha of maize and 3.5 q/ha yield of mungbean (Dhingra *et al.*, 1991). Fingermillet-mungbean intercropping system was found superior over sole crop under drought situations of southern states (Rao *et al.*, 1982).

Intercropping with oilseeds

In rainfed situations at Dipu, Assam, greengram-sesame (alternate rows), was the most profitable intercropping system compared to other systems tried which gave net return of Rs. 6586. 14/ha and benefit : cost ratio of 3.35 (Sharma and Kakti, 1991). At Sardar Krushinagar, castor-munbgean (1:3) system with 100 and 25% population, respectively, proved remunerative giving net return of Rs. 8819/ha.

Intercropping with pulses

Results reveal that blackgram intercropped with pigeonpea or clusterbean has been observed to cover more risk than the sole crop. In Bharauch region, a 3-year study confirmed that pigeonpea + blackgram system was most productive and recorded highest cost benefit ratio and net return (Goyal *et al*., 1991). In pigeonpea + blackgram (2:3) intercropping system both components proved complementary to each other, as a result both crops recovered by 61.7 and 81.5 per cent over their sole yield, respectively. Additional advantage of this system was : 43 per cent higher yield, 8.9 per cent energy output : input ratio and a net profit of Rs. 3776/ha, the system with 1.43 LER proved most productive and efficient in rainfed situations of Udayagiri (Padhi *et al.*, 1992).

Double Cropping

Rice followed by mungbean or blackgram has been considered the stable sequence under limited water availability situations of Tamil

Nadu. Safflower followed by blackgram was found the best cropping sequence for efficient utilization of conserved soil moisture when grown after fingermillet or rice in southern states (Rao *et al.*, 1982). Similarly, relay cropping of mungbean 130 days after sowing of rice has been considered feasible in upland rice growing belts of southern states. In Guntur area, paddy-blackgram has been observed a better production cropping sequence (Sudhakara Babu *et al.*, 1989). In central Uttar Pradesh, growing of Indian mustard following blackgram or mungbean was found a most productive system (Suraj Bhan and Khan, 1981). A 3-year study at Delhi revealed a gross return of Rs. 8352/ha and as high as Rs. 12000/ha in good rainfall year. Further, double cropping under rainfed situations was more feasible in case of rains occurring at the end of September or beginning of October (Giri and Gangasaran, 1984).

Horsegram

Depending on the monsoon rains, horsegram has wide range of planting time. For instance, it is raised as late monsoon crop in mid of September to October in Karnataka, Orissa, Andhra Pradesh and Maharashtra, whereas in Tamil Nadu it is grown in late October after the receipt of north-east monsoon rains.

Experiments at Bailhongal (Karnataka) have revealed that reasonably good yields of horsegram could be obtained in intercropping with hybrid sorghum and groundnut. Horsegram-fingermillet system of inter-cropping was found promising at Bangalore.

Mothbean

At Jodhpur, a 4-year study revealed that intercropping of mothbean and other *kharif* legumes like greengram, cowpea and groundnut did not affect the yield of main crop sunflower. In general, total productivity of intercropping system was 74 per cent higher than the sole crop (Singh and Singh, 1977). Mothbean is a most compatible legume for intercropping in arid areas. Pearlmillet + mothbean is a good combination compared to various crop mixtures in arid areas.

Lentils

Lentils are grown in subsistence mixed cropping but are hardly grown in monoculture. In India, lentils are planted after harvest of rice and grown as winter crop on residual soil moisture. A 4- year study at Pusa (Bihar) indicated that *paira* crop of lentils yielded significantly higher (12.7 q/ha) than the pure seeding on prepared lands after the harvest of paddy (Sharma *et al.*, 1984).

SOWING TIME

Several studies have established the significance of optimum sowing time of concerned crops for realization of reasonably good yields under specific situations and cropping systems.

Due to photosensitive nature of pigeonpea the seeding time has profound effect on its grain yield, hence timely sowing of pigeonpea is of great significance. In traditional Indian system, the sowing of late maturing pigeonpea is suggested to be done by last week of June to first week of July. Delay in sowing is known to cause heavy yield losses due to the reproductive phase coinciding with receding soil moisture. Short duration pigeonpea cultivars which are planted on assured soil moisture supply in pigeonpea-wheat rotation in north-western part of India, need to be planted at the optimum time so that fields are vacated by mid of November for sowing of wheat. In such system, the planting of pigeonpea by mid of June or at the most by third week of June has been observed optimum. Delayed sowing is unsuitable for wheat and for pigeonpea due to early flowering and slow growth in pigeonpea (Singh and Saxena, 1981). In areas with pigeonpea + mungbean - wheat rotation, the pigeonpea must be sown with mungbean in first fortnight of April. The extra early pigeonpea cultivars can be safely sown latest by mid of July. In *rabi* season, the pigeonpea sowing is suggested to be done by beginning of September in eastern zone, delayed sowing however, depressed yield at Ranchi (Srivastva, 1984).

Studies under All India Coordinated Project on Chickpea Improvement have, by and large, established the optimum sowing period of chickpea from mid of October to mid of November, and deviation caused heavy yield losses (Saxena and Singh, 1977; Sharma, 1978). However, in areas where winters are shorter as in eastern and southern zones of the country, the sowing time becomes still restricted. For instance, in Berhampore, first week of November was found to be the optimum sowing time (Sen, 1978) whereas, in Northern and Western India, the optimum sowing time of chickpea was found to be mid of October. Sowing of chickpea can however be done upto the end of November at Rewa. On black soils of Southern India, sowing of chickpea, is suggested to be advanced for efficient utilization of receding soil moisture.

Optimum sowing time for mungbean and urdbean under rainfed situations has been observed by mid of July (Ali *et al.*, 1981). Delayed sowing of these crops is known to depress yield severely due to unfavourable photoperiodicity and the moisture coinciding with critical reproductive growth period.

Summer Sowings

Planting of mungbean by February 15 to March 15 has been observed optimum at Sardar Krushinagar (Sharma *et al.*, 1989). At Delhi, however, optimum sowing time for mungbean was observed from April 5 to April 25 (Singh *et al.*, 1980). At Berhampore, mid of March was the optimum sowing time for summer mungbean. The data in Table 19 however indicates that the cultivars PDM-11, 11/395 and Pusa 105 were observed ideal under early and late planting conditions at Berhampore, Orissa (Padhi and Sahoo, 1991). Like mungbean, the optimum sowing time of summer uradbean has also been observed to be around mid of March. A 2-year study at Faizabad revealed that sowing of uradbean on March 23 yielded 8.5 and 20.6 per cent higher yield over Appril 2 and April 12 sowings, respectively (Singh and Dixit, 1989).

Planting of lens from October 15 to first week of November has been observed optimum period in India. For instance sowing of lens on November 6 yielded 23.7 q/ha at Kanpur, that was 24.5 per cent more over December 6 planting, cultivar p639 was however suitable for late planting also (Varshney 1989). At Sehore, Oct.15 was observed as the optimum sowing date, sowing delayed upto October 30, considerably depressed the production of lentils.

Ideal sowing time in respect of mothbean under rainfed situations in western Rajasthan has been observed around mid of July. However, under good rainfall situations sowing can be done safely upto the end of July.

Optimum sowing period in case of horsegram varies with the regions and the cropping systems. In Maharashtra, the optimum sowing period is postrainy season. In coastal areas of Orissa and Andhra Pradesh, the planting of horsegram from mid to last week of September has been found ideal. In Tamil Nadu, the planting has been observed optimum from mid of October to last October.

ESTABLISHMENT OF PLANT STAND

Optimum plant stand of a crop depends on sowing time, seed rate, inter- and inter-row spacings, soil moisture and other cultural practices. In pulses, the optimum plant population appears to differ with plant type of the cultivar and the environmental conditions. In case of chickpea, for instance, under favourable conditions the

Table 19. Grain yield of greengram as influenced by sowing dates

Date	Grain yield (q/ha)											
	PS 16		K851		PDM11		Pusa 105		11/395		Dhaub	
	1984	1985	1984	1985	1984	1985	1984	1985	1984	1985	1984	1985
1 March	5.9	2.9	7.0	2.4	10.8	9.8	7.5	7.0	8.8	6.0	5.4	4.0
15 March	9.1	8.8	11.1	7.7	13.7	12.0	13.6	12.0	10.5	11.5	7.3	8.6
30 March	7.2	6.5	8.6	5.8	12.3	7.8	8.9	6.6	9.5	7.2	4.8	5.7
15 April	2.8	6.4	2.0	4.1	3.3	7.4	2.6	4.2	2.4	5.2	1.4	3.7

Source : Padhi and Sahoo 1991.

genotype appear to show little variation in yield due to plant density (Saxena and Sheldrake, 1977). Under irrigated conditions, the yield of chickpea increased upto a density of 50 plants/m^2. The role of plant stand appears to have its impact only when the yield starts declining due to limited soil moisture. Under drought situation therefore, a stand of 24.8 plants/m^2 was found optimum. At Bellary, under dryland conditions, chickpea yields did not increase beyond a plant density of 185,000 ha and the required plant stand m^2 was still low to the extent of 18.5 m^2. At Faridkot, a plant population of 3.33 lakhs/ha following 30x10 cm spacing was found ideal.

Plant stand is an important agronomic factor in present day short duration pigeonpea cultivars which are sensitive to close spacings and are planted in assured moisture conditions. Optimum density for pigeonpea is 30 plants/m^2 which can be maintained by adopting 30x10 cm spacing (Chauhan et al., 1987 b). Plant population also depends on the cultivar. TPCL-87, for instance, required 25 plants/m^2 for exploration of its fullest yield potential. Similarly, cultivars Bahar and T 21, yielded higher at 25 cm spacing whereas Laxmi yielded maximum at 37.5 cm inter-row spacing at Ranchi (Srivastava, 1984). At Navsari, 90 cm spacing between pigeonpea rows yielded maximum (12 q/ha) and that was 40.5 and 19.1 per cent higher over 50 and 120 cm, respectively. On sandy loam soils of Agra, pigeonpea cultivars, Pusa Ageti, T 21 and Sharda, yielded significantly higher with 60 cm spacing compared to 40 and 80 cm spacings (Chauhan and Singh, 1981). Traditional and long duration varieties require less plant density (5 plants/m^2) for optimum yield potential. In such cultivars yields can be increased by adopting certain improved row proportions.

In Intercropping system, plant population has to be different from the sole cropping system. A plant population of 100 per cent (1,11,000 plants/ha) in respect of mungbean and 80 per cent (1,18,400 plants/ha) for pearlimillet were found optimum giving 10 q/ha of mungbean yield. Under sole cropping system a plant population of 2.5 to 3.0 lakhs/ha abtained from 40x10 cm spacing would be optimon for mungbean. Similarly, a plant population of 2.5 to 3.0 lakhs would be optimum for uradbean and pea under rainfed conditions. A planting pattern with 30x10 cm spacings which gave 2.2 lakhs plants/ha was found optimum for lentils at Sehore (Kalyan Singh et al., 1989). In case of horsegram also 2.5 to 3 lakhs plant population was, in general, optimum.

FERTILITY MANAGEMENT

Traditionally tall growing and long duration pigeonpea cultivars are not generally fertilized as they show little or no response to fertilization. Such cultivars are poor yielders and their fertilizer requirement is by and large met out by symbiotic nitrogen fixation. In short duration pigeonpea cultivars due to their intense management, the fertility requirement is increased compared to traditional cultivars (Rao, 1974). Information gathered so far show little response of pigeonpea to nitrogen application. Response to 'starter dose' of 15 to 20 kg N/ha is however, undisputed. Many studies conducted at various centres including those by Sadaphal (1988) have revealed that phosphorus was by and large a limiting nutrient for pigeonpea cultivation. Soils of rainfed areas are generally low to medium in phosphorus availability and pulses generally respond to 40 to 60 kg P_2O_5/ha (Pathak, 1970). Similarly, on sandy loam soils of Agra, a two-year study revealed that 80 kg P_2O_5/ha increased the yield of pigeonpea by 69.5 per cent over control (Chauhan and Singh, 1981). At Navsari, under rainfed conditions, a combined dose of N 20 and P 40 kg/ha was optimum for getting higher yields of pigeonpea (Patel *et al.,* 1984) Deficiency of Zn has been observed in some dryland soils. In vertisols of Andhra Pradesh, good response to zinc sulphate has been noticed.

In intercropping system, generally the nutrient requirement is more. Likewise, higher nutrient uptake in pigeonpea-mung and pigeonpea-maize system has been noticed over the sole cropping system. However, in a pigeonpea-sesame intercropping system at Delhi, the N dose from 0 to 50 kg/ha did not increase total productivity (Kumar and Ahlawat, 1986). The argument is that by applying more N to cereal component, it increased the vigour of cereal crop which in turn suppressed the growth and yield of pigeonpea. In pigeonpea cereal system, generally the requirement of phosphorus is 15 to 20 kg P_2O_5 /ha. However, in pigeonpea-urad system 40-60 kg P_2O_5/ha was found optimum (Singh *et al.*, 1986).

Like pigeonpea, in chickpea also positive response to starter nitrogen dose of 15 to 20 kg/ha has been observed, particularly in texturally poor soils (Sharma *et al.*, 1975; Rathi and Singh, 1976). Such responses have not however, been found in soils improved in texture (Dhingra, 1975). Response to N was in general more on the soils poor in nodulation. Response to phosphorus upto 50 to 75 kg P_2O_5/ha has been observed in the soils low in phosphrus availability as observed by Rathi and Singh (1976) at Kanpur. However, response to phosphorus was inconsistent in black soils due to high phosphorus fixing ability of these soils. On calcareous soils of Sirsa (Haryana), grain yield of chickpea significantly increased upto 40 kg P_2O_5/ha, whereas, there was no response to additional dose of nitrogen (Mudholkar and Ahlawat, 1979). In rainfed conditions of Delhi, Prabhakar and Saraf (1991), reported that irrespective of the source, the phosphorous dose upto 40 kg P_2O_5/ha increased the grain yield (22.5 q/ha) of chickpea with a significant margin over control (16.8 q/ha). Water-use-efficiency and consumptive water use were also more over control (Table 20). The phosphorus application, has registered very useful interaction with herbicides and hand weeding on sandy soils of Jobner (Table 23). Such interaction effect of two hand weeding with 60 kg P_2O_5/ha registered the highest yield of 16.1 q/ha compared to other treatments including control (Dadhich and Mali, 1991).

Application of phosphorus @ 40 to 60 kg P_2O_5/ha in mungbean during rainy season has been observed economic. Efficacy of applied P. can be increased by combined application with FYM. At Navsari on clayey soils, grain yield of uradbean significantly increased upto 20 kg N/ha (12.8 q/ha) and in case of P upto 40 kg P_2O_5/ha (11.4 q/ha) over control. On the contrary, response in case of summer *mung* was noticed upto 60 kg P_2O_5/ha at Faizabad (Singh and Dixit, 1989).

In case of blackgram, a basal dose of 15-20 kg N/ha and 40-60 kg P_2O_5/ha has been recommended optimum (Subramanian and Palanianppan, 1981). On problem soils, application of sulphur @ 30kg/ha in the form of gypsum or elemental sulphur has been recommended (Lal and Jaiswal, 1981). Fertility requirement of summer blackgram is like that of summer mungbean (N 20 and P 40-60 kg/ha).

Cowpea too responds to phosphorus fertilizer upto 25-50 kg P_2O_5/ha. Response, however, depends on the P-fixing ability of the soil (Bhat, 1981). Inoculation of cowpea with suitable rhizobium culture has been useful (Bhagyaraj and Hegde, 1978). However, at Navsari, summer cowpea responded upto 20 kg N and 80 kg P_2O_5/ha.

Grain yield of pea increased upto 40 kg P_2O_5/ha at Hisar. The water use efficiency also increased upto 40 kg P_2O_5/ha, whereas consumptive use increased upto 80 kg P_2O_5/ha (Chauhan *et al.*, 1991). On loamy soils of medium fertility at Jhansi, grain yield of pea and Lathyrus significantly increased upto 80 kg P_2O_5/ha

Table 20. Effect of phosphorus on grain yield, WUE and consumptive use of chickpea (mean of 2 years)

Treatment	Grain yield (q/ha)	Consumptive water use (mm)	Water use efficiency (kg/ha/mm water)
No P	16.4	246.9	6.80
PSP Inoculation	18.4	260.2	6.80
40 kg P_2O_5 /ha as SSP	22.6	276.0	8.17
40 kg P_2O_5 /ha as MRP +- PSB	21.1	272.3	7.73

Source : Prabhakar and Saraf (1991).

whereas, water use efficiency increased upto 40 kg P_2O_5/ha (Shukla and Lal, 1991). Response of these crops upto 80 kg P_2O_5/ha at Jhansi appeared to be due to poor phosphorus availability of the soil than Hisar soils. Horsegram and mothbean are also known to respond to 20-30 kg P_2O_5/ha. For instance, at Ambikapur (M.P.) in rainfed situation, 20 kg P_2O_5/ha was found an economic fertility dose in horsegram-niger intercropping system (Bajpai *et al.*, 1989).

SUPPLEMENTARY IRRIGATION

Medium to long duration pigeonpea cultivars may face terminal drought, particularly when grown on the alfisols. However, deep vertisols once fully charged can support a good pigeonpea crop. Water requirement of pigeonpea is, therefore, different on these soils. For instance, one irrigation on vertisols may give around 20 per cent additional yield, whereas 2-3 irrigations are required to double the yield of this crop on alfisols. Similarly, experiments at Badnapur, Ludhiana and Akola have revealed that one supplementary irrigation at most sensitive stage (bud/pod formation) increased the grain yield by 80-90 per cent. Bhan and Khan (1979) observed significant response to one or two supplementary irrigations at Kanpur (Table 21). On Rajasthan soils, one supplementary irrigation applied at branching stage increased yield upto 34 per cent (Makhanlal and Gupta, 1984). Many workers have not observed response of pigeonpea to supplementary irrigation. The logic is that the crops must be irrigated only when it experiences moisture stress. Irrigation prior to drought may prove wasterful and cause yield losses due to waterlogging condition. The crop responds well to limited irrigation in sandy soils and in low rainfall areas. Supplementary irrigation must therefore, be made at the sensitive stage only. Response to irrigation is more in post rainy season crop as it has to depend on stored moisture only. At ICRISAT, Hyderabad, application of 2-3 irrigations about one month of sowing increased grain yield by 150-160 per cent over the control.

Table 21. Effect of irrigation on grain yield (q/ha) of pigeonpea during rainy season 1976-77 at Kanpur

Stages of irrigation	Grain yield (q/ha)
Control	12.8
Flower initiation	13.0
Peak flowering	12.2
Pod filling	21.4
Flower initiation and pod filling	24.3
Irrigation after 80 mm CPE	22.7
Irrigation at 120 mm CPE	23.6
CD 5 %	0.49

Source : Bhan and Khan (1979).

Chickpea responds well to supplementary irrigation (Sharma *et al.*, 1974; Panwar, 1978; Rakesh Kumar and Tiwari, 1989). Irrigations at early preflowering stage on the soils with poor water holding capacity and at late vegetative stage and pod formation stage on deeper soils have shown promise. At Faizabad, chickpea responded very well in terms of yield increment to the two irrigations, one at seedling stage and another at flowering stage (Yusuf *et al.*, 1980), the WUE was more with one irrigation at seedling stage. In Hisar conditions, one irrigation applied at 45 DAS gave significantly higher yields over no irrigation or one irrigation at 75 DAS or at early pod filling stage (Singh *et al.*, 1980). At Raipur, pod initiation stage was very sensitive, irrigation at this stage gave highest grain yield (1401.8 kg/ha) over control and other stages (Rakesh Kumar and Tiwari, 1989); and the cultivar ICCC-37 yielded maximum (1908 kg/ha). Two year study at Faizabad revealed that the irrigation at 0.4 IW:CPE ratio gave 7 q more yield than no supplementary irrigation, moreover irrigation on the basis of this schedule gave high WUE and consumptive water use as well (Prabhakar and Saraf, 1991).

Urdbean generally grown with residual soil moisture in rice-fallow system during *rabi*, suffers from terminal drought which coincides with pod maturity stage. A 3-year study revealed that crop irrigated at least once during vegetative phase yielded satisfactorily (Rameshaiah *et al.*, 1989). On silty loam soils of Faizabad, the summer urdbean irrigated with 0.8 IW/CPE significantly increased yield, yield components and protein yield over 0.4 and 0.6 IW/CPE irrigation schedule. Urdbean plants are more sensitive to irrigation at branching and flowering stages, irrigation applied at these stages will therefore be more rewarding. Plants are also specially sensitive to flooding at the seedling stage (Vijai Kumar *et al.*, 1984).

Summer *mung* also responds well to applied irrigation. A 2-year study at Karnal revealed satisfactory yield with 150 mm CPE (IW:CPE = 0.4) applied and fertilized with 80 kg P_2O_5/ha (Sharma *et al.*, 1984). Cowpea responded well to irrigation when the available soil moisture receded below 35% at Hisar. Cowpea had maximum nodulation under well irrigated conditions when soil metric potential was -0.6 bar (Robert Zablotowicz *et al.*, 1981).

Field pea is susceptible to soil moisture stress at flowering stage and the grain yield depends on the availability of soil moisture at the reproductive stage. Application of one post sowing irrigation at pre-sowing stage to pea resulted in 62 per cent increase in yield over no post sowing irrigation, whereas high WUE was observed with one irrigation at the pre-flowering stage (Singh *et al.*, 1978). Similarly, at Hissar, pea responded to one post sowing irrigation with the application of 40 and 80 kg P_2O_5/ha, the irrigation resulted in more moisture extraction from 0 to 30 cm layer (Chauhan *et al.*, 1991).

Lens also responds very well to post-sowing irrigation (Table 22). For instance, on sandy loam soils of Agra, one irrigation at 45 DAS gave significantly higher yield (9.2 q/ha) over pre-sowing irrigation (7.3 q/ha) and pre-sowing + one irrigation at 90 DAS. One post sowing irrigation also increased grain protein content, nitrogen and phosphorus uptake (Tomar and Singh, 1991).

WEED CONTROL

Weeds pose a serious threat to pulse cultivation on drylands, particularly during rainy season where adequate soil moisture, suitable atmospheric temperature and poor compatibility of slow growing pulses create favourable conditions for their fast growth at the cost of critical growth period of pulses. The magnitude of crop-weed competition differs considerably in sole and inter-cropping systems. The weed flora in intercropping systems depends on a number of factors like, crop species, plant density, sowing geometry, growth stage etc. (Moody and Shetty, 1981). Effective weed control means have been briefed here.

Yield losses due to weed infestation in pigeonpea on rainfed areas have been estimated to be as high as 90 per cent (Saxena and Yadav, 1975). Hand weeding is very effective in case of pigeonpea also. A 4-year consecutive study at Bharauch (Gujarat) under rainfed conditions confirmed that consistently maximum yield (15.3 q/ha) of pigeonpea could be obtained with one hand weeding and with interculture 30 days of sowing (Goyal *et al.*, 1991). Many other workers (Prasad *et al.* 1985; Panwar *et al.*, 1988) have also reported high yield with hand weeding over herbicides. Amongst weedicides, use of Fluchloralin 1 kg ai/ha as pre-emergence followed by one interculture 60 DAS may be very effective in controlling weeds in pigeonpea. Besides, both Fluchloralin and Metolachlor weedicides were found equally effective at Rewa, either of these however when followed by one interculture resulted in much higher weed control and grain yield production under rainfed situation (Dwivedi *et al.*, 1991).

Yield losses due to weed infestation in chickpea have been estimated to range from 30 to 50 per cent (Panwar and Pandey 1977; Singh *et al.*, 1978). Hand weeding an age old practice in

Table 22. Grain yield, grain protein content, N and P uptake in lentil as affected by irrigation schedule.

Irrigation treatment	Grain Yield (q/ha)	Protein content (%)	N uptake (kg/ha)	P uptake (kg/ha)
Pre-sowing irrigation	7.3	22.2	25.6	3.1
Pre-sowing + irrigation at 45 DAS	9.2	22.6	33.4	3.8
Pre-sowing irrigation at 90 DAS	7.9	22.3	28.3	3.2
CD at 5%	0.13	–	–	–

Source : Tomar and Singh, 1991.

dryland areas has been very effective after 30 to 50 days of sowing in chickpea. However, in circumstances where hand weeding is not feasible, various herbicides can also be used. Pre-emergence application of 1.5 kg ai/ha of Nitrofen or 0.5 kg a.i/ha of prometryne were found very effective at Kanpur (Panwar and Pandey, 1977). On sandy loam soils of Ludhiana, preplant application of Basalin (48 EC) at a rate of 1 kg product/ha was found effective. Two hand weeding with 60 kg P_2O_5/ha gave maximum grain yield of chickpea and was better than other weedicides (Table 23).

Pre-sowing application of Basalin @ 1 kg ai/ha controlled the weeds effectively and was comparable to two-weeding in mungbean. A 2-year study at Kanpur indicated that nitrogen @ 1 kg ai/ha alone or mixed spray with Alachlor 1 kg ai/ha effectively controlled the weeds giving highest grain yield (8.7 q/ha) and maximum net return (Rs 1290/ha).

Pre-emergence application of Alachlor @ 1 kg ai/ha gave 18.4 q/ha grain yield of urdbean at Hisar (Singh and Faroda, 1977). Besides Nitrofen @ 1.5 kg ai/ha supplemented by one hand weeding was also found very effective treatment in controlling weeds in urdbean. Under rainfed situation of Gwalior, Fluchloralin @ 1 kg /ha was also found an effective weedicide (Jain *et al.*, 1989).

Yield losses due to weed infestation in pea have been estimated around 80 per cent (Singh *et al.*, 1986). Hand weeding 20 to 40 DAS gave higher yield (1263 kg/ha) with weed control efficiency (WCE) of 66.2 per cent, whereas Pendimethalin @ 1.5 kg ai/ha with 64.2 per cent WCE was almost at par to hand weeding and gave grain yield 1196 kg/ha under rainfed situations at Jorhat (Gogoi *et al.*, 1991).

Lentils are poor competitors with weed flora due to its slow growth during winter season. Inadequate weed control may cause yield losses to the tune of 75 per cent in this crop. The growth period with crop competition with weeds was most deleterious was observed 30 to 60 days of sowing (Hawtin *et al.*, 1980). Several herbicides are effective with lens (Summerfield *et al.*, 1982). Lentils are however, more sensitive to herbicides toxicity than other pulses.

From the vast literature on weed control in pulses it is evident that hand weeding is very effective in controlling weeds in dryland areas, however, it is time consuming, expensive and may be more feasible on small cropped areas only. Herbicides too have problems, as they have proved somewhat less inferior to hand weeding and are known to lose their efficacy with time. Thus, one hand weeding in herbicide treated plots may prove a sustainable, economic and most effective weed control measure.

Table 23. Interactive effect of herbicides and phosphorus application on grain yield of chickpea (q/ha)

Weed-control measures	P_2O_5 (kg/ha)		
	0	30	60
Weedy check	5.0	6.3	8.7
2-hand weedings 25 and 45 DAS	10.6	15.0	16.1
Fluchloralin @ 0.40 kg/ha	8.5	10.6	12.1
Fluchloralin @ 0.60 kg/ha	8.3	12.3	13.5
Pendimethalin @ 0.50 kg/ha	7.8	9.4	11.1
Pendimethalin @ 0.75 kg/ha	8.2	9.8	11.3
CD 5%		0.8	

Source : Dadhich and Mali 1991.

PEST MANAGEMENT

For intensification of pulse production, it is desired to stabilize and enhance the pulse productivity by preventing insect-pest damage occurring at various growth stages. Major insect-pests of different pulses along with their control measures have been briefed below.

Pigeonpea

Almost 200 insect species have been observed to infest pigeonpea crop. However, a dozen of them are of economic importance. The pests damaging buds, flowers, pods and grains are of serious concern.

Pod Borer (*Helicoverpa armigera Hiib*)

The pod borer is most important and widely distributed insect pest of polygamous nature. Its damage is most common in southern India and the extent of loss may even go as high as 90 per cent. The moths fly in the night and lay their eggs on a large range of plants. Adults are strongly attracted towards flowers. Large sized holes (3 to 4 cm) can be seen on damaged pods.

Pod borers can effectively be controlled by spraying endosulfan @ 0.07 per cent or monocrotophos @ 0.04 per cent, or dusting 5 per cent malathion or 2 per cent endosulfan @ 20-25 kg ha^{-1}.

Pod fly (*Melanagromyza oblusa*)

It is the second most damaging pest of pigeonpea in India. The small black flies lay their eggs insides the pods and the larvae feed on the seeds. This pest is considered most damaging in northern India and the estimated damage caused may be to the extent of 60 per cent during severe infection.

Spray of monocrotophos @ 0.04 per cent or dimethoate @ 0.03 per cent protects the infected plants quite effectively.

Plume moth (*Exelastis atmosa*)

This moth complex also includes three more pest species and is more prevalent in northern India. The damage caused by this moth may range from 5 to 10 per cent. The caterpillar feeds on the seed from the outside.

Spraying the crop with endosulfan @ 0.07 per cent, or dusting with 2 per cent methyl-parathion or with 5 per cent carbaryl/malathion dust @ 25 to 30 kg/ha can effectively control plume moth.

Spotted caterpillar (*Maruca lestulates*)

The extent of damage caused by these caterpillars range from 5 to 20%. Young larvae feed on the seed. The infested flowers and pods are webbed together and the holes on the pods are plugged with excreta.

Weevils (*Coleoptran borer*)

Weevils attack pigeonpea in the states of Andhra Pradesh, Karnataka, coastal areas of Orissa, Bihar and West Bengal.

Weevils and caterpillars can be controlled by spraying endosulfan @ 0.07 per cent or dusting with quinolphos 1.5 per cent or malathion 5 per cent.

Chickpea

Nearly 54 species of insects and other arthropoda are known to attack chickpea in India (Lal, 1989). However, gram pod borer (*Helicoverpa armigera*), black cut worm (*Agrotis tpsilon*), black aphid (*Aphis craccivora*), termite (*Odontotermes obesus*) and semilooper (*Autographa nigrisigna*) are chief pests of chickpea. The pod borer is the single most important pest (Reed *et al.*, 1979). The larvae feed voraciously on the crop during whole life cycle and cause the damage to the crop to the extent of 9-90 per cent (Ram Ujagar and Sehgal, 1989).

Pod borer can effectively be controlled by spraying the crop with endosulfan @ 0.07 per cent, *neem* seed kernel extract (5%). For controlling termites and cut worms, treating of seeds with aldrin EC @ 1 l/100 kg of seed is suggested.

A 4-year study at Kanpur revealed that mixed/intercropping of chickpea with linseed, wheat, barley or mustard was helpful in minimizing the pod borer damage compared to sole crop (Yadava *et al.*, 1989).

Over 5 year basis, screening of 109 varieties of chickpea against pod borer at Pantnagar, the varieties viz., PDG-84-16, PDG-84-5, HUG-183, ICCC-41, ICCC-14 and BG-3002, showed 41 to 61% reduced damage (Ram Ujagar and Sehgal, 1989).

***Vigna* species**

The *Vigna* group is infested by more than two-dozen pest species. However, important are : white fly, *Bemisia tabaci,*; leaf hopper, *Empoasca* spp., thrips, *Megalurothrips distalis*; bean stem fly,

Ophiomyia phaseoli; til caterpillar, *Acheronitia styx*; hairy caterpillars, *Spilosoma obiqua*; semilooper, *Anomis flava*; black aphids, *Aphis craccivora* and pod borer, *Maruca testulalis, Heliothis armigera, Lamphides boeticus.*

Estimated damage caused by these pests range from 5.2 to 47.6 per cent in mothbean, 13.9 to 50% in cowpea, 19.8 to 61.5 per cent in black-gram and 21.0 to 64.7% in mungbean.

For control of Jassid, aphids and white flies the spray of phosphamidon or monocrotophos (0.03%) is suggested. Cowpea stem fly can be controlled by dusting carbaryl (4%) or spraying with 0.05% monocrotophos. Spray of endosulfan (0.07%) or quinolphos (0.05%) against hairy caterpillar is suggested.

In order to avoid development of bruchids, it is advisable to fumigate seed material with ALP @ 1-2 tablets/t and EDB @ 30 ml/t.

Pest resistant varieties

Mungbean - ML 325 and ML 529 are promising against jassid.

Blackgram - LU-178, LU-190 LU-397 and Pant U-19 are promising against white fly.

Cowpea - Co-4 is promising against stemfly in Tamil Nadu.

DISEASE MANAGEMENT

Pigeon pea : Fusarium wilt (*Fusarium udum* Butler)

The wilt is a major problem in Bihar, Maharashtra, Uttar Pradesh, Madhya Pradesh and Tamil Nadu. It may cause 30 per cent yield losses when attack is at pre-harvesting stage and 100 per cent in case of attack at pre- poding stage (Kannaiyan and Nene, 1981). The foliage shows slight chlorosis and becomes bright-yellow before wilting. In later stage, branches may dry up and the plants wilt out.

Rotation/intercropping with sorghum may reduce wilt infection. Deep ploughing during summer and growing of green manuring crop can be helpful in reducing the inoculation of its pathogen. Seed treatment with 1.5 kg thiram + 1.5 kg brasicol/kg seeds may control wilt infection.

Phytophthora stem blight

It is very destructive in monsoon season due to water stagnation. Its incidence is more prevalent in short duration and closely spaced cultivars. Severe incidence at the early seedling stage may cause early seedling mortality. Disease shows the appearance of dark brown lesions on the stem slightly above the ground level. The stem breaks at the point of infection.

Chemical control of this disease is very difficult. However, seed treatment with metalxyl (4g ai/kg seed) can control the incidence upto 30 per cent at the early stage. Water stagnation in the field may be avoided.

Sterility mosaic virus

This viral disease is transmitted through a mite *Accria cajami* and is more severe in the states of Bihar, Uttar Pradesh and Tamil Nadu (Kannaiyan and Nene, 1981). Plants become pale green and show loaping of flowers and pods. Leaves are severely reduced in size and the plants show complete sterility.

Spray of Acaricides, Tedion$^{(R)}$ and Morestan$^{(R)}$ @ 0.1 per cent on infected plants has proved effective in killing more than 90 per cent mites.

Alternaria leaf spot (*Alternaria tenuissima)*

This fungal disease is important in post rainy season, particularly in north plain-east zone. Initially small necrotic spots appear on the leaves which enlarge in size to characterize lesions with dark and light brown rings. Severely infected leaves drop ultimately.

Spray of dithane M 45 @ 0.3 per cent or blitox @ 0.4 per cent as per disease intensity is suggested.

Yellow mosaic virus

It is transmitted through white flies. Disease intensity is relatively more in late sown crops. More than 40 per cent yield losses have been reported in India due to this disease (Beniwal *et al.*, 1983). Disease first appears in the form of yellow, diffused spot on leaf lamina, leaf size is reduced and the leaves turn yellow.

No effective control measures are available. White flies can however, be controlled by spraying the crop with 0.05% malathion or 0.03 per cent metasystox.

Chickpea

Out of 50 pathogens reported in India, 5 to 7 are more important which have been discussed below.

Wilt *(Fusarium oxysporium* **sp. ciceri)**

It causes 60 to 70 per cent yield losses at a time. Alkali soils appear to favour this disease. Sudden drop of leaves and petioles, black internal discoloration of xylem and pith and no external rotting are its important symptoms (Nene *et al.*, 1989).

Treating the seeds with benomyl and thiram mixture or with bavistin 25 per cent + TMTD 50 per cent @ 2.5 g/kg seed is suggested for its control.

Ascochyta blight *(Ascochyta rabiei/Phyllosticta rabiei)*

The blight in the beginning shows dark lesions on the stem and leaves and later on pods. In severe cases lesions surround the stems causing blighting the parts above ground level.

For control measures, the seed treatment with either thiram, calixin M, babytan or benomyl is suggested.

Dry Root Rot *(Rhizoctonia bataticola)*

It is more severe in central and southern India. Low soil moisture and temperature between 25 to 35°C favour the spread of this disease. Sudden drying of plants without dropping of leaves is important symptom of this diseases.

Seed treatment with captan or thiram @ 2.5 g kg/ha is suggested for its control. Early sowing and early maturing cultivars can minimise its incidence.

Root Rot *(Rhizoctonia solani)*

This disease may cause more damage at the seedling stage when soil moisture content is usually high. Disease occurs in temperature range of 18-30°C. Rotting of roots with discoloration extending above the ground level leading to wilting of plants gradually are the symptoms of this rot.

Treating of seeds with captan or thiram @ 2.5 g/kg seeds is suggested.

Botrytis Gray Mould *(Botrytis cinerea)*

This disease is more prevalent in prolonged cool and dry spells.

Treating of seeds with bavistin @ 2.5 g/kg seed or bavistin + thiram combination @ 2.5 g/kg seed is suggested for its effective control.

Vigna **Species**

Mungbean : yellow mosaic virus (MYMV)

It is destructive disease of mungbean, blackgram and mothbean and can cause 100 per cent yield losses in mungbean and blackgram. It is transmitted through white flies (*Bemisia tabaci*). Important symptoms are : leaves may turn yellow, maturity is generally delayed, and flowers and pods production is severely reduced.

White flies population can be reduced by use of aldicarb. Resistant cultivars against MYMV is the real check (Table 24).

Table 24. Disease resistant varieties of certain pulses crops.

Disease	*Varieties*
Pigeonpea	
Alternaria blight	ICPL 151, WB 20
Sterility mosaic virus	Pant A3, ICPL 151, Pusa 33, Bahar ICP 8863
Yellow mosaic virus	Bahar
Wilt	Pusa 33, R 60, BDN-1, ST-1, C 11, ICPL 87
Pod borer	Pusa 33
Fusarium wilt	*GAUT*-82-9, *GAUT*-82-74, *GAUT* 83-23
Chickpea	
Wilt	Pusa 209, BGM 404, BG-256, Pnule G-5, BGM 417, GNG-146

Table Contd...

Disease	Varieties
Botrytis grey mould	Pusa 209,
Ascochyta blight	BGM 404, BG-244, H75-35, GNG 146
Root rot	BGM 417, ICCC-32,
Stunt virus	BEG 417
Fusarium wilt	JG 315
Pod borer	ICCC-32
Mugbean	
Macrophomina blight	UPM 79-1-1, UPM 79-4--12, ML 131, Pusa 105
Yellow mosaic virus	ML 26-10-3, UPM 79-4--12, PDM 11, ML 131, KM1, ML5, Pusa 105, HYb. 12-4, ML1 +
Powdery mildew	UPM 79-4-12, PDM 11, ML 337, Pusa 105,
Leaf curl virus	UPM 79-4-12, Pant Mung-3 +
Cercospora leaf spot	PDM 11, ML5, ML 337,
Aphid, Jassid, white fly	Pusa 105
Pod borer	KM -1
Urbean	
Yellow mosaic virus	Pant U-19, Pant U-30,
Powdery mildew	Co-4
Stem fly	KM-2
Field pea	
Powdery mildew	Pant P-5
Cowpea	
Root rot	V 240, Co-4,
Bacterial blight	V 240, S 488
Tip blight	V 240, Co-4
Wilt	Co 4
Stem fly	Co 4
Lens	
Rust	Lens 4076

Source : Asthana *et al.* (1986).
* Patel *et al.* (1988).
* Shanmugasundaram (1988).

Powdery mildew (*Erysiphe polygoni* D.C.)

It is a common fungal disease of mungbean and black-gram and is common in cool and dry climate. In the beginning infected plants show gray spots on the upper surface of leaf and in the advanced stage older leaves defoliate.

Spray of wettable sulphur @ 0.04 per cent or karathan @ 0.1 per cent are suggested for its control.

17.3.iii. Cercospora leaf spot (*Cercospora canes cens* Ell and G. Martin) :

Disease is more severe in warm temperature (22.5 to 23.5°C) and high humidity (75 to 85%). Close and dense canopy may prove congenial for the spread of this disease (Sud and Singh, 1984b). Brown leaf spots with grayish white centres and reddish brown margins develop, causing premature leaf fall.

Foliar spray with mancozeb or zenab are known to reduce its incidence (Grewal *et al.*, 1980). The seed treatment (hot water, solar heat or streptomycin + captan) alone may give 40 per cent increase in yield over control (Jindal and Thind, 1987).

Charcoal rot (*Macrophomina phaseolina* (Tassi) Goidanich

Pulses like pigeonpea, mungbean, black-gram, cowpea and horsegram are infested by damping off, root rot and wilt caused by soil born diseases like *M. phaseolina*, *R. solani*, *F. udum*, respectively.

Leaf blight can be controlled by spray with benomyl and thiophanate- methyl @ 0.05 per cent (Taneja and Grover, 1978). Mungbean cultivars resistant to this diseases may be used (Table 24).

SCOPE FOR FURTHERING PULSE PRODUCTION IN INDIA

An annual production of 14.6 million tons with a productivity of 576 kg/ha is providing only 36.5 g pulse/day/adult in India against 70 g pulse/day/adult prescribed by WHO. Such a considerably poor availability of pulse therefore, justifies exploration of the prospects for furthering the pulse production in India. Since, the chickpea and pigeonpea collectively contribute nearly 60 per cent to the total pulse production thus, augmenting production, principally deserves realization from these crops.

The prospects of higher production of chickpea lies in northern and eastern zones of the country where assured winter precipitation promises a good harvest, provided, it is backed-up with sufficient fertility (particularly P). One or two supplementary irrigations may provide a good promise. On black soils in Southern India where chickpea suffers due to receding soil moisture and warm winters, early maturing cultivars with improved yield potential can be of great use in increasing chickpea production. One or two supplementary irrigations in M.P., where agroclimatic conditions are very congenial for chickpea, can boost yield substantially. In Northern India, for better adoption and expansion of chickpea cultivation, the cultivars having inbuilt potential for setting of pods in low temperature (5 to 10^{0}C) are essentially required.

Wilt, Ascochyta blight and Botrytis gray mould are the serious diseases of chickpea as they may cause complete losses during epidemic years. Similarly, pod borer is a serious pest. Control measures by chemical means or by use of tolerant cultivars may prove of great use in enhancing chickpea yields.

In subtropical environments, the early maturing pigeonpea cultivars grow tall and their maturity is delayed as a consequent delaying and adversely affecting the prospects of concerned sequentional crop. Remedy, therefore lies in breeding of extra-early cultivars (100 days) characterized with synchronous maturity and photoinsensitiveness. In arid areas, where rainy season is short, the prospects of such early maturing strains may be useful as a sole crop and intercropping with the crops of similar maturity group like clusterbean and mothbean.

Introduction of high yielding, determinate type early maturing (140 to 150 days) pigeonpea cultivars in irrigated double cropping system with wheat is a landmark technology for northern India. Dissemination of this system in more areas will bring further yield enhancement in pulses.

In rice-fallow system, introduction of short duration thermoinsensitive pigeonpea cultivars that can tolerate high soil moisture content have prospects of yield enhancement in eastern and southern zones.

Diseases viz., Phytophthora stem blight in high rainfall zones and water stagnated areas, Alternaria leaf blight and yellow mosaic virus in post rainy season areas and wilt in Uttar Pradesh, Bihar, Madhya Pradesh, Maharashtra and Tamil Nadu have greater degree of incidence in pigeonpea, hence effective chemical or cultural measures are required for boosting pigeonpea yields.

Susceptibility of pulses to abiotic factors is a major cause in lowering down their yields on real farm situations. For instance, the terminal drought occurring in Peninsular India reduces the yield of medium duration pigeonpea to about a half that could be achieved due to supplementary irrigation. Similarly, arid pulses show high degree of yield reduction in north-west part of the country due to water scarcity and high temperature coinciding with effective reproductive growth stage. Agronomic management and tailoring of thermoinsensitive and drought tolerant cultivars would be helpful in increased pulse production. Poor nodulation in waterlogged pulses in northeast and eastern zones lead to decreased yield of certain pulses. Soil and water salinity due to poor precipitation cause high degree of yield losses due to poor seedling emergence and their early mortality. Pulses tolerant to salinity particularly at the emergence stages are, therefore, desired for considerable salt prone areas.

SUMMARY

Compound growth rate in area of pulses has steadily decreased during last four decades over pre-green revolution era, whereas growth rates of production and productivity have considerably improved during the past decade. Amongst the major pulses, the growth rate of area and production in chickpea and that of productivity in pigeonpea have depicted negative trends during 1980-81 to 1990-91 period leading to stagnation in pulse production in the country. Considerably high population growth rate (2.2% per annum) and quite meagre growth rate in pulse production (0.4% per annum) have led to decreased pulse availability from 61 g in 1951 to as low as 36.5 g/day/adult in 1990 which is less than 50 per cent prescribed by ICMR and FAO. India, however, has a distinction of being the largest pulse producing (25%) and growing

(40%) country of the world. However, pulse productivity is least (576 kg/ha) in India and being maximum (5751 kg/ha) in Netherlands. World's 90, 62, 60 and 38 per cent pulse production in respect of pigeonpea, mungbean, chickpea and lentil, respectively is realized from India, hence these are the major pulse crops in Indian context. Constraints to pulse production in India are many however, chief of them are : agro-ecological, biological, technological, economical and adaptive constraints. Pusa ageti, Manak, and Bahar are the improved cultivars of pigeonpea suited to double cropping system. Prabhat, an extra early pigeonpea cultivar is also suited to pigeonpea-wheat double cropping system in northern zone. BGM 117 with multidiseases resistance potential is most suited to central dryland zones of chickpea. Pusa 209 chickpea cultivar is suited to both irrigated and rainfed situations. In mungbean, Jawahar-45 and ML 131 are suited to *kharif*, whereas K 851 and PDM-11 are suited to spring season. In urdbean UG-128 resistant to YMV is an ideal variety for dryland situations. In cowpea, FS-68 and Charodi are important cultivars. Pigeonpea-groundnut in central and southern zones and pigeonpea-sorghum on vertisols are the most efficient intercropping systems. Introduction of pigeonpea-wheat sequential system in assured soil moisture conditions of north-west zone is the landmark in pulse production. Chickpea-safflower in medium black soils and mustard-wheat in Haryana and Rajasthan are important intercropping systems. Similarly, mungbean-pearlmillet is an effective and productive system in arid areas.

Mid of October to end of October is the optimum sowing time of chickpea in Northern India, however, in Southern India sowing can be advanced. Sowing of traditional pigeonpea cultivars can be done by mid of June to first week of July, whereas sowing of short duration cultivars should be adjusted so that planting of wheat is not delayed. Optimum sowing time of urdbean, mungbean and cowpea has been observed from beginning to mid of July. Summer planting of mungbean and urdbean is suggested from February to mid of March. Optimum sowing dates of lentils are October 15 to first week of November, whereas the sowing time of horse-gram depends on the rainfall pattern in southern and coastal states.

Pulses show inconsistent and poor response to N-fertilization, however a starter dose of 15 to 20 kg N/ha has been observed responsive. Pulses invariably respond to P from 20 to 40 kg (urdbean, mungbean, cowpea) to 50 to 60 kg P_2O_5/ha (pigeonpea and chickpea). One or two supplementary irrigations at the most critical/sensitive growth stages have proved useful in pigeonpea, chickpea and mungbean, particularly. Hand weeding in appropriate herbicide treated fields is very effective weed control measure. Pod borer, semi-looper, hairy caterpillar and grain weevils are the major pests of pulses. Their effective control measures have been discussed. Amongst the diseases, fusarium wilt, yellow mosaic virus, dry root rot, powdery mildew, cercospora leaf spot and charcoal rot are the major diseases of pulses, their symptoms, chemical control measures and available disease resistant varieties have been highlighted.

REFERENCES

Ali, M. 1981. Effect of row spacing and phosphate levels on *arhar* production under dryland conditions. *Indian J. Agron.* 26: 346- 348.

Ali, M. 1988. Weed suppressing ability and productivity of short duration legumes intercropped with pigeonpea under rainfed conditions. *Trop.Pest Manag.* 34 : 384-387.

Anonymous 1976. *Annual Report of 1975 on Grain Legumes.* Tehran Univ. School Agric., Karaj, Iran.

Augustin, J. and Klein, B.P. 1989. Nutrient composition of raw, cooked and canned and sprouted legumes. In : *Legumes, Chemistry, Technology and Human Nutrition.* (Ed. R.H. Mathws), Marcel, Dekker, Inc., N.Y. pp. 187-217.

Aykroyd, W.R., Doughty, J. and Walker, A. 1982. *Legumes in Human Nutrition,* F.A.O., Food and Nutrition paper, 20.

Bajpai, R.P., Singh, V.K. and Dwivedi, R.K. 1989. Performance of horse gram of different dates of sowing, phosphorus levels and intercropping *Nat.Symp.New Frontiers in Pulses Res. and Dev* Kanpur, pp. 77.

Bashir, M., Zubair, M. and Malik, B.A. 1988. Disease resistant sources and utilization in breeding improved mungbean in Pakistan. *Mungbean Proc. Second Intern.Symp.* Asian Vegetable Res. and Dev. Cent., Shanhua, Taiwan.

Beniwal, S.P.S., Deena, E. and Nene, Y.L. 1983. Effect of yellow mosaic on yield and its components in post rainy-season pigeonpea. *Intern.Pigeonpea Newsl.* 2 : 48.

Bhagyaraj, D.J. and Hegde, S.V. 1978. Response of cowpea to rhizobium seed inoculation. *Curr.Sci.* 47 : 548-549.

Bhan, S. and Khan, S.A. 1979. Response of *kharif* crops to irrigation in light textured alluvium of Uttar Pradesh. *Indian J. Agron.* 24 : 410-416.

Bhat, S.N. 1981. Effect of row spacing and phosphorus levels on the growth and yield of cowpea genotypes and their residual effect on the yield of succeeding wheat. *Mysore J. Agric. Sci.* 15 : 348-349.

Bhatia, M.S. 1991. Economic constraints in increasing pulse production. *Agric. Situation in India* 46(5) : 279-284.

Brathwaite, C.W.D. 1981. Diseases of pigeonpea in the caribbean area. In : *Proc.Intern.Workshop on Pigeonpea, Vol. 1 ; ICRISAT Cent., India, A.P. pp.129-136.*

Bressani, R. 1973. Legumes in human diets. In : *Nutritional Improvement of Food Legumes by Breeding* (Ed. M. Milner), Protein Advisory Group of the United Nations, N.Y., pp. 15-42.

Brotonegoro, S., Radjit, B.S. and Laumans, Q.J. 1988. Mungbean research and development in East Java, Indonesia. *Mungbean: Proc. Second Intern. Symp.*, Asian Vegetable Res. and Dev.Cent., Shanhua, Iran.

Chainuvati, C., Potan, N. and Worasan, T. 1988. Mungbean and blackgram production and development in Thailand. *Mungbean: Proc.Second Intern.Symp.* Asian Veg. and Res. and Dev. Cent., Shanhua, Iran.

Chandra, S. and Ali, M. 1986. Recent achievements in pulses production. *Tech.Bull.* No.1. DPR, Kanpur, pp. 17-18.

Chauhan, R.S. and Singh, K.B. 1981. Response of pigeonpea varieties to levels of phosphorus and row spacing under rainfed conditions. *Indian J.Agron.* 26(1) : 49-52.

Chauhan, Y.S., Venkataratnam, N. and Sheldrake, A.R. 1987 b. Factors affecting growth and yield of short-duration pigeonpea and its potential for multiple harvests. *J.Agric.Sci.Cambridge* 109 : 591-593.

Chauhan, Y.S., Saxena, N.P. and Johanson, C. 1989. Abiotic factors limiting chickpea and pigeonpea production. *Nat.Symp.New Frontiers in Pulse Res. and Dev.*, Kanpur, p.3.

Chauhan, D.R., Panwar, K.S. and Balyan, R.S. 1991. Moisture extraction pattern and water use efficiency of field pea as influenced by irrigation, phosphorus and weed control practices. *Indian J.Agron.* 36(3) : 363-367

Dadhich, S.C. and Mali, A.L. 1991. Effect of herbicide and phosphorus on crop weed competition in chickpea. *Indian J.Agron.* 36(suppl.): 283-285.

Davies, D.R. 1976. In: *Evolution of Crop Plants.* (Ed. N.W. Simmonds) Longman, London, pp. 172-174.

Dhingra, K.L. 1975. Transmission of urdbean laeaf crinkle virus by two aphid species. *Indian Phytopathol.* 28: 80-82.

Dhingra, K.K., Dhillon, M.S., Grewal, D.S. and Sharma, K. 1991. Performance of maize and mungbean intercropping in different planting patterns and row orientations. *Indian J.Agron.* 36(2) : 207-212.

Dwivedi, V.D., Pandey, R.P. and Namdeo, K.N. 1991. Weed control in pigeonpea-sorghum intercropping system. *Indian J.Agron.* 36(2) : 234-238.

Eusebio, J.S. and Eusebio, P.S. 1984. Biological testing of protein quality of selected legume cultivars. *Philippine Agriculturist* 67 : 209-221.

Evans, R.J. and Bandemer, S.L. 1967. Nutritive value of legume seed proteins. *J.Agric. & Food Chem.* 15 : 439-443.

F.A.O. 1989. *FAO Production Yearbook*, 1988, 42, FAO, UN, Rome.

Giri, G. and De, R. 1978. Intercropping of pigenopea with other grain legumes under semi-arid rainfed conditions. *Indian J.Agric.Sci.* 48 : 659-665.

Giri, G. and Gangasaran, 1984. Studies on cropping systems under semi- arid rainfed conditions of North-West India. *Indian J. Agron.* 29(2) : 162-169.

GOI 1990. *Economic. Survey* 1990, Govt. of India, Ministry of Fianance, New Delhi.

Gogoi, A.K., Kalita, H., Pathak, A.K. and Deka, J. 1991. Chemical control of weeds in field pea *(P.sativum) Indian J. Agron.* 36 : 287-288.

Gonzales, J.M., Miranda, C.L., Brennisen, L.D. and Aguilar, C.G. 1972. Digestibility, biological value and net protein utilization of some beans. *Philippine J.Nutr.* 25 : 103-112.

Goyal, S.N., Tikka, S.B.S., Patel, N.L., Patel, N.M. and Ahlawat, I.P.S. 1991. Integrated weed management in pigeonpea. *Indian J.Agron.* 36(1) : 52-54.

Grewal, J.S., Pal, M. and Kulshrestha, D.D. 1980. Control of *Cercospora* leaf spot of greengram by spraying bavistin. *Indian J.Agric.Sci.* 50 : 707-709.

Gupta, J.P. 1987. Effect of tillage and mulch on soil and growth and yield of cowpea grown in the arid tropics. *Arid Soil Res.Rehab.* 1(3) : 161-172.

Gupta, J.P. and Gupta, G.K. 1986. Effect of tillage and mulching on soil environment and seedling growth under arid conditions. *Soil & Tillage Res.* 7(3) : 233-240.

Gupta, J.P. and Rao, A.V. 1989. Effect of mulching on nodulation, nitrogenase activity, growth and yield of greengram and cowpea grown in arid western Rajasthan. *Indian J.Agric.Sci.* 59 : 181-182.

Gupta, V.P. and Kapoor, A.C. 1980. Chemical evaluation of protein quality of various grain legumes. *Indian J.Agric.Sci.* 50 : 393- 398.

Gupta, Y.P. 1983. Nutritive value of food legumes. In : *Chemistry and Biochemistry of Legume.* (Ed. S.K Arorad, Edwar Arnold, London, pp. 287-327.

Hawtin, G.C., Singh, K.B. and Saxena, M.C. 1980. In : (Eds. Summerfield, R.J. and Bunting, A.H.) (Eds.) *Advances in Legumes Science*) London, HMSO, 613-624.

Haytowitz, D.B. and Matthews, R.H. 1986. *Composition of Foods : Legumes and Legume Products.* USDA, Agric. Handb. No. S-16.

ICRISAT 1990. *Agroclimatology of Asian Grain Legume Growing Areas.* Res.Bull.No.14, ICRISAT, A.P. India.

Jain, V.K., Chauhan, Y.S. and Khandekar, M.P. 1989. Effect of weed management on yield and yield attributes of urdbean. *Nat.Symp.*on New Frontiers *in Pulse Res. and Dev.,* Kanpur, pp. 58.

Jindal, K.K. and Thind, B.S. 1987. Control of *Cercospora* and bacterial leaf spots of greengram. *Indian J.Agric.Sci.* 57 : 372-375.

Kalyan Singh (Mrs), Singh, S., Jain, A. and Singh, P.P. 1989. Response of lentil varieties to planting time and row spacing. *Nat.Symp.New Frontiers in Pulses Res. and Dev.,* Kanpur, pp. 85.

Kambar, N.S. and Kadapa, S.N. 1983. Review of plant improvement and agronomic research on horsegram in north Karnataka. *Proc. First ICAR Group Meeting on Horsegram Res.Workers,* All India Coodinated Reseach Project for Dryland Agriculture, Hyderabad.

Kannaiyan, J. and Nene, Y.L. 1981. Influence of wilt at different growth stages on yield loss in pigeonpea. *Trop.Pest.Mangag.* 27 : 141.

Kannaiyan, J., Ribeiro, O.K., Erwin, D.C. and Nene, Y.L. 1980 b. Phytophthora blight of pigeonpea in India. *Mycol.* 72(1) : 169- 181.

Kaul, J.N., Sekhon, H.S., Dhingra, K.K. and Tripathi, H.P. 1980. Date of planting and row-spacing studies with early maturing pigeonpea varieties under irrigated conditions for double cropping. *Trop.Grain Legume Bull.* 19 : 20-23.

Kumar, A. and Ahlawat, I.P.S. 1986. Effect of planting geometry and nitrogen fertilization in pigeonpea based intercropping systems. *Indian J.Agron.* 31 : 112-114.

Lal, R.B. and Jaiswal, P.C. 1981. Influence of sulphur fertilizer on grain yield and economics of urd cultivation. *Indian J.Agron.* 26 : 323-330.

Lanka, D. and Satpathy, R.K. 1976. Response of pigeonpea varieties to levels of nitrogen and phosphorus in lateritc soil. *Indian J.Agron.* 21 : 217-220.

Ling, Y., Gu, H., Liu, C. and Sun, L. 1988. Mungbean production in China and AVRDC's variety performance in Jaingsu *Province. Mungbean : Proc.Second Intern.Symp.*Asian Veg.Res. and Dev.Cent., Shanhua, Taiwan.

Makhan Lal and Gupta, O.P. 1984. Effect of irrigation schedules on pigeonpea in Rajasthan, India. *Intern.Pigeonpea Newsl.* 3 : 25-26.

Mayouf, M.A. 1978. Food Legumes in Iraq. In : *Food Legume Imp. and Dev., Proc.Workshop,* (Eds. G.C. Hawtin and G.J. Chancellor) Aleppo (Syria), ICARDA Syria and IDRC (Canada), pp. 55-57.

Meekin, S., Troedron, R.J., Byth, D.E. and Wallis, E.S. 1988. Release of a quantum in Queensland, Australia. *Intern.Pigeonpea Newsl.* 7 : 6-8.

Moody, K. and Shetty, S.V.R. 1981. Weed management in intercropping systems. In : Proc.Intern. *Workshop on Intercropping,* ICRISAT, India, pp. 229-237.

Mudholkar, N.J. and Ahlawat, I.P.S. 1979. Response of bengal gram to nitrogen, phosphorus and molybdenum. *Indian J.Agron.* 24(1) : 61-65.

Muthoka, M.S. and Shakoor, A. 1988. Mungbean improvement and production in the semi-arid areas of Kenya. *Mungbean : Proc.Second Intern. Symp.* Asian Veg.Res.and Dev.Cent., Shanhua, Taiwan.

Narayanan, A. and Sheldrake, A.R. 1979. Pigeonpea as a winter crop in peninsular India. *Exp.Agric.* 15 : 91-95.

Navarro, R.S. 1988. Mungbean production and research in Philippines. *Mungbean : Proc.Second Intern. Wrokshop Symp.* Asian Veg.Res. & Dev.Cent., Shanhua, Taiwan.

Nene, Y.L., Sheila, V.K. and Sharma, S.B. 1989. *A World List of Chickpea and Pigeonpea Pathogens. Legumes Pathol.* Progress Report-7, ICRISAT, 23 p.

Nyabyenda, P. 1987. Grain legume production in Rwanda. In : *Grain Legumes in Eastern and Central Africa.* Summary Proc.Consultative Group Meeting for Eastern and Central African Reg.Res.on Grain Legumes, ILGA, Addis Ababa, Ethiopia, ICRISAT, pp. 57- 58.

Nygaard, D.F. and Hawtin, G.C. 1981. In : (eds.) *Lentils.* C. Webb and G.C. Hawtin Farnham Royal : Commonwealth Agricultural Bureau, 7-13.

Osiru, D.S.O. and Kibira, G.R. 1981. Sorghum/pigeonpea and finger millet/groundnut mixtures with special reference to plant population and crop arrangement. In : *Proc.Intern.Workshop on Intercropping,* ICRISAT, India, pp. 78-85.

Padhi, A.K. and Sahoo, P.N. 1991. Response of greengram varieties to sowing dates during spring season. *Indian J.Agron..* 36 : 115-118.

Padhi, A.K., Sahoo, B.K. and Das, K.C. 1992. Productivity of rainfed pigeonpea-based intercropping system. *Indian J. Agric.Sci.* 62(9) : 594-608.

Panwar, K.S. 1978. Latest technology for gram cultivation. *Seminar on Pulse Prod.Tech. (Rabi pulses),* Kanpur.

Panwar, K.S. and Pandey, K. 1977. Weed control studies in bengalgram. *Indian J.Agron.* 22(4) : 257-259.

Panwar, K.S. and Yadav, H.L. 1981. Response of short duration pigeonpea to early planting and phosphorus levels in different cropping systems. In : *Proc.Intern.Workshop on Pigeonpeas,* Vol.1, ICRISAT, pp. 37-44.

Panwar, R.S., Malik, R.K. and Bhan, V.M. 1988. Chemical weed control in cotton. *Indian J.Agron.* 33(2) : 205-207.

Patel, R.G., Patel, N.P., Patel, H.C. and Patel, R.B. 1984. Effect of graded levels of nitrogen and phosphorus on growth, yield and economics of summer mungbean. *Indian J.Agron.* 29(3) : 291-294.

Patel, B.R., Singh, D. and Gupta, M.L. 1991. Effect of irrigation and intercropping on gram and mustard. *Indian J.Agron.* 36(2) : 283- 284.

Pathak, G.N. 1970. Redgram-Pulse_crops of India. ICAR, New Delhi, pp. 14-53.

Patra, A.P. and Chatterjee, B.N. 1986. Intercropping of soybean with rice, maize and pigeonpea in the plains of West Bengal. *Indian J. Agric.Sci.* 56 : 413-417.

Patwardhan, V.N. 1962. Pulses and bean in human nutrition. *Amer.J.Clinical Nutr.* 11 : 12-30.

Poehlman, J.M. 1991. Quality and utilization. In: *The mungbean,* Oxford & IBH Publishing Company, New Delhi, Bombay and Calcutta, India, pp. 312-345.

Prabhakar, M. and Saraf, C.S. 1991. Influence of irrigation and phosphorus on growth, yield and water-use-efficiency of chickpea genotypes. *Indain J.Agron.* 36(3) : 357-362.

Prasad, K., Gautam, R.C. and Mahta, N.K. 1985. Effect of planting patterns and weed control methods on growth characters, yield and yield attributes of *arhar* intercropped with soybean. *Indian J.Agron.* 30 : 429-433.

Rafey, A. and Verma, U.K. 1988. Production potential of legumes, oilseeds and cereal in intercropping system with pigeonpea. *Indian J. Agric.Sci.* 58 : 433-436.

Rajaram, S. 1984. Effect of base crops and the growth and development of redgram SA-1 as an intercrop in rainfed black soils of Aruppukottai region. *Madras Agric.J.* 71 : 480-481.

Rakesh Kumar, D. and Tiwari, O.P. 1989. Low cost inputs in increasing chickpea production under paddy based cropping system in Chattishgarh. *Nat.Symp.New Frontiers in Pulses Res. & Dev.,* Kanpur, India, pp. 65-66.

Rameshaiah, K., Sudhakara Babu, K. and Prabhakara Rao, Y.P. 1989. Response of urdbean to irrigation in rice fallows. *Nat.Symp.New Frontiers in Pulse Res. & Dev.,* Kanpur, India, pp.59.

Ram Ujagir and Sehgal, V.K. 1989. Chickpea evaluation for *Heliothis armigera* resistance. Nat.Symp. New *Frontiers in Pulse Res. and Dev.,* Kanpur, India, pp. 102-103.

Rao, J.V. 1974. Studies on fertilizer management of wheat in maize-wheat and arhar-wheat cropping systems. unpublished Ph.D. thesis, I.A.R.I. New Delhi, India.

Rao, M.V., Jha, K.P., Murthy, B.T.S. and Mandal, B.K. 1982. Intercropping greengram and groundnut with rice and finger millet in the rainy season and the feasibility of a second crop in winter on rainfed uplands of coastal Orissa. *Indian J. Agric.Sci.* 52 : 657-664.

Rathi, S.S. and Singh, D. 1976. Effect of nitrogen and phosphate fertilization on the growth and yield of gram. *Indian J.Agron.* 21(3) : 305-306.

Reddy, D.S., Reddy, D.R. and Acharya, G.V.N. 1978. Studies on water intake and hydraulic conductivity in shallow and deep ploughing operating under rainfed agriculture in Anantpur. *Ann. Arid Zone* 17 : 291-297.

Reed, W., Lateef, S.S. and Sithanantham, S. 1979. Insect pest management on chickpea. *In : Proc. Intern.Workshop on Chickpea Improvement*, ICRISAT, Hyderabad, pp. 179-183.

Robert Zablotowicz, M., Focht Dennis, D. and Connell Glen, H.S. 1981. Nodulation and N_2 fixation of field grown California cowpeas as influenced by well irrigated and droughted conditions. *Agron.J.* 73 : 9-12.

Sadaphal, M.N. 1988. Agronomy of Pulses. In : *Pulse Crops (Grain Legumes)*, (Eds. B. Baldew, S. Ramamynujam) and H.K. Jain, Oxford and IBH Publishing Company, N.D., pp. 456-512.

Sandhu, T.S., Singh, K. and Singh, B. 1988. Mungbean germplasm sources : evaluation and utilization. *Mungbean : Proc.Second Intern.Symp.* Asian Veg. Res. & Dev.Cent., Shanhua, Taiwan.

Saraf, C.S., Singh, A. and Ahlawat, I.P.S. 1975. Intercropping of compatible crops with pigeonpea. *Indian J.Genet.* 35 : 248-252.

Saxena, M.C. and Yadav, D.S. 1975. Some agronomic considerations of pigeonpeas and chickpeas. In : *Inter.Workshop on Grain Legumes*, ICRISAT, A.P., Indian, pp. 31-61.

Saxena, M.C. and Singh, H.P. 1977. Studies on agronomic requirements of winter pulses, In : *Research on Winter Pulses, G.B.P.U. of Agric. & Tech., Pantnagar Exp.Sta.Bull.*, 101 : 23-42.

Saxena, N.P. and Sheldrake, A.R. 1977. Pulse physiology *Annual Report 1976-77, part II : Chickpea Physiology*, ICRISAT, Hyderabad, pp. 179.

Sen, S.N. 1978. Studies on suitability of the different improved types of gram under early and late sowing conditions. 1977-78 *Agron.Report of Pulse and Oilseeds Research Sta*, Berhampore, W.Bengal.

Sevilla-Eusebio, J., Mamaril, J.C., Eusebio, J.A. and Gonzales, R.R. 1968 b. Studies on Philippine leguminous seeds and protein foods 1. Evaluation of protein quality in some local beans based on their amino acid pattern. *Philippine Agrilt.* 52 : 211-217.

Sharma, J. 1978. *Report of the Agronomy Section of the pulses Improvement Project*. Univ. of Agric., Udaipur Res.Sta., Durgapura, 1977-78.

Sharma, D. and Green, J.M. 1975. Perspective of pigeonpea and ICRISAT's breeding programme. In : *Intern. Workshop on Grain legumes*, ICRISAT, Hyderabad, India, pp. 19-29.

Sharma, A.N., Paraya, R.P. and Patel, M.P. 1989. Effect of sowing dates and row spacings on leaf area, leaf efficiency, harvest index and yield of mungbean genotypes during summer season. *Nat. Symp.New Frontiers in Pulse Res. & Dev.*, Kanpur, India, pp. 85.

Sharma H.C., Singh, T. and Mohan, D.S.R. 1974. Response of gram varieties to irrigation. *H.A.U.Jour.Res.* 4 : 255-260.

Sharma, N.N. and Kakti, N.N. 1991. Compatibility of sesame as an intercrop or mixed crop with maize, greengram and blackgran. *Indian J.Agric. Sci.* 61(8) : 583-585.

Sharma R. R.P., Thakur, H.C., Sharma, H.M., Mishra, S.S. and Thakur, S.S. 1984. Effect of fertilization and inoculation of *paira* and late sown pure lentil. *Indian J.Agron.* 29(4) : 459-462.

Sharma, P.P., Singh, P. and Singh, P.P. 1975. Response of N and P in relation to method of application on the yield of gram. *J.N.K.V.V. Res.J.* 9(1/2) : 21-23.

Shukla, N.P. and Lal, M. 1991. Response of winter legumes to moisture regimes and phosphorus. *Indian J.Agron.* 36 : 282-283.

Singh, B.P. and Chauhan, C.P.S. 1991. Plant density relationship in pearlmillet + greengram under dryland conditions. *Indian J.Agron.* 36 : 100-103.

Singh, J.P. and Dixit, R.S. 1989. Effect of sowing dates and phosphorus levels on growth and yield of summer urdbean. *Nat.Symp.New Frontiers in Pulses Res. & Dev.*, Kanpur, India, pp. 67.

Singh, R.B. 1988. Trends and prospects for mungbean production in south and southeast Asia. *Mungbean : Proc.Second Intern.Symp.*Asian Veg. Res. & Dev.Cent., Shanhuan, Taiwan.

Singh, R.P. 1988. Soil-water and crop management practices for improving productivity of drylands of semi-arid tropics of India. In : *Curr. Practices in Dryland Resour. & Tech-4* (Eds. A. Singh, G.R. Chowdhary and D. Kumar) *Geo-Eviron. Acdemia* Jodhpur, pp. 113-131.

Singh, H.P. and Saxena, M C. 1981. Effect of date of planting on the time of flowering in greengram, blackgram, cowpea and pigeonpea. *Indian J. Agric. Sci.* 51 : 33-37.

Singh, H.P., Saxena, M.C., Yadav, D.S. and Sharma, R.P. 1980. Chemical and mechanical weed control in pigeonpea under humid-subtropical conditions of Pantnagar. *Legume Res.* 3 : 22-26.

Singh, R.P. and Das, S.K. 1987. Management of chickpea and pigeonpea under stress conditions with particular reference to drought. Adaptation of Chickpea and Pigeonpea to Abiotic Stresses. *Proc. 'Consultants' Workshop*, ICRISAT, Hyderabad, India, pp. 51-61.

Singh, R.P., Singh, K.C. and Ramakrishna, Y.S. 1978. Effect of systems of planting green gram on the yield, total productivity, moisture use and monetary returns. *Indian J. Agric.Sci.* 48 : 320-328.

Singh, R.C. and Faroda, A.S. 1977. Weed control in pigeonpea. *Indian J.Weed Sci.* 9 : 38-42.

Singh, R.C., Faroda, A.S. and Singh, H. 1986. Response of pigeonpea to phosphate fertilization under different cropping systems. *Indian J.Agron.* 31 : 244-247.

Singh, V.P., Chhabra, A. and Kharab, R.P.S. 1988. Production and utilization of mungbean in India. *Proc. 2nd Intern.Symp.* Asian Veg.Res. & Dev.Cent., Shanhua, Taiwan.

Singh, K.C. and Singh, R.P. 1977. Intercropping of grain legumes with sunflower. *Indian J.Agric.Sci.* 47 : 563-567.

Singh, C. and Yadav, B.S. 1978. Production potential of mungbean and gaps limiting its productivity in India. *Proc.First Intern.Mungbean Symp.* Asian Veg. & Res.Cent., Shanhua, Taiwan.

Solh, M. and Erskine, W. 1981. In : *Lentils.* (C. Webb and G.C. Hawtin eds.) Farnham Royal Commonwelath Agric.Bureau, 53-67.

Souci, S.W., Pachmann, W. and Kraut, H. 1986. *Food Composition and Nutrition Tables.* Wissenschaftliche Verlagsgesellschaft Mbh Stuttgart.

Srivastva, G.P. 1984. Effect of sowing dates and row spacing on pigenopea. *Indian J.Agron.* 29(3) : 317-321.

Srivastava, G.P. and Srivastava, V.C. 1991. Production potential of pigeonpea based cropping system under different planting pattern. *Indian J.Agron.* 36(2) : 173-176.

Subramanian, A. and Palaniappan, S.P. 1981. Effect of method of planting, plant density and phosphorus fertilization on yield of blackgram in irrigated system. *Madras Agric.J.* 68 : 96-99.

Sudhakara Babu, K., Prabhakara Rao, Y.P. and Ramaseshaiah, K. 1989. Performance of different pulse crops in a rice-based crop sequence. *Nat.Symp. New Frontiers in Pulse Res. & Dev.*, Kanpur, India, pp.60.

Summerfield, R.J., Muchlbauer, F.J. and Short, R.W. 1982. *Production Res.Report No.181.* Washington D.C. USDA.

Summerfield, R.J., Muchlbauer, F.J. and Roberts, E.H. 1985. In : Halevy, A.M. (eds.) *A Handbook of Flowering Plants.* Florida : CRC Press.

Suraj Bhan and Khan, S.A. 1981. Double cropping under rainfed conditions of central Uttar Pradesh. *Indian J.Agron.* 26(4) : 371-376.

Sud, V.K. and Singh, B.M. 1984 b. Effect of environmental factors on the development of leaf spot in urdbean. *Indian Phytopathol.* 37 : 511-515.

Taneja, M. and Grover, R.K. 1978. Differential responses of benimidazole fungitoxicants against isolates of *Rhizoctonia* species. *Indian Phytopathol.* 31 : 123.

Tickoo, J.L., Ahri, C.S., Chen, H.K. and Shanmugasundaram, S. 1988. Utilization of the genetic variability from AVRDC mungbean germplasm. *Mungbean : Proc. 2nd Intern.Symp.*, Asian Veg. Res. & Dev.Cent., Shanhua, Taiwan.

Tomar, S.P.S. and Singh, R.R. 1991. Effect of tillage, seed rates and irrigation on the growth, yield and quality of lentil. *Indian J.Agron.* 36(2) : 143-147.

USAID 1971. *Food Grain Legumes as a Major Means of Combating Malnutrition in LDCs.* US Agnecy for Inern.Dev., Tech.Series Bull.No.5.

Umrani, N.K., Shinde, S.H. and Dhonde, P.M. 1984. Studies on intercropping of pulses in *kharif* sorghum. *Indian J.Agron.*, 29 : 27-30.

van der Malsen, L.J.G. 1983. *World Distribution of Pigeonpea.* Inform.Bull.No.14, ICRISAT, India, pp.40.

van der Malsen, L.J.G. 1986. *Cajanus* DC and *Atylosia* W&A (Leguminosae). Agric.Univ.Wageningen Papers 85-4 (1985). Agric.Univ.Wageningen, The Netherlands, pp. 255.

Varshney, J.G. 1989. Performance of lentil cultivars at different row spacings under optimum and late sown conditions. *Nat.Symp.New Frontiers in Pulses Res. & Dev.*, Kanpur, India, pp. 76.

Ventakt Rao, S., Leela, R., Swaminathan, M. and Parpia, H.A.B. 1964. The nutritive value of the proteins of the leguminous seeds. *Indian J.Nutr. & Dietetics* 1 : 304-321.

Vijay Kumar, N.D., Kolandaiswamy, S., Muthukrishnan, P. and Thirumurugan, V. 1984. Effect of moistur stress and phosphorus levels on urd. *Pulse Crops Newsl.* 4 : 56.

Umrani, N.K., Shinde, S.H. and Dhonde, P.M. 1984. Studies on intercropping of pulses in kharif sorghum *Indian J. Agron.* 29 : 27-30.

Wallis, E.S., Byth, D.E. and Whiteman, P.C. 1988. Mechanized dry seed production of pigeonpea. In : *Proc.Intern.Workshop on Pigeonpea*, Vol.1, ICRISAT, Hyderabad, India, pp. 51-60.

Willey, R.W., Rao, M.R. and Natarajan, M. 1981. Traditional cropping system with pigeonpea and their improvement. In : *Proc.Intern.Workshop on Pigeonpea*. Vol.1, ICRISAT, India, pp. 117-127.

Yadav, C.P., Lal, S. and Sachan, J.N. 1989. Effect of mixed/inter- cropping on gram pod borer damage in chickpea. *Nat.Symp.New Frontiers in Pulse* Res. & Dev., Kanpur, India pp. 96.

Yusuf, M., Singh, N.P. and Dastane, N.G. 1980. Yield and water-use efficiency in relation to water supply in chickpea. *Indian J. Agron.* 25(2) : 235-238.

24

ALTERNATE LAND USE SYSTEMS FOR DRYLANDS

R.P. SINGH and MOHD. OSMAN

INTRODUCTION

In India, the per capita cropped land and forest land availability is less than 0.2 and 0.1 ha respectively, as against 1.3 and 1.9 ha in USA. Deforestation and overgrazing are the two principal factors responsible for soil degradation. The average plantation; deforestation ratio of 76 countries including India is 1:10 and the problem is more acute in Africa. The other factors for land degradation are : increased cropping on marginal lands, shifting cultivation, flooding and poor drainage, nutrient stress and acidity, intensive use of agricultural chemicals, acid precipitation, and surface mining.

Eighty five per cent of total cultivated rainfed area (100 m ha) in India is degraded. Only 6 m ha out of 67 m ha under forests are real forests as assessed by remote sensing techniques. About 94.0 m ha non- forest area in the country has been identified as wasteland, of which 64 per cent is drought prone. According to National Remote Sensing Agency, culturable wastelands occupy 32.8 m ha. These could be brought under tree or grass cover through improved technology. Area under different categories of land having different degrees of constraints to production is set out in Table 1. The atternate land use systems to overcome the production custraints in drylands are discussed in this chapter.

ALTERNATE LAND USE SYSTEMS

The term is applicable to all classes of land aiming at generation of assured income with minimum risk through efficient utilization of available resources. Agroforestry is a part of alternate land use systems. Agroforestry essentially involves a woody perennial, while alternate land use systems sometimes lack a woody component. Agroforestry is a combination of two or more components like tree with grass, but not tree or grass alone. Alternate land use systems can be defined as a perennial systems/practice adopted to replace or modify the traditional land use. Alternate land use systems possess more potentiality and flexibility in land use than the traditional crop production systems. Alternate land use systems aim at a suitable farming system matching the land capability class like tree farming, ley farming, pasture management etc.

In countries like India where land is scarce and labour is in plenty but less productive, coupled with scarce capital and high interest rates, alternate land use systems are an ideal option. There is predominance of marginal and small farmers representing 75 per cent of the total holdings of less than 2.0 ha. Small land holding have some inherent constraints to increased productivity, the principal being lack of resources and impoverished soils. Alternate land use systems are more appropriate in areas where subsistence farming is practiced in fragile ecosystems. These systems help in efficient utilization of resources like land and labour (Fig 1).

LAND USE ACCORDING TO ITS CAPABILITY

There are eight well recognized land capability classes. Classes I, II, III and IV lands

Table 1. Estimates of culturable wastelands in India

Type of Waste Land	State ranking and area (mha) (1)	State ranking and area (mha) (2)	Others	Total area (mha)
Salt-affected land	Gujarat (2.06)	Uttar Pradesh (1.23)	0.61	3.90
Gullied or ravinous land	Madhya Pradesh (1.24)	Uttar Pradesh (0.99)	2.10	4.33
Waterlogged or marshy land	West Bengal (0.23)	Uttar Pradesh (0.22)	0.43	0.88
Undulating upland	Maharashtra (2.32)	Andhra Pradesh (1.67)	6.81	10.80
Jhum or forest blank	N.E.Hilly (1.3)	Madhya Pradesh (0.29)	0.81	2.40
Sandy (Coastal or Desert)	Rajasthan (9.79)	Punjab (0.18)	0.53	10.50
Total	–	–	–	32.81

Figures in parenthesis indicate area in m ha.
Source : NRSA, 1985
Figures in parenthesis indicate area in m ha.

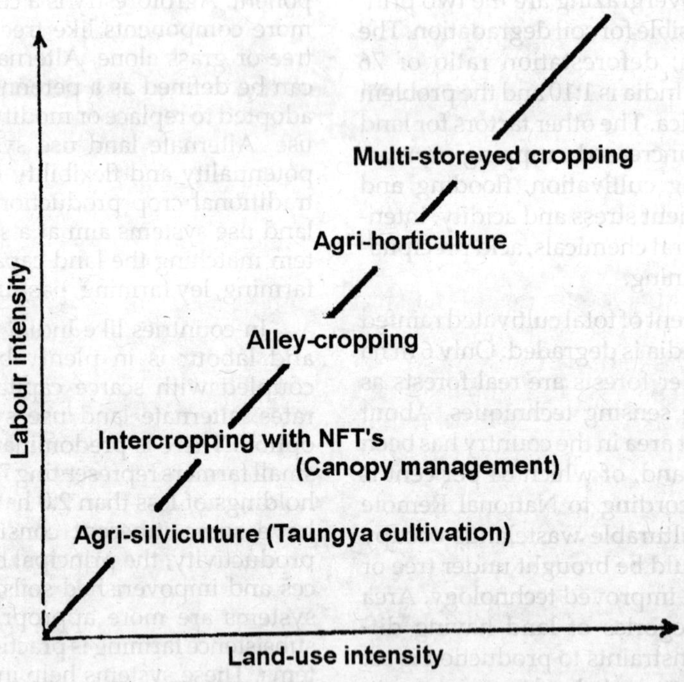

Fig. 1 : Agroforestry systems in relation to labour and land-use intensity

are fit for all agricultural uses including crop production, grassland/pasture development, forestry, etc. as per the need. Class I land is nearly level without any constraint to agricultural production. Class II to IV lands have progressively increasing hazards for cultivation such as more slope, and susceptibility to erosion, less deep soils, less efficient drainage and a slight to severe salt accumulation. Classes V, VI, VII and VIII lands are not suitable for agriculture. Land use according to its capability is set out in Table 2.

Table 2 Suitable farming systems for different classes

Land use capability classes	Sutiable farming system
I	Crop production, intensive cultivation
II & III	Moderately intensive cultivated-alley cropping, agro-horticulture, agri-silviculture
IV	Restricted cultivation-Ley farming
V & VI	Horti/silvi-pastoral system
VII	Low density long-rotation tree plantations
VIII	Recreation, wildlife, watershed protection

FARM PLANNING AND ALLOCATION OF RESOURCES

Food, fodder, fuel and timber are the primary requirements of farmers. The production levels of these products often fluctuate widely from year to year due to variation in rainfall. Improved alternate land use (ALU) systems not only meet the various requirements but also generate continuous and stable income.

The standard dietary requirement suggest that on an average there is need to produce 730, 137 and 182 kg of cereals, pulses and oilseeds, respectively to meet the requirements of a farm family having six members. A family having one milch animal requires approximately about 3.65 tons of grass and 1.83 tons of legume fodder. Larger area is required to be allocated for fodder system as only about 1 ha area can meet the food requirements of a farmer's family having six members. In drought years the scarcity of fodder is general and green fodder in particular is more felt than food grains.

NEED-BASED ALTERNATE LAND USE SYSTEMS

Among the several needs of a farmer, food always remains the first priority item, although fodder requirement is more as compared to food.

Alternate Land use Systems

Food (Arable land) II and III	Fodder (Non-arable land) IV and V	Fuel/Timber/ Fibre (Marginal degraded land VI and VII
–Alley Cropping	–Horti-pastoral	– Tree farming
–Agro-horticulture	–Silvi-pastoral	–Timber cum-fibre (TIMFIB)
–Intercropping with NFT's	–Ley farming	
	–Pasture management	

Some of the need-based alternate land use systems matching the land capability class are discussed below.

ALTERNATE LAND USE SYSTEMS FOR ARABLE LANDS :

Alley Cropping

Crop production in unassured rainfall areas is risky. Low and unstable yields are common and so are the incomes of dryland farmers. For imparting stability and providing sustainability to the farming system, a tree-cum-crop farming system will be the one most appropriate for such situations. One such system, called alley cropping-a version of agroforestry system, could meet the multiple requirements of food, fodder, fuel, fertilizer etc.

Alley cropping, also known as hedgerow intercropping, involves raising of food crops in alleys formed by hedgerows of trees or shrubs. The primary purpose of alley cropping is to maintain or increase crop yields by improvement of soil and micro climate. Hedge rows on contours also serve as vegetative barriers. Alley cropping is a suitable alternative to shifting cultivation.

Alley cropping is an agroforestry system in which food crops are grown in alleys formed by hedgerows of trees or shrubs. During the rainy season, the hedgerows are usually cut to about one metre at planting of crops and kept pruned to reduce competition with crops. Alley cropping has been recommended for the humid tropics primarily as an alternative to shifting cultivation and to improve soil fertility (Kang *et al.* 1985). In the semi-arid regions of India, the main benefit of alley cropping is fodder production during the dry season because mulching in alleys with hedgerow prunings does not consistently increase crop production (Singh *et al.* 1987).

There are several advantages of alley cropping. Some of the more important advantages are : (i) it provides higher total biomass per unit area than arable crops alone, (ii) it utilizes off-season precipitation which otherwise would go waste, (iii) it provides green fodder during the lean period of fodder availability, (iv) it provides additional employment opportunities during the off-season, and (v) when planted along the contours on a slopy land, it provides a barrier to run-off water, holds the silt, and conserves moisture.

Objectives of Alley Cropping

The main objective of alley cropping is to get green and palatable fodder from hedge rows in dry season and produce reasonable quantum of grain and stover in the alleys during the rainy/cropping season. This necessarily calls for cutting back (lopping) of hedge rows during the off-season so as to meet the dry season fodder requirements.

Depending on the requirements, one could have an arboreal alley cropping system where both fodder and poles (as co-product) could be obtained. At every 2 to 3 m in the hedge row a plant could be allowed to grow into a pole and the plants in between poles could be cut back at appropriate height for use as green fodder or mulch.

Three versions of alley cropping system, based on different objectives are :

1. Forage alley cropping,
2. Forage-cum-mulch alley cropping, and
3. Forage-cum-pole alley cropping.

In all the three systems, crops are grown in the alleys and forage is obtained from loppings of hedge rows. Whatever be the system, two components form an essential part of the system. These are : (a) the hedge row, and (b) the crop grown in the alley. So the agronomic management of both becomes extremely important to optimize production.

The hedge row species

Leucaena leucocephala has been the most popular tree species to serve as hedge row in the system. Admittedly, this is an extremely narrow choice. However, quite a few tree species like *Gliricidia, Calliandra, Sesbania* are available and could be tried with advantage. The tree species should meet certain basic criteria. One such criteria should be that it must be amenable to lopping management, besides being multi-purpose (including nitrogen fixing) and fast growing.

The research work on alley cropping has been in progress since 1981 in AICRPDA centres (All India Coordinated Research Project for Dryland Agriculture) and at CRIDA, Hyderabad.

The following four aspects of the hedgerow management were taken up in these studies.

* Choice of crops
* Alley width
* Cutting height
* Frequency/Interval of cutting

The experiences with regard to each on these are subjoined :

Choice of crops : In an experiment on alley cropping carried out at CRIDA, four crops, viz., sorghum, pearl millet, pigeonpea and castor were grown in the alleys of 3.6 and 7.8 m widths. Based on three years experimental results (Table 3), it is seen that yield reduction was higher (61 to 62 %) in pigeonpea and castor than sorghum (48%) and pearl millet (30%). In the year 1986, the grain yield of pearl millet was 14 per cent higher than the crop grown outside the alley.

Alley width : Alley widths varying from 2 m to 7.8 m were tried at different centres. There is ample evidence that alley widths between 3 to 4 m are severely competitive with crops, more so in a drought years. A comparison with pearl millet growtn in six alley widths at three sites has been presented in Table 4. Even at 2 years, the ratio of alley crop vs. sole crop, pearl millet yield

Table 3. Percent reduction in yield of arable crops in alley cropping system

Crop	1984		1985		1986		Percent mean reduction
	Grain yield (kg/ha)	% reduction	Grain yield (kg/ha)	% reduction	Grain yield (kg/ha)	% reduction	
Sorghum	548	49	636	68	909	11	48.0
Pearlmillet	720	53	997	51	2026	+14	30.0
Pigeonpea	227	61	9	65	–	–	61.0
Castor	284	56	149	77	211	47	62.0
Rainfall during crop growth (mm)							
Sorghum & pearl millet	208.6		145.1	–	280.0	–	
Castor & pigeonpea	329.4		286.9	–	300.9	–	–

Note : + 14 indicates increase in pearlmillet yield
Source : Korwar, unpublished data.

Table 4. Effect of alley row spacing on yield of millet

Alley Age = 2 Years

Location	Alley spacing (m)	Seasonal rainfall (mm)	Grain yield		
			Sole (t/ha)	Alley (t/ha)	Ratio (%)
Anantapur	2.0	326	0.24	0.10	40
Kovilpatti	3.0	396	0.73	0.59	81
Hyderabad	3.6	190	1.31	0.89	68
Anantapur	4.0	326	0.24	0.18	75
Kovilpatti	6.0	396	0.73	0.65	89
Hyderabad	7.8	190	1.31	1.22	93

Source : TNAU (1983)
APAU (1986)
Korwar, unpublished data.

decreased systematically with decreasing alley width.

Moisture is the main limiting factor in semi-arid tropics. In sub-humid and humid tropics narrow widths could be alright. However, in dry land situations wider row widths - 10, 15 or even 20 m could be tried with advantage. It would be preferable if the hedge rows are set up on contours.

The yearly comparison of competitive effects between *leucaena* hedges and crops grown in the alley is set out in Table 5. In the first year, pearl

Table 5. Effect of alley age on yield of pearlmillet, Hyderabad

Age (yr)	Alley width (m)	Seasonal Rainfall (mm)	Grain yield		Alley/sole ratio (%)
			Alley Crop (t/ha)	Sole Crop (t/ha)	
1	3.6	549	1.22	1.47	83
	7.8		1.24		84
2	3.6	190	0.89	1.31	68
	7.8		1.22		93
3	3.6	717	1.35	2.60	52
	7.8		1.54		59
4	3.6	209	0.47	1.54	31
	7.8		0.97		63
5	3.6	145	0.53	2.05	26
	7.8		1.36		66

Source : Korwar, unpublished data.

millet grown in 3.6 m wide alleys produced 83 per cent of the grain yield of sole pearl millet, however, by the fifth year this ratio decreased to just 26 per cent, a nearly linear trend. Yields in wider alleys (7.8 m) also decreased from a high of 93 per cent of sole crop in the second year to about 60 to 65 per cent in years 3-5. Three years (1984-1986) results are summarised in Table 6. Irrespective of the crop, the decline in yield was more in case of narrow (3.6 m) row widths than in wider (7.8 m) row widths, the per cent reduction being more in case of castor and pigeon pea.

Cutting height: In the beginning, varying cutting heights 50 cm, 100 cm, and 150 cm were tried. Hedge row heights of 100 and 150 cm are severely competitive with crops, in regard to light and moisture, even through light is not a limiting factor in semi-arid tropics. It is the shading effect of taller hedge row heights at anthesis/grain formation which effects crops in the alleys adversely.

Later, lower-cutting heights - 10, 20 and 30 cm were tried during the crop growth season, so as to eliminate the shading effect. In 1986 a cutting height of 15 cm was tried, which resulted in 14 per cent yield advantage in case of pearl millet. Effect of cutting height on crop and *Leucaena* yield is shown in Table 7.

Table 6. Effect of alley width (m) on grain yield (kg/ha) of arable crops

Alley width	1984		1985		1986		Percent mean reduction
	Grain yield	% reduction	Grain yield	% reduction	Grain yield	% reduction	
Sorghum							
3.6 m	406	62	391	80	839	18	60
7.8 m	690	36	881	55	979	4	37
Pearl millet							
3.6 m	474	69	608	70	1979	+12	43
7.8 m	966	37	1387	32	2072	+17	18
Pigeon pea							
3.6 m	137	76	8	69	–	–	76
7.8 m	317	45	10	62	–	–	46
Castor							
3.6 m	282	56	69	89	171	57	69
7.8 m	387	40	228	64	250	37	49
Rainfall during crop growth (mm)							
sorghum & pearl millet	208.6		145.1		280.0		
Castor & pigeonpea	329.4		286.9		300.9		

Note : + 12 and + 17 indicate per cent increase in pearl millet yield. Pigeonpea crop failed during 1986
Source : Korwar, unpublished data.

Table 7. Effect of cutting height on crop and *Leucaena* yield

System	Grain yield (q/ha)		*Leucaena* yield (DM, q/ha)	
	Cutting height		Cutting height	
	60 cm	15 cm	60 cm	15 cm
Pearl millet				
Sole	18.0	20.1	0	0
7.8 m mulch	11.9 (66)	23.7 (118)	18.4	30.5
7.8 m fodder	11.7 (65)	17.3 (86)	26.9	36.2
Sorghum				
Sole	15.2	18.5	0	0
7.8 m mulch	8.4 (55)	19.1 (103)	16.0	36.6
7.8 m fodder	7.3 (48)	16.7 (90)	23.1	46.5

Figures in parenthesis indicate grain yield per cent of sole crop.
Source : Korwar, unpublished data.

Cutting interval : At Bangalore, harvest intervals of 40 to 50 days and cutting heights of 75 and 150 cm yielded as much as 50 per cent more foliage than a 15 cm cutting height. Longer harvest intervals narrowed the yield difference between 15 and 75 cm cutting heights, but stumps of 150 cm produced more foliage at longer intervals.

Crop management

Crop management in the alleys deserves consideration. All improved agronomic practices that are recommended for sole crops have to be followed in case of alley crops as well. It is advisable that a crop rotation recommended for the area/region is followed so as to obviate the single crop ill-effects.

The results of an experiment on alley cropping of subabool *(Leucaena)* with groundnut and green gram carried out at Rajkot are given in Table 8. In Table 9 are set out the yield of four crops in mulched and fodder (unmulched) alleys of 3.6 and 7.8 m (Korwar - unpublished data). *Leucaena* prunings in mulched alleys were applied to the crop during the cropping season and removed for fodder during the off- season. The crops started responding to mulch application in the third year when pearl millet and sorghum gave an yield advantage of 31 and 4 per cent, respectively, over the crops grown outside the alleys without mulch.

Table 8. Alley cropping of subabool with groundnut and greengram at Rajkot

Treatment	Grain (q/ha)				Green fodder (q/ha)			
	1982-83	1983-84	1984-85	1986-87	1982-83	1983-84	1984-1985	1986-87
Subabool sole (paired row)	–	–	–	–	107.0	147.7	62.0	27.3
Ground nut sole	7.89	10.79	–	–	–	–	–	–
Greengram sole	4.52	6.81	–	0.58	–	–	–	–
Subabool + groundnut	1.25	1.42	–	–	66.4	109.5	52,4	36.0
Subabool + greengram	1.22	1.57	–	0.69	68.8	133.6	55.6	38.6

Note : Cutting height earlier - 90 cm and reduced to 25 cm
Source : AICRPDA, 1987

Table 9. Grain yield (kg/ha) as influenced by mulch and fodder alley cropping systems.

Crop	1984		1985		1986		Percent mean reduction
	Grain yield	% reduction	Grain yield	% reduction	Grain yield	% reduction	
Sorghum							
Mulch	587	45	736	63	1066	+4	41
Fodder	509	53	535	73	753	27	56
Pearl millet							
Mulch	721	53	1054	49	2320	+31	24
Fodder	719	53	941	54	1732	2	37
Pigeon pea							
Mulch	233	60	13	50	–	–	60
Fodder	220	62	5	81	–	–	63
Castor							
Mulch	295	54	180	72	232	42	58
Fodder	273	58	118	82	189	53	66
Rainfall during crop growth (mm)							
Sorghum & pearl millet	208.6		145.1		280.0		
Castor & pigeonpea	329.4		286.9		300.9		

Note : + 4 and + 31 indicate per cent increase in yield. Pigeonpea crop failed during 1986
Source : Korwar, unpublished data.

The effect of the system on crop growth and yield as here given in Table 10. As been given in Table 10 Reduction in alley crop yields when compared with sole crops is more severe in 1985 than in 1984, due to lower and more poorly distributed rainfall. The hedgerows, on the other hand, gave more green fodder in the poor rainfall year; 4.01 t/ha of green fodder in 1985 versus only 2.74 t/ha in 1984.

Table 10 : Grain yields of crops grown in 5 m alleys with arboreal Leucaena every 2m

Crop	Year	Grain yield Sole Crop (t/ha)	Alley Crop (t/ha)	Yield reduction (%)
Millet	1984	1.92	1.35	30
	1985	3.68	0.77	79
Sorghum	1984	1.98	1.34	32
	1985	3.19	0.84	74
Mungbean	1984	0.56	0.43	24
	1985	0.92	0.27	71
Cowpea	1984	0.84	0.45	46
	1985	1.46	0.30	80
Castor	1984	0.83	0.47	44
	1985	0.72	0.06	92
Groundnut	1984	1.80	0.88	51
	1985	0	0	–

Cropping season rainfall = 1984 – 431 mm and 1985 – 375 mm.
Source : Singh, R.P., unpublished data.

Yield reduction of grain crops in 1984 was about normal as compared with other alley crop experiments in CRIDA's nearby Hayatnagar Research Farm. Yield reduction during the bad year of 1985 was worst of all the sites. It is clear that allowing *Leucaena* trees to grow unmanaged in cropped fields is not advisable, particularly in areas with high probability of monsoon failure and soils with low water holding capacity.

Leucaena poles could be attractive source of income to a small farmer and the interactive effects of crop competition can be overcome if the tree canopies are lopped or better still, pollarded. In a related experiment at CRIDA, *Leucaena* trees were established at 7.5 m along the contours with 2 m spacing within rows. When the trees were about 4 years old and about 8 m tall, four treatments viz., unlopped, lightly lopped (33% canopy removal), heavily lopped (66% canopy removal) and pollarded at 2 m were established and sorghum grown underneath the trees in 1985. Pollarded plots faired much better in terms of sorghum grain and stover production, yielding 54 per cent and 68 per cent of control, respectively. Pollarded plots also produced over twice the amount of fodder than lopped treatments. In terms of total phytomass sorghum under pollarded *leucaena* was twice as productive as sorghum alone and 35 per cent more productive than *leucaena* alone.

The results of several studies revealed that short duration cereals are compatible with frequently pruned hedge rows. In general, wider alley width and paired rows of *Leucaena* were found suitable. *Leucaena* paired hedge rows yielded 2 tons of fodder and 1 ton of fuel/ha/yr when placed 7.8 m apart. Mulching of *Leucaena* leaves increased the sorghum and pearl millet yield by 9 and 16 per cent, respectively over control. This yield advantage was mainly due to mulching and monthly cutting of hedges close to the ground at 15 cm height.

It is noted that options are now available to meet different objectives, the choice depending on the situation and the needs. One very strong argument in favour of alley cropping is its ability to produce usable material even in years of severe drought. At Rajkot in 1985, rainfall received during the season was only 30 per cent of normal precipitation. There was total failure of grain production of three legume crops tried in the system. In sole crop plots, production was limited to 0.5 to 1.7 t/ha of green fodder. However, in alley cropped plots, *Leucaena* hedgerows produced over 5 t/ha of green fodder (Table 11). Similar was the experience at the Anantpur Centre in 1984. The cropping season rainfall was only 144 mm as against normal of 495 mm. All crops (groundnut, pigeonpea, sorghum) failed, and even stover production was severely affected. However, the *Leucaena* hedgerows produced 2 t/ha of dry leaf material. Thus, alley cropped systems when properly planned (*Leucaena* hedgerows) can remove a part of the risk facing the small farmer in India.

Table 11. Risk reduction in poor years through alley cropping at Rajkot

Teatment	Leucaena Fodder (wet) (t/ha)	Crop yield	
		Grain (t/ha)	Stover (t/ha)
Sole crops			
Leucaena *	5.75	–	
Groundnut	–	0	1.67
Mungbean	–	0	0.68
Urdbean	–	0	0.47
Alley crops			
Groundnut	5.24	0	1.26
Mungbean	5.56	0	0.25
Urdbean	6.03	0	0.22

* 1.5 m rows
Note : Rainfall 177 mm (120 mm in cropping season), vs. 625 mm average.
Source : Annual report, AICRPDA – Rajkot, 1985

An economic evaluation of the alley cropped systems has revealed that gross cash returns are higher in all cases than the sole crop systems. Alley-cropped sorghum yield nearly twice the income as sole sorghum, whereas the alley-cropped pigeonpea system yielded almost 7 times the income over sole pigeonpea, due to the value of the *Leucaena* co- products. Most of the value of the *Leucaena* is from its sale as a fodder which appears to be of greater economic benefit than its application to the crops as a green mulch.

Much of the work on alley cropping in India is confined to the investigation of agronomic treatments like cutting height and spacing on the yield of crops and trees and very little is known of the microclimate modification within the alleys. In contrast, microclimate studies in intercropping in the semi-arid tropics have demonstrated that shading by a tall crop eg. sorghum does indeed improve the efficiency of utilization for dry matter production e.g. groundnut (Marshall and Willey, 1983) and increas the yield of crops (Harris and Natarajan, 1987). In an study carried on microclimate and growth of sorghum and cowpea in alley cropping carried out at Gunegal Research Farm of CRIDA, Hyderabad, (Singh, Ong and Saharan, 1989) it was revealed that the severe reduction in crop yield observed in alley cropping in the semi-arid tropics is due to moisture competition. Although the crops adjacent to the *Leucaena* hedgerow experienced intense shading (30 to 85%) for much of the growing period the presence of shallow root barrier was sufficient to eliminate the reduction in cowpea and sorghum yield. For long duration crops like castor, the presence of a root barrier was not sufficient to fully compensate for the yield reduction because a large part of the reproductive period occurred in the dry season when water was drawn from depth below 0.5 m.

The *Leucaena* hedgerow was a relatively 'permeable' barrier to environmental factors and there was only a minor microclimate modification in the alleys. Such modifications are unlikely to influence the growth of crops as the windspeed, temperature, radiation and humidity are favourable during the cropping season. For example, studies in controlled environment have shown that the growth rate and leaf expansion of groundnut are relatively unaffected when saturation deficit varied from 1 to 2 kPa (Ong 1984, Ong *et al.* 1985). Similarly mean temperature in the range from 22 to 25°C has been found to be optimal for the canopy photosynthesis and grain yield of tropical crops e.g. millet (Ong and Monteith, 1985). However, if the temperature exceeds 36°C and the saturation deficit exceeds 2.5 kPa during the cropping season then modification of the microclimate by the trees would have a major influence on the crop growth and yield. Such conditions are rarely encountered during the cropping season in Peninsula India but in North-Western India and regions of the Sahel, conditions are more extreme for crop growth and microclimatic modification would be beneficial.

A similar conclusion was reached by Corlett *et al* (1987) when they examined the microclimate modification in 3.8 m wide alley cropping involving *Leucaena* and pearl millet near Hyderabad. In their system, the hedgerows were pruned to 1 m throughout the cropping season and analysis of both temperature and saturation deficit in the sole millet and alleys confirmed that the modifications have only a negligible effect on major crop processes. They also included a root barrier treatment which showed that alley millet produced the same biomass and grain yield as sole millet. However, the yield of alley millet without the root barrier was only 40 per cent of the sole millet. In the present study, the increased crop yield in the middle of the alleys was due to the residual effect of the *Leucaena* hedgerow which was removed in the previous year. The soil moisture record showed that infiltration rate had been increased but the main effect was probably due to the well documented residual effect of legumes in enriching soil nutrient status (Dart *et al*). Without this residual effect, the maximum yield of the crop in the middle of the alley would not be greater than in the sole crop since microclimate modification has a minor influence on crop yield. Therefore, the reduction in crop yield in the 10 m alley in the absence of a residual hedgerow would be even greater.

The importance of below ground interaction in alley cropping is in sharp contrast to the study on intercropping eg. millet and groundnut (Willey and Reddy 1981). They showed that the presence of a root barrier had no apparent influence on the yield and yield components of the intercrop and concluded that improvement in the utilization of solar radiation into dry matter was the main cause of increased productivity. A major difference between the intercropping and alley cropping system is that in intercropping the establishment of the root systems commences at the same time (Gregory and Reddy 1982) but in agroforestry the root system of the perennial species are well-established when the crop is sown. Consequently in a dry environment, the crops are less competitive than the trees.

Thus, the success of an agroforestry system in the semi-arid tropics mainly depends on combining a cropping system with a perennial species which avoids the competition for moisture. A classic example is the case of the *Acacia albida* or *Prosopis cineraria* in the arid region where the phenology of the trees allows the crop to complete a significant proportion of its life cycle before the canopy of the tree is fully developed (Felker 1978). An interesting analogue is the combination of a slow growing pigeonpea (*Cajanus cajan*) with sorghum which allows a full yield of sorghum to be harvested and a 70 per cent yield of pigeonpea (Natarajan and Willey 1980). A further possibility is to have very low population of trees amongst crops or have trees planted around the border of fields which is commonly observed in farmer's field.

Agri-silvicuture System (Intercropping with NFT'S)

Raising of nitrogen fixing trees (NFT) in arable lands not only provides rich organic matter and atmospheric nitrogen, but also improves soil structure, and prevents land degradation. Nitrogen fixing tree species offer immense possibility of supplementing the nitrogen requirement of crops through natural leaf fall, loppings, or green-leaf manuring. In semi-arid tropics, it is not possible to have a residual build-up of organic matter and, therefore, it is important that organic matter supplies are to be replenished every year through leaves, litter, crop residues, etc.

Suitable N-fixing genera

Intercropping of trees helps in incorporating litter and lopped leaves every year. This acts as green-leaf manuring *in situ* and woody sticks can be used as fuel. Inclusion of top-feed N-fixing genera helps in mitigating the fodder scarcity and life saving of livestock in years of drought. There are 640 species of nitrogen fixing trees known so far but their number is estimated to be around 4000. Not all genera belong to family leguminaceae, eg. *Casurina* and *Alnus.* Most of the common N-fixing genera are *Acacia, Albizia, Calliandra, Hardwickia, Dalbergia, Erythirina, Glyricidia, Inga, Leucaena, Parkinsonia, Pongamia, Prosopis, Sesbania,* etc.

Nutrient contributions by NFT's

NFTs contribute to the soil fertility enrichment by the *in situ* fixation of N in soil and the return of nutrients to the soil through leaf fall. Preliminary studies conducted on one year old unmanaged *Leucaena leucocephala* showed that they contribute 7.20 q total dry matter and 19 kg N/ha in one year through fall of different plant parts. Break-up of fallen parts into petiole, leaflets, podwalls and seeds showed that seeds alone contribute to about 70 per cent of the N. Some of these seeds decompose while majority germinate following monsoon and get incorporated into soil as seedlings. Similarly, the contribution from 8 years old trees was found to be 21.5 q of dry matter and 42 kg N/ha, respectively. (Table 12).

Tree-crop interaction

Studies on under cropping in 4 years old *Leucaena* trees planted at 7.5 m x 2 m with different intensity of loppings, showed that grain yield of sorghum in *kharif* 1985 and fodder yield of pear millet in year 1986 was high in pollarded treatment (53 per cent of sole) and was negligible in unlopped treatment (Table 13). These studies have showed that competition does occur when the trees and crops are raised together. The total biomass production was higher in tree + crop system when compared to sole tree or sole crops systems, thereby imparting stability to the production and risk reduction in poor rainfall years like 1986.

In another experiment of intercropping of arable crops with N-fixing trees, a marginal increase in yield of sorghum and castor was recorded with *Acacia albida, Acacia ferruginea,* and *Prosopis cineraria,* where as both the crops failed when inter-cropped with *Leucaena leucocephala* (Table 14). Crop failure in *Leucaena leucocephala*

Table 12. Cumulative annual dry matter and nitrogen addition as leaf fall by *Leucaena leucocephala*

Component	Dry matter (kg/ha)		Total nitrogen (kg/ha)	
	One year old trees	Eight year old trees	One year old trees	Eight year old trees
Pod wall	82	200	1.05	3.17
Petiole	241	533	2.46	6.81
Leaflet	84	1115	1.37	18.80
Seeds	313	297	14.45	13.70
Total	720	2145	19.33	42.48

Source : Venkateswarlu - unpublished.

Table 13. Yield and biomass (q/ha) of sorghum and pearl millet undercropped in *Leucaena leucocephala*

Treatments	1985			1986		
	Sorghum			Pearl millet		
	Grain	Stover	Total biomass	Grain	Stover	Total biomass
Leucaena + crop						
Unlopped	0	7.0	7.0	0	1.5	1.5
Lightly lopped	1.0	18.0	19.0	0	2.9	2.9
Heavily lopped	2.0	17.0	20.0	0	3.9	3.9
Pollarded	15.0	39.0	54.0	0	21.5	21.5
Sole crop	28.0	57.0	85.0	0	40.4	40.4

Source : Hocking and Rao (1985) and Osman, unpublished data.

shows the importance of canopy management for successful under cropping.

An investigation conducted by Peter Poschen (1986), in Eastern Ethiopia in farmers' field indicated a statistically significant increase in crop's yield by 56 per cent on an average for the crops under the tree canopies compared to those away from the trees. Maize yield under the trees was 3.4 t/ha as against 1.9 t/ha in the 'open', the figures for sorghum being 2.13 and 1.57 t/ha, respectively representing a 76 per cent increase for maize and 36 per cent increase for sorghum. This increase was caused by the improvement in 1000 grain weight and number of grains in plants 'under the tree', indicating that the trees enhanced the fertility status of the soil and improved its physical conditions in terms of crop growth (Table 15).

fruit plants develop full canopy after several years and some of them require regular pruning, thus permitting an intercrop. Intercropping provides weed-free area and yield obtained will be a bonus to the farmer. This systems mainly focusses on higher returns per unit area. The economic aspect of fruit culture is also no less important. Well maintained and established orchards bring better returns than field crops from the same piece of land. Fruit growing is a long-term venture and requires a high initial investment. The concept is to encourage small farmers to take up tree planting and ensure good returns through intercropping. This system also helps to generate more employment for the farmer especially during the off- season when the crops are not cultivated. It is very well recognised now that fruits should not be considered a luxury, since

Table 14. Growth parameters of 2 year old trees, yield of arable crops intercropped with NFT'S.

Tree species	Grain q/ha		Tree growth parameters	
	Sorghum	Castor	Height (m)	dbh (m)
Acacia albida	5.57	6.35	2.87	2.64
A. ferruginea *	5.30	6.06	1.40	1.65
Albizia lebbeck	4.67	5.07	3.08	3.71
Prosopis cineraria	5.20	5.34	0.60	1.29
Leucaena leucocephala	0.73	–	4.79	5.71
Control (no tree)	5.17	5.16	–	–
C.D. (0.05)	0.99	0.62	–	–

* One year late planted
Source : Osman, unpublished data.

Table 15. Estimates of annual gross returns * per hectare of sorghum and maize crops with and without *A. albida* cover

Output	Without *A. albida*	With 6 trees/ha (9 % cover)	With 20 trees/ha (31 % cover)	With 65 trees/ha (100% cover)
Crop	1685	1778	1992	2485
Wood	–	19	65	259
Fodder	–	30	100	250
Total	1685	1827	2157	2994
Percent increase over control (without trees)	–	8	28	78

Source : Peter Poshen, 1986.
* (1 Eth $ = 0.48 US $, 1983)

Agro-horticulture System

This system is also termed as food-cum-fruit system in which short duration arable crops are raised in the interspaces of fruit trees. Most of the

they are protective foods necessary for the maintenance of human health. Many fruits are also known to possess specific medicinal values. This is more important for a country like India where the majority of the population is vegetarian.

The system works best in medium to deep soils with good water holding capacity. Individual farm ponds and supplemental watering in the off- season will certainly improve the scope of fruit farming in drylands, if suitable species/varieties are chosen. *Jamum* (Syzigium cuminii), custard apple, guava, mango, *Sapota* are the suitable fruit species. The trees can be established in the contour furrows/saucer shaped pits. The rain water in the catchment could thus be efficiently used by concentrating it in the pits. Depending on the intercrop, fertilizer application (40-30-0 for cereals and 10-30-0 for pulse crops) has to be given. Generally, it is not desirable to interplant cereals and other exhaustive crops so as to avoid competition with fruit plants.

Studies involving *Ber* (*Ziziphus mauritiana*) and short duration legumes revealed higher returns than their respective sole crops. Horsegram and clusterbean gave and additional gross return of Rs. 3,000/- and Rs. 1,900/- per ha, respectively. The gross return from *Ber* was Rs. 7,320/- per ha, besides 1.2 tons of fuel wood obtained from pruning. (Table 16).

Table 16. Agro-horticulture system

System	Yield (kg/ha)		Gross returns (Rs/ha)		
	Ber *	Intercrop	*Ber*	Intercrop	Total
		1986-87			
Ber with cowpea	919	250	2757	750	3507
Ber with greengram	919	245	2757	1225	3982
Ber alone	920	–	2760	–	2760
		1987-88			
Ber with clusterbean (Vegetable)	1828	1918	7312	1918	9230
Ber with horsegram	1812	1025	7228	3075	10303
Ber alone	1830		7320	–	7320

* The marketable Ber yield account for only 1/3rd of total, since the rest was infested with fruitfly.

termed as non-arable lands. These are unfit for crop production. The topography of these lands is highly undulating. They lack vegetative cover. Run-off and soil loss is inevitable. The carrying capacity of these lands is less than 2 sheep/ha. These lands have to be brought under close canopy vegetative cover to prevent these lands from further degradation, improve the carrying capacity of native grass lands, and for stabilising the economy of farming community.

Grasses and legumes associated with hardy fruit trees/ top-feed trees is the panacea for non-arable lands. Horti-pastoral system is defined as, "One of the agroforestry systems which involves integration of fruit trees with pasture. When a fruit tree is replaced with a top-feed tree, it is called as silvi-pastoral system".

Selection of fruit trees

The tree species should be so chosen that they are complementary or less competitive for understorey plants. Slow growing and late bearing fruit tree species like tamarind, wood apple, *Jamun, aonla, bael,* cashew are highly suitable for

II. ALTERNATE LAND USE SYSTEMS FOR NON-ARABLE LANDS

Horti/silvi-pastoral System for Non-arable Land

There are four hundred million bovines in India, scarcity of green fodder is more felt than dry fodder. Livestock rearing is the mainstay of farmers in India. Class IV and above soils are

drylands. In a continuous grazing system, custard apple is an ideal option as it is not browsed by livestock and it is hardy too. Fruit trees like *ber* will go very well in combination with pasture, as the foliage is very nutritious and palatable. *Ber* yields good quantity of foliage during the lean period of April-May. In waterlogged areas and in alkali soils *aonla* is best suited.

Selection of top-feed tree species

The tree species should have the following features :

(i) fast growing, having multiple uses;

(ii) high palatability and digestability of foliage;

(iii) good coppicing ability;

(iv) ability to withstand browsing, trampling and intensive lopping; and

(v) resistant / tolerant to drought and extremes of climatic conditions.

A list of suitable top-feed tree species is given in Table 17

Table 17. Suitable top feed tree species

Common name	Botanical name
Babool	*Acacia lencophloea*
Babool	*Acacia nilotica*
Gum Arabic	*Acacia senegal*
Khairwal	*Bauhinea purpurea*
Mopane	*Colophospermum mopane*
Shisham	*Dalbergia sissoo*
Indian rose-wood	*Dalbergia latifolia*
Gumhar	*Gmelina arborea*
Dhaman	*Grewia optiva*
Anjan	*Hardwickia binata*
Subabool	*Leucaena leucocephala*
Khejri	*Prosopis cineraria*
Safeda, Fras	*Populus* sp.
Agati	*Sesbania grandiflora*
Farash	*Tamarix aphylla*

SELECTION OF SUITABLE GRASS AND LEGUME SPECIES

Studies carried out under the pasture scheme at All India Co-ordinated Research Project for Dryland Agriculture (AICRPDA) during 1975-83, revealed only a few species of grasses and legumes are potentially useful for dryland areas. Different grasses from genera *Dicanthium, Cenchrus, Lasiurus, Chloris, Urochola, Panicum, Pennisetum, Sehima, Chrysopogon, Paspalum, Digitaria, Heteropogon* have been evaluated. Of these, *Dicanthium, Urochloa, Lasiurus, Sehima* were found to be ideal for rainfed areas. *Cenchrus ciliaris* was found to be good in all areas.

In rainfed areas, different annual and perennial legumes from genera *Dolichos, Macroptilium, Clitoria, Centrosema, Glycine, Vigna, Alysicarpus* and *Stylosanthes* have been evaluated. The ideal legume found to be excellent with regard to persistence, nutritive value and palatability was *Stylosanthes* species.

In a grass-legume mixture five species of *Stylosanthes* were tested and *S. scabra* (40205) yielded the highest drymatter yield of 53.4 q/ha. The contribution of stylo to total yield was the highest with *S. hamata* (35.5 q/ha) and was closely followed by *S. scabra* (33.8 q/ha). The lowest yield was obtained with *S. guyanensis*. It indicates that *S. hamata* grows and spreads rapidly and does not allow grass to grow. More grass yield was obtained in *S. guyanensis* whose vegetative growth was poor (Table 18).

A list of suitable grasses and legumes for drought prone areas with different rainfall and soil types is set out in Table 19.

Table 18. Yield potentials of *Stylosanthes* species and grass

Species	Dry herbage yield (q/ha)		
	Stylo	Grass	Total
S. hamata	35.5	7.8	43.3
S. scabra (40205)	33.8	19.6	53.4
S. viscosa (34904)	18.1	30.2	48.3
S. humilis	15.8	19.8	35.6
S. guyanensis	6.2	27.3	33.5

Source : Reddy and Hampaiah, 1982

Table 19 : Suitable pasture species for drought prone areas

Pasture species		Minimum rainfall (mm)	Soil type	Seed rate (kg/ha)	Dry forage yield (t/ha)
Grass					
Sewan	*Lasiurus sindicus*	250	Light	3-5	3.5
Buffel (Dhaman or Anjan)	*Cenchrus ciliaris*	300	Versatile	5-6	4.0
Bird wood (Black dhaman)	*Cenchrus setigerus*	250	Versatile	8-10	3.0
Blue panic	*Panicum antidotale*	500	Versatile	2-4	3.0
Marvel	*Dicanthium annulatum*	375	Medium to heavy	4-5	2.5
Sain	*Sehima nervosum*	300	Mixed red and black	7-8	2.5
Dinanath	*Pennisetum pedicellatum*	400	Versatile	8-9	3.0
Bahia	*Paspalum notatum*	450	Light	11-17	3.5
Sabi	*Urochloa mosmabicensis*	450	Versatile	2-4	3.0
Napier (Elephant	*Pennisetum purpureum*	750	Medium to heavy	12,500 cuttings	4.0
Rhodes	*Chloris gayana*	600	Sodic	8-10	2.5
Legume					
Perennials					
Stylo	*Stylosanthes hamata*	450	Light to medium	5-7	3.5
Stylo	*Stylosanthes scabra*	325	Light to medium	4-6	2.5
Siratro	*Macroptilium atropurpureum*	500	Light to medium	7-8	2.8
Butterfly pea (Aparjit)	*Clitoria ternatea*	300	Light to medium	15-20	3.0
Bankulthi	*Atylosia scarabaeoides*	300	Light to medium	10-12	2.0
Annuals					
Cowpea	*Vigna unguiculata*	300	Light to medium	30-40	3.0
Guar	*Cyamopsis tetragonoloba*	300	Light to medium	25-30	3.0
Horsegram	*Dolichos biflorus*	300	Light to medium	40	2.0

Establishment techniques

A moderately good tilth seed bed is required for better establishment of pastures and must be sown at the beginning of the rainy season on a drizzling day. Seeds could be broadcasted. Since the seeds are small in size, seeding at shallow depth is important.

A seed rate 4 to 6 kg/ha mixed with wet soil or sand 10 times of its volume would ensure good establishment of *Stylosanthes*. Though it establishes and grows well in poor marginal soils but responds well to the application of single super phosphate (30 kg P_2O_5/ha should be broadcasted before sowing stylo). Weeding once in the first year improves the establishment. *S. scabra* is a shy germinator and scarification of seeds is a must, and it could be done by rubbing the seeds with coarse river sand.

For seeding *Cenchrus ciliaris*, a seed rate of 5 to 6 kg/ha is required. Broadcasting of seeds or planting of root slips at 50 cm x 50 cm on a rainy day would ensure good establishment. It responds well to fertilization (40-20-0 NPK/ha).

Horti-pastoral System

In a horti-pastoral system, fruit trees Guava (*Psidium guajava*) and custard apple (*Annona squamosa*) were planted and seeds of *Stylosanthes hamata* and *Cenchrus ciliaris* were broadcasted one year later to ensure good establishment of fruit trees. After 18 months of pasture establishment survival and growth of fruit was found to be poor with grass than legume association. In terms of herbage yield, *Cenchrus* grass outyielded the stylo legume but fruit plants suffered severely due to moisture stress (Table 20).

Table 20 : Growth parameters of fruit trees and forage yield in horti-pastoral system

System	Growth parameters		Dry forage yield (t/ha)	Survival (%)
	Heght (m)	Collar diameter (cm)		
System-I				
Guava + *S. hamata*	1.63	2.75	5.98	93
Guava + *C. ciliaris*	1.77	2.67	6.77	69
Guava alone	1.78	3.33	–	95
System-II				
C. apple + *S. hamata*	1.07	2.79	5.16	83
C. apple + *C. ciliaris*	1.50	2.66	6.52	23
Custard apple alone	1.22	3.65	–	80

Source : Osman, unpublished data.

Silvi-pastoral System

Three-month old seedling of *Leucaena leucocephala* were planted in contour furrows in 1981 at 7.5 m apart with plant to plant spacing of 2.0 m on a plot having about 2 per cent slope. As an undergrowth, improved pasture species *Stylosanthes hamata* in lower four strips and *Cenchrus ciliaris* in upper four strips were sown.

The performance of *Leucaena* + *Stylosanthes hamata* was found better (Stylo 4.2 t/ha/year DM) than *Leucaena* + *Cenchrus ciliaris* (Cenchrus 2.5 t/ha/year DM). The traditional sorghum crop in a similar soil (0-20 cm soil depth) seldom yielded a biomass of 2 t/ha/year. *Leucaena* trees were felled after eight years (700 trees/ha) and yielded 35 t/ha of biomass. The yield per tree, on an average, was 51.1 kg, of which stem, branch and twigs contributed 60, 24 and 16 per cent, respectively (Table 21).

Ley farming

A rotation in which two or more crops are grown in a fixed sequence. If the rotation included a period of pasture (a ley), which is used for grazing and conservation, the system is called "Ley farming".

Studies on ley farming were carried out with *Stylosanthes hamata* as the legume pasture ley and was rotated with the two main arable crops of the region, viz., sorghum and castor.

During year 1, year 2 and year 3 the arable crops received their recommended dose of fertilisers during each year and stylo received 10 kg N and 30 kg P_2O_5/ha during its first year only. During year 4, sorghum received only 30 kg P_2O_5/ha and no fertiliser nitrogen was applied.

The three years ley with stylo resulted in a nitrogen build up of 35 kg/ha over normal rotation. The two years ley resulted in a nitrogen build-up of 20 to 30 kg/ha. One year ley did not cause any appreciable soil nitrogen build-up. The sorghum grain yield also followed the same trend, i.e. maximum grain yield was obtained with 3 year stylo ley (Table 22).

Table 21 : Biomass from 8-year old *L. leucocephala* trees

dbh group (cm)	No. of plants	Mean dry wight (kg/tree)			
		Stem	Branch	Twigs	Total
< 8	8	12.16	3.90	3.61	19.67
8-10	15	21.76	8.38	5.58	35.72
10-12	31	28.63	11.36	7.13	47.12
> 12	35	39.55	18.90	9.86	68.31

Table 22 : Soil fertility and grain yield as influenced by ley farming

Treatment No.	Treatment particulars				Available nitrogen (kg/ha)		Grain yield of year 4 sorghum crop (kg/ha)
					Before the year 4 sorghum crop	After the year 4 sorghum crop	
T_1	Castor	Sorghum	Castor	Sorghum	109.76	81.54	1802
T_2	Castor	Sorghum	Stylo	Sorghum	112.89	97.21	2549
T_3	Castor	Stylo	Stylo	Sorghum	125.40	98.28	2899
T_4	Stylo	Stylo	Stylo	Sorghum	144.75	106.13	3912

Source : Korwar, 1984.

Pasture Management

Native pastures are highly degraded ones. Their improvement and management involves a set of technical and social interventions. The important technical interventions are : identification and introduction of suitable grass and legume species, using suitable establishment techniques, fertilisation of the pasture lands, regulating the grazing pressure and using an optimum stocking rate, use of rotational grazing system if feasible, and increasing the grazing period through introduction of top-feed tree species.

Grazing pressure and stocking rate

The access of livestock to pastures should be controlled so that grazing pressure could be managed. It is no exaggeration to say that production from properly managed pastures is 5 to 10 times than the land which is constantly under excessive random grazing pressure.

Studies on stocking rate carried out at CRIDA, Hyderabad with *Decanni* sheep have shown that the improved pastures can carry upto 6 sheep/ha in a continuous grazing system, without any supplemental feed. The native pastures did not carry even 2 sheep/ha. Grazing had to be withdrawn from these pastures because of lack of forage on them. The effect of the quality of pasture was evident in terms of number of lambings and the lamb weight at birth (Table 23). The improved pastures produced more number of lambs than the natural pastures which was because of the better quantity and quality of forage available in the former case.

Rotational grazing system is one in which the pasture is divided into a number of compartments, but it requires additional investments. This system has several advantages. There is no wastage and the pastures get enough of regrowth period.

Increasing grazing period

A major shortcoming of most of the common pastures is lack of production during hot summer lean period. A traditional way to over come this is to use tree leaf fodder during this period. In the pasture lands lot of trees are seen growing. If these are replaced with top-feed tree species, that would prolong the grazing period and improve the carrying capacity of the pasture.

Top-feed tree species were introduced in improved pastures maintained at CRIDA farm. Initial observations on survival and height indicate that some species hold good promise. Survival was high in case of *Acacia senegal, Acacia aneura, Leucaena leucocephala* and *Acacia nilotica*. The survival was poor in case of *Sesbania grandiflora, Prosopis cineraria, Acacia albida* and *Acacia bavinosa*.

Various options are now available for adoption, according to production objectives and agro-climatic conditions. Some of these options are listed in Table 24 and depicted in Figs. 2(a), 2(b), and 3.

Tree Farming

It is said that agriculture is for the plains having assured rainfall, while trees are for hills, degraded soils and drylands, where greater risk of crop failure exists. Trees can flourish and yield abundantly in many places where arable crops would fail to grow. Another comparative ad-

Table 23 : Number of lambings (in 20 month period) and lamb weight (kg) at birth

Treatment	No. of lambings	Lamb weight (kg) at birth
Natural pasture		
2 sheep/ha	1.89	2.25
4 sheep/ha	1.56	2.23
6 sheep/ha	2.00	2.24
Improved pasture		
2 sheep/ha	2.78	2.66
4 sheep/ha	2.78	2.68
6 sheep/ha	2.44	2.39

Source : Korwar, 1984.

Table 24 : Farming systems options for different agro-climatic conditions

Annual rainfall (mm)	Soil type	Farming systems	Suitable tree / grass / legume species
A) Less than 500	Shallow (0-30 cm)	Tree farming	*Prosopis cineraria, P. juliflora, Acacia aneura, A. nilotica, A. tortilis, Pithecellobium dulce*
	Medium (0-45 cm)	Pasture management	*Lasiurus sindicus* (light textured soils). *Cenchrus setigerus, Sehima nervosum, Stylosanthes scabra, Clitoria ternatea*
B) 500-750	Shallow (0-30 cm)	Silvi-pastoral system	*Acacia nilotica, Colophospermum mopane, Dalbergia sissoo, Hardwickia binata, Cassia sturti, Albizia amara, Leucaena*, sp. *Cenchrus ciliaris, C. setigerus, Dicanthium annulatum, Panicum antidotale, Stylosanthes hamata, Macroptilium atropurpureum*
	Medium (0-45 cm)	Horti-pastoral system	Custard apple, *Ber, Jamun Aonla*, Tamarind, Woodapple, *Bael, Cenchrus ciliaris, Panicum antidotale, Urochloa mosambicensis, Stylosanthes hamata, Macroptilium atropurpureum, Clitoria ternatea*
C) More than 750	Shallow (0-30 cm)	Ley farming or silvi-pastoral system	3 years *Stylosanthes hamata* and 4th year arable crop (sorghum on heavier soils; pearl millet on lighter soils) Silvi-pastoral system as above
	Medium (0-45 cm)	Ley farming or Horti-pastoral system	Ley farming as in B above. Mango, *Sapota*, Guava, *Aonla, Stylosanthes hamata/ Macroptilium atropurpureum*

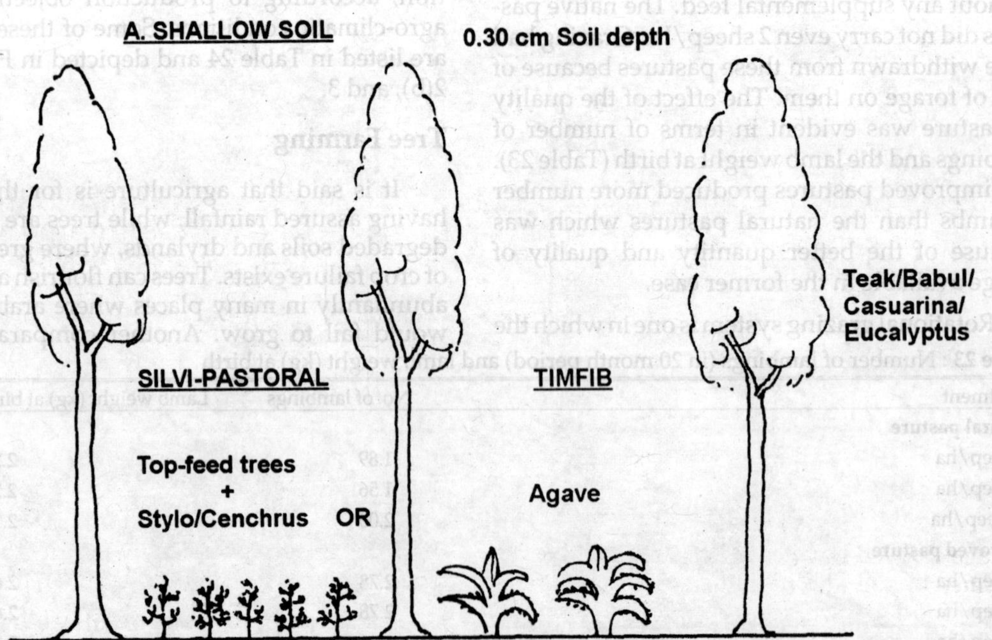

Fig. 2(a). Two Tier cropping models for SAT (A. Shallow soil)

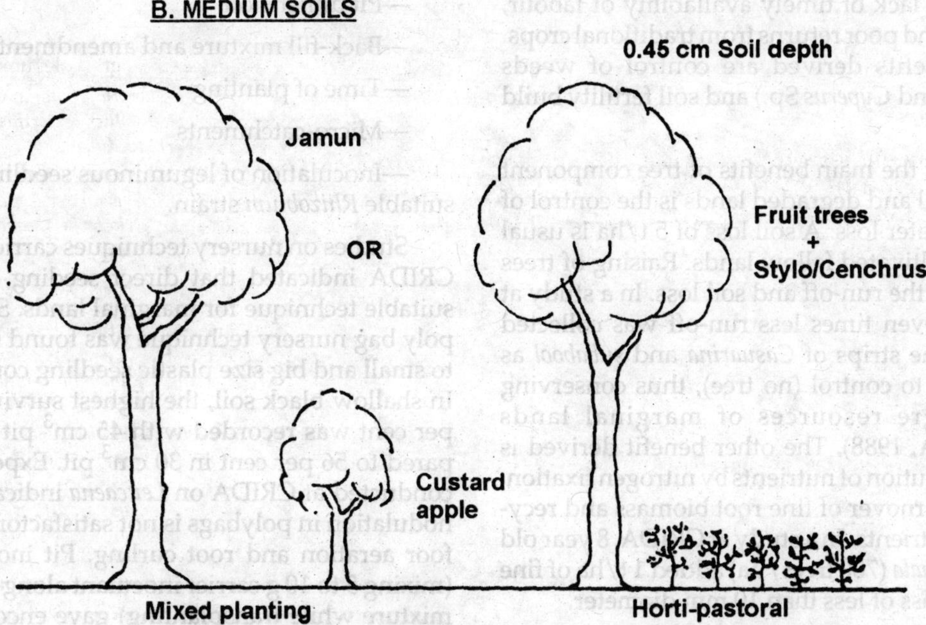

Fig. 2(b). Two Tier cropping models for SAT (Medium soils)

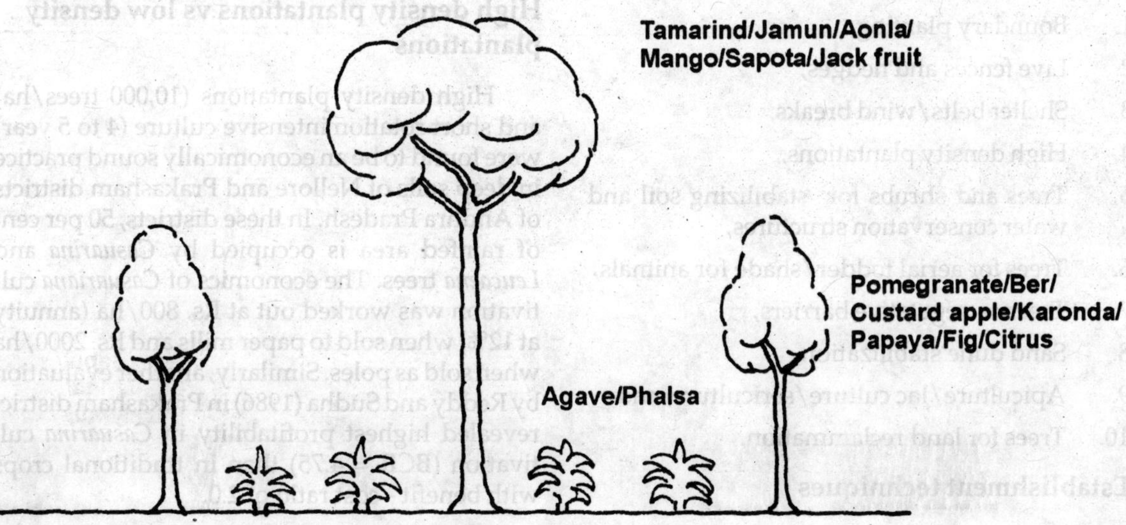

Fig. 3. Three tier or multistoreyed cropping models for SAT.

vantage enjoyed by trees is that they can tolerate extremes of soil and climatic conditions under which food production would be difficult and/or impossible. India is blessed with a variety of trees, unsurpassed anywhere in the world. There are trees for almost all human needs.

Farmers are resorting to tree farming, mainly

because of lack of timely availability of labour, high cost and poor returns from traditional crops. Other benefits derived are control of weeds (*Cynodon* and *Cyperus* Sp.) and soil fertility build up.

One of the main benefits of tree component in marginal and degraded lands is the control of soil and water loss. A soil loss of 5 t/ha is usual from uncultivated fallow lands. Raising of trees minimises the run-off and soil loss. In a study at Bijapur, seven times less run-off was collected through the strips of *Casuarina* and *subabool* as compared to control (no tree), thus conserving the meagre resources of marginal lands (AICRPDA, 1988). The other benefit derived is the contribution of nutrients by nitrogen fixation, leaf fall, turnover of fine root biomass and recycling of nutrients. In a study at CRIDA, 8 year old *L. leucocephala* (700 trees/ha) added 1 t/ha of fine root biomass of less than 10 mm diameter.

Tree farming practices

Tree raising practices that can be applied are as follows :

1. Boundary planting,
2. Live fences and hedges,
3. Shelter belts/wind breaks,
4. High density plantations,
5. Trees and shrubs for stabilizing soil and water conservation structures,
6. Trees for aerial fodder/shade for animals,
7. Trees as vegetative barriers,
8. Sand dune stabilization,
9. Apiculture/lac culture/sericulture, and
10. Trees for land reclammation.

Establishment techniques

Survival and growth of tree seedlings depend on nursery and establishment techniques. These include :

—Good site preparation
—Size of seedling containers
—Age of seedling
—Pit dimension
—Back-fill mixture and amendments
—Time of planting
—Micro-catchments
—Inoculation of leguminous seedlings with suitable *Rhizobium* strain.

Studies on nursery techniques carried out at CRIDA indicated that direct seeding is not a suitable technique for marginal lands. Standard poly bag nursery technique was found superior to small and big size plastic seedling containers. In shallow black soil, the highest survival of 72 per cent was recorded with 45 cm^3 pit as compared to 56 per cent in 30 cm^3 pit. Experiments conducted at CRIDA on *Leucaena* indicated that nodulation in polybags is not satisfactory due to foor aeration and root curling. Pit inoculation (mixing 5 to 10 g carrier inoculant along with pit mixture while transplanting) gave encouraging results. In shallow alfisols one year old seedlings of *neem* and *babool* recorded 99 per cent survival as compared to 80 per cent (3 month old).

High density plantations vs low density plantations

High density plantations (10,000 trees/ha) and short rotation intensive culture (4 to 5 year) were found to be an economically sound practice in deep soils of Nellore and Prakasham districts of Andhra Pradesh. In these districts, 50 per cent of rainfed area is occupied by *Casuarina* and *Leucaena* trees. The economics of *Casuariana* cultivation was worked out at Rs. 800/ha (annuity at 12%) when sold to paper mills and Rs. 2000/ha when sold as poles. Similarly, another evaluation by Reddy and Sudha (1986) in Prakasham district revealed highest profitability in *Casuarina* cultivation (BCR = 4.75) than in traditional crops with benefit - cost ratio of 2.0.

High density plantations (HDP) and short rotation intensive cultures (SRIC) will not be viable in marginal and degraded lands. The focus in marginal lands should be on survival than on number of seedlings. Low density, long rotation should be practiced and intensive care must be given in the first two years. This is possible only when few seedlings (500/ha) are planted.

Suitable MPTS for marginal lands

At Hayathanagar Research Station of CRIDA, several MPTS were planted in different sites and their growth was monitored regularly. All the species planted recorded more than eighty per cent survival without supplemental watering and were found suitable for marginal lands. *Leucaena leucocephala* recorded the highest growth rate in monoculture and also in association with crops and *Stylosanthes hamata*, compared to other species tested, viz., *Acacia albida, Albizia lebbeck, Acacia nilotica* var *cupressiformis* etc. Tree species suitable for areas having different annual rainfall are given in Table 25.

Table 25. Tree species suitable for areas having different annual rainfall

< 500 mm	500-750 mm	> 750 mm
Acacia nilotica	*Acacia nilotica*	*Acacia nilotica*
Acacia aneura	*Acacia ferruginea*	*Albizia lebbek*
Acacia tortilis	*Albizia lebbek*	*Azadirachta indica*
Acacia albida	*Albizia amara*	*Casuarina equisetifolia*
Colophospermum mopane	*Azadirachta indica*	*Dalbergia sissoo*
Prosopis cineraria	*Casuarina equisetifolia*	*Dalbergia latifolia*
Prosopis juliflora	*Cassia sturti*	*Gmelina arborea*
Pithecellobium dulce	*Dalbergia sissoo*	*Grewia optiva*
	D. latifolia	*Grevillea robusta*
	Leucaena leucocephala	*Hardwickia binata*
	Tamarindus indica	*Melia azadarach*
	Terminalia sp.	*Populus sp*
		Sesbania sp.
		Terminalia sp.

Tree farming in open grazing lands

Neem (*Azadirachta indica*) and Australian babool (*Acacia auriculiformis*) are two tree species, which are damaged the least by livestock. According to FAO report, *'neem'* has been coined as 'mobile dispensary of Indian villages'. Every part of the tree is full of different therapeutical properties. There are about 4 crore *neem* trees in India. *Neem* fruiting starts after third year of planting and yields about 30 to 35 kg fruits / tree. Seeds are a good source of non- edible oil and the oil content ranges between 40 to 45 per cent. *Neem* timber is valuable and is used for furniture as it resists insect attack. One cubic metre of timber weighs about 925 kg.

Keeping in view the above uses, neem was plated at 4m x 4m spacing in the year 1984 and the growth was compared with *Acacia auriculiformis*. Survival and growth of neem was found to be superior to *Acacia auriculiformis* (Table 26).

Pole/fruit/timber tree options for degraded land

Three multipurpose species viz., *Acacia anerua, Syzygium cuminni* and *Dalbergia sisoo* were planted in July, 1986 at a spacing of 5 m x 5 m. These species were planted on contour furrows and crescent bunds of 0.3 m^2 section were made to concentrate run-off. The land is undulating with more than 2 per cent slope and mostly gravelly. All the species recorded 100 per cent survival, indicating their suitability for degraded lands. They also recorded good growth (Table 27).

Timber-Cum-Fibre System (TIMFIB)

This system was initiated at Bijapur, Karnataka. Intercropping of *subabool* with agave was found to be more remunerative than arable crops on marginal shallow soils. Fibre yield was increased with an increase in plant population and had no significant effect on the subabool plant growth (Table 28). However, intercropping of

Table 26 : Survival and growth of *neem* and *Acacia auriculiformis*.

Tree species	ht (m)	dbh (cm)	Survival (%)
Azadirachta indica	4.21	8.69	85
Acacia auriculiformis	5.07	8.23	68

Date of planting 15.7.84, spacing 4 m x 4m. Date of observation : 20.7.89.

Table 27 : **Growth parameters of tree species suitable for marginal and degraded lands**

Tree species	ht (cm)	dbh (cm)	Remarks
Acacia aneura	3.03	3.07	Pole-cum-timber
Syzygium cuminii	2.53	5.25	Fruit, timber, fodder,medicinal
Dalbergia sissoo	3.56	4.49	Timber-cum-fodder

Date of planting 16.7.86 Spacing 5m x 5m
Date of observation 16.7.89

subabool with agave has reduced the fibre yield to the extent of 100 kg/ha as compared to sole crop. The annual fibre yield obtained was 371 kg/ha from second year onwards.

Table 28 : TIMFIB : Permanent intercropping of *subabool* and *Agave* - Bijapur.

Plant population/ha		Mean fibre yield (kg/ha) of 3 years
Agave	*Subabool*	
10000	6944	516
6666	3888	334
5000	3472	311
3333	2083	236
2966	1889	199
2500	1250	189

Source : AICRPDA, 1988.

Sisal *(Agave)* is widely distributed throughout the drier regions of India, largest area being in Orissa state. The economics of cultivation appears to be encouraging with a yearly profit of about Rs. 6000/ha, averaged over six year period (Theagarajan, 1986). *Agave* fibre is strong, coarse, superior and more flexible than Manila hemp and it ranks next to cotton in U.S.A. Each plant yields about 250 to 300 leaves during its life span (7-8.5 years) and leaves yield 27 to 35 kg of dry fibre. Only about 2000 tons of sisal fibre is being produced annually in India against a demand for about 23000 tons.

ECONOMICS OF ALTERNATE LAND USE SYSTEMS

A detailed comparative economics of various, "Alternate Land Use Systems" tried at Central Research Institute for Dryland Agriculture, at Hyderabad is set out in Table 29. The choice among them lies with the farmer's production objective in relation to land capability class, rainfall and market considerations.

Alley cropping and agro-horticulture systems are an ideal option for better sites. Returns will be higher with timely management and adequate plant protection measures.

Horti-pastoral and silvi-pastoral systems seem to be more paying than sheep rearing on non-arable lands. More income could be generated if it is integrated with improved breeds of milch animals. Horti-pastoral system could be a more stable system on medium deep soils if right type of fruit tree and forage species is selected. The interspaces in the initial years could be efficiently utilized by raising *stylo* or mixture of *stylo* and *cenchrus* grass. For silvi-pastoral system to be successful, suitable selection of a multi- purpose top-feed species is necesary. It could be mounted on shallow soils.

Ley farming may be advantageous in areas prone to severe soil erosion, endemic diseases and perennial weeds. In this system there is possibility of continuous rotational grazing with an arable crop every fourth year without any additional input of fertilizer.

Tree farming appears to be more paying in areas located nearby paper mills and high density and short rotation intensive culture is possible only in better sites.

Table 29 : Comparative economics of different alternate land use systems

Farming system	Average cost/(Rs/ha)				Average returns/ year (Rs/ha)				Value cost ratio	Annuity at 12 % (Rs/ ha)
	1	2	3	4	1	2	3	4		
Alley cropping										
Fodder system	2000	1800	1800	1800	6000	6425	6425	6425	3.39	4441
Mulch system	2000	1800	1800	1800	6000	6350	6350	6350	3.36	4388
Agro-horticulture										
Ber-cluster bean/ horsegram	5811	1600	2500	2500	2500	2500	3500	12000	2.69	5146
Horti-pastoral system										
Custard apple /Guava Stylosanthes	5240	1739	1739	750	–	875	875	5000	2.28**	2475**
Silvi-pastoral system										
Leucaena-stylosanthes	2500	750	750	2500 (8th year)	–	875	875	14000 (8th year)	1.50 **	582 **
Ley-farming										
Stylo-sorghum (IVth year)	900	300	300	1500	–	1250	1250	6000	2.22	1150
Pasture management										
Sheep rearing										
Tree farming	875	2776	976	976	–	1800	1800	1800	1.28*	335
Leucaena sp.	2500	500	500	10,000 (10th year)	–	–	–	40,000 (10th year)	2.10	1181

* Highly fluctuating market price.

** Returns will be higher if integrated with livestock.

NOTES : 1) No. of lambs 1.7/sheep/year. Cost of sheep Rs. 350/- having body weight of 29 kg.

2) Rs. 400/t of fuelwood.

3) Cost of stylo legume Rs. 0.25/kg dry matter.

REFERENCES

AICRPDA, 1987. *Annual Report 1986-87*. All India Coordinated Research Project for Dryland Agriculture (ICAR). Main Centre, Rajkot, Gujarat Agricultural University. pp. SA-8.

AICRPDA, 1988. *Annual Report, 1987-88*. All Inida Coodinated Research Project for Dryland Agriculture (ICAR). Main Centre, Bijapur, University of Agril. Sciences, Dharwad, pp. 59-69.

Corlett, J. Ong, C.K., and Black, C.K. 1987. Comparison of microclimate in intercropping and alley cropping. In : *Meteorology and Agroforestry* (Ed. w.S. Reifsnyder and T.O. Application of meteorology and Agroforestry Systems Planning and Management. February 9-13, 1987. Nairobi, Kenya, pp 419-430.

Felker, P. 1978. State of the art : Acacia albida as a complementry permanent intercrop with annual crops. Prepared under USAID grant No. : AIF/Afr-C-1361, Riverside. Department of Soil and Environmental Sciences, University of California, 133 P (Mimeo).

Filius, A.M. 1982. Economic aspects of agroforestry. *Agorforestry Systems*, 1 : 29-30.

Gregory, P.J. and Reddy, M.S. 1982. Root growth in an intercop of pearmilet/groundnut. *Field Crops* Res. 5 : 241-252.

Harris, D. and Natarajan, M. 1987. The physiological bassis for yield advantage in a sorghum : groundnut intercrop exposed to drought. 2. Plant temperature, water stress and components of yield. *Field Crop Res.* 17 : 273-288.

Hocking, D. and Rao, D.G. 1985 *Annual Report*, Central Research Institute dor Dryland Agriculture, Hyderabad, P. 23

Kang, B.T., Grimma H. and Lawson, T.L. 1985. Alley Cropping sequentially cropped maize and cowpea with *leueaena* on a sandy soil in southern Nigeria. *Plant and Soil 85* : 267-277.

Korwar G.R. 1984. *Annual Report*. All India Coordinated Research Project for Dryland Agriculture,. pp. 55.

Marshall, B. and Willey, R.W. 1983. Radiation interception and growth in an intercrop of pearl millet/groundnut. *Field Crop Res.* 7 : 141-160.

Natrajan, M. and Willey, R.W. 1980. Sorghum-Pigeonpea intercropping and the effect of plant population density. II Resource Use. *J. Agric. Sci.*, Camb. 95 : 59-65.

Ong, C.K. 1984. The influence of temperature and water deficit on the partitioning of dry matter in groundnut (*Arachis hypogaea* L.) *Journal of Experimental Botany. 35* : 746-755.

Ong, C.K., Black, C.R., Simmonds, L.P. and Saffell, R.A. 1985. Influence of saturation deficit on leaf production and expansion in stands of groundnut (*Arachis hypogea* L.) grown without irrigation, *Annals of Botany 56* : 523-536.

Ong, C.K. and Monteith , J.L. 1985. Response of pearl millet to light and temperature. *Field Crops Res. 11* : 141-160.

Peter Poschen, 1986. An evaluation of the *Acacia albida* based agroforestry practices in Hararghe highlands of Eastern Ethiopia. *Agroforestry Systems*, 4 : 129-143.

Reddy, N.V. and R. Hampaiah. 1982. *Alternate land* use in dryland agriculture In *"A Decade of Dryland Agricultural Research in India, 1971-80*, AICRPDA, Hyderabad.

Reddy, Y.V.R. and M.Sudha, 1987. Economic of alternate land use systems. In : *Annual Report*. Central Research Institute for Dryland Agriculture, Hyderabad, pp. 83.

Singh, R.P. 1986. Alley cropping in semi-arid tropics of India - The CRIDA experiences. *ICAR-ICRAF Training-cum-workshop on Agroforestry*, CRIDA, Hyderabad, pp. 1-13.

Singh, R.P., Vanden Beldt, Hocking, D., Korwar, G.R. 1987.Alley cropping in the semi-arid regions of India. *Pro. of Workshop on Alley Cropping. International Institute of Tropical Agriculture* (IITA), Ibadan, Nigeria.

Singh, R.P. and Korwar, G.R. 1986. Agroforestry options for drylands of India. *Indian J.Dryland Agri.Res.&Dev.*, 1(1); 1-10.

Singh, R.P. and Osman, Mohd. 1987. *Agroforestry systems* for small hodlings. *Paper Presented in Workshop on 'Agroforesty for Rural Needs'*, Indian National Science Academy, New Delhi.

Singh, R.P. Ong, C.K. and Saharan, Neelam 1989 Above and below ground interactions in alley - cropping in semi-arid India. *Agroforestry Systems. 9 : 259-274.*

Singh, R.P., Ong, C.K. and Saharan, N. 1988. Microclimate and Growth of Sorghum and cowpea in Alley - cropping in semi-arid India. *Challenges in Dryland agriculture : A Global Perspective Proceedings of the International Conference on Dryland Farming*, Amarillo/Bushland, 15 to 19 pp. 163-169.

Theagarajan, K.S. 1986. Cultivation and utilization of Sisal (Agave), *Van Vigyan* 24 (1&2) : 38-39.

Willey, R.W. and Reddy, M.S. 1981. A field technique for separating above and below ground interactions in intercropping : an experiment with pearlmillet/groundnut, *Experimental Agriculture* 17 : 257-264.

25

AGROFORESTRY — PRINCIPLES AND PRACTICES

R. DEB ROY

INTRODUCTION

Agroforestry is neither a new concept nor being used only recently. Agroforestry systems have been in existence since very early times. In our country many Agroforestry systems are age old and traditional viz., shifting cultivation (*jhum*), *taungya*, growing plantation crops like tea, coffee under shade and nitrogen fixing trees, intercropping under coconut, palm, alder (Alnus), *Prosopis* etc.

In Burma, raising of teak plantation through *taungya* was available as early as 1806 (Blandford, 1958; King, 1987) but became popular since 1887 (Hailey, 1957). First probable *taungya* cultivation in India was in Chitagong (now Bangladesh) in 1890 and in Bengal in 1896 (Raghavan 1960) and reintroducing in 1920's in other parts of India. Shifting cultivation is believed to have originated in the neolithic period long long back in 7000 BC (Singh, 1992). Trees and shurbs and other woody perennials including fruit trees, crops and animals have been raised traditionally altogether all over the world specially in the developing countries on small farms. Initially shifting cultivation was a low input land use system which was sustainable due to low population pressure and availability of large tract of undisturbed forests. But with the rising population pressure this system of cultivation is not only unsustainable now but it promotes heavy soil erosion and further land degradation because of the drastic reduction of the gestation period (*jhum* cycle). With the advancement of scientific knowledge and specialisation in the individual scientific field, trees, crops and animals are managed separately under modern agriculture, animal husbandry and forestry. In the developing countries like India which has got about 2 per cent of the world's geographical area but she has to provide various needs of its ever increasing human and livestock population of 875 million and 465 m which is likely to exceed 1000 m and 600 m respectively by the end of the century (Deb Roy, 1990). Although the so called forest area has been estimated to be around 75 m ha but the present productive and effective forest area would be around 12 to 15 per cent of the total land area giving a per capita forest area of 0.06 ha against the world average of 1.6 ha (Deb Roy and Gill 1990). Looking to the present availability and future requirement by 2000 AD the deficit in respect of food, fuelwood and fodder has been estimated to be to the tune of 75, 140 and 1104 m metric tons, respectively (Deb Roy 1992). In addition, the requirement of timber would be to the tune of 65 m^3. Increasing thrust given for food production as a result of green revolution more and more of the marginal, submarginal and other wasted land has been put to arable farming creating more problem of soil erosion and further land degradation than solving the problem of food deficits. India is also passing through the energy crisis and acute shortage of fuel and the resultant high price leads to the practice of burning cowdung cakes @ 80 to 100 mt annually representing 400 to 500 m t of wet dung which could have increased the agricultural and other biomass production substantially had it been diverted to the agricultural fields.

Under the above circumstances, the principles of integrated agroforestry systems using multipurpose trees including nitrogen fixing trees and shrubs and fruit trees alongwith crops/pasture in different cultivated, marginal, submarginal and other wasted land to solve the pressing needs besides meeting the national target to bring 33 per cent area (60% in the plains and 20% in the hills) under various tree/pasture cover have significant importance. In this chapter, an attempt is made to discuss briefly the principles of agroforestry and practices under various agroforestry systems in vogue in India and being improved upon through research.

What is Agroforestry ?

The word Agroforestry is so commonly used now a days that it hardly needs any further elucidation as it has already got distinct identity of its own all over the world. It may be defined as an integrated self sustained land management system which involves deliberate introduction/retension of woody components including trees, shrubs, bamboo etc. with agricultural crops including pasture/livestock simultaneously or sequentially on the same unit of land meeting the ecological as well as socio-economic needs of the people. One of the earliest definition of Agroforestry is that of Bene *et al.* (1977) as well as King and Chandler (1978) which defines Agroforestry as "a sustainable land management system which gives the yield of the land component with the production of crops including tree crops and forest land and/or animal simultaneously or sequentially on the same unit of land and applied management practices that are compatible with the cultural practices of the local population". The definition of Agroforestry has been discussed in great details by Nair (1989), Mac Dicken and Vergera (1990). Of the various definitions given by number of authors one of the simple definition of Agroforestry is that of Lundgren and Raintree (1982) which define Agroforestry as "a collective name of land use systems and technologies where woody perennials (trees, shrubs, palms, bamboos etc.) are deliberately used from the same land management units as agricultural crops and/or animals in some form of spatial arrangement or temoral sequence. In agroforestry systems, there are both ecological and economic interaction between different components".

In agroforestry, the three important things should be kept in mind is that the land management system is sustainable, there is direct or indirect influence of woody perennials with crops including pasture crops/animals and it has to meet the ecological as well as socio- economic needs of the local populace.

Agroforestry Principles

With the rising population pressure of both human as well as the livestock with the limited land resources available especially in the developing countries to meet the multifarious needs of the people we should utilize the concept of vertical expansion of plant components to capture maximum solar energy for increasing biomass production through various agroforestry systems. Under agroforestry systems benefits of sharing solar energy, moisture and nutrients can be taken for optimising biomass production, taking advantage of certain plants having shade tolerance/diffused light requirement. In India and other developing countries, land is limited and as such there is hardly any scope for horizontal expansion but there is ample scope for expansion vertically for growing multiple plant species for higher biomass production.

Preliminary results from research that was being conducted in different parts of the world had indicated that in intercropping system more effective use was made of the natural resources of sunlight, land and water; that intercropping system might exercise beneficial effect on pest and disease problems; that there are advantages in growing legume and non-legume in mixture; and that as a result of all these, higher yield were being obtained per unit area even when multi-cropping systems were compared to sole cropping systems (King 1987). Agroforestry is fast becoming recognised as an integrated system which is capable of yielding both wood and food, besides conserving and rehabilitating ecosystems.

CLASSIFICATION OF AGROFORESTRY SYSTEMS

ICRAF's inventory on agroforestry systems (AFSI) during 1982-87 revealed that agroforestry systems are widespread in the developing countries in various ecological zones (Nair 1987). Based on the kind of associated agricultural products, major function of the tree component, spatial arrangement of trees and duration of the combination, Cambe (1982) suggested 24 agroforestry systems. Nair's (1985) classification of the agroforestry systems included agro-ecological zone as well as socio-economic aspects and gave a detailed account of the various agroforestry systems. According to Tejwani (1987) there could be 48 agroforestry systems if Cambe's system of classification is followed. Based on the nature of the components common agroforestry system in India was classified by Deb Roy (1989) as *Agrisilviculture* (crops + trees); *Silvipasture* (trees + pasture / + animals); *Agri-horticulture* (crops + fruit trees); *Agrisilvipasture* (crops + trees + pasture/ + animals); *Hortipasture* (fruit trees + MPTS + Pasture/animal). *Agri-horti-silviculture* (crops + fruit tress + MPTS); Homestead agroforestry (multiple combination of various components) and other. Other common agroforestry system could be *Agro-pisci- silviculture* (crops + fish + trees); *Silvi-apiculture* (trees + honey bees) and *Pisci-silvicuture* (fish + trees). In the similar process by combining different components there could be many more agroforestry systems.

Classification of various agroforestry systems has been dealt in great details by Nair (1985). He has also given some examples of important agroforestry systems and practices in the developing countries.

Agroforestry practices

The agroforestry practice usually denotes specific land management operation of an agroforestry nature on a farm or other management unit and consists usually of arrangement of agroforestry components (Nair 1985). Several such practices will be involved in the constitution and maintenance of an agroforestry system. Commonly, this practice includes the sustainable system management in space and time vis-a-vis the major function of the tree component. Although several agroforestry systems have been accounted from different parts of the world the distinctive agroforesty practices that constitute this system for various eco-regions are only a few.

Alley cropping (hedge row intercropping); boundary planting of trees; trees and shrubs as shelter belts and wind breaks; use of woody perennial for soil conservation; tree gardens; wood lots on agricultural lands etc. are commonly found agroforestry practices in most agroforestry systems. These commonly found agroforestry practices and essential characteristics are given in details by Nair (1985). Any of these agroforestry practices become an agroforestry system and it is developed or spread to such an extent on a specific area so as to form a definite land utilization type in that area. The agroforestry system, sub-system and agroforestry practices are used rather promiscuously and even sometimes synonymously in common agroforestry usage. As such, it is often difficult to differentiate practices from one another.

Traditional Agroforestry Systems in India

In our country, agroforestry in some form or other is being practised by our farmers since ancient times. Common agroforestry systems in vogue in our country are enumerated by Deb Roy (1989). Some common agroforestry practices and the systems under which they are grouped are given below.

Shifting cultivation, taungya cultivation, plantation crops, commercial crops with trees, trees grown in field boundaries, shelter belts and wind breaks, alley cropping or hedge row intercropping. All these common practices can be grouped under the agroforestry system-Agrisilviculture. Trees and shrubs on pasture and introduction of pasture in forest plantation can be grouped under agroforestry-Silvipasture-often called as Forage Forestry. Similarly, growing crops including fodder crops in association with fruit trees can be grouped under Agro-horticulture. In homestead Agroforestry, there can be many agroforestry systems depending on the mixture of the components raised in a particular system.

Tejwani (1987) tried to differentiate between Agrosilviculture and Silviagriculture depending on the principal component in the system. If the agriculture is the main component it may be agrisilviculture, whereas if the tree is the main component it could be silviagriculture. Similarly, he said agrihorticulture and hortiagriculture, silvipasture and pastoral silviculture but author feels these differentiations are artificial, as one can merge with an other. For example, in agrisilvicultural system one can start with large number of trees per unit area and gradually pruning and thinning them out and ultimately run the system with small number of trees. The main idea of various agroforestry systems is to improve the overall land productivity by using various agroforestry components. This is only a variation in the management and should not be confused. Some people are of the opinion that if the forest department is growing crops under trees then it should be called as silviagriculture and if the farmers grow trees in their agricultural fields it should be called agrisilviculture. There is often a confusion about some people writing agri-silviculture, agrihorticulture, silvipasture etc. whereas others write it as agrosilviculture, agrohorticulture, silvopasture etc. Even Nair (1985) classified the system as agrosilvicultural system, silvopastoral system, agrosilvopastural system, agrosilvofishery system etc. Author feels that it should be written as agrisilvicultoral system, silvipastoral system, agrisilvipastoral system, agri-silvi-fishery system etc.

Traditional Agroforestry Practices and their Modifications

Shifting cultivation

This type of agroforestry practice is quite extensive in north-eastern hill regions and in Andhra Pradesh, Bihar, Madhya Pradesh, Orissa and Karnataka states (Tejwani 1987). This type of agroforestry practice is in vogue in our country since long and very popular among the tribals. The entire way of life, training to youth, ceremonies and festivals and the philosophy are associated with this type of cultivation (Sharma, 1981). In earlier years, the farmers used to return to the same plot after about 40 years but with the rising population now the cycle has come down to as low as 3 to 5 years, thereby the system has lost its sustainability. It has been reported (Anonymous 1982) that out of 43,43,000 tribals in 1971 shifting cultivation was practised by half of them in this region. Tejwani (1987) has dealt in greater details about the shifting cultivation. It has been mentioned by Mishra and Ramakrishnan (1981) that in the shorter cycle (5 years) main emphasis is on growing maize, leafy vegetables and tuber crops, whereas in the larger cycle (15 years) varieties of crops have been grown. The chores as performed by males and females are well defined. Singh and Singh (1981) estimated the annual loss due to soil erosion in the early state of shifting cultivation which may vary from 3.3 to as high as 201 t/ha. Borthakur *et al.* (1981) suggested a model land use plan for the north-eastern hill region as an alternative to shifting cultivation. They suggested the top 1/3 of the hill slopes be put under forest, the middle 1/3 under horticultural use and the bottom 1/3 under bench terraced cultivation, but the author is of the opinion that the top 1/3 should be under silvipasture (trees with pasture), middle 1/3 under hortipasture and the bottom 1/3 under agrisilviculture. Silvipasture and hortipasture will provide necessary cover to the steep land and the bottom 1/3 will get the benefit from the run off and the nutrients received from the upper slopes under silvipasture and hortipastoral system. This model needs to be tried widely to make a suitable technology alternative to shifting cultivation. Author is of the opinion that if it is not possible to change the shifting cultivators to adopt this system they may atleast be helped to put fast growing MPTS and perennial grasses before leaving the area bare so that it will be able to recoup to a reasonable extent when they return to have a sustainable production.

Taungya cultivation

Taungya cultivation is a system of establishing forest plantations of commercial species found most economical by the Forest Department with the help of farmers who are allowed to grow agricultural crops on temporary basis in between rows of trees. *Taungya* is a Burmese word meaning hill cultivation. The shifting cultivation in north eastern hill regions of India is dealt in greater details by Sharma (1981) and Ingty and Goswami (1981). Seth (1981) reported that taungya cultivation was introduced in India by Brandis in 1956 and the first *taungya* was raised in 1963 in north Bengal. *Taungya* is practised in the states of Kerala, West Bengal, Uttar Pradesh, Tamil Nadu, Andhra Pradesh, Orissa, Karnataka, Assam and other north eastern hill regions. The different trees and associated agricultural crops grown under *taungya* system in different states have been given by Tejwani (1987). *Tectona grandis* (teak), *Shorea robusta* (sal) and *Eucalyptus* species are the most common among the tree species used. Most commonly grown crops are paddy, maize, ginger, turmeric and vegetables. Trees and associated crops grown in various states in India are given in Table 1. In South India, it is popularly known as *'cumri'* and is usually practised in assured high rainfall (over 12000 mm) region. Agricultural crops are normally grown for 1-3 years after which the forest department does not allow them as according to them it affects the tree growth and as such the farmers are compelled to move to some other area. This can be also compared to shifting cultivation as the farmers are shifted from one place to another after a duration varying from 2 to 3 years but is different from the practice in that the land is given by the forest department. It is practised both in plains as well as in hill region. In certain areas where there is shortage of land like Kerala, Andhra Pradesh etc. the forest department realises lease money from the farmers (Reddy 1981 and Nair 1981), whereas in other states the farmers are allowed to cultivate from 1 to 3 years freely but they have to do necessary operations as directed by the forest department.

It has been observed that the tree growth was much faster under the *taungya* cultivation than under pure forest plantation, thereby the important commercial plants like teak, sal and eucalyptus could be grown most economically. The *taungya* plantation was initiated to economise growing commercial trees in shortest period and in most economic way but it could form an important agroforestry system if it is properly managed. It is felt that this could be a suitable agrisilvicultural system for specific trees.

The research conducted at the National Research Centre for Agroforestry, Jhansi U.P., showed that it is possible to grow as many as 1250 MPTS including *Eucalyptus tereticornis, Casuarina equisetifolia, Dalbergia sissoo, Acacia cupressiformis, Leucaena leucocephala* (Subabool), *Hardwickia binata* etc. in association with crops in the initial year of establishment of the MPTS. By annual pruning and thinning (removing 40% of the plants) the trees in the fourth year it has been observed that there is not much variation in the crop yield even in the fourth year and tree growth is also better (Deb Roy and Gill, 1991-92, 1992 and 1992-93). It is proposed to remove another 40 per cent of the MPTS in the seventh year to reduce light as well as root competition so as to minimise reduction in crop production.

Teak, sal, sissoo, eucalyptus etc. are initially grown by the forest department with a large plant population and then there is a practice of thinning after a specific period. There is ample scope for allowing cultivation of crops for a longer period than the normal period assigned by the forest department under *taungya* cultivation of a number of important tree species. There is enough scope for doing systematic research work on this line to make suitable model for further improvement.

Intercropping with Fruit Trees

Fruit trees normally take a period of 4 to 5 years to mature and give cash returns. In some cases they take even 8 to 10 years or so. These trees are normally grown wide apart and there is scope for growing agricultural crops in the initial

Table 1. Crops grown in taungya system in India

State	Tree crop	Associated agricultural crops.
Uttar Pradesh (UP)	*Shorea robusta* *Tectona grandis* *Acacia catechu* *Dalbergia sissoo* *Eucalyptus* spp. *Populus* spp.	Maize, paddy, sorghum, pigeonpea, soybean, wheat, barley, chickpea, rape seed and miscellaneous
Andhra Pradesh (AP)	*Anacardium occidentale* (cashew) *Tectona grandis* *Bombax ceiba* *Bamboo* *Eucalyptus* spp.	Hill paddy, groundnut, sweet, potato
Kerala	*Tectona grandis* *Bombexc ceiba* *Eucalyptus* spp.	Paddy, tapioca, ginger, turmeric
Assam	*Shorea robusta*	Paddy
Tamil Nadu	*Tectona grandis* *Bamboo* *Santalum album* *Tamarindus indica* *Acacia nilotica* *Acacia mearnsil* *Ceiba pentandra* Cashew Rubber	Millets, pulses, groundnut, cotton, tapioca, potato
West Bengal (WB)	*Tectona grandis* *Shorea robusta* *Schima wallichii* *Cryptomeria japonica* *Quercus* spp. *michelia doltsopa*	Paddy, maize, millets, turmeric, ginger, ladys finger, pineapple, hemp
Maharashtra		Sunhemp, jute, mesta, sunflower, castor, etc.
Andaman and	*Pterocarpus dalbergoides*	Sugarcance, maize

Source : Tejwani (1987).

stage of the growth of the fruit trees. It has been estimated that in 1982 there were more than 205 m ha under different fruit trees in India. Out of different fruit trees, mango occupied maximum (above 40%) of the total area under fruit cultivation and is distributed in various states especially in Uttar Pradesh, Kerala, Andhra Pradesh, West Bengal, Bihar etç. Other commmon fruit trees are banana, orange, apple, guava, pineapple which altogether occupy an area of more than 22 per cent (MOA 1983).

Although farmers/orchard owners grow intercrops in the early stage to get some economic return from such plantations but not much research results are available on this aspect. The crops normally grown vary from fodder crops like berseem to rainy season vegetables and also irrigated winter vegetables, beans etc. Farmers normally grow 60 to 100 trees/ha of improved varieties of mango but with the availability of dwarf varieties like Mallika, Amrapali etc. 500-600 trees/ha are not uncommon today. In an agrihorti-silvicultural system at NRCAF, Jhansi it has been observed that in addition to growing 400 fruit trees like pomegranate, guava, *Citrus sinensis* var. kinnow, indian plum (*Zizyphus mauritiana*) it is possible to grow 200 to 300 fast growing MPTS like *subabool* in between the fruit trees in lines alongwith crops and the *subabool* could be harvested annually to provide annual return of nutritious leaf fodder and fuel to the tune of more than 1.13 to 1.78 t/ha respectively (Gill and Deb Roy 1990-91, Gill *et al.* 1991). It has also been observed that the relative crop yield was sometimes higher or at par with control (without fruit trees and MPTS) especially with 200 fruit trees/ha even in the third or fourth year (Gill and Deb Roy 1990-91, 1991-92). It has been observed that production of fruit trees especially indian plum and guava was not significantly affected by growing crops as well as *subabool.* This gives an indication that fruit trees and MPTS could be profitably grown alongwith suitable agricultural crops.

Four varieties of mango viz., Amrapali, Mallika, Dasheri and Langra at 6 x 6 m and 6 x 12 m spacing are being studied in association with food, fodder and vegetable crops and also without crop. In between the fruit trees in lines two plants of fast growing NFTS like *subabool* are also planted which are being harvested annually for fodder and fuel production (Gill and Deb Roy; 1991-92, Gill *et al.* 1992).

Intercropping with Plantation Crops

Most common plantation crops in our country are coconut and arecanut which are popularly grown in Kerala, West Bengal, Assam, Tamil Nadu, Karnataka and Andhra Pradesh. Ninety per cent of the total area and also the total production of coconut were confined in south India especially in Kerala, Tamil Nadu, Karnataka and Andhra Pradesh. According to Menon and Pandalai (1960) the mean annual temperature of 25°C with a diurnal variation of 6 to 7°C and a well distributed rainfall between 1500 to 2250 mm is optimal for the growth of coconut. As the trees are grown wide (8-10 m) apart there is ample scope of growing annual, biennial and perennial crops under its canopy. From the intercropping point of view, the first eight years and from 25 years onwards there is very good scope of growing intercrops but the period from 8 to 25 years when the canopy covers 80 per cent of the ground there is not much scope of growing other crops except shade loving crops like turmeric, ginger etc. in the interspaces during this period. Roots of adult coconut plant are concentrated within 2 m radius around the tree and top 30 cm of soil are practically devoid of functional roots (Kushwaha *et al.* 1973). Coconut is mostly grown as homestead agroforestry alongwith various intercrops by the farmers but not much information is available. Most of the information available on the intercrops is mostly from the Research Institutes/Agricultural Universities like Central Plantation Crops Research Institute, Kasargod; Kerala Agricultural University; Central Tuber Crops Research Institute (Varghese *et al.* 1978a, & b, Nair, 1983; Nair and Shreedharan, 1986 etc.). Coconut and cocoa are quite compatible and contribute to the increase in yield of both the crops. It has been observed that the farmers are growing banana, tapioca and other root crops, guinea grass etc. under the canopy of coconut. Although some work has been done by scientists of CPCRI (Kannan and Nambiar; 1976; Nelliat and Bhat, 1979, Verghese *et al.* 1978a & b) there is lot of scope for doing further research as some of the farmers grow very many crops in association under the canopy of coconut. With the availability of dwarf varieties of coconut suitable studies in intercropping should be carried out to identity compatible crops separately for the tall varieties and also for the newly evolved dwarf varieties. Retinam (1991) suggested that in a coconut plantation with a spacing of 7.5 x 7.5 m, the available space of 56.25 sq.m.(of which roots are concentrated only on 12.57 sq.m.) and thus the remaining area is available for arable crops with 56 per cent light. He suggested pepper, cocoa, ginger, turmeric and pineapple in association with coconut is a viable multi-storied cropping system. This system provides similarity to the ecosystem with that of the forest.

Out of the arecanut producing countries in the world, India is producing almost three fourths (about 1,5,000 tonnes) of the total world production covering a land area of about 1,80,000 ha and most popular in the states like Assam, Tripura, West Bengal, Kerala, Tamil Nadu, Karnataka and Andhra Pradesh (Bavappa 1980). The climatic prameters suitable for arecanut are more or less similar to that of coconut. For better growth, well distributed high rainfall of 2250 mm and the temperature varying from 18 to 38°C without extreme variation is necessary. At young stage arecanut plant needs to be protected from scorching sun which is normally done by growing a shade crop of banana. As it is very tall plant giving minimum light obstruction it is possible to grow crops like cocoa, pineapple, black pepper, banana, beetle wine, tapioca, turmeric, ginger, guinea grass etc. but there is very little information available regarding the interaction of the crops with arecanut. In North Bengal only one crop is taken by the farmers. The intercrop taken varies from state to state, although banana is a common crop grown in association with arecanut (Singh *et al*. 1982). Production of arecanut often increases when intercropped with ginger, beetle vine, black pepper, elephant foot, yam etc. Although some of the farmers do grow intercrop in association with arecanut but there is lack of information about the long term management of arecanut- based agroforestry system. Like coconut, arecanut alongwith intercrops is an important and profitable homestead agroforestry system which needs further improvement.

Shade Loving Commercial Plantation crops:

Tea, coffee and cardamon are the three most important commercial crops in India which love shade. As such these are planted in association with number of leguminous and other trees for their optimum growth and production.

Tea (*Camellia sinensis* var. **Assamica**) *is most* common in Assam, West Bengal and part of Tamil Nadu. Small scale plantation of tea is also being initiated in other states including Himachal Pradesh. In Assam and Bengal, the most commonly grown leguminous trees are Albizia species preferably *A.chinensis, Delbergia serica, Derris robusta, Gravillea species.* In some places, *Leucaena leucocephala* is also being grown to provide shade and also ameliorate soil. In South India, *Acacia* species, *Gravillea* species and *Erythrina lithosperma* are common (Tejwani 1987). Some information on the effect of shading, use of fertilizer on yield of tea has been investigated by Wight (1959). It had been observed that in tea plantation instead of choosing the single leguminous trees a number of trees are choosen in order to provide shade, ameliorate soil through nitrogen fixation and also through litter fall. These trees are also used in due course for providing fuelwood. It has been reported by Wight (1959) that the increased yield of tea due to planting *Albizia stipulata* is equivalent to about 90 kg N/ha. He has also stated that if it is not provided with shade the yield could be obtained by using high dose of nitrogen fertilizers. There is no doubt that the shade is essential to increase the production and quality of the tea leaf. It is very important to find out the proper shade (diffused light) for optimum production of quality and quantity of tea leaf. Although some research work on the fertilizer and shade effect was studied by Tocklai Expt. Station, Jorhat; Tea Research Station, Cinchona, Coimbatore but there is an ample scope to find out number and also the type of trees to be planted to provide the required diffused loght as well as the effect of shade tree on the increased production and quality of tea leaf.

The total area under tea cultivation in our country is about 3,58,000 ha and a production of 468 million kg is obtained. A considerable amount of foreign exchange is obtained by exporting tea to other countries (Venkataramni and Venkattram, 1987).

Coffee (*Coffea arabica* and *Coffea canephora* (Robusta). These two important species extensively cultivated in India. The cultivation of coffee is restricted to the southern states of Karnataka, Tamil Nadu, Kerala and Andhra Pradesh. Recently attempts are being made to have coffee plantation in noth eastern hill regions. Coffee plant grows better if the temperature does not go below 12°C and higher than 36°C with an elevation varying from 900 to 1200 m although robusta could grow at much lower elevation. Although both the coffee species are shade tolerant but *C.arabica* is more tolerant to shade. Most common leguminous trees grown to provide shade to coffee are *Erythrina lithosperma* and number of *Albizia* species including *A.stipulata* besides Ficus glomerata and F.nervosa (Tejwani 1987). Other shade trees are *Grevillea robusta, Dalbergia latifolia* etc.

Coffee is often grown under the shade of coconut as well as arecanut. It has been reported by Central Coffee Research Institute (1987) that the shade trees required for coffee plantation needs constant attention by way of prunning and lopping to furnish the required shade. Initially large number of shade trees are planted which are gradually thinned out as they grow and spread out so that canopy is 10 to 15 m above coffee plants. Permanent shade trees are generally planted at 12 to 14 m apart. The total production of coffee is about 85,000 to 1,00,000/annum covering an area of 1,55,257 ha.

Cardamom

There are two types of cardamom; small (*Eletaria cardamomum*) and large. (*Ammomum subulatum* and *A.aromaticum*). The small cardamom is cultivated in India in an area of 93,947 ha of which about 90 per cent is confined in Kerala and Karnataka and about 10 per cent in Tamil Nadu. The plants are 2 to 4 m in height grown in the evergreen forest of western ghat at an elevation varying from 600 to 1200 m with annual rainfall of 1500 to 4000 mm and the average temperature 10 to 35°C (Cardamom Board 1986). Cardamom plants require light shade for proper growth in the early stage. Most common shade trees used are *Mesopsis emini* commonly known as African shade tree, a fast growing tree which can be grown as temporary shade tree and *Cedrella tooa* and *Artocarpus heterophyllus* can be grown as permanent shade tree. For providing shade tall trees with wide branches and small leaves are preferred. The spacing commonly adopted vary from 1.5 x 2.2 m to 3 x 3 m. The yield of small cardamom is about 100 kg/ha with economic yield reaching at 12 to 15 years (Tejwani 1987).

Large cardamom is cultivated in the north eastern Himalaya region, Sikkim, Darjeeling district of West Bengal and Arunachal Pradesh. It is grown under the shade of natural forest in flat as well as slopy land by planting shade tree. The common shade trees are *Alnus nepalensis* (Utis) and *Schima wallichii* (Chillowane). The shade trees are lopped during September- October. When harvesting of cardamom capsule completes, farmers require forage in dry winter months (Singh *et al.* 1989). It takes 4 to 5 years to give production and is reported to yield 250 to 300 kg of cardamom/ha (Gupta 1982, Upadhyay and Ghosh 1983). Although farmers grow cardamom under the shade of *A.nepalensis* and *S.wallichii* but the number of trees and the method of its management, amount of shade/diffused light has not been standardised. There is lot of scope for improving cardamom based agroforestry system as it has got very good export market.

Tea, coffee and cardamom are very important plantation crops which require diffused light (shade loving plants) for their quality production. There is very good scope for identifying the shade trees and also the quantum of diffused light required for these plantation crops.

It is a common agroforestry practice by the farmers to grow or allow MPTS on the boundaries of the field in almost all the agro-ecological regions of our country. The species and also their intensity differs in different regions. In arid regions, common tree/shrub species are *Prosopis cineraria, Zizyphus* species, *Acacia nilotica, Azadirachta indica, Prosopis juliflora, Ailanthus excelsa, Tecomella undulata, Eucalyptus camadulensis, Acacia senegal, Capparis decidua* etc. In semi-arid region many species are grown viz. *Acacia nilotica, A.cupressiformis* (Ramkanti), *Zizyphus mauritiana, Dalbergia sissoo, Leucaena leucocephala, Azadirachta indica, Albizia* species, *Eucalyptus tereticornis, Anogeissus* species, *Casuarina equisetifolia, Madhuca latifolia, Melia azedarach, Hardwickia binata, Sesbania* spp., *Glyricidia sepium, Borassus flabellifer, Phonix sylvestris* etc.

For hill region at an elevation from 500 to 2300 m having subtropical to temperate climate the most common species grown around the field boundary are *Grewia oppositifolia, Celtis autralis, Morus serrata, Toona ciliata, Alnus nepalensis, Ficus* species, *Schima wallichii, Bauthinea* species, *Eudcalyptus globulus, Pinus khasea* etc.

For humid and sub-humid regions the common spcies grown in the field boundaries are *Acacia nilotica, Albizia stipulata* and other *Albizia* species, *Leucaena leucocephala, Sesbania* species, *Casuarina equisetifolia, Bombax ceiba, Albizia falcateria, Artocarpus heterophyllus, Lanea coromondelica, Glyricidea sepium, Gravellia robusta, Moringa oleifera, Erythrina lithosperma* etc. Although farmers grow or allow some of these species on the field boundaries for multipurpose uses but there is very little information on the effect of these trees on the crops or biomass production of these trees. Some of the fast growing species like *Leucaena leucocephala, Acacia auriculiformis, Albizia falcataria, Eucalyptus tereticornis, Sesbania sesban* etc. to which more attention is paid to study their introduction and growth on the production of crops. Except *Eucalyptus*, other species are leguminous and nitrogen fixing. *Leucaena, Albizia* and *Sesbania* species have good fodder value.

These trees besides being grown for specific purpose also serve as a demarcation of the field and also serve as wind breaks and shelter belts. *Leucaena leucocephala* is widely used both in the field boundaries and also inside the field in rows or in alleys. In Bundelkhand *Leucaena* grown in the field boundaries as single row at 2 m apart produce at 4th year 11 t and 0.5 t/row km of dry fuelwood and top feed, respectively (Pathak et al. 1981). In the second cycle (coppice) the yield was 27.4 and 2.5 t/row km, respectively.

In an agrisilvicultural study, Deb Roy and Gill (1990-91) reported maximum fuel and fodder biomass of 7.91 and 1.85 q/ha by annual pruning from *Leucaena* when grown at 10 x 2 m spacing at NRCAF, Jhansi. Under sorghum, maximum grain production (28.34 q/ha) was also recorded with *Leucaena*. In the second year maximum ralative grain yield was 78.8 per cent which was next only to *Casuarina* with 81.6 per cent yield (Deb Roy and Gill 1991). It has been observed that by manipulating the canopy through pruning and thinning, crop yield reduction could be further minimised and at the same time nutritious leaf fodder and fuelwood requirement could be met.

Acacia nilotica and *A. cupressiformis* are grown or allowed to grow in the field boundaries as well as inside the agricultural field especially the latter species because of its special upright branching habit. Both these species are pruned every alternate year to have leaf fodder for one year and pod production in the alternate year for fodder purpose. Nambiar (1990) reported the single row boundary planting of *L.leucocephala* and *E.tereticornis* at Vasad resulted in reduction of tobacco and pearl millet yield to the tune of 37 and 13.8 per cent respectively compared to 38 and 36 per cent by the latter tree species for 5 year average. The farmers inspite of the higher loss in crops preferred *Eucalyptus* because of its good monitory return by selling the poles. In view of good fodder value and improved soil fertility by *Leucaena*, farmers are gradually turning to grow *Leucaena* instead of *Eucalyptus* under various agroforestry systems including boundary planting.

Sesbania sesban is another fast growing nitrogen fixing fuel- cum-fodder tree which has got high potential for growing in the field boundaries. Gupta *et al.* (1983) reported biomass production of single row 2 m apart boundary planting of *S.sesban* varying from 6 to 17.7 kg/tree in different habitats. Studies in progress at NRCAF, Jhansi revealed that this tree can be planted in the field boundary 2 m apart in single row to provide quick fuel and fodder production besides acting as a wind break to protect the crops especially fruit crops from hot as well as cold wind. Under partially irrigated field it could be harvested after third/fourth year to produce biomass varying from 15 to 40 kg/tree. *Sesbania* is a poor coppicer compared to *Leucaena* especially when cut in 3rd or 4th year.

In arid region, *Prosopis cineraria* is grown both on the field boundary as well as in the agricultural field. Each and every part of this tree is utilised by the people to meet their demand (Mann and Saxena 1980). Trees are lopped in winter and a moderate sized tree produces 25 to

30 kg of dry leaves per tree for use by livestock and about 5 kg dry pods for human consumption.

Grassland with Trees

Livestock is an integral part of Indian agriculture. We have got a huge livestock population (465 m) and most of the livestock depend on grazing in the village grazing lands or in the adjoining forest land with scattered trees. In dryland agriculture zone where only one crop is taken, livestock are also allowed to graze in the agricultural field after the harvesting of the crop. There are five grass covers of India (Dabadghao and Shankarnarayan 1973) and by virtue of their agro- ecological and edaphic conditions trees are also available in the grassland some of which are often lopped and given to the animals. Out of five grass covers, two namely *Sehima-Dichanthium* and *Lasiurus-Cenchrus* are two main grass covers which contribute significantly to livestock production in our country. As trees are an inevitable component of the grasslands, similarly grasses are also invariably found in the forest. Arid and semi-arid regions of our country occupy 54 per cent of our total geographical area and 72 per cent of the total cropped area (Deb Roy 1990). Livestock rearing is one of the major subsidiary occupation which imparts stability to the farmers income. Silvipastoral systems for arid and semi-arid areas and their problems and prospects have been discussed in details recently by Deb Roy (1990).

In arid regions, silvipasture is the most important land use system. *Prosopis cineraria* and *Zizyphus nummularia* are grown both in association with crops as well as on field boundary. Gupta and Mohan (1982) worked out the economics of trees vs. annual crops for marginal agricultural land and recommended that silvipastoral system is a most economic system in arid Rajasthan. Muthanna *et al.* 1985 and Shankarnarayan *et al.* (1987) reported that when *Acacia tortilis* is grown in association with *Cenchrus* the growth of the trees was effected due to grass in the initial 2 to 3 years and thereafter there was better growth with grasses. They reported better return from silvipastoral system of *A.tortilis* grown at 10 x 5 m spacing with *C.ciliaris* compared to growing pure tree or pure grass. Muthana *et al.* (1980) also reported that the grass yield of *C.ciliaris* was not effected when grown in association with *A.tortilis* at 5 x 5 m spacing and got an average forage yield of 8.8 q/ha/annum during 1979-80 but the better forage production was obtained by them in association with *Hardwickia binata* (9.8 q/ha) and *Colospermum mopane* (8.7 q/ha) at wider spacing of 10 x 10 m. Deb Roy (1990) pointed out that the growth of the tree seedlings in the initial years could be enhanced by selecting one year old tree seedlings and sowing of grasses atleast 50 cm away from the tree Seedlings. He pointed out that in addition to indigenous trees like *P.cineraria*, *Acacia senegal*, *Z.nummularia* and *T.undulata*, the introduced trees and shrubs like *A.tortilis*, *Dichrostachys cineraria* and *Colospermum mopane* in the silvipastoral system showed very good promise. Both *Dichrostachys cineraria* and *C.mopane* species has very good self regenerating capacity through root sucker (*D.cineraria*) and through self sowing seed (*C.mopane*). The leaves as well as the pods of *D.cineraria* are highly palatable to the livestock especially goat and sheep. The most important perennial grasses of the arid tract are *L.sindicus*, *Panicum turgidum* and *C.Ciliaris*.

The grass production under about 15 years old plantation of 4 desert trees was reported by Ahuja *et al.* (1978) and found that over a period of 4 years green forage yield was maximum (1.54 t/ha) in *P.cineraria* and minimum (0.69 t/ha) in *A.senegal*. Deb Roy and Rai (1990) and Deb Roy et al. (1983) have discussed in details the top feed resources and their management for sheep and goat for arid and semi-arid areas.

As there are good population of *Zizyphus nummularia* (Bordi) in arid and semi-arid regions, it could be allowed to grow alongwith *C.ciliaris* and *L.sindicus* both under silvipastoral system as well as under horti-silvipastoral system. By *in situ* budding of some of the *Z.nummularia* plants with improved varieties of *Z.mauritiana* var. Gola, Seb etc. (developed by CAZRI, Jodhpur) which could be an efficient horti-silvipastoral system for improving the production of fruit and pasture and livestock besides, combating the desertification process.

For semi-arid regions, lot of work has been done on various aspects of silvipastoral system. While reviewing the silvipastoral research and development of the country Deb Roy and Pathak, (1983) emphasised the need for further research on this system and monitoring various aspects associated with this system for long term studies. Deb Roy et al. (1980) reported that in association with *A.tortilis, Leucaena leucocephala, C.ciliaris* could produce 3.5 t/ha/annum of dry forage without significantly affecting the growth of the tree species. They have also reported wood production of 7 to 8 t/ha/annum of *A.tortilis* at a rotation of 13 years in the silvipastoral system (Deb Roy *et al.* 1982).

Biomass production potential of *Albizia lebbek* and *Albizia procera* under silvipastoral system was discussed in detail by Deb Roy (1988a and 1988b). The mean woody biomass at 9th year of *A.procera* was 150.26 kg/tree or 4.67 t/ha/annum and the total aerial biomass was 173.86 kg/tree (5.58 t/ha/annum) in semi-arid Bundelkhand region. The forage production from this system varies from 4 to 6 t/ha/annum depending on the rainfall. Deb Roy *et al.* (1987) reported biomass production of *H.binata* and *A.amara* from silvipastoral system in Bundelkhand region. In addition to getting 3 to 5 t/ha/annum of dry forage it was possible to get aerial biomass production of 66 and 52 kg, respectively at 4 x 4 m spacing. They have also reported higher forage production under *H.binata* compared to under *A.amara* because of the inherent branching habit from the base of the latter species thereby obstructing light.

Under ravinous situation around Agra, *C.ciliaris* and *Dichanthium annulatum* have been successfully grown under the canopy of *Acacia nilotica* giving a forage production of 2.5 t/ha/annum (Prajapati 1979). In an experiment at Dehradun *Chrysopogon fulvus* was grown in association with *Eucalyptus* hybrid and it has been reported that the grass production for a period of 4 years varied from 3.4 to 3.9 t/ha/yr. at 5 x 5 m spacing. It has also been stated that in association with *Dalbergia sissoo* the *C.fulvus* gave much higher production (Puri and Joshi 1979). From alkali soil, the fresh forage production of Karnal grass (*Diplachne fusca*) in association with *A.nilotica* varied from 7.7 to 16.5 t/ha (Grewal and Abrol 1987).

The detailed management strategy of silvipastoral system for production of sheep and goat was discussed recently by Deb Roy (1989) and Bhattacharyya and Sharma (1989). Recently Rai and Rao (1990-91) developed an improved silvipastoral system with woody component of *Albizia amara, D.cineraria* and *L.leucocephala* in association with *C.fulvus* and a mixture of *Stylosanthes hamata* and *S.scabra* giving a forage production of 6 to 8 t/ha and introducing small livestock like sheep and goat in the second year. It has been observed that the livestock could be maintained in the silvipastoral system even during the dry summer months without giving any additional concentrate.

Although there are some information available on the livestock performance on some of the silvipastoral systems but there is an ample scope for improving the management of the silvipastoral system and evaluating them through livestock grazing. Contribution of fodder trees in the system is of paramount importance and accordingly lopping studies and the contribution of leaf fodder and fuelwood production of the associated fodder trees needs to be standardised. Information on the lopping studies are very meagre (Roy, and Deb Roy, 1983; 1986; Srivastava, 1988; Deb Roy *et al.* 1982 and Bhimaya et al. 1964). Because of the dearth of moisture growth of pasture during summer months is restricted. It is very important to increase the leaf fodder production potential of the woody component of the silvipastoral system so that the system could become productive and popular among the farmers.

The major portion of our vast wasted land could be reclaimed by developing the silvipastoral systems which will improve these lands and ameliorate soil and reduce loss of huge quantity of soil and water which could be recycled for better phytomass and livestock production. The fuelwood produced from the system will be able to partially meet the fuel requirement of the farmers thereby helping them to divert the major portion of the cowdung to the agricultural field for increasing crop productivity. It has been pointed out by Hari Kant (1981) that by diverting 1 ton of cowdung the grain production could be increased by 50 kg. As such improved silvipastoral system should be well demonstrated and popularised amongst the farmers.

REFERENCES

Anonymous 1982. Shifting cultivation in north eastern regions. *North Eastern Council Publication No. 17*, pp. 1-24, Shillong, India.

Ahuja, L.D., Verma C.M., Sharma S.K. and Lamba, T.R. 1978. Range management studies on the contribution of ground storey (grass) in afforested areas in arid region. *Ann. Arid Zone* 17(3): 304-331.

Bavappa, K.V.A. 1980. Plantation crops - Arecanut. In : *Hand Book of Agriculture*. pp. 865-920, ICAR, New Delhi, India.

Bane, J.G., Beall H.W.and Cote, A. 1977. Trees, Food and People, IDRC, Ottowa.

Bhattacharyya, N.K. and Sharma, K. 1989. Forage for goat production. *Nat. Symp. on Strategy for Forage Production and Improvement by 2000 AD.* KAU, Vellayani, Trivandrum, Sept. 21-23.

Bhimaya, O.P., Kaul, R.N. and Ganguli, B.N. 1964. Studies on lopping intensities of *Prosopis cineraria*. *Ind. Forester*, 90:19-23.

Blandford, H.R. 1958. Highlights of 100 years of forestry in Burma. *Empire Forestry Review*, 37(1):33-42.

Borthakur, D.N., Prasad, R.N. Ghosh, Singh S.P., A. Awasthi, Rai R.P., Verma R.N. A., Datta H.H., Sachan J.N., and Singh, M.D. 1981. Agroforestry based farming system as an alternative to jhumming. *Proc. of the Agroforestry Seminar*, ICAR, New Delhi. pp. 109-131.

Cambe,J. 1982. Agroforestry techniques in tropical countries - potential and limitations. *Agroforestry System* 1:13-28.

Cardamom Board 1986. *Cardamom Cultivation*. Cardamom Board, Cochin, India, 38 p.

C.C.R.Station 1980. Plantation Crops - Coffee. In: *Hand Book of Agriculture*, ICAR, New Delhi, pp.871-886.

Dabadghao, P.M. and Shankarnarayan K.A. 1973. *Grass Cover of India*, ICAR, New Delhi.

Deb Roy, R. 1988a. Biomass production potential of *Albizia lebbek* under silvipastoral system. In: *Advances in Forestry Research in India*, (Ram Prakash), Intl. Book Distributors, Dehradun, Vol. II: 33-46.

Deb Roy, R, 1988b. Tree growth and aerial biomass production of *Albizia procera* under silvipastoral system. In: *Rangeland Resources and Management*. (Panjab Singh and P.S.Pathak) (Eds.), RMSI, IGFRI, Jhansi.

Deb Roy, R. 1989. Basic research needs in Agroforestry - Indian context. *Nat. Sym. Agroforestry Systems in India*, CRIDA, Hyderabad.

Deb Roy, R. 1990. Silvipasture in arid and semi-arid tracts of India. *Symp. Agroforestry systems and their Management*, GAU, Navsari, Jan. 8-11.

Deb Roy, R. 1992. Role of MPTS in agroforestry systems. In: *Multipurpose Tree Species for Agroforestry systems*. P.S.Pathak, R. Deb Roy, and Panjab Singh, (Eds.), RMSI, IGFRI, Jhansi.

Deb Roy, R., Patil, B.D. and Pathak P.S., 1980. Forage production of *Cenchrus ciliaris* and *C.setigerus* under silvipastoral system of management. *Ind. J. Range Mgmt*. 1(2):113-119.

Deb Roy, R., Patil, B.D. Pathak, P.S. and S.K.Gupta 1982. Growth of *A.tortilis* under silvipastoral system. Ind. J.Range Mgmt. 3(1 & 2): 45-52.

Deb Roy, R., Pathak, P.S. and M.M. Roy, 1983. Lopping of fodder trees-a scientific approach. In: *Proc. Top Feed Resources, their Production, Utilisation and the constraints*. Manohar Singh) (Ed.), CSWRI, Avikanagar.

Deb Roy, R. and Pathak, P.S. 1983. Silvipastoral research and development in India. *Ind. Rev. Life Sci.* 3: 247-264.

Deb Roy, R., Pathak, P.S. and Gupta, S.K. 1987. Production potential of wasteland through silvipastoral system. *Annual* Rep., IGFRI, Jhansi.

Deb Roy, R. and Rai, P.1990. Development of top feed resources and their management for sheep and goat. *Nat. Sem. Sheep and Goat Nutrition and Pasture Legumes*, CSWRI, Avikanagar.

Deb Roy, R. and .Gill, A.S 1990, Agroforestry Research in India. *Int. Symp. on Natural Resources Management for a Sustainable Agriculture*, New Delhi, Feb. 6-10.

Deb Roy, R. and Gill, A.S. 1991-92. Agri-silvicultural studies. *Annual Report*, NRCAF, Jhansi.

Deb Roy, R. and Gill, A.S. 1991. Tree growth and crop production under agri-silvicultural system. *Ind.J. Range Mgmt. & Agroforestry* 12(1):69-78.

Deb Roy, R. and Gill, A.S. 1991-92. *Annual Report*, NRCAF, Jhansi (Unpublished).

Deb Roy, R. and Gill, A.S.1992. *Agroforestry Newsletter*. NRCAF, Jhansi.

Deb Roy, R. and GIll, A.S. 1992-93. Annual Report, NRCAF, Jhansi (Unpublished).

Gill, A.S. and Deb Roy, R. 1990-91. *Annual Report*, NRCAF, Jhansi.

Gill, A.S. and Deb Roy, R. 1991-92. *Annual Report*, NRCAF, Jhansi.

Gill, A.S., Deb Roy, R. and Bajpai, C.K. 1991. Performance of *Leucaena* in agri-silvi-horticultural system. *Leucaena Res. Rep.* 12:34-36.

Gill, A.S., Deb Roy, R. and Bajpai, C.K. 1992. Growth of *Leucaena leucocephala* in a mango based intercropped agroforestry system. *Leucaena Res. Rep. 13:22-23.*

Grewal, S.S. and Abrol, I.P. 1987. Effect of rain water conservation in alkali soil on trees and grass growth and soil amelioration. In : *Afforestation of salt Affected soils.* Vol. 3:87-104.

Gupta, P.N. 1982. Export potential in large cardamom. *Indian Horticulture*, 26:21-25.

Gupta, S.K., Pathak, P.S. and Deb Roy, R. 1983. Biomass production of *Sesbania sesban* on different habitats. *Myforest* 19(3):145-155.

Gupta, T. and Mohan, D. 1982. *Economics of trees* vs. *Annual Crops on marginal Agricultural Land.* IBH Pub. Co., Oxford, New Delhi.

Hailey, Lord 1957. *An African Survey*, OUP oxford.

Hari Kant 1981. Forest - a renewable energy sources of forests. *Ind. J. Range Mgmt.* 2(1 & 2):57-64.

Ingty, P.S. and Goswami, N. 1981. Shifting cultivation in the north eastern hill- a new approach to the problem. In: *Proc. of the Agroforestry Seminar, ICAR, New Delhi.* pp. 138-146.

Kannan, R. and K.P.P.Nambiar 1976. Studies on intercropping coconut garden with annual crops. *Coconut Bulletin*, 5:1-3.

King, K.F.S. 1987. History of Agroforestry. *In: Agroforestry : A Decade of Development.* H.A.Steppler & P.K.R.Nair (eds.), ICRAF, Nairobi.

King, I.F.S. and Chancler, M.T. 1978. *The Wasted Land*, ICRAF, Nairobi.

Kushwaha, B.L., Nelliath, E.B. Marcose, B.T. and Sunny, A.S. 1973. Rooting pattern of coconut. *Ind. J. Agronomy*, 18:71-74.

Lundgren, B.O. and Raintree, J.B. 1982. Sustained agroforestry. In: *Agricultural Research and development - Potential and challenges in Asia*, Bnestel (ed.), ISNAR, The Hague, pp. 37-49.

MacDicken, K.G. and Vergera, N.T. 1990. Introduction to Agroforestry. In: *Agroforestry Classification and Management.* (Eds. MacDicken and Vergera), John Willey & Sons, New York/Singapore, pp. 1-30.

Mann, H.S. and Saxena, S.K. 1980. *Khejri* (*Prosopis cineraria*) in the Indian Desert. Monograph No. 11, CAZRI, Jodhpur, India, 78. pp.

Menon, K.P.B. and K.M.Pandalai 1960. T*he Coconut Palm - A Monogram.* Central Coconut Committee, Ernakulam, India.

Mishra, B.K. and Ramakrishnan, P.S. 1981. The economic yield and energy efficiency of hill ecosystems at higher elevation of Meghalaya in North Eastern India. Acta Oecologica, *Oecol Applic.* 2: 369-389.

MOA 1983. *Statewise Area and Production of Important fruits in India from 1976-77 to 1978-79 and Projection upto 1982-83* Ministry of Agriculture, New Delhi. (Memio).

Muthana, K.D., Sharma, S.K. Raina, A.K. and Meena G.L. 1980. Silvipastoral studies. *Annual Report*, CAZRI, Jodhpur.

Muthana, K.D., Sharma, S.K. and Harsh L.N.1985. Studies of silvipastoral system in arid zone. *Myforest* 21(3): 233-238.

Nair, K.K. 1981. Agroforestry in the Southern hills. In: *Proc. Agroforestry Seminar*, ICAR, New Delhi, pp. 234-241,

Nair, P.K.R. 1983. Agroforestry with coconuts and other tropical plantation crops. In: *Plant Research and Agroforestry*, (Ed. Huxley, P.A.), pp. 79-102, ICRAF, Nairobi.

Nair, P.K.R. 1985. Classification of Agroforestry System. *Agroforestry System* 3: 97-128.

Nair, P.K.R. 1987. Agroforestry and biomass energy firewood production. In: *Agroforestry Systems in the Tropics.* (Ed. P.K.R.Nair), Kluwer Academic Publishers in Cooperation with ICRAF, pp. 591-597.

Nair, P.K.R. 1989. Agroforestry defined. In: *Agroforestry Systems in the tropics.* (Ed. P.K.R.Nair), Kluwer Academic Publisher in Cooperation with ICRAF.

Nair, M.A. and C.Sreedharan 1986. Agroforestry farming planting of *Eucalyptus tereticornis* and *Leucaena leucocephala* in the tobacco-bajra cropping system at *Vasad. Agroforestry Newsletter* 2(1):3.

Nelliatr, E.B. and Bhat, E.S. 1979. Multiple cropping coconut and arecanut garden.*Tech. Bulletin* No.3, CPCRI, Kasargod, India.

Pathak, P.S., Gupta, S.K. and Deb Roy, R. 1981. Production of aerial biomass in *Leucaena leucocephala. Indian For.* 107 (7) : 416-419.

Prajapati, R.C. 1979. Silvipastoral system of productions from ravines. *Proc. Summer Instt. on Silvipastoral production,* IGFRI, Jhansi, pp. 28-40.

Puri, D.N. and P. Joshi 1979. Economic utilization of class V, VI and VII lands for raising fuel and fodder. *Annual Report,* CS & WC & RTI, Dehradun.

Raghavan, M.S. 1960. Genesis and history of cumri system of cultivation. In : *Proceeding of the 9th Silviculture Conference, Deadun,* F.R.I. Dehradum, India.

Rai, P. and *Rao, G.R. 1990-91. Annual Report, NRCAF,* Jhansi,

Reddy, C.K. 1981. Agroforestry in coastal Andhra Pradesh. In : *Proceeding Agroforestry Seminar. ICAR, New Delhi.* pp. 77-82

Retinam, P. 1991. Planatation crop based agroforestry systems for humid.tropics. *Agroforestry Newsletter,* NRCAF, Jhansi, Vol. 3 No. 1.

Roy, M.M. and Deb Roy, R. 1983. Lopping studies of *Bauhinea purpurea. Myforest* 19 : 193-196.

Roy, M.M. and Deb Roy, R. 1986. Leaf fodder and fuelwood production from *Albizia amara* on wasteland through lopping management. *Nat. Sem. Silvipastoal System - land use for Wasteland,* IGFRI, Jhansi, Nov. 22-26.

Seth, S.K. 1981. *India and Sri Lanka Agroforestry Report for local Community Development Programme.* UN/FAO Rome, Italy.

Shankarnrayan, K.A., Harsh, L.N. and Kathju, S. 1987. Agroforestry in the arid zone of India. *Agroforestry Systems* 5(8) : 69-88.

Sharma B.S. 1981. Shiftying cultivation in Manipur and Nagaland. In : *Proceeding Agroforestry Seminar.* ICAR, New Delhi. pp. 132-137.

Sharma, T.C. 1980. Pre-history document of shifting cultivation. In : *Shifting cultivation in north east India.* Shillong, India. 1-24. pp.

Singh, A and M.D. Singh 1981. Soil earosion hazard in north eastern hill region. Bulletin No. 10, ICAR Research Complex, North Eastern Hill Region Shillong, India, pp. 20.

Singh, K.A., Rai, R.N. Patiram and Bhutia, D.T. 1989. Large cardamom (*Ammomum subulatum*)- an age old agroforestry system in Himalayas. *Agroforestry System* 9 : 1241-1261.

Singh, Punjab 1992. Multipurpose Trees for agroforestry systems. In : *Multipurpose tree species for Agroforestry Systems.* (Eds. P.S. Pathak, R.Deb Roy and Panjab Singh), RMSI, IGFRI, Jhansi.

Singh, R.K., N. Yadukumar; Roy, K.N. Burman and Roy, A.C. 1982. Intercropping in arecanut garden in north Bengal. *Ind. Farming* 32 : 13-15.

Srivastava, J.P. 1988. Lopping studies of *Prosopis cineraria. Ind. Forester,* 104 : 269-274.

Tejwani, K.G. 1987. Agroforestry practices and research in India. In : *Agroforestry Realities, Possibilities and Potential.* (Ed. H.L.Gholz), Martinus Nijhoss Publishers, Lancaster/Boston/Dordrecht.

Upadhyay, R.C. and S.P. Ghosh 1983. Wild cardamom of Arunachal Pradesh. *Indian Horticulture* 27 : 25-27.

Varghese, P.T. et al. 1978a. Beneficial interaction of coconut- cocoa crop combination. In : Proceedings *Plantation Crop Symposia.* pp. 383-392.

Varghese, P.T. et al. 1978b. Intercropping with tuber crops in cocunut gardens, *Ibid,* pp. 399-405.

Venkataramani, K.S. and Venkatram, C.S. 1987. Plantation crops - Tea. In : *Hand Book of Agriculture,* ICAR New Delhi. pp. 865-870.

Wight, W. 1959. The shade tree tradition in the tea gardens of north India value of shade. *Report Indian Tea Association,* Tocklai Expt. Station, Jorhat, pp. 75-98.

26

MANAGEMENT OF PASTURE LANDS IN THE ARID REGIONS.

K.C. SINGH

INTRODUCTION

Crop production in the hot desert of Rajasthan is unstable due to unfavourable climatic conditions. Pasture development for livestock raising has a very important role to play in the economy of this region. But due to heavy pressure of livestock on pasture lands the productivity has decreased. In order to evolve scientific technology for the improvement of arid pasture lands, studies have been conducted at different rangelands on various aspects viz., moisture conservation, methods of reseeding, fertilizer use, grass-legume mixture, silvi- pastures, agri-pastures and utilization of pastures. The results of these studies are discussed in this chapter.

MOISTURE CONSERVATION

Contour furrows (60 cm wide and 22 cm deep) were constructed on different rangelands at a distance of 8 to 10 m across the slope of land. Reseeding of perennial grass species was done in furrowed and non-furrowed plots. Results revealed that contour furrowing increased forage production of perennial grass species by 130.6 per cent over non-furrowed plots. The herbage yield of annual species was not influenced significantly by contour furrowing compared to control (non-furrowed plot).

Inter-row water harvesting (IRWH) system in which 30 cm wide ditch alternated with 70 cm wide raised bed was tried on different rangelands. Two rows of pasture legumes or grasses were sown on the edges of the ditch. IRWH system increased fodder yield of legumes and grasses by 124 and 66 per cent, respectively over conventional system of planting.

IMPROVED GRASSES

Lasiurus sindicus produced high forage yield (4 t/ha) on sandy soils with annual rainfall of 250 mm and below. *L.sindicus* also showed good performance on sand dunes. *Cenchrus ciliaris* and *Cenchrus setigerus* gave forage yield 2 to 3 t/ha on light to medium soils with annual rainfall 300 mm and above. *Dichanthium annulatum* gave high yield (3.3 t/ha) on heavy textured soil and annual rainfall above 400 mm. *Sporobolus* performed well (1.5 t/ha) on saline soils with rainfall of 300 mm and above, while *Panicum antidotale* showed good performance on sandy and sandy loam soils. Some improved strains for high productivity have been identified and these are CAZRI Nos. 75 (*Marwar Anjan*), 357 and 358 of *Cenchrus ciliaris*, CAZARI Nos. 76 (*Marwar Dhaman*), 175 and 296 of *Cenchrus setigerus*. These strains have shown both stability of production as well as better persistence over the years at different locations in arid zone. The crude protein content of perennial grasses varied from 2.8 to 9.2 per cent in different seasons of the year and it gradually decreased from September onwards (Chakravarty *et al.* 1970).

RESEEDING

Due to poor germination of seeds of *Lasiurus sindicus* and *Dichanthium annulatum*, the sowing of the mixture of seeds of *Cenchrus* species and

Lasiurus sindicus and of *Cenchrus* species and *Dichanthium annulatum* was found advantageous for large scale development of pasture in arid region of Rajasthan.

The establishment of *Lasiurus sindicus* (Fig. 3) and *Dichanthium annulatum* was found more assured by transplanting of rooted slips or seedlings compared to direct seeding.

In order to have better establishment of *L.sindicus, D.annulatum* and *Cenchrus* Species pastures on large scale, the seeds, should be pelletted. Pellets are prepared by mixing grass seed with cow dung, clay and sand in proportion of 1:1:3:1 using sufficient quantity of water for preparing round pellets of the size of about 0.5 cm diameter.

Cenchrus ciliaris seed sown in dry soil before onset of rain gave 36 per cent higher forage yield over monsoon sowing. Seeds of *Cenchrus* species should be mixed with moist soil three to four times the volume of seed and drilled uniformly in lines 50 cm apart in 8 to 10 cm deep furrows with a soil cover of less than 2 cm. Such a sown pasture (Fig. 4) needs 1 to 2 weedings during first year of its establishment (Singh, 1988).

The increase in forage production due to reseedig under different grasslands were from 30 to 122 per cent in 'Poor' and 29 to 107 per cent in 'fair' grasslands. The increase in forage production due to reseeding at Bhopalgarh rangeland was 47.2 per cent.

FERTILIZER USE IN PASSTURE GRASSES

Application of fertilizer to pasture grasses increased forage as well as seed yield (Singh and Singh, 1973). Application of 20 kg N/ha to *Cenchrus ciliaris* (358) pasture resulted in 83.1 per cent increase in dry forage yield and 112.6 per cent in seed yield. Application of phosphorus did not increase the yield of grass (Figs. 1 & 2). The protein content of fertilized pastures was higher than that of unfertilized ones.

GRASS LEGUME MIXTURE

Introduction of *Stylosanthes hamata* after light soil working in the established pasture of *Dichanthium annulatum* proved beneficial in increasing productivity of grassland in rainfall zone of more than 400 mm. The forage yield of *Stylosanthes hamata* obtained from pure and mixed stands was 14.1 and 8.6 q/ha, respectively. Yield of *D.annulatum* from pure stand was 33.1 q/ha compared to 33.3 q/ha from mixed stand with *Stylosanthes hamata* and therefore total productivity of *D.annulatum + S.hamata* was highest (42.0 q/ha). The crude protein content in *S.hamata* was the highest i.e. 12.53 per cent in pure stands followed by 11.42 per cent in mixed stands. Crude protein in *D.annulatum* was 4.18 per cent in association with *S.hamata* and 3.81 per cent under pure stand (Shankarnarayan & Singh, 1990).

AGRI-PASTURES

Studies were conducted at Central Arid Zone Research Instittute (CAZRI), Jodhpur on agri-pastoral approach i.e. cropping between grass strips laid out against the prevalent wind direction. (Fig. 5.) In a study it was revealed that the average yields of *Vigna radiata* (Mung) and *Vigna aconitifolia* (Moth) were 272 and 257 kg/ha, respectively obtained under unprotected plots increased to 484 and 372 kg/ha, respectively when the cropped strips were protected by perennial protective strips of *Lasiurus sindicus* established at right angles to the general direction of the prevailing winds. In addition to the grain yield in the protected plots, 1650 kg/ha dry forage yield of *L.sindicus* was also obtained (Mishra, 1964).

In another study on agri-pastoral system carried out at CAZRI, Jodhpur and Bikaner, the average seed yield of clusterbean at Jodhpur was 418 kg/ha under unprotected plot, which increased to 503 kg/ha in the protected plot. Out of 6 years (1981-86), clusterbean failed to produce seed in three low rainfall years (1984-86), while grass produced dry forage yield (726 kg/ha). Also, in severe drought year of 1987, clusterbean did not succeed but grass component produced forage. The average yield of *Cenchrus ciliaris* and *Lasiurus sindicus* was 2 and 4 t/ha, respectively. The amount of soil blown during summer months from unprotected plot was higher than that of protected plot.

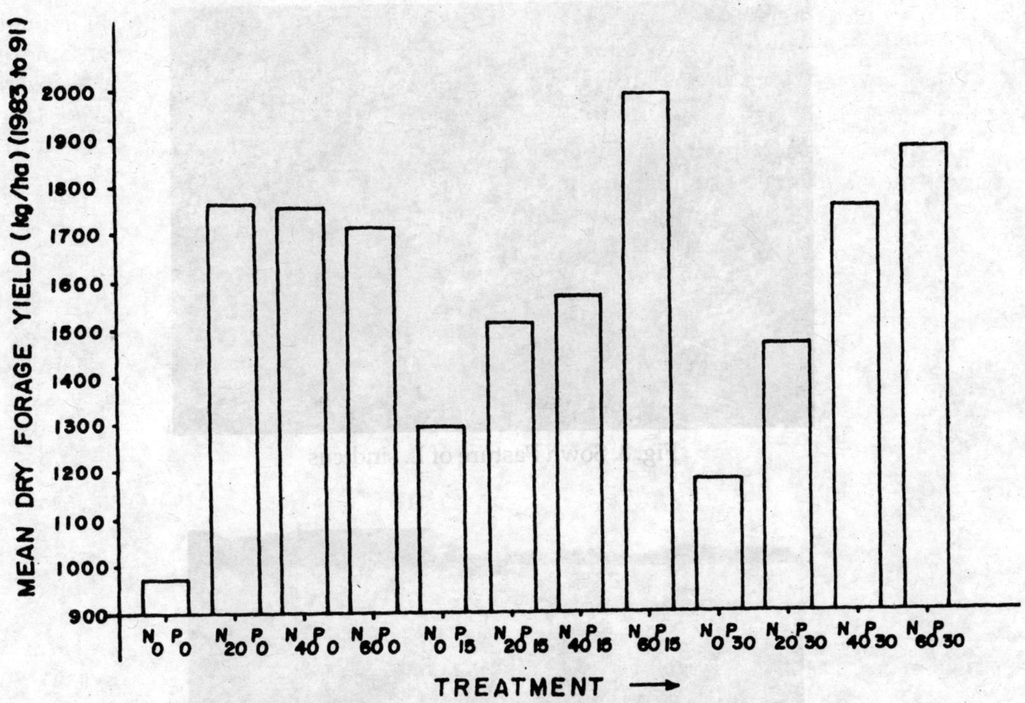

Fig. 1 Effect of fertilizer on dry forage yield of Cenchrus ciliaris (358).

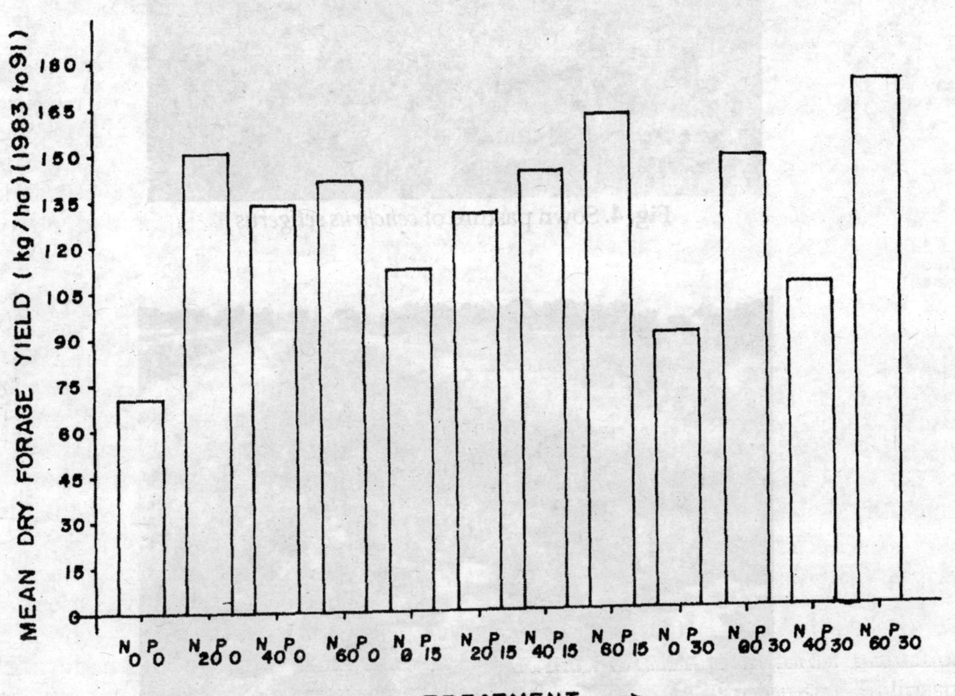

Fig. 2. Effect of fertilizer on Seed yield of cenchrus ciliaris (358).

Fig. 3. Sown Pasture of L. sindicus

Fig. 4. Sown pasture of *cenchrus setigerus*

Fig. 5. Wind strip cropping grass and clusterbean

At Jodhpur, pearl millet was grown in 1988 and 1989 in the plots which were under grass and clusterbean for 7 years (1981-87). The mean grain yield (1988 and 1989) of pearl millet after grass was 8.50 q/ha as compared to 5.65 q/ha after cluster bean (Singh, 1989).

Organic carbon content of crop plots in 0 to 15 and 15 to 30 cm soil layers were 0.16 and 0.14 percent respectively, whereas it was 0.19 and 0.17 per cent respectively under grass plots. Grass plots were also found to contain more soil moisture than that of crop plots.

These results reveal that grass component can create substantial stability to crop production in the arid region.

CHOICE OF TREE SPECIES FOR ARID GRASSLANDS

Prosopis cineraria is important drought hardy top feed tree which grows on pasture lands without any detrimental effect on grass grown in association with it. The tree is lopped in winter for its protein rich (about 14%) leaves.

Zizyphus nummularia, a bush, provides very valuable nutritive leaf forage containing 14 to 16 per cent protein. It grows well on cultivated fields and grasslands. *Acacia tortilis* grows very well on wastelands and is good for fixation of sand dunes. Its pods provide nutritive feed for sheep and goats. *Calligonum polygonoides* is a sand dune fixing plant. Its tender stems provide good top feed for animals. *Acacia senegal* grows well on rocky sites and provides leaves and pods for sheep and goats. *Prosopis juliflora* performs well on all types of soils. It is good for live fencing of pasture lands. It also provides nutritious seed from pods for sheep and goats. *Acacia nilotica* grows well on medium to heavy textured soils in rainfall areas above 400 mm and provides nutritive top feed through its leaves and pods. In extremely dry areas, animals eat the green shoots of *Capparis aphyll*. *Colophospermum mopane*, *Dichrostachys nutan* and *Hardwickia binata* are performing well on arid grasslands.

SILVI-PASTURES

In a study carried out at Jodhpur on forage production under *Prosopis cineraria* trees in sown pasture of *Cenchrus ciliaris*, the forage yield, of grass reduced significantly only upto one metre distance from trunk (Fig. 6). The difference in forage yield from 2 to 5 m distance from trunk and in open area at 15 m distance was found non-significant. The dry forage yield of *C.ciliaris* obtained upto 1 metre distance from trunk was low (11.6 q/ha). But forage yield under tree canopy at 2 to 5 m distance was 27.8 q/ha as compared to 32.8 q/ha from open area at 15 m distance from *Prosopis cineraria* which did not differ significantly. In general, grass seed yield increased with the increase in distance from the tree. Soil moisture and organic carbon content were higher under tree canopy compared to open area (Jain *et al.* 1989).

To evaluate the forage production under different tree species, a study was undertaken in 10 years old tree plantation of *Acacia tortilis, Dischrostachys nutan, Colophospermum mopane* and *Azadirachta indica* at RM&SC (Range Management and Soil conservation) area, Borunda (Jodhpur). There were three rows of each tree species having 14 trees per row at a spacing of 6 x 6 m. *Cenchrus setigerus* was sown in interspaces after establishment of trees.

Results revealed non-significant differences in mean dry forage yield of *Cenchrus setigerus* in association with *C.mopane* (18.43 q/ha); *D.nutan* (20.73 q/ha) (Fig. 7) compared with the yield of pure grassland without tree (22.87 q/ha). But forage yield of grass in association with *A.tortilis* (8.39 q/ha) and *A.indica* (14.96 q/ha) was reduced significantly compared to the grass yields with other tree species and pure grassland. *A.tortilis* showed 97.6 per cent survival, followed by *C.mopane* (90.5%), *A.indica* (39.3%) and *D.nutan* (40.5%). Maximum height was recorded in *A.indica* (425 cm) followed by *A.tortilis* (393 cm), *C.mopane* (301 cm) and *D.nutan* (261 cm) (Table 1).

In another trial conducted at rangeland Bhopalgarh (Jodhpur) involving 9 tree species planted in 1984 at a spacing of 10 m row to row

Table 1. Performance of different tree species at rangeland, Borunda (1990)

Tree species	Survival %	Plant height (cm)	Collar diameter (cm)	Mean forage yield of *C. setigerus* grown with trees (q/ha)
Acacia tortilis	97.6	393 (210 to 565)	11.6 (4.6 to 16.5)	8.39
Dichrostachys nutan	40.5	261 (150 to 300)	6.6 (4.8 to 8.5)	20.73
Colophospermum mopane	90.5	301 (100 to 450)	9.3 (3.0 to 14.0)	18.43
Azadirachta indica	39.3	425 (380 to 510)	16.0 (12.5 to 20.8)	14.96
Control (without tree)	–	–	–	22.87

Year of tree planting 1976.
Figures in parenthesis indicate the range.

and 4 m plant to plant. *Cenchrus ciliaris* was sown in interspaces after establishment of tree species. Data in respect of survival percentage, height and collar diameter of different tree species recorded in 1991 showed 98 per cent survival of *A.tortilis* and *D.nutan*, followed by *A.senegal* (88%), *C.mopane* (82%), *A.indica* (75%) and *Z.nummularia* (65.0%). *A.tortilis* showed maximum height (258.5 cm), followed by *A.indica* (188.8 cm) and *A.senegal* (143.7 cm). *A.tortilis* also showed maximum collar diameter (6.1 cm) (Table 2). The dry forage yield of grass obtained from different plots of tree species did not differ significantly in any year of the experimentation (1984 to 1991). But forage production varied considerably from year to year and was closely related with rainfall. The mean dry forage yield of *Cenchrus ciliaris* (Table 2) was 25.56 q/ha as compared to pure pasture without tree (32.2 q/ha). The annual rainfall received during 1984, 1985, 1986, 1987, 1988, 1989, 1990 and 1991 was 221.9, 188.2, 164.3, 78.3, 127.3, 295.3, 286.7 and 331.3 mm, respectively. There was low and erratic rainfall varying from 78.3 mm in 1987 to 331.3 mm in 1991. About 80 per cent of rainfall was received from July to September (Table 3).

Table 2. Performance of different tree species at rangeland, Bhopalgarh (1991)

Tree species	Survival %	Plant height (cm)	Collar diameter (cm)	Mean forage yield of *C. ciliaris* grown with trees (q/ha)
Prosopis cineraria	10	74.17 (38 to 118)	1.47 (0.2 to 2.6)	24.67
Acacia tortilis	98	258.51 (85 to 670)	6.14 (1.5 to 13.4)	11.83
Albizzia lebbek	5	118.00 (30 to 124)	1.43 (0.1 to 2.4)	36.00
Zizyphus nummularia	65	100.96 (7 to 280)	1.59 (0.1 to 4.8)	27.00
Dichrostachys nutan	98	127.23 (23 to 230)	2.53 (0.1 to 6.4)	23.67
Colophospermum mopane	82	58.45 (21 to 148)	0.64 (0.1 to 4.3)	26.83
Hardwickia binata	53	47.20 (14 to 122)	0.66 (0.1 to 2.6)	26.33
Azadirachta indica	75	188.78 (18 to 390)	3.86 (0.1 to 10.9)	22.17
Acacia senegal	88	143.71 (70 to 215)	4.24 (1.1 to 22.8)	31.50
Control (without tree)	–	–	–	32.16

Year of tree planting 1984
Figures in parenthesis indicate the range.

Fig. 6. Sown pasture of *C. ciliaris* associated with *Prosopis cineraria* trees

Fig. 7. Silvi-pasture *cenchrus* + *C. mopane, D. nutan.*

Table 3. Rainfall (mm) at rangeland, Bhopalgarh (Mean rainfall 368 mm)

Month	1984	1985	1986	1987	1988	1989	1990	1991
Jan.	–	–	–	6.4	4.3	2.4	–	–
Feb.	–	–	–	1.0	–	–	34.6	2.0
Mar.	–	8.2	–	–	0.4	3.6	1.2	–
Apr.	–	9.0	–	–	–	3.0	–	–
May	–	16.0	46.0	22.4	1.8	–	14.0	–
June	9.7	–	–	12.8	33.6	25.6	20.0	40.2
July	54.6	43.0	107.1	19.0	51.4	70.9	116.0	111.4
Aug.	50.8	37.6	11.2	11.7	27.8	183.7	63.7	151.9
Sept.	106.8	64.0	–	–	7.4	6.1	28.2	25.2
Oct.	–	–	–	–	0.6	–	9.0	–
Nov.	–	–	–	–	–	–	–	–
Dec.	–	10.4	–	5.0	–	–	–	0.6
Total	221.9	188.2	164.3	78.3	127.3	295.3	286.7	331.3

Table 4. Forage production under continuous and deferred rotational systems of grazing (dry forage yield, q/ha)

Year	Continuous grazing				Deferred rotational grazing			
	High perennials	Low perennials	Annuals	Total	High perennials	Low perennials	Annuals	Total
1979	3.29	1.74	0.04	5.07	6.10	2.53	0.08	8.71
1980	6.98	5.12	0.15	12.25	10.57	5.46	0.67	16.70
1981	3.00	2.43	0.68	6.11	4.20	4.34	0.61	9.15
1982	7.15	5.40	0.63	13.18	9.86	4.47	0.88	15.21
1983	15.88	13.59	0.94	30.41	21.38	8.46	0.38	30.22
1984	9.40	6.00	2.15	17.55	15.34	5.39	1.55	22.28
Mean	7.62	5.71	0.76	14.09	11.24	5.11	0.70	17.04

The results of these silvi-pasture studies indicated that apart from the benefits to be achieved from the fuel and fodder trees, it is also possible to derive an extra dry forage yield of perennial grasses through proper silvi-pastoral management.

FORAGE PRODUCTION AS INFLUENCED BY GRAZING MANAGEMENT SYSTEMS

An experiment involving two systems of grazing viz., continuous and deferred rotational with sheep and heifers was conducted at different rangelands. Results revealed that forage production increased when the forage was utilized through deferred rotational system compared to continuous system of grazing. The per cent, increase in the yield of high perennials was 47.5 per cent over continuous grazing. There was no significant difference in the yield of annual species under two type of grazing management system (Table 4). In another trial forage production in *L.sindicus* and *Cenchrus* species pastures increased by 62 and 20 per cent, respectively when grasses were harvested from half portion of paddocks for stall feeding compared to grazing alone (Shankarnarayan and Singh, 1990).

REFERENCES

Chakravarty, A.K., Ram Ratan and Singh, K.C. 1970. Utilization of vegetation cover, grazing behaviour of sheep and seasonal variation of crude protein content of plants in different pastures. *Ann. Arid Zone* 9(1) : 10-16.

Jain, B.L., Singh, K.C., Rathore, S.S. and Prasad, R.N. 1989. Influence of tree canopy on characteristics and productivity of loamy sand soils in arid region. *International Symposium on Managing Sandy Soils,* pp. 598-601.

Misra, D.K. 1964. Agronomic investigations in arid zone. *Proceedings* of the *Symposium on Problems of*

Indian Arid Zone, pp. 165-169.

Shankarnarayan, K.A. and Singh, K.C. 1990. Rehabilitation of degraded pasture land in the arid region of Rajasthan. In : *Technologies for* Wasteland Development, (Eds. I.P. Abrol and V.V. Dhruvanaryana), ICAR Publication, pp. 21-31.

Singh, K.C. 1988. Management of arid rangelands of Rajasthan. In : *Wastelands Development and their Utilization*. Scientific Publishers, pp. 287-293.

Singh, K.C. 1989. Importance of grasses in the management of sandy soils - Strip cropping of grasses. *International Symposium on Managing Sandy Soils*, pp. 561-564.

Singh, R.P. and Singh, K.C. 1973. A note on seed production in some promising desert grasses grown on drylands as influenced by fertilizer use. *Ind. J. Agron.* 18(4) :

27

CHANGING SCENARIO OF WEEDS AND THEIR MANAGEMENT IN DRYLAND CROPS AND CROPPING SYSTEMS

A. RAMAKRISHNA and B. TRIPATHI

INTRODUCTION

Weeds are no strangers to man. They have been present ever since crops began to be cultivated around 10,000 BC (Hay,1974). When man first started to grow crops for food and fibre, he soon learned that yields were higher when weeds were removed. Thus, the concept of weed control is as old as agriculture itself. However, the universal occurrence of weeds as constant companion of the agricultural environment as opposed to the epidemic nature of other pests has delayed recognition of the importance of weed science in crop production. From the beginning of agriculture through the middle of the twentieth century, the plow and hoe have been the only widely employed means of weed control.

Of the total losses caused by pests (insects, diseases, mematodes and weeds) in world agriculture, weeds alone account for one-third of the loss, amounting to about 25 billion dollars (Joshi, 1987). Losses due to weeds in India are considerable and are sufficiently alarming as to warrant urgent attention (Mitra, 1962). In 1975 Parker and Fryer reported that about 10 per cent of the losses caused by weeds to the principal agricultural produce amount to approximately Rs. 4200 million per annum. Presently the estimated losses due to weeds in India amount to Rs. 16500 millions (Joshi, 1987).

Approximately, 70 per cent of the area in the country is cultivated under rainfed conditions but these areas contribute only 42 per cent of the total food grains production. Immense potentiality exists for increasing productivity in dryland areas by adapting available improved weed management technology and by developing economically viable technology for managing troublesome weeds in dryland crops and cropping systems. In this chapter, an attempt is made to review the available weed management technology for dryland crops and cropping systems and identify future research priorities.

Weeds of dryland crops and cropping systems

A large part of the loss in agricultural production due to weeds is caused by few weed species. Of the 250,000 plant species that exist, less than 250 have become important weeds (Holm, 1971). Although the number of serious weeds is decreasing steadily as a result of changes in agricultural production, for those that remain, the scope for expansion is increasing. Based on life span, weed species are classified into annuals, biennials and perennials. The greatest difference in growth behaviour between different weed species is between annuals and perennials. Annuals tend to have a fast and vigorous early growth with a high proportion of biomass accumulating in above-ground plant parts. These plants prepare for the generative phase that ends in seed setting. Perennials, on the other hand, invest more of their energy in the growth of underground plant parts, slowly but steadily expanding their vegetative propagation system.

Although the classification of weeds is based on their life span, the duration of a particular weed is at times determined by climatic factors. Many weeds that are annuals or biennials in a harsh climate may act as biennials or perennials in milder climates. Majority of weeds are perennials; a smaller portion are annuals. Biennials comprise only a small percentage of weeds (Rao, 1983). Weed problems in dryland crops and cropping systems vary -- from one vegetation zone to another. Even within the same vegetation zone, variation in weed flora as well as intensity of weed infestation occurs because of differences in cropping systems, intensity of land use, rainfall, soil fertility and method of weed control used in individual crops. Difference in plant growth, architecture, morphology, growth cycle and canopy development influence what weed species will be most competitive in a given crop or cropping system.

Some perennial weeds (eg. *Cyperus rotundus* L., *Cynodon dactylon* (L.) Pers.; *Sorghum halepense* (L.) Pers.) pose a serious problem because they are very competitive and difficult to control when routine crop husbandry practices are employed. The broad leaved perennials (eg. *Convolvulus arvensis* (L.) and *Solanum elaegnifolium* Cav.) are also some of the most pernicious, ubiquitous weeds which are difficult to control. Among the parasitic weeds, *Orobanche* spp. are fairly wide spread in semi-arid regions damaging a few crops including certain legumes, vegetables and tobacco, while *Striga* spp. are restricted to hotter arid regions affecting summer cereals including maize and sorghum apart from sugarcane and sudan grass. *Cuscuta* spp.; a stem parasitic weed, is also becoming a major problem in crops like green gram and black gram grown after rice. It is also posing a serious problem to niger and certain vegetables.

The *Asphodelus tenuifolius* (L.) Cav. has also invaded large areas in the country and has become a major problem in northern belt of the country (more than 50%) where pearl millet is grown. The incident of *Ageratum conyzoides* L. is recently reported to be a problem weed in maize and many other crops of dryland farming. A comprehensive list of major weeds of dryland crops is presented in Table 1.

NATURE OF WEEDS FLORA AND FACTORS AFFECTING

Climatic, edaphic and biotic factors determine the distribution, prevalence, competing ability, behaviour, and survival of weeds. Man also plays an important role in changing the environment by crop husbandry practices.

Climatic Factors

Important climatic factors in the environment that affect weed flora are light, temperature, rainfall and humidity. Light intensity, quality, and duration are important in influencing growth, reproduction, and distribution of weeds. Tolerance to shading is a major adaptation that enables weeds to persist. Atmospheric and soil temperature affects the latitudinal and altitudinal distribution of weeds. Soil temperature affects seed germination and dormancy, which is a major survival mechanism of weeds. Rainfall and water have a significant effect on weed flora. The weed species found in desert conditions are different from those of aquatic environments. Weeds of temperate regions are rarely found in the tropics.

Soil Factors

Soil-water aeration, temperature, pH, fertility level and cropping systems influence the weed flora. Some weed species are characteristically 'alkali' plants known as 'basophiles' (pH range 7.4 to 8.5) and grow well in alkali soils. Alkaligrass (*Puccinalla* spp.) and quackgrass (*Agropyron repens* Beauv.) are the best examples of basophiles. Similarly, *Digitaria sanguinalis* (Retz.) Koeler and *Borreria* spp. inhabit only acidic soils and are termed as 'acidophiles', which have a pH range of 4.5 to 6.5. Many weeds can adapt and grow well in soils having poor fertility. For example, thatch grass, *Imperata cylindrica* Beauv, grows well in low fertility soils and can also adapt well to soils of high fertility.

Table I. Major weeds of major dryland crops and cropping systems

Weeds	Cereals	Millet	Oilseed crops **	Pulses ***	Fiber crop (Cotton)
Ageratum conyzoides L.	X	X	–	X	–
Alysicarpus vaginalis (L.) DC.	X	X	–	–	–
Alysicarpus rugosus (Willd) DC.	X	X	–	–	–
Amaranthus viridis L.	X	X	X	X	X
Amischophacelus axillaris (L.) Rolla Rao & Kamathy	X	X	–	–	X
Aristolochia bracteata (L.) Retz.	X	X	X	X	X
Asphodelus tenuifolius (L.) Cav.	–	–	–	–	X
Brachiaria ssp.	–	X	–	X	X
Celosia argentea L.	X	X	X	X	X
Convolvulus arvensis L.	X	–	–	X	X
Commelina benghalensis L.	X	X	X	–	X
Corchorus trilocularis L.	–	–	X	–	–
Cyanotis cucullata (L.) Kunth	X	X	X	X	X
Cynodon dactylon (L.) Pers.	X	X	X	X	X
Cyperus iria L.	X	X	X	X	X
Cyperus rotundus L.	X	X	X	X	X
Dactyloctenium aegyptim (L.) Willd.	X	X	X	X	X
Desmodium dichtomium (Willd) DC.	X	–	X	–	X
Digera muricata (L.) Mart.	X	–	–	X	X
Digitaria ciliaris (Retz.) Koel.	X	–	X	X	X
Echinochloa crusgalli (L.) P. Beauv.	X	X	–	–	–
Echinochloa colona (L.) Link.	X	X	–	–	X
Eclipta prostrata (L.) L.	X	–	X	X	–
Eleusine indica (L.) Gaertn.	X	X	X	X	X
Eragrostis spp.	X	–	–	–	X
Euphorbia spp.	–	–	X	X	X
Hedyotis corymbosa (L.) Lam.	X	X	X	–	–
Mollugo pentaphylla L.	X	X	–	–	–
Panicum spp.	X	X	X	X	X
Pennisetum glaucum (L.) R.Br.	X	–	X	X	X
Rottboellia cochinchinensis (Lour) W.D.	X	–	–	–	–
Saccharum spontaneum L.	X	X	–	–	X
Sorghum halepense (L.) Pers.	X	X	X	X	X
Sonchus arvensis L.	–	–	X	X	X
Trianthema portulacastrum L.	X	X	X	–	X
Trichodesma indicum (L.) R. Br.	X	X	X	X	X
Tridax procumbens L.	X	X	X	X	–
Vicia sativa L.	–	–	X	X	–
Xanthium strumarium L.	X	–	X	X	X

* Cereals represented by rice (up land), maize and sorghum.
** Oilseed crops include groundnut, sunflower, safflower, castor, sesamum.
*** Pulse crops comprise of soyabean, mungbean, blackgram, chickpea, pigeonpea.
Source : Annual Report, AICRPWC, (1988) and Shetty et. al., (1983).

Similarly, *Commelina benghalensis* L. thrives in both moist and dry soil conditions.

Biotic Factors

Plants and animals modify the weed flora in a variety of ways that affect weed persistence directly and indirectly. Cropping systems and agricultural practices associated with cultivation of a crop encourage or discourage the weeds.

Concept of competition and critical period

In dryland areas, during the early stages of crop growth, weeds compete for nutrients. Later, competition for light and moisture becomes important, as weeds produce abundant foliage by then. Weeds may absorb nutrients from the soil more rapidly and in relatively larger amount than does the crop. NPK contents of common weeds and crops are compiled in Table 2. Research has established that competition for moisture usually occurs along with other forms of competition. It becomes critical with increasing soil moisture stress, as found in arid and semi-arid areas. In general, for producing equal amounts of dry matter weeds transpire more water than do most of crop plants. Therefore, in dryland agriculture, the actual evapotranspiration from weedy crop fields is much more than evapotranspiration from a weed-free crop field (Kanitkar *et al.* 1960). Transpiration coefficient of weeds ranged from 221 (*Tribulus terrestris*) to 1402 (*Tridax procumbens*) while that of maize and

Table 2. Nutrient composition in different crop and weed species

Plant species	Nutrients (on dry weight basis) %		
	N	P_2O_5	K_2O
Crops :			
Paddy (*Oryza sativa* L.)	1.13	0.33	1.15
Wheat (*Triticum aestivum* L.)	1.34	0.66	1.50
Maize (*Zea mays* L.)	1.21	0.20	1.18
Sorghum (*Sorghum vulgare* Pers.)	0.41	0.24	2.16
Pearlmillet (*Pennisetum typhoides* Burm.)	0.64	0.74	2.50
Blackgram (*Vigna mungo* (L.) Hepper.	1.35	0.52	1.22
Greengram (*Vigna radiata* (L.) Wilezek	1.24	0.49	1.21
Pea (*Pisum sativum* L.)	1.25	0.54	1.31
Mustard (*Brassica juncea* (L.) (Czern and coss)	0.76	0.30	0.62
Gram (*Cicer arietinum* L.)	1.18	0.43	1.10
Weeds :			
Echinochla colonum (L.) Link.	2.95	0.62	2.50
Amaranthus viridis L.	1.90	1.52	3.12
Digera arvensis forsk.	3.25	1.65	3.15
Xanthium strumarium L.	2.50	0.70	2.50
Parthenium hysterophorus L.	2.10	1.34	2.42
Chenopodium album L.	3.50	1.40	3.45
Melilotus indica (L.) All.	2.35	1.45	2.20
Vicia hirsuta (L.) S.F. Gray	2.95	1.55	2.04
Fumaria parviflora lamk.	3.25	1.50	2.12
Anagallis arvensis L.	1.58	1.55	1.90
Asphodelus tenufolius (L.) Cavan.	2.50	1.75	2.40
Commelina benghalensis L.	2.12	1.50	1.90
Phalaris minor Retz.	1.78	0.84	1.90
Cynodon dactylon (L.) Pers.	2.05	1.45	1.25
Cyperus rotundus L.	1.65	1.50	1.24

Source : Rao and Agarwal, (1984).

sorghum was 352 and 394 respectively (Kanitkar *et al.* 1960).

Critical Period

Critical period is defined as a time span before and after which weed competition does not reduce the crop yield. The existence of this critical period was verified and quantified. Critical period vary from crop to crop depending upon competitiveness of weeds, environment, location, plant density, soil fertility and crop cultivar etc. In general, 4 to 6 weeks period is more crucial from competition point of view in almost all the major crops. Kasasian and Seeyave (1969) proposed that all crops need a weed-free period for one-fourths to one-thirds of their growing season. The critical period of important dryland crops is given in Table 3.

'Kandiari' (*Carthamus oxyacantha* Bieb.) and prickly pear (*Opuntia* spp). In Uttar Pradesh and Madhya Pradesh, *'Kans'* (*Saccharum spontaneum* L.) occupies a large area.

Weed Flora Shifts as a Result of Spread of Plants between Regions and Continents

Exotic weeds can become very aggressive in a new habitat if the ecological conditions are favourable. In a short time, they may become dominant in new areas, successfully repressing earlier established species of a plant community. Some of the important weed species introduced to India are discussed below.

Eichhornia crassipes (Mart.) Solns. (Water hyacinth) is one of the most troublesome weeds. Since its introduction in Bengal from South America around 1890, it has spread throughout India and has heavily infested eastern India, particularly the Bengal delta, the plains around the river Ganga and the Brahmaputra (Joshi, 1987). It blocks water bodies and irrigation systems. It also interferes directly with rice growth on many deep water paddy fields and indirectly with the growth of a large number of crops because the

Table 3. Critical periods of competition and percentage yield reduction

Crop	Critical period	% yield reduction	Remarks
Rice (Up land)	10-40 DAS	60-80	Approximate % losses caused by weeds.
Maize	15-40 DAS	40-60	However, the extent of reduction in yield
Sorghum & Millets	15-35 DAS	30-40	mainly depends upon
Cotton	30-60 DAS	45-85	the type of weeds present and other factors.
Groundnut	25-55 DAS	55-70	Erect
		20-25	Spreading
Sunflower & Safflower	30-60 DAS	30-50	
Sesamum	15-30 DAS	45-50	
Castor	30-60 DAS	35-40	
Soyabean	20-45 DAS	25-35	
Mungbean (Greengram)	20-45 DAS	50-55	
Pigeonpea	25-60 DAS	40-60	
Blackgram	25-35 DAS	45-60	

* Days after sowing

Source : AICRPWC comprehensive report for the period of 1978-84 and Singh and Saraswat, (1986)

Weed flora shifts

Agriculture in India, which is undergoing a transition from extensive to intensive cultivation, a very important dimension is the shift in weed flora. In Uttar Pradesh alone, about 400,000 ha of cropped area are infested with the obnoxious weed known as *'baisure'* (*Pluchca lanceolata* Oliv.). In Madhya Pradesh large tracts are infested with

water surfaces covered with this weed have much higher evapotranspiration rates than open water surfaces (Gopalakrishnan and Joy, 1977).

Salvinia molesta Roxb. (Water fern) is a native of southeastern Brazil and has spread rapidly throughout India (Holm, 1977). The plant is very well adapted and colonizes vast areas of water mainly by virtue of its plasticity. In the coastal state of Kerala, it competes successfully with water hyacinth and has largely replaced the previously dominant aquatic weeds.

Partnenium hysterophorus L. is mainly a wasteland weed and is considered a native of Central and North America (Mani *et al.* 1975). It first appeared in India in 1956 near Pune, Maharastra (Parihar and Kanodia, 1987). By 1987, it had spread throughout India and now occupies more than 5 million ha of urban waste land and village grazing land (Parihar and Kanodia, 1987).

Phalaris Minor L. is of mediterranean origin and is reported to be a major weed in Mexico. Its incidence has increased dramatically in areas cultivated to wheat in recent decades in Northern India (Kundra, 1983).

Weed Flora Shifts as a Result of Changes in Cropping Systems

Several factors contribute to unfavourable and unintentional shifts in weed flora composition due to changes in cropping systems. Imported seed lots of improved varieties and food grain may bring unwanted weed seed into the country. Changes in agronomic practices may also be conducive to creating opportunities for imported weed seeds to create major infestations.

With the introduction of new high-yielding crop varieties of wheat in India, the wild oat (*Avena fatua* L.) problem has recently become acute. This weed has reduced the yield of new wheat varieties in large areas. Previously it was not considered a serious problem in traditional tall wheat cultivars (Yaduraju and Mani, 1980). New cultivation practices were adopted with the introduction of high- yielding varieties of wheat. Irrigation, nitrogenous fertilizers, and insecticides became important inputs. The change in the cropping system resulted in the shift of weed flora from dicots to grassy (monocotyledonous) weeds (Hooda and Malik, 1980).

Another weed, *Phalaris minor* L. was introduced to India during large-scale import of wheat in the 1960s (Rao, 1983). It become a major weed of wheat in the wheat-growing regions of Northern India when rice-wheat rotation became common with the introduction of canal irrigation (Sen, 1981). It grows well in compacted soil (in wetland rice cultivation) and has a competitive advantage over many other weeds. Previously, indigenous tall wheat cultivars were sown from the end of October to the first week of November and wheat would grow sufficiently tall before *P.minor* could offer sufficient competition. The incidence of *P.minor* was less in traditional wheat cultivation than in late-sown HYV wheat, which is sown when the temperature is ideal (27°C) for emergence of this weed. The development of the rice-wheat system is the main reason for the occurrence of *P.minor* as the major species. Rapid spread of *A.fatua* and *P.minor* is because of their morphological similarity with the wheat plant up to flowering, their high reproductive potential, shedding of seeds two or three weeks before wheat harvest, and seed dormancy.

Weed Flora Shifts as a Result of Routine Use of Herbicides

Introduction of herbicides and their routine use is an important example of a change in agronomic practices that may cause serious changes in weed flora in an unfavourable direction. *Cyperus rotundus* has become increasingly troublesome in recent years. This weed is currently under control in mixed-plant communities since it is sensitive to shade and other weeds are stronger competetors. However, with the introduction of selective herbicides, more easily - controllable weeds were eliminated from the mixed plant cummunity and have given ample scope for weed shifts. Thus, a temporary successful attempt at weed control can become a long-term failure in weed management.

WEED CONTROL IN IMPORTANT DRYLAND CROPS

The semi-arid/arid environment is ecologically fragile, and farmers in this region generally have low resource-base. Weed control options should be those seeking to develop effective and economically viable weed management practices that will minimize drudgery and be environmentally neutral or favourable. It is not only the method *per se* which affects the weed control but several others associated factors like timeliness and precision of operation. Irrespective of method of weeding, timeliness is considered to be more important than even number of operations involved.

Rice (Upland)

Weed competition is higher in dwarf and slow growing rice HYVs like Japonicas. Unchecked weed growth causes yield loss of 50-70 per cent (Vaishya *et al.* 1992). Traditional weeding is still a common practice to remove weeds. Hoeing with narrow bladed tools is commonly employed. Frequent weeding (3-4 times) is often required to control weeds. Use of herbicides offers the most practical and economical solution. Some of the commonly used herbicides are: Oxadiazon, Butachlor, Thiobencarb, Pendimethalin and 2, 4-D. The details of doses, time of application and weeds controlled are given in Table 4. Under upland rice conditions, integration of the above herbicides with one hand weeding at 30 days after sowing provides a better weed control than herbicides used alone (Pandey *et al.*, 1991)

Maize

Maize is very sensitive to weed competition during the first 3 to 4 weeks after sowing due to intial slow growth (Sandhu and Gill, 1973). Uncontrolled weed growth causes yield loss of 40 to 60 per cent. Some weeds like *Rottboellia cochinchinensis* and *Striga* spp., can cause total crop loss. The extent of infestation of different weeds varies from one zone to another.

Manual weeding and inter cultivation are common practices in dryland areas. Herbicides that have proved to be effective are presented in Table 4. Integration of herbicides with one hand weeding at 30 to 45 days after sowing was found to be promising.

Sorghum

Sorghum is very susceptible to weed competition, as the seedling starts off weak and frail under dry conditions (Ramakrishna *et al.* 1991). While annual and perennial weeds are common problems, sorghum is particularly vulnerable to damage by parasitic weeds like *Striga asiatica* and *Striga hermonthica.* Two hand weedings are normally required, the first can be carried at thinning and another one at 35 to 45 days after sowing (Subbareddy *et al.* 1976). Sorghum roots are easily damaged by cultivation, so careful hoeing is essential. Animal drawn blade harrow is normally used by farmers to control inter-row weeds in row-planted sorghum. Herbicides that can be used for effective weed control are given in Table-4.

Striga is a serious weed in some parts of semi-arid regions. The promising herbicides for *Striga* control are 2,4-D and MCPA. The best control of *Striga* can be obtained with 2,4-D (2 kg/ha) applied at 30 days after planting without any crop injury (Korwar and Friesen, 1984). In addition to chemical weed control crop rotation can also be practiced to suppress striga.

Pearl Millet

Delayed weeding could reduce the pearl millet yields by 44 per cent (Ramakrishna *et al.* 1992). Significant yield increase was noticed by timely weeding in moisture deficit condition than when soil moisture was adequate (Umrani and Bhoi, 1982).

Apart from other weeds, *Striga* also pose a serious problem to pearl millet. There appears to be differences in the strains of *Striga hermonthica* and *Striga asiatica* that will parasitize pearl millet. Herbicides which are more promising for weed control in pearl millet are listed in Table 4. An effective control of *Striga asiatica* could be obtained by spraying 2,4-D at 30 days after sowing of millet. This corresponds to the date at which the parasitic weed starts germinating. Other good crop husbandry practices, like tillage, crop rotation, manipulation of time of sowing also reduce the infestation to a great extent.

Table 4. Selective herbicides for weed control in major dryland crops

Crops	Herbicides combination	Rate kg/ha	Time of application	Remarks
Rice (Upland)	Butachlor	1-1.5	PE (6-10 DAS)	Annual grasses & some broad leaved weeds.
	Thiobencarb	1-1.5	PE (6-10 DAS	
	Oxadiazon	0.5-1.0	PE	
	2, 4-D	1-1.5	POST-E	Broad leaved weeds
	Oxadiazon fb One HW	0.5	PE	Annual grasses
Maize	Atrazine	0.5-1.0	PE	Broad spectrum control of broad leaved weed and annual weeds
	Pendimethalin	1-1.5	PE	
	Simazine	0.75-.1.0	PE	
	Alachlor	1-2.0	PE	
	Atrazine + Pendimethalin	0.5+0.5	PE	Many annuals including Rottboellia spp.
	Fluchloralin	0.75-1.0	PPI	
Sorghum	Atrazine fb 1 HW	0.25-0.5	PE	Broad spectrum control of annual grassy and broad leaved weeds
	Simazine	0.5-1.0	PE	
	Alachlor	1-1.5	PE	
	Isoproturon	0.5-0.6	PE	
		0.2-0.5	PE	
		0.5-1.0	PE	
Millets	Atrazine	0.2-0.5	PE	Mainly grassy weeds
	2, 4-D	0.75-1.0	POST-E	To control *Striga* spp.
Cotton	Fluchloralin	0.75-1.0	PPI	Broad spectrum control of annuals
	Pendimethalin	1-1.5	PE	
	Diuron	0.5	PE	
	Alachlor	1-1.5	PE	
	Oxadiazon	0.5-0.75	PE	
Groundnut	Alachlor	1-1.5	PE	Broad spectrum control of grassy and broadleaf weeds.
	Simazine	1-1.5	PE	
	Fluchloralin	1-1.25	PPI	
	Pendimethalin	1-1.5	PE	
	Butachlor	1-1.5	PE	
Sunflower	Alachlor	1.0	PE	Broad spectrum control of annuals and grassy weeds
Safflower	Fluchloralin	0.75-1.0	PPI	
	Pendimethalin	1.0	PE	
Sesamum	Pendimethalin	1.0	PE	Control grassy and broad leaved weeds
	Fluchloralin	0.75-1.0	PPI	
	Alachlor	1.0	PE	

Table 4 Contd. . . . . .

Crops	Herbicides combination	Rate (kg/ha)	Time of application	Remarks
Castor	Alachlor	1-1.5	PE	Broad spectrum control of annual and grassy weeds
	Fluchloralin	1.0	PPI	
	Pendimethalin	1-1.5	PE	
Soybean	Fluchloralin	0.75-1.0	PPI	Annual grassy and broad leaved weeds
	Alachlor	1-1.5	PE	
	Pendimethalin	0.75-1.0	PE	
	Metribuzin	0.25-0.5	PE	
Pigeonpea	Alachlor	1.0	PE	Broad spectrum control of annuals
	Pendimethalin	1.0-1.5	PE	
	Fluchloralin	0.75-.10	PPI	
Mungbean (Greengram)	Alachlor	0.75-1.0	PE	Broad spectrum control of annuals
	Pendimethalin	0.5-1.0	PE	
	Fluchloralin	0.75-1.0	PPI	
Blackgram	Pendimethalin	0.5-1.0	PE	Broad spectrum control of annuals
	Fluchloralin	0.5-1.0	PPI	
	Alachlor	0.75-1.0	PE	

PE = Pre-emergence; POST E = POST-emergence; PPI = Pre plant incorporated; HW = Hand weeding; DAS = Days after sowing., f.b. = followed by.

Cotton

Young cotton will be severely affected by delayed or inadequate weeding. Wider row spacing and high nutrient requirement of high yielding varieties provide an ample scope for weeds to grow vigorously and rapidly (Singh *et al.* 1988). Cotton is particularly susceptible to weed competition when soil fertility and moisture are low. Weed competition causes 45 to 80 per cent reduction in yield. Inter-row weeding about two weeks after emergence of cotton and intra-row weeding one week later when the crop is thinned provide an adequate weed control in the beginning and 2 to 3 subsequent weedings would take care of remaining weeds till the crop flowers and produces good canopy (Tiwana and Brar, 1990). Herbicides that can be used successfully in cotton are listed in Table 4.

Groundnut

Groundnut is very sensitive to weed competition because of its initial slow growth, short stature and prostrate growth habit (Naidu, *et al.*, 1983). Weeds also interfere with pegging, pesticide application and harvesting. Hand weeding and interrow cultivation are generally done after the emergence of weeds and before the peg formation. Pre-emergence application of herbicides (Table 4) provide season long control of a wide spectrum of weeds. Integration of herbicide with other cultural practices resulted in more efficient control of weeds than use of herbicides alone (Ramakrishna *et al.* 1991b).

Sunflower and Safflower

Weed competition is more severe and damaging during the early stage of crop growth. Singh and Singh (1978) recorded highest yield when weed- free condition was maintained for the first 60 days after sowing. The effective pre-emergence herbicides for sunflower and safflower are given in Table 4.

Castor

Castor is generally grown on poor and less fertile soils. Wider row spacing provides an opportunity for luxuriant weed growth. Herbicides like Alachlor, Pendimethalin and Fluchloralin can be used safely (Rao and Shetty, 1984).

Soybean

Weeds pose a serious threat to soybean cultivation during the early phase of crop growth till 45 days after sowing. Reduction in yield may vary from 27.71 per cent depending upon the type and intensity of weeds, and time of their occurrance (Muniappa *et al.* 1986). Weeding thrice at 20 and 40 days after sowing or pre-plant incorporation of Fluchloralin (1.0 kg) or Alachlor (1.5-2 kg) provides a sufficient long duration suppression of weeds (Singh and Sharma, 1990). Some other promising herbicides which are commonly used in soybean crop are given in Table 4.

Sesamum

The critical period of competition in sesamum is between 15-30 days after sowing (Kondap and Chandrasekhar, 1978). Crop yield reduction to an extent of 49 per cent was recorded by Subramanian and Sankaran (1978). Most promising herbicides for weed control in sesamum are listed in Table 4. Herbicides in combination with one hand weeding provide better control as compared to herbicide alone.

Pigeonpea

The first 25 to 33 per cent of the life cycle of pigeonpea is critical for weed competition, but when crop is well developed, it will effectively suppress weeds (Masood Ali, 1991). Dwarf cultivars of pigeonpea are more susceptible to weed competition than tall cultivars. Herbicides like Pendimethalin, Alachlor and Fluchloralin provide season long weed suppression and results in increased crop yield (Rao and Shetty, 1984).

Mungbean/Blackgram

Two weedings during the first 35 days after sowing of mungbean provide effective control of weeds. Pre-emergence herbicides like Alachlor, Pendimethalin and Fluchloralin are useful for weed control in mungbean and blackgram (Kundra *et al.* 1991).

HERBICIDE EFFICACY AND GENETIC IMPROVEMENT FOR CROP TOLERANCE TO HERBICIDE

Herbicide efficacy Time and method of application determine the efficacy of the applied herbicide. Herbicide efficacy can also be improved by selecting suitable formulation and mixing with suitable adjuvants. Rethinam and Sankaran (1976) reported that in cotton, E.C. formulation of Alachlor was having better selectivity than granular form and bed system of planting was superior to ridge system. The pH of spray solution, the addition of detergents and sucrose were found to effect the penetration and translocation of foliar applied 2,4-D in purple nutsedge (Veerabhadraiah *et al.* 1980) and *Solanum elaegnifolium* Cav. (Devendra *et al.* 1986). The rate of symoplast translocated herbicide was enhanced by additives like sucrose (Shastry *et al.*, 1973).

The scope for reducing the dose of isoproturon and metoxuron by 20 per cent when applied in combination with 0.1 per cent Sandovit and 0.25 per cent Hyoxid without any loss in the bioefficacy to control *Phalaris minor* was reported by Walia and Gill (1987). Exploration of such possibilities in different dryland agro-ecosystems would be helpful in reducing the weed menace with low-cost.

Genetic Improvement for Crop Tolerance to herbicides

Differential tolerance to herbicides among crop varieties was reported (Sahi *et al.* 1978, Lall and Chaudhary, 1974, ICRISAT, 1977). Selectivity of herbicides depends on factors such as genetic, morphological characteristics and metabolising capacity of a crop variety. Tolerant crop cultivars would permit use of higher rates to reduce the selection for tolerance within some susceptible weed species. It would also permit the use of herbicides which were previously non-selective. There is need for intensifying biotechnology research efforts to improve herbicide tolerance in crop cultivars as development of herbicide tolerance in crop cultivars would be more efficient and economical way to control unmanageable weeds than consistently screening new herbicides.

Weed Management in Cropping Systems

Though there are a number of terms coined in which crop production/unit area can be increased, the cropping systems -- the spatial and temporal combinations of cultivars in any one plot in one year -- presently practiced may be grouped into two major categories viz., intercropping and sequential cropping.

Intercropping

Control of weeds may be a greater problem in intercropping than when the component crops are grown alone. Mechanical weeding may be difficult or even impossible in certain spatial arrangements or when the row spacings of the component crops are too close to each other. Comparing the ability of intercrops and sole crops in competing with weeds, it has been reported that intercrops are better than either or both of the component crops as sole crops in weed suppression. Experiments carried out at ICRISAT, Hyderabad, indicated 50 to 70 per cent reduction in weed infestation by intercropping (Shetty and Rao, 1977). Although intercropping may reduce weed infestation and growth, there is still a need for some weeding in most cases. But the labour input for weeding is greatly reduced because of limited weed growth. The results obtained by Rao and Shetty (1976) in sorghum/pigeonpea intercropping and by Moody (1980) and Shetty and Rao (1981) in sorghum/cowpea and sorghum/mungbean intercropping systems indicated that one weeding would be sufficient to get yields equivalent to weed- free control.

Most herbicides are crop specific. A herbicide that does not harm both the component crops usually does not control a large number of weeds. Further, the more complex the intercrop system, the less the likelihood of finding suitable herbicides. A limited number of experiments conducted have identified some herbicides for different crop combinations (Table 5). Weed management research in intercropping is still in its infancy and systematic and coordinated research efforts in this area are to be intensified to improve the efficiency and productivity of intercropping.

Relay cropping of short duration crops, particularly grain legumes, either by broadcasting or row sowing into the main crop of cereal- rice, maize, sorghum before their harvest is a common practice. Weeds can be a major obstacle to relay planting. Weeds and crop stubble may hinder

Table 5. Promising herbicides identified for managing weeds in intercropping systems

Intercropping System	Promising herbicide	Reference
Maize/mungbean	Butachlor	Bantilan, et al. 1974
Maize/cowpea	Butachlor	Pamplona and Imlan, 1976
Maize/cowpea Maize/lablab Sorghum/cowpea Sorghum/lablab	Alachlor	Damodaran and Sankaran, 1974
Maize/groundnut	Trifluralin	Damodaran and Sankaran, 1974
Sorghum/blackgram Sorghum/cowpea	Alachlor	Bhoopathi Babu, 1978
Sorghum/lablab	Dinitramine	Bhoopathi Babu, 1978
Maize/mungbean	Alachlor and Linuron	CIAT, 1976
Sorghum/pigeonpea	Ametryn, Prometryn, Terbutryn	ICRISAT, 1977
Pigeonpea/groundnut	Pendimethalin Fluchloralin	Vijaya Kumar *et al*. 1988.

relay crop establishment. Weeds can cause considerable damage to relay cropped chickpea or safflower due to frequent rains during the latter part of monsoon (ICRISAT, 1979).

Sequential Cropping

Sequential cropping aims at growing two or more crops in sequence on the same field in a year. The problems of weed management in sequential cropping are different from those in intercropping. Continuous presence of crop cover, residual toxicity of herbicides applied to the previous crop and change in weed flora need a different approach in weed management. Apart from optimum weed control, agronomic considerations which would determine the success of sequential cropping include early crop establishment, time and method of tillage, date and method of planting and harvesting, crop rotation and the length of the turn around time (Moody, 1976).

Minimum and Zero Tillage

Frequent cultivation of the field is not possible because of the time factor and minimum and zero tillage techniques favour reduction in turn around time thereby result in early crop. establishment. (Shetty, 1980; Moody, 1982.) Rangaiah (1981) observed a small increase in weed dry weight under minimum tillage (the land was tilled only once) compared to conventional tillage in a cotton-sorghum-ragi system. Weeds were effectively checked by pre-emergent herbicides followed by a late weeding. This technique reduced the turn around time, cost of land preparation and labour requirement. In zero tillage system, the ratoon from the first crop will act as weed and may offer severe competition to the second crop (ICRISAT, 1980, 1981). Herbicides which kill weeds in the planted crop and at the same time also kill the previous ratooning crop are needed. Further research is needed on the role of herbicides in minimum and zero tillage, and the physiological basis for yield increase due to reduction in turn around time and reduced tillage needs to be further understood.

Dry Season Tillage

Ploughing the field well ahead of the planting seasons, either at the end of the rainy season or after summer rains, helps to kill most of the weeds and reduces the weed problem in the ensuing season. Tillage during the dry season keeps the soil free of weeds and helps in conserving soil moisture (Krantz, 1979).

Crop Rotation

Rotation of crops having drastically dissimilar life cycles or cultural conditions is an effective and practical weed control method. By rotating crops, the possibility of build-up of certain weed species is usually reduced. When an upland crop is rotated with a low-land crop, the total number of weeds is either lower than under continuous lowland or upland culture (Moody, 1977). Gill and Brar (1972) observed *Phalaris minor* L., a relatively minor weed, when tall traditional wheat was grown as single crop is increasing at alarming rates in an intensive rice-wheat crop rotation system.

Weed Control in Component Crops and its impact on Subsequent Crops

The severity of weed problem in a crop sequence is determined by the weed control measures practiced in the component crops. Saraswat (1976) observed that in jute-oat rotation, the number of weeds growing in association with the oat crop was directly related to the weeds growing in the jute crop. The better the weed control in jute, the less weeds in the oat.

Selective herbicides are available for sole crops and effective weed control can be obtained by using herbicides. But the residual effect of these herbicides have to be carefully evaluated before using them in crop sequence. There are reports on the hazards of herbicide residues in the succeeding crop especially if legumes are planted after cereal crops. Atrazine applied @ 0.5 kg ai. /ha as pre-emergence gave effective weed control in sorghum but the establishment of legumes such as mungbean or groundnut which followed sorghum was poor but following cotton was not affected (Palaniappan and Ramaswamy,

1976). However, Atrazine, Diuron and Amiben applied to maize had no harmful residual effect on the succeeding crop of soybean at lower doses but increased dosage reduced the emergence of soybean (Hugar and Hosmani, 1977). Atrazine applied at 1,2 and 4 kg/ha and alachlor at 1,2 and 4 kg/ha as pre-emergence spray to maize gave good control of weeds in maize and left no harmful residues that could effect emergence, growth or yield of peas, chickpea and lentil (Singh and Mani, 1981).

In sequential cropping systems, the weeds allowed to grow later in the season in the first crop may cause a serious threat to the second crop, apart from the difficulty they impose during the harvest of the first crop and land preparation for the second crop. Studies to determine the critical period of crop-weed competition on a cropping systems basis are lacking and therefore, should be initiated.

SCOPE OF BIOLOGICAL WEED CONTROL IN DRYLAND CROPS AND CROPPING SYSTEMS

Biological Control

The need to reduce residual toxicity and increased resistance of weeds to herbicides has encouraged research into biological control. Historically, biological control has worked best on large infestation of a single weed species: Examples of successful biological weed control in India include the control of *Opuntia vulgaris* (Miller) by *Dactylopius ceylonieus* (Green). After this first success, seed fly *Ophiomyia lantanae* (Frogg), a native of Mexico was introduced in 1809, from Hawaii and tingid bug *Teleonemeia scrupulosa* Stal. imported from Australia in 1952 for *Lantana* control have established but failed to provide economic control. Recent finding has established that *Zygogramma bicolorate* do well on *Parthenium hysterophors* L. which has become a nuisance weed in crops like maize, sorghum and pulses in many parts of the country (Visalakschy 1989). A comprehensive list of some successful bio-agents is listed in Table 6.

Microbial Control (Mycoherbicides)

The concept of using fungi, bacteria and even viruses as bioherbicides is biologically feasible with several host-pathogen combinations. Today, however, only fungal pathogens have been exploited commercially as mycoherbicide, marketed for practical application in developed countries.

The mycoherbicides devine (R), a product of fungus *Phytophthora palmivora* and collego (R), a formulation from *Colletotrichum gloeosporoides* (Penz) Sacc; F. sp. aeschynomene, are registered in USA for the control of stranger vine (*Morrenia odorata* (Lindl) and *Aeschynomene virginica* in citrus groves and rice and soybean crops, respectively.

Biological control of weeds with pathogens offered opportunities for overcoming several of the drawbacks imposed by chemical herbicides like injury to non-targets from spray drifts, residual toxicity and alteration of environment beyond tolerance limits. Hence, there is an urgent need for intensifying research in identifying biological weed control agent with attributes such as specificity of action, coincidence of distribution with the host, high virulence and production of abundant inoculum.

ECO-PHYSIOLOGY AND WEED MANAGEMENT IN CROPPING SYSTEMS

An understanding of the ecology and physiology of weeds associated with any cropping system is of utmost importance so as to create an altered environment, that is no longer favourable to weeds. Some of the biological aspects of crop mangement that have strong bearing in shifting the crop weed balance to the advantage of crop are: crop species, variety, plant population; and in case of intercropping, nature of crops used, relative proportion of crops in the mixture; and geometry of planting (Rao and Shetty, 1976; Rao and Shetty, 1983a; Rao, 1980).

Crops differ in their relative growth rates, spreading habit, height, canopy structure and inherent competitive character and accordingly differ in their weed suppressing ability. Pearl millet and maize showed high initial weed

Table 6. Potential for biological control of some terrestrial weed with bio-agents in India

Weeds	Pest status	Promising bio-agent
Opuntia vulgaris (Miller) (Green)	Occupies grazing land	*Dactylopius ceylonicus*
Opuntia dillent (Ker. Gawler) (Lichtenstein)	Occupies grazing land	*Dactylopius opuntiae*
Opuntia elatior (Miller) (Stone)	Occupies grazing land	*Proceidocharus utilis*
Eyoatiruyn adenophorum (Sprengel)	Occupies grazing land hilly areas	
Cyperus rotundus (L.) Bactra	Crop weed	*Athesapeuta cyperi* (Mehl), *minima minima* (Meyr) *B. Venosana* (Zeller)
Parthenium hysterophorus (L.) (Pallister)	Found in maize, sorghum and many pulses	*Zygogramma bicolorate*
Orobanche spp. Kaltenbach	Invades crops like tobacco, brinjal, tomato, sunflower etc.	*Phytomyza orobanchiae*
Striga spp.	Semi parasitic or parasitic on graminicious plants	*Smicrony umbrinus* *Smicronyx albovarieagatus* (Faust) *Eulocastra argentisparsa* (Hamps) *E. undulata (stellen)* *Ophiomyia strangalis* (Spencer)
Cuscuta spp.	Parasitic on plants, transmit virus and microplasma diseases.	*Melangromyra* spp.
Chromolaena odorata	Crops like rubber, pea, coffee, carlomon, teak and citrus orchards	*Pareuchaetas pseudoinsulata*
Loranthus spp.	Perennial partial parasitic on fruit trees, Fiscus & Memm	*Delias hypareta*

Source : Visalakchy, (1989) and Singh, (1989).

smothering ability followed by cowpea and groundnut. Sorghum progressively increased its competitive ability with time. Pigeonpea and castor are poor competitors with weeds. Weed growth in compact pigeonpea genotypes was higher than that observed in spreading types (Rao and Shetty, 1976). In sole cropping systems, Rao and Shetty (1976) recorded decreased weed dry weight with increased crop population.

The spread of high yielding wheat varieties is a contributing factor towards dissemination of *Phalaris minor* Retz. in wheat growing states, causing major weed problems in wheat and other rabi crops. *P.minor* has become the dominant weed suppressing the native weed species. Tiwari and Bisen (1985) observed that for effective suppression of *Phalaris* in the fields where it is dominating and for expression of greater crop yields, pea and wheat crops will be more suitable as compared to chickpea, linseed, or safflower. The tall wheat genotypes exerted a strong suppressing effect on weed intensity and growth of *Phalaris minor* Retz. *Chenopodium album* L. and *Melilotus indica* (L.) All. (Gill and Mehra, 1981). Thus it may be advisable to go in for medium dwarf varieties which combine a reasonably good level of yield and also adequately smother the weeds.

The competitiveness of weeds was attributed to the characteristics of weeds such as prolific seed production, greater root/shoot ratio, presence of more stomata and accumulation of greater quantities of N,P and K than crop species (Rao and Agarwal, 1984).

In the post rainy season crops, less weed growth occurs in fields that had been planted during the rainy season than in fields left fallow during the rainy season (Shetty, 1978). Rao and

Shetty (1983) observed increase in weed dry weight in sorghum field with the increase in irrigation. Higher reduction in sorghum grain yield due to weeds was recorded under limited irrigations, than under optimum irrigation, thereby revealing better competitive ability of sorghum under optimal irrigation.

The greater reduction in finger millet yield was recorded due to grassy weeds than by dicot weeds (Nanjappa and Hosmani, 1985), as grassy weeds and fingermillet belong to the same taxanomic family and have similar morphological characteristics. Krishnamurthy *et al.* (1977) screened forty two crop plants in pots and found that crop plants like greengram, sorghum, cowpea, chilies, deccan hemp, blackgram and wild mung etc., induce abundant germination of *Orbanche* seeds, thus acting as promising trap crops for *Orbanche* control. These trap crops would be of immense use as rotation crops with tobacco for eliminating *Orbanche* incidence in fields.

Intercropping was proved to be superior to component crops in weed suppressing ability (Shetty and Rao, 1977, 1981) and thus, it provided an opportunity to utilize crops themselves as tools of weed management. The weed suppressing ability of intercrops is dependent upon many factors such as the component crops selected, the genotypes used, plant density adopted, proportions of component-crops, their spatial arrangement and the fertility and moisture status of the soil (Moody, 1980; Moody and Shetty, 1981). Rao and Shetty (1976) observed that competitive ability of intercropping was enhanced by high plant population pressure provided by the component species together. If the combined density is greater than the optimum density of either of the sole crop, there is likelihood of reduction in weed infestation. Rao and Shetty (1976) and Shetty and Rao (1977,1981) reported that the weight of weeds growing in association with intercrops such as maize/mungbean, maize/soybean, sorghum/pigeonpea, sorghum/cowpea and sorghum/mungbean is as low or lower than that growing in association with the sole crops. Maize, sorghum, millet, cowpea and mungbean are considered to be more competitive crops than pigeonpea and groundnut. Shetty (1978) reported that the critical period of weed competition in a sorghum/pigeonpea intercrop was longer than in either of the sole crops. Such studies on other crop combinations are lacking. Rao and Shetty (1981) recorded that weed biomass accumulating in sorghum/cowpea and sorghum/mungean smother cropping system with one hand weeding was observed to be less than that observed in sorghum sole situation with two hand weedings. Light interception pattern and leaf area index (LAI) observations revealed that inclusion of smother crop viz. cowpea and mungbean resulted in quicker and earlier attenuation of maximum LAI and maximum percentage of light interception by component crops. Significant positive correlation was observed between LAI and percentage ligh interception and significant negative correlation was observed between percentage light interception by component crops and weed biomass accumulation.

The growth habit of the genotype used in intercropping has a great influence on weed suppression. Early canopy coverage seems to be more important than rapid increase in height. Mungbean cv. MG_50 intercropped in maize competed with weeds better than CES-14, since the former had a better early lateral growth (Bantilan *et al.*, 1974). Spreading genotype of pigeonpea, ST-1 was found to be more competitve than a compact genotype, HY-3C against weeds under sole cropping. However, when intercropped with sorghum, the genotype differences in weed suppressing ability were not significant (Rao and Shetty, 1976). Weed competition usually increases with increase in soil fertility. Bantilan *et al.* (1974) observed that intercrops were less effective in controlling weeds at high nitrogen levels than at low levels. Similarly in a naturally fertile Vertisol more weed growth was noticed than in a comparatively low fertile Alfisol (ICRISAT, 1977). Tripathi and Singh (1983) verified the density of soybean intercropping with maize and found that across fertilizer levels, the total weight of crop seeds produced and competitive suppression of weeds were positively correlated. Sandhu and Gill (1973) showed that weedy maize yield decreased as the competition period increased at high fertility, but not at low fertility.

Besides the direct competition, some of the weeds exert chemical stress on crops by their phytotoxic root exudates and plant leachates secreted or exerted into the soil. Due to these phytotoxic substance other plants growing in proximity are adversely effected. The term allelopathy generally refers to the detrimental effects of one species on germination, growth and development of another species. Lall and Savongdy (1981) reported that root exudates of growing plants and leachates of plant residues of purple nutsedge caused significant reduction in plant height and dry matter production of pearl millet. Allelopathic impact of *Parthenium hysterophorus* has been studied by many (Kanchan, 1975; Kanchan and Jayachandra, 1976; Rao *et al.* 1977; Vallaiappan and Towers, 1988). Roots of *Lolium temulentum* and roots of *Phalaris minor* and *Avena fatua* had inhibitory effect on growth characteristics of wheat (Bansel and Singh, 1986). On the contrary, Bhatia et al., 1984 observed that a few weeds such as *Chenopodium album*, *Celosia argentea*, *Rumex dentatus* and *Phalaris minor* had promotory effects on growth and development of wheat and such effect are species specific. Thus allelopathic interactions could be beneficial and further research is needed to ascertain particular weed crop interactions and identifying compounds responsible for such behaviour in different cropping systems. Alleopathy poses two major chllenges to agricultural investigators viz. i) minimise the negative impacts of allelopathy on crop growth and yield and (ii) to exploit allelopathic mechanisms to control weeds in plant growth regulations strategies.

A well directed weed research strategy for managing weeds in cropping systems should include both ecophysiological and agronomic studies. Ecophysiological studies should aim at manipulating the cropping systems to obtain better weed management and agronomic studies would involve evaluation of various weed control measures for selected cropping systems of the region.

RESEARCH NEEDS FOR EFFECTIVE TACKLING OF WEED PROBLEMS

Agricultural environment in the semi-arid tropics is very complex. It is impossible to obtain the high-yield potential of dwarf, insect and disease-resistant varieties unless weeds are controlled adequately, even more so when high amounts of fertilizers are applied. Likewise, the increasing emphasis on multiple cropping and intercropping systems have pointed to the role of better weed control as part of the systems. The primary objective of weed research should be to help utilize the available resources and crop inputs, and to harness the maximum benefical effects of herbicide technology. Weed management is a changing technology and new technology must always be developed to suit the changing agricultural and economic situation of the farmer and the country. Following are some of the weed research areas that should be given due importance.

Weed Biology and Competition

Inadequate knowledge of weeds is a major limitation in the development of suitable weed control technology. If we knew as much about weeds as we knew about crops, vastly improved weed control methods and more effective herbicides could be developed. Research on weed biology must probe for a better understanding of the survival mechanisms, dormancy, apical dominance, the relationships between buds and shoots in perennial weeds, the effect of habitat change and soil disturbance on seed dormancy and germination, the influence of various climatic and soil factors on weed populations, population shifts, growth, life cycles etc. This knowledge would lead to application of weed control practices at times when weeds are most vulnerable, to develop habitat-management practices that would induce all weed seeds in the soil to germinate at the same time for complete control, and to develop and identify chemcials that induce seed germination, extend the dormancy of weed seeds until they lose viability or prevent the fertilization of weed flowers. Weed competition is maxlimized during the active growth stage and this critical period varies from one crop to another and from one weed species to another. Identification of these critical periods and economic threshold levels is helpful in choosing appropriate herbicides (in mixture or rotations) and/or weed control practices and in economizing the cost of weed control.

Herbicide Technology

Herbicide technology includes research on basic and applied aspects. Basic research is needed on absorption, translocation and action mechanisms of herbicides as well tranformation and degradation pathways in plants and soil. It will aid in understanding these mechanisms and help identification of the proper dosage, placement methods, application time and effective use of herbicides.

Another area of great importance is related to herbicide persistence, residues and residual effects. Effects of herbicides on soil microbes and vice-versa is also of major concern as usage of herbicides is increasing rapidly. An important area of applied research in herbicide technology is studies related to the enhancement of herbicide activity by adjuvants, surfactants and solubilizers for possible reductions in rate of application. Proper herbicide mixtures and rotations need to be determined for each crop and cropping system for economic weed control, to widen the spectrum of weeds controlled, and to prevent emergence of tolerant weed species through weed shifts. There is a great paucity of sutitable herbicide application equipment. Such equipment includes; sprayers, spray nozzles, granule applicators, herbicide incorporators, etc. Suitable and inexpensive sprayers need to be developed for large-scale as well as small scale application of herbicides.

Weed Management Technology

Traditional technology has involved mainly manual, mechanical and cultural methods that have been the mainstay of weed control for centuries. There is an immediate need for research on weed management systems that will provide guidelines to select the best combinations of techniques for a given weed situation. Research needs to look into various alternative agronomical and mechanical methods in combination with herbicides and then choose from these combinations the most viable and economically feasible system. The most economical and effective methods or combination of methods is that which destroys all weed seeds/propagules in the soil and prevents weed multiplication.

Habitat Management Technology

Continuous use of any method of weed control may lead to possible undesirable side effects. There is a continuing need to determine long-term effects of different weed control programs, crop production practices, and habitat management systems on weed problems. Specific information needs to be generated on the effects of habitat management on weed resistance, the relationship between herbicides and soil microbial activities, and the effects of soil-applied herbicides on the physical, chemical and biological characteristics of the soil. There is already growing evidence of second generation problems from shifts in weed flora due to continuous use of the same or closely-related herbicides. It is necessary to prevent these problems before they become serious.

Biocontrol Technology

Introducing natural encemies of weed species to new environments has proved to be one of the most spectacular means of weed control, although its protential has not yet been fully explored. Research needs to be focused on understanding the nature of host- plant specificity, the identification of accessory substances, and those conditions that lead to host-plant acceptance of plant enemies. Research is also needed to continue the identification of specific organisms that attack plants and groups of organisms that possess suitable specialization in their diet to control weeds.

Economic Analysis

Farmers will respond to a new recommendation if it is economically beneficial. Weed research reports and communications should describe the relative costs and benefits associated with new and existing practices, so that the farmer can choose the one most appropriate to him. Many research stations are located on the best soils and usually have good access to irrigation facilities. Dryland farmers, on the other hand, are forced to work with degraded, erosion-prone lands without irrigation. Very few research stations dealing with weed research conduct trials on such lands A successful weed management program from the perspective of farmer welfare requires involvement of the farmer in planning and execution of weed research programs.

CONCLUSIONS

Very little work has been done on developing weed management strategies for cropping system as a whole and on a year round basis. The role that weeds play in the determination of cropping patterns is not well documented. Little information is available on the implications of weeds on cropping systems and vice-versa. Weed control in cropping systems needs to be considered on a different basis from that in monoculture. Weed management strategies for the small farmers should involve combinations of crop production practices and specific weed control technologies aimed at reducing weed competition and should focus on the entire cropping system with emphasis on year round and multi-year management of weed populations.

Research in cropping systems requires the evaluation of several crops, not separately but as a package following a prescribed arrangement either in time or space. There is a need to evaluate specific weed management factors in a crop sequence as a whole. In addition, the evaluation needs to be done in both agronomic and economic terms. Experiments should, therefore, be planned in such a way as to provide adequate agronomic and economic data.

The subsistence dryland farmer, who is slow to accept weed control technology and herbicide use, can be made to accept those technologies if losses caused by weeds, timeliness of weed control and cost-benefit ratios of various weed control practices are demonstrated to him effectively. The fact that he understands the importance of insect and disease control and accepts recommendations to control them with little effort and persuation indicates that he would be receptive to new production technology. There is a definite need to develp weed control technology suitable to his farm and farming conditions. For this, we must learn more about the ecology of weeds, their association with the surrounding environment, their cycle, morphology, and physiology. Emphasis should be placed on developing weed control strategies for the farming systems as whole.

REFERENCES

AICRPWC 1984. *Annual Report* 1978-84.

AICRPWC 1989. *Annal Report 1988.*

Bhatia, R.K. Gill, H.S. Bhandari, S.L. and Khurana, A.S. 1984. Allelopathic interactions of some tropical weeds. *Indian J. Weed Sci. 16 : 182-189.*

Bansel, G.L. and Singh, C.M. 1986. Allelopathic effect of different plant parts of grassy weeds of wheat on the germination and growth of rice. *Indian J. Weed Sci. 18 : 108-110.*

Bantilan, R.T., Palada, M.C., and Harwood, R.R.1974. Integrated weed management, 1. Key factors affecting crop-weed balance. *Philipp. Weed Sci. Bull* 1 : 14-36.

Bhoopathi Babu, D. 1978. Studies on Weed Control in Sorghum Based Intercropping Systems M.Sc (Ag) Unpublished Thesis, TNAU, Coimbatore.

CIAT 1976. Bean Production Systems. *Annual Report* for 1975. Cali, Colombia, 58p.

Damodaran, A. and Sankaran, S. 1974. A note on evaluation of herbicides in cereal-legume mixtures. *Madras Agri. J.* 61 : 924-925.

Devendra, R., Prasad, T.G., Rama Rao, S., Uday Kumar, U. and Krishna Murthy, K. 1986. Translocation of foliar applied 2,4 D as affected by additives and pH in *Solanum elaegnifolium* Cav. *Indian J. Weed Sci. 18 : 153-158.*

Gill, H.S., and Mehra, S.P. 1981. Growth and development of *Phalaris minor, Chenopodium album* and *Melilotus indica* in wheat crop ecosystem. In : *Proc 8th APWSS Conf.*, Bangalore, India pp. 175-179.

Gill H.S., and Brar L.S. 1972. *Punjab Agric. Univ. Res. J. 9 : 287-294.*

Gopalakrishnan, R. and Joy, P.J. 1977. Aquatic weeds - *Salvinia* and *Eichhornia. India Farming* 26 : 39-46.

Hay, J.R. 1974. Gains to the grower from weed science. *Weed Science 22 : 439-442.*

Holm, L.G. 1971. The role of weeds in human affairs. *Weed Science* 19 : 485-490.

Hooda, I.S. and Malik, D.S. 1980. Note on selective action of herbicides for control of wild canary grass (*Phalaris minor* Retz.) in wheat. *Indian J. Agric. Res.* 14 : 259- 261.

Hugar, P.V. and Hosmani, M.M. 1977. Chemical and cultural methods of weed control in sorghum. *Mysore J. Agric. Sci.* 11 : 165-167.

ICRISAT 1977. *Annual Report* 1976-1977 : *Agronomy and Weed Science.* Hyderbabad, India. 239p.

ICRISAT 1979. Annual Report 1977-78 : *Agronomy and Weed Science,* Hyderbad, India 288p.

ICRISAT 1980. *Annual Report 1979-79 ; Agronomy* and *Weed Science,* Hyderabad, India, 288p.

ICRISAT 1981. *Annual Report 1979-80, Agronomy and Weed Science.* Hyderabad, India. 304p.

Joshi, N.C. 1987. *Manual of Weed Control.* Researchco Publication, Delhi, India. 545 pp.

Kanchan, S.D. 1975. Growth inhibitors from *Parthenium hysterophorus* Linn. *Curr Sci.* 44 : 358-359.

Kanchan, S.D., and Jayanchandra. 1976. *Parthenium - a positive danger.* Bciro, U.S.A., Bangalore, India. p11.

Kanitkar, N.V., Sirur, S.S., and Gokhale, D.H. 1960. *Dry Farming in India.* I.C.A.R., Pub., 2nd ed., New Delhi.

Kansasian, L. and Seeyave, J. 1969. *Pans.* 15 : 208.

Kondap, S.M. and Chandrasekhar, J. 1978. Weed control in sesamum (*Sesamum indicum*L.) *All India Weed Sci. Conf. Abst* No. 41.

Korwar, G.R. and Friesen, G.H. 1984. Control of witch weed (*Striga asiatica*) in sorghum. *Trop. Pest Mang. 30 : 14-17.*

Krantz, B.A. 1979. Intercropping on an operational scale in an improved farming system *Proc. International Workshop on Intercropping, Hyderabad, India. pp319-327.*

Krishnamurthy, G.V.G., Nagarajan, K. and Ramji Lal. 1977. Some studies on *Orobanche cernua* Loefl, a parasitic weed on tobacco in India. *Indian J. Weed Sci.* 9 : 95-106.

Kundra, H.A. 1983. *Control of Phalaris minor Retz. in Wheat.* Unpublished Ph. D Thesis, Indian Agricultural Research Institute, Delhi, India.

Kundra, H.C., Gosal, K.S. and Brar, H.S., 1991. Effect of weed management practices on nutrition uptake by summer mung and associated weeds. *Indian J. Weed Sci.* 23 : 31-35.

Lall, M. and Chaudhary P.N. 1974. Varietal tolerance of sugarcane to atrazine. *Indian J. Pl. Protection. 1 : 53-60.*

Lall, M. and Savongdy. 1981. Allelopathic effect of purple nutsedge on the growth of pearl millet. *Proc. 8th Asian Pacific Weed Science Soc. Conf.* Bangalore, India pp.

Mani, V.S., Gautam, K.C. and Arora R.K. 1975. Beware of the pertidy of *Parthenium. Indian Farming* 25 : 15-16.

Masood Ali, 1991. Weed suppressing ability and productivity of short duration legumes intercropped with pigeonpea under rainfed conditions. *Trop. Pest Mang. 37 : 384-387.*

Mitra, N. 1962. Losses caused by weeds - a retrospect. *Sci. and Cul.* 28 : 160-162.

Moody, K. 1976. Weed control in Asian Soybeans. In : *Expanding the Use of Soybeans Ser. No. 10.* (Ed. R. M. Goodman), pp. 69-73.

Moody, K. 1977. *Philli. Weed Sci. Bull.* 4:27-38.

Moody, K. 1980. Weed control in intercropping in tropical Asia. p101-108. In Akobundu, I.O. (ed). Weeds and their control in the humid and subhumid tropics. *Int. Inst. Trop. Agric. Proc. Ser. No.3* Ibadan, Nigeria.

Moody, K. 1982. Weed control in dryland crops after rice. p75-84. In *Int. Rice Res. Inst. Workshop on Cropping Systems Research in Asia.* Los Banos Laguna, Philippines.

Moody, K., and Shetty, S.V.R. 1981. Weed Management in Intercropping Systems. pp.229-237. In *Proc. International Workshop on Intercropping.* 10-13 Jan 1979. ICRISAT, Hyderabad.

Muniappa, T.V., Ramachandra Prasad and Krishna Murthy, K. 1986. Critical stages of crop-weed competition and chemical weed control in soybean. *Indian J. Weed Sci.* 18: 34-38.

Naidu, G.L.K., Rajan, M.S.S., Rao, R.S., and Reddy, G.H.S. 1983. Correlation studies in crop weed competion in rainfed groundnut. *Indian J. Agron.* 28: 479-480.

Nanjappa, H.V., and Hosmani, H.M. 1985. Effect of weed density on crop growth and yields of drill-sown finger millet. *Indian J. Weed Sci.* 17: 53-56.

Palaniappan, S.P., and Ramaswamy, R. 1976. Effect of atrazine applied to sorghum on the succeeding crops. *Madras Agric. J.* 63: 196-197.

Pamplona, P.P. and J.S. Imlan, 1976. MIT Res. J.6:1-9.

Pandey, J., Singh, R.P., and Shukla, K. 1991. Study on the chemical weed control in upland rice. *Indian J. Weed Sci.* 23:7-9.

Parihar, S.S. and Kanodia, K.C. 1987. Beware of the exotic weeds. *Indian Farming* 37: 24-27.

Parker, C. and Fryer, J.D. 1975. Weed control problems causing major reductions in world food supplies. *FAO Plant Protection Bulletin* 23: 83-95.

Ramakrishna, A., Ong, C.K., and Reddy, S.L.N. 1991a. Integrated weed management for rainfed groundnut. *J. Plant Protec. in the Trop.* 8: 111-119.

Ramakrishna, A., Ong, C.K., and Reddy, S.L.N. 1991b. Studies on integrated weed management in sorghum. *Trop. pest mang.* 37: 159-161.

Ramakrishna, A., Flower, D.J., and Reddy, S.L.N. 1992. Crop-weed competion studies in pearl millet. *Indian J. Dryland Agric. Res. and Dev.* 7: 102-107.

Rangaiah, P.K. 1981. Studies on weed control in a cropping system of irrigated ragi, cotton and sorghum. Ph.D. Thesis, TNAU, Coimbatore.

Rao, M.R., and Shetty, S.V.R. 1976. Some biological aspects of intercropping systems on crop-weed balance. *Indian J. Weed. Sci.* 8: 32-42. p73

Rao, A.N. 1980. Some ecophysiological aspects of weed management in different cropping systems. Post-doctoral report submitted to ICRISAT. 119p.

Rao, A.N., Dagar, J.C. and Dubey, P.S. 1977. Phytoallelopathic potentials of *Parthenium hysterophorus* Linn. *Indian J. Weed Sci.* 9: 24-30.

Rao, A.N., and Shetty, S.V.R. 1981. Investigation on weed suppressing ability of smother cropping system in relation to canopy development and light interception. p357-364. In Proc. 8th Asian Pacific Weed Sci. Soc. Conference, Bangalore, India.

Rao, A.N., and Shetty, S.V.R. 1983a. Some observations on the ecological approach for managing weeds in Indian semi arid tropical agro-ecosystems. p.181-191. In. *Proc. Ninth Asian Pacific Weed Science Society Conference,* Manila, Philippines.

Rao, A.N. and Shetty, S.V.R. 1983b. Weed occurrence and competitiveness and grain yield of sorghum under different moisture regimes. p.261-270. In *Proc. Ninth Asian Pacific Weed Science Soc. Conf.* Manila, Philippines.

Rao, A.N., and Shetty, S.V.R. 1984. Possible approches for managing weeds in three tropical oil seed crops. *Proc. First Tropical Weed Sci.* Conf. Vol. 2:367-377, Hatyai, Thailand.

Rao, A.S., and Agarwal, J.P. 1984. Seed production, root/shoot ratio, stomata number and nutrient uptake by different crop and weed species. *Indian J. Weed Sci.* 16: 250-254.

Rao, V.S. 1983. *Principles of Weed Science.* Oxford IBH Publishing Co., New Delhi, India. 540 pp.

Rethinam, P., and Sankaran, S. 1976. Studies on the selectivity of alachlor formulations in cotton under two systems of planting and adsorbent seed treatment. *Indian J. Weed Sci.* 8: 74- 76.

Sahi, B.K., Lall, M., and Chaudhary, P.N. 1978. Differential tolerance of sugarcane varieties to herbicides. *Indian J. Weed Sci.*10: 54-62.

Sandhu, M.S., and Gill, G.S. 1973. Studies on critical period of weed competion. *Indian J. Weed. Sci.* 5(1): 1-5.

Saraswat, V.N. 1976. Sodium 2,2,3,3 - tetrafluropropionate, a new herbicide for jute field weeds. *PANS.* 22: 391-398.

Sen, D.N. 1981. *Ecological Approaches to Indian Weeds.* Geobios International. Jodhpur, India. 380 pp.

Shastry, K.S.K., Rama Rao, S., Miniyappa, T.V., and Vishwanath, S.R. 1973. Studies on the use of weedicide in the control of nutsedge and prevention of regeneration of the tubers. Final Technical Report. U.A.S. Bangalore.

Shetty, S.V.R. 1978. Some observations on weed intensity and crop management systems. p.145-155 In Sen. D.N. (ed) *Environmental Physiology and Ecology of Plants,* Dehradun, India.

Shetty, S.V.R. 1980. Some agroeconomical aspects of improved weed management systems in Indian semi-arid tropics. p.899-909 In *Proc. 1980 British Crop Protec. Conf. Weeds,* Brighton, England.

Shetty, S.V.R., and Rao, M.R. 1977. Weed Management studies in pigeonpea based intercropping. p.655-672 In *Proc. 6th Asian-Pacific Weed Sci. Soc. Conf.,* Jakarta, Indonesia.

Shetty, S.V.R., Maiti, R.K., and Sharma, M.M. 1983. Prevalent weeds of semi-arid tropics. Agronomy and Weed Science. Progress Report. No. 8 ICRISAT, Hyderabad, India.

Shetty, S.V.R., and Rao, A.N. 1981. Weed management studies in sorghum/pigeonpea and pearl millet/groundnut intercropping systems - some observations. p238-248. *Proc. International Workshop on intercropping,* ICRISAT, Hyderabad, India.

Singh, Dheer, and Sharma, K.C. 1990. Economics of soybean cultivation under different methods of weed control. *Indian J. Weed Sci.* 22: 7-11.

Singh, G.B., and Saraswat, V.N. 1986. Paper presented at the joint Indo-Pakistan Weed Control Workshop and 7th USDA/ICAR- AICRPWC workshop at Durgapur, Agric. Research Station, Jaipur held on 22-24 March, 1986.

Singh, G., and Singh, V. 1978. Requirement of weedfree period in sunflower. *Indian J. Weed Sci.*10:79-82.

Singh, S.P. 1989. Introduction to biological suppression of weeds in India. In Proceeding of Training-cum-workshop on Biological suppression of weeds, held at ICAR Biological Centre, NCIPM, Bangalore from 25-27 July, 1989.

Singh, R., and Mani, V.S. 1981. Effect of atrazine and alachlor in maize-pulses (peas, gram and lentil) rotation. pp. 127-130. In *Proc. 8th Asian-Pacific Weed Science Soc. Conf.* Bangalore, India.

Singh, Vireshwar., Verma, S.S., and Kairon, M.S. 1988. Effect of weed control and nitrogen on weed growth and yield of cotton. *Indian J. Agron.* 33: 376-379.

Subba Reddy, G., Venkateshwarlu, J., and Dryden, R.D. 1976. Weed management in rainfed sorghum. *Indian J. Weed Sci.* 8: 95-100.

Subramanian, A., and Sankaran, S. 1978. Effect of herbicides on nutrient removal of weeds in sesamum and its impact on seed production. *All India Weed Sci. Conf. Abst.* No.42

Tiwana, U.S. and Brar, L.S. 1990. Effect of herbicides on weed control efficiency and production potential of American cotton. *Indian J. Weed Sci. 22: 6-10.*

Tiwari, J.P., and Bisen, C.R. 1985. Competive impact of *Phalaris*- weed ecosystem on link coefficient of some rabi crops. *Indian J. Weed Sci.* 17: 21-27.

Tripathi, B., and Singh, C.M. 1983. Weed and fertility management using maize/soybean intercropping in the north-western Himalayas. *Trop. Pest Mang.* 29: 267-270.

Umrani, N.K., and Bhoi, P.G. 1982. Studies on weed control in bajra under dryland conditions. *J. Maharashtra Agric. Univ.* 7: 145-147.

Vaishya, R.D., Singh, V.K., and Amal Saxena 1992. Mechanical and chemical weed control in upland direct seeded rice. *Indian J. Weed Sci.* 24: 11-16.

Valliappan, K., and Towers, G.H.N. 1988. Allelopathic effect of root exudates from the obnoxious weed *Parthenium hysterophorus* L. *Indian J. Weed Sci.* 20: 18-22.

Veerabhadraiah, G.P., Rama Rao, S., Uday Kumar, M., and Krishna Sastry, K.S. 1980. Increase of penetration and translocation of foliar applied 2,4-D in to the under ground parts of nutsedge. *Indian J. Agric. Sci.* 50: 352-356.

Vijaya Kumar, B., Shiv Shankar, and Rao, A.N. 1988. Weed control in pigeonpea/groundnut intercropping system. *Indian J. Weed Sci.* 20: 96-97.

Visalakschy, N.G. 1989. Biological suppression of *Chromolaena odorata* Inc. Proceeding of Training-cum- workshop on Biological suppression of weeds, held at ICAR Biological Centre NCIPM, Bangalore from 25-27 July, 1989.

Walia, U.S., and Gill, H.S. 1987. Efficiency of substituted urea herbicides as influenced by surfactants and their influence on wheat quality and uptake of nutrients. *Indian J. Weed Science.* 19: 37-46.

Yaduraju, N.T. and Mani, N.T. 1980. Watch for grass weeds in wheat. *Indian Farming* 29: 16-18 & 25.

28

PRINCIPAL DISEASES OF DRYLAND CROPS AND THEIR MANAGEMENT

S.K. SINGH

INTRODUCTION

Predominent rainfed crops of the Indian subcontinent are coarse cereals, cotton, oilseeds and pulses. Higher levels of yield can be achieved through improved agronomic practices including improved cultivars, adequate fertilization and effective crop protection measures. The agricultural significance of diseases is that the damage they cause reduces the *quality* and *quantity* of yield. Often presence of a disease is evidenced by particular type of disease symptoms it exhibits.

Nevertheless, it is of vital importance that the causal organism be correctly identified so that the economic significance of the disease can be correctly assessed and that appropriate control measures can be applied. To achieve this important objective characteristic symptoms and appropriate management strategies of principal diseases of dryland crops with special reference to Indian conditions are described in this chapter.

CEREALS

Most important dryland cereal crops of India are pearl millet, sorghum and maize.

Pearl millet (*Pennisetum glaucum* (L.) R. Br.)

Most drought tolerant of all the cereals, it is grown widely as a rainfed crop in semi-arid regions of Africa and India. It is grown both as food and forage crop. An excellent identification handbook of the internationally important pearl millet and sorghum diseases has been compiled by ICRISAT (Williams *et al.* 1978). Four major diseases of pearl millet have been dealt with in detail.

Downy mildew or green ear disease (*Sclerospora graminicola* (Sacc.) Schroet.)

Systemically infected plants remain stunted with excessive tillering and deformed chlorotic leaves. The amount of chlorosis and deformation depends upon reaction of a particular cultivar. Typical symptoms are general yellowing of leaves beginning from the base; under humid conditions, downy white growth mainly on under surface and transformation of inflorescence (partially or totally) into a leafy stucture. Phytosanitary practices like destruction of crop residues and long term crop rotation reduces the primary inoculum. Hybrids 'PHB-10', 'PHB-14' and 'HB-1' are resistant. Seed treatment with Apron 35 SD at 6 g/kg seed can control the disease (Muthuswami and Narayanaswamay, 1981).

Ergot (*Claviceps fusiformis* Loveless)

In India, this disease is on the increase during the past ten years in Maharashtra and Karnataka states. Infected florets exude a cream to red sticky liquid (honey dew) which develops into cream to black hard structures called sclerotia instead of grains. The seed set is poor or remains completely inhibited. The fungus perpetuates through the seed-borne or soil-borne sclerotia. The contaminated grains, if fed to cattle or used by human beings, can cause poisoning. Sclerotia can be removed from grain by immersion in 10 per cent NaCl solution in which the grains sink but the sclerotia float. Deep ploughing can reduce the incidence of disease.

Smut (*Tolyposporium penicillariae* Bref.)

Smut is more prevalent in Tamil Nadu, Andhra Pradesh and Maharashtra. The pathogen infects the florets and transforms them into large green smut sori which later turn brown and black smut spores. Such spores serve as a source for secondary spread. Removal and destruction of diseased ears early in the season and treatment of seeds with 3 g captan/thiram / kg of seed can reduce the disease incidence.

Rust (*Puccinia penniseti* Zimm.)

Symptoms first appear on the lower older leaves as typical eruptive pustules containing a reddish brown powder (uredospores).

As the infection advances brown teliospores are produced. Symptoms can occur on both upper and lower surfaces on the leaves. The leaves dry and grain yield is reduced. Host plant resistance is now available.

Sorghum (*Sorghum bicolor* (L.) Moench)

Sorghum, great millet or *jowar*, is an important food and fodder crop of India. About 50 diseases of sorghum are known, about 30 of which are found in India. Only a few of them are of major importance. (Williams *et al.* 1978).

Sorghum downy mildew (*Sclerospora sorgi*, Weston and Uppal)

Systemically infected young plants show stunting, clorosis and produce sterile ears similar to that of pearlmillet downy mildew. Under humid conditions, downy white sporulation of the conidial stage occurs on the underside of leaves. Chlorotic stripping is a common symptom. Diseased leaves become necrotic as they mature and shredding releases the oospores. Oospores of the fungus remain viable in soil for several years. Transmission can also occur through seeds. Phytosanitary measures such as destruction of crop residues and deep ploughing reduce oospores survival in soil. Systemic fungicides effective against oomycetes such as Metalaxyl and Al-ethyl phosphonate can reduce the disease incidence. Resistant cultivars such as 'CS 3541' 'CL 168', 'CSH 5' and 'CSH 6' are now available.

Ergot or sugary disease (*Sphacelia soogi* Mc Rae)

Ergot infected florets show secretion of a creamy sticky liquid (honey dew). Under favourable conditions of infection straight or curved, cream to light brown, hard sclerotia develop and gives blackened appearance due to associated growth of sooty moulds. Use of 10 per cent solution of common salt enables removal of floating sclerotia mechanically. Wash the seeds thoroughly, dry and treat with thiram or Agrosan GN. Three sprays of mencozeb or captafol starting from the emergence of ear and subsequent sprays at 5 days interval reduced the incidence of disease (Anahosur, 1979).

Head smut (*Sphacelotheca reiliana* (Kuhn) Clinton)

In India head smut is reported from Tamil Nadu, Karnataka, Andhra Pradesh, Maharashtra, Gujarat, Madhya Pradesh, Uttar Pradesh and Punjab. The head is completely or particularly replaced by a large whitish gall which ruptures before the head emerges from the boot and expose a brown-black powder (smut spores). Infected heads are completely destroyed. The spores remain viable for over 18 months in soil and gain entry into the host through the root hairs and younger roots before becoming systemic. Periodical removal of smutted plants avoiding smut spores mixing with the soil prevents the spread of the diseases. Resistant varieties are available.

Grain smut or Covered kernel smut (*Sphacelotheca sorghi* (Link.) Clinton)

Destructive of all the smuts, it is prevalent in Tamil Nadu, Karnataka, Andhra Pradesh, Maharashtra, Gujarat and Uttar Pradesh. The individual grains are replaced by small conical cream to light brown sacs that persist upto threshing and contain black powder (Smut sori). The disease is seed-borne. In a field trial seed treatment with Panoctine (0.4%), Panoran (0.4%) or Vitavax gave 100% control. (Shah and Mariappam, 1988).

Loose kernel smut (*Sphacelotheca cruenta* (Kuhn) Potter)

It occurs in Tamil Nadu, Karnataka, Andhra Pradesh and Maharashtra, but less common than grain smut. Individual grain converted to sac containing black powder, which is released prior to or soon after head emergence, with conspicuous long black, often covered, pointed structures embedded in the powder. Treating the seeds with sulphur, copper carbonate, or organo mercurials, is recommended.

Anthracnose and Red rot (*Colletotrichum graminicola* (Cesati) Wilson)

The fungus causes both leaf spot disease (anthracnose) and a stalk rot (red rot) in sorghum. Small spots straw coloured centres and wide purple, red, or tan margin appear. Under favourable conditions fruiting bodies (acervuli) with *setae* can be seen with hand lens. Mid- rib infection show elongated red or purple lesions on which the black acervuli can be clearly seen. Red rot infected stem when split open show discontinuous discoloration which gives morbled appearance. Seed treatment with carbendazim + thiram or captafol gave the best control of *C. graminicola* including other seed-borne fungi (Valarini *et al.* 1988).

Charcoal rot (*Macrophomina phaseolina* (Tassi) Goid.)

It occurs in some parts of Tamil Nadu and Maharashtra states. Upon systemic infection, lower stem becomes soft and weak resulting in lodging, usually at second or third internode. Infected stem if split open, the vascular fibers appear black because of the sclerotial bodies of the fungus. Therefore, it is called charcoal rot. Drought stress and high temperatures at grain-filling stage increases the infection. The fungus is soil-borne and survives saprophytically on crop residues; hence difficult to control.

Maize (*Zea mays* L.)

In India it is grown for both grain and fodder purposes. Where the climatic conditions are fairly warm it can be grown throughout the year. With the introduction of hybrid maize varieties the cultivation practices have become very intense. This has favoured several diseases. About 20 diseases of maize have been reported from India. At present few are of major importance, causing severe damage to the crop (Rangaswami, 1988).

Philippine Downy Mildew (*Peronosclerospora philippinensis*)

The infected plants are conspicuous in appearance due to the presence of chlorotic stripes on young developing leaves and pronounced whitish downy growth on all infected tissues. Such plants show yellow discoloration, browning and necrosis of the blade and stunting. Young plants show systemic infection whereas older plants are resistant. Management includes destruction of the weed host, *kans* grass (*Saccharum spontaneum*), in the fields and protection of young seedlings by spraying 0.2 per cent mancozeb.

Stalk and Ear rot (*Diplodia maydis* (Bert.) Sacc.)

The pathogen infects the lower nodes which become yellow brown in colour, soft and brittle in texture; the pith disintegrates. Early ear infection results in premature bleaching of the grain husk and the entire rotting of the ear. The pathogen also causes seedling blight due to infected mesocotyles which turn pale grey. Cultural practices are important such as balanced fertilizer regimes and crop rotation. 'Flint' varieties are more resistant than 'dent' varieties. Thiram seed dressing can control seed-borne infection.

Stalk rot (*Pythium aphanidermatum* (Edson) Fitz.)

The disease is favoured by high atmospheric humidity and appear in devastating proportions on young plants soon after heavy rains. The symptoms are broadly similar to those caused by *D. maydis*.

Rust (*Puccinia sorghi* Schw.)

Rust caused by *P. sorgi* is more important than *P. polyspora* Underw. The fungus produces uredial and telial stages on maize. Minute uredia occur on both leaf surfaces and sometimes on the husk and other floral parts. The rust postules are yellowish in the early stages but later become brown, and are surrounded by chlorotic haloes. In severe infections, the entire plant becomes pale and looks unhealthy, even from a distance. Among the available varieties 'Cuzco' is resistant to all biotypes of the rust in India.

Common smut (*Ustilago maydis* (DC) Cda.)

The fungus produces gall on the ears, axillary buds, tassels, stalks, and more rarely, on the leaves. The epidermal tissues of the gall are dull-white or grey in colour, much bulged from the surface and shiny. They rupture to expose the black powdery mass of spores. The spores are mixed up with gelatinous matrix when young, but become dry with maturity. Cultivation of smut resistant varieties provides most effective control. Seed treatment with one percent organo-mercurial fungicide at the rate of 2.5 g / Kg of seed can control the disease.

Head smut (*Sphacelotheca reiliana* (Kuhn) Clinton)

The fungus attacks the pistillate and staminate inflorescences, forming tumours in place of flowers and the floral parts get converted into black powder mass of smut spores. The sori vary in mainly through soil-borne spores. The disease is checked by careful sanitation. Seed treatment with Baytan 19.5 (Triadimenol) gave the best results against both soil-borne and seed-borne infections by *S. reiliana* (Pelmus and Pelmus, 1988).

Charcoal rot (*Macrophomina phaseolina* (Tassi) Goid.)

The disease may be troublesome in drought-stressed crops. The fungus produces a greyish dry rot of the crown and stem base. Black sclerotial bodies of the fungus can be seen in affected stems. *Trichoderma harzianum* can provide biological control under field conditions (Elad *et al.* 1986).

CASH CROP

Cotton (*Gossypium* spp.).

The world's most important fibre crop is extensively grown in USA, China, USSR, Egypt, Mexico, Uganda, Nigeria, Tanzania, Sudan, India and West Indies. The production of cotton in India is greater than that of all other fibres put together. This important cash crop is attacked by a number of diseases and insect pests (Hill and Waller, 1988). Four major diseases are as follows :

Black arm or Bacterial blight (*Xanthomonas compestris* pr. *malvacearum*)

This disease is of major concern in Maharashtra, Karnataka, Andhra Pradesh, Tamil Nadu and Madhya Pradesh. The disease is both seed-borne and soil-brone. Lesions on the leaves appear as minute, watersoaked, angular spots, which subsequently become brown and finally black angular dead lesions on both sides of the leaf. On balls, it causes small, water-soaked spots. Shedding of leaves and balls cause heavy damage. Black elongated lesions on young stems and branches cause girdling and death known as black-arm. Sulfuric acid, mercuric chloride or Agrosan GN as seed dressers and terramycin, penicillin or streptomycin as sprays can prevent infection (Akhtar and Khan, 1988).

Wilt (*Fusarium oxysporum* f. sp. *Vasinfectum* (Atk.) Snyder and Hansen)

Wilt is predominent in black cotton soils of Karnataka, Tamil Nadu, Madhya Pradesh and Western Uttar Pradesh. The plants are affected during all stages of plant growth. Infected young seedlings show vein- clearing of the leaves followed by inter-veinal tissue necrosis and soon wilt and die. If the plants are attacked a little late, the leaves become flaccid which result in drooping and wilting, starting from the oldest leaves at the base and proceed upward. Often there is complete defoliation, leaving the stem alone standing in the field. The base of affected stem turns black and the plant eventually wilts. The characteristic symptom is the brown discoloration of the vascular system, when the stem is cut open. Varieties 'Jayadhar', 'Jarila', 'Vijay', 'Pratap' and 'BDS' are wilt resistant. Good drainage, soil tillage, and adequate application of potash are some of the more important precautionary measures.

Rhizoctonia root rot (*Rhizoctonia bataticola* (Taub.) Butter and *Rhizoctonia solani* Kuhn)

Root rot of cotton is caused by *R. bataticola* and *R. solani* in Punjab and Haryana and only by *R. bataticola* in Rajasthan. The chief symptom is a sudden and complete wilting of the plants. The disease spreads in the field in concentric circles. Affected plants when pulled come out easily, remain devoid of lateral roots and show shredding of wood. Minute sclerotial bodies can be seen on the surface of the wood. Mixed cropping has been recommended for checking the disease intensity (Rajpurohit, 1983). Host plant resistance is also available.

Anthracnose (*Glomerella gossypii* (Ascomycete) imperfect stage = *Colletotrichum gossypii* Southw.)

Typical reddish brown sunken anthracnose lesions can occur on leaves, stems and bolls. On seedlings, it causes post-emergence blight. Young bolls are susceptible. Circular sunken lesions on bolls expand to envelop the whole organ and lint inside is ruined. Cultivar 'Stoneville' is resistant. Destruction of crop residue and crop rotation help to restrict disease development. Spraying of plants after boll formation with 1 per cent Bordeaux mixture and other recommended chemicals can check the disease.

OILSEED CROPS

India is the largest producer of oilseeds in the world. A preliminary appraisal of pathogenic fungi showed that major diseases of oil seed crops are leaf spots, blights, rusts, powdery mildews, and downy mildews (Chohan, 1978). Important rainfed oilseed crops grown in India are: groundnut, castor, sesamum, rapeseed mustard, linseed, safflower, sunflower and soyabean.

Groundnut (*Arachis hypogaea* L.)

It is the most important oilseed crop of India. It is grown both under rainfed and irrigated conditions. Symptoms, factors affecting disease development and control measures of important diseases are as follows :

Tikka : Combination of early leaf spot (*Cercospora arachidicola* Hori) and late leaf spot (*Phaeoisariopsis personata* (Berk. and Curt) V. Arx)

This disease has been referred to as mycosphaerella leaf spot, cercospora leaf spot, brown leaf spot, peanut cercosporiosis, viruela and tikka. Early and late leaf spots, combinedly, cause yield losses ranging from 10 per cent to over 50 per cent, but vary considerably from place to place and between seasons (McDonald *et al.* 1985).

For both early and late leaf spots small chlorotic spots appear on leaflets 10 days after incubation. The spots then develop in about 5 days into mature, sporulating lesions. Lesions caused by *C. arachidicola* are subcircular and vary from 1 to 10 mm in diameter. They are dark brown in the adaxial leaflet surface where most sporulation occurs, and a lighter shade of brown on the abaxial leaflet surface. Lesions caused by *P. personata* are smaller, more nearly circular, and darker in colour than those of *C. archidiocola.*

Sporulations that occur on abaxial surface lesions are rough in appearance and black in colour.

Integrated control strategy should include all available and compatible disease control measures. Fungicides should be used to combat both leaf spots and rust in combination with insecticides where insect pests are also a problem. The International Crops Research Institute for the Semi-Arid Tropics has developed resistant cultivars.

Rust (*Puccinia arachidis* Speg.)

In conjunction with leaf spots caused by *C. arachidiocola* and *P. personata*, it is devastating (Subrahmanium *et al.* 1984) Symptoms are readily recognized as orange coloured postules (Uredia) on the abaxial surface of the leaves and then rupture to expose masses from 0.3 to 1.0 mm in diameter. Rust damaged leaves tend to remain on the plant. Strict plant quarantine regulations should be enforced to avoid movement of rust spores on pods or seeds to areas where the disease is not present. Economic benefits of spraying fungicides to groundnut foliar disease have been worked out (Kannaiyan and Haciwa, 1990).

Stem rot (*Sclerotium rolfsii* Sacc.)

It is found in all major peanut growing areas of the world. Initial symptoms are wilting of a branch or main stem. Leaves turn dark brown and shed prematurely. While mycelial growth of the fungus can be seen at or near the soil line around affected plant parts. Infected pods are rotten. Deep - plowing and seed treatment with thiram at the rate of 2g/kg seeds have been reported to be effective in the management of this disease.

Charcoal rot (*Macrophomina phaseolina* (Tassi.) Goid.)

Distributed throughout the world, it causes disease in a wide range of crops (Singh, *et al.*, 1990). In peanut, it is responsible for seed and seedling rots, wilt, root and stem rots, leaf spot, rotting of developing pods and seeds, and concealed damage. In India, it is most common as peanut seedling disease. The occurrence of water-soaked lesions of the hypocotyl near the soil surface is a characteristic sign of the disease. After the hypocotyl is girdled, the seedling dies. Older plants wilt and dead tissues turn black, as sclerotia of the fungus develop profusely. Carbendazim (lg/kg seed) is effective against pre- and post-emergence seedling rots caused by the fungus (Sharma and Bhowmik, 1982).

Crown rot (*Aspergillus niger* Van Tiegh.)

An important peanut disease. Seedling and young plants are most susceptible. Seed rot and pre-emergence damping-off are common components of the disease which result in sudden wilting of young plants within 30 days after planting. Protectant fungicides such as thiram and captan at the rate of 2 g/kg seeds can reduce the incidence (Ravindranath, 1974).

Yellow mold (*Aspergillus flavus* and *A. parasiticus*)

Produces aflatoxin contamination in the stems of peanut seedlings and within peanut pods and seeds. Toxin produced by the fungi are translocated throughout the seedling. Such seedlings lack a secondary root system, a condition known as 'aflaroot'. Avoid moisture build-up during transport. Destroy contaminated lots. Formaldehyde, Sodium hypochlorite, or ammonia reduces aflatoxin levels.

Rape Seed-Mustard (*Brassica compestris, B. juncea* and *Eruca sativa*)

Alternaria leaf blight *Alternaria brassicae* (Berk.) Sacc.

Most widespread and destructive disease of rapeseed and mustard in India. Both leaves and pods are damaged by the disease. infected seeds. become deformed and discoloured. Heavy losses occur under conditions of late winter rains. Three sprays of mixture of Mancozeb and Metasystox at fortnightly intervals not only reduce disease incidence and aphid infestation but also give higher yields (Munde and Bhowmik, 1984).

White rust (*Albugo condida*)

The fungus appears on the undersurface of the lower leaves and later on both sides. Almost simultaneously, the crop is attacked by Alternaria and aphids. The combined effect is serious. Metalaxyl alone or with Captafol or Metasystox with Captafol provided significant reduction in intensity of white rust, leaf blight and aphid infestation (Bhowmik and Singh, 1984).

Castor (*Ricinus communis* L.)

Castor is an important oilseed crop of India. Rust, leaf spots and charcoal rot are important diseases.

Rust (*Melampsora ricini* (Biv.) Pass.)

It is a serious disease. The rust pustules occur mostly on the lower surface of the leaf. Lower leaf surface is of orange-yellow colour with corresponding areas on the upper surface of the leaves as yellow. Sori are grouped in concentric rings and many coalesce and cause drying up of leaves. Resistant varieties are now available.

Leaf spot (*Cercospora ricinella* Sacc. and Berless)

It is common in all castor growing regions of India. Minute water soaked lesions, which enlarge to turn brown and irregular with deep brown margins and greyish white centres, 2 to 5 mm in diameter, coalesce to form large patches of the leaf black. Older infected leaves become blighted and withered.

Charcoal rot (*Macrophomina phaseolina* (Tassi.) Goid.)

It is an important disease in rainfed areas. Plants at all ages are susceptible to the disease. Affected plants show drooping of leaves and top of the plant. Infected stem bear minute pycnidial bodies of the fungus. Plants remain stunted.

Sesamum (*Sesamum indicum* L.)

Grown extensively in tropical and sub-tropical climates. Of several diseases, phyllody, charcoal rot and bacterial blight are economically important.

Sesamum phyllody (*Orosium albicinctus* (*Vasudeva* and *Sahambi*)

Transmitted by a Mycoplasma Like Organism (MLO). Vein cleaning of the leaves, phylloid flowers, profuse branching and shortening of internodes are the characteristic symptoms of the disease. Late sowing and insecticidal sprays can prevent effective transmission. Sesamum lines, 'Sl-289', '951', '1535', '2194', '2201', '2373' and '2635' possess resistance to phyllody.

Charcoal rot (*Macrophomina phaseolina* (Tassi) Goid.)

The germinating seeds and seedlings may become brown and ultimately rot. The older plants are affected at the base of the stem, causing drooping of leaves and top of the plants. Plants remain stunted. The mycelium of the fungus progress upward in the stem, and as the stem dries, pycnidia are formed as minute black dots. Affected capsules open prematurely, exposing shrivelled and discolored seeds. The fungus is both seed-and soil- borne. Seed treatment with captafol, thiram or carbendazim gives complete control of seed-borne infection (Vyas 1981). Varieties 'Ajinio Attar 55', 'C 50', 'RT-1', 'G-5' and 'NP 6' are reported to be resistant.

Bacterial blight (*Xanthomonas compestris* (Pamel) Donosn Pv. *sesami* (Sabet and Donson) Dye)

It is serious in the states of Madhya Pradesh, Rajasthan and Uttar Pradesh. About 20 per cent loss in yield has been reported from Jabalpur area of Madhya Pradesh. Small water soaked light brown lesions develop on the margin of the cotyledonary leaf about 1 to 12 days after sowing and cover entire cotyledons. In older plants water soaked spots turn dark olive green, increase in size (2 to 3 mm), and show a dark red-brown to black colour. The spots are angular and coalesce to form large blotches and result in death of the plant. High temperature and humidity favour the disease. Urdaneta and Mazzani, 1976 found Copper oxychloride useful in reducing the severity of the disease.

Linseed (*Linum usitatissimum* L.)

It is grown extensively as a *rabi* crop in Madhya Pradesh, Uttar Pradesh, Rajasthan, Bihar, Gujarat and Maharashtra. Rust, blight and wilt are economically important diseases.

Rust (*Melampsora lini* (Pers.) Lev.)

It is common in North India. Characteristic bright orange coloured, minute pustules (Uredosori) are produced on the leaves and stem. With the disease advancement, pustules turn black (telial stage). Excessive nitrogenous fertilizers favour the disease development. Tridemorph (0.05%) provided best control (Thakure *et al.*, 1987).

Alternaria blight (*Alternaria lini* Dey)

Severe in parts of Uttar Pradesh and Madhya Pradesh. It produces dark brown, irregular patches on the leaves, which curl and dry up when infection is severe. The fungus attacks floral buds, which shrink and fail to open. The disease is both seed-and air-borne. Seed treatment with Agrosan GN reduces seed-borne infection. Early sowing reduces the disease incidence.

Wilt (*Fusarium oxysporum* f. sp. *lini* (Bolley) Snyder and Hensen)

Prevalent in North India and Rajasthan. The fungus is soil-borne. It causes wilt within 3-4 days of the appearance of first symptoms. Infected older plants survive longer. A black discoloration of the vascular tissues can be seen. Low soil moisture, and light sandy soils are reported to favour the disease.

Safflower (*Carthamus tintorius* L.)

Extensively grown in Madhya Pradesh, Andhra Pradesh, Karnataka, Tamil Nadu and other states. Two major diseases are described in details.

Wilt (*Fusarium oxysporum* Schlecht. f. sp. *Carthami* Klisiewicz Houston)

It is a threat to the cultivation of safflower in India. Symptoms of the disease are manifested at all stages of growth. In seedlings, cotyledonary leaves show small brown spots which tend to become rolled and curved. Golden yellow discoloration of the leaf, followed by wilting, epinasty and vescular browning are characteristic symptoms. The pathogen perpetuates through seed and soil. Wilt is less severe where paddy or millets precede safflower.

Alternaria blight (*Alternaria carthami* Choudhary)

Small brown to dark-brown concentric spots of 1 to 2 cm diameter appear on leaves. The centre of the mature spots is usually lighter in colour and develop shot holes and in severe infections irregular cracking of the leaf blade occurs. Flower heads are attacked and flower buds shrivel without opening. Thiram seed treatment and foliar Dithiocarbamate fungicidal sprays reduce the disease incidence (Anon. 1982; Basavarasaiah *et al.* 1977).

Sunflower (*Helianthus annus* L. Var. *marcocarpus* (DC) CK.

With continuous cultivation of the crop in the same field, the crop suffers from diseases resulting in large losses in yields. The flowering behavior facilitates attack by insects. Alternaria blight and charcoal rot are major diseases.

Alternaria blight (*Alternaria helianthi* (Hansf.) Tubaki and Nishihara.)

Characteristic symptoms are development of dark-brown to black circular-to-oval spots, varying from 0.2 to 5.0 mm in diameter. The spots are surrounded by a necrotic chlorotic zone with grey-white necrotic centre marked with concentric rings. At the later stages, elongated spots are formed on petioles, stems, and ray florets. Under high humidity conditions, the spots enlarge in size and coalesce resulting in blighting of leaves and sometimes rotting of flower heads.

Kolte *et al.* 1979 demonstrated that 0.3 per cent mancozeb, when sprayed four times at an interval of 7 to 10 days, gives effective control. Use of resistant cultivars and adjustment in planting dates provide effective management of the disease (Kolte and Tewari, 1977; Agrawal *et al.* 1980).

Charcoal rot (*Macrophomina phaseolina* (Tassi) Goid.)

The disease is of great economic importance in arid areas of the world. The common symptom is sudden wilting of plants after pollination. Early symptoms are not visible on infected plants, but the plants become weak, mature early, and when dry they exhibit black ashy discoloration of the stem and, therefore, called 'charcoal' rot. In affected stem, black microsclerotia are formed abundantly in the pith. Thiram seed treatment has been reported to be effective in control of seed-borne infection of the fungus (Sivaprakasam *et al.*, 1975).

Soybean (*Glycine max* (L.) Merrill)

A primary source of vegetable oil and protein. Common diseases are :

Downy mildew (*Peronospora manshurica* (Naum.) Syd. exhaum.)

It appears on the upper surface of young leaves as pale green to light yellow spots which enlarge and turn grayish brown to dark brown with a yellowish green margin and finally turn entirely brown. On the lower surface of leaves, particularly in moist weather, lesions are covered with tufts of grayish to purple sporangiophores. Pod infection may occur without external symptoms. Infected seeds can produce systemically infected seedlings. Seed treatment, crop rotation and use of resistant cultivars can control the disease.

Bacterial blight (*Pseudomonas syringae* pv. *glycinea* (Coerper) young, Dye & wilkie)

Blight lesions are most conspicuous on leaves but also occur on stems, petioles, and pods. Small, angular, translucent, water-soaked yellow to light brown spots appear on leaves, and are surrounded by a water-soaked margin bordered by a yellowish green halo. Young leaves are most susceptible. Infected leaves are distorted, stunted, and chlorotic. Pod lesions turn dark brown to black with age. Use of disease free seeds and crop rotation can reduce the disease incidence.

PULSES

Pulses are a major source of protein in the predominantly vegetarian diet of the people of India. India produces about 90 per cent of the pigeonpea and 70 per cent of the chickpea grown in the world (Nene, 1986). The major obstacles in the way of increased pulse production are the diseases and insect pests, which are responsible for reduction and uncertainty in pulse yields, besides other stress factors (Grewal, 1982). Principal dryland pulses are chickpea, pigeonpea, black and green gram, lentil, peas, cowpea, mothbean and rajmash.

Chickpea (*Cicer arietinun* L.)

More than 50 pathogens and 60 insect pests have been reported from different parts of the world (Nene, 1980). Only few of them have the potential to devastate the crop. Blight, grey mould, wilt and dry root rot are responsible for significant losses in this crop, (Nene and Reddy, 1987).

Blight (*Ascochyta rabiei* (Pass.) Labr.)

The disease is a limiting factor to chickpea production in India and Pakistan. Repeated epiphytotics have been reported from important chickpea growing countries, viz., India, Pakistan, Spain, Syria and Tunisia (Nene *et al.* 1984).

The pathogen attacks all above-ground parts of the plant. Necrotic lesions girdle the stems or the base of the branches and the plant parts above the lesions rapidly die. On the leaflets, the lesions are initially round and watersoaked. Later they develop into well defined, round or elongated brown spots with sunken tissues and dark margins. The lesions on the pods are round with dark margins. Pycnidia as dark black small bodies can be observed on all the infected plant parts. Primary inoculum is seed-and soil-borne. Secondary spread is air- borne. To control seed-borne infection, Grewal (1982) recommended seed treatment with Bavistin 25% + TMTD (Thiram 50% at the rate of 250 g/q of seed. Foliar sprays are generally ineffective under epidemic situations. Use of resistant cultivars like 'Gaurva' is the best method, since the other methods are either impractical or uneconomical.

Grey mould (*Botrytis cinerea* Pers. ex Fr.)

The disease was responsible for heavy losses in the Indo-Gangetic plains of India (Grewal and Laha, 1983). The symptoms appear on stems, leaves, inflorescences and pods as gray to dark brown lesions covered with mouldy sporophores under high humidity conditions. Growing tips and flowers are particularly susceptible to infection. The plants show yellowing and there is shedding of bark. Seed treatment with Bavistin 50 WP or Bavistin + Thiram at the rate of 2.5g/kg of seed not only eliminated seed-borne infection but also protected the seedling upto 8 weeks after sowing against aerial infection by *B. cinerea* (Grewal and Laha 1983). Resistant lines, viz. 'ICCV 5', 'BG267' and 'ICCC 32' are now available.

Wilt (*Fusarium oxysporum* f. sp. *ciceri* (Padwick) Synd. & Hans.)

Fusarium wilt is a serious disease of chickpea in India, Iran, Pakistan, Nepal, Burma, Spain and Tunisia (Nene *et al.* 1984). Affected plants show drooping of the entire plant. There is a slow change of the green colour and the plants look dull green. Roots of the wilted plants show no external rotting, drying or discoloration (Nene *et al.* 1978). Roots and stems of wilted plants when split vertically, show internal discoloration of the pith and Xylem. The seed-borne inoculum can be eradicated by seed dressing with Benlate T (benomyl 30% + thiram 30%) at 0.15% rate (Haware *et al.*, 1978). Varieties 'ICCV 2', 'ICCV 5' and 'ICCC 32' are resistant.

Dry root rot (*Rhizoctonia bataticola* (Taub.) Butler; *Macrophomina phaseolina* (Mauble.) Ashby)

Dry root rot is an important disease in Central and Southern India (Vidhyasekharan *et al.* 1977). The tap root when pulled out is found to be dry and devoid of lateral and finer roots. The dead root shows shredding of bark and is brittle towards tip. Minute dark brown fungal sclerotia can be seen below the bark. Drying of plants appear suddenly in the field where affected plants are scattered (Nene *et al.* 1978). Multiple disease resistant cultivars viz., 'ICCV 2' 'ICCV 5' and 'ICCC 32' are now available.

Pigeonpea (*Cajanus cajan* (L.) Millsp.)

Commonly known as red gram or *arhar* is cultivated in the states of Maharashtra, Uttar Pradesh, Madhya Pradesh, Karnataka, Gujarat, Andhra Pradesh and Tamil Nadu. Over 90 per cent of the world pigeonpea is produced in India. Wilt, sterility mosaic and phytophthora blight are the major diseases of this crop (Nene *et al.* 1985).

Wilt (*Fusarium udum* Butler)

The annual pigeonpea crop loss due to wilt in India alone has been estimated at US $ 36 million, while in eastern Africa annual losses were estimated at 5 million (Kannaiyan *et al.* 1984). Symptoms are yellowing of the leaves from the base upwards. Brown or dark purple bands appear on the stem surface of partially wilted plants extending upwards from the base. Inter-cropping with sorghum reduces wilt incidence. Varieties 'C-11', 'BDN-1', Maruthi '(ICP 8863)' are resistant to wilt.

Sterility mosaic (Viral)

The disease causes an estimated annual loss of 205, 000 tonnes of grains in India alone (Kannaiyan *et al.* 1984). The symptoms are profuse upright vegetative growth and the absence of flowering branches (Nene, 1972). The disease is mainly transmitted through Eriophyid mite (*Aceria cajani*). Three lines 'ICP 786', '10976' and 'ICP 10977' have been identified by ICRISAT as resistant or tolerant to the disease. Acaricides like tedion, morestan, and kelthane, at 0.01 per cent concentration can kill more than 90 per cent of the mites.

Phytophthora blight (*Phytophthora drechsleri* var. *cajani,* Pal Grewal and Sorbhoy)

The disease causes severe production losses in the states of Uttar Pradesh, Andhra Pradesh, Tamil Nadu and Maharashtra. Symptoms are curling and drying of infected leaves, leading to the infection of the stem at the point of leaf attachment. The lesions on stems girdle the stem causing portions of the plant above the lesions to dry out but remain attached to the point. Longitudinal cuts into newly formed lesions show brown-to-black discoloration of the bark and cambium, but not the older xylem. The roots of infected plants are healthy. Application of potassium fertilizers reduces the incidence of disease. The genotypes 'ICP 5097', '6344', '7200', '8564', '8610', '12749' and 'KPBR 80-2-1' showed field tolerance. Disease tolerant varieties 'BDN-1', 'Bori-1' and 'Pusa Ageti' should be cultivated in disease prone areas.

Black & green gram : Black gram or urdbean (*Vigna mungo* (L.) Hepper) and green gram or mungbean (*V.radiata* L. Wilezek)

Mainly grown in Maharashtra, Madhya Pradesh, Orissa, Andhra Pradesh, Tamil Nadu, Rajasthan and Uttar Pradesh. The pathogens attacking both the crops are mostly common. Yellow mosaic, Cercospora leaf spots, dry root rot and bacterial leaf spots are predominent diseases of these two crops.

Yellow mosaic (Viral)

It is a most serious viral disease of mungbean and urdbean. Typical symptoms are yellow specks which expand and cover the entire leaf area. Irregular yellow and green patches can be observed on the leaves. Reduced infected pods bear immature seeds. The disease is transmitted by white fly (*Bemasia tabaci*) and has a wide host range. Black gram varieties 'Pant U-19' and 'Pant U-30' are moderately resistant and green gram variety, 'Pusa 101', is resistant to yellow mosaic disease.

Cercospora leaf spots (*Cercospora canescence* Ellis and Martin, *C. cruenta* Sacc., *C. dolichi* Ellis & Everlast, *C. kikuchi* Matsumoto & Tomoyasu)

The losses due to cercospora leaf spots can be as high as 47 per cent during warm and wet seasons (Amin and Singh, 1987). The lesions are brown with reddish-brown margins on leaves on susceptible genotypes. If the infection starts early in the season, it causes reduction in the yield. Spraying with captan or zineb @ 2g/litre or Bavistin 0.5 kg in 1,000 litres of water / ha controls the disease incidence (Grewal *et al.*, 1980). Urdbean variety '4-5-2' and mungbean varieties 'ML-5', 'ML-65' and 'L-24-2' have been reported resistant of cercospora leaf spots.

Dry root rot (*Rhizoctonia bataticola* (Taub.) Butler *Macrophomina phaseolina* (Tass) Goid.)

Rhizoctonia bataticola causes seedling rot, collar rot and dry root rot. Dark brown lesions appear on roots and shredding of bark occurs. Small circular brown spots appear on young leaves. Dark brown lesions on stem kill the branches. Pycnidial bodies are produced on leaves and stems. The pathogen is both seed-and soil-borne. Foliar spray with Benomyl is recommended (Kotasthane and Agarwal, 1976). The cultivar 'PIMS-1 to 4' are tolerant to the disease.

Anthracnose (*Glomerella lindemuthianum* (Sacc. & Magn. Scrib.)

This is a common disease of black gram (*Vigna mungo* (Linn.) Hepper, green gram (*Vigna radiata* (Linn.) Wilczek, Frenchbean (*Phaseolus vulgaris* Linn. and cowpea (*Vigna unguiculata* (Linn.) walp. The fungus attacks all plant parts above ground level and at any stage of plant growth. Prominent symptoms appear as water-soaked lesion on pods, which later become brown, and enlarge to form circular spots of varying size. The spots are usually depressed with dark centres, with bright-red, yellow or orange margins. Lesions may occur more often on the under surface of the leaf, leaf petiole and stems. Disease-free seeds and crop rotation can reduce the disease incidence.

Lentil (*Lens culinaris* M. = *Lens esculenta* M.)

Cultivated mostly in the central, northern and eastern regions of India either as sole or as a inter-crop, it is attacked by wilt, collar rot and rust diseases.

Wilt (*Fusarium oxysporum* f. sp. *lentis* Synder & Hansen)

Losses upto 50 per cent were recorded in some fields (Khare *et al.* 1979). The fungus is both seed- and soil-borne. The wilted plants become prominent at flowering and pod formation stage. The plant growth is checked, leaves curl and dry up. Lentil varieties 'Pusa 3', 'Pusa 234', NP-II and 'Vidishah' are resistant. Seed treatment with Carbendazim + thiram @ 250 g/q of seed is recommended.

Collar rot (*Sclerotium rolfsii* (Sacc.) Curzi)

In humid weather, the pathogen spreads on soil and attacks plants at ground level. Brown coloured mustard size sclerotia of the fungus are produced on the white mycelium strands. To avoid collar rot, cropping system, lentil following maize should be avoided. Seed treatment with Carbendazim + thiram @ 250 g/q of seed reduces the incidence.

Rust (*Uromyces fabae* (Pers.) de Bary)

The disease appears in January-February. Brown disease patches can be seen in low lying areas of the field under high humidity conditions towards the end of January. Aecial cups are formed on leaves, pods and stem followed by uredo and teluto postules. Dark brown to black telia of the fungus are formed late in the season mostly on stems and branches. Severely infected plants dry up even before seed setting. Clean seeds can reduce the incidence. Resistant varieties like 'LG 60', 'LG 120', 'JL 279', 'Pant 406' and 'Pant 639' need to be popularized.

Pea (*Pisum sativum* Linn.)

Cultivated both as vegetable and pulse crop. Powdery mildew, rust, and downy mildew are important pea diseases.

Powdery mildew (*Erysiphe polygoni* DC.)

The fungus produces a white powdery growth on the leaves, stems and other plant parts. Lesions first appear on lowest and oldest leaves. These lesions appear as white powdery areas, and severely infected foliage looks blue-white. Tissues beneath these infected areas may turn purplish, after which small black fruiting structures (Cleistothecia) form in the mature lesions. Severe pod infection may result in "hollow" peas. Seed infection may impart gray-brown discoloration to the seeds. The disease was effectively controlled by 3 sprays of Karathane (dinocap), Bavistin (carbendazim) or Benlate (benomyl). 'Rachana' and 'DMR 11' are resistant varieties.

Downy mildew (*Peronospora pisi* Syd. emend. Campbell)

Systemically infected plants show downy growth on the under surface of leaves. As the disease advances, the affected tissues turn gray, dry up and the leaves drop off prematurely. Seed treatment with Metalaxyl (Apron 70SD) at 200 g/100 kg of seed provides effective control (Miller & Whalley, 1981). Deep tillage, removal of crop debris and extended crop rotations are useful techniques for combating this pathogen.

Rust (*Uromyces fabae* (pers.) de Bary)

Light to brown uredosori of the fungus appear on both surfaces of the leaves. Severely infected leaves wither and fall off prematurely. This rust is autoecious, completing its life cycle on peas. Two mancozeb (0.25%) sprays soon after the appearance of rust can provide effective control.

Moth Bean (*Vigna aconitifolia* (Jacq.) Marechal)

Mothbean is an important pulse crop of the desert region. Alternaria blight and dry root rot are two important diseases of this crop (Singh and Srivastava, 1984 and 1989).

Alternaria leaf spot (*Alternaria alternata* (fr.) Keissler)

This disease was first reported by Singh and Srivastava, (1984a). The infection first appears as pale yellow spots either on tips or on the margins of the leaves which gradually advances towards the petiole. In rare cases, small patches in middle of the leaf lamina can also be seen. During initial stages of infection, the diseased spots are separated from the healthy tissues by a dark coloured band, but later on, the lesions spread over the entire leaf and become irregular, brittle, dark olive green to black in colour due to fungal growth.

Dry root rot (*Macrophomina phaseolina* (Tassi) Goid.)

The leaves of affected plants show yellowing and drooping. Dark lesions can be seen on the bark of the stem at the ground level. Root tissues become weak and break off easily. The roots of such affected plants first become brown and then black. In advanced infections, sclerotial bodies can be seen scattered on the affected tissues. All plant stages are susceptible.

Cowpea (*Vigna unguiculata* (Linn.) Walp.)

Anthracnose (*Glomerella lindemuthianum* Shear)

The fungus attacks all aerial parts. Orange to brown elongated stem lesions can occur resulting in a die-back of leaves and branches if the stem is girdled. The centre of the lesions on pods develop a reddish colour, in which are embedded dark acervuli of the fungus, with a brown to purple margin. Since pathogenicity of *C. lindemuthianum* is variable, resistance is often only temporary.

Bacterial blight (*Xanthomonas campestris* pv. *vignicola*)

The initial symptoms appear as tiny water-soaked dots on leaves. These dots remain small and the surrounding tissues die, developing a tan to orange coloration with a yellow halo. On heavily infected leaves, the dead spots merge so that large areas of leaf are affected. The pathogen also infects the stem, causing cracking (stem canker), and pathogen enters the seed. Methods of control include the use of clean seed and of resistant varieties, such as 'S 488', 'K 11', 'T 26' and 'T 5826'.

Cercospora leaf spot (*Cercospora canescence* Ell. & Mart. *Cercospora cruenta* Sacc.)

C. canescens produces circular to irregular cherry-red to reddish brown lesions, upto 10 mm diameter. *C. cruenta* spots begin as a chlorosis (yellowing) on the leaf upper surface which become dotted with spots of dead tissue that enlarge until the whole lesion area is necrotic. Both fungi are controlled by use of clean seed, resistant varieties and application of Benomyl sprays.

Rajmash (*Phaseolus vulgaris* L.)

Attacked by a large number of plant pathogens and insect pests, many of which can reduce yields drastically (Howard and Guillermo, 1980). Four major diseases of rajmash are dealt here.

Angular leaf spot (*Isariopsis griseola* Sacc.)

Initial symptoms are gray to brown lesions surrounded by a chlorotic halo with indefinite margins. Later become necrotic and well-defined with a typical angular shape. Such lesions may increase in size, coalesce and cause partial necrosis and yellowing of leaves, followed by premature defoliation. Lesion size is usually inversely related to lesion number per leaf or leaflet. On pods, lesions are oval to circular with reddish-brown centres surrounded by darker coloured borders. A characteristic symptom of the fungus is the production of dark gray to black synnemeta and conidia in lesions on the lower leaf surface, stems, branches and pods during long periods of high humidity or free moisture. The pathogen can be seed-borne. Crop rotation for 2 years, planting pathogen-free seed in well-drained soils, and removal of previously infected crop debris are advised control procedures. Gonzalez *et al.*, (1977) obtained eoconomical control by applying mancozeb, captafol or Metiram 20, 30 and 40 days after planting.

Anthracnose (*Colletotrichum lindimuthianum* (Sacc. & Magn.) Sarib)

Lesions may develop initially on leaf petioles and the lower surface of leaves and leaf veins as small, angular, brick red to purple spots which become dark brown to black. Pod infections appear as flesh to rust-coloured lesions which develop into sunken cankers (1-10 mm in diameter) delimiting by a slightly raised black ring surrounded by a reddish brown border. Carbendazim and benomyl have been used to control anthracnose (CIAT. 1976 Annual Report).

Rhizoctonia root rot (*Rhizoctonia solani* Kuhn)

The fungus may induce damping off, stem canker, root rot and pod rot. During initial hypocotyl and root infection, the fungus causes dark circular to oblong sunken cankers delimited by brown margins. As infection progresses the sunken cankers enlarge, become red, rough, dry, pithy and retard plant growth. Infected seedlings show girdling of the stem which may cause damping off. Fungicides which control *R. solani* include PCNB, Benomyl, vitavax, thiram, zineb, demosan and captan (1- 3g a.I/Kg seed).

Bean yellow mosaic virus (BYMV)

BYMV particles are transmitted mechanically and by insect vectors such as aphids. Initial symptoms of BYMV systemic infection appear as small chlorotic spots one to three mm in diameter surrounded by yellow halo. These spots enlarge and coalesce to produce general chlorosis. Leaves become brittle, concave and glossy on the upper surface. Yellow and green mottling becomes more intense on leaves as they age. Affected plants remain stunted. Plant resistance appears to be the most reliable control measure.

REFERENCES

Agrawal, J.M., Chhipa, H.P., and Mathur, S.J. 1980. *Indian J. Mycol. Plant Pathol.* 9(1) : 85.

Akhtar, M.A., and Khan, I.U. 1988. *Pak. J. of Agri. Res.* 9(2) : 202-204.

Amin, K.S., and Singh, R.A. 1987. *Natl. Seminar on Pl. Prot. in Field crops.* CPPTI, Hyderabad.

Anahosur, K.H. 1979. *Indian Phytopathol.* 32 : 487-489.

Anon. 1982. *Annual Progress Report,* 1981-82, Directorate of Oilseeds Research, ICAR, Rajendranagar, Hyderabad.

Basavarasaiah, A.B., Kulkarni, S., and Siddaramaiah, A.L. 1977. *Curr. Rev.* 8(4) : 71.

Bhowmik, T.P., and Singh, A. 1984. *Symp.* on *Chem. Control* of *Pl. Dis.*, 1984. Indian Phytopathological Society, New Delhi (Absts).

Chohan, J.S. 1978. *Indian Phytopathol.* 31 : 1-15.

CIAT, 1976. *Annual Report.* Centre Internacional de Agricultura Tropical, Cali, Colombia.

Elad, Y., Zvieli, Y., and Chet, I. 1986. *Crop Protection* 5(4) : 288-292.

Gonzalez, L.C., Gutierrez, R., Cascante, F., and Pertilla, E. 1977. *Agron. Cost.* 1 : 107-118.

Grewal, J.S., 1982. *Indian J. Gen. & Plant Breeding* 42 : 393- 398.

Grewal, J.S., and Laha, S.K. 1983. *Indian Phytopathol.* 36(3) : 516-520.

Grewal, J.S., Pal, Mahendra. and Kulshrestra, D.D. 1980. *Indian J. agric. Sci.* 50(9) : 707-709.

Haware, M.P., Nene, Y.L., and Rajeshwari, R. 1978. *Phytopathol.* 68 : 1364-1367.

Hill, D.S. and Waller, J.M. 1988. *Pests and Diseases of Tropical Crops,* Vol.2 - Field Handbook. Longman Scientific and Technical, Longman Group UK Ltd., England.

Howard, F.S. and Guillermo, E.G. 1980. *Bean Production Problems.* CIAT publication, Cali, Colombia, 424 pp.

Kannaiyan, J. and Haciwa, H.C. 1990. *Tropical Pest Management* 36(1) : 21-22.

Kannaiyan, J., Nene, Y.L., Reddy, M.V., Ryan, J.G. and Raju, T.N. 1984. *Tropical Pest Management* 30(1) : 62-71.

Khare, M.N., Agarwal, S.C., and Jain, A.C. 1979. *Tech. Bull. Deptt. Plant Pathol.*, JNKVV, Jabalpur, M.P., India, 29 pp.

Kolte, J.S., Balasubrahmanyam, N., Tewari, A.N. and Ashwani, R.P. 1979. *Indian J. Agric. Sci.* 49(7) : 555.

Kolte, J.S. and Tewari, A.N. 1977. *Pantnagar J. Res.* 2(2) : 236.

Kotasthane, S.R. and Agarwal, S.C. 1976. *Pesticides* 10 : 35- 36.

McDonald, D., Subrahmanyam, P., Gibbons, R.W., and Smith, D.H. 1985. *Information Bull. No. 21.* ICRISAT, Patancheru, Andhra Pradesh, India.

Miller, M.W. and Whalley, C.V. 1981. *Proc. Br. Crop Prot. Conf.* 1. 341-348.

Munde, P.N. and Bhowmik, T.P. 1984. *Symp.* on *Chem. Cont.* of *Pl. Dis.*, Feb. 16-18, 1984. Indian Phytopathological Society, New Delhi.

Muthuswamy, M. and Narayanaswamy, P. 1981. *Indian Phytopathol.* 34 : 418-420.

Nene, Y.L. 1972. *Res. Bull. No. 4.* GB Pant Univ. of Agri. and Tech., Pantnagar, U.P., India.

Nene, Y.L. 1980. *Proc. Intl. Workshop on Chickpea Improvement.* ICRISAT, 28 Feb.-2 Mar. 1979, Hyderabad, A.P., India.

Nene, Y.L. 1986. *Indian Phytopathol.* 39(3) : 33-342.

Nene, Y.L., Haware, MP. and Reddy, M.V. 1978. *Information Bull. No.3 ICRISAT, Patancheru, Andhra Pradesh, India.*

Nene, Y.L., Kaiser, W.J., Grewal, J.S., Kannaiyan, J. and Beniwal, S.P.S. 1985. *An Annotated Bibliography of Pigeonpea Diseases* 1906-81. ICRISAT, Patancheru, A.P., India.

Nene, Y.L. and Reddy, M.V. 1987. *The Chickpea* (Saxena, M.C. and Singh, K.B., Eds), 233-270, CAB Publication, London.

Nene, Y.L., Sheila, V.K. and Sharma, S.B. 1984. *Pulse Pathol. Progr. Rep.* 32., ICRISAT, Patancheru, Andhra Pradesh, India.

Pelmus, V., and Pelmus, A. 1988. *Probleme de Protectia Plantelor* 16(4) : 253-258.

Rajpurohit, T.S. 1983. *Madras Agric. Journal* 70(11) : 751-752.

Rangaswami, G. 1988. *Diseases of Crop Plants in India.* Prentice Hall of India Pvt. Ltd., New Delhi.

Ravindranath, V. 1974. *Ph.D. Thesis.* Banaras Hindu University, Varanasi, U.P., India.

Shah, S.E., and Mariappam, V. 1988. *Pesticides* 22(10) : 41-42.

Sharma, R.C. and Bhowmik, T.P. 1982. *Pesticides* 12 : 31-32.

Singh, S.K., Nene, Y.L. and Reddy, M.V. 1990. *Plant Disease* 74(10) : 828.

Singh, S.K. and Srivastava, H.P. 1984. *Natl. Acad. Sci. Let.* 7(7) : 211-212.

Singh, S.K. and Srivastava, H.P. 1984a. *Natl. Acad. Sci. Let.* 78(8) : 243.

Singh, S.K., and Srivastava, H.P. 1989. *Indian Phytopathol.* 42 : 164-167.

Sivaprakasam, K., Pillayarsamy, K., and Ayyamperumal, S. 1975. *Madras Agric. J.* 62 & 73.

Subrahmanyam, P., Williams, J.H., Mc Donald, D., and Gibbons, R.W. 1984. *Annals of Appl. Biol.* 104:467-476.

Thakore, B.B.L., Mathur, S. and Singh, R.B. 1987. *Indian J. of Mycol. and Pl. Pathol.* 17(3) : 328-329.

Urdaneta, U.R. and Mazzani, B. 1976. *Agron. Trop.* 26(1) : 47.

Valarini, P.J., Lasca, C.C., Vachiato, M.H., Schmidt, J.R., Dian, P., Chiba, S. 1988. *Fitopatologia Brasileira* 13(3) : 238-243.

Vidhyasekharan, P., Arjunan, G., Mariappam, V., Raganath, K. and Kolandisamy, S. 1977. *Indian Phytopathol.* 30 : 537-538.

Vyas, S.C. 1981. *Pesticides* 15(9) : 10.

Williams, R.J., Fredesiksen, R.A. and Girard, J.C. 1978. *Information Bull. No.2,* ICRISAT, Patancheru, Andhra Pradesh, India.

29

POST HARVEST TECHNOLOGY FOR DRY LAND CROPS

T.P. OJHA

INTRODUCTION

Post harvest technology studies in respect of food grains, orchard crops and vegetables and other food materials aim at determining the gains derived through value addition activities and preventing losses to make more of food materials available for human consumption, animal production as well as to the advantage of the society at large. In the most traditional manner, the post harvest activities involve the following operations which may be improved to take better advantage by the rural producers of the raw materials. The activities are listed in Table 1.

POST HARVEST LOSSES

In the food grains production and utilization system, the post harvest losses occur at all stages of operations. These losses occur right in the field with standing crops at all stages namely, transport, threshing and so on. But these losses can be avoided by adopting improved post harvest practices evolved through researches. The reductions of post harvest losses, especially those caused by insects, micro-organisms, rodents, birds and animals can increase available food

Table 1. Post harvest operations

SNo.	Activities	Traditional methods	Improved methods
1.	Threshing	Mannual beating or Animal treading	Mechanical threshing with important designs of these machines
2.	Cleaning	Mannual action with ordinary baskets	Mechanical cleaning with mannual or mechanical power
3.	Final cleaning	Mannually operated SUPA simple devices	Manual cleaner cum graders or power operated machines
4.	Drying	Open yard sun drying	By means of solar dryers or heated air dryers using mechanical power
5.	Storage	Earthen storage bins or bag storage	Metalic bins, stone bins or brick bins of improved designs
6.	Milling	Traditional methods using hand pounding devices or Stone wheels (*chakkis*) or oil *ghanis*	Modern rice or flouroilmills of different capacities
7.	Utilization of by-products	Direct feeding to animals	Using solvent extraction for separating oil from cakes or making other food products
8.	Marketing	Selling the raw material to middleman at low prices	Selling the graded products or processed materials to Super Bazar with good profits.

supplies, particularly in developing countries and more so in backward pockets of the countries where the losses are very high and the need is the greatest. It has been reported that the post harvest losses amount to the extent of 9.33 per cent in India. However, in certain unfavourable pockets these losses have been much more than the average value. The overall scenario of harvest losses taking place in respect of paddy crops has been presented in Table 2.

Table-2. Grain losses at different harvesting time

Variety	Stage of harvesting	Percent loss
	One week before maturity	0.77
IR-8	Two weeks after maturity	3.35
	Three weeks after maturity	40.70
	Four weeks after maturity	60.46

The avoidable losses occurring in the field, at the threshing yard, during the storage and while converting the raw material (food grains) to a ready to cook food or process products, must be controlled by adapting modern methods and improved technologies. It has also been observed that the small farmers of different regions and more so of the poor strata are not aware of the modern technologies in respect of post harvest activities. Adequate training and demonstration programmes need to be organized to push the technologies to them. Both men and women have to be educated in this respect.

Losses vary by crop, variety, year, pest and pest combination, length of storage, methods of threshing, drying, handling, storage, processing, transportation and distribution rate of consumption and according to both the climate and the culture in which the food is produced and consumed.

MOST POPULAR FOOD GRAINS AND FRUITS OF ARID ZONES

Kharif or rainy season crops are the most predominant food grains of the arid regions where rainfed agriculture is the main support of the farmers. *Bajra*(Pearlmillet), *arhar*(Pigeon pea), Millets, *guar* (Clusterbean) constitute the main crops of the region. Among the fruits, the arid fruits like *ber*, *bel*, mango and *Anar* are the most popular fruits which are produced in the region. However, vegetables of all types can be grown in the region.

Since the commencement of green revolution or even prior to that period, much needed efforts were made in increasing the productivitiy of different crops. But, the maximum efforts were made mainly for the crops like paddy and wheat. Not so much attention was paid towards improving the productivitiy of other crops i.e. coarse grains, oil seeds and pulses. Perhaps it would not be incorrect to say that hardly any significant work was done in improving the varieties of vegetables, fruits and plantation crops which are grown under severe climatic conditions of deserts. As a consequence, no post harvest attention was paid to the crops grown in Arid regions of India. The farmers of these regions have not been able to take advantage of the modern technologies developed in respect of post harvest activities relating to food grains, fruits and vegetable.

Storage moisture content is an important parameter which determines the shelf life of food grains and other food materials. The values of moisture contents for the safe storage of food grains are presented Table-3.

POST HARVEST OPERATIONS

Post Harvest operations are those activities which are carried out from the stage immediately after harvesting to the final stage of utilization of the products. Briefly, these activities include threshing, drying, cleaning, grading, transport, storage, processing, separating, by-product utilization and so on. In simple words, these activities are termed as "Value Addition" activities so as to improve the quality, storability and utilization efficiencies. Thus, the product (raw material) is transformed into different grades, shapes and sizes which provide opportunities to the producers to earn more income than what he gets by selling the raw products without value addition. Let us take two products, one from cereal group and another from fruits to understand the utility of post harvest operations

Table 3. Moisture content during harvesting and storage :

Crop	Per cent (W. B*) Optimum during harvesting	Usual humidity when harvested in central part of the country	Required for safe storage	
			For 1 Year	For 5 Years
Paddy	22-25	16-24	14	12
Wheat	18.5-20	9.17	12-14	11-12
Corn	24-30	14-30	13	12
Barley	20-25	14-25	13	12
Sorghum	20-25	14-25	12	11
Oats	20-25	14-20	13	12

Source : Hall, C.W. : Drying Farm Crops.
* W.B. Wet Basis. The moisture content on wet basis is obtained by dividing the weight of water present in the material by the total weight of material with moisture.

which are carried out to convert the raw material into value added finished products suitable for human consumption.

Paddy - Rice System in a Rice Mill

In order to simplify the diagram, it has been assumed that the dry, clean and graded paddy is received at the rice mill. It is also worth mentioning that value addition activities for converting paddy into rice involves paddy shelling by rubber roll shellers and rice polishers or metallec hullns or central fugal shellers. (Fig. 1). Similarly, the value addition activities on bran and husk can be done to produce more valuable products like cooking oil, soap oil, gums, waxes, deoiled cakes etc. Also, the husk can be utilized to produce energy, white ash, ordinary ash, ceramics, silicon etc. All such finished products are obtained by means of value addition activities performed on the raw materials and also on their by products.

Processing operations as stated earlier aim at improving the raw materials by 'Primary Processing' and transforming into the new products by 'Secondary and Tertiary Processing' activities. For proper storage of paddy or polished rice, the Pusa Bin Fig. 2 or Metal bins can be used satisfactorily. These bins are suitable for small and medium farmers.

Processing of Mango Fruits

The freshly harvested raw or ripe mangoes are subjected to primary and secondary processing activities to transform the raw materials into value added useful finished products for human and animal consumption. Unripe mangoes are graded, washed and then used for making pickles or powders. The operations like washing, boiling, drying, size reductions, packaging etc. are performed to convert raw mangoes into use-

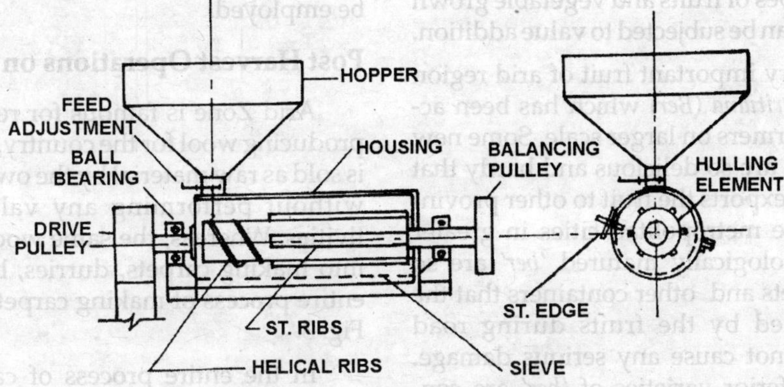

Fig. 1(a) Engleberg Rice Huller

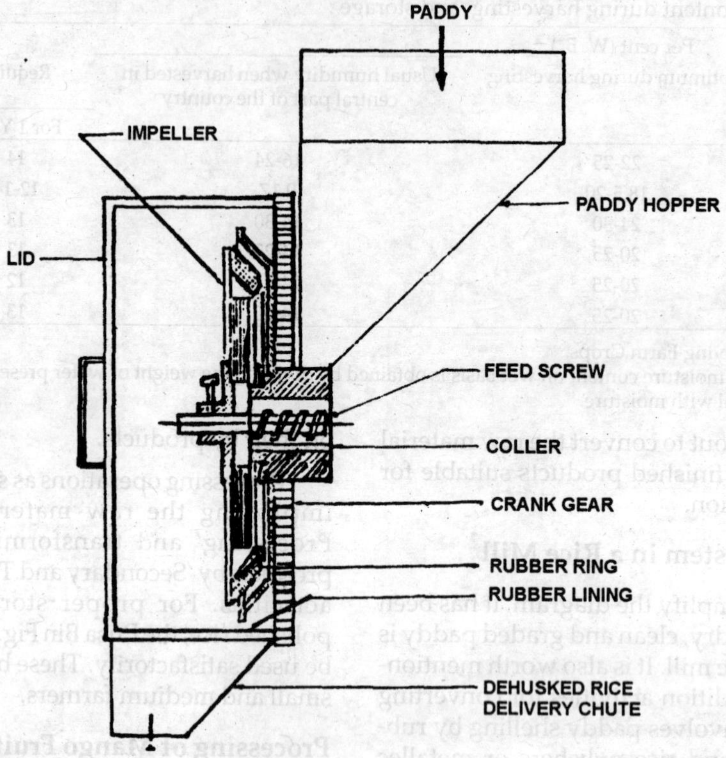

Fig. 1(b) Centrifugal Sheller

ful products.

Similarly, ripe mangoes are subjected to washing, grading, deskining, crushing, filtrations, boiling, drying or dehydration to make jam, squash, fruit mat, nectar, mango powder etc. Also, the mango stone is crushed and used for preparing animal feed or extraction of mango fat.

Thus, all types of fruits and vegetable grown in arid regions can be subjected to value addition.

Another very important fruit of arid region is *Zizyphus mauritiana* (*Ber*) which has been accepted by the farmers on larger scale. Some new varieties of *'ber'* are so delicious and hardy that Rajasthan State exports the fruit to other provinces or large size metropolitan cities in greater volumes. Physiologically matured *'ber'* are so packed in baskets and other containers that the impacts received by the fruits during road transport does not cause any serious damage. Some of the inferior varieties of *'ber'* are consumed locally or processed to make *'ber'* powder after dehydration by open yard drying or by means of solar dryers. Solar Cabinet dryers or polyethylene chambers can be successfully employed for this operation. The most common features of the devices which can be used to produce superior quality dried products are shown in Fig. 2. However, for making the powders, traditional hammer mills or burr mills can be employed.

Post Harvest Operations on Wool

Arid Zone is famous for rearing sheep and producing wool for the country. The sheep wool, is sold as raw material by the owners of the herds without performing any value addition activities. Whereas, the same wool is transformed into making carpets, durries, blankets etc. The entire process of making carpets is presented in Fig. 3.

In the entire process of carpet making so many value addition operations are performed

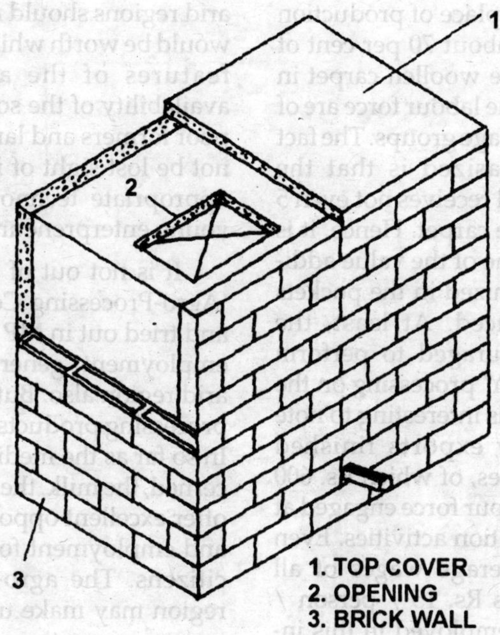

Fig. 2. Pusa Bin

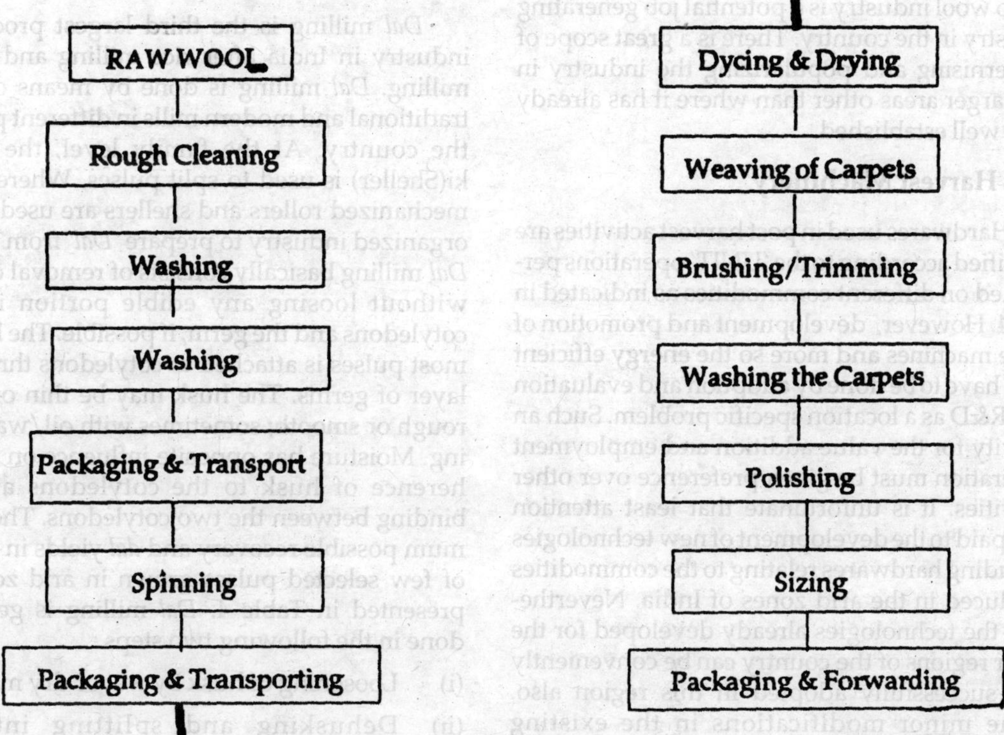

Fig. 3. Wool Processing System

generally far away from the place of production of wool. It is reported that about 70 per cent of the total market value of the woollen carpet in India is the labour charge. The labour force are of various skill groups and also age groups. The fact which needs to be emphasized is that the producer of the raw material receives not even 5 per cent of the value of the carpet. Hence, it is worth emphasizing that some of the value addition activities must be organized in the pockets where the wool is produced. At least, the producers should be encouraged to perform what is called the 'PRIMARY' processing on the raw and uncleaned wool. It is interesting to note that the Carpet industry exports finished products worth Rs. 800 crores, of which Rs. 600 crores or so is paid to the labour force engaged at various stages of value addition activities. Even if we assume that the average wages of all categories of labour force is Rs. 15 / person / hour, the total man-hours employed in this industry would be (600x1/5) i.e. about 40 crore man-hours. Hence, post harvest activities relating to wool industry is a potential job generating industry in the country. There is a great scope of modernising and popularizing the industry in still larger areas other than where it has already been well established.

Post Harvest Machinery

Hardwares used in post harvest activities are classified according to the 'UNIT' operations performed on different commodities as indicated in Fig. 4. However, development and promotion of these machines and more so the energy efficient ones have to be done by adoption and evaluation and R&D as a location specific problem. Such an activity for the value addition and employment generation must be given preference over other activities. It is unfortunate that least attention was paid to the development of new technologies including hardwares relating to the commodities produced in the arid zones of India. Nevertheless, the technologies already developed for the other regions of the country can be conveniently and successfully adopted in this region also. Some minor modifications in the existing hardwares to suit the needs of the products of the arid regions should not be a very difficult task. It would be worth while to mention that the special features of the arid zone' i.e. abundant availability of the solar energy, large number of poor farmers and landless labourers etc., should not be lost sight of in adapting and developing appropriate technologies for the farmers and young enterpreneurs of the region.

It is not out of context to mention that the 'Agro-Processing Centres' as depicted in Fig. 4 and tried out in MP State for value addition and employment generation can be adopted in the arid region also. But these centres should aim at processing products which are locally produced. In so far as the medium scale industries are concerned, the milk, the meat, the wool, the fruits etc. offer excellent opportunities to generate income and employment for the local youths and other citizens. The agro-processing centres of this region may make use of these commodities for income generation.

DAL MILLING TECHNOLOGY

Dal milling is the third largest processing industry in India after rice milling and wheat milling. *Dal* milling is done by means of both traditional and modern mills in different parts of the country. At the family level, the Chakki(Sheller) is used to split pulses. Whereas, the mechanized rollers and shellers are used by the organized industry to prepare '*Dal*' from pulses. *Dal* milling basically consists of removal of husk without loosing any edible portion i.e. the cotyledons and the germ, if possible. The husk of most pulses is attached to cotyledons through a layer of germs. The husk may be thin or thick; rough or smooth; sometimes with oil/wax coating. Moisture has opposite influence on the adherence of husk to the cotyledons and the binding between the two cotyledons. The maximum possible recovery and *dal* yields in respect of few selected pulses grown in arid zone are presented in Table 4. *Dal* milling is generally done in the following two steps :

(i) Loosening of husk by wet or dry methods.

(ii) Dehusking and splitting into the cotyledons using appropriate machines.

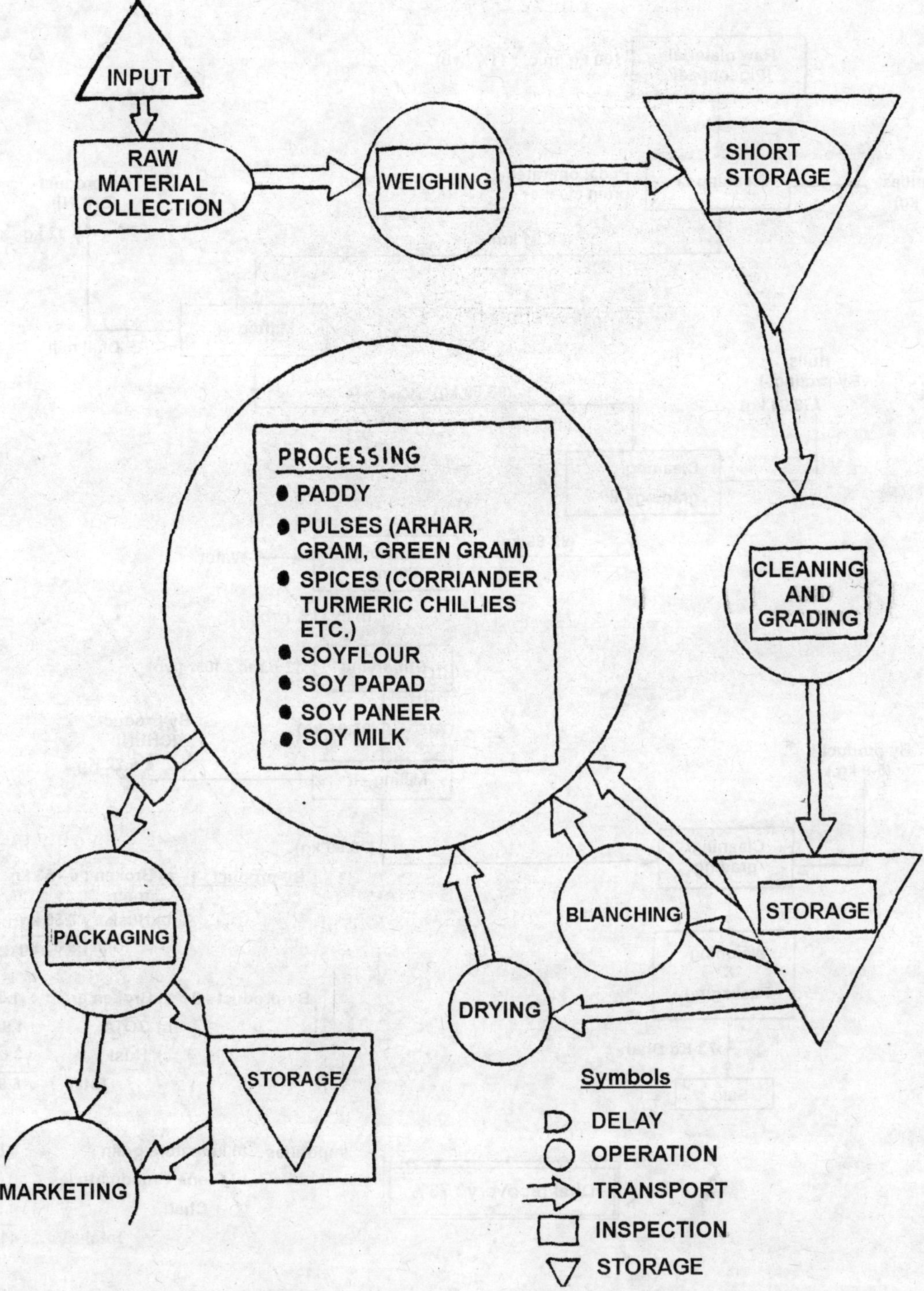

Fig. 4. Process flow digram of Agro Processing centre at village Islam Nagar

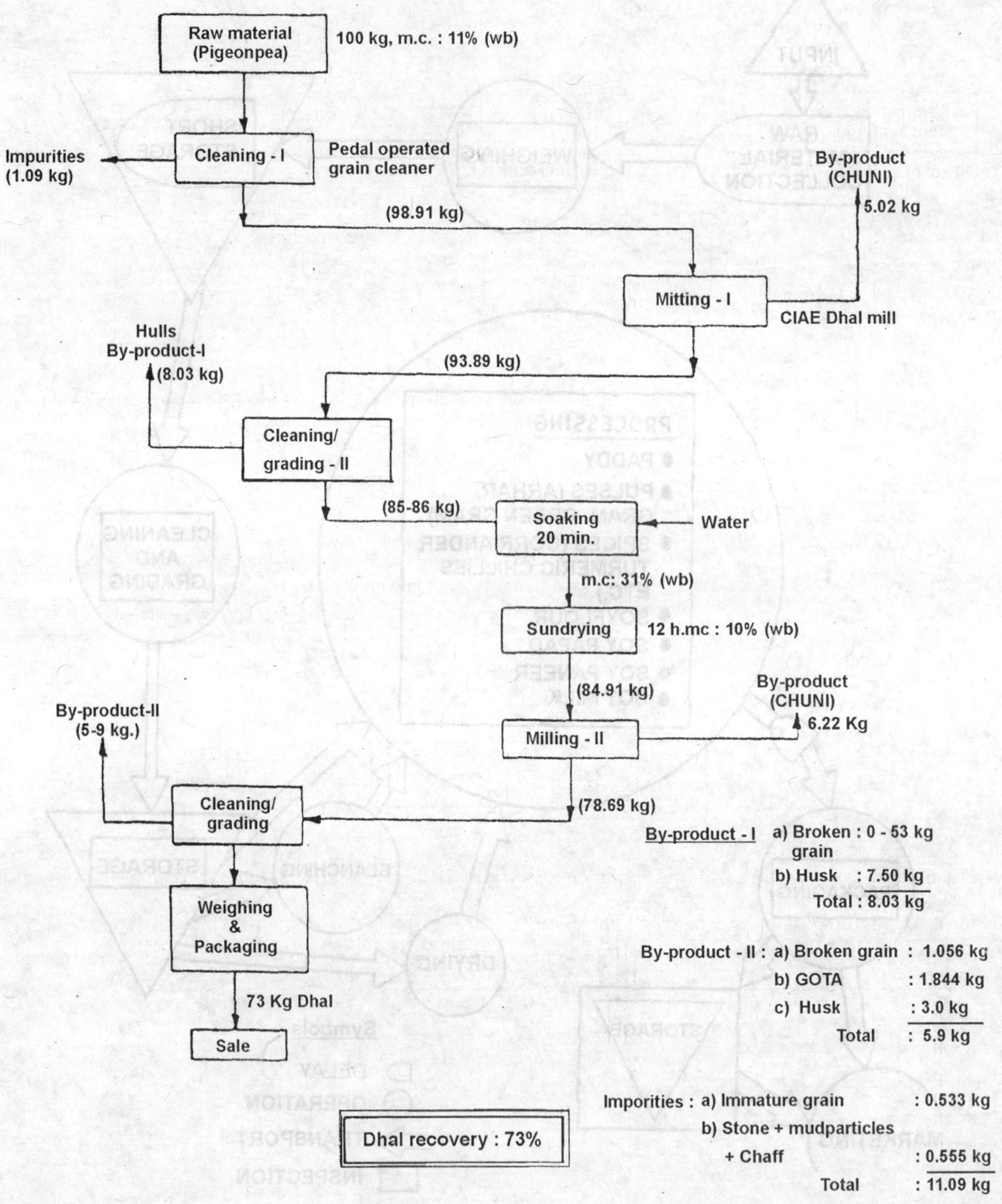

Fig. 5. Process flow chart and mass balance for making pigeonpea *dhal* at Agro-Processing Centre

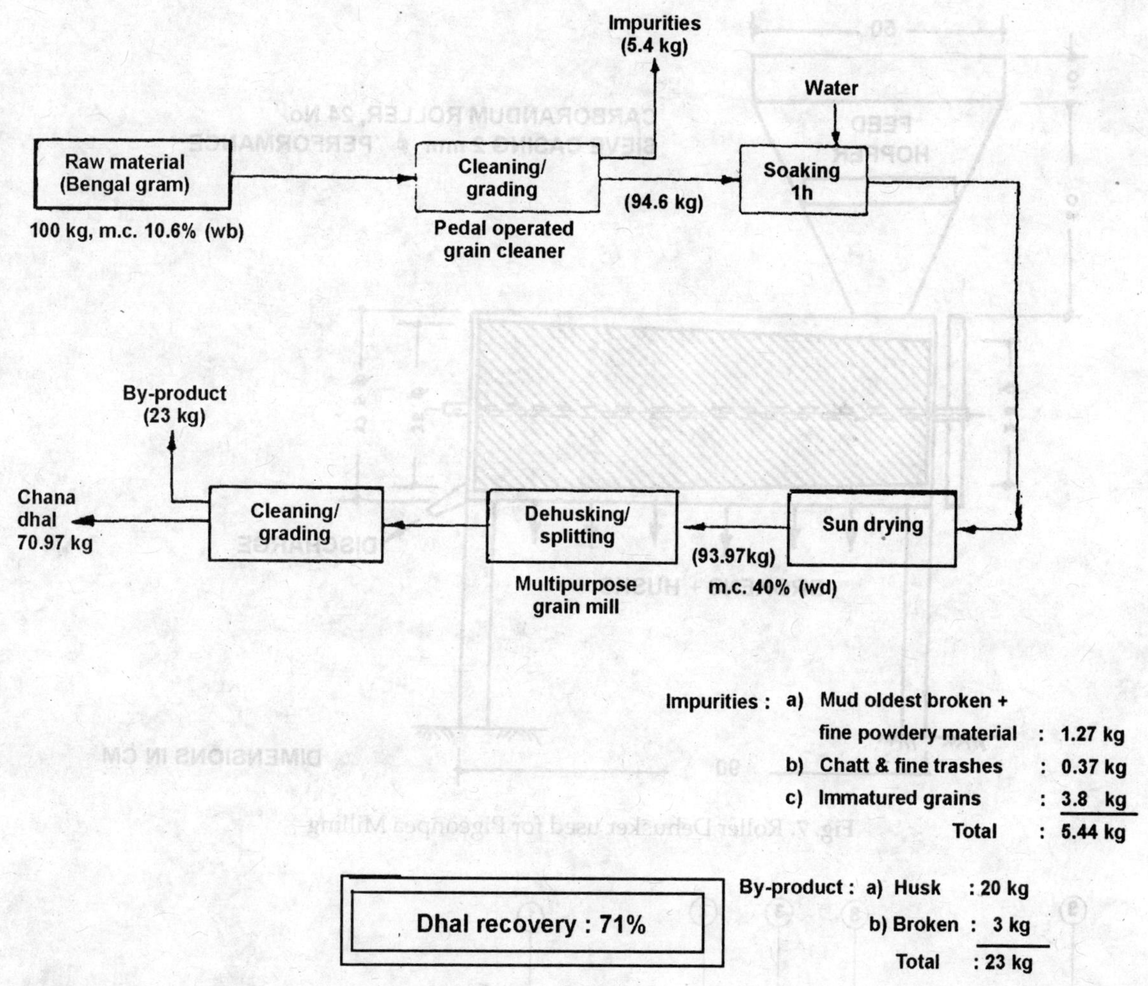

Fig. 6. Process flow chart and mass balance for making Bengal gram dhal at Agro-Processing Central

Table 4. Maximum possible recovery and dal yield

Legume	Content of kernel after husk removal (%)	Yield of head dal (%)	Yield by traditional method (%)
Bengalgram			
a) Big grains	88.50	84.00	74.00
b) Small grains	87.50	83.00	72.00
Pigeonpea			
a) Big grains	88.50	85.00	75.00
b) Small grains	85.00	82.00	68.00
Greengram	89.50	84.00	65.00
Blackgram	87.50	83.00	71.00
Soybean	88.50	84.00	–
Cowpea	89.50	86.00	–
Kidney beans	90.30	86.00	–
Horsegram	87.50	81.00	–

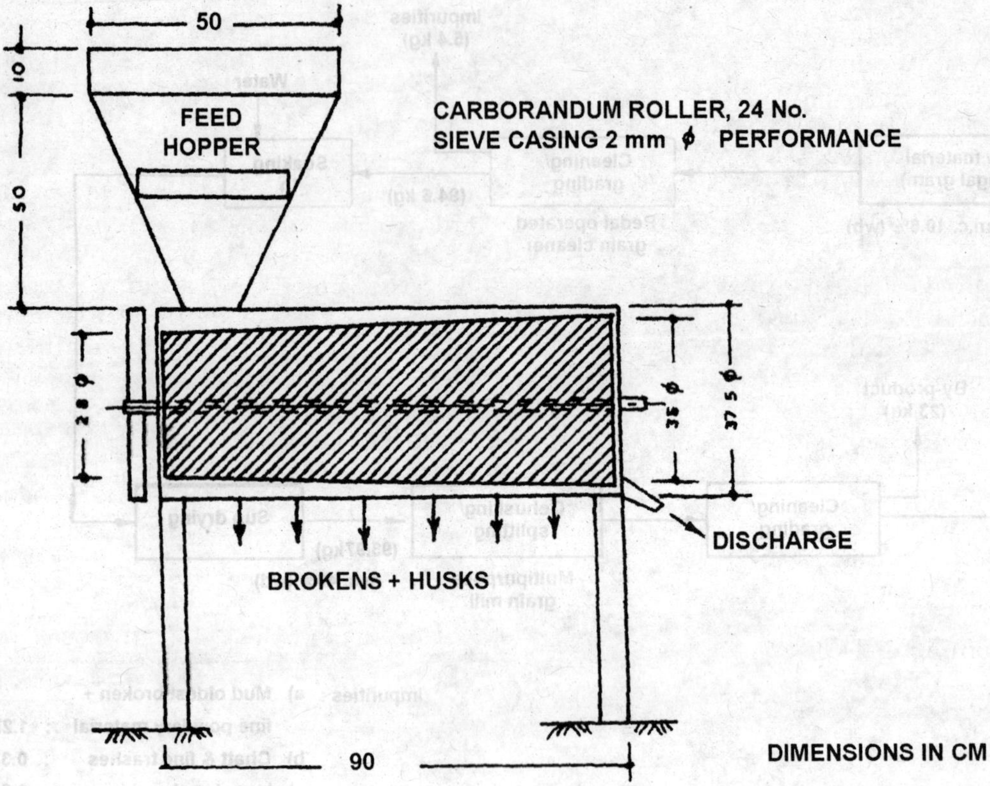

Fig. 7. Roller Dehusker used for Pigeonpea Milling

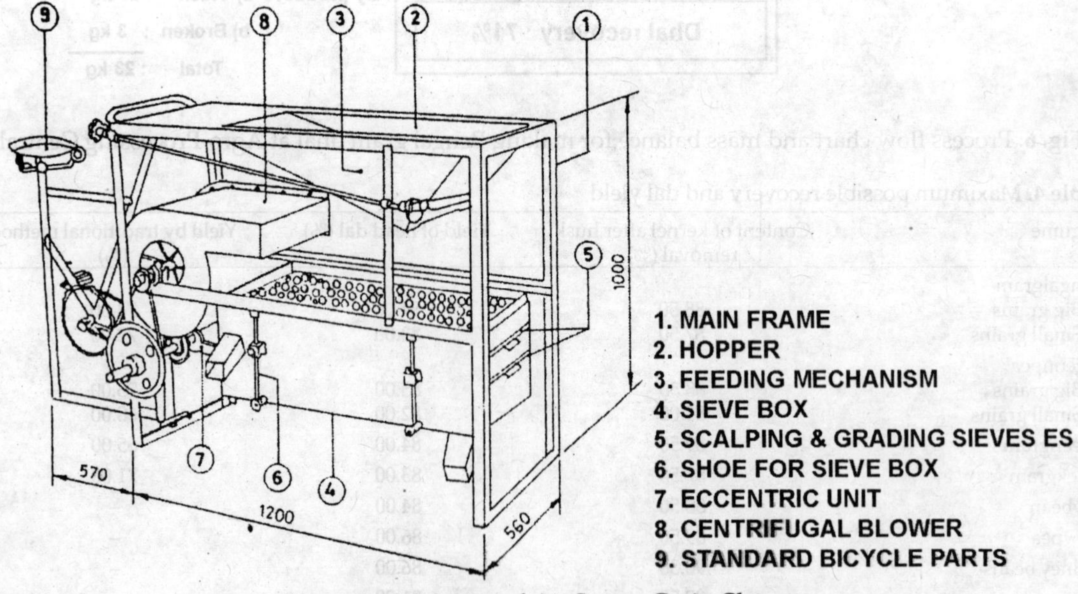

Fig. 8. Pedal Operated Air Screen Grain Cleaner

The amount of oil mixed with pulses vary from place to place from 50g/q. of grain to 400 g/q. Similarly, addition of water also varies from 4 to 20 kg/q. of grain. For losening of husk and its complete removal, 3 to 7 passes through the roller in case of pigeon pea are given to the conditioned grain.

Dal from Pigeon Pea

For production of pigeon pea dal, the process and machinery developed by the Central Institute of Agricultural Engineering (CIAE). Bhopal is shown in Fig. 5. Some of the important machines and equipments used during the process of handling 100 kg pulse are :

(i) Dal Mill - Emery carborandum roller dehusker (Fig. 7).

(ii) Pedal-operated grain cleaner (Fig. 8).

(iii) Dal Polisher (Fig. 9).

Dal from Bengal Gram

The actual process flow and mass balance for the production of bengal gram *dal* is presented in Fig. 6. The vertical stone burr grinder and the pedal operated grain cleaner having the capacity of 50 kg/hour can be successfully used. At the current price, the cost of the machines would be Rs. 5000 approximately.

OILSEED STORAGE AND EXPELLING TECHNOLOGY

In India, about 20 million hectare area is under oil seed crops. Most of these crops are grown under unirrigated and on marginal lands with low productivity status. Nutritional oils are the richest source of energy (9 cal/g) and provide essential fatty acids and fat soluble vitamins. Due to high demand and low production, the oil seed and oil are being imported every year. It is essential not only to increase the production of oil seed but also to reduce the losses during harvesting and post-harvesting operations of oil crops. The losses which can be minimized by adopting improved technology already available in the market are presented in Table 5.

Table 5. Improved technology for minimising losses.

Operation	Losses (%)
Harvesting	5-15
Threshing	2-5
Cleaning	2-3
Handling & Storage	10-15

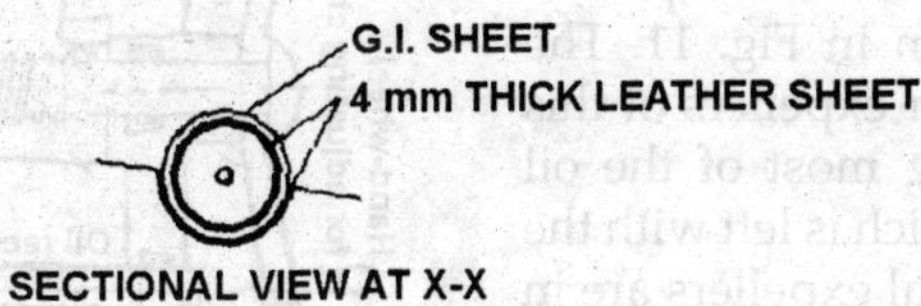

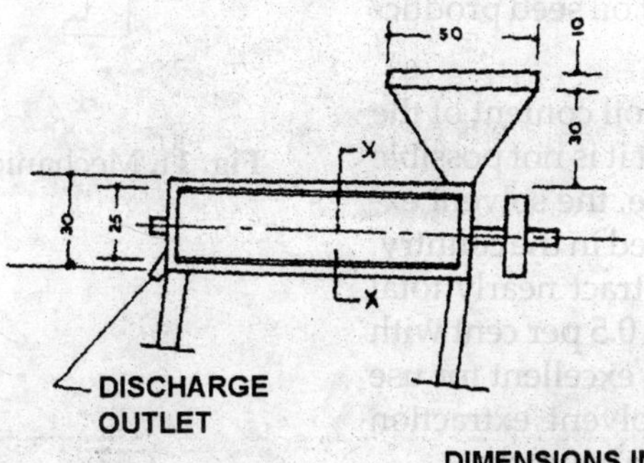

Fig. 9. Pigeonpea Dal Polisher (Nylon Polisher)

Oil Seed Storage

After harvesting and drying the oil seed crops, the low moisture oil seeds can be stored in one of the following improved storage structures depending on the quantity and environmental conditions. In case of small quantity of such materials, the metal bins, Pusa bin and stone bins can be successfully used. However, for the large quantities of such stocks, the water proof godowns or warehouses are preferred to stock, bags full of oil seeds. Even concrete silo or storage structure made of hollow cement blocks or bricks laid in cement sand mortar should be preferred. Storage of oil seeds under high humid conditions would not be desirable and must be avoided. Arid regions are best suited for storage food materials without much of precautions except fumigation and moisture entry in the godown or structure.

Oil Extraction

Traditional oil extraction equipment of India is the bullock-drawn 'ghani' which leaves 10-15 per cent oil with cake. The mechanical power *ghanis* have now been introduced in the country but they also leave 8-12% oil with cake. A simple power *ghani* has been shown in Fig. 10. These *ghanis* are phasing out in favour of power operated oil expellers as sown in Fig. 11. The single and the double chamber expellers of this type are capable of extracting most of the oil except about 6 to 7 per cent which is left with the cake. About 155,000 mechanical expellers are in use in India. Quite a large number of these units are operating in the dry land areas which contribute to the maximum share of oil seed production in the country.

As far as possible, the total oil content of the the seed should be extracted but it is not possible with mechanical devices. Hence, the solvent extraction plants have been installed in the country. These plants have ability to extract nearly total quantity of oil leaving less than 0.5 per cent with the cake. The deoiled cakes are excellent for use as animal feeds. About 230 solvent extraction plants have been installed in the country.

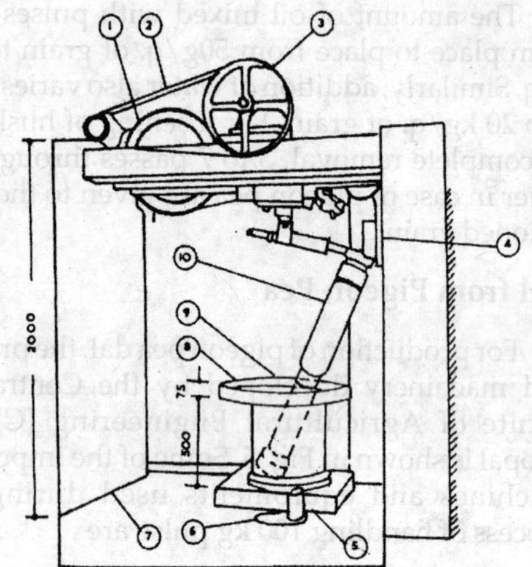

Fig. 10. Power Oil Ghani

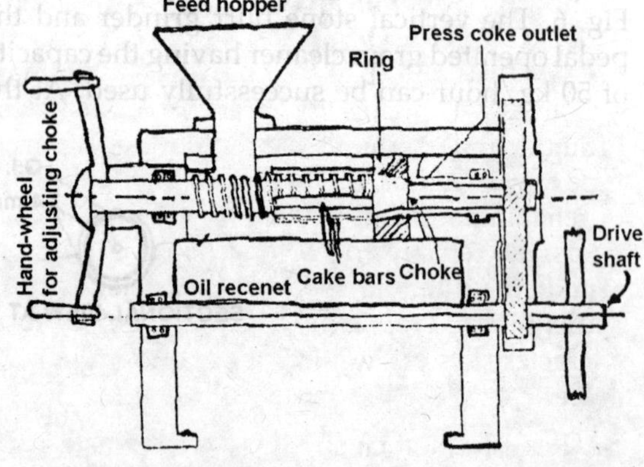

Fig. 11. Mechanical Oil Expeller– Screw Press

30

WASTELANDS AND THEIR REHABILITATION IN INDIA

SURENDRA SINGH and R.P. SINGH

INTRODUCTION

In India, human and livestock population has increased manifold in the last two to three decades. The human population from 548 million in 1977 increased to 685 million in 1981 and 844 million in 1991. It is estimated that by 2000 AD; it will be around 1000 million, as the population is increasing at the rate of 2.4 per cent per year. The livestock population which was around 400 million in 1982 will also increase manifold by 2000 AD. In order to meet the demand of food, fodder, fuel and fibre of the increasing human and livestock population more land has to be brought under plough by reclaiming different type of wastelands. But before taking up the task of rehabilitation and development of the wastelands, the reliable and comprehensive information on the different aspects of wastelands such as definition, type, extent, genesis, biophysical characteristics etc. will be of paramount importance.

Realising the importance of the wastelands in catering to the demand of food, fodder and fuel of the increasing population, some studies on the above aspects of wastelands have been done under different morpho- climatic zones. The different workers, have defined the term in different ways. Simply stated, wasteland is one which is unfit for use, that is uncultivable, barren or without vegetation. Sir James (1928) defines it as an uncultivable land, producing a little or nothing, a wild desolate region, a desert and a barren land. Dudley Stamp (1948) has defined the wasteland as that land which has been previously used but which has been abandoned and for which no further use has been found. Wasteland Survey and Reclamation Committee has defined the term wasteland as those lands which are either not available for cultivation or left out without being cultivated like fallow and culturable waste.

Wastelands are basically degraded lands which denote dimunition of the biological productivity expected of a given tract of land. Degradation of drylands (arid, semi-arid and sub-humid lands) is called 'deseritification' because desert like conditions appear where none existed before (World Resources, 1988-89).

According to the Technical Task Group constituted by Planning Commission in 1986, the wasteland is that land which is degraded and is presently lying unutilised except current fallows due to different constraints. The member Planning Commission Prof. M.G.K.Menon in 1987 constituted an International Working Group under the Chairmanship of Mr. R. Venkatanaryan, Member Secretary NWDB. This group prepared a revised booklet and provided a working description of wasteland and 13 wasteland categories identified by Technical Task Group were reduced to 7 categories. As per this document, the wasteland means degraded land which can be brought under vegetative cover with reasonable effort, and which is currently unutilised and land which is deteriorating for lack of appropriate water and soil management or on account of natural causes. Wasteland can result from inherent imposed disabilities such as by location, environment, chemical and physical properties of the soil or financial or management constraints (NWDB, 1987).

In this chapter, the type, extent, processes and factors responsible for the development of wastelands, their biophysical characteristics and the rehabilitation strategies have been discussed. In the end, the economics of rehabilitation of different type of wastelands has also been highlighted.

TYPE AND EXTENT OF WASTELANDS

Based on the above definitions of wastelands, the different type of wastelands have been identified and classified by various workers and their extent has also been worked out. The different estimates of the total area of the wasteland in India has been given by various organisations (Table 1). The total area of wasteland varies from as low as 36.90 million hectares to as high as 175 million hectares (m ha).

Table 1. Reported area under wastelands in India

S.No.	Organisations	Area under waste-lands (m ha)	Percentage of total geographical area
1.	National Committee on Agriculture Ministry of Agriculture and Irrigation, Govt. of India, New Dehli.	45.0	13.7
2.	Directorate of Economics and Statistics, Govt. of India New Dehli (1978-79)	38.4	11.7
3.	Directorate of Economics and Statistics, Govt. of India, New Delhi (1980-81)	36.9	11.2
4.	Department of Environment, Govt. of India (B.B. Vohra, 1980)	95.0	28.9
5.	Ministry of Agricutlure & Cooperation, New Delhi (1982)	175.0	53.2
6.	National Remote Sensing Agency, Dept. of Space, Govt. of India, Hyderabad (1985), based on Satellite remote sensing data of 1980-82	53.0	16.2
7.	Society for Promotion of Wastelands Development (D.R. Bhumbla and Arvind Khare, 1984-85)	93.7	28.5
8.	National Wastelands Development Board Ministry of Environment & Forest New Delhi, 1985.	123.0	37.5

Source : Gautam and Narayan (1988).

The major categories of wasteland and their extent identified by Ministry of Agriculture, Govt. of India (1986) are presented in Table 2.

Table 2. Major categories of wastelands in India

S.No.	Land categories	Extent of area (m ha)
1.	Serious water and wind erosion	150.00
2.	Shifting cultivation	3.00
3.	Waterlogging	6.00
4.	Saline soils	4.50
5.	Alkali soils	2.50
6.	Riverine (Diara) lands	2.40
7.	Other culturable wastelands fit for reclamation	6.60
	Total	175.00

Source : Ministry of Agriculture (1986).

The statistics given in Table 1 does not account for the area affected by annual floods (50 m ha) and irrigated area threatened by waterlogging and salinity (10 m ha) each year. Thus, about 75 per cent of the effective land area (226 m ha) requires urgent remedial measures to bring them back to their inherent productive capacity.

The categories of degraded lands or wastelands have been identified and their approximate extent (based on NCA, 1976; INCOR 1986) and NRSA wasteland map in 1:3.5 million scale) is given in Table 3.

Table 3. Extent of degraded lands in India

Category	Area (m ha)	Occurrence
Total area subject to serious water and wind erosion of which 69 m ha are critically affected	150.0	Water erosion areas in all climatic regions specially in sub-humid and humid areas Wind erosion areas mostly in arid region of Gujarat, Rajasthan and Haryana
Riverine land, coastal sandy areas, and high altitude and steeply slopy lands	11.3	Coastal region, interior mountaneous regions and steep slopes in Himalayas
Salt affected lands	7.0	Semi-arid and Indo-Gangetic plain and other semi-arid areas in Peninsular region
Waterlogged areas	6.0	Low-lying areas in high rainfall regions and irrigated areas
Ravine land	4.0	Gujarat, Rajasthan, Madhya Pradesh and Uttar Pradesh
Shifting cultivation	3.0	Mostly in north eastern region
Barren hill ridges or rock outcrops	2.7	Distributed in several States

Source : Hirekerur, L. R. *et al.* (1990).

The broad sub-divisions of degraded land resources, according to Ram Prasad (1988) are : in agriculture, out of 143 m ha atleast 40 m ha or so are degraded, in forestry sector, out of 75 m ha, 30 m ha are without tree cover and 10 m ha have some shrubs only; and the pasture and grazing lands and other cultivated lands are largely degraded.

Bhumbla and Khare (1987) have categorised wastelands on the basis of problems afflicting various regions. These categories are : (i) sheet erosion, (ii) ravines, (iii) waterlogging, (iv) riverine lands, (v) shifting cultivation, (vi) gully erosion, stream bank erosion, (vii) salinity and alkalinity, (viii) wind erosion, and (ix) coastal sand dunes. Ram Prasad (1988), however, has classified the wastelands into seven categories : (i) ravinous lands, (ii) mined out areas and overburdens, (iii) lateritic soils and 'Bhata' lands (iv) saline and alkali soils, (v) arid and semi-arid areas, (vi) degraded forests, and (vii) other wastelands including waterlogged areas and periodically inundated areas.

Realising the importance of rehabilitation and development of wastelands, National Remote Sensing Agency (NRSA), Hyderabad by developing eight fold classification based on Landsat MSS images, has completed the wasteland mapping of entire country. The eight categories of culturable and unculturable wastelands identified and mapped (Fig. 1) by NRSA (1985) and their extent are presented in Table 4.

Table 4. Category and extent of wastelands in India

Major categories	Sub-categories	Extent (m ha)
Culturable wastelands	Salt affected land	3.90
	Gullied or ravinous land	4.32
	Waterlogged or marshy land	0.89
	Upland with or without scrub	10.79
	Thorn or forest blank	2.42
	Sandy area	10.53
Unculturable wastelands	Barren hill ridge or rock outcrop	2.75
	Snow covered	17.70
	Total	53.30

Source : NRSA Wasteland Map of India (1985).

It may be seen from the above Table that out of total wasteland area (53.30 m ha); 32.85 m ha or 61.5 per cent is under culturable wastelands and 20.45 m ha or 38.5 per cent under unculturable wastelands.

Based on wasteland mapping of each State and Union Territories, its areal extent was computed and is presented in Table 5.

Table 5. Areal extent of wasteland in India

S.No.	States and Union Territories	Area (m ha)	Percentage
1.	Andhra Pradesh	2.64	8.91
2.	Assam	0.19	2.43
3.	Bihar	1.00	5.79
4.	Gujarat	3.30	16.89
5.	Haryana	0.33	7.54
6.	Himachal Pradesh	2.04	36.67
7.	Jammu & Kashmir	13.36	60.14
8.	Karnataka	1.64	8.57
9.	Kerala	0.22	5.82
10.	Madhya Pradesh	2.91	6.57
11.	Maharashtra	2.98	9.69
12.	Manipur	0.32	14.35
13.	Meghalaya	0.16	7.46
14.	Nagaland	0.15	9.59
15.	Orissa	1.15	7.41
16.	Punjab	0.35	6.99
17.	Rajasthan	12.94	37.81
18.	Sikkim	0.27	36.94
19.	Tamil Nadu	1.01	7.77
20.	Tripura	0.09	8.94
21.	Uttar Pradesh	4.31	14.67
22.	West Bengal	0.56	6.92
23.	Arunachal Pradesh	1.21	14.48
24	Mizoram	0.21	9.98
25.	Union Territories	0.05	3.10
	Total	53.30	16.21

Source : NRSA Wasteland Map of India (1985).

The highest percentage (60.14) of wasteland in the country is in Jammu and Kashmir, followed by Rajasthan (37.81%), Sikkim (36.94%) and Himachal Pradesh (36.67%). The simple reason of the high percentage of wasteland in Rajasthan State is large extent of sandy and stony/rocky areas. Whereas in other above three states, permanent snow and/or glacial cover has contributed to the large extent of wasteland.

However, the wasteland maps of India prepared on 1:1000,000 scale could not be proved useful for the rehabilitation and development of wastelands due to the limitation of scale. The National Wasteland Development Board, therefore, felt the necessity of mapping wastelands on a large scale (1:50,000) which would enable the identification of the wastelands with respect to all categories developed by the Technical Task Group and to rehabilitate and develop them. Accordingly, the mapping of following 13 wasteland categories on 1:50,000 scale in critically affected districts was completed by using Landsat TM and IRS LISS-II False Colour Composites.

1. Gullied and/or ravinous land
2. Land with or without scrub
3. Waterlogged and marshy lands
4. Land affected by salinity/alkalinity
5. Shifting cultivation
6. Sandy (desertic/coastal)
7. Mining/industrial wastelands
8. Under utilized/degraded notified forest land
9. Degraded pastures/grazing land
10. Degraded and non-forest plantation land
11. Barren rocky/stony waste/sheet rock area
12. Steep sloping area
13. Snow covered and/or glacial area

The salient features of the above different type of wastelands identified and mapped by NRSA (1985) has been discussed by Gautam and Narayan (1988). The salient features involve the description, classification and nomenclature of wastelands problems of land degradation, methodology for wastelands mapping, spatial distribution of wastelands in India and their reclamation.

GENESIS AND CHARACTERISTICS OF WASTELANDS

Factors and processes responsible for the genesis of different type of wastelands are both, natural and anthropogenic such as water erosion in the form of sheet, rill and gully, wind erosion/deposition, salinity/alkalinity, waterlogging, mining, industrial effluent and surface crust etc. The factors and processes of soil degradation of the wastelands are given in Table 6. Singh and Kar (1988) have reported that the genesis and distribution of different types of

Table 6. Factors and processes of soil degradation

Processes (actions & interactions)			Factors (agents and catalysts)		
Physical	Chemical	Biological	Agricultural activities	Industrial activities	Urbanisation
A. Deterioration of soil structure leading to :	Leaching Fertility depletion	Decline in biomass carbon	Deforestation	Waste disposal	Application of city waste
(i) compaction (ii) crusting	Sodication Laterization Toxification (Al^{+3} Mn^{+3}) heavy metals	Reduction in organic matter content	Excessive & untimely ploughing	Acid-rain	Conversion of land to non-agricultural uses
(iii) accelerated erosion (iv) hard setting		Decrease in population activity and species diversity of soil fauna and flora.	Intensive row cropping and mono-culture	–	–
B. Imbalance in water/air ratio :	–		Indiscriminate and excessive use of chemicals	–	–
(i) wetness (ii) drought		Alterations in biological processes from favourable to unfavourable trends	Excessive grazing and high stocking rate	–	
C. Extremes of temperature : (i) permafrost (ii) supra-optimal	–				

Source : Lal, Hall, and Miller (1989).

wastelands are associated with various landforms and geomorphic processes responsible for their formation. The genesis and salient characteristics of wastelands are discussed below.

Gullied/Ravinous Land

Ravines meaning deep gorges, comprise of a system of gullies, running more or less parallel to each other, in deep alluvium and draining into a major river or its tributary which is flowing much lower than the surrounding table lands. These wastelands are mostly associated with the flood plains of the major rivers like the Yamuna, the Chambal, the Krishna, the Narbada and the Thungbhadra etc.

Ravines represent the last stage of water erosion, some of the causes of ravine formation are :

1. Clearance of forest growth and putting the land under agriculture,
2. Unscientific agricultural methods of soil working,
3. Burning of ground cover and uncontrolled grazing,
4. Low annual rainfall, with the down pour frequency of greater intensity,
5. Soft alluvial liable to easy scouring
6. Pronounced vertical difference between the general ground level and the river bed level, and
7. Extreme dryness of the soil due to easy drainage and high soil porosity.

These gullies are the first step of excessive land dissection followed by their net working, leading to the development of ravinous land. (Fig. 2).

Land With or Without Scrub

The wastelands occupy (relatively) higher topography like uplands or high grounds with or without scrub. These lands are generally prone to degradation or erosion. These exclude hilly and mountaneous terrain. Genetically, these wastelands are associated with pediments, plateaus, lateritic formations and *Bhata* lands and largely occur in Maharashtra, Andhra Pradesh and Rajasthan. Such wastelands are characterised by rocky, gravelly and stony surfaces, shallow deposits and presence of drainage channels along the joints (Singh *et al.* 1989).

The wastelands associated with lateritic formation have been formed under conditions of high rainfall wherein all basic constituents in the parent material have been leached out and only a matrix rich in iron and aluminium oxides and hydroxides with some silica is left behind. Kaolinite is the predominant clay mineral. These soils show typical properties of clay-plasticity, cohesion, shrinkage, expansion, etc.

Bhata lands are generally found in the Chhatisgarh region of Madhya Pradesh. The climate of the region is hot and dry. The topography is undulating, interspersed by a series of convergent streams. The soils of these lands are the poorest soils of the tract.

In contrast to the poor and nutrient deficient *Bhata* soils found on the ridges, the *Kanhar* soil available in valleys is very rich and fertile. These soils produce good agricultural crops even without irrigation. A third type of soils is also found on the slopes, joining ridges to the adjoining valleys. These soils are yellow in colour and locally called *Matari* and *Dorsa*. Crops can be grown on these soils with irrigation and fertilization.

Waterlogged and Marshy Lands

These wastelands have developed due to canal seepage, high water table and improper drainage and largely occur in West Bengal, Bihar and Uttar Pradesh. Canals flow above the general ground level to provide irrigation water to the fields by gravitation. In certain types of soils this water seeps and as a result the water table of the surrounding areas tend to rise. The rise in water table increases the concentration of inorganic salts and new saline soils get formed and areas are waterlogged.

Simply stated, waterlogged land is one where the water is at/or near the surface and water stands for most of the year. Such lands usually occupy low lying areas in the toposequence. Land with surface water bodies like lakes, ponds and tanks are precluded from this definition.

The littoral salty marshes and mud flats are confined to the coastal belt notably in the estuaries of rivers, in creeks and lagoons and along the lower reaches of tidal streams.

Marshy land is characterised by permanent or periodic inundation by water and one which carries grasses and weeds as the predominant vegetation.

Land Affected by Salinity/Alkalinity

These wastelands are found almost in all parts of the country but largely occur in Uttar Pradesh, Gujarat, Punjab, Harayana and Rajasthan. The factors responsible for the development of such wastelands are : (i) arid and semi-arid climate, (ii) hard impervious pan in the sub-soil, (iii) high water table, (iv) basin shaped topography, (v) impeded drainage conditions, (vi) salt bearing sub- strata, (vii) excessive canal irrigation leading to seepage from non- lined canals and consequent waterlogging, (viii) use of saline and brackish water for irrigation, (ix) ingress of sea water, (x) salt impregnated winds in coastal areas, and (xi) construction of tanks/reservoirs at the unfavourable geomorphic sites (Fig. 3).

The saline/alkali wastelands are also known as *Usar* lands and could be classified into four classes viz. (i) saline, (ii) saline- alkali, (iii) non-saline alkali, and (iv) degraded alkali. The soils of such wastelands contain excess of both soluble salts and exchangeable sodium. The soils of saline wastelands contain different soluble salt mainly sodium and sodium sulphate. Whereas, the soils of alkaline wastelands constitute dissolved sodium carbonate and sodium bicarbonate. Such wastelands are characterised by white or grey efflorescence on surface, salt crust, heavy textured highly saline/alkali soils underlain by hard *kankar* or clay (Sharma *et al.* 1990). They are generally characterised as the land that has adverse effects on plant growth due to the action or pressure of excess soluble or high exchangeable sodium. Alkali land has an exchangeable sodium percentage (ESP) of about 15 which is generally considered as the limit between normal and alkali soils.

The soils of these wastelands present numerous problems in management on account of their inherent physico-chemical and biological properties. Some of these characteristics are as follows :

1. Physical Soil Characteristics

(i) Unfavourable soil structure and consistance;

(ii) Dispersed conditions and bad aeration;

(iii) High osmotic pressure;

(iv) Poor soil permeability;

(v) Hard subsoil and impeded drainage.

2. Chemical Properties

(i) High pH;

(ii) Low availability of nutrients;

(iii) Excess of sodium.

3. Biological Properties

(i) Reduction in number of micro-organisms.,

(ii) Reduction in wild fauna.

Shifting Cultivation

The cyclic landuse practice of felling trees and burning of forest areas with the purpose of growing crops lead to shifting cultivation. These lands are also known as *Jhum* lands. The practice leads to excessive loss of soil and development of wasteland. Since this practice of cultivation is still continuing by tribal people in northeastern region comprising of Assam, Arunachal Pradesh, Mizoram, Manipur, Meghalaya and Nagaland and new wastelands are being created.

The shallow to moderately deep soil, irregular slope, presence of drainage channels, severe to very severe water erosion and constant human pressure are the major limitations for the development of these lands.

Sandy Area (Coastal or Desertic)

The sand dunes and sandy undulating plains created by aeolian action along the coastal region and in the Thar desert of India are responsible for the creation of such type of wasteland (Singh *et al.* 1989). The new sandy wastelands are being created due to increasing anthropogenic activities like cultivation, cutting and lopping of trees and overgrazing. The fresh sands are deposited on the cultivated lands in the form of barchan and shrub-coppice dunes and sand sheets resulting in depletion of their biological productivity. (Fig. 4). The riverine sands transported by unprecedented floods and reworked by aeolian action are also creating new wastelands in many parts of the country (Saxena and Singh, 1990).

These wastelands are characterised by high relief of 5 to 60 m, gentle to moderately steep slopes varying from 2° to 20°, compact and loose structureless sands, poor fertility and water holding capacity, presence of alternate layers of calcium carbonate, degraded surfaces and high vulnerability to wind erosion (Singh, 1977; Singh and Sharma, 1984). The soils of such wastelands are single grained, non- coherent and structureless. They are characterised by large sized pores, high infiltration rate as high as 30 to 60 cm/hr, moisture status varies from 1.5 to 3.5 per cent, very low water retention capacity, low carbon content (0.04 to 0.05%), low nitrogen content (0.005 to 0.006%). However, they contain relatively higher proportion of mineral and hydrolysable form of nitrogen. The fertility status as such was found sufficient to sustain plant growth for the rehabilitation of such wastelands. The genesis and salient characteristics of the sandy wastelands of the Indian arid zone have been discussed in detail by Singh (1990).

Mining/Industrial Wastelands

As a result of mining of minerals or building materials etc. by unscruplous methods and without following mining regulations properly, large areas have developed as mining wastelands. Improved mining technology and continuously increasing demand of mined minerals i.e. building material, gypsum, soapstones etc. for local and industrial use have created new wastelands and more and more areas will be turned into wastelands in near future (Saxena and Singh, 1990).

These wastelands are characterised by undulating topography, presence of pits and debris heaps of various dimensions, water and wind erosion and irregular slopes. The mining of limestone creates deep and hollow structures as permanent scars in the surrounding areas. The mining activities have created several environmental problems like loss of surface top soil, surface water pollution, ground water pollution, fire hazard, air pollution etc. besides creating adverse soil conditions (Fig. 5).

In different parts of the country, the development of various types of industries generate industrial effluents which cause ill effects to the agricultural lands finally results into the industrial wastelands. By and large, the industrial effluents are diverted to the major rivers which affect the surrounding areas and pollute the surface and ground water.

These wastelands are characterised by salt encrustations, surface crust, poor soil fertility, stunted growth of crop and subsurface clay pan etc.

Under-utilised/Degraded Notified Forest Land

The main cause of degradation of forests are intensification of pressure on forests for meeting the growing demands of fuel, fodder and timber for the increasing human and bovine population. Some times, faulty management practices have contributed to degradation of forests. Some socio-economic and political reasons are also responsible for large scale degradation of forests.

Degradation of forests has many ecological implications. The structure and composition of natural forest community changes into wastelands, scrubs and then into grazing lands. The degraded forests associated with hills, mountains, plateaus, uplands, rocky/gravelly areas are characterised by eroded and undulating topography, steep to very steep slope, poor growth of trees and shrubs with density less than 20 per cent, wind and water erosion and their soil cover.

Degraded Pastures/Grazing Land

This type of wasteland has developed on varied type of topography like rocky/gravelly pediments, sandy undulating and flat alluvial surfaces, saline alluvial plains, eroded uplands due to the increasing anthropogenetic stress like overgrazing and cutting of trees (Singh *et al.* 1989).

The stunted growth of plants, presence of wind, water and salinity hazards, irregular slopes, very shallow to moderately deep and very deep gravelly sand to sand and loamy sand to sandy loam soils and poor soil moisture status are the major characteristics of this category of wastelands.

Degraded Non-Forest Plantation Land

These lands occur outside the notified forest areas under different climatic zones. The increasing natural and human activities like wind and water erosion, cutting of trees, overgrazing are responsible for the degradation of these lands.

The arid zone degraded lands are characterised by ill defined profile development, presence of *kankar* pan at varying depth, poor moisture status, poor growth of vegetation, salinity and fluvial hazards.

In the semi-arid tropics, the soil erosion over the years had degraded these lands to such an extent that soil cover has become thinner making them impoverished and unproductive. Shallow with limited moisture storage capacity and undulating topography has made the situation still worse. The flooding, salt encrustations and surface crust are also responsible for the degradation of these lands and pose the serious problems for their rehabilitations.

Barren Rocky/Stony Waste/Sheet Rock

The isolated hills, rock outcrops, sheet rock and lateritic areas form this category of wasteland. These wastelands have resulted due to the accelerated and natural erosion and weathering processes. The steep slope, rocky surfaces, rills and gullies, shallow soil cover, rocky substrata and presence of rock fragments and gritty material are the major limitations of this category of wastelands for their rehabilitation.

Steep Sloping Area

This type of wastelands have developed along the slopes of the mountains due to natural and accelerated erosion and weathering processes. These wastelands are characterised by steep to very steep slope, debris accumulation in the form of 'talus', stunted stand of vegetation and severe degradation due to water erosion and biotic impact.

Snow Covered and/or Glacial Areas

This type of wastelands are created by the glacial and fluvial glacial activities in the mountainous areas due to the accumulation of snow. They occur on mountain peaks and slopes in association with high relief and glaciers. (Fig. 6).

Such wastelands are characterised by large and extensive size, thick snow deposits, irregular and discontinuous shape, smooth texture and contiguous pattern.

REHABILITATION STRATEGIES OF WASTELANDS

GULLIED/RAVINOUS LAND

The reclamation of ravines involves : (i) preventive measures to stop further spread of gullies, and (ii) curative measures. Protection of table lands is essential. Contour bunds and graded bunds in alluvial soils and black soils, with cross section of 0.9 to 1.3 m^2, and 0.7 to 1.05 m^2, respectively, were found efficaceous for retention of rain water and for reducing soil loss. Bunds need be stabilized by planting adaptable grasses.

Small gullies (upto 3 m deep and 18 m or more wide) may be reclaimed by levelling the bed and construction of diversion-cum-check bunds (1.5 m cross section) at 30-45 m intervals and providing grass ramps and composite check dam at the end of the gully. Medium gullies (3 to 9 m deep, 18 m or more wide with side slope 8 to 15%) may be reclaimed by levelling the bed and constructing a series of composite earth and brick masonary check dams at vertical intervals of 1.2 m and terracing the side slopes (Fig. 7). The terraced faces, outlets and earthen dams need to be stabilized by suitable grasses like *Dichanthium annulatum* and *Cynodon dactylon* (Hirekerur *et al.* 1990).

The deep and narrow gullies are best controlled by putting them to permanent vegetation after closure to grazing. Afforestation with suitable tree species like *Acacia nilotica, Azadirachta indica, Dalbergia sissoo, Tecona grandis, Eucalyptus* species and other adaptable species has been found suitable. Besides the above species, following species of trees, shrubs and grasses have been found suitable for rehabilitation of ravinous lands. The suitable tree species are : *Leucaena latisiliqua, Dendrocalamus strictus, Albizzia lebbeck, Prosopis chilensis* (Syn. *Prosopis juliflora*), *Syzygium cuminii, Terminalia arjuna.* Among grasses and legumes, *Chrysopogon fulvous, Cenchrus ciliaris, Pennisetum pedicellatum, Saccharum spontaneum, Dichanthium annulatum* and *Phaseolus atropurpureous, Stylosanthes* species, besides Kudzu (*Pueraria* species) have been found suitable for such lands (Singh, 1990).

For the treatment and rehabilitation of ravine (gullied) areas, the whole gullied watershed is to be considered. A gullied watershed is a hydrological unit having two important components : (a) table and marginal lands contributing run-off to the gullies, and (b) gullies proper. Rehabilitation of gullies involves treatment of these two components (Verma and Chinnamani, 1990).

Control of run-off from table and marginal lands to gullies involves the following measures : (i) contour cultivation, (ii) crop rotations including cover crops during rainy season, (iii) intercropping, including legumes to cover more effective vegetative cover, (iv) contour, graded bunding, broad base terracing, bench terracing, (v) run-off harvesting, and (vi) earthen marginal bund (Section 2.0 -2.5 m^2) between adjacent area and gully heads. Graded bunds, bench-terrace rises and marginal bunds, are stabilized with vegetation. Waterways are stabilized with grasses (Verma and Chinnamani, 1990).

Strategy for gully rehabilitation or reclamation is based on 'Gully Rehabilitation Classification'. Six reclamability classes have been identified by Verma (1981) based on inherent soil properties, like texture and gully features (depth, bed width, and side slopes). Reclamability classes 1, 2, 3 gullies may be reclaimed for agriculture, class 4 for horticulture/agri-horticulture, whereas classes 5, 6 may be planted under perennial vegetation like fuel-fodder or silvi-pasture. Forests can be developed by closure, planting indigenous tree species or introducing more economical species like bamboo (*Dendrocalamus strictus* (Verma and Chinnamani, 1990).

Gullied lands could be rehabilitated by adopting the following measures :

1. Retention of run-off in the drainage area.
2. Diversion of run-off in the drainage area.
3. Conveyance of run-off through a gully. The strategy comprising of establishing a good vegetation cover and construction of gully control structure.

The gully formation could be controlled by constructing temporary check dams like woven-wire, brush, loose rock, plank and log and pole. For gully plugging where no run-off is expected from the top, earthen gully plugs of 1.1 sq.m cross section with grassed ramp 22.5 cm below the top and spaced at 45 to 60 m horizontal intervals have been recommended. On the other hand, where runoff is expected from the top, earthen gully plugs of 2.2 m cross section with a pipe outlet are recommended (Tejwani *et al.* 1960).

The gullied lands could be developed into silvipastures by using suitable plant species after proper protection from biotic interference by fencing. The grasses can be raised through direct seeding but for the trees seedlings raised in polyethlene bags should be used. The spacing between plant to plant and row to row should be 5 m.

Land with or without Scrub

The root cause of the present degradation of these lands was deforestation. In order to reverse the process of degradation or to bring it back to the original productive state, afforestation alone can be the solution of this malady. The afforestation technique has been standardized (Ram Prasad, 1988) and the same is briefly described below :

1. The area to be planted must be properly fenced, one year before taking up plantation.
2. Pits of size 45 cm^3 should be dug at the desired spacement which should normally be 2.0 x 2.0 m. Fertile soil, preferably tank silt, should be used for refilling the pits. Aldrex 30 EC solution should be sprayed in pits before refilling. Polypotted plants, not less than 45 cm in height, should be planted after the onset of monsoon.
3. Three weedings in the first year, two in the second year and one in the third year, around each pit upto a radius of 50 cm should be done. Hoeing should also be done along with weeding.
4. Ammonium sulphate @ 25 g and an equal quantity of superphosphate should be applied around each plant after 1-2 months of planting, and should be repeated every year, atleast for the first three years, during rainy season.
5. Among the species suitable for *Bhata* soils, *Eucalyptus tereticornis, Eucalyptus camaldulensis, Eucalyptus* hybrid, *Hardwickia, binnata, Cleistanthus collinus, Gmelina arborea, Emblica officinalis, Dalbergia sisoo, Acacia auriculiformis, Anogeissus pendula, Albizzia proceera* and *Dendrocalamus strictus* should be selected for afforestation. Inter planting of *Agave sislana* and palatable grasses would be desirable to protect soil from further erosion.

Waterlogged and Marshy Land

Remedial measures for rehabilitating waterlogged lands are : (a) drainage either surface or sub-surface or both to remove surplus water, (b) lining of canals to prevent seepage and rise of water table, and (c) sinking tubewell and utilizing water for irrigation thereby lowering water-table. Waterlogged lands contiguous to the sea-coast may be ameliorated for crop cultivation or used for fish culture (NCA, 1976).

Once the land has been drained, it is advisable to provide a cover with suitable tree species tolerant to high water table. Tree species found suitable for waterlogged areas are : *Acacia auriculiformis, Acacia nilotica, Dalbergia sissoo, Eucalyptus tereticornis, Eucalyptus robusta, Pongamia pinnata, syzigium cumini, Albizia procera.*

In waterlogged and salinized gully beds of Chambal (Kota), para grass (*Brachuria mutica*), a forage grass has been found to be successful. The waterlogged lands could be reclaimed by adopting following reclamation measures (Khan, 1990).

1. The waterlogged area should be reclaimed by introduction of improved surface and sub-surface drainage and shallow tubewells. These systems link down the water table and reduce the waterlogged area. Seepage is also reduced by canal lining.
2. Afforestation is another potent measure to reclaim the waterlogged land. The afforestation with selected species of trees i.e. *Eucalyptus* helps in lowering the water table on account of increased transpiration from new vegetation.
3. Some crops like rice, jute, sugarcane, dhaincha etc. and selective grasses like Dalis, Napier, Guinea, Karai and Murta, which can with stand waterlogging conditions, should be grown on such area.

Land Affected by Salinity/Alkalinity

The soils of this category of wasteland are broadly grouped under two broad classes viz. saline and alkali. These have distinct management and reclamation requirements. As a general rule, saline soils contain neutral salts (chlorides and sulphates of sodium, magnesium and calcium) whereas saline alkali soils contain salts capable of alkaline hydrolysis i.e. sodium carbonate, bicarbonate and silicate. Based on climatic, environmental and soil characteristics, the salt affected soils have been further grouped under 9 broad classes (Bharagava and Abrol, 1990). The alkali soils of the north western part of the country need simple alkali neutralization, whereas those in parts of Uttar Pradesh need vertical drainage too. Saline soils are largely responsive to horizontal drainage in the inland regions and surface flushing or a combination of surface flushing and prevention of sea water inundation in coastal and deltaic regions.

Reclamation of alkali soils requires replacement of excess sodium on the soil-exchange complex by calcium through amendments and leaching out of the exchanged sodium from the root zone. For any lasting reclamation, provision of adequate drainage is important. Vertical drainage has proved to be an effective measure for controlling groundwater table in alkali areas. It involves the making of a bore hole through which underlying water is extracted by pumping (Abrol and Gupta, 1990). Besides control of groundwater table, management of monsoon rains (averaging 600-700 mm) during the rainy season poses a major problem in alkali areas.

As an amendment, gypsum ($CaSO_42H_2O$) has proved by far the most effective and cheapest for reclamation of alkali soils of wastelands. The quantity of gypsum required for the improvement of alkali soils depends on the quantity of the exchangeable sodium that must be replaced and the depth to which soil improvement is desired. If a proper choice of crops is made, it is sufficient to add gypsum to improve only the surface 15 cm soil. For highly deteorated illitic soils, 14-15 tonnes of gypsum per hectare is sufficient to start cropping with rice and wheat in a rotation (Abrol and Gupta, 1990). Rice culture enjoys a favoured place in any cropping sequence recommended for adoption during reclamation of alkali soils. Proper choice of crops during reclamation of alkali soils is important. In general, crops that are able to withstand excess moisture conditions are also more tolerant to sodic conditions. Besides the tolerant crops (rice, barely, sesbania, *raya,* wheat (Fig. 8), sunflower, linseed, garlic, *guar*, groundnut, cowpea, onion, pearl millet, bengal gram, soybean, safflower, etc. in that order), grasses such as *Diplachne fusca* (Karnal grass), *Chloris gayana (Rhodes grass) and Brachiaria mutica* (Para grass) can also be grown. Regarding nutrient management of alkali soils, a crop sequence requires,on an average, 20 to 25 per cent more nitrogenous fertiliser than a calcareous soil. Large gains in yields by the zinc fertilization of alkali soils have been reported.

Management of waterlogged saline soils basically involves lowering of water table below the root zone of crop and leaching of excess salts (Rao and Singh, 1990).

There is good scope for afforestation of salt affected soils, provided soils are ameliorated to desired level by adopting appropriate technique. It basically involves soil replacement technique. A new technique commonly called posthole method of planting tree species has been evolved (Gill, 1985). Following post hole technique of planting trees in alkali soils, several species were evaluated for their tolerance to highly alkali environment and the tree species found suitable are given in Table 7.

Table 7 : Tree species and their relative tolerance to soil alkali

Tolerant	Moderately tolerant	Sensitive
Prosopis chilensis (Syn. *Prosopis juliflora)*	Azadirachta indica Parkinsonia aculeata	*Melia azadarach* *Morus species*
Acacia nilotica	Eucalyptus umbellata	*Populus species*
Casuarina equisetifolia	(Syn. E. tereticornis)	*Syzygium cumini*
Albizia lebbeck	Dalbergia sissoo	*Sesbania grandiflora*
Termapalia arjuna	Leucaena latisiliqua	*Grewia optivia* (Syn.
Tamarix aphylla	(Syn. L. leucocephala)	*G. oppositifolia)*
(Syn. *T. articulata)*	Grevillea robusta	*Dendrocalamus strictus*
Sesbania sesban	Syzgium fruticosum	*Citrus species*
Ailanthus excelsa	Acacia auriculae *formis*	–
Acacia catachu	*Tamarindus indica*	–
Zizyphus jujuba	*Moringa oleifera*	
Butea monosperma	*Bombax ceiba*	
Capparis decidua (Syn. *C. aphylla)*	*Pithecellobium dulce*	

Source : Gill *et al.* (1990).

The species tolerant to soil salinity are given in Table 8.

Table 8. Tree species tolerant to soil salinity

Tolerant	Moderate lytolerant	Sensitive
Prosopis chilensis	Eucalyptus umbellata	*Leucaena latisiliqua*
Tamarix troupii	Tamarinda arjuna	*Morux species*
Tamarix aphylla	Pongamia pinnata	*Salix balyloncia*
Acacia tortilis	Dalbergia sissoo	*Populus species*
Acacia nilotica	Acacia catechu	*Tamarindus indica*
Casuarina obesa	Acacia senegal	*Melia azadarach*
Casuarina equisetifolia	Prosopis cineraria	*Azadirachta indica*
Callistemon lanceolatus	–	*Albizzia lebbeck*
Eucalyptus camaldulensis	–	–
Acacia auriculae-formis	–	–

Source : Gill *et al.* (1990).

Saline / alkali lands also known as *usar* lands could be rehabilitated by leaching the salts and also by growing manure plants such as moong (*Vigna radiata*) and dhaincha (*Sesbania aculeata*). The Central Soil Salinity Research Institute, Karnal has evolved an effective technology for the reclamation of this category of wasteland. The main components of the technology are as under:

1. Land levelling and bunding to ensure uniform application of water and leaching.
2. Application of amendments like gypsum and pyrite. Gypsum is the most effective amendment for alkali soil, especially in Indo-Gangetic plain.
3. Growing of suitable varieties of rice and wheat in a rotation and this cropping sequence should be used at least for 3 to 4 years of reclamation.

4. Adoption of best suited agronomic cultural and water management practices.
5. Growing of Karnal (*Deplachne fusca*) and para (*Brachiaria mutica*) grass which are very tolerant to alkalinity and can be grown in highly alkali soils.

Shifting Cultivation

Some alternative to shifting cultivation found suitable of northeastern hills region are : agricultural land use, horticultural land use, livestock based land use, agri-horti-silvipastoral land use, and fish production.

Agricultural land use can be adopted on hill slopes upto 50 per cent gradient where soil depth is greater than 1.0 m. Contour bunding at 0.5 to 1.0 m vertical interval draining into a common grassed waterway is an essential requirement. It is advisable to construct few bench terraces at the foot hill portion of the hill slope (Prasad and Singh, 1990). For horticultural land use, slope of the land should not be more than 100 per cent. Soil depth must be minimum 1.0 m. Land upto 100 per cent slope with minimum 0.5 m soil depth can be utilised for livestock farming. Carrying capacity of such a land use has been estimated to be 4 to 5 livestock unit per ha with setaria and stylo (1:1 mixture) of fodder production. Land upto 100 per cent slope having soil depth greater than 1.0 m can be used for agri-horti-silvipastoral system. The system comprises agricultural land use towards the foothills, horticulture in the mid-portion of the hill slope and silvipastoral land use towards the top of the hill. Dug out-cum-embankment type of ponds can be created as seasonal or perennial water bodies and utilised for fish production.

Sandy Areas (Coastal or Desertic)

India has a long sea-coast line (about 5600 km). Sand dunes occupy large areas in the coastal districts of Tamil Nadu, Maharashtra, Gujarat and Orissa. During the monsoon and cyclonic periods, there is blowing and shifting of sands causing damage to the standing crops in the neighbouring areas. Hence, plantation of *Casuarina* on the sandy foreshore backed by plantation of other adaptable tree species further inland can minimize this problem (Hirekerur *et al.* 1990).

The technique for rehabilitation of coastal areas consists of planting *Casuarina* upto 10 m width along the coast and inner areas with cashewnut interspersed with coconut. The plantation gives economic returns after 5 years. The tree species suggested for the regions are : *Casuarina equisetifolia, Cocos nucifera, Ficus* spp., *Calophylum inophyllum, Anaeardium occidentale, Mangifera indica, Thespesia populnea* etc. and among grasses - *Sporobolus* spp., *Cynodon dactylon, Panicum maximum, Cenchrus* spp., *Chloris* spp., *Spinifex squamosus* etc. (Ram Mohan Rao, 1990).

The Indian or the Thar desert, a hot season rainfall region, situated in the north-west, covers about 2,86,000 km^2. More than 68% of this area is covered with stabilised and active sand dunes of different types, shape and pattern and they fall under category of sandy wasteland (Singh and Shankarnarayan, 1986).

In the successful sand dune/sandy wastelands rehabilitation with vegetation there are three most important processes viz. (i) protection against biotic interference by fencing the area with barbed wire or thorny bushes, (ii) sand dune fixation to check the movement of sand, and (iii) afforestation by grasses and trees. Some of the more important options for sand dune fixation are : checker board and parallel hedge system, the foredune system mulching, and chemical mulching (Harsh and Tewari, 1993). In areas with wind blowing from different directions a checker board system (2-3 m^2) has been found to be very effective in arresting the movement of sand (Gupta, 1990). The distance between the strips depends upon the wind velocity, steepness of slopes and the form of sand dunes. Generally these should be closer together on the tops and windward sides of the dunes and widely spaced on low dunes or depressions (Fig. 9).

After fixation, sand dunes can be permanently stabilised with vegetation - trees, shrubs and grasses like *Prosopis chilensis* (Syn. *Prosopis juliflora*), *Prosopis cineraria, Acacia tortilis, A. radiana, Zizyphus mauritiana,* among tree speices, *Calligonum polygonoides, Crotalaria burhia, Aerva javanca, Zizyphus numularia,* among shrubs; and *Lasiurus sindicus, Panicum turgidum, Panicum antidotale* among grasses, were found most suitable species for sand dune stabilisation in the arid zone of western Rajasthan.

The sandy wastelands in the form of highly reactivated stabilised sand dunes may be rehabilitated by planting top feed species in a phased manner. The plantation should be done after due enclosures of large blocks of 1000 ha or more. Reactivated and highly unstable crest and flanks or the sand dunes should be managed by seeding and root slip plantation of suitable grasses like *Panicum turgidum, Lasiurus sindicus, Cenchrus ciliaris* and *Aristida funiculata.*

The locally adapted species viz. *Acacia senegal* (Kumut); *Calligonum polygonodes* (Phog), *Prosopis cineraria* (Khejri) and *Maytenus emarginatus* (Kangkera) may be most suitable for the rehabilitation of sanddunes/sandy wastelands. The dunes thus reclaimed must be kept protected against grazing by all kind of livestock and should be managed by cut fodder (Saxena and Singh, 1976).

Mining/Industrial Wastelands

The site preparation of such mined out areas is a pre-requisite. It involves levelling of the surface, digging of pits of appropriate dimensions at variable spacings and finally filling these pits with a fertile soil. The site preparation is followed by the right choice of species suitable to the site, from a number of exotic and indigenous plant species. Interplanting of *sal* (the original species) in bauxite and coal mine areas was found to give encouraging results (Ram Prasad, 1988). Natural regeneration in *Acacia auriculiformis,* production of bulk litter and natural succession of other plant species, prove that the sites are progressing effectively towards rehabilitation.

Rehabilitation of mine spoils (Chatterji, 1988) can be undertaken in the following three ways :

(i) Establishment of vegetation i.e. forest or grassland, orchards or crops, either alone or in combination i.e. agroforestry or silvipastoral or crop.

(ii) Development of water resources for recreation, drinking, irrigation, fishery, recharge to aquifer and / or artificial aquifer.

(iii) Miscellaneous uses like development of biogas and manures.

Adopting any of the above method would involve one or more of the following :

(i) Preservation of top soil;

(ii) Selection of proper site for disposal;

(iii) Proper configuration of the disposal dump;

(iv) Soil conservation measures;

(v) Soil amendment;

(vi) Planting techniques; and

(vii) Water bodies.

The bunding, contour furrowing, mulching, levelling and soil compaction will help to restore productivity of the more spoiled areas by reducing erosion and retention of moisture. It is proposed to construct half moon terraces of 5 x 5 m distance for planting 425 plants.

The following measures for the rehabilitation and management of industrial effluent responsible for the development of industrial wasteland are suggested (Singh, 1992; Aggarwal and Kumar, 1990).

1. The polluted water discharged from the textile industries and other industries be treated at the spot by chemical and biological methods before it is used for industrial and irrigation purpose.

2. The untreated polluted water may be used for irrigated farming once in three years, in second year, the land may be left fallow followed by irrigation farming in third year to avoid the development of hard surface crust, to enhance the infiltration rate of water and for better growth of the plants.

3. The suitable limits of total dissolved salts (TDS), biological oxygen demand (BOD), sodium adsorption ratio (SAR) in the polluted water if it is discharged in the rivers and used for irrigation should be imposed.
4. The dry *Calotropis procera* leaf powder, latex of *Calotropis procera, Euphorbia caducifolia* activated charcoal and alum may be used to reduce pH and EC in the polluted water before using it for irrigation to improve the productivity of industrial wasteland.
5. The flyash produced from thermal plants may be used for removing the colour from the polluted water to reduce its harmful effect on land, soil and land use to upgrade the production of wasteland and to check the development of new wastelands.

Under-utilized/Degraded Notified Forest Land

The degraded forest could be rehabilitated by protecting these lands from increasing biotic activities like cutting of trees, mining, control of degrading factors like sheet, rill and gully erosion by contour furrows, contour bunding and trenching etc. The forest blank areas could be revegetated by suitable plant species given below.

Acacia albida	*Crotalaria burhia*
Acacia nilotica	*Commiphora wightii*
Agathis damkuara	*Cordia myxa*
Agave americana	*Cynodon dactylon*
Albizia lebbeck	*Lantana camara*
	Panicum antidotale
Anogeissus pendula	*Saccharum bengalense*
Aristida depressa	*Poa compressa*
Arundo donax	*Portulaca oleracea*
Boswellia serrata	*Tecomella undulata*
Azadirachta indica	*Simmondsia chinensis*
Butea monosperma	*Sesbania bisponosa*
Calligonum polygonoides	*Tussilago farfava*
Cenchrus biflorus	
Cenchrus setigerus	
Citrullus colocynthis	
Clerodendrum phlomoides	

Degraded Pastures/Grazing Land

The increasing impact of natural and accelerated factors like wind erosion/deposition, water erosion, salinity, overgrazing and lopping and cutting of plants should be stopped for rehabilitation of this type of wastelands. The soil and water conservation measures like wind breaks, shelterbelts, mulching, contour bunding, contour furrows, leaching of salts and use of gypsum are recommended to upgrade the productivity of these lands. The degraded areas should be reseeded and planted with suitable plant species. The deferred control grazing should be done according to the carrying capacity of these pastures/grazing land. Deferred grazing in a three compartment scheme is a suitable management technique. In this scheme, the third compartment is deferred first for two-thirds of the season and the animals are grazed rotationally in first and second compartments until 70 per cent utilization is achieved. The aeolian landforms could be developed into silvipastures by using suitable plant species. The rocky/gravelly pediments and shallow eroded alluvial plains could be developed into pastures. The trees should be planted in staggered contour ridges cum trenches 3 m in length and 60 x 60 cm in cross section. The suitable plant species for reclamation of degraded pastures/grazing land are as under:

Acacia tortilis	*Albizzia lebbeck*
Acacia senegal	*Zizyphus mummularia*
Arundo donax	*Indigofera cordifolia*
Butea monosperma	*Heteropogon contortus*
Cymbopogon caesius	*Sporobolus aciocides*
Dactyloctenium aegyptium	*Suaeda martima*
Crotalaria burhia	*Vicia villosa*
Magnifera indica	*Vigna hosei*

Salvadora oleoides	*Zysia matrella*
Saccharum bengalense	

Degraded Non-Forest Plantation Land

This type of wasteland, particularly in the arid and semi-arid regions, could be reclaimed by using mechanical and agronomic measures.

The mechanical measures involve contour cultivation, ploughing, harrowing, sowing and interculture, contour bunding, terracing, graded trenching and grassed water ways. Soil conservation, forestry, agrostology and horticulture involving tree, bushes and grasses reduce run-off by providing a good soil cover. These eroded lands could be economically put under the above uses for sustained production. Agronomic measures such as strip cropping, mixed cropping and intercropping will also help to rehabilitate such wastelands. The suitable plant species for growing on such wastelands are given below.

Albizzia lebbeck	*Maytenus emarginata*
Albizzia amara	*Heteropogon contortus*
Anogeissus latifolia	*Ficus glomerata*
Apluda mutica	*Grewia tenax*
Capparis decidua	*Leptadenia protechnia*
Chloris virgata	*Populus alba*
Parkinsonia aculeata	*Tamarindus indica*

Barren Rocky/Stony Waste/Sheet Rock Area

Barren hill ridges or rock outcrops are observed in all the hill ranges in the country which serve as catchment for water supply. Intermittent patches with shallow skeleton soils on sloping lands may be put to grasses and biotic interferences controlled, wherever possible, depending on the land conditions, afforestation may be initiated, establishment of which will check the flow or run off (Hirekerur *et al.* 1990). Stony and gravelly lands, normally found on rolling topography, should be closed to grazing till vegetative cover is established.

The run-off from steep slopes and rocky areas is required to be safely drained through diversion drain or diversion bund at non-erosive velocity to avoid damage to flat lands. Contour trenches of 4.0 m x 0.5 m x 0.5 m at 10 m horizontal interval are to be made for conserving rain water with an equilizer every 4 m apart. The end of these trenches are to be hooked up. In between the trenches, pits of 0.5 m x 0.5 m x 0.5 m are to be made in a row at a spacing of 4.0 m. In trenches and pits seedling may be planted after filling half of them diagonally with soil and utilizing the remaining soil to form a level mound or the downhill side to encourage water collection (Ram Mohan Rao, 1990).

Lateritic areas appear like barren mounds devoid of vegetation. Some of these areas can be put under plantation crops by trenching and filling with fertile soil from neighbouring places.

This category of wasteland also provide the ideal sites for harvesting surface runoff in the village tanks, particularly in arid and semi-arid regions. The water thus harvested could be used for drinking purpose.

Based on the present status of different types of wastelands, the suitable soil and water conservation measures and plant species have been suggested (Table 9) for their rehabilitation and development (Saxena and Singh, 1990).

ECONOMIC EVALUATION FOR REHABILIATION OF WASTELANDS

Reclamation of Alkali Wastelands

The capital requirement for reclamation of soils of alkali wastelands was worked out by Joshi (1985). Total capital requirement at 1985 prices came to Rs. 4365 / ha at 50 per cent subsidy of amendment and Rs. 3225/ha at 75 per cent subsidy. The capital requirement includes the cost of bunding, (Rs. 435/ha), levelling (Rs. 1085./ha), and irrigation (Rs. 200/ha).

Cost of production and returns from alkali soils reclamation are set out in Table 10.

Table 9. Rehabilitation measures for the development and management of wastelands in the Indian Arid Zone

Type of wastelands	Present status and type of degradation	Developmental Technology		Utilization	Grazing management	Other development plans
		Reclamation measures	Suitable species			
Gullied and ravinous land	Poor; water erosion in the form of rills and gullies	Gullies plugging, levelling	*Cenchrus, Dichanthium, Acacia, Prosopis*	Silvipastoral/ Hortipastoral	rotational eferred grazing	Agro-industry, medicinal
Land with or without scrub	Poor to moderately fair; water erosion.	Contour furrowing and trenching	*Cenchrus, Dichanthium Sehima, Acacia, Salvadora, Prosopis,*	Grasslands/ silvipastoral /hortipastoral	Deferred rotational grazing	Medicinal, energy plantation
Land affected by salinity/ alkalinity	Poor to very poor; salinity/alkalinity	Trenching, raised bed, mixing FYM + gypsum + normal soil, auger hole method	*Sporobolus, Dichanthium Cenchrus, Salvadora, Capparis*	Grassland	Deferred rotational grazing	Energy plantation
Sand-desertic	Poor to moderately fair; wind erosion/deposition/ water erosion	Micro-wind break, mulching, aerial seeding etc,	*Panicum, Lasiurus, Cenchrus, Calligonum Prosopis, Acacia*	Silvipastoral	Cutting and stall feeding	Economic, agro-industry and energy plantation
Mining/industrial wasteland	Poor; water erosion/ wind erosion/ denudation	Land levelling, filling and other suitable method needed for specific site	*Cenchrus, Dichanthium Lasiurus, Panicum, Acacia, Albizzia, Zizyphus, Prosopis, Salvadora*	Hortipastoral/ silvipastoral	rotational eferred grazing	Agro-industry and economic, medicinal plantation
Underdutilized degraded notified forest land	Moderately fair to fair; water erosion/salinity, denudation	Contour furrowing, bunding, trenching	*Lasiurus, Cenchrus, Sehima, Prosopis, Acacia, Albizzia Zizyphus*	Forests	Cutting and stall feeding	Energy and medicinal plantation
Degraded pasture /grazing land	Poor to moderately fair; water/ wind erosion/ salinity	Bunding, trenching and other operation according to need	*Lasiurus, Panicum, Cenchrus, Dichanthium Prosopis, Zizyphus, Acacia*	Silvipastoral	Deferred rotational grazing	Exclusive for animal husbandry
Baren rocky/ stony waste/ sheet rock area	Very poor; water erosion	Contour trenching channelization of surface water	–	Water harvesting and stone mining	Watering points	
Steep sloping areas	Very poor; water erosion	Terracing	–	Water harvesting	Watering points	

Source : Saxena and Singh (1990).

* All wastelands need protection by fencing.

Table 10. Cost of production and returns (Rs/ha)

Item	Cost	Returns
	First Year	
Tubewell installation	2,000	–
Land preparation and amendment	4,365	–
Paddy production	4,270	6,300
Wheat production	2,890	3,140
Sub-total	13,575	9,440
	Second Year	
Paddy production	4,270	7,000
Wheat production	2,890	3,925
Sub-total	7,160	10,925
	Third Year	
Paddy production	4,270	8,400
Wheat production	2,890	4,710
Sub-total	7,160	13,110
Total	27,845	33,475
Net return in 3 years		5,630

Source : Joshi (1985).

Reclamation of Gullied/Ravinous Land

The economic evaluation of denuded ravine lands has two facets : (i) for agricultural purposes and (ii) for non-agricultural purposes.

Seckler (1978) considering 1977 price level worked out the benefit : cost ratio and economics of reclamation or ravine lands in Gujarat (Table 11).

Table 11. Benefit : Cost ratio and economics of reclamation of ravine lands in Gujarat

Treatment	Capital cost (Rs/ha)	Average annual cost (Rs/ha)	Present value (Rs/ha)	Total cost (Rs/ha)	Average annual benefits (Rs/ha)	Present values (Rs/ha)	B : C ratio
Humps having a slope upto 6% reclaimed							
(a) with manual labour	426	409	3482	3844	861	7330	1.86
(b) with bulldozer + manual labour	346	320	2724	3070	681	5798	1.89
(c) a system of shallow and medium gullies	2955	288	2435	5390	543	4623	0.86

Source : Seckler (1978).

Economic evaluation of a ravine watershed (30.2 ha) was undertaken. The reclamation consisted of construction of contour and peripheral bunds in the tablelands, bench terraces in medium gullies) and gully plugs, composite check dams, grassed water ways, chute structures and pipe outlets, etc. in deep gullies during 1958-59. The period of analysis considered was 25 years and 12 per cent annual interest rate. The B : C ratio of the project on the given site and climatic conditions worked out to be 1.37:1 (Vimal kishore *et al.* 1982).

Puri and Khebri (1975) had carried out studies on afforestation in Chambal ravines, on lands suited to non-agricultural uses. The benefit cost ratio at 6 per cent rate of interest with 20 years rotation period was worked out to be 1.55 and the annual net income that could be obtained from such lands was worked out to be Rs. 349/ha.

Economic evaluation of mixed afforestation, consisting of fuel and fodder species in deep and narrow ravines at Vasad has been carried out. A fifteen year project life and 12 per cent discount rate for calculation of B : C ratio was considered in the analysis. The B:C ratio obtained is 1.38:1 and IRR/(Internal rate of return) obtained is 15 per cent. The internal rate of return fully supports the economic feasibility of such social projects (Vimal Kishore *et al.* 1982).

Economics of tree crops for hot arid zones of Rajasthan was worked out by Gupta *et al.* (1982) and the same is given in Table 12.

Table 12. Economics of tree crops for hot arid zone of Rajasthan (at 11% discount rate)

Tree species	Year of life cycle	Annuity value (Rs/ha)
Acacia tortilis	8	361.60
Albizia lebbeck	15	547.90
Prosopis cineraria	20	1610.10
Prosopis chilensis (Syn. *P.juliflora*)	6	950.00
Leucaena latisiliqua (Syn. *L.leucocephala*)	15	2909.00
Zizyphus species (Grafted)	25	3267.50

Source : Gupta *et al.* (1982).

Dhruvanarayana and Ram Babu (1990) have worked out the benefit : cost ratio of fruit trees grown at Dehradun (Table 13).

Table 13. Benefit : Cost ratio and payback period of fruit trees grown on wasteland at Dehradun

Discount rate	Lemon				Mandarin (Kinnow)			
	Cost	Benefit	B:C ratio	Pay back period	Cost	Benefit	B:C ratio	Pay-back period
10	8371	11745	1.40	12	7,511	16,140	2.13	10
15	6397	7631	1.20	—	5,890	10,652	1.79	—

* Present value in rupees
Source : Dhruvanarayana and Ram Babu (1990).

Economic Viability of Alternative Land use Technologies for wastelands

Kalla (1988) has compiled information on economic viability of various alternative land uses suited for different wastelands and the same is presented in Table 14.

Table 14. Economic viability of alternative land use technologies for wastelands in arid areas of western Rajasthan

Technology	Suitability	Attributes				
		NPV (RBS)	DB(-)CR ratio	PBP (yrs)	IRR %	Annuity (Rs/ha)
1 Sand dune stabilisation	Duny wastelands	–	1.85:1	12	12.50	702.88
2 Improved pasture and grazing management	Duny plains	–	1.79:1	5	37.70	948.54
3 Cultivable pastures (mixed stands)	Saline wastelands	–	–	–	–	–
(a) CC + LS	Below 300 mm annual rainfall zone	1122.59	1.18:1	–	16.42	80.20
(b) CC + CS		651.93	1.81:1	–	13.78	-46.57
(c) CC + CS +LS		2247.42	1.30:1	–	15.82	160.57
(d) Natural		254.16	1.03:1	–	15.70	9.07
4 Fodder grasses	Below 300 mm annual rainfall zone					
(a) *Cenchrus ciliaris*	–	–	1.56:1	–	–	–
(b) *C. setigerus*	–	–	3.30:1	–	–	
(c) *Panicum antidotale*	–	–	1.95:1	–	–	
(d) *Lasiurus sindicus*	–	–	1.53:1	–	1	1
5 Fuel tree species	Arid wastelands					
(a) *Albizzia lebbeck*		–	1.12:1	–	–	–
(b) *Azadirachta indica*		–	0.99:1	–	–	–
(c) *Prosopis juliflora*		–	1.06:1	–	–	–
(d) *Acacia tortilis*		–	1.71:1	–	–	–
6 Afforestation	Alkali wastelands	6581.32	1.63:1	–	–	1580.00
7 Sown pastures (in terms of alternative uses)	Sandy wastes	–	–	–	–	–
(a) Sheep grazing		10753.30	1.02:1	14	16.80	1715.80
(b) Cattle grazing		60889.30	1.14:1	12	24.02	9715.80
(c) Mixed herds		35821.30	1.08:1	13	21.01	5715.80
(d) Seed collection		414974.80	1.79:1	8	31.60	66215.86

Surce : Kalla (1988).

Economic Evaluation of Agroforestry Systems

The economic evaluation of this system was carried out by Singh and Reddy (1986) and the same is set out in Table 15.

Table 15. Economic evaluation of agroforestry systems at 11% and 20% discount rates (Rs/ha)

Agroforestry systems	Life span (years)	Annuity value 11 %	Annuity value 20 %	B : C ratio 11 %	B : C ratio 20 %	IRR %
Farm Forestry						
(i) Casuarina	10	3000	2145	11.08	7.52	90
(ii) Eucalyptus	30	1741	1337	18.60	10.64	98
Agroforestry	10	1117	853	1.73	1.33	32
(i) Agri-horticulture	20	17881	14638	5.78	5.08	59
(ii) Silvi-agriculture	10	1204	1141	1.80	1.74	35
(iii) Silvi-pasture	10	1115	529	2.56	1.68	34

Source : Singh and Reddy (1986).

Based on the experience gained in working with the Model Watersheds Development of the ICAR, Singh and Reddy (1986) worked out the cost of some of the important soil and moisture conservation measures, which are given in Table 16.

Table 16. Cost (Rs/ha) of some of the important soil and moisture conservation measures

Item	Cost estimates (Rs/ha) at 1986 prices
Field bunding	200
Contour cultivation	Nominal
Contour bunding	500-800
Graded border strip	1,200-1,500
Ridge and furrow system	300-400
Broad bed and furrow system	1,000-1,200
Deep ploughing	400-500
Residue management	200-300
Application of surface mulches	200
Dug out pond	6,000-18,000*

*Varies with the size of the pond and lining.
Source : Singh and Reddy (1986).

Integrated Wasteland Development

At the instance of District Rural Development Agency (DRDA) a micro plan for integrated wasteland development of 5000 ha area in the vicinity of Jodhpur located in the district of Jodhpur (Rajasthan) was prepared by the Central Arid Zone Research Institute, Jodhpur. The cost to be incurred on the rehabilitation of different types of wasteland has also been estimated on the basis of the rate of PWD BSR 1991 (Table 17). The items of work involved in estimating the cost of wasteland rehabilitation are as under (CAZRI, 1991).

(i) Surveying and layout
(ii) Stone wall fencing
(iii) Cost of raising plants
(iv) Preparation of terraces
(v) Digging of pits
(vi) Construction of soil conservation devices
(vii) Cost of seeds
(viii) Cost of watering
(ix) Cost of weeding and hoeing
(x) Purchase and application of insecticides, pesticides, fertilizers and manure
(xi) Construction of water storage tank
(xii) Construction of guard hut
(xiii) Cost of watch and ward
(xiv) Amenities to labourers, and
(xv) Miscellaneous expenses

Table 17. Cost estimates for development of different wasteland types in Kailana area, Jodhpur (Rajasthan)

S.No.	Type of wasteland	Development activity	Area (ha)	Cost (Rs/ha)	Total cost (Rs)
1.	Gullied/sandy undulating plains	(i) Silvipastoral (ii) Agroforestry	60 82	10,512 200	6,30,720 16,400
2.	Sandy hummocky plains	(i) Silvipastoral revegetation	48	10,512	5,04,576
		(ii) Agroforestry	308	10,512	61,600
3.	Rocky plateau / Sheet rock with soil pockets	Regeneration and ecological management	2295	11,450	2,62,77,750
4.	Foot hills zone, moderately sloping	Regeneration and ecological management	1360	11,450	1,55,72,000
5.	Valley fill region with deep soil	(i) Horti-agriculture	348	750	2,61,000
		(ii) Agroforestry	348	200	69,000
6.	Stream bed with deep coarse and sandy soil	Vegetable cultivation	26	–	–
7.	Mine spoil area with boulder and stones	Rehabilitation forestry	54	16,000	8,64,000
8.	Hills	Regeneration and ecological management for resource conservation	487	11,45,055	55,76,150
9.	Obstacle sand dune	Silvipastoral revegetation	10	10,512	1,05,120
	Entire area	Road side and channel side plantation on 61 running km	–		5,22,648
Water storage reservoirs 6 numbers @ 6.5 lac each					
	Total		5,426		543,61,564

Cost of development Rs. 10,018/ha
Source : CAZRI–Micro Plan for Integrated Wasteland Development near Kailana (1991).

REFERENCES

Abrol, I.P. and Gupta, R.K. 1990. Alkali soils and their management. In : *Technologies for Wasteland Development* (Eds. I.P. Abrol and V.V. Dhruvanarayana), ICAR, New Delhi, pp. 317-334.

Aggarwal, R.K. and Kumar, Praveen. 1990. Textile industrial effluent : Implication and possible use for afforestation in western Rajasthan. *Ann. Arid Zone* 29(4) : 295-300.

Bhargava, G.P. and Abrol, I.P. 1990. Nature and extent of salinity and waterlogging problem in India. In : *Technologies for Wasteland Development* (Eds. I.P. Abrol and V.V. Dhruvanarayana), ICAR, New Delhi, pp. 307-316.

Bhumbla, D.R. and Khare, A. 1984. *Estimate of Wastelands in India.* Society for Promotion of Wasteland Development, New Delhi.

Bhumbla, D.R. and Khare, A. 1987. Estimate of wastelands in India. In : *Wastelands Diagnosis and Treatment* (Ed. H.R. Yadav), Concept Publ. Co., New Delhi, pp. 29-48.

CAZRI 1991. *Micro-Plan for Intergrated Wasteland Development near Kailana.* Resource Survey and Monitoring Division, CAZRI, Jodhpur, 46 p.

Chatterji, P.C. 1988. Mine spoils and their rehabilitation. In : *Mining and Environment in India* (Eds. S.C. Joshi and G. Bhattacharya), H.R. Publishers, Nainital, pp. 394-409.

Dhruvanarayana, V.V. and Ram Babu. 1990. Social and economic constraints in the management of the soils of hilly regions. In : *Technologies for Wasteland Development* (Eds. I.P. Abrol and V.V. Dhruvanarayana); ICAR, New Delhi, pp. 273-284.

Gautam, N.C. and Narayan, L.R.A. 1988. *Wastelands in India.* Pink Publishing House, Mathura, 96 p.

Gill, H.S. 1985. *Studies on the Evaluation of Selected Tree Species for Their Tolerance to Sodicity and Mechanical Impedence in a Highly Sodic Soil with particular References to Root Growth Behaviour.* Unpublished Ph.D. Thesis, CSSRI, Karnal, 271 p.

Gill, H.S. Abrol, I.P. and Gupta, R.K. 1990. Afforestation of salt affected soils. In : *Technologies for Wasteland Development* (Eds. I.P. Abrol and V.V. Dhruvanarayana), ICAR, New Delhi, pp. 355-380.

Gupta, J.P. 1990. Sand dunes and their stablilisation. In : *Technologies for Wasteland Development* (Eds. I.P.

Abrol and V.V. Dhruvanarayana), ICAR, New Delhi, pp. 59-68.

Gupta, Tirath and Deepinder Mohan. 1982. Economics of Trees Versus Annual crops on Marginal Agricutural Lands. Oxford and IBH Publishing Co., New Delhi, pp. 38-69.

Harsh, L.N. and Tewari, J.C. Sand dune stabilisation, shelterbelts and silvipastoral plantation in dry zones. In : *Desertification and its Control in the Thar, Sahara and Sahel regions* (Eds. A.K. Sen and Amal Kar). Scientific Publishers, Jodhpur, pp. 269-279.

Hirekerur, L.R., Lal S.and Pofali, R.M. 1990. Approaches to development planning for degraded lands. In : *Technologies for Wastelands Development* (Eds. I.P. Abrol and V.V. Dhruvanarayana), ICAR, New Delhi, pp. 43-58.

INCOR 1986. *Area Estimation of Wastelands in India.* Institute for Coastal and Offshore Research, Visakhapatnam.

Joshi, R.K. 1985. Alkali soil reclamation - costs and benefits. In : *Ann. Report, CSSRI, Karnal,* pp. 6-11.

Kalla, J.C. 1988. Criteria of economic evaluation of wasteland development programmes. In : *Wastelands Development and their Utilization* (Ed. K.A. Shankarnarayan), Scientific Publihsers, Jodhpur, pp. 356-363.

Khan, N. 1990. Reclamation of wastelands in Faizabad district, U.P. In : *Utilization of Wastelands for Sub-tainable Development in India* (Eds. S.C. Sharma, R.B. Chaturvedi and O.P. Misra), Concept Publ. Co., New Delhi, pp. 419-430.

Lal, R. Hall, C.F. and Miller, F.P. 1989. Soil degradation : I. Basic processes. *Land Degradation and Rehabilitation* 1(1) : 51-69.

Malhotra, S.P. and Kalla, J.C. 1990. Social and economic constraints in optimal land use under arid conditions In : *Technologies for Wasteland. Development* (Eds. I.P. Abrol and V.V. Dhruvanarayana), ICAR, New Delhi. pp. 32-42.

NCA 1976. *Report of the National Commission on Agriculture.* Ministry of Agriculture and Irrigation, Govt. of India, New Delhi.

NRSA 1985. *Mapping of Wastelands in India from Satellite Imagery.* Department of Space, Govt. of India, Hyderabad.

NWDB 1986. *Wastelands Definition and Classification.* Technical Task Group Report, 12 p.

NWDB 1987. *Description and Classification of Wastelands.* Ministry of Environment and Forest, Govt. of India, p. 5.

NWDB 1985. *Wastlands Development Programme : Guidelines for Action.* Ministry of Environment and Forest, New Delhi.

Prasad, R.N. and Singh, A. 1990. Shifting cultivation and land degradaton. In : *Technologies for Wasteland Development* (Eds. I.P. Abrol and V.V. Dhravanarayan), ICAR, New Delhi, pp. 214-222.

Prasad, Ram 1988. *Technology of Wasteland Development.* Associated Publ. Co. 356 p.

Puri, D.N. and Khebri, M.L. 1975. Economics of Chambal ravines afforestation. *Ind. For.* 10(8) : 448-451.

Rama Mohan Rao, M.S. 1990. Approaches to the management of denuded lands in the Peninsular India. In : *Technologies for Wasteland Development* (Eds. I.P. Abrol and V.V. Dhravanarayana), ICAR, New Delhi, pp. 232-244.

Rao, K.V.G.K. and Singh, O.P. 1990. Waterlogging and salinity problems - The management alternatives. In : *Technologies for Wasteland Development* (Eds. I.P. Abrol and V.V. Dhruvanarayana), ICAR, New Delhi, pp. 335-344.

Saxena, S.K. and Singh, S. 1976. Some observations on the sand dunes and vegetation of Bikaner district in western Rajasthan, *Ann. Arid Zone* 15(4) : 313-322.

Saxena, S.K. and Singh, S. 1990. Wastelands in making, measures to mitigate the effects of factors responsible. In : *Dry Farming and Wasteland Development - Part II.* Afro-Asian Rural Reconstruction Organisation, New Delhi, pp. 149-154.

Secklar, D. 1978. *A Guide to the Economic Evaluation of Soil and Water Conservation Projects.* CSWCRTI, Dehradun.

Sharma, S.C., Tripathi, D.N. and Sharma, M.L. 1990. Usar land reclamation in India with special reclamation to Sultanpur block, district Sultanpur (U.P.) In : *Utilization of Wastelands for Sustainable Development in India* (Eds. S.C. Sharma, P.B. Chaturvedi and O.P. Misra) Concept Publ. Co., New Delhi, pp. 287-301.

Singh, P. 1990. Forage and fuel production from degraded and wastelands. In : *Technologies for Wasteland Development* (Eds. I.P. Abrol and V.V. Dhruvanarayana), ICAR, New Delhi, pp. 108-124.

Singh, R.P. and Reddy, Y.V.R. 1986. Degradation and rehabilitation of drylands - Some technical and economic aspects. *Paper Presented in the Regional Workshop for Nine Countries of South and Southeast Asia on Economics of Dryland Degradation and Rehabilitation, New Delhi.*

Singh, R.P. 1990. Land degradation problems and their management in the semi-arid tropics. In : *Technologies for Wasteland Development* (Eds. I.P. Abrol and V.V. Dhruvanarayana), ICAR, New Delhi, pp. 125-136.

Singh, S. and Sharma, B.K. 1984. Some observations on the morphological and physio-chemical aspects of the sand dunes in Jodhpur district (Rajasthan). *Ann. Arid Zones* 23(1) : 31-39.

Singh, S. 1977. Sand dunes and palaeoclimate in Jodhpur district, western Rajasthan (India). *Man & Environ.* 1 : 7-15.

Singh, S. and Shankarnarayan, K.A. 1986. Dynamics, morphology and management of sand dunes in the Great Indian desert. In : *Current Practices in Geotechnical Engineering - Science and Engineering of Desert Soils* (Eds. Alam Singh and M.L. Ohri) Int. Book Traders, Delhi and Geo-Environ Academia, Jodhpur, pp. 23-48.

Singh, S. and Kar, A. 1988. Geomorphological appraisal of arid wastelands by using remote sensing techniques. In : *Wastelands Development and Their Utilization* (Ed. Shankarnarayan, K.A.) Scientific Publishers, Jodhpur, pp. 62-68.

Singh, S., Sharma, K.D., Singh, N. and Vats, P.C. 1989. Wastelands and rural development in the Rajasthan desert. *The Geographer* 36(1) : 37-49.

Singh, S. 1990. Genesis, distribution and utilization of sandy wastelands in the Rajasthan desert. In : *Utilization of Wastelands for Sustainable Development in India* (Eds. S.C. Sharma, R.B. Chaturvedi and O.P. Misra), Concept Publ. Co., New Delhi, pp. 131-146.

Singh, S. 1992. Water pollution and its effect on ecosystems in an arid envirohment. In : *Environmental Pollution-Consequences and Measures* (Ed. B.P. Chaurasia), Chugh Publications, Allahabad, pp. 192-206.

Sir James 1928. Quoted in M.Sc. Thesis of Hadimani, A.S. (1960). *Studies on the Wastelands of Delhi State.* Associate, Indian Agricultural Research Institute, New Delhi.

Stamp, L.D. 1948. *The Land of Britain : Its Use and Misuse.* London, 433 p.

Tejwani, K.G., Dhruvanarâyana, V.V. and Satyanarayan, T.J. 1960. *Soil Water Conservation in India.* ICAR Publication, 65 p.

Verma, B. and Chinnamani, S. 1990. Rehabilitation of ravine lands. In : *Technologies for Wasteland Development* (Eds. I.P. Abrol and V.V. Dhruvanarayana), ICAR, New Delhi, pp. 263-272.

Verma, B. 1981. *Classification of Ravines for Optimum Utilization in India.* CSWCRTI Research Centre, Kota, Bulletin No.R-11/K-1.

Vimal Kishore, Nambiar, K.T.N. and Subhash Chandra 1982. *Economics of Fuel-Cum-Fodder Plantation in Mahi Ravines of Gujarat State.* Ann. Rep. CSWCRT, Dehradun, 185 p.

Vohra, B.B. 1980. A policy for land and water. In : *Patel Memorial Lectures.* Publication Division, Ministry of Information and Broadcasting, Govt. of India, New Delhi, 5 p.

World Resources 1988-89. *A Report by the World Resources Institute and Institute for Environment and Development.* Basic Books, Inc. New York, 372 p.

31

WATERSHED MANAGEMENT IN DRYLAND AREAS—PRINCIPLES AND PRACTICES

SHRINIWAS SHARMA and P.K. MISHRA

INTRODUCTION

The Indian subcontinent continuously faced grim food situation in the pre-independence days; the trend did not show much improvement after attaining the political freedom. However, there has been some sigh of relief from the 60's, as the national food grain production has increased manifolds, mainly due to Green Revolution concentrated particularly in the irrigated tracts. But this additional grain production is still not sufficient in right sense for feeding the increasing population. Even, this increasing trend of grain harvest will stabilize soon, unless a breakthrough in production technology takes place particularly in drylands, as these lands account for about 70 per cent of the total cultivable area. And this increase in production in dryland areas is directly linked with soil resource management along with the water resource development. Apparently, soils of most drylands are degraded due to erosion, extended periods of aridity, poor soil fertility and low moisture retention; moreover uncontrolled grazing further makes things worse. Likewise, due to the erosion problem the productivity of the lands in higher elevations is getting reduced and at the same time disastrous floods create havoc in the lower regions as a result of sediment accumulation in the water courses and reservoirs. Land and water resources in dryland areas, therefore, need to be managed on watershed basis as the country still depends on drylands for nearly half of its food grain production.

In general, watershed-based planning of resource-management has generated wide appreciation, particularly for assured dividends. Recently, concept of maintaining an ecological balance embedded in the watershed programme has also started getting major attention in different sections of society. But all this does not lower down the importance of scientific irrigated farming. At national level, 10 to 15 million hectares, out of the total 55 million hectares of drought prone area (spread over 99 districts), may be conveniently covered under watershed programme; but still 40-45 million hectares are likely to depend on irrigation water having the source beyond the watershed. Though the history of land and water management is as old as that of agriculture itself, the issue of providing substantial Government support to the programme has always remained debatable. However, national consensus in respect of watersheds emerged in the early eighties. Presently, the watershed management programme in the rainfed areas is getting momentum in a big way through National Watershed Development Project for Rainfed Agriculture. The watershed management and its principles and parctices have been highlighted in this chapter.

WHY MANAGEMENT ON WATERSHED BASIS ?

The concept of watershed management has

developed from the definition of watershed - a convenient unit for planning. Hydrologically, a watershed is an independent entity by itself, which allows all its run-off to reach a particular point, called the outlet of the watershed. Thus, the boundary of the watershed demarcates the water harvesting boundary (catchment) of the outlet point. As the entire process of agricultural development depends on the status of water resources, the watershed with the distinct hydrological boundary is considered ideal for taking developmental programme. In brief, planning and design of soil and water conservation structures such as bunds, waterways, overflow hydraulic structures, water harvesting structures etc. are carried out considering the expected rate and amount of run-off water. In this context, the watershed concept is of practical significance in estimating the rate of run-off and flood volumes to be handled by these structures. Thus, it is essential to have various developmental programmes on watershed basis in conjunction with the required basic soil and water conservation measures as hardware treatments. And to be on safer side, the entire development needs to be taken up on topographic considerations - from the ridge line to the outlet point.

Many-a-time, all causes of the land and water problems are not confined at the place where the problem is noticed; deposition of infertile sands in lower flat lands may be due to unscientific land management in upper reaches; rapid rill formation in the fields situated along deep gullies may be due to mis-management of vegetation and lack of mechanical protection at vulnerable places in the gully system. Accordingly, by adopting sizeable watershed, the cause and effect relationship can be conveniently established. In watershed approach, major efforts are made to find solutions of watershed problems from the watershed area itself.

Basically, the purpose of integrated watershed management is to practice improved and scientific land management programme for improving the hydrologic behaviour of the watershed by reducing run- off and soil loss. In other words, the watershed development programme aims at preventing degradation of land and increasing land productivity; and simultaneously it maintains ecological balance between land, water, plant and animal kingdom in the area. Thus, watershed management particularly in drylands is regarded as holistic approach towards optimising the integrated use of land, water and vegetation in an area for providing an answer to alleviate drought, moderate floods, prevent soil erosion, improve water availability and increase fuel, fodder, fibre and agricultural production on a sustained basis. More specifically in the watershed management, soil conservation work is enmeshed with crop management and alternate land use practices and other allied agricultural activities like animal husbandry, pisci-culture, sericulture, etc. for increasing and stabilizing the farm production and income.

PRINCIPLES OF WATERSHED MANAGEMENT

The main principles of watershed management based on resource conservation, resource generation and resource utilization, are :

(i) Utilising the land according to its capability;

(ii) Protecting productive top soil;

(iii) Reducing siltation hazards in storage tanks and reservoirs and lower fertile lands;

(iv) Maintaining adequate vegetation cover on soil surface throughout the year;

(v) *In-situ* conservation of rain water;

(vi) Safe diversion of excess water to storage points through vegetative waterways;

(vii) Stabilisation of gullies by providing checks at specified intervals and thereby increasing ground water recharge;

(viii) Increasing cropping intensity and land equivalent ratio through intercropping and sequence cropping

(ix) Safe utilisation of marginal lands through alternate land use systems with agriculture - horticulture - forestry - pasture systems with varied options and combinations;

(x) Water harvesting for supplemental and off-season irrigation;

(xi) Maximising agricultural productivity per unit area per unit time and per unit of water;

(xii) Ensuring sustainability of the eco-system befitting the man- animal-plant-water complex;

(xiii) Maximising the combined income from the inter related and dynamic crop-livestock-tree-labour complex over years;

(xiv) Stabilising total income and to cut down risks during aberrant weather situations;

(xv) Improving infrastructural facilities with regard to storage, transportation and marketing of the agricultural produce;

(xvi) Setting up of small scale agro-industries; and

(xvii) Improving the socio-economic status of the farmers.

OBJECTIVES OF WATERSHED MANAGEMENT

As mentioned earlier, watershed management is enshrined with the concept of sustainability meeting the needs of present population without compromising the interests of future generations. It is multi-pronged approach for steady uplift of masses living in the area. The main objectives of this multi-purpose programme can be described in symbolic form by the expression : "POWER". Here the letters symbolize the following :

P = Production of food -fodder - fuel - fruit - fibres - fish - milk combine on a sustained basis

– Pollution control

– Prevention of floods

O = Over-exploitation of resources to be minimised by controlling excessive biotic interferences like over-grazing

– Operational practicability of all on-farm operations and follow- up programmes including easy approachability to different locations in watershed

W = Water storage at convenient locations for different purposes

– Wild animal and indigenous plant life conservation at selected places.

E = Erosion control

– Eco - system safety

– Economic stability

– Employment generation

R = Recharge of ground water

– Reduction of drought hazards

– Reduction of siltation in multi - purpose reservoirs

– Recreation

The above objectives can be achieved by planning and implementing the programme in a systematic way with active participation of farmers including constitution of Cooperative Watershed Management Societies. Rather, as a convention - not as a rule - following guidelines may be followed :

1. The implementation programme should start from the ridge line of the watershed to the valley - not on piecemeal basis in isolated patches.

2. Development of both arable and non-arable lands should be done together.

3. Forest, pasture, cultivable land and waste-lands should be treated as inter-linked units of hydrological entity. The condition of all lands has to be improved to meet the demands of increasing man and animal population.

4. Essentially, all developmental activities are to be carried out on watershed basis - whole watershed area needs to be covered - may be in planned phases. However, treatment of some lands lying outside the watershed - but geographically contiguous - can also be taken up based on socio-economic considerations.

COMPONENTS OF WATERSHED MANAGEMENT PROGRAMME

The main components of watershed programme are :

1. Soil-water conservation and water harvesting.
2. Crop management
3. Alternate land use systems

Successful implementation of the above three components in the watersheds would lead to the development of livestock, poultry, pisciculture and other associated and allied activities; finally resulting in improved standard of living of the farmers and rural masses.

Soil-Water Conservation and Water Harvesting

Soil and water conservation measures coupled with water harvesting help to improve the moisture availability in the soil profile and surface water availability for supplemental and off-season irrigation. The interventions through soil conservation measures have greater role to play in transferring a part of surface water to ground water by recharge. Based on the nature and type of hydraulic barriers and their cost, the conservation measures in arable lands can be divided into following three categories.

1. Hardware treatments
2. Medium software treatments
3. Software treatments

Hardware treatments

Hardware treatments are generally of permanent type, provided for improvement of relief, physiography and drainage features of watershed. These are erected with major Government support with the purpose to check soil erosion, regulate overland flow and reduce peak flow. At times, they are imposed to completely divert the upland run-off from running into downstream fertile lands. Different hardware treatment are explained below.

Waterways : Quite often, high run-off volumes are observed, even in drylands, particularly in watersheds located in slopy areas. Such run- off water should, therefore, be channelised through a few waterways; as far as possible some of the existing waterways may be developed. Waterways draining larger areas may be designed on hydrologic and hydraulic considerations. No doubt, at places, sizeable land strip may be occupied by such waterways, but this may be found highly advantageous for the survival of many multi-purpose useful trees when planted along the waterways. Depending upon land situations alignment of small size waterway may be adjusted with field boundaries; at times, such arrangements are specially useful for safe-guarding crops against over-stagnation of rain water in black soil areas. Such water courses may also be safeguarded against erosion by providing mechanical checks and also by raising vegetation which may be useful fodder for animals. In some situations, waterways may also have a small section side bund with necessary openings at water-entry points.

Bunds : These are low height earthen embankments constructed across the slope in cultivated lands - after deciding location of waterways. The bunds function to intercept run-off, increase infiltration opportunity time and dispose excess rainfall safely. Such bunds can be either contour bunds or graded bunds; many-a-time, these bunds are adjusted with field boundaries, if deviation from grade or contour is not too much; no doubt, spacing depends on many engineering considerations.

Contour bunds are recommended in dry farming areas with light textured soils of slopes upto 6 per cent and where annual rainfall does not exceed 600 mm. They are designed for an expected run-off of 24 hours duration and 10 years frequency. Surplussing arrangements (waste weirs) are provided to dispose of the excess run-off beyond the design storage. The cross section area of contour bunds follows the depth and type of soil; however, $0.5m^2$ is minimum (Bali, 1988).

Graded bunds are constructed in medium to high rainfall areas in poor permeable soils (vertisol), having 2 to 6 per cent slope. They are also quite suitable for the soils having crust formation tendency like red *'chalka'* soils of Telengana region of Andhra Pradesh. By and large, graded bunds with 0.3 to 0.5 m^2 section are constructed with longitudinal gradient of 0.2 to 0.4 per cent depending on the site condition (Singh et al. 1990). Graded bunds particularly are made with the following objectives :

(i) To reduce run-off ;

(ii) To reduce soil loss; and

(iii) To divert run-off.

Thus, graded bunds along with waterways and water harvesting structures not only check soil erosion, but they also provide an ideal rain water management system for many watershed situations.

For making bunds, required soil may generally be taken from lower side. Sometimes, excavations and pits for making bunds - in deep black soil areas particularly are made on upper side. But these may be got levelled or ploughed up as early as possible after bund construction. Invariably - barring a few situations - borrow pits should be made on lower side at the places where the bunds drain into waterways. In real field situations, bunds are covered with natural grass and in most cases, these are dependable sources of forage. In addition, quite often some trees also comp up in their close vicinity, to the liking of cultivators. In some places loose stones are easily available, hence stone-bunds are quite advisable in such pockets.

Terracing : Cultivated lands having land slopes above 10 per cent particularly in hilly areas - should be put under bench terracing by converting the lands into series of platforms. The width of bench terrace depends on the land slope and the permissible depth of cut. Bench terracing is very much effective in reducing soil erosion in hilly areas. At Ootacamund, it has helped in bringing down the annual erosion rate from 39 t/ha to less than 1.0 t/ha on 25 per cent sloping lands (Jayakumar et al 1982). At places where scattered stones are available, loose stone walls can be made to act as risers for bench terrace construction.

Zingg terrace consists of a contributing (donor) area with mild slope and a receiving levelled area at lower side - half to one third of the donor area. They are particularly suited for areas having medium to deep soils (Joshi 1976).

Medium software treatments

Medium softwares are also provided particularly as inter-bund treatments, where field sizes are large and conventional bunds are constructed along field boundaries. Such treatments are usually of semi-permanent type and are adopted to minimise the velocity of overland flow. They may need major initiative from cultivators in addition to some grants from the Government side. Such measures may last 2 to 10 years; vegetation component and land configurations at times may also provide some direct returns on short term basis. However, they need to be modified at times to maintain effectiveness for erosion control and moisture conservation, and also to minimise risk of providing shelter to harmful pests in and around these measures and in certain situations like taking up corrective measures to avoid too much spread of introduced vegetation and being counter-productive.

Small section/key line bunds : A small section bund may be created across the slope at half of the vertical bund spacing. Such bunds can be nearly 0.1 m^2 in section; may be renovated at an interval of 2 to 3 years.

Strip levelling : About 4 to 5 m strip of land above the bund across the major land slope may be levelled for the purpose. Similarly, one or more strips can be created at midlength of slope. Such strips can be created by running blade harrows after ploughing the field with mould board or disc plough. Such minor rough levelling programme may be taken up after every 2 to 4 years.

Live beds : One or two live beds (2-3 m wide) can be created either on contour or on grade in the inter bund space. The vegetation on the beds may be according to the liking of cultivator; it can be annual or perennial or a combination of both.

Vegetative/live barriers : One or two barriers of close growing grasses or legumes can be created along bunds as well as at the mid - length of slope to filter the run-off water or slow down overland flow. *Khus* could be one of such vegetations. Several other promising grass species that serve as valuable fodder for cattle are being explored as an alternative to *khus* grass. A miniature bund at lower side of barrier is recommended to help in the development of live barriers, particularly in the initial stages.

Software treatments

A mention was made that hardware type land treatments are useful for safe run-off disposal and similarly, medium soft-wares are essential to slow down the velocity of overland flow in cultivated fields. However, on several occasions, these are found inadequate in attaining equitable moisture distribution for crop growth. In such cases software treatments are taken up for ensuring fairly uniform soil moisture for satisfactory crop performances. By and large, software treatments are temporary in nature; in that case these are required to be remade or renovated every year. The entire cost of applying such treatments is to be met by the farmers. Because of favourable economics, a few of these treatments have gained wide acceptance in the recent years.

Contour farming : Contour farming is one of the easiest and most effective and low cost method of controlling erosion and conserving moisture. With contour farming, tillage operations are carried out along contours. It creates numerous ridges and furrows for harvesting sezieable amount of run off inside the field itself.

Compartmental bunds : Compartmental bunds, converting the area into square/rectangular parcels - are useful for temporary impounding of water for improving moisture status of the soil. These are made using bund formers. In medium deep black soil, they are found advantageous in storing the rainfall received during the rainy season in the soil profile - thereby augmenting the soil moisture for use by *rabi* crops. The size of the compartments may be fixed considering the size, of the inter bunded land.

Broad bed and furrows : Broad bed and furrow system implies shaping alternate bed and furrows. This technique is especially suited to black soils, where crops are sown on pre-formed beds. This system is made before the season and is maintained year after year. The planting is done on the bed. Generally, the depth of each furrow is kept 0.15 m and the inter furrow spacing is maintained at 1.5 m (ICRISAT, 1983).

Dead furrows : Dead furrows are laid across the land slope in rolling lands, to intercept the run-off. The spacing between dead furrows varies between 2 to 5 m or 4 to 7 crop rows. This system works well in alfisols (Singh and Rao, 1988).

Tillage : Tillage operations help in rain water infiltration. Off- season tillage, in particular, has been found quite useful in most rainfed areas. The tillage operations make the soil receptive to rainfall. This practice is very useful in light soils often prone to crusting.

Mulching : Surface mulching protects soil against beating action of rain drops and it also increases water infiltration into the soil. Further it helps in minimising water evaporation from soil surface. Sometimes dry soil mulch created simply by stirring the soil has been found effective for good performance of crops.

Soil and water conservation in non - arable lands

Control grazing is one of the simplest approach in reducing soil loss from denuded sloping lands. As seen from the experiences of Operational Research Projects, soil and moisture conservation measures are required even in the non - agricultural lands. These practices include contour and staggered trenches and contour furrows. They help in accelerating vegetation establishment, encouraging natural regeneration of species and combating soil erosion. Contour and staggered trenches are usually spaced 10 to 20 m apart across the slope for raising forest species. In general, 0.45 m wide trenches with 0.45 m depth are made at regular intervals. Sometimes in place of trenches, contour furrows are formed.

at a spacing of 2 to 10 m (Singh *et al.* 1990). In rocky areas, crescent bunds of loose boulders are constructed in place of trenches. All the above stated measures are useful in slowly regaining the lost forest cover and fertility of most degraded grasslands; community efforts in the right direction are likely to make major dent in re-greening such areas.

Water harvesting measures

Since time immemorial, several kinds of water harvesting structures have been in practice in our country. Till recently, local experiences and precedences were utilised for designing these structures. In order to bring cost effectiveness, efforts are also necessary to harmonize the designing of water harvesting structures based upon traditional wisdom and scientific contemplations. In fact, farm ponds, surface water tanks and percolation tanks need to be viewed as conservation structures in addition to sources for supplemental irrigation. In real watershed situations, ponds may be located in conjunction with waterways - some may be of a few hundred cubic metre size, whereas couple of structures may be of bigger size. In selected pockets, integrated catchment - cum - command approach should be followed. To ensure proper storage in light soils, some of the ponds can be lined with sealant materials. All large ponds have to be constructed at Government cost. All such structures are likely to recharge ground water reservoir. When located in proper eco-cultural hydrologic matrix, in conjunction with other factors, they also help in reducing load on ground water aquifer. It may be advisable to work out design details of all the required structures in the first phase of the project, no matter construction of a few of these may have to be deferred - may be planned to be taken up in drought years to provide employment to the rural folk. The objectives of water harvesting structures are :

(i) To store water for supplemental and off season irrigation;

(ii) To act as silt detention structure;

(iii) To recharge the ground water;

(iv) To raise aquaculture species; and

(v) Recreation and allied agricultural uses.

Thus in most situations, in addition to reviving traditional water harvesting systems, other measures are also needed. Some of such typical water harvesting structures adopted in watersheds are as follows :

Minor irrigation tanks : Minor irrigation tanks are constructed across the major streams with canal system for irrigation purpose by constructing low earthen dams. A narrow gorge should be preferred for making the dam in order to keep the ratio of earthwork to storage as minimum. The height of the dam may vary from 5 to 15 m. The tanks are provided with well designed regular and emergency spill-ways for safety against side cutting. In micro-watersheds, water harvesting *bundhies,* similar to small scale irrigation tanks are also recommended. By and large, the water harvesting *bundhies* are not integrated with extensive canal system.

Farm ponds : Farm ponds are typical water harvesting structures constructed by raising an embankment across the flow direction or by excavating a pit or a combination of both. Dug out ponds in light soil require lining of the sides as well as the bottom with suitable sealants. This is not the case with heavy black soil. Considering the benefit of the harvested water, different type of linings, viz., brick, concrete, HDPE, etc. may be used.

Nala bunds and percolation tanks : Nala bunds and percolation tanks are located in the nalas having permeable formations with the primary objective of recharging the ground water. A strict regulation on the silt load entering the downstream reservoirs is an additional advantage of percolation tanks. The percolation tanks encourage the digging up of wells downstream of recharged area for irrigation purposes. The percolation tanks are provided with emergency spillways for safe disposal of flow during floods.

Stop dams : Stop dams are permanent engineering structures constructed for raising the water level in the *nala* for the purpose of providing life saving irrigation during drought periods. These are located over flat *nalas* at narrow gorges carrying high discharge of long durations. They are created over stable foundation conditions where hard murrum or rocks are encountered. For the stop dam, a site with larger water storage capacity would be desirable.

Crop Management

The rainfed areas of our country contribute about 45 per cent of the total food grain production. In the watersheds there is scope of increasing the overall productivity by adopting the integrated farming systems approach. For achieving sustainability in crop production in the watersheds due importance is given to : (i) evolve and follow up simple and low cost crop production technology including improved varieties; (ii) provide alternate crop production technologies to match weather aberrations; and (iii) optimise the use of natural resources like land and water.

Considering the availability of rain water over space and time, different crops have been identified as given in Table 1 for different rainfall zones.

Many drylands of the country are not only thirsty; these are hungry as well. Optimum doses of fertilizers are, therefore, recommended for the crops depending upon physico - chemical properties of soil. And farmers are advised for proper crop residue management - stubbles of the plants being left in the field for decomposition and building up the fertility. However, major breakthrough has yet to be achieved in cropping systems in the watersheds by changing over to organic farming with tree-based cropping programmes. The research work is being initiated on these lines through alternate land use systems.

Alternate Land Use Systems

In the watersheds, most of the uplands are degraded and these are also marginal lands with low productivity level. Due to the population pressure, more and more marginal and sub-marginal lands are being brought under cultivation. Apart from being uneconomical in the long run (with more inputs), cultivation of such lands, can lead to serious imbalances in the eco-systems. For

Table 1. Agro-ecological conditions in different rainfall zones of India

Mean annual rainfall (mm)	Predominant soil types	Mean annual temperature (°C)	Length of growing season under rainfed conditions (days)	Predominant rainfed crops
< 400	Sandy Soils (Aridisols)	20 - 30	30 - 80	Grasses Pearlmillet
		0 - 10	30 - 90	Short duration pulses
400 - 1000	Sandy soils (Aridisols); Red soils (Alfisols and related soils);	20 - 30	80 - 200	Pearlmillet, Sorghum Maize, Oilseeds, Pulses
	Black soils (Vertisols and Vertic Incepti sols)	–	–	Cotton
1000 - 1800	Alluvial soils (Entisols); Laterite soils (Alfisols and related soils); Black soils (Vertic and Vertic Inceptisols)	20 - 30	200 - 300	Maize, Paddy, Wheat, Barley, Mustard and Soybean
> 1800	Sub - montane soils (Inceptisols)	15 - 20	> 300	Paddy, plantation crops

Source : Singh and Das (1989).

Depending on the water availability periods, examples of some selected crops for rainy season and post rainy season are given in Table-2.

such lands, some alternate efficient land use system other than arable cropping would be desirable. In alternate land use system, by en-

Table 2. Major efficient food crop-based double cropping systems for different rain dependent regions of India (based on water availability periods)

Water availability period (days)	Rainy season crops	Post rainy season crops
110 - 150	Cowpea/blackgram	Safflower/chickpea
	Soybean	Mustard/safflower
	Greengram	Mustard
	Pearlmillet	Chickpea/barley
150 - 175	Greengram	Sorghum
175 - 200	Cowpea	Sorghum
	Greengram	Safflower
	Blackgram	Barley/mustard
	Pearlmillet	Chickpea
	Maize	Wheat/wheat + Chickpea /mustard
	Rice	Chickpea + wheat
	Sesame	Chickpea
200 - 250	Sorghum/groundnut/maize	Safflower
	Soybean	Wheat
		Safflower/chickpea
		Soybean/maize
> 250	Rice/maize/fingermillet/groundnut	Wheat/chickpea/linseed/lentil/horsegram/barley
	Soybean	Fingermillet
	Pearlmillet	Wheat

Source : Singh and Das, (1989).

couraging tree and grass components, the demand for food and fodder can be solved to some extent. Thus, stability in production will be achieved alongwith the safety of environment. As seen from some model watersheds, alternate land use systems not only help in generating much needed off-season employment but they also help in utilising off-season rain which may otherwise go waste.

In general, land capability classes are the determinant factors for deciding different alternate land use options. By and large, in land classes I to III, arable farming can be practised and beyond that lands should be utilised for forestry and grassland management. But from economic considerations, agroforestry options - integrating farming with other land uses like horticulture, silviculture and pasture may find acceptance from the farmers. The evaluation of all these systems are under research investigations. The tree component of the system should be selected according to the soil and climatic conditions. The trees should preferably have the characteristics like fast growing with high palatability, good coppicing ability, and drought resistance - also having properties like adding organic manure in soil profile. Some of the tree species for different places in different rainfall zones are presented in Table 3.

The pasture component in such systems should have the characteristics like growing well even under shade, drought resistance, easy propagation, high palatability, conserving soil and moisture, withstanding over-grazing etc. Different legumes like stylo species and grasses including Cenchrus species have been found performing well in many situations.

Dryland horticulture is quite promising enterprise in slightly favourable soil conditions - especially where there is provision of water harvesting. In this, hardy fruit species such as *ber*, promogranate, custard apple, guava and *aonla* may have to be preferred over others.

EXPERIENCES FROM WATERSHED PROGRAMME

The problems of dryland watersheds are quite complex; perceptible progress is, therefore, necessary on two fronts - a short term programme to provide immediate benefit to the inhabitants living in the area and a long term

Table 3. Tree species found suitable at different dryland research centres

Centres with good annual rainfall distribution	Centres with moderate annual rainfall distribution	Centres with poor annual rainfall distribution
Rakh Dhiansar	Indore Akola, Solapur, Dantiwada, Jhansi,	Hoshiarpur,
Dehra Dun	Hyderabad, Anantapur	Hissar, Jodhpur
Varanasi, Rewa	Bangalore and Bellary	Udaipur and
Ranchi,		Rajkot
Bhubaneswar and		
Kovilpatti		
Alnus nepalensis	*Acacia nilotica*	*Acacia nilotica*
Bauhinia purpurea	*Ailanthus excelsa*	*A. catechu*
Casuarina equisetifolia	*Albizia lebbeck*	Faidherbia albida
Dalbergia sissoo	*Azadirachta indica*	*A. aneura*
Emblica officinalis	*Butea monosperma*	*A. senegal*
Eucalyptus sp.	*Cassia fistula*	A. tortilis
Gmelia aborea	*Casuarina equisetifolia*	Colophospermum mopane
Grewia optiva	*Dalbergia sissoo*	Eucalyptus camaldulensis
Grevillea robusta	Eucalyptus sp.	*Parkinsonia aculeata*
Hardiwickia binata	*Leucaena leucocephala*	Prosopis cineraria
Melia azedarach	Tamarindus indica	*P. juliflora*
Morus alba		
Populus sp.		
Sesbania sp.		
Terminalia sp.		

Source : Singh *et al.,* (1987)

approach for full assessment of resource conservation and management problems and providing practical and permanent solutions to various land and water problems. Model Watershed Programme on moderate scale was therefore undertaken by Government of India in the eighties for testing the integrated Watershed approach. The action plans of the watersheds were prepared by the two ICAR Institutes : (i) Central Soil and Water Conservation Research & Training Institute, Dehradun; and (ii) Central Research Institute for Dryland Agriculture, Hyderbad. The programme implementation was carried out through respective state governments.

Fortunately the success story of projects-like Sukhomajhri in Haryana has magnified the concept of watershed management. It has set the ball of watershed development rolling in the country; presently the programme is going on in a big way in different parts of our country. In Sukhomajhri watershed - with the hardware treatments - the average yield of wheat and maize increased from 10 to 30 q/ha and 10 to 20 q/ha, respectively with supplemental irrigation, improved seed and fertilizer. Due to imposition of soil conservation measures sediment rate was brought down from 300 to 30 tonnes/ha/year.

Due to the pasture development programme in Nartora (M.P.) watershed, where milk availability in pre-programme period was negligible, the daily milk production after watershed treatment has gone up to 2000 litres. In Mittemari (Karnataka) watershed, the yield of different crops have improved over a period of time (1984-1988) due to adoption of better management practices in the watershed programme (Table 4).

Similarly, in Bajar - Ganjar and Siha (Haryana), complete change has come after the watershed programme. The irrigated area has doubled; the number of wells and sprinkler sets has gone up by four times; crop yields have increased substantially; and income of farmers has increased by as much as 166 per cent. Likewise,

Table 4. Yield improvement of crops in Mittemari watershed

Crops	Benchmark yield q/ha	Mean yield (1984-88) q/ha)
Ragi + pigeonpea (10 : 2)	6.50 + 0.00	10.70 + 2.38
Groundnut + Pigeonpea (7 : 2)	3.00 + 0.00	8.65 + 2.67
Cowpea	2.50	6.06
Pigeonpea	–	8.43
Sunflower	–	7.05

Source : Hegde *et al.* (1989).

in Dhanwal and Patihwa watersheds in Mirzapur district of U.P., due to the watershed programme the crop yields have gone up by 25 per cent in pulses and 116 per cent in cereals. In Tejpura watershed (U.P.), the productivity of different crops increased by 220 and 300 percent during *Kharif* and *rabi* season respectively in a period of five years. In Chevella (A.P.) watershed the positive effect of soil and water conservation has been reflected on crop yields. (Table 5). Further, the groundwater status of the watershed has increased.

Likewise, after the inception of model watershed programme, there has been considerable increase in the number of wells in most of the watersheds due to the improvement of groundwater status (Table 6). The increased water availability in wells and bundhis has enabled farmers to provide life saving irrigation to additional areas (Table 7).

Table 5. Effect of adoption of package of practices on crop yields at Chevella watershed (1986-88).

Technology	Crop yield (q/ha)			
	Greengram	Blackgram	Sunflower	Sorghum
Traditional	2.46	0.74	2.52	4.37
Recommeded crop management without soil and water conservation	2.69	1.73	3.01	5.33
Recommended crop management with soil and water conservation	6.60	4.65	7.30	10.68

Source Singh *et al.*, (1989)

Table 6. Increase in number of wells in model watersheds

Watershed	1984-85	1985-86	1986-87	1987-88
		Open wells		
Mittemari	28	29	30	30
Naranka	72	82	97	97
Kolhewadi	38	38	45	45
Rendhar	12	12	20	25
Tejpura	6	6	12	20
Aril	33	33	66	80
		Tube wells		
Naratora	4	–	9	10
Chevella	–	–	–	2
Rendhar	–	–	–	2

Source : Dhruvanarayana *et al.*, (1989).

Table 7. Increase in area under supplemental irrigation in model watersheds

Watershed	Area (ha) in base year (1984-85)	Increased Area (ha) (1987-88)
Kolhewadi	2.7	11.0
Yernal	3.0	18.0
Paddalsinghi	0	15.0
Rendhar	0	45.0
Tejpura	28.0	342.0
Aril	29.5	208.0
Mittemari	66.0	87.0
Naratora	32.0	98.0

Source : Dhruvanarayana *et al.*, (1989)

From a review of the model watershed programme, it was revealed that the watershed approach has helped in mitigating the effect of 1987 - drought to a considerable extent as a result of cumulative improvement in water resources and groundwater status due to construction of conservation structures during relatively favourable monsoon seasons of earlier years. About 80 to 90 per cent of the area could be seeded in the watersheds in the drought year, compared to 30 to 50 per cent outside. As a consequence of the overall improvement of hydrology and cropping programme including judicious fertilizer application, significant gain in crop yield is noticeable in the model watersheds (Table 8).

when action plans were prepared; the data base is still not adequate. Further improvements in such master plans are definitely possible in future if requisite hydrological information is collected in respect of different areas and same is utilised for formulation of Watershed Master Plans. Similarly, still better quality can be expected in field structures if implementation is carried out by trained persons in the line. To sum up, if these things also start receiving necessary priority, the future of watershed programme in the country is going to be very bright.

EPILOGUE

By following the watershed approach, several tangible and intangible benefits are likely to accrue. Integrating soil and water conservation including water harvesting, with alternate land uses and improved crop and animal production programmes would also ensure proper health of watershed. Different land-use options—approaches when capped as per land capabilities in a systematic manner would no doubt improve the socio-economic status of the farmers. The full concept of watershed management in drylands can be summarised as follows :

1. The entire programme should be planned and implemented on watershed basis, on the principle of ridge-to-valley. Major structures need to be planned on the basis of adequate hydrologic data.

Table 8. Productivity gains in dryland crops in model watersheds

Crop	Yield (q/ha)				
	1983-84	1984-85	1985-86	1986-87	1987-88
Sorghum (10)	6.63	6.83	9.09	10.08	17.08
Greengram (11)	5.80	2.62	5.74	4.07	5.17
Groundnut (10)	4.14	5.07	9.81	9.81	11.94

Source : Singh and Sharma 1989)
Note : Figure in parentheses indicate number of watersheds considered for averaging the yield.

In addition to these visible advantages, other community needs like fodder and fuel are also being met in most cases. And as a result of such major gains, income of farmers and village people has significantly gone up in a span of 4 to 5 years. All these achievements are specially significant, considering the fact that all the hydrological data were not available at the time

2. For effective soil and moisture conservation, both mechanical and vegetative measures should go hand in hand. For relatively flat areas, the medium software treatment may be preferred over the hardware treatments.

3. For sustainable crop production on long term basis, due priority should be given to soil conservation work with active participation of Government machineries and beneficiaries. In fact, it is the primary role of the Government to maintain country's basic resources of soil and water.

4. Merits of traditional systems should also be kept in view while planning the watershed programme particularly at micro-level. The new management techniques should not be a burden to the farmers. In general, creation of awareness among the farmers regarding the programme is a pre-requisite. Peoples active participation is also vital for greening most of denuded lands and degraded slopy areas.

5. The constitution of cooperative societies in the watersheds would be useful in equitable distribution of available and generated resources.

6. With necessary efforts from concerned agencies, the watershed areas can be developed into self-sustaining, self-motivating, self-financing and self-perpetuating socially acceptable and viable units of growth in the long run; by conserving and managing land and water resources and by optimum utilization of man and animal power.

REFERENCES

Bali, Y.P. 1988. A review of soil and water conservation strategies. *Proc. 1st Meeting of working Group on Soil and Water Conservation for Formulation of Eighth Plan.*

Bhardwaj, S.P., Dhruvanarayana, V.V., Sharma, A.K. and Singh J.P. 1989. Fortunes take a turn in Aravali Watershed. *Indian farming* 39(9) : 11-13.

Das, D.C. 1977. Soil conservation practices and erosion control in India - A case study. FAO Soils Bulletin *No.33,* Rome, pp. 11-15.

Dhruvanarayana, V.V., Singh, R.P., Bharadwaj, S.P., Sharma, S.N. Sikka, A.K. and Vittal K.P.R. 1987. *Watershed management for drought mitigation.* Indian Council of Agricultural Research, New Delhi, p. 38.

Dhruvanarayana, V.V., Singh, R.P., Bharadwaj, S.P. and Sharma, S.N. 1989. *Achievements and Performances of Model Watersheds* (Eds : I.P. Abrol and N.S. Randhawa). ICAR, New Delhi.

Dhruvanarayana, V.V., Sastry, G. and Patnaik, U.S. 1990. *Watershed Management.* ICAR, New Delhi.

Gil, N. 1985. Watershed Development with special reference to Soil and Water Conservation. *FAO Soils Bulletin No.* 44, Rome.

Hazra, C.R. 1989. Crop yields rise dramatically in Tejpura watershed. *Indian farming.* 39(9) : 37-39.

Hudson, N.1971. *Soil Conservation.* B.T. Batsford Ltd., London, U.K.

Hedge, B.R. and Pandurangaiah, K. 1989. In Karnataka Mittemari prospers amidst scarcity. *Indian farming* 39(9) : 7-10.

ICRISAT (International Crops Reserach Institute for the Semi-Arid Tropics). 1983. *Annual* Report 1982, 262-264.

Jayakumar, M., Padmanabhan, M.V. and Jeevarathnam, K. 1982. *25 Years Research on Soil and Water Conservation in Southern Hilly Regions.* Monograph No.6. CS & WCR & TI, Dehara Dun.

Joshi, P.G. 1976. *Soil and Water Conservation Techniques.* Government Text Book Press, Mysore, India, 1-221.

Katyal, J.C., Sharma, S., Mishra, P.K. and Padmanabhan, M.V. 1992. Cost effective technologies for soil and moisture conservation in rainfed areas. *Rainfed Agriculture Research News Letter, CRIDA, Hyderabad,* 1 : 15-21.

Miranda, S.M., Pathak, P., and Srivastava, K.L. 1983. Runoff management on small agricultural watersheds : the ICRISAT experience. In *Dryland Agricultural Research in India., Thrust in the Eighties.* AICRPDA, ICAR, Hyderabad 88-97.

Padmanabhan, M.V., Mishra, P.K. and Sharma, S. 1992. Soil and water conservation research in rainfed areas. In *Dryland Agriculture Research in India - State of Art of Research in India.* (Eds. Soman; L.L., Vittal, K.P.R. and Venkateswarlu, B.), Scientific Publishers, Jodhpur.

Rai, L., Singh, U.N. and Singh, S.R. 1989. Drylands of Mirzapur spring to life. *Indian farming* 39(9) : 21-23.

Rao, C.S., Sivamohan, M.V.K., Chandy, S and Rao, R.S. 1990. *Pilot Project for Watershed Development in Rainfed Area, Maheswaram.* Administrative Staff College of India, Hyderabad.

Sharma, S., Rao, V.K. and Vijayalakshmi, K. 1982. Soil

and moisture conservation for drylands. A Decade *of Dryland Agricultural Research in India 1971-80.* AICRPDA, ICAR, Hyderabad pp. 90-102.

Sharma, S.N. and Singh, R.P. 1988. Pay offs from hydrologic improvements in watersheds. *Indian J. Dryland Agri. Res. & Dev.*, 3(3) : 102-111.

Sharma, S and Mishra, P.K. 1992. Watershed development for drought proofing. In *Dryland Agriculture in India - State of the Art of Research in India* (Eds. Somani *et al.*), Scientific Publishers, Jodhpur.

Sharma, S., Katyal, J.C. and Mishra, P.K. 1992. Soil and water resource management for sustainable crop productivity in rainfed regions. *Paper Presented in the Colloquium on Sustainable Development and Information Technology, CESS, Hyderabad.*

Singh, S.P. and Rao, U.M.B. (1988). Rain water management for stabilising productivity of drylands,. *Indian J.Drylands Agri. Res. & Dev.*, 3(3) : 203-214.

Singh, R.P., Vijayalakshmi, K., Korwar, G.R. and Osman, M. 1987. Alternate land use systems for drylands of India, *Research Bulletin No.6*, CRIDA, Hyderabad.

Singh R.P. and Das, S.K. 1989. Agronomic aspects of plant nutrient management in rain dependent food crop system in India. *Paper Presented in FAO/FIAC.* Technical Sub-Committee Meeting, Rome.

Singh, R.P. and Sharma, S. 1989. Managing rain water to better farmers' lot. *Indian farmin*, 39(9) : 61-63.

Singh, R.P., Vittal, K.P.R., Das, S.K. and Sanghi, N.K. 1989. Watershed technology stabilises yields in Andhra Pradesh. *Indian Farming* 39(9) : 34-36.

Singh, R.P., Sharma, S., Padmanabhan, M.V., Das, S.K. and Mishra, P.K. 1990. *Field Manual on Watershed Management*. CRIDA, Hyderabad.

Vijayalakshmi, K. 1987. Rain water management in drylands. In *Technological Advances in Dryland Agriculture*, CRIDA, Hyderabad, pp. 19-45.

32

ECONOMICS OF RECOMMENDED DRYLAND TECHNOLOGY IN INDIA

Y.V.R.REDDY, J.C. KATYAL, SHAIK HAFFIS and S.K.DAS

INTRODUCTION

Drylands occupy an important place in Indian agriculture. They constitute about 68 per cent of total cultivated area of 143.8 m ha and contribute to 44 per cent of total food grains production as also more than 90 per cent of pulses and oilseeds production in India. The productivity levels are very low and unstable in drylands due to the vicissitudes of monsoon rainfall which is erratic, unpredictable and also highly fluctuating over years under different agro-climatic conditions. The situation calls for an adverse effect on the production and income distribution which finally hampers the economic growth of the country (Lipton, 1970). Unless there is transformation of dryland agriculture, the country will continue to face the shortage of pulses and oilseeds in particular and other food grains in general. It is evident that though India registered phenomenal increase in agricultural production on irrigated lands due to the emergence of Green Revolution in the mid sixties, yet, planners and irrigation experts feel that increasing agricultural production either through area expansion or irrigation development has reached a saturation point. The projected increase in population has necessitated the country to set a food grains target of 240 million tonnes at the present level of consumption and fire wood at 1 billion tonnes by 2000 A.D. as against the present availability of less than 60% in India. Keeping in view the ever increasing population and increasing demand for food, fodder and fuel/firewood, there is dire need to exploit production potential of drylands by adopting recommended/appropriate technologies in order to meet the projected needs of food, fodder and fuel-wood.

The recommended dryland farming technology consists of improved varieties of crops, appropriate agronomic practices, appropriate seeding time and method, optimum plant population, fertilizer use, timely weeding, application of plant protection chemicals; mid season corrections alternative uses of land involving different agroforestry systems and also watershed-based system of farming. Consequent to the establishment of All-India Coordinated Research Project for Dryland Agriculture (AICRPDA) in 1971 with 23 cooperating centres in major agro-climatic zones and after its full-fledged conversion into Central Research Institute for Dryland Agriculture, Hyderabad (ICAR) in 1985, concerted efforts have been made to increase and stabilise production on dryland through development of technologies. It is gratifying to note that today, dry farming research is better equipped to generate technologies to suit different farming situations and hence the dryland farmer at present, has a wide range of crop options to choose a cost effective low-cost and high yield technology from the spectrum of technology options suiting to his resource's position and agro-climatic conditions. Therefore, an emphasis has been placed in this chapter to evaluate the economic viability of recommended dryland farming technology under different resources utilisation, crop options/farming systems and agro-climatic conditions in India.

DATA BASE

The detailed data recorded in prescribed proforma at different agro- climatic regions/centres in India were used for this study to evaluate economics of recommended dryland technology. For the purpose of comparison, control plots (traditional technology) representing real situations with resources and managerial skills of a farmer and demonstration plots with recommended practices were included for this study. Demonstration and control plots belonging to the same farmer were preferred and monitored from 1976-77 through 1979-80.

To evaluate the economics of new dryland technology for watershed based system of farming, a comparative study was undertaken inside as well as outside watershed at Chevella-Vertisol region of Medak district in Andhra Pradesh and Mittemari-Alfisal region of Kolar district in Karnataka. The data were collected in prescribed proforma from about 20 per cent farmers in both the watershed and non-watershed villages by using stratified random sampling with proportion to stratum size. The data were pertaining to the agricultural year 1987- 88.

To evaluate the economics of sustainable alternate land use pattern (agro forestry systems) prevalent in drylands, a couple of pilot studies carried out during 1983 at Bangalore and Hyderabad as also detailed study made during 1985-86 and 1986-87 at Cuddapah, Chittoor, Prakasham and Range Reddy districts of Andhra Pradesh were used for this study.

RESULTS AND DISCUSSION

Efficiency of Different Crop Varieties

To evaluate the economics of recommended technology under dryland conditions, selection of a suitable short duration, drought tolerant crop variety serves as a predominant part of the cropping strategy. Hence, economic efficiency of different crops/varieties under similar levels of inputs were worked out and the results are presented in Table 1.

It is seen that out of the four varieties tested in case of *guar* at Jodhpur, Durgapur Safed excelled yield levels than other varieties with a net profit of Rs. 170/ha with a 50 per cent return on investment. However, hybrids registered higher yields compared to local variety. Similarly, variety T-2 of mothbean and S-8 of greengram were found to be more promising than local variety from the point of view of profitability.

At Hissar, out of the three varieties of chickpea tested, H-208 variety gave the highest yield. Even with the higher cost of cultivation, the net profits derived from this variety were relatively higher compared to the other two varieties. It is interesting to note that in case of *rabi* sorghum at Solapur, M-35-1 was the most profitable, though CSH-8R variety yields higher than other varieties. This was attributed mainly due to the higher fodder value of the former and higher cost of cultivation incurred in case of the latter, which drifted the profits at a low level. As regards safflower, S-4 variety gave the highest yield with a net profit of Rs. 764/ha.

Out of the three varieties tested, CSH-5 was the best for Hyderabad region. Though, the cost of cultivation/ha in Hyderabad was 2-3 fold more that the local variety, the profits were relatively higher.

It was noticed that though consumers, preference for the local variety of maize was more and also the cost of cultivation was lower by Rs. 112/ha compared to Deccan-101 variety at Banglore, profits accrued per ha were more by Rs. 400 in case of the latter.

At Akola, sorghum hybrid CSH-5 produced the highest yield, but in terms of net profits, variety 370 was better. It was due to fetching better price compared to CSH-5. Regarding the cotton varieties, the SRI-I was found to be the most superior to other three varieties, viz., H-4, Varalaxmi, and DHY-286 from yield point of view and quality of lint.

Table 1. Economic efficiency of different crops/varieties under similar levels of inputs at different dryland centres

Centre	Crop and variety		Yield (q/ha)	Gross value (Rs/ha)	Working cost (Rs/ha)	Net returns (Rs/ha)
1	2		3	4	5	6
Jodhpur	Guar					
	Durgapur Safed	(2,4)	1.7	509	338	171
	KVS-2	(2,4)	1.3	369	346	23
	FS-277	(2,4)	1.4	420	348	72
	Local	(2,4)	0.6	259	292	-33
	Mothbean					
	T-2	(1,1)	3.7	982	793	189
	T-18	(1,1)	2.9	740	793	-53
	Jadia	(1,1)	2.4	615	811	-196
	Local	(1,1)	3.5	869	719	150
	Greengram					
	S-8	(1,2)	10.3	2790	553	2237
	Local	(1,2)	8.9	2412	676	1736
Hissar	Chickpea					
	H-208	(1,32)	19.3	3566	802	2764
	C-214	(1,21)	15.1	2791	753	2038
	Local	(1,45)	13.8	2551	554	1997
Solapur	*Sorghum*					
	CSH-8R	(1,2)	22.0	2051	846	1205
	M-35-1	(1,2)	14.5	2904	448	2456
	Local	(1,2)	8.2	1428	360	1068
	Safflower					
	N-62-8	(1,5)	3.9	1098	657	441
	Tara	(1,3)	4.0	1114	617	497
	S-4	(1,3)	4.9	1378	614	764
	Local	(1,3)	2.7	750	607	143
Hyderabad	Sorghum					
	CSH-5	(1,5)	11.4	1151	762	389
	CSH-6	(1,2)	7.5	1150	883	267
	Local	(1,4)	3.4	502	298	204
Bangalore	Maize					
	Daccan-101	(1,3)	28.3	3120	1973	1147
	Local	(1,3)	23.3	2630	1881	749
Akola	Sorghum					
	CSH-5	(4,18)	22.6	2650	1280	1370
	CSH-6	(2,9)	15.6	1516	820	696
	CS-3541	(2,NA)	15.8	2517	1504	1013
	370	(2,NA)	19.4	2960	1445	1514
	SPV-106	(1,NA)	7.8	740	454	286
	SPV-102	(1,NA)	11.2	1066	463	603
	Local	(1,NA)	5.4	513	445	68

Table 1 Contd . . . .

Centre	Crop and variety		Yield (q/ha)	Gross values (Rs/ha)	Working cost (Rs/ha)	Net returns (Rs/ha)
1	2		3	4	5	6
	Cotton					
	DHY-286	(4,NA)	5.1	1909	1131	778
	Varalaxmi	(1,5)	6.7	2155	1892	263
	H-4	(1,5)	8.1	2573	1868	705
	SRI-I	(3,NA)	8.5	3518	1290	2228
Varanasi	**Wheat**					
	C-306	(1,8)	10.0	1577	1019	558
	S-308	(1,8)	9.7	1479	1019	460
	K-65	(1,8)	8.2	1245	1019	226
	RR-21	(1,8)	11.0	1705	1019	686
	K-68	(1,8)	12.5	1937	1019	918
	H.D.1981	(1,8)	9.6	1408	1019	389
	Barley					
	Jyoti	(1,2)	15.8	1920	991	929
	Ratna	(1,2)	9.2	1052	991	61
	Local	(1,2)	5.4	574	964	-390
	Mustard					
	T-59	(1,2)	2.6	829	613	216
	Local	(1,2)	1.3	375	531	-156
	Chickpea					
	T-3	(1,5)	8.8	2033	887	1146
	T-6	(1,5)	7.7	1784	887	897
	Local	(1,5)	7.2	1502	869	633
	Linseed					
	Mukka	(1,2)	2.7	810	662	148
	Neelam	(1,2)	2.6	780	662	118
	Local	(1,2)	1.3	364	529	-165

(Figures in parenthesis indicate the number of year and number of trials respectively).

Out of the 8 varieties tested for wheat at Varanasi, K-68, RR-21 were the most suitable varieties in that order, in terms of yield potential and profits. Jyoti variety of barley fared well among the three varieties tested in this region but, local variety had poor yield potential which did not cover half of the total cost of cultivation. T-59 mustard gave more than double the yield advantage of local variety. Hybrid gram varieties T-3 and T-6 in order were found to be superior over local variety in terms of yield and profits. Similarly, Mukka and Neelam varieties of linseed were found to be more suitable and superior varieties for this region over local variety in terms of yield and profits. However, research results from several studies (Mehra, 1981 and Hazell, 1982) suggest that variability in Indian foodgrains production is increasing. Walker (1985) found that greater adoption of hybrids increased inter-regional yield covariances in both sorghum and peralmillét production.

Thus, hybrid seed varieties of different crops at different centres established their superiority.

Fertilizer Use in Drylands

Fertilizer use is an important component of recommended technology. Therefore, economic evaluation of fertilizer use at recommended level for different crops like sorghum, pearl millet, finger millet, wheat, lentil, chickpea, etc., was carried out at different agro-climatic regions and the results are presented in Table 2.

Table 2. Economics of fertilizer use

Centre	Crop	Level of fertilizer (kg/ha)	Yield (q/ha)	Gross value (Rs/ha)	Working cost (Rs/ha)	Net returns based on working cost (Rs/ha)	(Rs/Re invested
1	2	3	4	5	6	7	8
Nagaur	Hybrid pearlmillet	(3,80) 40 N + 20 P_2O_5	8.2	1180	495	685	1.38
	Hybrid pearlmillet	Unfertilised	3.9	586	210	376	1.79
	–	Difference	4.3	594	285	309	1.08
	Local pearlmillet	(3,44) 40 N+ 20 P_2O_5	4.3	586	420	166	0.40
	Local pearlmillet	Unfertilised	1.9	265	162	103	0.64
	–	Difference	2.4	321	258	63	0.24
Bangalore	Finger millet	(3,39) 45 N + 30 P_2O_2 + 10 K_2O+ 28 cartloads FYM	19.8	2659	1545	1114	0.72
	Finger millet	30 N + 20 P_2O_5 + 5 K_2O +20 cartloads FYM	15.1	2100	1401	699	0.50
	–	Difference	4.7	559	144	415	2.88
Akola	Hybrid sorghum	(3,10) 60 N	26.6	3305	1569	1736	1.11
	Hybrid sorghum	30 N	17.7	2193	1178	1015	0.86
	–	Difference	8.9	1112	391	721	1.84
Varanasi	Wheat	(1,8) 40 N	19.6	2961	1032	1929	1.87
	Wheat	Unfertilised	9.9	1380	505	875	1.73
	–	Difference	9.6	1581	527	1054	2.00
	Chickpea T-3	(1,8) 10 N + 25 P_2O_5	12.1	2786	887	1899	2.14
	Chickpea T-3	Unfertilised	5.7	1160	474	686	1.45
	–	Difference	6.4	1626	413	1213	2.94
	Lentil T-36	(1,1) 60 N + 30 P_2O_5	8.5	2150	718	1432	1.99
	Lentil T-36	Unfertilised	3.5	887	405	482	1.19
	–	Difference	5.0	1263	313	950	3.04
Hyerabad	Sorghum hybrid	(2,24) 40 N + 30 P_2O_5	13.3	1370	589	781	1.33
	Sorghum hybrid	Unfertilised	2.0	320	383	-63	-0.16
	–	Difference	11.3	1050	206	844	4.10
	Pearl millet BJ-104	(1,5) 40 N + 30 P_2O_5	16.3	1432	456	976	2.14
	Pearl millet BJ-104	Unfertilised	2.1	224	298	-7	-0.25
	–	Difference	14.2	1208	158	1050	6.65
Bellary	Sorghum M-35-1	(1) 30 N + 40 P_2O_5	7.8	1354	588	766	1.30
	Sorghum M-35-1	Unfertilised	4.2	604	241	363	1.51
	–	Difference	3.7	750	347	403	1.16

(Figures in parenthesis indicate the number of years and number of trials respectively).

It is clear that hybrids responded better at the same level of fertilizer application (40 kg N/ha and 20 kg P_2O_5/ha) than local in trials conducted at Nagaur in Rajasthan which is an arid region. About 73 per cent of the total working cost (Rs. 495) was spent on fertilizer use in case of hybrids and 78% (Rs. 420) in case of local variety. However, the response of hybrid pearlmillet to the same level of N and P was significantly higher at Hyderabad, a SAT region. It was found that though, the working cost being more or less the same both at Nagaur and Hyderabad, net profit accrued per ha (76%) due to fertilizer use was greater at Hyderabad.

With a fertilizer dose of 40 kg N and 30 kg P_2O_5/ha the yields of hybrid sorghum increased by 2 to 13 q/ha in different trials at Hyderabad region with an additional investment of Rs. 200 and a net profit of Rs. 800/ha. But, without using fertilizer, even the cost of hybrid seeds could not be repaid from the gross returns obtained from this crop.

In case of finger millet, out of the two trials conducted at Banglore, the higher dose of fertilizer use (45 kg N + 30 kg P_2O_5 + 10 kg K_2O/ha) to the the same level of FYM use (28 cart loads or 280 q) yielded higher.

At Akola, the use of sub-optimal dose of N (30 kg) did not pay for the extra expenditure incurred but with full dose (60 kg) substantial net profits could be derived substantially. At Bellary, for *rabi* sorghum genotype M-35-1, the net additional profit accrued per ha was more than Rs. 400 and the return on additional investment was 116 per cent with the use of 30 kg N + 40 kg P_2O_5/ha.

In Varanasi region, application of 40 kg N/ha to wheat recorded 9 q increase in yield. The net return accruing from the additional yield could cover the expenditure incurred in fertilizer use. In case of chickpea, with 10 kg N and 25 kg K_2O, the yield was significantly higher (more than twice) compared to underutilized crop resulting in substantial rise in net profits. Similar results were obtained in case of lentil for higher dose of N (60 kg) and P_2O_5 (30 kg) application.

Thus, the contribution of fertilizer use made a mark for increasing yield levels and stabilizing farm returns on drylands.

Economic Analysis of Recommended Technology Package

As mentioned in mateials and methods, a great deal of demonstrations on improved practices viz., hybrid/improved seed, recommended dose of fertilizers, intra-row weeding and plant protection measures were conducted at farmers' fields and the results are compared with the farmers traditional practices. The data on economic analysis of a package of recommended practices vis-a-vis traditional practices are presented in Table 3.

It is clear from Table 3 that the additional net benefit-cost ratio was more than unity in many test crops indicating a fairly higher return on every additional rupee invested on these crops. Exceptions were *kharif* sorghum and pearlmillet at Kovilpatti; maize, paddy and finger millet at Bhubaneswar, pearlmillet at Nagaur, Jodhpur, Hissar, Anantapur, Bijapur and Solapur. Similar results were obtained in case of greengram at Nagaur and Hissar, with chickpea at Bellary. In case of oilseeds, the additional benefit-cost ratio recorded to be more than unity in safflower at Bijapur. Similar results were noticed in mustard at Hissar and Varanasi. This indicated that recommended technology package if adopted provided higher gross income for efficient crops suitable for a given region and it can not hold true for all crops at all centres/regions.

Economics of Intercropping Versus Sole Crop Systems

Intercropping systems provide stable income and serve to minimize risk in arid and semi-arid regions in India. Therefore, economics of intercropping vis-a-vis sole crops were worked out and the results are set out in Table 4.

It is evident from Table 4 that the per ha total cost of cultivation incurred was higher in case of intercropping system at all the centres under study. It was also found that the gross returns accrued per ha was relatively higher in this system. The benefit-cost ratios were on higher side

Table 3. Costs, returns and benefit-cost ratios for major cereal pulse and oilseed crops grown at different locations in India

Name of the crop	Centre	Cost (Rs/ha) Traditional	Improved	% increase over traditional	Net Returns (Rs/ha) Traditional	Improved	% increase over traditional	Net additional benefit-cost ratio
1	2	3	4	5	6	7	8	9
				Cereals				
Sorghum *(Kharif)*	Nagaur	357	797	123	306	865	153	1.7
	Rajkot	1045	1229	18	1011	1681	66	3.64
	Kovilpatti	799	1253	57	364	810	123	0.98
	Hyderabad	574	898	56	147	774	427	1.94
	Akola	1164	1554	34	604	1185	96	1.49
Sorghum (Rabi)	Bellary	389	714	84	127	334	163	0.94
	Rahuri	381	617	62	277	559	102	1.19
	Bijapur	236	583	147	214	758	254	1.57
	Solapur	394	640	62	454	1014	123	2.28
Maize	Bangalore	1969	2057	4	661	1063	61	4.57
	Bhubaneswar	734	1150	57	54	127	135	0.18
Upland Paddy	Varanasi	624	1029	65	408	1339	70	2.30
	Bhubaneswar	923	1615	75	200	635	218	3.60
Wheat	Varanasi	579	1123	94	721	1758	144	1.91
	Ranchi	1023	1910	87	1323	3050	131	1.95
Barley	Varanasi	567	1062	87	315	1141	262	1.67
	Ranchi	572	1004	76	24	823	3329	1.85
				Millets				
Pearlmillet	Nagaur	298	604	103	218	369	69	0.49
	Anantapur	361	532	47	-49	36	173	0.50
	Hissar	757	1329	76	-44	631	1534	0.18
	Rajkot	1044	1292	24	166	383	331	2.21
	Kovilpatti	595	1076	81	33	55	67	0.05
	Solapur	267	659	147	87	275	416	0.92
	Hyderabad	396	62	58	-172	532	409	3.09
Fingermillet	Bangalore	1262	1553	23	-25	634	2636	2.26
	Bhubaneswar	753	1009	34	-88	97	210	0.72
				Pulses				
Greengram	Nagpaur	270	518	92	261	285	9	0.10
	Jodhpur	465	722	55	1603	2030	27	1.66
	Hissar	698	1163	67	-124	191	254	0.68
	Bijapur	285	639	124	207	1038	401	2.35
Chickpea	Hissar	777	1034	33	916	1563	71	2.52
	Bellary	408	802	97	72	325	351	0.64
	Varanasi	518	1011	95	677	1769	161	2.22
	Ranchi	627	927	48	438	2300	425	6.21

Table 3 Contd . . . .

1	2	3	4	5	6	7	8	9
				Oilseeds				
Groundnut	Rajkot	1562	1517	-5	263	133	331	–
Safflower	Bellary	323	667	107	60	345	475	0.83
	Bijapur	273	532	95	143	813	462	2.59
	Solapur	521	687	32	133	239	110	0.88
Mustard	Hissar	903	1141	26	1077	2060	48	4.13
	Varanasi	378	889	135	363	1274	251	1.78
Til	Nagaur	239	428	79	156	277	78	0.64
	Jodhpur	530	783	48	1311	1491	14	0.71
Linseed	Varanasi	281	793	182	219	653	198	0.85

Table 4. Additional costs and returns involved in case of intercropping systems over sole crop and income equivalent ratio (IER)

Centre	Crop	Inter crop			Sole crop			Additional			IER
		Total cost (Rs/ha)	Gross returns (Rs/ha)	B.C ratio	Total cost (Rs/ ha)	Gross returns (Rs/ ha)	B.C. ratio	Cost (Rs/ ha)	Returns (Rs/ ha)	B.C ratio	
Udaipur	Sorghum+pigeonpea	3277	15540	4.82	2897	9864	3.40	330	5676	17.20	1.58
Indore	Sorghum+pigeonpea	4586	6377	1.39	3762	4128	1.10	824	2449	2.79	1.54
Jhansi	Sorghum+pigeonpea	2922	6030	2.06	2731	4975	1.82	191	1055	5.52	1.21
Kovilpatti	Sorghum+pigeonpea	2088	7900	3.78	1708	4308	3.52	380	1884	4.96	1.31
Kovilpatti	Sorghum+cowpea	2088	5065	3.43	1708	6016	3.52	380	1137	2.99	1.19
Rajkot	Pearlmillet+pigeonpea	3211	5967	1.83	3060	3696	1.21	151	2171	14.38	1.59
Solapur	Pearlmillet+pigeonpea	1927	6334	3.29	1794	2780	1.55	133	3554	26.72	2.28
Bijapur	Pearlmillet+pigeonpea	983	4906	3.00	848	1623	1.91	135	3283	24.32	3.02
Hyderabad	Pearlmillet+pigeonpea	1508	1780	1.18	1493	1385	0.93	15	395	26.33	1.29
Bijapur -	Groundnut+pigeonpea	2545	12257	4.82	2052	9298	4.53	493	2959	6.00	1.32
Rajkot	Groundnut+castor	3722	8879	2.39	3657	7396	2.02	65	1483	22.83	1.20
Indore	Maize+soyabean	4335	4826	1.12	3397	4554	1.34	938	272	0.29	1.60
Dhiansar	Maize+blackgram	5840	5970	1.02	5526	4920	0.89	314	1050	3.34	1.21
Bhubane-swar	Pigeonpea+maize	5067	7865	1.55	3761	4100	1.09	1306	3765	2.88	1.92
Bhubane swar	Pigeonpea+maize	3916	4644	1.19	3761	4100	1.09	155	544	3.51	1.14
Bijapur	Chickpea+safflower	1499	5836	3.89	1055	4095	3.88	444	1741	3.92	1.43
Agra	Chickpea+mustard	1713	6591	3.86	1589	6129	3.86	124	465	3.75	1.08
Ranchi	Rice+pigeonpea	2339	3283	1.40	1742	2395	1.37	597	888	1.49	1.37
Bangalore	Fingermillet+soyabean	2758	5299	1.92	2554	5009	1.96	204	290	1.42	1.06
Bellary	Safflower+coriander	988	3815	3.86	683	1860	2.72	205	1955	6.41	2.05

Note : Income Eqnivalant Ratio (IER) = $\dfrac{\text{Income intscropSystem/ha}}{\text{Income sole cropsystem/ha}}$

in case of intercropping system compared to sole crops at all locations, except in case of finger millet + soybean intercropping system at Banglore. However, additional benefit-cost ratios that indicate the real extra investment made and returns accrued thereon due to intercropping system varied from 1.42 in finger millet + pigeonpea at Bangalore region to 26.72 in pearlmillet + pigeonpea intercropping system at Solapur region. Thus, intercropping system in drylands emerged as a promising system in India to fetch higher returns as reflected in higher additional benefit-cost ratios.

Economics of Sequence Crop

Sequence cropping systems can be practised in high rainfall years utilising short duration varieties. The economic analysis of sequence crop versus single crop was attempted in this study and the results are presented in Table-5.

Table 5. Economics of sequence cropping versus single crop at different centres in India

Centre	Crop	Sequence crop			First crop		
		Total cost (R/ha)	Gross returns (Rs/ha)	B.C ratio	Total cost (Rs/ha)	Gross returns (Rs/ha)	B.C ratio
Indore	Maize-Gram	5552	7886	1.42	3398	4542	1.34
Dehra Dun	Maize-Gram	5538	7965	1.44	3091	4075	1.32
Hoshiarpur	Maize-Wheat	6845	10575	1.54	3382	5400	1.59
Dehra Dun	Wheat-Maize	6207	9625	1.55	3116	5560	1.78
Agra	Greengram-Mustard	2549	5736	2.25	1186	1316	1.11
Anantapur	Greengram-Horsegram	2236	3016	1.35	1143	2050	1.79
Udaipur	Blackgram-Safflower	4836	12850	2.66	2189	2200	1.01
Bangalore	Cowpea-Fingermillet	4293	5024	1.17	1897	2720	1.43

It was found that generally the cost of cultivation/ha involved and gross returns accrued were higher in case of sequence cropping system compared to single crop. The net returns derived/ha. were on higher side in sequence cropping system at all centres, except Anantapur and Bangalore. The benefit-cost ratios were higher in case of sequence cropping system at Udaipur, Agra, Dehradun, Hoshiarpur and Indore in that order but, low at other centres which was due to the variation in cost structure and returns. Thus, sequence cropping systems are feasible and viable system of high rainfall regions in India.

Economics of Agroforestry Systems vis-a-vis Arable Crops

Agroforestry systems which include perennial component of trees/shurbs/grasses in arable crops are basically meant for stabilising farm income of dryland farmers during drought years. Annuity values and benefit-cost ratios were worked out for different regions to know economic viability of agroforestry systems and the results are presented in Table 6.

A perusal of the data presented in Table 6 indicate that income accrued in case of agroforestry system was substantially higher compared to arable crops in all the regions under study. However, annuity values and benefit-cost ratios showed that this system is a paying proposition in increasing and stabilising income from an unit land in dryland farming regions in India. In addition to these tangible benefits, there are numerous intangible benefits from agroforestry systems like maintenance of balanced eco-system, minimising losses from weather aberrations in drylands etc.. Venkateswarlu (1989) suggested the need for growing *Prosopis cineraria, Zizyphus nummularia* and *Acacia nilotica* as also promising tree species like *Acacia tortilis* and *Acacia senegal* to bring about sustained production and risk minimization in the desert areas of Rajasthan in India.

Table 6. Economics of agroforestry systems as compared to arable cropping at Bangalore and Hyderabad

System		Period (Years)	Annuity values (Rs/ha)	Benefit-cost ratio
	Bangalore region			
Arable crop	Fingermillet	Annual	1095	1.53
Alternate	Casuarina	10	3000	11.08
	Eucalyptus	30	1741	18.50
	Hyderabad region			
Arable Crop	Sorghum	Annual	683	1.39
	Castor	Annual	575	1.24
	Sorghum + pigeonpea	Annual	1373	1.72
Alternate	a) Agro-forestry system (Acacia tortilis) + Sorghum/Horsegram	10	1117	1.73
	b) Agri-horticulture : *Ber* + Sorghum/Castor	30	17881	5.78
	c) Silvi-agriculture : Subabool + Sorghum/Castor	10	1204	1.80
	d) Silvi-pastoral : Subabool + Stylosanthes hamata	10	1115	2.56
	Chittoor region			
Mango	With arable inter crop	4	4228	3.21
	Without arable inter crop	–	3896	2.27
	Cuddapah region			
Acidlime	With arable inter crop	–	9645	3.16
	Without arable inter crop	–	9851	3.04
Sweetlime	With arable inter crop	–	3307	2.89
	Without arable inter crop	–	3255	1.49
	Ranga Reddy region			
Guava	With arable inter crop	–	4526	2.18
	Without arable inter crop	–	3722	2.08

Impact of Watershed Programme on Economics of Dryland Crops

Benefit-cost ratios

Benefit-cost ratios help us to know the profitability of a crop. Hence, an attempt was made to analyse the benefit-cost ratio for different dryland crops to judge the impact of watershed programme on rupee invested. The results are presented in Table 7.

It was clear that the benefit-cost ratios varied from 2.03 in sorghum to 4.06 in blackgram + pigeonpea in watershed as against 1.40 in sorghum + safflower (in post-rainy season) to 3.39 in sunflower + pigeonpea (in rainy season) in non-watershed villages at Chevella.

At Mittemari, the B-C ratios in all the crops under study were higher in watershed villages in addition to higher gross returns per unit area compared to non-watershed villages. Thus, the B-C ratios of all crops were on higher side in watershed compared to non-watershed villages at Chevella and Mittemari indicating higher profitability.

Additional benefit-cost ratios

Additional benefit-cost ratio is the real indicator to decide the economic efficiency of new dry farming technology due to watershed programme. The additional benefit-cost ratios were compared by taking into consideration the real additional benefits accrued and additional cost involved/ha in watershed compared to non-watershed villages. Additional benefit-cost ratios of different crops indicated that an additional investment on crop production fetched additional returns due to watershed programme

Table 7. Impact of watershed programme on economics of recommended dryland technology

Crop	Watershed			Non-watershed			Additional B.C ratio		
	Total cost (Rs/ha)	Gross returns (Rs/ha)	B.C ratio	Total cost (Rs/ha)	Gross returns (Rs/ ha)	B.C. ratio	Cost (Rs/ ha)	Benefit (Rs/ ha)	B.C. ratio
1	2	3	4	5	6	7	8	9	10
Chevella									
Rainy season									
Sorghum (L)	1001	2031	2.03	896	1315	1.47	105	716	6.82
Sorghum (L) +pigeonpea (L)	1017	2744	2.61	939	2166	2.31	78	578	7.41
Greengram (I)	1165	2924	2.51	888	1332	1.50	277	1592	5.75
Greengram (I)+pigeonpea (H)	1515	3712	2.45	1198	2037	1.70	317	1675	5.28
Blackgram (I)	993	3416	3.44	810	1332	1.64	183	2084	11.39
Blackgram (I)+pigeonpea (H)	1052	4268	4.06	950	3162	3.33	102	1106	10.84
Sunflower (I)	1091	3124	2.86	926	1824	1.97	165	1300	7.88
Sunflower (I)+pigeonpea (H)	1095	3730	3.41	1014	3441	3.39	81	289	3.57
Post-rainy season									
Sorghum (L)	806	1721	2.14	712	1222	1.70	114	499	4.38
Sorghum (L) +Safflower (L)	939	2288	2.44	866	1213	1.40	73	1075	14.73
Mittemari									
Rainy season									
Groundnut (I)	2047	5137	2.51	1836	4090	2.23	211	1047	4.96
Fingermillet (I)	1860	2777	1.49	1386	1802	1.30	474	975	2.06
Groundnut (L)+pigeonpea (L)	3035	6798	2.24	2646	4242	1.60	389	2556	6.57
Groundnut (I)+pigeonpea (H)	3104	8354	2.69	2584	4558	1.76	520	3796	7.30
Fingermillet (L)+pigeonpea (L)	2080	2666	1.28	1682	1934	1.15	398	732	1.84
Fingermillet (I)+pigeonpea (H)	2404	5713	2.38	1725	3337	1.93	679	2376	3.50

(Table-7). Additional B-C ratios were higher in all crops in the two seasons at Chevella and in groundnut + pigeonpea at Mittemari.

SUMMARY AND CONCLUSIONS

Adoption of recommended technology package, though cost-intensive gave higher yields and income even at farmers, fields for efficient crops suitable for the region. Adoption of HYV seed and recommended level of fertiliser use derived higher net returns compared to local/traditional system of farming in different regions in the country.

The cost of cultivation involved and returns accrued registered higher in intercropping system to sole cropping system. The benefit-cost ratios of intercropping systems showed greater than unity at all centres under study indicating higher net returns. Thus, intercropping system is a viable and feasible system for increasing yields and stablising income of farmers in drylands.

Sequence cropping system is an economically viable and profitable system under dryland conditions as established by higher B-C ratios, but it is limited to those places where the rainfall is relatively higher and assured.

Agroforestry is an alternative land use system in drylands for stablising income of the farmer. This system registered phenomenal increase in terms of B-C ratios and annuity values indicating that the income accrued in agroforetry system was higher compared to arable competitive crops which suggests best scope for stabilising income on drylands.

Resource Centred Technology (watershed management), an area approach, showed clear impact on economics of new dry farming technology. This was evident by higher values of additional B-C ratios of different crops which indicate that an additional investment made on crop production increased returns in watershed compared to non-watershed villages at Chevella and Mittemari.

REFERENCES

Hazell P.B.R. 1982 *Instability in Indian foodgrain production*. Research Report 30, International Food Policy Research Institute, IFPRI, Washington, D.C.

Lipton, M. 1970 Farm Price Stabilisation in underdeveloped agriculture : Some effect on income Stability and income distribution In P.Streeeton (Ed.) : *Unfashoinable Economics, Weidenfeld and Nicholson*, London : 3-22.

Mehra, S. 1981 Instability in Indian Agricultre in the context of the New Technology Research Project 25, International Food Policy Research Institute, IFPRI, Washington, D.C.

Venkateswarlu, J. 1989. Strategies for harnessing natural resources in arid West Rajasthan. *Indian Farming* : 19-31.

Walker, T.S. 1985. *HYVs and Variability in Sorghum and Pearlmillet Production in India* Research Report, ICRISAT, Patacheru P.O., Andhra Pradesh, India.

33

EXTENSION STRATEGY FOR RAINFED AGRICULTURE

C. PRASAD

INTRODUCTION

India was one of the leading developing nations where Green Revolution was brought about in late 1960s. The pace of agricultural production in the second phase of Green Revolution is now staggering. Therefore, challenges for production are being examined from different angles. Issue number one being discussed is whether or not we have the appropriate technologies for rainfed farming? And secondly, if we have technologies, how effective is our delivery system for reaching the dryland farmers and promoting technologies best suited in their socio-economic conditions? Scientists do agree that we have lagged behind in evolving improved technologies for rainfed farming vis-a-vis irrigated technologies. But still we have farm technologies for the rainfed farmers which can raise production upto 200 per cent. The productivity of rainfed crops has remained, on an average, a little over half a tonne per hectare.

The issue in respect of technology dissemination has been discussed in various forums in 1980s. The International Seminar on the Generation and Transfer of Technology for Small Farmers held in Republic of Korea in 1988, besides other things, highlighted the lack of appropriate extension strategies for rainfed agriculture and stressed the need for developing innovative technology transfer mechanisms for such farmers (Kesseba, 1989).

The first phase of Green Revolution has been marked by more general and blanket recommendations which, by and large, benefited the farmers in irrigated - the rich resource users. Similarly, extension system(s) also approached every one alike whether rich or poor, big farmer or small farmer. In this process, the farmers in irrigated areas benefitted the most, being resource rich and influencial.

In technology generation as well as technology promotion and adoption process, socio-economic considerations are of special significance, which, somehow, could not catch the imagination of the scientists and the extension personnel in the past. Appropriate technology and suitable extension approach and methodology are two important dimensions of the transfer of technology system. Both must be compatible for accelerating agricultural and allied production and have, therfor, been highlighted in this chapter.

Features of Rainfed Farming

Of the world's 1,474,100 thousand ha agricultural land, 227140 thousand ha are irrigated (15.4%) - remaining 84.6 per cent are rainfed; in Asia- Pacific regions, 1,27,018 thousand ha are irrigated (30.4%), 69.6 per cent are rainfed, and in India 44,350 thousand ha (26.3%) are irrigated - 73.7 per cent are rainfed. Of the total arable land of 143.8 million hectares m.ha 100.0 m.ha are non-irrigated / rainfed - 34.5 mha rainfed dryland area and 65.5 m ha rainfed wetland area. In rainfed / dryland farming, the rainfall is half the potential evapotranspiration - moisture conservation is the main concern; in the rainfed wetland farming, rains are more or less adequate, but drainage of excessive water is the major problem.

At present, in India 70 per cent rainfed farming produces only 42 per cent of the foodgrains. The farmers are mostly small and marginal and resource-poor. Institutional support is weak; people are illiterate, less enlightened, and rather easy-going. The rainfed areas also represent the bulk of the weaker sections of the society; and the weaker sectors in agriculture are producers of oilseeds and pulses, coarse grains, and feeds and fodder. Rainfed farmers do not have much access to modern technologies as well as tailor-made methodologies. The intensity of population in rainfed area is less than in irrigated areas; thus, the population/village are rather scattered. Communication and transportation systems are generally not well- developed.

The rainfall in India is rather erratic - sometime too much and sometimes none; and hence floods and droughts occur accordingly. Therefore, in rainfed farming, the management is required to handle the extremes - deep water to dry uplands. In scientific terms, it is the management of arid regions - rainfall 500 mm and below - good for pasture- livestock system only; semi-arid regions - rainfall between 500 mm and 1,100 mm - can practise farming with limited soil moisture; and humid or wet regions - rainfall above 1,100 mm - farming requiring skillful management of water.

The tenancy reforms and implementation of land ceilings in rainfed areas have yet to deliver the goods. The consolidation of holdings is yet another unaccomplished area. Checking fragmentation by suitable law, as in Japan, has not been thought of. Common community lands have not been developed and utilized. Further, arresting degradation and retrieving the ones already degraded should be The priority areas.

The potentiality of production exists even at the present level of technological advancement in rainfed agriculture. The productivity shown in various crops in rainfed condition under the National Demonstrations (ND) programmes for 1986-87 is given below in Table-1.

For R & D sectors, it is very important to remember that, even when irrigation potential is fully utilized in the country, there will still be about 50 per cent land under rainfed agriculture. Implications are obvious - no complacency is expected at any stage and on any account.

Table 1. Yields (q/ha) of different crops in National Demonstrations vis-a-vis national averages, the highest yields in ND and increase in ND average over the country average in rainfed areas

Crop	National average (a)	ND average (b)	ND highest (c)	Increase (%) (b) over (a)
Maize	12.8	29.2	45.0	128
Sorghum	5.7	32.7	70.5	469
Pearl-millet	4.0	17.0	45.0	326
Groundnut	8.4	19.8	32.3	136
Mustard	7.0	8.91	19.3	27
Gram	6.8	15.5	26.6	129
Soybean	6.0	14.5	25.0	141

SUSTAINABLE AGRICULTURAL DEVELOPMENT

Sustainability concept is universally true, may it be agriculture or industry, research or extension. In case of sustainable agricultural development, the concept is skewed in favour of research or development, depending upon the perceptions of the individuals. But a balanced view of the subject entails that there are four discrete functional areas - four pillars in a way, for sustaining agricultural development : (1) research, (2) education, (3) extension and (4) training (Fig. 1). We are generally used to first three functional areas; 'training' has been a diffused area - an integral part of extension or any other discipline. This is why, the area of training could not develop as much as it should. Training is so vital at all levels, starting from farmers vocational training to advanced level training of profes-

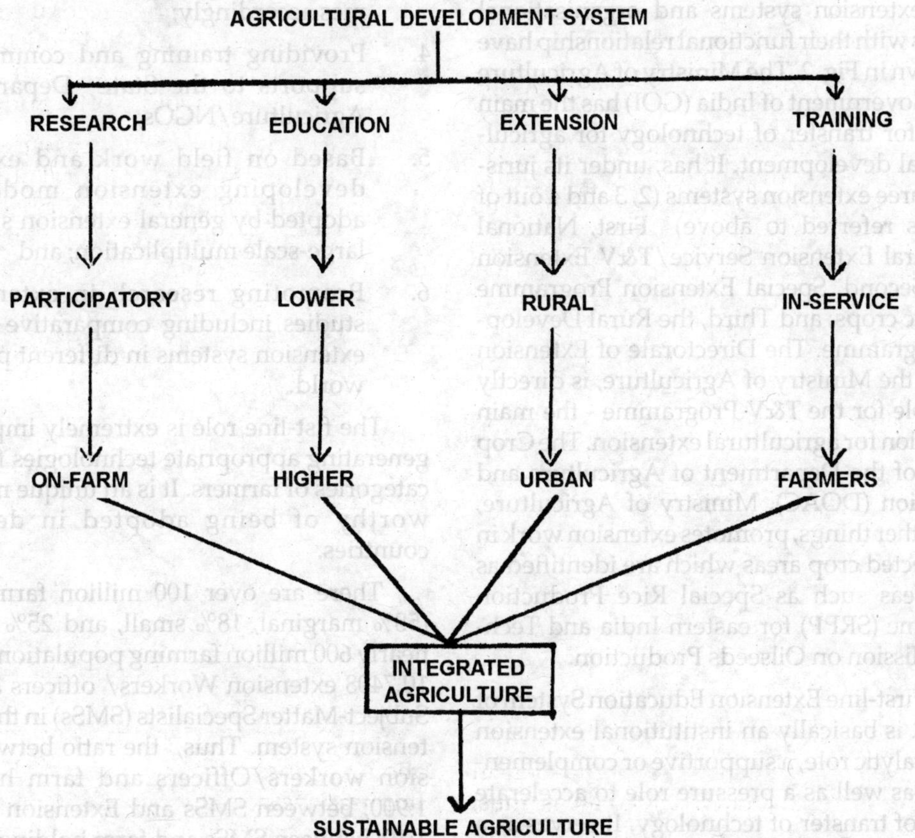

Fig. 1. Development system for sustainable agriculture

sionals. Millions of people in agriculture cannot afford to go for formal certificates, diplomas or degrees, but, as a matter of right, they must have access to training, short or long as the case may be, for upgrading their knowledge, understanding and skills. Therefore, training must be given its rightful place - status of an independent functional area, and thus adequate resources should be available earmarked for this purpose. (Prasad, 1990).

Rainfed agriculture must be treated in a special way in all these four functional areas for generating appropriate technologies, developing manpower most suited to dryland/rainfed agriculture, following extension approach and methods most befitting to the rainfed farmers, and organizing training programmes for all shades of people for promoting rainfed technologies.

AGRICULTURAL EXTENSION SYSTEMS

At present, five extension education/transfer of technology systems/sub-systems are in vogue in the country, each having its own mandate, approach and thrust with ultimate attack on freedom from hunger and poverty. These systems are :

1. First-line Extension Education System.
2. National Agricultural Extension Service (NAES)/Training & Visit (T&V) System;
3. Special Extension Programme on specific crops;
4. Rural Development Programme; and
5. Extension efforts of the Non-government Organizations (NGOs).

The extension systems and organisational structures with their functional relationship have been shown in Fig. 2. The Ministry of Agriculture (MOA), Government of India (GOI) has the main mandate for transfer of technology for agricultural/rural development. It has, under its jurisdiction, three extension systems (2, 3 and 4 out of 5 systems referred to above) First, National Agricultural Extension Service/T&V Extension System; Second, Special Extension Programme on specific crops; and Third, the Rural Development Programme. The Directorate of Extension (DOE) of the Ministry of Agriculture, is directly responsible for the T&V Programme - the main organisation for agricultural extension. The Crop Division of the Department of Agriculture and Cooperation (DOAC), Ministry of Agriculture, besides other things, promotes extension work in some selected crop areas which are identified as thrust areas such as Special Rice Production Programme (SRPP) for eastern India and Technology Mission on Oilseeds Production.

The First-line Extension Education System of the ICAR is basically an institutional extension role, a catalytic role, a supportive or complementary role as well as a pressure role to accelerate the pace of transfer of technology. It serves as a window of educational and research institutions through which the extension agencies as well as the farmers can gaze the latest agricultural technologies. In specific terms, the objectives of the ICAR extension education systems are :

1. Promptly demonstrating the latest agricultural technologies to the farmers as well as the extension workers of the State Departments of Agriculture/Non-government Organizations with a view to reduce the time-lag between the technology generation and its adoption;
2. Testing and verifying the technologies in the socio-economic conditions of the farmers and identifying the constraints;
3. Getting the first-hand specific feed-back by the scientists, in order to keep abreast with the performance of the technologies and the farming problems, and reorient their research, education and training programmes accordingly;
4. Providing training and communication supports to the State Departments of Agriculture/NGOs;
5. Based on field work and experience, developing extension models to be adopted by general extension systems for large-scale multiplication; and
6. Promoting research in extension and studies including comparative studies of extension systems in different parts of the world.

The fist-line role is extremely important for generating appropriate technologies for various categories of farmers. It is an unique mechanism worthy of being adopted in developing countries.

There are over 100 million farm holdings (58% marginal, 18% small, and 25% big) - and nearly 600 million farming population. There are 10,7408 extension Workers/ officers and 10,195 Subject-Matter Specialists (SMSs) in the main Extension system. Thus, the ratio between extension workers/Officers and farm holdings is 1:900; between SMSs and Extension Officers is 1:10; between SMSs and farm holdings is 1:9600; and between extension workers and farmers is 1:3000 (effective farmers). In addition, the first-line extension system has about 1100 SMSs and 600 technical assistants.

These extension systems/sub-systems must be differently geared to rainfed farming and the farmers.

Approaches to Extension Strategy

There are 3 basic approaches in extension strategy, viz., consultancy approach, promotional approach and participlative approach (Dexter, 1986). In the consultancy approach, the Extension Agency provides advice on request by farmers, whereas in the promotional approach the agency takes its own decision on what to do and how to do it on the basis of views and needs of the farmers. In the participative approach, it is a partnership between the farmers and the Extension Agency and together they decide on what

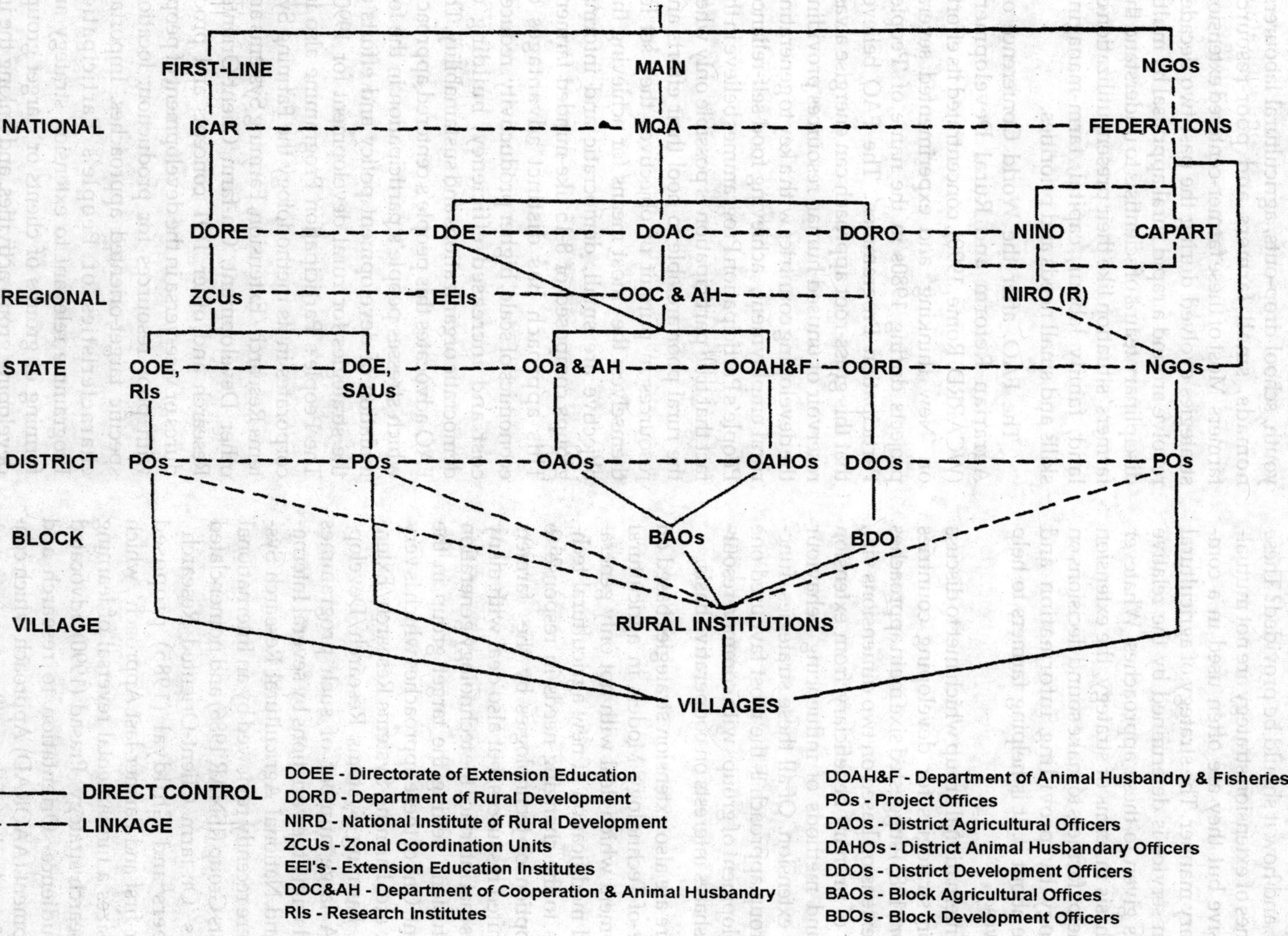

Fig. 2. TOT Systems in Agriculture

technologies are important, what information is required, and how it should be provided? These approaches of extension strategy are not mutually exclusive but they are often used in a complementary manner. The strategy of agricultural extension service is determined by the relative emphasis given to these approaches. Whatever the emphasis in terms of strategy, the extension service helps farmers to make sound decisions on technology by providing information and guidance about it. It is helping farmers to help themselves.

An International group which met to discuss extension strategies for developing countries (*Haverkort*, 1983) suggested sixe main approaches to rural extension based on two dimensions viz., the unit of primary beneficiary from extension efforts and methods of influencing behaviour used by extension. Of all these strategies, functional group approach is the most favoured one for development of groups with common resources and shares interests on cooperative basis.

There are also extension strategies based on Transfer-of-Technology Models in agricultural development which deal with not only generation and modification of new agricultural technologies to the changing needs in response to non-adoption of technologies by the farmers with limited resources, but also deal with many concepts that interface the technology generation and diffusion, setting the target group in the fore-front. One of these approaches which is very popular are: Farming Systems Research/Extension or Farming Systems Research/Development. A large number of such programmes adapted with little variations by several International and National Agricultural Research Systems were recently reviewed by an International Scientists Group (ISNAR 1986) and nomenclated them as 'On-farm Client-Oriented Research'. Chambers and Ghildyal (1985) proposed 'Farmer-first and Farmer-last Approach', which emphasizes a fundamental reversal of learning and research strategy. Prasad (1990) advocated Area-audience- Orientation to research and development (AAORAD) Approach which considers 2 basic dimensions: (i) area irrigated, rainfed, hilly, coastal, remote or inaccessible; and (ii) audience - male farmers, farm women, farm youth, school drop-outs, agricultural labourers, nomads, small farmers and poor-resource farmers. Most of these farmer-centered extension strategies evolved during the past two decades revolve around a rapid rural appraisal by multidisciplinary teams of scientists to understand the farmers' situation like their present utilization of land, family labour, capital, farm-managing skills and small needs and priorities.

The FAO, after the World Conference on Agrarian Reform and Rural Development (WCARRD, Rome, 1990), concentrated its efforts on 'New thinking' and experimented several projects during 1980s in the name of 'Peoples Participation Programmes'. The FAO believes that this grass-root approach can energise a vast reservoir of unused human resources providing the developing countries with a key to generating rural employment, achieving food self-reliance. People's Participation Programmes believe in the fact that true participation is possible only when the rural poor are able to pool their efforts and resources in pursuit of objectives they set for themselves; the most means for achieving this objective are small, democratic and informal groups composed of 8-15 like-minded farmers. This approach has distinct advantages of economics of scale, higher productivity, reduced cost and increased efficiency, building of democratic organizations and sustainability. The FAO advocates this people's centered approach, which stresses people's participation in the formulation of development policy and efforts as the strategies for rural development for 1990s. The People's Participation Programme also incorporates in its methodology the Farming Systems Research/Extension, Farming Systems and other Development, On-farm Client-Oriented Research and other TOT concepts and procedures or vice-versa in the development of people with poor resources for production; location-specific target-oriented approaches. Important characteristics of People's Participation Programme relevant to extension strategy are: forming of groups of clients or target groups, developing group activities, and linking the implementing agencies with the groups. Important means and methods used in development process are: group promoters, participatory training, participatory research, monitoring and evaluation.

Thus, gradually, the development path has been moving from a generalized (extensive) approach to specific (intensive) problem or area approach. And of late, area-audience-oriented research and development for specific target groups has been emphasized. The demand is for both location-specific technology as well as appropriate extension pedagogy or methodology.

EXTENSION STRATEGY FOR RAINFED FARMING

Strategy is a planned design to attack or tackle a problem or related problems concerning an individual or organization for achieving targetted goal within a specified time. The inherent feature of any strategy is that it would vary from situation to situation for precision of approach and implementation. Therefore, flexibility in such a planning is an expected principle, based on reality of circumstances and resources. - both men and material. But once a work plan/strategy is developed after a meticulous planning and exercises, its implementation is mission-oriented - do or die type of attack or effort to achieve that goal within specified period. During implementation only limited flexibility is allowed for unforeseen situations.

The development is not a simple process unlike ancient days; societies have become more complex, large and demanding, population is ever- growing; there are trained manpower; and there are institutional framework. The resources are scarce vis-a-vis demands; and, though, there is an army of trained manpower, they have yet to develop the needed devotion, discipline and commitment. Development is a function of several related important factors: (i) clientele - rural, urban, male, female; (ii) coordination; (iii) infrastructures/institutions; (iv) manpower - trained, disciplined, extension, subject-matter specialists; (v) methodology; (vi) materials; (vii) monitoring; (viii) evaluation; (ix) technology; (x) training; (xi) planning; (xii) approach; (xiii) linkage; (xiv) area; (xv) feed-back; (xvi) outcomes; and (xvii) situations. These factors may be remembered as CCIMMMMETT - PALAFOS. This CCIMMMMETT-PALAFOS model can be utilized for studying, analysing and understanding any organization or systems - it is a system-analysis model comprising all the relevant elements of extension. These 17 elements or components of a system must work and interact in a right mix and balance and in a proper harmony and relations for synergistic effects and outcomes. The transfer of Technology (TOT) System in irrigated or rainfed agriculture must be analysed on these factors in order to understand their different requirements.

The CCIMMMMETT-PALAFOS model (Figure-3 cuts across all the extension approaches and presents a most practical way of studying, analysing and understanding a system or subsystem. A group of subject-matter specialists with long experience of field extension work were used to score on the elements of this model on a four-point scale (Excellent, Very Good, Good and Poor) both in irrigated and rainfed conditions. The results are presented in Table-2.

It is obvious that the extension work in the rainfed areas is a neglected sector. The 17 elements of the TOT System is now briefly discussed in the context of rainfed agriculture.

1. *Clientele:*

The profile of the rainfed farmers, broadly speaking, reveals that they are small, marginal or agricultural labourers with poor-resource base, very limited education and limited access to modernization and modern technologies and methodologies. Because of the harshness of the economic condition particularly, they have subdued living and life-style - less prone to change, fatelistic attitude, leaving things at the mercy of the almighty and pessimistic value system. In the struggle to acquire information, technologies and means to adopt technologies, they are left behind; the survival of the fittest prevail - the progressive, rich and innovators get the most, even at the cost of unfairness. The first charge of the extension agents is to prepare the farmers - small and poor for newer technologies for more

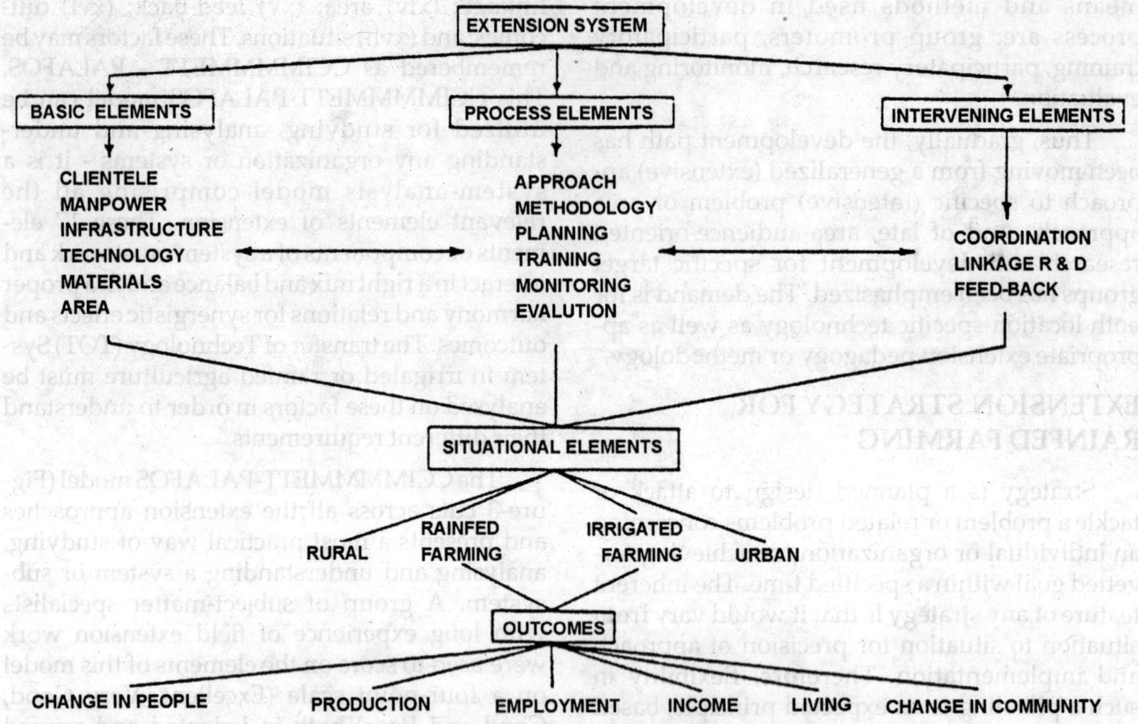

Fig. 3. A Model for analysing an Extension System

employment, production and income; and help rich and progressive to assist, guide and help the poorer. A very conducive atmosphere for change has to be brought about before technologies could be adopted by all especially the small and poor rainfed farmers. The adoption process follows the normal curve but more subdued than the irrigated farmers. We must cater for these clientele as per their specific needs - farmers (male), farm women, young farmers (boys and girls), and agricultural labourers.

Coordination :

Agricultural development system presents a complex coordination phenomenon - inter-institutional and inter-disciplinary. Coordination is required within the disciplines/specializations, between institutions and departments, and in functional areas like research, extension and training. The old concept of people's participation and the new thrust on participatory research and development bring clientele/farmers also in the framework of interactions at all levels. Coordination is a skilled job; it is management of people of a system and between allied systems in the realm of technological as well as sociological perspectives. More allied agencies have to be brought to gether to serve the rainfed farmers on the line of farming systems approach.

Infrastructures/Institutions :

It is apparent from Table-2, that infrastructure/institutionwise, the dry belts are more than half weak as compared to irrigated areas - weakness is more in functional domain - in the past, attention has been more towards the richer and potential. Keeping in view the local conditions, the VEW Village Extension Workers and farmers ratio of 1:700 has to be reduced to 1:500 for covering larger areas and scattered population as also for more personalized approach. Rural institutions especially village cooperatives must function to meet the local needs of inputs - small and marginal farmers cannot afford to spend a lot of time in searching for the inputs. 'Village Resources Development Societies', being promoted in

Table-2 Relative ranking of 17 elements of CCIMMMMETT-PALAFOS Extension model by subject-matter specialists in irrigated (I) and rainfed (R) situations (1990)

Elements of TOT System	Excellent		Very Good		Good		Poor	
	I	R	I	R	I	R	I	R
Clientele	10	5	20	10	45	35	25	50
Coordination	10	–	20	5	40	15	30	80
Infrastructures/ Institutions	20	10	35	25	25	30	20	35
Manpower	10	5	20	15	40	35	30	45
Methodology	5	–	20	15	20	20	55	65
Materials	10	–	30	20	40	25	20	55
Monitoring	5	–	15	10	25	20	55	70
Evaluation	–	–	25	15	25	20	50	65
Technology	25	5	30	20	20	15	25	60
Training	5	–	20	15	25	20	50	65
Planning	–	–	15	10	25	15	60	75
Approach	5	–	10	10	15	15	70	75
Linkage	10	5	20	15	30	20	40	60
Area	10	5	20	15	20	20	50	60
Feed-back	–	–	10	5	15	10	75	85
Outcomes	10	–	15	10	30	25	45	65
Situations	5	–	15	15	25	20	55	65

watershed areas under the Operational Research Projects of the ICAR involving local institutions and people in planning and implementation needs promotion and strengthening for sustainable rainfed agriculture. Krishi Vigyan Kendras may be established in rainfed districts on priority basis.

Manpower:

Keeping in view the need of rainfed farmers, the number of field staff need to be increased in such areas - the ratio between extension workers and the contact farmers has to be narrowed. If possible, more women extension workers have to be employed. Both subject-Matter specialists and extension professionals have to be thoroughly geared to and trained in the precision and intricacies of rainfed farming - deep water paddy, hill agriculture, water technology, watershed approach, etc. The subject-matter specialists would need insight and training in extension methodology and social sciences, whereas the extension specialists would need training in subject matters relevant to rainfed farming.

Methodology:

Both technologies and extension methodologies are equally important, the latter, somehow, have been given lower priority. The educational nature and the educational processes involved in extension have to be practised and followed. The participatory approach in extension has yet to be fully introduced. In case of rainfed farmers, more personal and group approaches will help: field demonstrations, personal visits and advices, audio-visual aids made on local conditions will be effective; published materials will have limited value in view of mass illiteracy. For making the limited resources of the farmers viable, a farming-systems methodology has to be followed - promoting more than one enterprise in right combination - diversification in farming. Cooperative or the group methods being followed in Kerala now require promotion in dryland areas for collective efforts in selected activities; loan distribution, training, supply of inputs, etc. The concept of adopting small and marginal farmers and agricultural labourers and providing 'community inputs' followed under the Lab-to- Land Scheme have proved very practical and helpful.

Watershed Operational Research and Extension Approach (WOREA) has been found The right mechanism for promoting S&T inputs in farming. Linking such watershed, with Krishi Vigyan Kendras for guidance and training support is still another good approach.

Materials:

Mobility is the first requirement for the extension work in such areas. To economize on travels, motor-cycles may be used instead of jeeps, station wagons, etc. Low-cost indigenous audio-visual aids may be preferred over expensive communication gadgets Nevertheless, the farmers once in a while may have the thrill of sophisticated audio-visual aids, i.e. 16 mm films, video-tapes, etc. Village library may be promoted where extension publications in local language could be available for at least those who can read and write, the latter can function as extension agents for others. Farmers' fair is proving a good mass-communication device, where farmers besides knowing and seeing new technologies, have occasion to interact with the scientists and can buy improved seeds, plants, chemicals, etc. Availability of transportation for rainfed farmers would be necessary to avail of such opportunities. In rainfed areas, the concept of mini-farmers' fair in villages can be promoted. Mobile exhibition van and soil- testing kits are proving useful and effective in such areas. Loans, perhaps, long-term have to be made available for initial investments in land development including irrigation structures and water harvesting devices.

Community inputs, specially the farm equipments or implements may be provided free of cost to be shared among the poor-resource farmers. Providing a 'bull' for serving the animals in the area; and bucks in the goat and sheep and herds as community inputs were found rewarding.

Extension staff should not suffer for want of simple common requirements like papers, pencils, pens, typing facility and other stationeries for effective bottom-up communication and reporting. The local extension staff are commonly handicapped in these aspects.

Monitoring

This is a regular and continuous activity through personal visits, meetings, reports, occasional seminars and conferences. In this process, two-way communication, often talked about (bottom-up and top- down) must be ensured. Action taken review and follow-up on all accounts will improve monitoring. Intervention of senior officers on one hand and progressive farmers and local leaders including farm women on the other, should figure in the monitoring system. For instance, presence of the Directors of Agriculture, Agriculture Production Commissioners, Vice-Chancellors, or Directors of Research and Extension of SAUs in monthly workshop, once in a while, will bring prestige to such interactions as well as good feed-back. Care may also be taken that reporting should not become a burden - a full-time job; it should be short, pin-pointed and fast. A good manual for monitoring will be helpful for extension personnel.

Evaluation

If monitoring is regular and effective, frequent evaluation is not necessary. However, more informal and day-to-day self-evaluation by the extension staff should be encouraged. This is not happening; this would depend on the commitment and motivation of the staff. All in- service training courses should emphasize on this aspect. Annual evaluation by the Units and five yearly evaluation by the External Agencies may be encouraged. Recommendations of such evaluations should be strictly followed up-latter is the lacunae almost everywhere.

Technology

For meeting the food requirement of the country, scientists put in all their best for high potential farm technologies, with the result, the Green Revolution was brought about in 1960s; self-sufficiency in food was almost achieved - production was increased over 300 per cent. Such a dramatic change came about owing to production technologies in irrigated agriculture; the rainfed agriculture did not get as much attention. The second phase of Green Revolution has posed this problem - how to improve rainfed agriculture, almost 70 per cent of the farming and it may continue so in 50 per cent of the cultivated area even after realizing the full potential of irrigation in the country. The present available improved rainfed technologies can increase production upto 200 per cent if well adopted and all the farming practices are followed. Even this has not yet been tapped fully. Of course, scientists have to work hard for still more productive technologies for arid, semi-arid and wet agriculture.

There is a general complaint from the extension systems that technologies are not very appropriate. This is partly true keeping in rainfed technologies in view, and partly such observation comes when the technologies are not being adopted. But non-adoption of technologies are also partly due to ineffective extension operations as well as socio- economic circumstances. We have to work on all fronts to realize the production potential of rainfed agriculture. Participatory research and inter-disciplinary approach in farming system context have to be followed; on-farm research should find priority.

Training

Poor-resource farmers of the rainfed areas have to be trained more and more, for they are less enlightened, mostly illiterate and neglected from the mainstream of development. Because of their background, trainings have to be based on the principles of 'training by doing' and 'learning by doing' like being followed by the Krishi Vigyan Kendras. Most of the trainings have to be *in situ*, off- campus/village-based for such farmers cannot afford to come for residential courses; they have to be one-day or two-day courses; many times they could be sandwitch courses (mixed courses). The training strategy should be to reach farmers (male), farm women, and young farmers including boys and girls with special priority to school drop- outs. We must build training infrastructures/institutions nearer to the farming situations as best as possible. The Krishi Vigyan Kendras (Farm Science Centres) being promoted in India for each district is a good example in this respect.

Planning

Grass-root level or bottom-up planning, needs to be vigorously promoted. This has rarely been done with the study of full data and precision. In case of rainfed farmers, meticulous planning is necessary, as their resources are poor as well as their risk taking capacity is very low. Planning is all the more important in view of diversification in farming which requires right combination of enterprises. Cooperative or group farming recommended for such farmers as well as close involvement of village cooperative societies and rural banks in their work require a lot of planned effort. Also, farm management, an important dimension of planning has not yet been promoted even among big and well-to-do farmers, what to say of resource- poor farmers? Farmers in India so far have hardly used paper and pencil in farming - management aspect must come, an important area which should be introduced by the extension agencies. Systematic documentation of events and results is another vital aspect of planning - data base on dryland farming has to be created.

Approach

People's participation in extension work has been emphasized since its inception; likewise, inter-disciplinary work in research. And in both the cases, we have far lagged behind. Firstly, because of individualist approach -self-centered rather than group-centered; and secondly, lack of appreciation of the intricacies of technology generation and the transfer. This is being focussed now but we are far away from the goal yet. We need a lot of effort in this direction, more so in rainfed farming.

Linkage

Research and development linkage is again not a new subject. We have met quite a success in this respect but we can do better, no doubt. The weakness is not in the formal mechanisms for linking and interactions, but in their functionality and follow-up. Scientific group functioning has to be promoted. Linkage, in the first instance, is a mental function - individuals must appreciate research and development alike. The process and methods of linking is more a social process requiring human dealings and approach; it is difficult in that sense - a challenge really. Many designs and approaches have been adopted by different countries in this respect. In case of rainfed farming, linkage among the scientists, extension workers and farmers becomes all the more delicate and important.

Area

Regional imbalances in development is found everywhere; some areas could get more attention than others owing to various reasons and factors. Rainfed areas are obviously neglected ones; so are the hill areas, tribal areas, coastal areas, etc. Therefore, both in technology generation and transfer of technology, this angle is necessary. We have more problems in the latter (TOT), for we are not able to employ and retain extension staff in such areas basically because of lack of basic amenities. This demands special consideration and approach in all respects.

Feed-back

This is the weakest area - scientific feed-back is not readily forthcoming. This is precisely because the field-level extension is not equipped and trained to do it; the feed-backs are mostly mechanical - shortage of seeds, fertilizers, etc. Therefore, the introduction of first-line extension system as an integral part of the research/educational institutions in India has been rewarding. We have a scientific class of extension workers in between research and general extension system - a strong bridge. The testing and assessing technologies in socio-economic conditions of the farmers is one of their main mandate. Such a system is worth introducing everywhere so that institutional extension role can be encouraged and highlighted. Involvement of farm scientists, even on a humble scale, in field extension work, would make the research, education and training more relevant and meaningful.

Outcomes

The expected outcomes will affect the total process of development. The general tendency is to go through the process and watch for the results. But a planned development would have well considered goals or targets both quantitative and qualitative, which will govern the process, methods, efforts and investments. For instance, getting more job opportunities and income is no guarantee for a healthy living - pattern of investments/spending, therefore, becomes a pointer toward proper or improper use of resources. Extension job is not over only after production; its wise use requires a lot of guidance and education - the post-production extension.

Situations

All the elements and processes of Research and Development will be affected by the situational variations - rainfed or irrigated agriculture for example. Both technology as well extension methodology may vary depending upon these situational variables. The situational theory of leadership, technology and methodology must be practised in a planned development and economy. This is very much wanting; we have been too generalized in our approach and recommendations. Therefore, adoption of technologies have been slow especially in rainfed areas.

CONCLUSIONS

Planned development for over four decades in India has resulted in economic growth and green revolution, but not in the reduction of the regional imbalances, disparities and inequalities among people and the societies. In fact, the modernization through Science and Technology (S&T) has widened the gap between the rich and the poor, the rural and the urban, the higher and the lower castes, the plains and the valleys, and the advantaged and the vulnerable sections of the society. In spite of excellent institutional infrastructures, philosophical development designs and laudable extension programmes, effective implementation of the programmes could have contributed much more. Poorer sections of the society are getting disillusioned, oppressed and frustrated. This is resulting in social disorder, upheavals and unrest. The commitment, discipline and hard-working habits, both among officials and non- officials, are far from desired - a real challenge for the human resource development machinery. Mission-oriented development strategies with strict discipline, management and monitoring are earnestly needed - the Government policy has to be harsh in this respect. Without rightful reward and punishment institution and organisations do not function.

Research and development in rainfed agriculture the world over have been the neglected sectors. The future of second phase of Green Revolution in India lies in raising the productivity of over 100 million hectares of such lands. This subject will be a continued challenge, for even after the exploitation of full irrigation potential, 50 per cent of Indian agriculture will remain rainfed. Therefore, added attention and investment in research and development for arid, semi-arid and wet land agriculture have to be made as a long- term agricultural policy in the country. It has not become the priority area of the government, not only for improving farm production but also for serving the resource-poor farmers. The demand for diversification in farming, farming system approach in research and development, and not mass production but production by masses, present a formidable task ahead.

There is no short-cut to development. No general rule of thumb would work. For sustainable agriculture, the four pillars of development - research, education, extension and training - must get equal support and treatment. Specificity in terms of area, audience, technology, methodology, and time is the crying need of today. Varied approaches and models of extension have been advocated but promoted least in the rainfed areas; a lot remains to be done. To bring objectivity and workability to the extension design, the CCIMMMMETT-PALAFOS model comprising 17 specific elements need careful consideration. For professionalism in extension, all these elements must work in proper harmony and relationship.

The watershed resources management approach being tried in India for the development of rainfed areas has emerged as an important tested device. The integrated agricultural development, leading to diversified agriculture and agro-industries with community participation in the areas, has shown considerable desired changes in the farming communities. As a policy, the Governments, Centre and States, must support the poor-resource farmers for land development, soil conservation and water harvesting.

REFERENCES

Anonymous, 1990. *Participation Practice - Lesson from the FAO People's Participation Programme.* Participatory Training. FAO.

Anonymous. 1990. *Annual Report - National Demonstrations Projects.* ICAR, New Delhi.

Burtus, Haverkort, 1983. Inventory of Problems and Possible solutions for Rural Extension Activities in the third World Countries. Proc. *Seminar on 'Strategies for Agriculture Extension in the 3rd World'.* Focus Recommendation Discussion Papers; International Agriculture Centre, Wageningen, Netherlands.

Chambers, R. 1983. *Rural Development; Putting the Last First.* Longman Group Ltd., London, U.K.

Chambers, R. and Jiggins, J. 1986/1987. Agricultural research for resource- poor farmers- A Parsimonious paradigm. Brighton (Sussex): Discussion Paper number 20 (1986). Later published in *Agricultrual Administration and Extension,* 27.

Chambers, Robert and Gildyal, B.P. 1985. Agricultral research for resource-pooer farmers: the farmer-first and -last model, *Agricultural Administration,* Vol.20, No.1.

Dexter, E. 1986. Strategies in the transfer of agriculture technologies with reference to Northern Europe. In Gwyn E. Jones (Ed.) *Investing in Rural Extension: Strategies and Goals.* Enslevier Applied Science Publishers, London and New York.

FAO, 1990. *Participation and Practice.* Lessons from the FAO Peoples' Participation Project, FAO, Rome.

ISNAR, 1986. *A Report of Scientific Panel on Evaluation of FAR/OFCOR Programmes in National and International Agricultural Research System ISNAR.* An Evaluation Report.

Kesseba, Abbas M.Ed. 1989. *Technology systems for Small Farmers - Issues and Options,* Boulder,

Colorado, Westview Press, Published in Cooperation with IFAD, the International Fund for Agricultural Development.

Prasad, C. 1990. Developing appropraite technologies for specific target groups of Farmers - Indian experience and approach. *International Seminar, National Productivity Council, New Delhi, India.*

Prasad, C. 1991. Extension management for watershed development. *Paper Presented in International Training Course organized by World Bank & Ministry of Agriculture, Govt. of India.*

34

SUSTAINABLE DRYLAND AGRICULTURE

R.P. SINGH

INTRODUCTION

A wide array of critical environmental threats is posing a challenge to mankind all the world over. Some of these threats include degradation of soil, water, forest, and marine resources essential to increased production of food, fodder, fibre and timber etc. The problem of degradation is compounded in dry areas because of fragile ecosystem. Common to all areas is the health- threatening pollution which is wide spread. Stratospheric ozone depletion, global climate change, and loss of biodiversity, are some of the concerns receiving the attention of all developed and developing countries alikec.

Deterioration of natural resources is the main issue threatening sustainable development of dryland agriculture, more so in the Third World Countries. Three main reasons are attributed to this state of affairs (FAO 1989):

1. lack of access to land, inputs and other productive resources by the majority of rural households and communities,
2. lack of awareness among policy makers of the economic costs involved, and
3. lack of environmentally sound alternative technologies acceptable to farmers."

Subsistence dryland farmers are forced to over exploit natural resources by way of mining soil nutrients, cultivating steep slopes, overgrazing rangeland, excessive collection of fuelwood, inorder to survive and feed themselves. The government policies should, therefore, strive to create favourable circumstances, whereby the farmers conserve natural resources, and are not forced to destroy them. It must be realized that the cost and level of inputs needed to restore the productivity of degraded lands are much greater, and returns much lower, in harsh environment of semi-arid tropics than in the humid temperate climates (Parr et al. 1990).

In this chapter the concept of sustainability, sustainable agriculture, and sustainable development has been explained. Sustainable agricultural practices and systems have been highhighted. In the end, some policy issues have been posed.

SUSTAINABILITY DEFINED

The concept of sustainability has an unidentified time dimension. The magnitude of the time dimension depends on one's objectives, being shorter for economic factors and longer for concerns pertaining to environment soil productivity and land degradation." (Lal and Miller,1990). In the definition of sustainability adopted by the Technical Advisory Committee of the Consultative Group on International Agricultural Research is, however, clearly discernible. "Successful management of resources for agriculture to satisfy changing human needs while maintaining or enhancing the quality of the environment and conserving natural resoures" (TAC, 1989), Almost all definitions of sustainability reflect a subjective rather than an objective concept. In other words, they donot project the quantifiable entity of sustainability.

"From an agroecological perspective, sustainability may be defined as a measure of productivity over time per unit input of non-renewable or a limiting resource, per unit amount of soil erosion, per unit decline in soil organic matter or pH, per unit energy input from fossil fuel, per unit efflux of Co_2 from soil, per unit increase in No_3 concentration in the ground water, per unit of economic cost and value added, etc." (Lal and Miller, 1990).

II Singh et al. 1990 have put a quantitative value to sustainability of dryland agricultural practices (Soil and Moisture conservation, cropping systems, etc.) and have documented the same in the publication, "Towards Sustainable Dryland Agricultural Practices"II, brought out by Central Research Institute for Dryland Agriculture, Hyderabad, A.P. India.

SUSTAINABLE AGRICULTURE

Sustainable agriculture must meet four important criteria. These are (1) it must produce adequate feed of high quality, (2) be environmentally safe, (3) protect the resource base, and (4) be profitable. "The goal of sustainable agriculture is to feed expanding populations, while farming in an ecologically-sound, regeneeative way." (Reganold et al. 1990). Economically riable systems that minimize the purchase of off-farm inputs such as pesticides and fertilizers, and rely on on-farm renewable resources, form the erux of sustainable agriculture. Farming methods that emphasize soil-building practices (like crop residues, animal manures, green manures), natural pest controls, crop and crop/livestock diversity, and crop rotations with nitrogen fixing legumes, all constitute systems that ensure substitution of external resources by internal resources, reduce production costs and are ecologically sound. (Francis and King, 1988; Rodale 1988).

SUSTAINABLE DEVELOPMENT - THE CONCEPT AND MEANING

The concept of sustainable development has been conceived differently by different authorities/authors. Some are more specific than others, some authors have given a narrow definition of sustainable development, focussing on physical aspects - using renewable natural resources only, while some confine to economic difinition, focussing on optimal resource management. In fact, in economic terms, "the use of resources today should not reduce real incomes in the future" (Markandya and Pearce, 1988). Human development is central to sustainable development. "Men, women, and children must be the centre of attention - with development woven around people, not people around development" (UNDP, 1991). 'Equity' is one aspect which cuts across all definitions of sustainable development. Increasingly, sustainable development encompasses several aspects or dimensions. The World Conservation Union, in *Caning for the Earth,* 1991, defines sustainable development as "improving the quality of human life while living within the carrying capacity of supporting ecosystems." Earlier, the World Commission on Environment and Development (1987), defined sustainable development as " development that meets the needs of the present without compromising the ability of future generations to meet their own needs."

There are several dimensions of sustainable development-the economic, human, environmental and technological dimensions being most common, each interacting with one another. In essence, sustainable development means protecting the natural resources needed for food production and cooking fuels - from soil to woodlots to fisheries - while expanding production to meet the needs of growing population." In other words, "sustainable development means more efficient use of arable lands and water supplies, as well as development and adoption of improved agricultural practices and technologies to increase yields" (World Resources, 1992-93). It implies avoiding overuse of chemical fertilizers, careful use of irrigation, and avoiding the expansion of agriculture into steep slopes or marginal soils.

DESERTIFICATION OF DRYLANDS

The United Nations Conference on Desertification held in Nairobi, Kenya, in 1977 defined desertification as the "dimunition or destruction of the biological potential of the land, and can lead ultimately to desert like conditions." Drought was considered as just one cause of desertification. More importantly, human misuse of the land or mismanagement practices like overgrazing, tree cutting, improper tillage methods, poor water distribution systems, and overexploitation, are now regarded as the direct cause of desertification.

The principal desertification processes are (1) degradation of vegetative cover, (2) water erosion, (3) wind erosion, (4) salinization, (5) soil crusting and compaction, (6) reduction in soil organic matter, and (7) excess of toxic substances in the soil (Dregne, 1990). The list suggests that the word land degradation in an arid land context, is nothing but desertification. The dominat problem varies from over grazing in rangelands to water and wind erosion in rainfed crop lands, to salinization in irrigated land. In drylands, however, the most serious long term land degradation problem is water and wind erosion.

"Degradation on drylands (arid, semiarid, and subhumid) is called *desertification* because desert-like conditions appear where none existed before. Desertified drylands are characterized by low primary productivity (less than 400 kilograms of dry matter per hectare per year) and poor rain use efficiency (usually less than 1 kilogram of dry matter produced per hectare per year for each millimeter of rainfall."(World Resources-1988-89).

Estimates of soil erosion world wide are rough. Crop lands lose far more top soil each year than is replaced naturally. About six to seven million hectares of crop land are rendered unproductive because of soil erosion each year. Sodication (water logging, salinization and alkalinization) is responsible for reducing the productivity of upto 1.5 million hectares of crop land each year.

Drylands cover 18 per cent of the landarea in developing countries (Table 1) highest proportion of drylands are in Central America (28 per cent) and Africa (25 per cent). More than 300 million people in developing countries live in drylands.

RECOVERING DEGRADED LANDS

The International Union for Conservation of Nature and Natural Resources (IVEN) adopted guidelines for rehabilitating or restoring degraded lands. The recommendations were developed at a workshop on restoration ecology held in Varanasi, India, in December, 1987. The source of important guidelines are :

➢ "Critical sites - such as those especially liable to erosion, sources of saline water in the upper reaches of a watershed, or habitats of threatened and endangered species - should

Table 1. Drylands (million hectares) and human populations (millions) in developing regions.

Growth days	Africa	South west Asia	South east Asia	Central America	South America	East Asia	Total
A. Drylands							
1–74	488.0 (17%)[6]	72.6 (11%)	47.4 (5%)	62.3 (23%)	114.6 (6%)	27.7 (3%)	812.6 (11%)
75–119	231.8 (8%)	61.8 (9%)	54.9 (4%)	14.5 (5%)	142.8 (8%)	70.7 (7%)	586.2 (7%)
Total Drylands	719.8 (25%)	134.4 (20%)	102.3 (11%)	76.8 (28%)	257.4 (14%)	98.1 (10%)	1388.8 (18%)
B. Human Populations							
1–74	33.1 (9%)[6]	15.9 (12%)	46.1 (4%)	8.5 (8%)	9.3 (4%)	x	112.9 (6%)
75–119	37.8 (10%)	24.3 (18%)	101.1 (9%)	3.7 (3%)	20.4 (9%)	x	187.3 (10%)
Total (All lands)	70.9 (19%)	40.2 (29%)	147.2 (13%)	12.2 (11%)	29.7 (14%)	x	300.2 (15%)

Source : FAO, Improving Productivity of Dryland Areas (FAO, Rome, 1987)

Notes : a. Excludes South Africa.

b. Parentheses indicate percentage of total land / population for each region.

c. Not available.

be rehabilitated or restored first.

- For the initial re-vegefation, plants should come from different successional stages to lessen the chances of natural succession being thwarted and to let slow-dispersing species be included.
- Plants that facilitate succession or maintain nutrient and water cycles should be included. Nitrogen fixes are an important example, as are plants whose roots can penetrate beyond infertile upper soil layers.
- Local people should be involved, some of whom may be able to contribute traditional knowledge about local plants and soils.
- Rehabilitation programmes, should provide economic benefits to the participants.
- Laws and customs affecting the area under recovery should be reviewed periodically.
- Tree plantations should be established on degraded lands rather than on undisturbed or otherwise productive tracts.
- As a corollary to recovery projects, protected areas should be set aside to couserve plant and animal wild life species that may be needed for future rehabilitations or restorations."

SUSTAINABLE AGRICULTURAL SYSTEMS

Dry regions are characterized by a highly fragile natural resource base. As such, these areas are extremely susceptible to wind and water erosion. According to Parr *et al* 1990, very few farming systems are immune to breakdown as a result of the deterioration of the natural resources. "Arable farming systems in the fringe semi-arid regions are Indangered by the extreme variability of precipitation which can result in former praisies and savannahs becoming dust bowls and deserts following cultivation" (Arnon, 1992). So the dangers are inherent in the system itself. In areas where irrigation is used to increase production and avoid the risks inherent in dryland production, the potential for soil degradation are increased manifold. This is particularly true when injudicious use of irrigation water is made.

Modern agricultural systems are capital intensive. Economic returns require use of high levels of inputs. Injudicious use of inputs leads to environmental pollution. Such systems do not endure for long. A farming system to be sustainable should have the capacity to endure indefinitely. The ultimate goal of sustainable agriculture is : "to develop farming systems that are productive and profitable, conserve the natural resource base, protect the environment, and enhance health and safety, and to do so over the long term" (Parr *et al*., 1990).

Two farming systems have been proposed for ensuring sustainability. These are : Low - input sustainable agriculture (LISA), and Organic farming.

Low -input sustainable agriculture (LISA)

In this system, minimal use of external production inputs is made. The production costs are obviously lower. The over-all risk of the farmer is considerably reduced. Besides the above advantages, pollution of surface and groundwater is avoided and healthy food with very little or no pesticide residues is ensured. These systems hold promise for both short and long term profitability. However, the system suffers from one serious drawback-continuation of low external input agriculture will perpetuate the vicious circle of low inputs-low yields, which the third World Countries with ever increasing population pressure can ill afford.

High input systems, on the other hand, will fail sooner or later, as they are not economically and environmentally sustainable. What is the solution then ? The third alternative, i.e. optimal-input farming system (Wagner, 1990) will meet the requirement of sustainability. The optimal-input farming system has the premise of low input per unit of output and lays emphasis on law of diminishing returns.

Organic farming

This system merits consideration on the

There are problems associated with Conventional Model, having on-farm and off-farm impacts. The same are highlighted below :

Conventional agricultural practices	On-farm impacts	Off-farm impacts
1. Deep ploughing and removing crop residus	1. Reduce the amount of organic matter 2. Reduce soil fertility and rooting depth	—
2. Growing shallow rooted crops, using heavy machinery or removing organic matter from the soil.	Encourage soil compaction which restricts root growth, water retention, and air exchange	—
3. Row-crop monocultures	1. Aggravàte soil erosion	
4. Over irrigation	1. Lead to salinization	
5. Over fertilization, particularly excessive use/of nitrogen fertilizer.	1. Contaminates farm wells as well as water down stream.	1. Impact on water quality, nitrate pollution from nitrogen fertilizers. 2. Inereasing nutrient levels in estuartes.
6. Use of pesticides	—	1. Unsafe concentrations in drinking water 2. High levels of DDT led to dramatic decline in predatory bind population.

[Adapted from World Resources, 1992-93].

ground that most of the ills of modern-day agriculture are avoided. Use of agro-chemicals is forebidden. There is emphasis on build-up or organic matter in the soil, thereby activating biological activity. Soil is treated as a living organism. Maintenance of favourable soil structure, and development and use of crop rotations that improve soil fertility, control pests and diseases, and prevent soil erosion, is emphasized, Biological control measures for control of diseases, pests, and weeds are adopted. The major difference between organic and conventional farming is the almost exclusive reliance of the organic farmer on organic matter for supply of nutrients.

As a consequence of conventional agricultural practices, soil erosion and air pollution, may occur. Eroded soils run into reservoirs, elogging water ways, etc., thereby existing an off-farm impact. In areas where soils are heavily fertilized for raising high value crops, air pollution is common. Fertilized soils emit 2 to 10 times as much nitrous oxide as unfertilized soils and pasture.

Conversion of wild life habitat and rural landscapes to agricultural use may lead to loss of biodiversity and degradation of landscapes an off-farm impact having long-term consequences on productivity and sustainability.

In view of the impacts that conventional agricultural practices have, what is needed is an economically and environmentally efficient agriculture.

Role of Rainfed Agriculture

Although irrigation is known to boost yields dramatically, India could enhance food production with more efficient farming. Rainfed agriculture, which accounts for roughly two thirds (68 percent) of the total cultivated area in Indian is crucial to its food production. Managing the small, fragmented holdings where this type of farming is practiced calls for a cooperative approach if economics of scale are to be utilized (Bottral, 1991; Singh, 1988).

Areas depending on rainfed agriculture are diverse both with regard to soils and rainfall availability. 'Improved farming practices in rainfed areas could enhance production otherwise cubed by poor soils, in particular the black and red types found in Central and South India.'

ENVIRONMENTALLY SUSTAINABLE DRYLAND FARMING SYSTEMS

Principles or relevant components of environmentally sustainable farming systems should include the following (Lal *et al*, 1988)

- Reducing soil erosion and improving water conservation,
- Inclusion of legumes and cover-crops in crop rotations,
- Agroforestry as an alternate land use system,
- Judicious use of organic wastes.

Environmentally sustainable dryland farming systems emphasize on conservation and efficient utilization of natural resources. "Agronomic practices of conservation, tillage and mulch farming, rotational cropping, use of legumes and cover-crops to grow nitrogen and suppress weeds, and efficiently utilizing cattle manure and sewage sludge", (Lal *et al*, 1988), are some of the ingredients of sustainable farming systems source of the experimental truths are : (1) No till with residue mulch is an effective eroson control measure even on steep slops (Lal, 1984), (2) Adverse effects of crust can be minimized by mulching, (3) One of the several useful measures for water conservation is the practice of tied-ridging, (4) ridge cropping is yet another practice widely adopted in rainfed farming, (5) In addition to supplying nitrogen, legume cover-crops suppress weeds-*Stylosanthes* sp. suppressing perennial weeds spear grass-*Imperata cylindrica*, in the tropics and, (6) cover and grain crops can also be grown together as live mulches-a desirable practice so long as the host crop does not compete with the grain crop.

As compared with monoculture systems, perennials used in the agroforestry systems impart ecological stability, besides adding biologically fixed nitrogen. "Perennial hedge rows established on contour may reduce run-off and erosion and provide additional mulch with readily mineralizable N.'

Scientifically proven components that go into developing sustainable farming systems, and sustainable farming systems for different ecological regions, are given in Tables 2 and 3, respectively.

Table 2 : Scientifically proven components for developing sustainable farming systems.

Objective	Recommended component/such-systems
A. Controlling erosion	Conservation tillage, strip cropping, cover crops, mixed cropping and agroforestory, mulch farming, terracing, tied ridging, shelter belts, wind breaks.
B. Conserving water	Mulch farming, eco-fallow, weed control, tied ridges, irrigation.
C. Alleviating compaction	Sowing deep legumes, chisel ploughing, reducing vehicular traffic, guided traffic.
D. Enhancing soil fertility	Legume based rotations, cover crop, organic amendments, chemical fertilizer.
E. Improving soil fauna, maintaining soil tilth	Mulch farming, crop diversification, reduced tillage, organic amendment, planted fallows.
F. Post management	Rotations, crop diversification residue management, pesticides use based on threshold needs

Source : Lal and Miller, (1990).

Table 3. Sustainable farming systems for different ecological regions

Soils	Sustainable Systems
Structurally active soils	(a) 1. Grain crop-cover crop rotation using improved cultivatr
	2. No till or minimum tillage methods
	3. Residue mulch
	4. Chemical amendments
	5. Planted fallow
	(b) 'a' with agroforestry system using suitable woody perennial or shrubes
	(c) 'b' in association with cattle at low stocking rate.
	(d) Strip cropping with growing alternate strips of woody shrubes/cover crops with grain crops.
Structurally inert soils	(a) 1. Mixed cropping with compatible mixtures

2. Tied ridges, following ploughing at the end of rains.
3. Organic compost or farm yard manure.
4. Chemical amendments
5. Planted fallow
6. Irrigation

(b) 'a' above with contour hedge rows of woody perennials at 1 to 2 m vertical interval.

(c) 'b' above and integration with livestock, based on light grazing.

Source : Lal and Miller (1990).

POLICY ISSUES

Sustainability related issues are many. The technical aspects of sustainability, sustainable agriculture, and sustainable development of dryland agriculture, have been discussed in the preceding paragraphs. The ill of all ills is the problem of land degradation. In the context of India, demographic pressure is yet another issue causing concern.

In order to tackle the problem of deteriorating natural resources, in particular land and water, and land degradation, a sound land use policy is called for. Land use according to land capability, and putting human and livestock population pressure on land in accordance with its carrying capacity, should form the central theme of land use policy. The mismanagement of land should be halted through people support and strong legislative measures. Agriculture on steep slopes and marginal lands should be given a go-bye. Alternate land use systems, is place of arable cropping, should be given incentives necessary to popularise them among small and marginal farmers. The technologies for rehabilitating marginal and wasted lands are available. Concerted efforts on the part of the Government are needed to push through programmes of rehabilitation and restoration of such lands. A strong political will and commitment is needed to combat forces leading to land degradation.

Farming systems that could ensure sustainability need to be developed for each of the agro-ecological regions, so as to conserve soil, water, vegetation, wild life, etc. Farming systems that include multi-purpose nitrogen fixing tree species, animals (livestock), and leguminouns crops, which are efficient utilizes of land and water, should be given a premium in the scheme of things. Minimum tillage and use of crop residues as mulch should be the guiding principles for farming systems.

In arid regions, introduction of tractors has done irreversible damage to the ecosystem. Whereas carrying out timely operations it extremely important in dry regions, is should not be at the cost of soil structure and stability. As a policy, tractorization should be prohibited in fragile ecosystem, particularly in arid regions.

Human resource development is vital. Infact, environment development programme and human resource development programmes should go together. An education policy which emphasizes on conservation and management of natural resources, and family planning, will go a long way in achieving sustainability.

REFERENCES

Arnon, I. 1992. Agriculture in Drylands-Principles and Practice. Developments in Agriculture and Managed Forest Ecology, 26, Elsevier, Amsterdam - London, New York - Tokyo.

Bottrall, Anthony. 1991, "Overview in *Future Directions for Indian Agriculture,* Ruth Meinzen Dick and Mark Svendsen, eds. (International Food Policy Research Institute, Washington, D.C. p. 150.

Dregne, H, 1990. Desertification of Drylands, in *Challenges in Dryland Agriculture - A Global perspective. Proc. of the Intt. Conf. on Dryland Fmg.,* Amarillo/Bushland, Texas, USA, 15 - 18 August 1988.

FAO, 1989. Sustainable development and natural Resource Management, In : *The State of Food and Agriculture* 1988, FAO, Rome, pp 84-99.

FAO, 1987, *Improving Productivity of Dryland Areas* (FAO Rome, 1987) p. 3.

Francis, C.A. and King, J.W. 1988. Cropping systems based on farm derived renewable resources. *Agricultural systems,* 27 : 67 - 75.

IUCN 1991, The World Conservation Union, United Nations Environment Programme (UNEP), and

World Wide Fund for Nature (WWF), *Caring for the Earth* (IUCN, UNEP and WWF) Gland, Switzerland, 1991) p. 10.

Lal, Ratan, 1984. Soil erosion from tropical arable lands and its control. *Adv. Agron.* 37 : 183-247.

Lal, R. And Miller, F.P. 1990, Sustainable Farming Systems, for the Tropies, in *Sustainable Agriculture - Issue, Perspectives and Prospects in Semi Arid Tropics, Proc. Intl. Symp. on Natural Resources Management for a Sustainable Agriculture,* Vol. 1. p. 69-89, Ind. Soc. Agron.

Lal, R. Eckert, D.J., Logan T.J., and Myers, D. 1988., Environmentally sustainable Dryland Farming Systems, In : *Challenges in Dryland Agriculture, Proc. of the Intl. Conf. on Dryland Fneg.,* Amarillo/Bushland, Texa, USA, 15 - 28 August, 1988.

Markandaya, Anil and Pearce, D.W., 1988. "Natural Environments and the Social Rate of Discount.", *Project Appraisal,* Vol. 3, No. 1, p. 11.

Parr, J.F., Stewart, B.A., Hornick, S.B., and Singh, R.P. 1990. Improving the sustainability of dryland farming systems : A global perspective. *Adv. Soil Sci.,* 13, 1-8.

Reganold, J.P., Papendick, R.I. and Parr, J.F. 1990. Sustainable Agriculture in the United States. An overview, In : *Sustainable Agriculture - Issues,* Perspective and prospects in Semi Arid Tropics, *Proc. of Int. Symp. On Natural Resources Management for a Sustainable Agriculture,* Ind. Soc. Agron. 444-459.

Rodale, R. 1988, Sustainability - An opportunity for leadership, presented at the *Intl. Conf. on Sustainable Agricultural Systems,* Columbus, Ohio. 19-23 Sept. 1988.

Singh, R.P., Das, S.K., Bhaskara Rao, UM., and Narain Reddy, M. 1990. Towards sustainable Dryland Agricultural Practices, Central Research Institute for Dryland Agriculture Hyderabad, p. 106.

Singh R.P. 1992. "Problems and Prospects of Dryland Agriculture in India" In : *Land and Soils,* T.N. Khoshoo and B.L. Deekashatulu, eds. (Har - Anand, New Delhi 1992). pp. 261-263.

Singh, Inderjit, 1988. *Land and Labour in South Asia,,* (The World Bank, Washington, D.C. 1988) pp. 13-14.

TAC, Secretariat, 1989. *Sustainable agricultural production : Implications for International Agricultural Research, LGIAR,* Washington, D.C. 1989.

UNDP, 1991. United Nations Development Programme, *Human Development Report* 1991 (Oxford University Press, New York, 1991) pp. 1-2.

World Resources, 1988-89. A Report by the World Resources Institute and the International Institute for Environment and Development in Collaboration with the United Nations Environment Programme, Basic Books, Inc. New York. pp. 372.

World Resources, 1990-91. A Report by the World Resources Institute in collaboration with the United Nations Environment Programme, and The United Nations Development Programme Oxford University Press, New York, Oxford 1990. pp. 383.

World Resources, 1992-93. A Report by the World Resources Institute in Collaborations with UNEP and UNDP, Oxford University Press, New York - Oxford. 1992. pp. 385.

World Resources, 1994-95. A Report by the World Resource Institute in collaboration with UNEP and UNDP, Oxford University Press, New York Oxford, 1994. pp. 400.